NOBEL LECTURES IN PHYSIOLOGY OR MEDICINE 1971–1980

NOBEL LECTURES

Including Presentation Speeches
And Laureates' Biographies

PHYSICS

CHEMISTRY

PHYSIOLOGY OR MEDICINE

LITERATURE

PEACE

ECONOMIC SCIENCES

NOBEL LECTURES

INCLUDING PRESENTATION SPEECHES
AND LAUREATES' BIOGRAPHIES

PHYSIOLOGY
OR
MEDICINE

1971–1980

EDITOR

JAN LINDSTEN

The Karolinska Medico-Chirurgical Institute
Stockholm, Sweden

World Scientific
Singapore • New Jersey • London • Hong Kong

Published for the Nobel Foundation in 1992 by

World Scientific Publishing Co. Pte. Ltd.
5 Toh Tuck Link, Singapore 596224
USA office: 27 Warren Street, Suite 401-402, Hackensack, NJ 07601
UK office: 57 Shelton Street, Covent Garden, London WC2H 9HE

British Library Cataloguing-in-Publication Data
A catalogue record for this book is available from the British Library.

NOBEL LECTURES IN PHYSIOLOGY OR MEDICINE (1971–1980)

ISBN-13 978-981-02-0790-8
ISBN-10 981-02-0790-5
ISBN-13 978-981-02-0791-5 (pbk)
ISBN-10 981-02-0791-3 (pbk)

Foreword

Since 1901 the Nobel Foundation has published annually "Les Prix Nobel" with reports from the Nobel Award Ceremonies in Stockholm and Oslo as well as the biographies and Nobel lectures of the laureates. In order to make the lectures available to people with special interests in the different prize fields the Foundation gave Elsevier Publishing Company the right to publish in English the lectures for 1910–1970, which were published in 1964–1972 through the following volumes:

Physics 1901–1970	4 vols.
Chemistry 1901–1970	4 vols.
Physiology or Medicine 1901–1970	4 vols.
Literature 1910–1967	1 vol.
Peace 1910–1970	3 vols.

Elsevier decided later not to continue the Nobel project. It is therefore with great satisfaction that the Nobel Foundation has given World Scientific Publishing Company the right to bring the series up to date.

The Nobel Foundation is very pleased that the intellectual and spiritual message to the world laid down in the laureates' lectures will, thanks to the efforts of World Scientific, reach new readers all over the world.

Lars Gyllensten
Chairman of the Board

Stig Ramel
Executive Director

Stockholm, June 1991

Preface

The Nobel Week is an annual, international event the highlight of which is the Nobel Prize Award Ceremony in the Concert Hall of Stockholm on December 10th, that is on the death-day of Alfred Nobel. It is a fascinating occurrence from both the scientific and social points of view, mainly because it puts the significance of science for society into bright focus. Only one item on the entire agenda is compulsory for the Laureates during this week, the Nobel Lecture. Thus, each Laureate has to deliver a lecture on the topic for which the prize has been awarded. The lectures are generally given on December 8th, that is before the Prize Award Ceremony, so that the Laureates can enjoy the festivities in a more relaxed way.

The Laureates in Physiology or Medicine present their lectures at the Karolinska Institute, that is on the ground of the Prize Awarding Institution. This gives the scientists and students at the Institute a unique possibility to enjoy presentations of some of the most significant contributions to biomedical science and also to meet the Laureates personally. That is why these lectures have such a special atmosphere (video tape recordings of the lectures are kept in the Nobel Archives at the Karolinska Institute).

The Nobel Lectures are published in an annual book series, "Les Prix Nobel". Each volume in this series contains, in addition to the lectures given by all Laureates for a specific year, short biographical notes and portraits of the Laureates as well as reports on the prize presentation ceremonies in Stockholm and Oslo.

In the present two volumes the Nobel Lectures given by the Laureates in Physiology or Medicine for the years 1971–1990 have been reprinted. Since these lectures are time bound documents, only minor changes, for instance of printing errors, have been made.

As a member of the Nobel Assembly and secretary-general of the Nobel Assembly and the Nobel Committee at the Karolinska Institute during the years 1979–1990, I have had the pleasure of meeting most of the Laureates from the time period covered in these two volumes. It has indeed been a pleasure to reread their lectures with the perspective that time now has given them.

Jan Lindsten
M.D., Professor
September 1992

Contents

Foreword v

Preface vii

1971 EARL W. SUTHERLAND 1
Presentation Speech by Peter Reichard 3

Studies on the Mechanisms of Hormone Action 5

1972 GERALD M. EDELMAN AND RODNEY R. PORTER 23
Presentation Speech by Sven Gard 25

Biography of Gerald M. Edelman 29
Antibody Structure and Molecular Immunology 31

Biography of Rodney R. Porter 57
Structural Studies of Immuhoglobulins 58

1973 KARL VON FRISCH, KONRAD LORENZ and NIKOLAAS TINBERGEN 69
Presentation Speech by Börje Cronholm 71

Biography of Karl von Frisch 75
Decoding the Language of the Bee 76

Biography of Konrad Lorenz 89
Analogy as a Source of Knowledge 97

Biography of Nikolaas Tinbergen 109
Ethology and Stress Diseases 113

1974 ALBERT CLAUDE, CHRISTIAN DE DUVE and GEORGE E. PALADE 131
Presentation Speech by Jan-Erik Edström 133

Biography of Albert Claude 137
The Coming of Age of the Cell 143

Biography of Christian de Duve 149
Exploring Cells with a Centrifuge 152

	Biography of George E. Palade	173
	Intracellular Aspects of the Process of Protein Secretion	177
1975	DAVID BALTIMORE, RENATO DULBECCO and HOWARD M. TEMIN	207
	Presentation Speech by Peter Reichard	209
	Biography of David Baltimore	213
	Viruses, Polymerases and Cancer	215
	Biography of Renato Dulbecco	229
	From the Molecular Biology of Oncogenic DNA Viruses to Cancer	232
	Biography of Howard M. Temin	243
	The DNA Provirus Hypothesis	245
1976	BARUCH S. BLUMBERG and D. CARLETON GAJDUSEK	265
	Presentation Speech by Erling Norby	267
	Biography of Baruch S. Blumberg	271
	Australia Antigen and the Biology of Hepatitis B	275
	Biography of D. Carleton Gajdusek	299
	Unconventional Viruses and the Origin and Disappearance of Kuru	305
1977	ROGER GUILLEMIN, ANDREW V. SCHALLY and ROSALYN S. YALOW	355
	Presentation Speech by Rolf Luft	357
	Biography of Roger Guilleman	361
	Peptides in the Brain. The New Endocrinology of the Neuron	364
	Biography of Andrew V. Schally	399
	Aspects of Hypothalamic Regulation of the Pituitary Gland with Major Emphasis on its Implications for the Control of Reproductive Processes	405
	Biography of Rosalyn S. Yalow	441
	Radioimmunoassay: A Probe for Fine Structure of Biologic Systems	447
1978	WERNER ARBER, DANIEL NATHANS and HAMILTON O. SMITH	469
	Presentation Speeches by Peter Reichard	471

	Biography of Werner Arber	475
	Promotion and Limitation of Genetic Exchange	481
	Biography of Daniel Nathans	495
	Restriction Endonucleases, Simian Virus 40, and the New Genetics	498
	Biography of Hamilton O. Smith	519
	Nucleotide Sequence Specificity of Restriction Endonucleases	523
1979	ALAN M. CORMACK and GODFREY N. HOUNSFIELD	543
	Presentation Speech by Torgny Greitz	545
	Biography of Allan Cormack	549
	Early Two-Dimensional Reconstruction and Recent Topics Stemming From It	551
	Biography of Godfrey N. Hounsfield	565
	Computed Medical Imaging	568
1980	BARUJ BENACERRAF, JEAN DAUSSET and GEORGE D. SNELL	587
	Presentation Speech by George Klein	589
	Biography of Baruj Benacerraf	593
	The Role of MHC Gene Products in Immune Regulation and Its Relevance to Alloreactivity	597
	Biography of Jean Dausset	625
	The Major Histocompatibility Complex in Man — Past, Present, and Future Concepts	628
	Biography of George D. Snell	643
	Studies in Histocompatability	645

1971

Physiology or Medicine

EARL W. SUTHERLAND

"for his discoveries concerning the mechanisms of the action of hormones"

Earl W. Sutherland

THE NOBEL PRIZE FOR PHYSIOLOGY OR MEDICINE

Speech by Professor PETER REICHARD, the Karolinska Medico-Chirurgical Institute

Translation

Your Majesty, Your Royal Highnesses, Ladies and Gentlemen,

What applies to bacteria also applies to elephants. This free quotation after the French Nobel prize winner, Jacques Monod, illustrates with some exaggeration one important principle of biology: that of the identity of the fundamental life processes.

Yet one need not be a Nobel prize winner to know the difference between bacteria and an elephant. The latter is not only much larger. The decisive difference lies in the fact that bacteria are unicellular organisms and that all the functions of life are contained in a single cell. In higher organisms on the other hand, there occurs a division of labor between different types of highly specialized cells. Nevertheless, the elephant must function as an integrated unity. The cells in the different organs must be coordinated in such a way that they rapidly adapt to the changing requirements of the environment.

The hormones form part of such a coordinating system. Among other things, the difference between a bacterium and an elephant lies in the fact that the latter—as well as all of us here—for the sustainment of his life is completely dependent of the proper function of hormones, while bacteria can do without them.

What then is the function of hormones? Ever since the first hormone was discovered about 70 years ago this has been a central theme of research for many scientists. This question is also of considerable medical importance. Many diseases are hormone diseases, amongst them diabetes. In spite of this the mechanism of hormone action remained a complete mystery until recently. The answer did not come until Earl Sutherland started his investigations on the function of the hormone epinephrine.

This hormone is produced in the adrenal glands and is transported to different organs of the body by the blood. It is formed in increased amounts during stress and adapts the individual to new situations. One of its important functions lies in the liberation of glucose inside the cells for the production of energy. Epinephrine serves as a chemical signal, as a messenger, which is sent out from the adrenals to activate different organs essential for the defense of the individual.

Sutherland investigated the effect of epinephrine on the formation of glucose in liver and muscle cells. He discovered a new chemical substance which serves as an intermediate during the function of the hormone. This substance is called cyclic AMP. It transmits the signal from epinephrine to the machinery of the cell, and Sutherland therefore called it a "second messenger". Furthermore, Sutherland made the important discovery that cyclic AMP is formed in the cell membrane. This means that epinephrine never enters the cell. We may visualize the hormone as a messenger which arrives at the door of the house and there rings the bell. The messenger is not allowed to enter the house. Instead the message is given to a servant, cyclic AMP, which then carries it to the interior of the house.

Sutherland suggested already around 1960 that cyclic AMP participates as a second messenger in many hormone mediated reactions, and that its effect thus is not limited to the action of epinephrine. First this generalization was not willingly accepted by the scientific community, since it was difficult to visualize how a single chemical substance could give rise to all the diverse effects mediated by various hormones. By now Sutherland and many other scientists have provided convincing evidence, however, that many hormones exert their effects by giving rise to the formation of cyclic AMP in the cell membrane. Sutherland had discovered a new biological principle, a general mechanism for the action of many hormones.

How can one then explain the specificity of different hormones? A good part of the explanation lies in the fact that different cells in their membranes possess specific receptors for various hormones. The different messengers thus must find their way to the right door in order to deliver their messages.

Cyclic AMP was discovered in connection with investigations concerning the function of hormones. It came therefore as a big surprise when Sutherland in 1965 reported that cyclic AMP also occurred in bacteria which apparently had no use for hormones. It was soon found that cyclic AMP was produced by other unicellular organisms, too. In all these cases cyclic AMP was shown to have important regulatory functions which aid the cells in their adaptation to the environment. Maybe we can look upon cyclic AMP as the first primitive hormone, regulating the behaviour of unicellular organisms. We then may look upon the true hormones of higher organisms as components of an overriding principle which was added during the course of evolution. Thus the difference between uni- and multicellular organisms does not, after all, appear to be so great, and with respect to cyclic AMP we can turn around Monod's dictum and say that what applies to elephants also applies to bacteria.

Dr. Sutherland,

Hormones were known in biology and medicine for a long time. The mechanism for hormone action remained a mystery, however, until you discovered cyclic AMP and its function as a second messenger. In recent years it has become apparent that cyclic AMP also serves as an important regulatory signal in microorganisms, and that its action thus is not limited to the function of hormones. When you discovered cyclic AMP you discovered one of the fundamental principles involved in the regulation of essentially all life processes. For this you have been awarded this year's Nobel prize in physiology or medicine. On behalf of the Karolinska Institute I wish to convey to you our warmest congratulations, and I now ask you to receive the prize from the hands of his Majesty the King.

STUDIES ON THE MECHANISM OF HORMONE ACTION

Nobel Lecture, December 11, 1971

by

EARL W. SUTHERLAND†
Vanderbilt University, Nashville, Tennessee, U.S.A.

When I first entered the study of hormone action, some 25 years ago, there was a widespread feeling among biologists that hormone action could not be studied meaningfully in the absence of organized cell structure. However, as I reflected upon the history of biochemistry, it seemed to me there was a real possibility that hormones might act at the molecular level. For example, recalling how the biosynthesis of urea was elucidated, we find a period when urea synthesis could be studied only in the whole animal; later its formation was studied in perfused livers, then in liver slices, and was finally obtained in cell extracts. By analogy, it seemed that a systematic analysis of hormone action might also proceed from studies of the intact animal to isolated tissues and finally to soluble systems. In any event, that was the thought I had in mind when I began studying epinephrine and glucagon. At that time we referred to glucagon as the hyperglycemic-glycogenolytic factor (HGF) of the pancreas, not realizing that Burger and his colleagues had studied it many years earlier in Germany.

At this point I should acknowledge a large debt of gratitude to Professor Carl Cori. When I returned to St. Louis after medical service in World War II, I was undecided as to whether I should enter medical practice or go into research. Cori convinced me, not so much by anything he said so much as by his example, that research was the right direction for me to take. Although I have occasionally felt an urge to see patients more often, I have never really regretted this decision to stay in the laboratory.

Of course the intellectual environment in Cori's laboratory was highly conducive to this decision. I believe that kind of stimulating environment, with the necessary "critical mass" of young and talented investigators, with the opportunity for the free exchange of ideas, is an important ingredient in the making of scientific progress. Such an environment existed at Washington University then, just as it exists at Vanderbilt now, but it has always been rare. I regret that it threatens to become rarer still, at least in the United States, as a result of the continuing decline in the federal support of basic research.

Returning now to epinephrine and glucagon, studies on the glycogenolytic action of these hormones were attractive for several reasons. Their effects on glycogen breakdown and glucose production in the liver were rapid, large, and reproducible. Liver slices could be used and a number of slices could be prepared from the liver of one animal. In addition, the basic biochem-

† Dr Sutherland died in 1974.

istry of glycogen breakdown had been established through the classical work of the Cori's and others, with phosphorylase, phosphoglucomutase, and glucose 6-phosphatase being the basic enzymes involved (1).

However, other enzymes were known to participate in glycogen metabolism, and at this stage even the possible hydrolysis of glycogen was thought to deserve consideration. Another basic point that had to be settled was whether or not the release of glucose might be a more primary event than glycogenolysis, since it seemed possible that glucose might be pumped out of the cell in response to the hormones, rather than overflowing as a result of increased glucose production. Measurement of glucose levels convinced us, however, that the increased glucose output resulted from an overflow phenomenon and not from an extrusion process in which glucose was actively transported from inside the cell (2).

Measurement of labelled intermediates, formed during incubation of rabbit liver slices in the presence of inorganic phosphate, led us to the conclusion that phosphorylase, rather than phosphoglucomutase or glucose 6-phosphatase, was rate-limiting in the conversion of glycogen to glucose, and that the hormones were acting to increase the activity of this enzyme (3). Glycogen loss was associated not only with an increase in glucose but with increased levels of both glucose 1-phosphate and glucose 6-phosphate. An important problem complicating analysis at this stage was the accepted theory or assumption that phosphorylase catalyzed the *in vivo* synthesis of glycogen as well as its degradation. We knew that the hormones always stimulated glycogen breakdown rather than synthesis, so that even after phosphorylase activation had been demonstrated in response to the hormones (Fig. 1), an additional factor was suspected. One factor preventing phosphorylase from catalyzing glycogen synthesis *in vivo* is the high ratio of inorganic phosphate to hexose phosphate that normally prevails inside cells. Glycogen synthetase was not discovered until later (5), and we now know that its activity is decreased while phosphorylase activity is increased (6) as a result of the same basic reaction (7).

What was the nature of this reaction? Although phosphorylase activation by epinephrine or glucagon could be shown easily if intact cell preparations were studied, as in Fig. 1, virtually all response to the hormones was lost if the cells were broken. It was not clear, therefore, that a better understanding of phosphorylase activation would lead us very directly to a better understanding of how the hormones were acting. We nevertheless decided that this was our best approach. Wosilait and I were able to purify active phosphorylase from dog liver extracts, and then found another enzyme which would catalyze the inactivation of phosphorylase. Since the molecular weight did not seem to change during inactivation, we suspected that a relatively minor change might be responsible for the loss of activity, the loss of a phosphate group being only one of many possibilities. Had one assay been as easy to carry out as any other, it is probable that we would have done many more experiments than we did before discovering that the inactivation of phosphorylase was indeed accompanied by the loss of inorganic phosphate (Fig. 2). Thus our inactivating enzyme was shown to be a phosphatase.

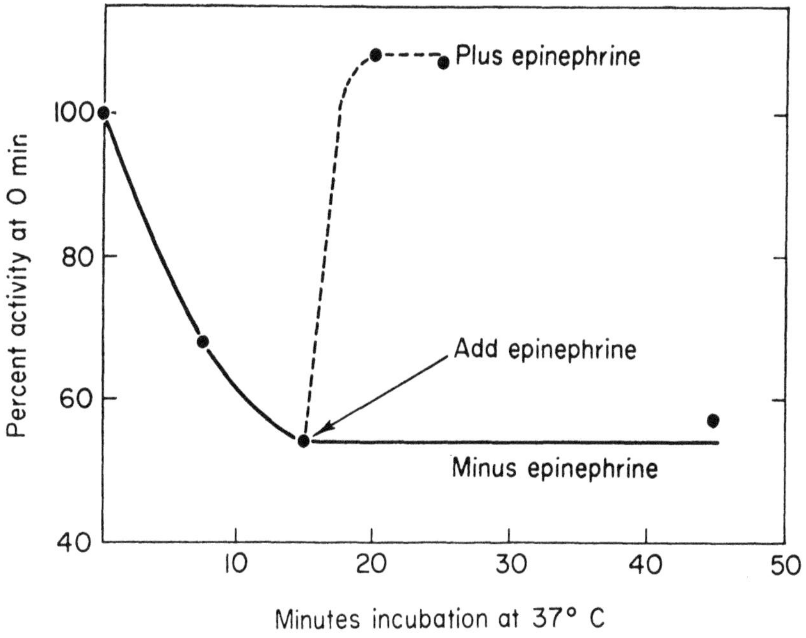

Fig. 1.
Effect of epinephrine on phosphorylase activity in dog liver slices. After incubating in a glycylglycine-phosphate buffer (pH 7.4) for the indicated times, the slices were homogenized and phosphorylase activity determined. From (4).

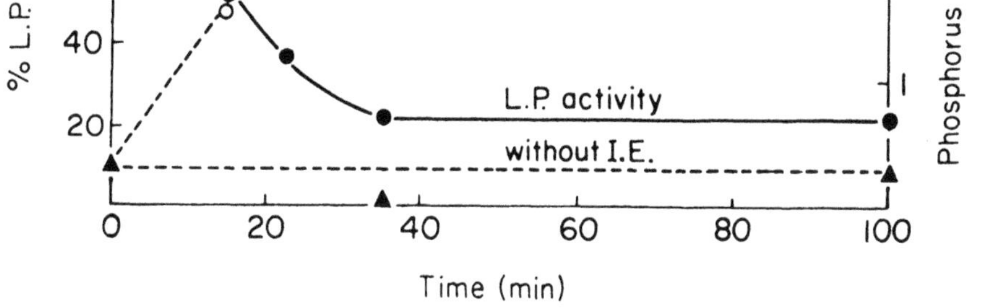

Fig. 2.
Effect of incubating with "inactivating enzyme" (I.E.) on liver phosphorylase (L.P.) activity and on the release of inorganic phosphate. Phosphate was measured in the supernatant fluid after TCA precipitation. From (8).

At about this time Ted Rall joined me, marking the beginning of a long and fruitful period of collaboration between us. The inactivation experiments had suggested that phosphorylase activation might be associated with phosphorylation of the enzyme molecule, and we were able to demonstrate this easily when liver slices were incubated in the presence of a phosphate buffer containing ^{32}P. Phosphate was rapidly incorporated into phosphorylase, and this was increased by the hormones, roughly in proportion to the degree of activation (4). It thus became clear that the concentration of active phosphorylase in liver represented a balance between inactivation by a phosphatase and reactivation by a process in which phosphate was donated to the protein. Krebs and Fischer (9) had meanwhile been studying phosphorylase activation in rabbit muscle extracts, and had demonstrated the requirement for ATP and Mg^{++}.

With this information in hand, we began to add hormones to our liver extracts in the presence of ATP and Mg^{++}. Progress was delayed for a time because we usually added fluoride to our homogenates, along with purified inactive phosphorylase. This seemed logical because we knew that fluoride would inhibit the phosphatase, and we wanted to inhibit this enzyme so as to preserve as much of the active phosphorylase that might be formed as possible. We now know that fluoride has another important effect in broken cell preparations—the stimulation of adenyl cyclase—and that this tends to mask the effect of hormones. Fortunately, we often incubated homogenates in the absence of fluoride, and in these experiments we began to see pronounced effects of epinephrine and glucagon (10). Our earlier discoveries—phosphorylase activation in slices, and the nature of the associated chemical change—were important landmarks, but not as exciting as finally establishing hormone effects in broken cell systems. Of course, we and others had often noted effects of hormone in this or that cell-free system, but, when pursued in greater depth, these effects had invariably proved to be non-specific. By contrast, the effects on phosphorylase activation appeared to be physiologically significant.

We then found that if we centrifuged our homogenates, to remove what we then thought of as cellular debris, the hormonal response was lost. It could be restored, however, by recombining the particulate fraction with the original supernatant (Fig. 3). We also found that if we incubated the particulate fraction with the hormones, a heat-stable factor was produced which could in turn activate phosphorylase when added to the supernatant fraction (10). The hormonal response was thus separated into two separate phases.

The next step was to identify the heat-stable factor. We showed that it was an adenine ribonucleotide (11) that could be produced from ATP by particulate preparations not only from liver, but also from heart, skeletal muscle, and brain (12). Lipkin and his colleagues had isolated the same compound from a barium hydroxide digest of ATP, and they later established its structure as adenosine 3', 5'-monophosphate (13), now commonly referred to as cyclic AMP or cAMP (Fig. 4).

Although cyclic AMP was not affected by a number of known phosphatases or diesterases, we did find an enzyme in animal tissues that rapidly

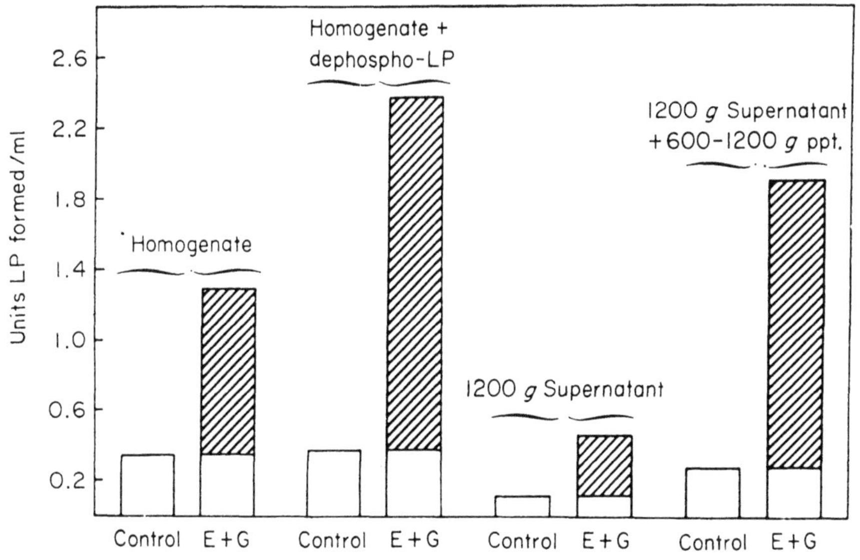

Fig. 3.
Effect of epinephrine and glucagon (E+) on phoshporylase activation in whole and fractionated cat liver homogenates. Inactive liver phosphorylase (dephospho-LP) was added where indicated, and the mixtures incubated with ATP and Mg^{++} in the presence and absence of the hormones. Phosphorylase activity was measured before and after 10 min incubation at 30° C. Each bar represents the amount of LP formed during this period. The effect of the hormones is indicated by the crosshatched areas. From (10).

inactivated it, with adenosine 5′-phosphate (5′-AMP) as the product of the reaction. It began to appear, therefore, that the concentration of cyclic AMP in tissues probably reflected a balance between two opposing reactions, formation from ATP, on the one side, and conversion to 5′-AMP, on the other. Our next series of studies were concerned with the enzymes responsible for these reactions.

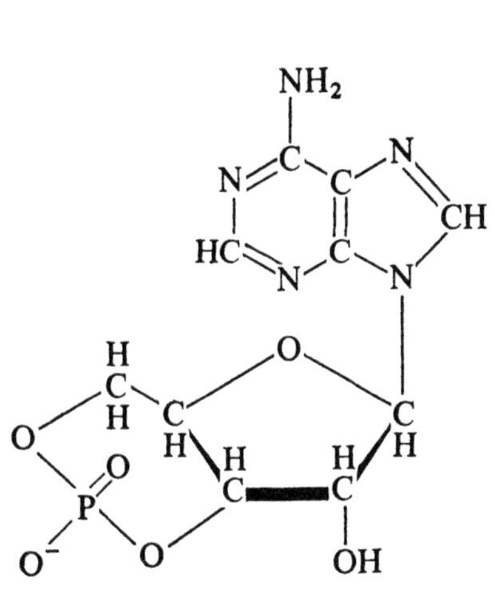

Fig. 4.
Structural formula of adenosine 3′, 5′-monophosphate (cyclic AMP).

The activity in particulate fractions responsible for converting ATP to cyclic AMP was originally named adenyl cyclase, although a chemically more correct designation might have been adenylyl cyclase or perhaps adenylate cyclase. We established that this enzyme was widely distributed, not only in mammalian tissues but in a variety of other animal phyla as well (14), and pyrophosphate was shown to be another product of the reaction (15). In the meantime, Bill Butcher studied the phosphodiesterase responsible for catalyzing the breakdown of cyclic AMP, and he used this enzyme to establish the widespread occurrence of cyclic AMP in cells and fluids (16).

It had become clear by this point that epinephrine and glucagon increased the accumulation of cyclic AMP in liver homogenates by stimulating adenylyl cyclase, rather than by inhibiting the phosphodiesterase. Glucagon was shown to be more potent than epinephrine in this system (17), as was later shown to be the case in the intact liver as well (18). Epinephrine and other catecholamines were also shown to be effective in preparation of cardiac muscle (19). Then evidence began to accumulate to suggest that some of the effects of other hormones might be mediated by this same mechanism. For example, Haynes (20) showed that ACTH (but not glucagon or epinephrine) was capable of stimulating adenylyl cyclase in adrenal preparations, whereas ACTH was *not* active in the liver. In collaboration with Mansour and Bueding, we found that serotonin but not catecholamines was active in preparations of the liver fluke, *Fasciola hepatica* (21). Then TSH was found to stimulate cyclase in thyroid preparations (22). Developments to this stage were summarized in more detail in several review articles (23).

Although the cell membrane seemed the most likely source of adenylyl cyclase, our finding that the enzyme could not be detected in dog erythrocytes, which lacked nuclei, whereas it did occur in the nucleated erythrocytes of birds, forced us to consider the possibility that the nucleus might be the major source of the enzyme. Davoren thereupon designed and built a pressure homogenizer which caused extensive fragmentation of cell membranes with little damage to the nuclei. When this procedure was followed by density gradient centrifugation, it was possible to show that most of the adenylyl cyclase was not associated with nuclear material, but rather fractionated with what appeared to be fragments of the cell membrane (24).

As a result of these and other developments, we gradually began to think of cyclic AMP as a second messenger in hormone action, with the hormones themselves acting as first messengers (25). This concept is illustrated in Fig. 5. It would appear that different cells contain receptors for different hormones, and that an important result of the hormone-receptor interaction in some cells is to stimulate adenylyl cyclase, leading to increased levels of cyclic AMP. The cyclic AMP then acts intracellularly to alter the rate of one or more cellular processes. Since different cells contain different enzymes, the end result of the change in cyclic AMP will differ from one type of cell to the next, e. g., phosphorylase activation in hepatic cells, steroidogenesis in the adrenal cortex, and so on. This concept seemed to clarify, in our own minds at least, many of the phenomena which we and others had observed.

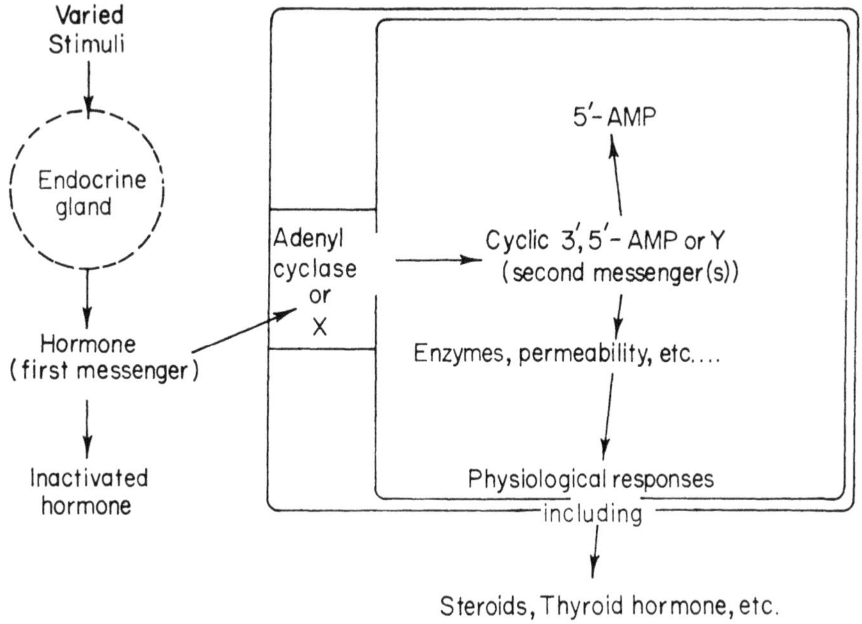

Fig. 5.
Schematic representation of the second messengers concept. From (26).

To further evaluate this hypothesis, and to help us determine which hormones might exert which of their effects by this mechanism, we decided to embark on a series of studies of different biological systems, with the following criteria in mind. First, if a given response to a given hormone is mediated by this mechanism, then adenylyl cyclase in the appropriate target cells should be stimulated by the hormone; conversely, hormones not producing the response should not stimulate the enzyme. Second, the level of cyclic AMP in intact tissues should change appropriately in response to hormonal stimulation; here we were thinking primarily of dose-response and temporal relationships. Third, we felt that drugs which inhibit phosphodiesterase, such as theophylline (16), should act synergistically with hormones that acted by stimulating adenylyl cyclase. Finally, it should be possible, at least in theory, to mimic the hormone by the application of exogenous cyclic AMP or one of its acyl derivatives. Of course we knew that most organic phosphate compounds such as cyclic AMP penetrated cells poorly if all, and the derivatives were synthesized by Theo Posternak in hopes that their more lipophilic nature might enable them to penetrate cell membranes more readily. Many of these derivatives were in fact more active than cyclic AMP, when applied to intact cells and tissues (27), but whether this can be related to better penetration remains to be established. These criteria have all had to be qualified, in the light of subsequent experience, but they did serve as useful guides for research. I will just mention briefly a few of the results obtained.

The positive inotropic response to epinephrine was of interest for many reasons, and was especially attractive to us because the heart contains only a limited number of different types of cells, with myocardial cells pre-

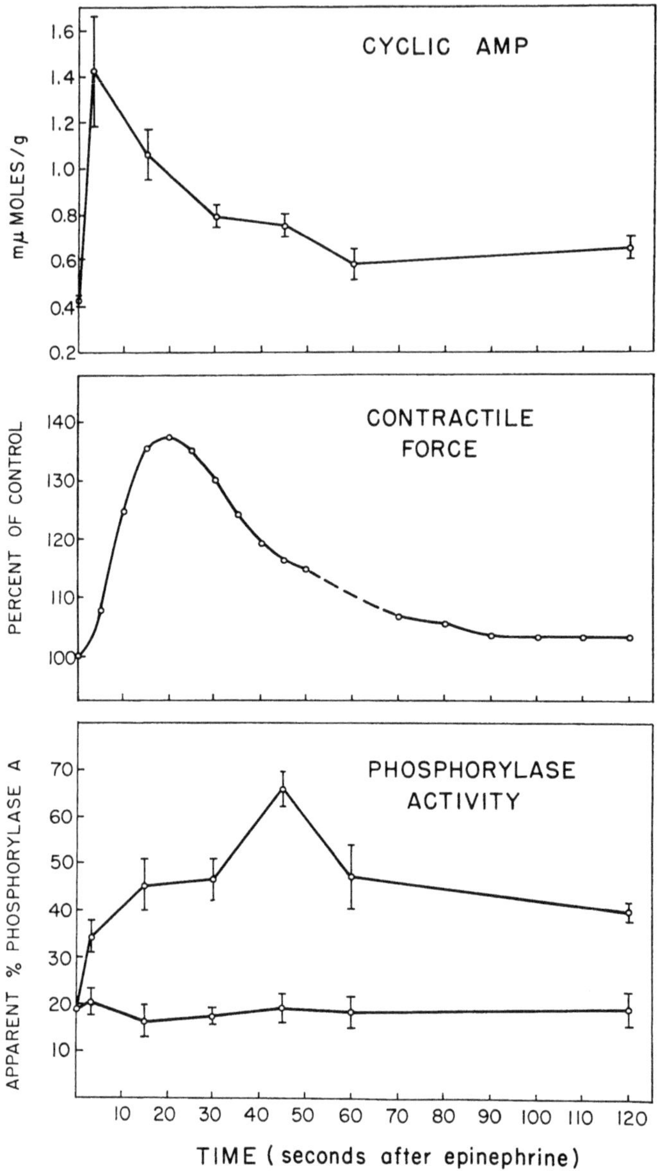

Fig. 6.
Effect of a single dose of epinephrine on cyclic AMP — contractile force, and phosphorylase activity in the isolated perfused working rat heart. Hearts were frozen at different times after injection of epinephrine. Lower curve in bottom panel shows lack of response to injected saline. From (28).

dominating. Hence we felt that biochemical studies of the heart would not be completely uninterpretable. The results of these studies have been summarized in more detail elsewhere (28—30), but an especially important result is shown in Fig. 6. We had previously found that the order of potency of several catecholamines in stimulating dog heart adenylyl cyclase was the same as others had reported for the inotropic response (19). Now we wondered if cyclic AMP in the intact heart could possibly increase with sufficient speed to account for this response. As shown in Fig. 6, we found that cyclic AMP increased with extraordinary rapidity, reaching a 4-fold elevation within 3 seconds of a single injection of epinephrine, clearly fast enough to account for the increased force of contraction. We also found that the ability of several adrenergic blocking agents to prevent cyclic AMP accumulation in response to catecholamines correlated with their relative potencies as antagonists of the inotropic response. Our experiments with the rat heart were important because they represented our first attempt to measure tissue cyclic AMP levels while simultaneously monitoring a mechanical or functional response which could not be measured chemically. They also led us to a more serious consideration of the possibility that beta adrenergic effects in general might be mediated by cyclic AMP (29). Subsequent studies by ourselves and others have tended to support the view that the positive inotropic response (30, 31) and other beta adrenergic effects (32) are mediated by cyclic AMP. A related development has been the evidence that some alpha adrenergic effects result from a fall in the level of cyclic AMP (32, 33). However, the mechanism by which alpha receptors mediate this effect, and how applicable it will be to alpha adrenergic effects in general, are questions that remain for future research.

Lipolysis by rat adipose tissue was of special interest for other reasons, one being the large number of hormones capable of stimulating this response. Epinephrine, ACTH, and glucagon, for example, were known to be capable of stimulating adenylyl cyclase in other cells, and it seemed reasonable to suppose that they might also do this in adipocytes. Confirming this expectation, Fig. 7 shows that all three hormones caused a pronounced rise in cyclic AMP when added to isolated fat cells; this figure also shows the selective blockade of epinephrine by the beta adrenergic blocking agent pronethalol. Combinations of supramaximal doses of two or more hormones were never additive, leading to the conclusion that the hormones were all affecting the same cells, and presumably the same adenylyl cyclase. The result with pronethalol, however, showed clearly that they were interacting with separate receptors. Other results, summarized elsewhere (30, 34—36), disclosed that phosphodiesterase inhibitors acted synergistically with the hormones to increase cyclic AMP and stimulate lipolysis, and that exogenous cyclic AMP (or at least the acyl derivatives) would also stimulate lipolysis. We also found that insulin and certain prostaglandins were capable of suppressing the accumulation of cyclic AMP in fat cells, and presumably these substances owe at least part of their antilipolytic activity to this effect. More recently, we established that the relative potencies of a series of xanthine derivatives to enhance lipolysis correlated well with their ability to inhibit phosphodiesterase in fat cells (37). Our most recent finding in this area has been that fat cells produce an antagonist in response to hormones that stimulate cyclic AMP formation, and that this agent prevents further response to the hormones (38). Perhaps this antagonist plays an importnt role in the negative feedback regulation of many cells.

In collaboration with others, we also helped investigate the effects of vasopressin on the toad bladder (39) and the steroidogenic effects of LH and

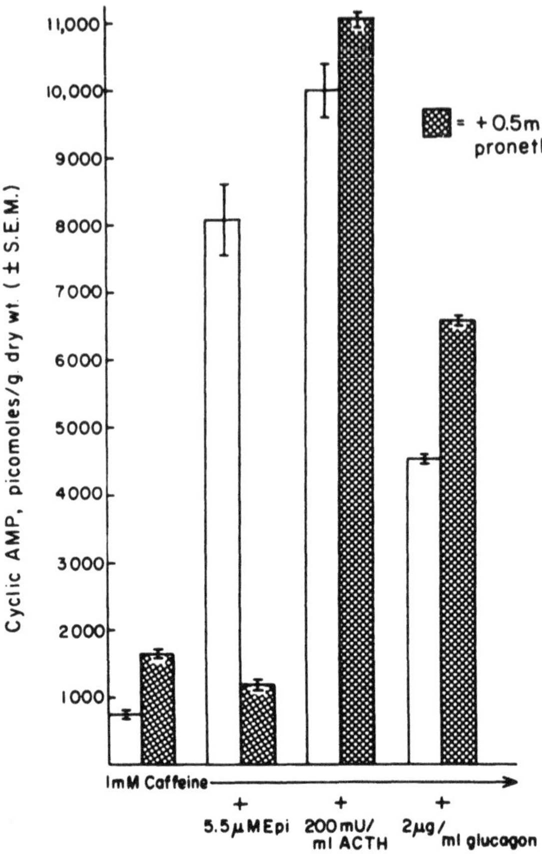

Fig. 7.
Effects of epinephrine (epi), ACTH, and glucagon, in the presence of caffeine, on cyclic AMP levels in isolated rat fat cells. Cells were incubated for 10 min in the presence of the hormones at the indicated concentrations, with and without pronethanol. From (34).

ACTH (40). It will not be possible here to discuss these and other results in detail. During recent years numerous investigators outside my own group have been active in this field, and progress has been correspondingly rapid. We attempted to summarize some of this progress in a recent monograph (41), and I might attempt a very brief recapitulation here.

Among hormones, we now know that vasopressin, ACTH, MSH, LH, parathyroid hormone, and TSH, besides epinephrine and glucagon, produce at least some of their effects by way of cyclic AMP. The production of certain other hormones, such as the steroid hormones, appears to be regulated by cyclic AMP. Thus the involvement of this nucleotide in endocrinology seems far-reaching. With reference to Fig. 5, we still do not understand how the hormone-receptor interaction leads to the stimulation of adenylyl cyclase. Robison developed an interesting model, according to which the protein component of the membrane adenylyl cyclase system was envisioned as being composed of two types of subunits (29): a regulatory subunit of which the receptors were a part, facing the extracellular space, and a catalytic subunit with its active center in contact with the cytoplasm. However, this has been a difficult model to test experimentally. Hirata and Hayaishi (42) obtained a soluble cyclase from bacteria, but the system in eukaryotic cells may be more complex. Phospholipids appear to be important (43), and recently Rodbell and his colleagues (44) have implicated guanyl nucleotides in the regulation of adenylyl cyclase in liver. Fluoride, which we had used initially to inhibit phosphorylase phosphatase, stimulates cyclase activity in broken cell preparations of most tissues (14, 45), but we still do not understand the mechanism of this effect. There is now evidence to suggest that cyclase is held in the membrane in a restrained or inhibited state (46), and perhaps the hormones and fluoride act by relieving this inhibition. Some interesting developments have also been reported concerning the phosphodiesterase, which may turn out to be a more interesting enzyme (or perhaps I should say enzymes) than it originally appeared to be (47).

Cyclic AMP has been found to produce a large number of effects since our early studies on phosphorylase activation, and I have listed some of them in Table 1. We have not ourselves studied the mechanism of action of cyclic AMP within recent years, but substantial progress has been made by others. Walsh, Perkins, and Krebs (48) discovered a protein kinase which was stimulated by cyclic AMP, and which is responsible not only for activating phosphorylase but also for inactivating glycogen synthetase (7). This enzyme is composed of two subunits, a regulatory subunit to which cyclic AMP binds, and a catalytic subunit which is inhibited by the regulatory subunit (49). When cyclic AMP binds to the regulatory subunit, a dissociation occurs such that the catalytic subunit becomes freely active. Protein kinases are widely distributed in nature (50), and their activation (or deinhibition) by cyclic AMP may account for many of the known effects of this nucleotide.

Of course the various relations in the endocrine system cannot all be understood in terms of the simple concept illustrated in Fig. 5. I have already mentioned that some hormones, such as insulin, produce some of their effects by lowering the level of cyclic AMP, although we still do not understand the mechanism of this effect. The primary action of the steroid hormones appears to follow an entirely different pattern (51), although some interesting relationships between these hormones and those that act through cyclic AMP will probably be discovered. An interesting recent development has been the evidence that prolactin affects the mammary gland by increasing the amount of protein kinase; this protein, rather than cyclic AMP, appears to limit the rate of casein production by this tissue (52).

Cyclic AMP has also been found to play an important role in non-endocrine regulatory mechanisms. Makman (53) found that glucose could suppress the formation of cyclic AMP by *E. coli,* and we now know, from the work of Pastan and Perlman and others (54), that cyclic AMP plays an important role in regulating bacterial metabolism. Bonner and Konijn and their colleagues (55) found that cyclic AMP was needed for the aggregation and differentiation of certain species of cellular slime molds; in these organisms, interestingly

Table 1. Some Known Effects of Cyclic AMP[1]

Enzyme or Process Affected	Tissue Organism	Change in Activity or Rate
Protein kinase[2]	Several	Increased
Phosphorylase	Several	"
Glycogen synthetase	Several	Decreased
Phosphofructokinase	Liver fluke	Increased
Lipolysis	Adipose	"
Clearing factor lipase	Adipose	Decreased
Amino acid uptake	Adipose	"
Amino acid uptake	Liver and uterus	Increased
Synthesis of several enzymes	Liver	"
Net protein synthesis	Liver	Decreased
Gluconeogenesis	Liver	Increased
Ketogenesis	Liver	"
Steroidogenesis	Several	"
Water permeability	Epithelial	"
Ion permeability	"	"
Calcium resorption	Bone	"
Renin production	Kidney	"
Discharge frequency	Cerebellar Purkinje	Decreased
Membrane potential	Smooth muscle	Increased
Tension	" "	Decreased
Contractility	Cardiac muscle	Increased
HCl secretion	Gastric mucosa	"
Fluid secretion	Insect salivary glands	"
Amylase release	Patorid gland	"
Insulin release	Pancrèas	"
Throid hormone release	Thyroid	"
Calcitonin release	Thyroid	"
Research of other hormones	Anterior pituitary	"
Histamine release	Mast cells	Decreased
Melanin granule dispersion	Melanocytes	Increased
Aggregation	Platelets	Decreased
Aggregation	Cellular slime molds	Increased
mRNA synthesis	Bacteria	"
Synthesis of several enzymes	"	"
Proliferation	Thymocytes	"
Cell growth	Tumor cells	Decreased

[1] References to most of these effects can be found in reference 41.

[2] Stimulation of protein kinase is known to mediate the effects of cyclic AMP on several systems, such as the glycogen synthetase and phosphorylase systems, and may be involved in many or even most of the other effects of cyclic AMP.

enough, cyclic AMP seems to function more as a first messenger than a second messenger. It is conceivable that cyclic AMP may at times act extracellularly in mammals too, since the proliferation of thymic lymphocytes was stimulated by exogenous cyclic AMP in concentrations similar to those that normally exist in plasma (56). Other evidence implicating cyclic AMP in the control of the immune response has been presented (57). Still another non-endocrine function of cyclic AMP may involve the regulation of vision (58) and possibly other senses. Our present understanding of the biological role of cyclic AMP is probably very small compared to what it will be in the future.

Most of our own research during recent years has centered around cyclic guanylic acid (cyclic GMP). Dr. Paul G. Hardman has been most active in this area. First found in urine by Price and his colleagues (59), it is still the only other 3′, 5′-mononucleotide known to occur in nature. Our approach here has been the opposite, in a sense, of what it was when I began my studies of epinephrine and glucagon. Then we had a function, and found a nucleotide; now we have a nucleotide, and are trying to discover its function. We still seem a long way from success. Cyclic GMP has been found in all mammalian tissues studied and in many lower phyla (60); in most cells its level is about an order of magnitude lower than that of cyclic AMP, although some insects contain more cyclic GMP than cyclic AMP. Guanyl cyclase has been studied (61) and found to differ in several respects from mammalian adenylyl cyclase: it is partially soluble in most tissues (in contrast to the totally particulate adenylyl cyclase) it requires Mn^{++} for activity and it is not stimulated by fluoride. More importantly, from the standpoint of hormone action, we have so far been unable to detect an effect of any hormone on the activity of this enzyme.

Although cyclic AMP and cyclic GMP seem clearly to be produced by separate enzymes, it is less certain that they are metabolized separately. Both nucleotides can serve as substrates for the same enzyme, under some conditions (47), although there is evidence that in vivo they may be hydrolyzed largely by separate enzymes (62). An intriguing observation has been that physiological concentrations of cyclic GMP are capable of stimulating (by several fold) the hydrolysis of cyclic AMP by liver phosphodiesterase (63).

We have also studied the excretion of cyclic nucleotides, and here we have made some interesting observations (64, 65). Normal plasma levels of these nucleotides are low, in the order of 10^{-8} M, but urine ordinarily contains levels which are several orders of magnitude greater. Normal humans excrete about 2 to 9 μmoles of cyclic AMP per day, with the rate of excretion of cyclic GMP being about 30 % of this. About 60 % of the urinary cyclic AMP is derived from plasma by glomerular filtration with the balance coming from the kidney itself. Essentially 100 % of urinary cyclic GMP is produced by glomerular filtration. We are still uncertain of which cells contribute how much to these extracellular values, but parathyroid hormone (PTH) seems to have an important influence in the case of cyclic AMP. Very striking increases in urinary cyclic AMP are seen after the injection of PTH in man, while the infusion of calcium chloride, which suppresses PTH secretion, leads to the opposite effect. This is shown for four patients in Fig. 8, which also illustrates the unexpected finding that urinary cyclic GMP levels were *increased* in response to calcium infusions. Other differential effects on these nucleotides have been noted. In rats, for example, hypophysectomy lowered the excretion of cyclic GMP to less than half of normal while reducing only slightly the excretion of cyclic AMP (64); conversely, the injection of glucagon greatly in-

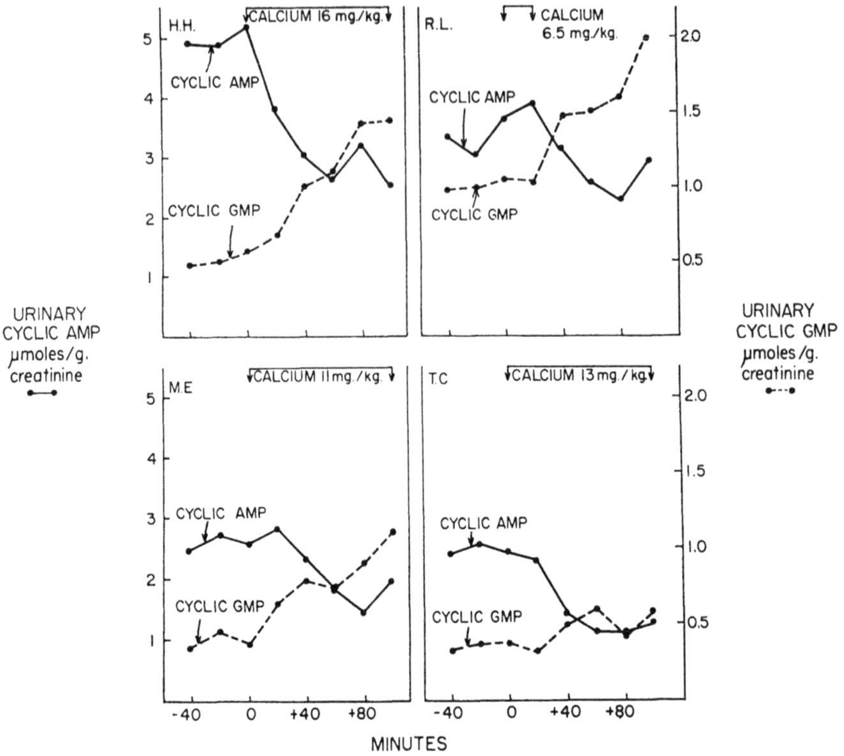

Fig. 8.
Comparison of urinary cyclic AMP and cyclic GMP in response to calcium infusion. Subjects received from 0.33 to 0.50 mg/kg/min of Ca^{++} during the first 15 min of the infusion. From (66).

creased cyclic AMP excretion with no effect on cyclic GMP. Another interesting finding was made by Goldberg and his colleagues (67); acetylcholine increased the level of cyclic GMP while reducing cyclic AMP in the isolated perfused rat heart.

These and other findings have led to the conclusion that the biological role of cyclic GMP, if it has such a role, must be quite different from that of cyclic AMP. An observation which seemed initially at variance with this conclusion was that exogenous cyclic GMP was almost as potent as cyclic AMP in stimulating glucose production by the isolated perfused rat liver. This was surprising because we had previously found (15) that cyclic GMP was less effective than cyclic AMP by several orders of magnitude in stimulating phosphorylase activation in dog liver extracts. Our first thought, that cyclic GMP might be inhibiting phosphodiesterase in the rat liver, thereby allowing cyclic AMP to accumulate, was found to be incorrect. We then found that exogenous cyclic GMP accumulated in the liver to a much greater extent than cyclic AMP, and this appears to account for its unexpected potency (68). We are continuing our studies to define a role for cyclic GMP. Many possibilities have yet to be explored, and whether cyclic GMP will prove to be as interesting as cyclic AMP remains to be seen.

I might turn briefly to the role of cyclic AMP in human disease. Because cyclic AMP serves primarily a regulatory function, it seems likely that a great variety of disorders may be related to defects in the formation or action of this nucleotide. One of the first examples to be studied was pseudohypoparathyroidism, a rare hereditary disease in which PTH appears incapable of stimulating cyclic AMP formation in target tissues in vivo (69); other hereditary disorders in which cyclic AMP has been implicated would include bronchial asthma (70), diabetes mellitus (71), and certain affective disorders (72). Some bacterial infections may involve cyclic AMP, and an especially interesting example has been cholera. The toxin responsible for this disease produces an apparently irreversible increase in adenyl cyclase activity, leading to prolonged high levels of cyclic AMP in intestinal epithelia; this in turn leads to the debilitating loss of fluids and electrolytes characteristic of this disease (73). As a final and possibly very important example, there is now evidence that defective cyclic AMP formation may be involved in the growth of tumors (74). To date, our knowledge that cyclic AMP may be involved in certain diseases has not lead to improved methods of therapy. It seems likely, however, that such methods will evolve as more detailed information accumulates. Several drugs already in use appear to act through cyclic AMP mechanisms.

In this lecture I have attempted to summarize some of my own research on the mechanism of hormone action, together with some results by others which have led to our present understanding of the role of cyclic AMP.

In conclusion I wish to suggest, or plead, that all of us exert a small amount of effort to stimulate interest in biological and medical research. A life in research can be a most enjoyable life with many frontiers to explore. In addition we need research to understand man and his ailments. I believe we are reaching a stage where research will be more and more helpful to man.

These points seem very obvious to me, and until the last few years they have appeared so obvious and simple that they were not worthy of mention. In these times, however, we may need to state and restate such simple points.

References

1. Cori, C. F., Physiol. Rev. **11**, 1943 (1931); Cori, C. F. and Welch, A. D., J.A.M.A., **116**, 2590 (1941); Cori, G. T. and Cori, C. F., J. Biol. Chem., **158**, 321 (1945).
2. Sutherland, E. W., Recent Progr. Horm. Res., **5**, 441 (1950).
3. Sutherland, E. W. and Cori, C. F., J. Biol. Chem., **188**, 531 (1951).
4. Rall, T. W., Sutherland, E. W. and Wosilait, W. D., J. biol. Chem., **218**, 483 (1956).
5. Leloir, L. F. and Cardini, C. E., J. Am. Chem. Soc., **79**, 6340 (1957).
6. Hers, H. G., De Wulf, H. and Stalmans, W., FEBS Lett., **12**, 72 (1970).
7. Soderling, T. R., Hickenbottom, J. P., Reimann, E. M., Hunkeler, F. L., Walsh, D. A. and Krebs, E. G., J. Biol. Chem., **245**, 6317 (1970)
8. Wosilait, W. D. and Sutherland, E. W., J. Biol. Chem., **218**, 469 (1956).
9. Krebs, E. G. and Fischer, E. H., Biochim. biophys. Acta, **20**, 150 (1956).
10. Rall, T. W., Sutherland, E. W. and Berthet, J., J. Biol. Chem., **224**, 463 (1957).

11. Sutherland, E. W. and Rall, T. W., J. Am. chem. Soc., **79,** 3608 (1957); ibid., J. Biol. Chem., **232,** 1077 (1958).
12. Rall, T. W. and Sutherland, E. W., J. Biol. Chem. **232,** 1065 (1958).
13. Lipkin, D., Cool, W. H. and Markham, R., J. Am. chem. Soc. **81,** 6198 (1959).
14. Sutherland, E. W., Rall, T. W. and Menon, T., J. Biol. Chem., **237,** 1220 (1962).
15. Rall, T. W. and Sutherland, E. W., ibid., **237,** 1228 (1962).
16. Butcher, R. W. and Sutherland, E. W., ibid., **237,** 1244 (1962).
17. Makman, M. H. and Sutherland, E. W., Endocrinology, **75,** 127 (1964).
18. Robison, G. A., Exton, J. H., Park, C. R. and Sutherland, E. W., Fed. Proc., **26,** 257 (1967); Exton, J. H., Robison, G. A., Sutherland, E. W. and Park, C. R., J. Biol. Chem., **246,** 6166 (1971)
19. Murad, F., Chi, Y. M., Rall, T. W. and Sutherland, E. W., J. Biol. Chem., **237,** 1233 (1962).
20. Haynes, R. C., ibid., **233,** 1220 (1958).
21. Mansour, T. E., Sutherland, E. W., Rall, T. W. and Bueding, E., ibid., **235,** 466 (1960).
22. Klainer, L. M., Chi, Y. M., Friedberg, S. L., Rall, T. W. and Sutherland, E. W., ibid., **237,** 1239 (1962).
23. Sutherland, E. W. and Rall, T. W., Pharmac. Rev., **12,** 265 (1960); Haynes, R. C., Sutherland, E. W. and Rall, T. W., Recent Progr. Horm. Res., **16,** 121 (1960); Rall, T. W. and Sutherland, E. W., Cold Spring Harb. Symp. Quant. Biol., **26,** 347 (1961); Sutherland, E. W., Harvey Lect., 57: 17 (1962).
24. Davoren, P. R. and Sutherland, E. W., J. Biol. Chem., **238,** 3016 (1963).
25. Sutherland, E. W., Øye, I. and Butcher, R. W., Recent Progr. Horm. Res., **21,** 623 (1965); Sutherland, E. W. and Robison, G. A., Pharmac. Rev., **18,** 145 (1966).
26. Sutherland, E. W. and Robison, G. A., Pharmac. Rev., **18,** 145 (1966).
27. Posternak, Th., Sutherland, E. W. and Henion, W. F., Biochim. Biophys. Acta, **65,** 558 (1962); Henion, W. F., Sutherland, E. W. and Posternak, Th., ibid., **148,** 106 (1967).
28. Robison, G. A., Butcher, R. W., Øye, I., Morgan, H. E. and Sutherland, E. W., Mol. Pharmacol., **1,** 168 (1965).
29. Robison, G. A., Butcher, R. W. and Sutherland, E. W., Ann. N.Y. Acad. Sci., **139,** 703 (1967).
30. Sutherland, E. W., Robison, G. A. and Butcher, R. W., Circulation, **37,** 279 (1968).
31. Epstein, S. E., Levey, G. S. and Skeleton, C. L., ibid., **43,** 437 (1971); LaRaia, P. J. and Sonnenblick, E. H., Circulation Res., **28,** 377 (1971).
32. Robison, G. A., Butcher, R. W. and Sutherland, E. W., in Fundamental Concepts in Drug-Receptor Interactions, Danielli, J. F., Moran, J. F. and Triggle, D. J., Eds. (Academic Press, London, 1969), pp. 59—91; Robison, G. A. and Sutherland, E. W., Circulation Res., **26** (Supp. I), 147 (1970).
33. Turtle, J. R. and Kipnis, D. M., Biochem. Biophys. Res. Commun., **28,** 797 (1967); Handler, J. S., Bensinger, R. and Orloff, J., Am. J. Physiol., **215,** 1024 (1968); Abe, K., Robison, G. A., Liddle, G. W., Butcher, R. W., Nicholson, W. E. and Baird, C. E., Endocrinology, **85,** 674 (1969); Robison, G. A., Arnold, A. and Hartmann, R. C., Pharmacol. Res. Commun., **1,** 325 (1969); Avioli, L. V., Shieber, W. and Kipnis, D. M., Endocrinology, **88,** 1337 (1971); Robison, G. A., Langley, P. E. and Burns, T. W., Biochem. Pharmacol., **21**, 589—92 (1972).
34. Butcher, R. W., Baird, C. E. and Sutherland, E. W., J. Biol. Chem., **243,** 1705 (1968)
35. Butcher, R. W., Ho, R. J., Meng, H. C. and Sutherland, E. W., ibid., **240,** 4515 (1965).
36. Butcher, R. W., Robison, G. A., Hardman, J. G. and Sutherland, E. W., Adv. Enzyme Regul. **6,** 357 (1968).
37. Beavo, J. A., Rogers, N. L., Crofford, O. B., Hardman, J. G., Sutherland, E. W.

and Newman, E. V., Mol. Pharmacol., **6,** 597 (1970).
38. Ho, R. J. and Sutherland, E. W., J. Biol. Chem., **246,** 6822 (1971).
39. Handler, J., Butcher, R. W., Sutherland, E. W. and Orloff, J., J. Biol. Chem., **240,** 4524 (1965).
40. Marsh, J. M., Butcher, R. W., Savard, K. and Sutherland, E. W., ibid., **241,** 5436 (1965); Grahame-Smith, D. G., Butcher, R. W., Ney, R. L. and Sutherland, E. W., ibid., **242,** 5535 (1967).
41. Robison, G. A., Butcher, R. W. and Sutherland, E. W., "Cyclic AMP" (Academic Press, New York, 1971).
42. Hirata, M. and Hayaishi, O., Biochim. Biophys. Acta, **149,** 1 (1967).
43. Levey, G. S., Biochem. Biophys. Res. Commun., **43,** 108 (1971).
44. Rodbell, M., Birnbaumer, L., Pohl, S. L. and Krans, H. M. J., J. Biol. Chem., **246,** 1877 (1971).
45. Øye, I. and Sutherland, E. W., Biochim. Biophys. Acta, **127,** 347 (1966); Johnson, R. A. and Sutherland, E. W., Fed. Proc., **30,** 220 Abs (1971).
46. Schramm, M. and Naim, E., J. Biol. Chem., **245,** 3225 (1970); M. J. Schmidt, Palmer, E. C., Dettbarn, W. D. and Robison, G. A., Dev. Psychobiol., **3,** 53 (1970); Perkins, J. P. and Moore, M. M., J. Biol. Chem., **246,** 62 (1971).
47. Beavo, J. A., Hardman, J. G. and Sutherland, E. W., J. Biol. Chem., **245,** 5649 (1970); Cheung, W. Y., ibid., **246,** 2859 (1971); Thompson, W. J. and Appleman, M. M., ibid., **246,** 3145 (1971); S. Kakiuchi, Yamazaki, R. and Teshima, Y., Biochem. Biophys. Res. Commun., **42,** 968 (1971).
48. Walsh, D. A., Perkins, J. P. and Krebs, E. G., J. Biol. Chem., **243,** 3763 (1968).
49. Gill, G. N. and Garren, L. D., Proc. Nat. Acad. Sci., U.S., **68,** 786 (1971); Brostrom, C. O., Corbin, J. D., King, C. A. and Krebs, E. G., ibid., **68,** 2444 (1971).
50. Kuo, J. F. and Greengard, P., ibid., **64,** 1349 (1969).
51. Jensen, E., Numata, N., Brecher, P. and Desombre, E., in "The Biochemistry of Steroid Action", Smellie, R. M. S., Ed. (Academic Press, London, 1971), pp. 133—159; O'Malley, B. W., Metabolism, **20**, 981 (1971).
52. Majumder, G. C. and Turkington, R. W., J. Biol. Chem., **246,** 5545 (1971).
53. Makman, R. S. and Sutherland, E. W., ibid., **240,** 1309 (1965).
54. Pastan, I. and Perlman, R., Science, **169,** 339 (1970); Zubay, G., Schwartz, D. and Beckwith, J., Cold Spring Harb. Symp. Quant. Biol., 35, 433, (1970); Yokota T. and Gots, J. S., J. Bact., **103,** 513 (1970); Nissley, S. P., Anderson, W. B., Gottesman, M. E., Perlman, R. L. and Pastan, I., J. Biol. Chem., **246,** 4671 (1971); Hong, J. S., Smith, G. R. and Ames, B. N., Proc. Nat. Acad. Sci. U.S., **68,** 2258 (1971).
55. Bonner, J. T., Barkley, D. S., Hall, E. M., Konijn, T. M., Mason, J. W., O'Keefe, G. and Wolfe, P. B., Develop. Biol., **20,** 72 (1969); Konijn, T. M., Chang, Y. Y. and Bonner, J. T., Nature, **224,** 1211 (1969).
56. MacManus, J. P. and Whitfield, J. F., Exptl. Cell Res., **58,** 188 (1969).
57. Parker, C. W., Smith, J. W. and Steiner, A. L., Int. Arch. Allergy, **41,** 40 (1971); Cross, M. E. and Ord, M. G., Biochem. J., **124,** 241 (1971)» Brun, W. and Ishizuka, M., J. Immunol., **107,** 1036 (1971).
58. Bitensky, M. W., Gorman, R. E. and Miller, W. H., Proc. Nat. Acad. Sci., U.S., **68,** 561 (1971).
59. Ashman, D. F., Lipton, R., Melicow, M. M. and Price, T. D., Biochem. Biophys. Res. Comm., **11,** 330 (1963); Price, T. D., Ashman, D. F. and Melicow, M. M., Biochim. Biophys. Acta, **138,** 452 (1967).
60. Goldberg, N. D., Dietz, S. B. and O'Toole, A. G., J. Biol. Chem., **244,** 4458 (1969); Ishikawa, E., Ishikawa, S., Davis, J. W. and Sutherland, E. W., ibid., **244,** 6371 (1969).
61. Hardman, J. G. and Sutherland, E. W., J. Biol. Chem., **244,** 6363 (1969); White, A. A. and Aurbach, G. D., Biochim. Biophys. Acta, **191,** 686 (1969); Schultz, G., Bohme, E. and Munske, K., Life Sci., Pt. **8,** 1323 (1969).

62. Thompson, W. J. and Appleman, M. M., Biochemistry, **10,** 311 (1971).
63. Beavo, J. A., Hardman, J. G. and Sutherland, E. W., J. Biol. Chem., **246,** 3841 (1971).
64. Hardman, J. G., Davis, J. W. and Sutherland, E. W., J. Biol. Chem., **241,** 4812 (1966); ibid, **244,** 6354 (1969).
65. Broadus, A., Hardman, J. G., Kaminsky, N. I., Ball, J. H., Sutherland, E. W. and Liddle, G. W., Ann N.Y. Acad. Sci., **185**, 50—66 (1971).
66. Kaminsky, N. I., Broadus, A. E., Hardman, J. G., Jones, D. J., Ball, J. H., Sutherland, E. W. and Liddle, G. W., J. Clin. Invest., **49,** 2387 (1970).
67. George, W. J., Polson, J. B., O'Toole, A. B. and Goldberg, N. D., Proc. Nat. Acad. Sci., U.S., **66,** 398 (1970).
68. Exton, J. H., Hardman, J. G., Williams, T. F., Sutherland, E. W. and Park, C. R., J. Biol. Chem. **246,** 2658 (1971).
69. Chase, L. R., Melson, G. L. and Aurbach, G. D., J. Clin. Invest., **48,** 1832 (1969); Marcus, R., Wilber, J. F. and Aurbach, G. D., J. Clin. Endocrinol., **33,** 537 (1971).
70. Szentivanyi, A., J. Allergy, **42,** 203 (1968).
71. Cerasi, E. and Luft, R., in "Pathogenesis of Diabetes", Cerasi, E. and Luft, R., Eds. (Interscience, New York 1970) pp. 17—40.
72. Paul, M. I., Cramer, H. and Bunney, W. E., Science, **171,** 300 (1971).
73. Shafer, D. E., Lust, W. D., Sircar, B. and Goldberg, N. D., Proc. Nat. Acad. Sci. U.S., **67,** 851 (1970); Sharp, G. W. G. and Hynie, S., Nature, **229,** 266 (1971); Kimberg, D. V., Field, M., Johnson, J., Hendersen, A. and Gershon, E., J. Clin. Invest., **50,** 1218 (1971).
74. Johnson, G. S., Friedman, R. M. and Pastan, I., Proc. Nat. Acad. Sci., U.S., **68,** 425 (1971); Sheppard, J. R., Ibid, **68,** 1316 (1971); Makman, M. H., ibid., **68,** 2127 (1971); Heidrick, M. L. and Ryan, W. L., Cancer Res., **31,** 1313 (1971); Otten, J., Johnson, G. S. and Pastan, I., Biochem. Biophys. Res. Commun., **44,** 1192 (1971).
75. Smith, H. W., "Man and His Gods" (Little, Brown and Co., Boston, 1952).
76. Meadows, D. L., The Futurist, **5,** 137 (1971); Roland, J. D., ibid., **5,** 145 (1971).
77. Although the food supply is probably not the most important factor limiting the populations of most species. See, for example, Christian, J. J., Biology of Reproduction 4, 248, (1971).
78. I am grateful to the many colleagues who helped in this research. James W. Davis has worked with me for over 15 years. Other important contributions were made by Drs. W. D. Wosilait, T. W. Rall, R. W. Butcher, G. A. Robison and J. G. Hardman, among others. I am further grateful to the National Institutes of Health and to the American Heart Association for their support over the years. I am indebted to Rollo Park not only for his friendship but for this creation of an ideal research environment.

1972
Physiology or Medicine

GERALD M. EDELMAN and RODNEY R. PORTER

"for their discoveries concerning the chemical structure of antibodies"

THE NOBEL PRIZE FOR PHYSIOLOGY OR MEDICINE

Speech by Professor SVEN GARD of the Karolinska Medico-Chirurgical Institute
Translation from the Swedish text

Your Royal Highnesses, Ladies and Gentlemen,
Immunebodies or antibodies is the designation of a group of proteins in the blood, that play an important part in the defense against infections and in the development of many different diseases. Their perhaps most characteristic property is the capacity to react and combine with substances, foreign to the organism, so-called antigens and to do so in a highly specific manner. There probably exist more than 50,000 different antibodies in the blood, each of them reactive against one particular antigen. Their main features are similar but they show individual characteristics and constitute, therefore, an extremely heterogeneous group. Since, in addition, they appear as very large molecules of a complex structure, it is understandable that the study of their chemistry for a long time offered great difficulties.

Up to 1959 the knowledge about their nature and mechanism of action was rather incomplete. That same year, however, Edelman and Porter separately and independently published their fundamental studies of the molecular structure of antibodies. Both of them had aimed at splitting the giant molecule into smaller, well defined fragments that might be more easily analysed than would the whole complex.

Porter's aim was to separate those parts of the antibody which are responsible for their specific reactivity. He hoped by this means to obtain a preparation lacking most of the biologic functions of the antibody but, on account of its capacity of combination, capable of competing with the antibody for the binding sites of the antigen. He succeeded in achieving this by means of treatment of the antibody, under strictly controlled conditions, with a protein-splitting enzyme called papain. By this treatment the antibody split into three parts. Two of these could combine specifically with the antigen and they were almost identical in other respects as well. The third fragment differed distinctly from the others, lacked binding capacity but possessed certain other biologic characteristics of the intact molecule.

Edelman for his part assumed the molecule, like those of many other proteins, to be composed of two or more separate chain structures held together by cross links of some kind, most probably so-called sulphide bonds. His assumption turned out to be correct. By means of a fairly rough treatment he was able to sever the cross bonds and release a number of separate chain molecules. Both he and Porter could later show that the antibody was in fact composed of four chains, one pair of identical, "light" chains and one pair of likewise identical, "heavy" chains.

On the basis of the collected evidence Porter built a model of the molecule which has later, with overwhelming probability, been proven correct.

Accordingly the antibody molecule appears in the shape of the letter Y, with a stem and two angled branches. Each branch is composed of one light and one half of a heavy chain in side by side arrangement. The stem is made up of the remaining halves of the heavy chains. The specific combining capacity is accounted for by the structure of the free tips of the branches and in like measure by the light and the heavy chain; separately they are inactive. Porter's papain treatment attacks the molecule exactly at the point of branching and splits off the branches from the stem.

These discoveries incited an intense activity in laboratories in the four corners of the world. Apparently there existed a latent need for immunochemical research that could not be satisfied until today's prize winners had opened the way and provided the means. During the two decades that have since past our knowledge about the processes of immunity has broadened and deepened to an extent that perhaps has not yet been fully appreciated, even by some specialists in closely related fields. Many novel and fascinating aspects on problems in the fields of molecular biology and genetics have grown out of the immunochemical studies. We have now a new and firmer grasp of the question of the role of immunity as defense against and as cause of disease. Our possibilities to make use of immune reactions for diagnostic and therapeutic purposes have improved. It is, thus, a very important pioneer contribution that has been rewarded with this year's prize in physiology or medicine.

Gerald Edelman, Rodney Porter,

By clarifying the principal chemical structure of immunoglobulins you achieved an extremely important break-through in the field of immunochemistry. You, so to speak, opened the sluice-gates and gave impetus to the flood of research that soon started gushing forth, irrigating previously arid land, making it fertile and producing rich harvests. By awarding you the prize in physiology or medicine the Karolinska Institute has recognized the great significance of your accomplishments for biology in general and medicine in particular. On behalf of the Institute I wish to express our admiration and extend to you our heart-felt felicitations.

Now I ask you to proceed to receive your prize from the hands of His Royal Highness the Crown Prince.

GERALD M. EDELMAN

Dr. Gerald M. Edelman was born on July 1, 1929 in New York City to Edward Edelman and Anna Freedman Edelman. His father is a practicing physician in New York.

After his education in New York public schools, Edelman attended Ursinus College in Pennsylvania and received the B.S. degree, magna cum laude, in 1950. He then attended the Medical School of the University of Pennsylvania where he received the M.D. degree in 1954. In the succeeding year, he was a Medical House Officer at the Massachusetts General Hospital. He became a Captain in the U.S. Army Medical Corps in 1955 and practiced general medicine at a Station Hospital connected with the American Hospital in Paris, France. In 1957, he joined the Rockefeller Institute as a graduate fellow in the laboratory of Dr. Henry G. Kunkel.

After receiving the Ph.D. degree in 1960, he remained at the Rockefeller Institute as Assistant Dean of Graduate Studies and started work in his own laboratory. In 1963, he became Associate Dean of Graduate Studies, a position from which he retired in 1966. From that time to the present, he has been a Professor of the Rockefeller University.

Edelman is a member of the National Academy of Sciences, the American Academy of Arts and Sciences, the American Society of Biological Chemists and the American Association of Immunologists, as well as a number of other scientific societies. He was a member of the Biophysics and Biophysical Chemistry Study Section of the National Institutes of Health from 1964 to 1967. Presently, he is an Associate of the Neurosciences Research Program at Massachusetts Institute of Technology, a member of the Board of Governors of the Weizmann Institute of Science and a member of the Advisory Board of the Basel Institute for Immunology.

He has given the Carter-Wallace Lectures at Princeton University in 1965, the National Institutes of Health Biophysics and Bioorganic Chemistry Lectureship at Cornell University in 1971, and delivered the Darwin Centennial Lectures at the Rockefeller University in 1971. In 1972, he was the first Felton Bequest Visiting Professor at the Walter and Eliza Hall Institute for Medical Research in Melbourne, Australia.

Edelman has received the Spencer Morris Award of the University of Pennsylvania in 1954, the Eli Lilly Award in Biological Chemistry given by the American Chemical Society in 1965, and the Annual Alumni Award of Ursinus College in 1969. In addition to his studies of antibody structure, his research interests have included the application of fluorescence spectroscopy and fluorescent probes to the study of proteins and the development of new methods of fractionation of both molecules and cells. His present research interests include work on the primary and three-dimensional structures of proteins, experiments on the structure and function of plant mitogens and studies of the cell surface.

In 1950, Edelman married Maxine M. Morrison. They have two sons, Eric and David and one daughter, Judith.

ANTIBODY STRUCTURE AND MOLECULAR IMMUNOLOGY

Nobel Lecture, December 12, 1972,

by

GERALD M. EDELMAN

The Rockefeller University, New York, N.Y., U.S.A.

Some sciences are exciting because of their generality and some because of their predictive power. Immunology is particularly exciting, however, because it provokes unusual ideas, some of which are not easily come upon through other fields of study. Indeed, many immunologists believe that for this reason, immunology will have a great impact on other branches of biology and medicine. On an occasion such as this in which a very great honor is being bestowed, I feel all the more privileged to be able to talk about some of the fundamental ideas in immunology and particularly about their relationship to the structure of antibodies.

Work on the structure of antibodies has allied immunology to molecular biology in much the same way as previous work on hapten antigens allied immunology to chemistry. This structural work can be considered the first of the projects of molecular immunology, the task of which is to interpret the properties of the immune system in terms of molecular structures. In this lecture, I should like to discuss some of the implications of the structural analysis of antibodies. Rather than review the subject, which has been amply done (1—4), I shall emphasize several ideas that have emerged from the structural approach. Within the context of these ideas, I shall then consider the related but less well explored subject of antibodies on the surfaces of lymphoid cells, and describe some recently developed experimental efforts of my colleagues and myself to understand the molecular mechanisms by which the binding of antigens induces clonal proliferation of these cells.

Antibodies occupy a central place in the science of immunology for an obvious reason: they are the protein molecules responsible for the recognition of foreign molecules or antigens. It is, therefore, perhaps not a very penetrating insight to suppose that a study of their structure would be valuable to an understanding of immunity. But what has emerged from that study has resulted in both surprises and conceptual reformulations.

These reformulations provided a molecular basis for the selective theories of immunity first expounded by Niels Jerne (5) and MacFarlane Burnet (6) and therefore helped to bring about a virtual revolution of immunological thought. The fundamental idea of these theories is now the central dogma of modern immunology: molecular recognition of antigens occurs by selection among clones of cells already committed to producing the appropriate antibodies, each of different specificity (Figure 1). The results of many studies by

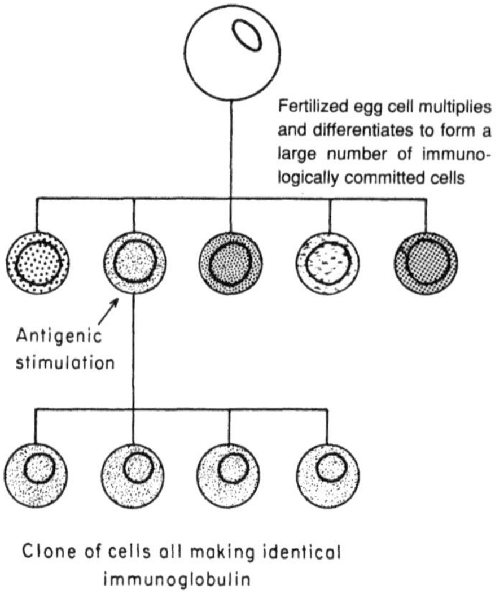

Fig. 1.
A diagram illustrating the basic features of the clonal selection theory. The stippling and shading indicate that different cells have antibody receptors of different specificities, although the specificity of all receptors on a given cell is the same. Stimulation by an antigen results in clonal expansion (maturation, mitosis and antibody production) of those cells having receptors complementary to the antigen.

cellular immunologists (see references 1 and 2) strongly suggest that each cell makes antibodies of only one kind, that stimulation of cell division and antibody synthesis occurs after interaction of an antigen with receptor antibodies at the cell surface, and that the specificity of these antibodies is the same as that of the antibodies produced by daughter cells. Several fundamental questions are raised by these conclusions and by the theory of clonal selection. How can a sufficient diversity of antibodies be synthesized by the lymphoid system? What is the mechanism by which the lymphocyte is stimulated after interaction with an antigen?

In the late 1950's, at the beginnings of the intensive work on antibody structure, these questions were not so well defined. The classic work of Landsteiner on hapten antigens (7) had provided strong evidence that immunological specificity resulted from molecular complementarity between the determinant groups of the antigen molecule and the antigen-combining site of the antibody molecule. In addition, there was good evidence that most antibodies were bivalent (8) as well as some indication that antibodies of different classes existed (9). The physico-chemical studies of Tiselius (10) had established that antibodies were proteins that were extraordinarily heterogeneous in charge. Moreover, a number of workers had shown the existence of heterogeneity in the binding constants of antibodies capable of binding a single hapten antigen (11). Despite the value of all of this information, however, little was known of the detailed chemical structure of antibodies or of what are now called the immunoglobulins.

The Multichain Structure of Antibodies: Problems of Size and Heterogeneity

If the need for a structural analysis of antibodies was great, so were the experimental difficulties: antibodies are very large proteins (mol. wt. 150,000 or greater) and they are extraordinarily heterogeneous. Two means were adopted around 1958 in an effort to avoid the first difficulty. Following the work of Petermann (12) and others, Rodney Porter (13) applied proteolytic enzymes, notably papain, to achieve a limited cleavage of the gamma globulin fraction of serum into fragments. He then successfully fractionated the digest, obtaining antigen binding (Fab) and crystallizable (Fc) fragments. Subsequently, other enzymes such as pepsin were used in a similar fashion by Nisonoff and his colleagues (14). I took another approach, in an attempt to cleave molecules of immunoglobulin G and immunoglobulin M into polypeptide chains by reduction of their disulfide bonds and exposure to dissociating solvents such as 6 M urea (15). This procedure resulted in a significant drop in molecular weight, demonstrating that the immunoglobulin G molecule was a multichain structure rather than a single chain as had been believed before. Moreover, corresponding chains obtained from both immunoglobulins had about the same size. The polypeptide chains (16) were of two kinds (now called light and heavy chains) but were obviously not the same as the fragments obtained by proteolytic cleavage and therefore the results of the two cleavage procedures complemented each other. Ultracentrifugal analyses indicated that one of the polypeptide chains had a molecular weight in the vicinity of 20,000, a reasonable size for determination of the amino acid sequence by the methods available in the early 1960s.

Nevertheless, the main obstruction to a direct analysis of antibody structure was the chemical heterogeneity of antibodies and their antigen binding fragments. Two challenging questions confronted those attempting chemical analyses of antibody molecules at that time. First, did the observed heterogeneity of antibodies reside only in the conformation of their polypeptide chains as was then widely assumed, or did this heterogeneity reflect differences in the primary structures of these chains, as required implicitly by the clonal selection theory? Second, if the heterogeneity did imply a large population of molecules with different primary structures, how could one obtain the homogeneous material needed for carrying out a detailed structural analysis?

These challenges were met simultaneously by taking advantage of an accident of nature rather than by direct physicochemical assault. It had been known for some time that tumors of lymphoid cells called myelomas produced homogeneous serum proteins that resembled the normal heterogeneous immunoglobulins. In 1961, M. D. Poulik and I showed that the homogeneity of these proteins was reflected in the starch gel electrophoretic patterns of their dissociated chains (16). Some patients with multiple myeloma excrete urinary proteins which are antigenically related to immunoglobulins but whose nature remained obscure since their first decsription by Henry Bence Jones in 1847. These Bence Jones proteins were most interesting, for they could be readily obtained from the urine in large quantities, were homogeneous, and had low molecular weights. It seemed reasonable to suggest (16) that Bence Jones proteins represented one of the chains of the immunoglobulin molecule that was synthesized by the myeloma tumor but not incorporated into the homogeneous myeloma protein and therefore excreted into the urine.

This hypothesis was corroborated one exciting afternoon when my student

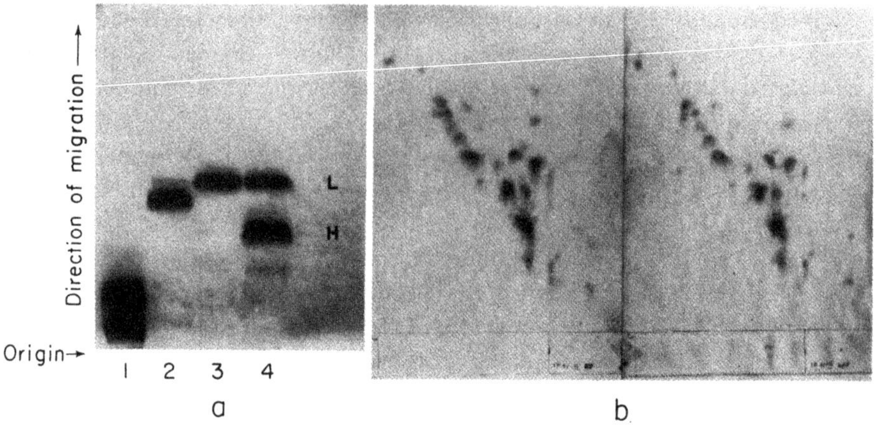

Fig. 2.
Comparisons of light chains isolated from serum IgG myeloma proteins with urinary Bence Jones proteins frcm the same patient. (a) Starch gel electrophoresis in urea. 1) serum myeloma globulin, 2) urinary Bence Jones protein, 3) Bence Jones protein reduced and alkylated, 4) myeloma protein reduced and alkylated. L — light chain; H — heavy chain. (b) Two-dimensional high voltage electrophoresis of tryptic hydrolysates. Pattern on left is of urinary Bence Jones protein; that on right is of light chain isolated from the serum myeloma protein of the same patient.

Joseph Gally and I (17) heated solutions of light chains isolated from our own serum immunoglobulins in the classical test for Bence Jones proteinuria. They behaved as Bence Jones proteins, the solution first becoming turbid, then clearing upon further heating. A comparison of light chains of myeloma proteins with Bence Jones proteins by starch gel electrophoresis in urea (17) and by peptide mapping (18) confirmed the hypothesis (Figure 2). Indeed, Berggård and I later found (19) that in normal urine there were counterparts to Bence Jones proteins that shared their properties but were chemically heterogeneous.

No physical means was known at the time that was capable of fractionating antibodies to yield homogeneous proteins. It was possible, however, to prepare specifically reactive antibodies by using the antigen to form antigen-antibody aggregates and then dissociating the complex with free hapten. Although we knew that these specifically prepared antibodies were still heterogeneous in their electrophoretic properties, it seemed possible that antibodies to different haptens might show differences in their polypeptide chains. Baruj Benacerraf had prepared a collection of these antibodies, and together with our colleagues (20) we decided to compare their chains, using the same methods that we had used for Bence Jones proteins. The results were striking: purified antibodies showed from 3 to 5 sharp bands in the Bence Jones or light chain region and antibodies of different specificities showed different patterns. In sharp contrast, normal immunoglobulin showed a diffuse zone extending over the entire range of mobilities of these bands. These experiments showed not only that antibodies of different specificities were structurally different but also that their heterogeneity was limited.

The results of these experiments on Bence Jones proteins and purified antibodies had a number of significant implications. Because different Bence Jones proteins had different amino acid compositions, it was clear that immunoglobulins must vary in their primary structures. This deduction, confirmed later by Koshland (21) for specifically purified antibodies, lent strong support to selective theories of antibody formation. Moreover, it opened the possibility of beginning a direct analysis of the primary structure of an immunoglobulin molecule, for not only were the Bence Jones proteins composed of homogeneous light chains, but their subunit molecular weight was only 23,000. The first report by Hilschmann and Craig (22) on partial sequences of several different Bence Jones proteins indicated that the structural heterogeneity of the light chains was confined to the amino terminal (variable) region, whereas the carboxyl terminal half of the chain (the constant region) was the same in all chains of the same type. This finding was soon extended by studies of other Bence Jones proteins (23).

Although some work had also been done on the heavy chains of immunoglobulins, there was much less information on their structure. For instance, it was suspected but not known that they also had variable regions resembling those of light chains. Comparisons of heavy chains and light chains even at this early stage did, however, clarify the nature of another source of antibody heterogeneity: the existence of immunoglobulin classes (24).

Antibodies within a particular class have similar molecular weights, carbohydrate content, amino acid compositions and physiological functions (Table 1) but still possess heterogeneity in their net charge and antigen binding affinities. Studies of classes in various animal species indicated that both the multichain structure and the heterogeneity are ubiquitous properties of immunoglobulins. The different classes apparently emerged during evolution (25) to carry out various physiologically important activities that have been named effector functions in order to distinguish them from the antigen-binding or recognition function. The various manifestations of humoral immune responses as well as their prophylactic, therapeutic and pathological consequences can now be generally explained in terms of the properties of the particular class of antibody mediating that response. As a result of comparing their chain structure, it became clear that although immunoglobulins of all classes contain similar kinds of light chains (Table 1), the distinctive class character (24) is conferred by structural differences in the heavy chains, specifically in their constant regions, as I shall discuss later.

With the clarification of the nature of the heterogeneity of immunoglobulin chains and classes, attention could be turned to the problem of relating the structure and evolution of antibodies within a given class to their antigen-binding and effector functions. We chose to concentrate on immunoglobulin G, for this was the most prevalent class in mammals and the work on chain structure suggested that it would be sufficiently representative.

Table 1. *Human Immunoglobulin Classes.*

Class	Physiological Properties	Heavy Chain[1]	Light Chains	Molecular[2] Formula	Molecular weight ($\times 10^{-3}$) and sedimentation constant	Carbohydrate Content
IgG	Complement fixation; placental transfer	γ	$\varkappa$ or λ	$(\gamma_2\varkappa_2)$ or $(\gamma_2\lambda_2)$	143 – 149; 6.7S	2.5 %
IgA	Localized protection in external secretions	α	$\varkappa$ or λ	$(\alpha_2\varkappa_2)$ or $(\alpha_2\lambda_2)$	158 – 162; 6.8 – 11.4S	5 – 10 %
IgM	Complement fixation; early immune response	μ	$\varkappa$ or λ	$(\mu_2\varkappa_2)$ or $(\mu_2\lambda_2)$	800 – 950; 19.0S	5 – 10 %
IgD	Unknown	δ	$\varkappa$ or λ	$(\delta_2\varkappa_2)$ or $(\delta_2\lambda_2)$	175 – 180; 6.6S	10 %
IgE	Reagin activity; mast cell fixation,	ε	$\varkappa$ or λ	$(\varepsilon_2\varkappa_2)$ or $(\varepsilon_2\lambda_2)$	185 – 190; 8.0S	12 %

(1) The class distinctive features of these chains are in their constant regions.
(2) IgA can have additional unrelated chains called SC and J. J chains are also found in IgM.

The Complete Covalent Structure and the Domain Hypothesis

An understanding of the chain structure and its relation to the proteolytic fragments (26, 27) made feasible an attempt to determine the complete structure of an immunoglobulin G molecule. My colleagues and I started this project in 1965, and before it was completed in 1969 (28) seven of us had spent a good portion of our waking hours on the technical details. One of our main objectives was to provide a complete and definitive reference structure against which partial structures of other immunoglobulins could be compared. In particular, we wished to compare the detailed structure of a heavy chain and a light chain from the same molecule.

Another objective was to examine in detail the regional differentiation of the structure that had been evolved to carry out different physiological functions in the immune response. The work of Porter (13) had shown that the so-called Fab fragment of immunoglobulin G was univalent and bound antigens whereas the Fc fragment did not. This provided an early hint that immunoglobulin molecules were organized into separate regions, each mediating different functions. In accord with selective theories of immunity, it was logical to suppose that V regions from both the light and the heavy chains mediated the antigen binding functions. Early evidence that some of the C regions were concerned with physiologically significant effector functions was obtained by showing that Fc fragments would bind components of the complement system (29), a complex group of proteins responsible for immunologically induced cell lysis. A more detailed assignment of structure to function required a knowledge of the total structure.

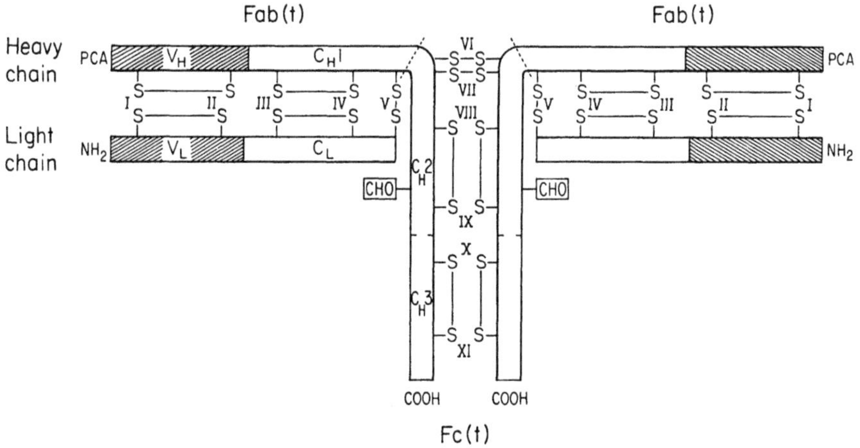

Fig. 3.
Overall arrangement of chains and disulfide bonds of the human γG_1 immunoglobulin *Eu*. Half-cystinyl residues are I–XI; I–V designates corresponding half-cystinyl residues in light and heavy chains. PCA, pyrrolidonecarboxylic acid; CHO, carbohydrate. Fab(t) and Fc(t) refer to fragments produced by trypsin, which cleaves the heavy chain as indicated by dashed lines above half-cystinyl residues VI. Variable regions, V_H and V_L, are homologous. The constant region of the heavy chain (C_H) is divided into three regions, C_H1, C_H2 and C_H3, that are homologous to each other and to the C region of the light chain. The variable regions carry out antigen-binding functions and the constant regions the effector functions of the molecule.

Amino acid sequence analysis of the Fc region of normal rabbit γ chains by Hill and his colleagues (30) demonstrated that the carboxyl terminal portion of heavy chains was homogeneous. On the basis of internal homologies in this region, Hill (30) and Singer and Doolittle (31) proposed the hypothesis that the genes for immunoglobulin chains evolved by duplication of a gene of sufficient size to specify a precursor protein of about 100 amino acids in length. Although direct confirmation of this hypothesis is obviously not possible, it was strongly supported by the results of our analysis (28) of the complete amino acid sequence and arrangement of the disulfide bonds of an entire IgG myeloma protein.

Comparisons of the amino acid sequences of the heavy chain of this protein with others studied in Porter's laboratory (32) and by Bruce Cunningham and his colleagues in our laboratory (33) showed that heavy chains had variable (V_H) regions, i.e., regions that differed from one another in the sequences of the 110—120 residues beginning with the amino terminus (Figure 3).

Examination of the amino acid sequences (Figures 4 and 5) allowed us to draw the following additional conclusions:

1) The variable (V) regions of light and heavy chains are homologous to each other, but they are not obviously homologous to the constant regions of these chains. V regions from the same molecule appear to be no more closely related than V regions from different molecules.

2) The constant (C) region of γ chains consists of three homology regions,

SEQUENCE HOMOLOGY IN EU VARIABLE REGIONS

										1									10
EU V_L (RESIDUES 1-108)										ASP	ILE	GLN	MET	THR	GLN	SER	PRO	SER	THR
EU V_H (RESIDUES 1-114)										PCA	VAL	GLN	LEU	VAL	GLN	SER	GLY	-	ALA
									20										30
LEU	SER	ALA	SER	VAL	GLY	ASP	ARG	VAL	THR	ILE	THR	CYS	ARG	ALA	SER	GLN	SER	ILE	ASN
GLU	VAL	LYS	LYS	PRO	GLY	SER	SER	VAL	LYS	VAL	SER	CYS	LYS	ALA	SER	GLY	GLY	THR	PHE
											40								
THR	-	-	TRP	LEU	ALA	TRP	TYR	GLN	GLN	LYS	PRO	GLY	LYS	ALA	PRO	LYS	LEU	LEU	MET
SER	ARG	SER	ALA	ILE	ILE	TRP	VAL	ARG	GLN	ALA	PRO	GLY	GLN	GLY	LEU	GLU	TRP	MET	GLY
	50											60							
TYR	LYS	ALA	SER	SER	-	LEU	GLU	SER	GLY	VAL	PRO	SER	ARG	PHE	ILE	GLY	SER	GLY	SER
GLY	ILE	VAL	PRO	MET	PHE	GLY	PRO	PRO	ASN	TYR	ALA	GLN	LYS	PHE	GLN	GLY	-	ARG	VAL
		70																	80
GLY	THR	GLU	PHE	THR	-	-	-	-	-	-	-	LEU	THR	ILE	SER	SER	LEU	GLN	PRO
THR	ILE	THR	ALA	ASP	GLU	SER	THR	ASN	THR	ALA	TYR	MET	GLU	LEU	SER	SER	LEU	ARG	SER
									90										
ASP	ASP	PHE	ALA	THR	TYR	TYR	CYS	GLN	GLN	-	TYR	ASN	SER	ASP	SER	LYS	MET	PHE	GLY
GLU	ASP	THR	ALA	PHE	TYR	PHE	CYS	ALA	GLY	GLY	TYR	GLY	ILE	TYR	SER	PRO	GLU	GLU	TYR
100																			
GLN	GLY	THR	LYS	VAL	GLU	VAL	LYS	GLY											
ASN	GLY	GLY	LEU	VAL	THR														

Fig. 4.
Comparison of the amino acid sequences of the V_H and V_L regions of protein Eu. Identical residues are shaded. Deletions indicated by dashes are introduced to maximize the homology.

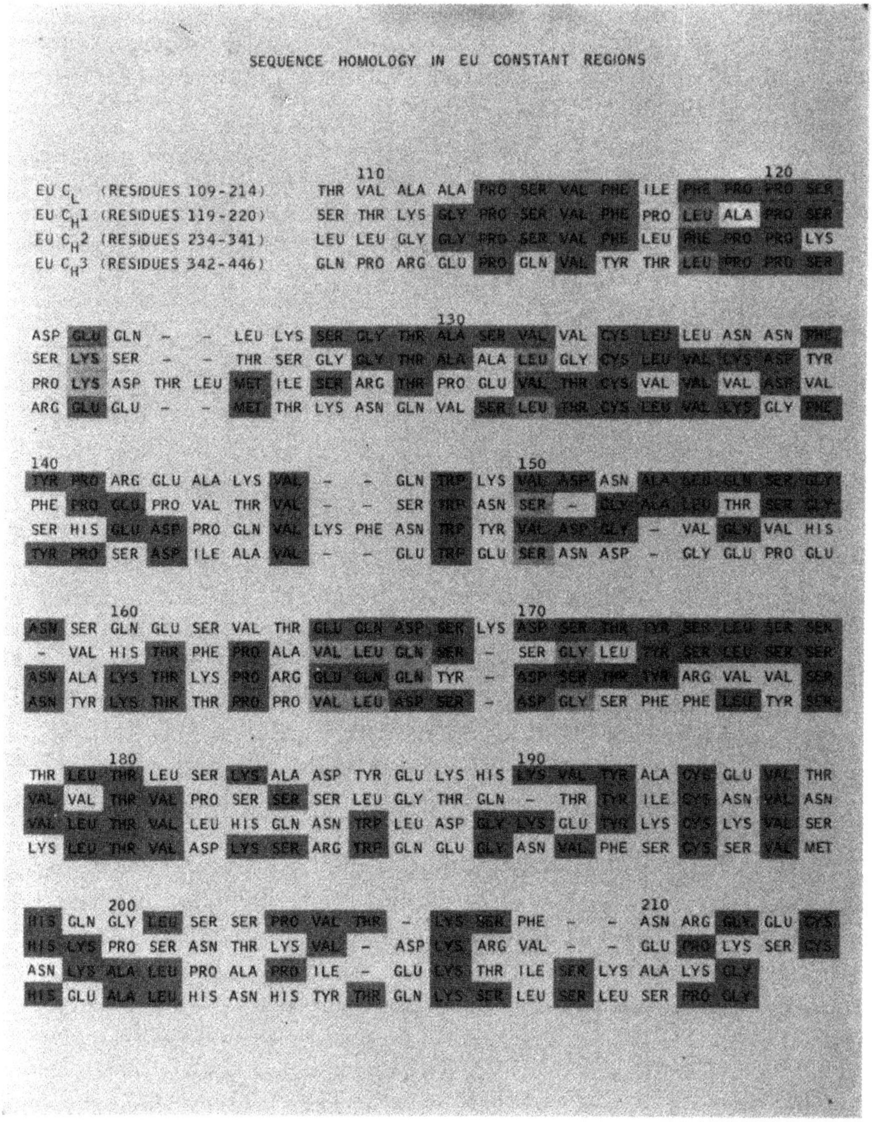

Fig. 5.
Comparison of the amino acid sequences of C_L, C_H1, C_L2 and C_L3 regions. Deletions, indicated by dashes, have been introduced to maximize homologies. Identical residues are darkly shaded; both light and dark shadings are used to indicate identities which occur in pairs in the same position.

C_H1, C_H2 and C_H3, each of which is closely homologous to the others and to the constant regions of the light chains.

3) Each variable region and each constant homology region contains one disulfide bond, with the result that the intrachain disulfide bonds are linearly and periodically distributed in the structure.

4) The region containing all of the interchain disulfide bonds is at the center of the linear sequence of the heavy chain and has no homologous counterpart in other portions of the heavy or light chains.

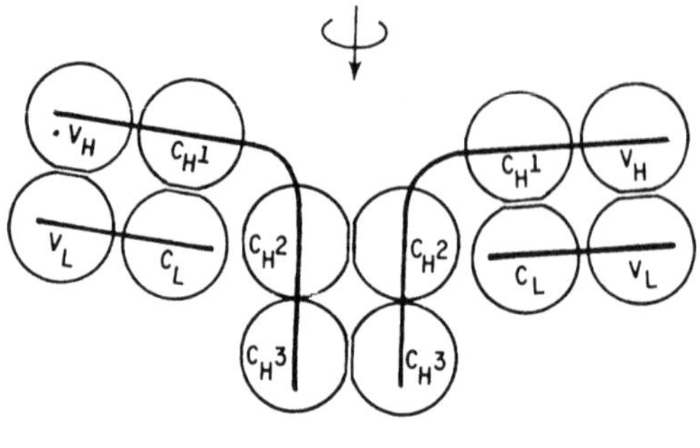

Fig. 6.
The domain hypothesis. Diagrammatic arrangement of domains in the free immunoglobulin G molecule. The arrow refers to a dyad axis of symmetry. Homology regions (see Figures 3, 4 and 5) which constitute each domain are indicated: V_L, V_H — domains made up of variable homology regions; C_L, C_H1, C_H2, and C_H3 — domains made up of constant homology regions. Within each of these groups, domains are assumed to have similar three-dimensional structures and each is assumed to contributed to an active site. The V domain sites contribute to antigen recognition functions and the C domain sites to effector functions.

These conclusions prompted us to suggest that the molecule is folded in a congeries of compact domains (28, 33) each formed by separate V homology regions or C homology regions (Figure 6). In such an arrangement, each domain is stabilized by a single intrachain disulfide bond and is linked to neighboring domains by less tightly folded stretches of the polypeptide chains. A twofold pseudosymmetry axis relates the V_LC_L to the V_HC_H1 domains and a true dyad axis through the disulfide bonds connecting the heavy chains relates the C_H2-C_H3 domains. The tertiary structure within each of the homologous domains is assumed to be quite similar. Moreover, each domain is assumed to contribute to at least one active site mediating a function of the immunoglobulin molecule.

This last supposition is nicely demonstrated by the interaction of V region domains. The reconstitution of active antibody molecules by recombining their isolated heavy and light chains (34, 35, 36) as well as affinity labelling experiments (31) confirmed our early hypothesis that the V regions of both heavy and light chains contributed to the antigen-combining sites. Moreover, the experiments of Haber (37) provided the first indication that Fab fragments of specific antibodies could be unfolded after reduction of their disulfide bonds and refolded in the absence of antigen to regain most of their antigen binding activity. This clearly indicated that the information for the combining site was contained entirely in the amino acid sequences of the chains. That this information is contained completely in the variable regions is strikingly shown by the recent isolation of antigen-binding fragments consisting only of V_L and V_H (38). The chain recombination experiments suggested an hypothesis to account in part for antibody diversity: the various combinations of different heavy and light chains expressed in different lymphocytes allow the formation of a large number of different antigen-combining sites from a relatively small number of V regions.

One of the remaining structural tasks of molecular immunology is to obtain a direct picture of antigen-binding sites by X-ray crystallography of V domains at atomic resolution. Although crystals of the appropriate molecule or fragment yielding diffraction patterns that extend beyond Bragg spacings of 3.0 Å have not yet been obtained, it is likely that continued searching will provide them. The details of a particular antigen-antibody interaction revealed by such a study will be of enormous interest. For example, certain sequence positions of V regions are hypervariable (39) and are very good candidates for direct contribution to the site. It will be particularly important to understand how the basic three-dimensional structure can accomodate so many amino acid substitutions. X-ray crystallographic work may also show in detail how the disulfide bonds in each of the V domains provide essential stability to the site (28, 33, 40).

The proposed similarities in tertiary structures among C domains have not been established nor have the functions of the various C domains been fully determined. There is a suggestion that C_H2 may play a role in complement fixation (41). A good candidate for binding to the lymphocyte cell membrane is C_H3, the function of which may be concerned with the mechanism of lymphocyte triggering following the binding of antigen by V domains. The C_H3 domain has already been shown to bind to macrophage membranes (42) and there is now some evidence that lymphocytes can synthesize isolated domains (43, 44, 45) similar to C_H3 as separate molecules.

Although many details are still lacking, the gross structural aspects of the domain hypothesis have received direct support from X-ray crystallographic analyses of Fab fragments (46) and whole molecules (47) in which separate domains were clearly discerned. Indirect support for the hypothesis has also come from experiments (38, 48) on proteolytic cleavage of regions between domains.

It is not completely obvious why the domain structure was so strictly preserved during evolution. One reasonable hypothesis is that although there was a functional need for association of V and C domains in the same molecule, there was also a need to prevent allosteric interactions among these domains. Whatever the selective advantages of this arrangement, it is clear that immunoglobulin evolution by gene duplication permitted the possibility of modular alteration of immunological function by addition or deletion of domains.

Translocons: Proposed Units of Evolution and Genetic Function

The evolution by gene duplication of both the domain structure and the immunoglobulin classes raises several questions about the number and arrangement of the structural genes specifying immunoglobulins. Although time does not permit me to discuss this complex subject in detail, I should like to suggest how structural work has sharpened these questions.

According to the theory of clonal selection, it is necessary that there preexist in each individual a large number of different antibodies with the capacity to bind different antigens. One of the most satisfying conclusions that emerged from structural analysis is that the diversity of the V regions of antibody chains is sufficient to satisfy this requirement. This diversity arises at three levels of structural or genetic organization, two of which are now reasonably well understood:

1) V regions from both heavy and light chains contribute to the antigen-

binding site and therefore the number of possible antibodies may be as great as the product of the number of different V_L and V_H regions.

2) Analyses of the amino acid sequences of V regions of light chains by Hood (49) and Milstein (50) and later of heavy chains from myeloma proteins (32, 33) indicated that V regions fall into subgroups of sequences which must be specified by separate genes or groups of genes. Within a subgroup, the amino acid replacements at a particular position are of a conservative type consistent with single base changes in codons of the structural genes. Variable regions of different subgroups differ much more from each other than do variable regions within a subgroup.

Although different V region subgroups are specified by a number of non-allelic genes (50), the analysis of genetic or allotypic markers suggests that C regions of a given immunoglobulin class are specified by no more than one or two genes. These allotypic markers, first described by Grubb (51) and Oudin (52) provide a means in addition to sequence analysis for understanding the genetic basis of immunoglobulin synthesis (4). V regions specified by a number of different genes can occur in chains each of which may have the same C region specified by a single gene. It therefore appears that each immunoglobulin chain is specified by two genes, a V gene and a C gene (4, 49, 50).

Work in a number of laboratories (reviewed in reference 4) has shown that the genetic markers on the two types of light chains are not linked to those of the heavy chains or to each other. These findings and the conclusion that there are separate V and C genes led Gally and me to suggest (4) that immunoglobulins are specified by three unlinked gene clusters (Figure 7). The clusters have been named translocons (4) to emphasize the fact that some mechanism must be provided to combine genetic information from V region loci with information from C region loci to make complete V—C structural genes. According to this hypothesis, the translocon is the basic unit of immuno-

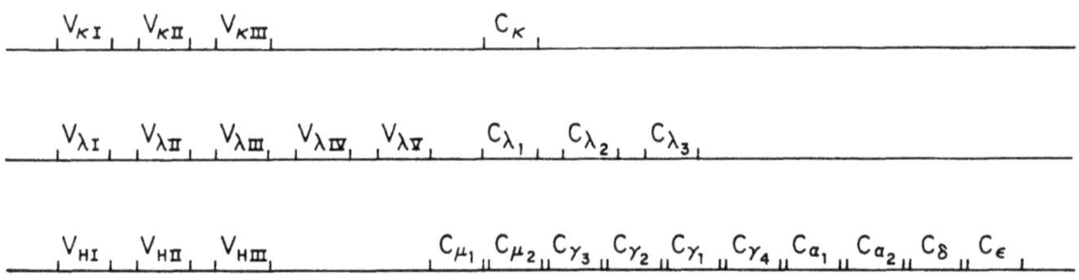

Fig. 7.
A diagrammatic representation of the proposed arrangement in mammalian germ cells of antibody genes in three unlinked clusters termed translocons. κ and λ chains are each specified by different translocons and heavy chains are specified by a third translocon. The exact number and arrangement of V and C genes within a translocon is not known. Each variable region subgroup (designed by a subscript corresponding to chain group and subgroup) must be coded by at least one separate germ line V gene. The number of V genes within each subgroup is unknown, however, as is the origin of intrasubgroup diversity of V regions. A special event is required to link the information from a particular V gene to that of a given C gene. The properties of the classes and subclasses (see Table 1) are conferred on the constant regions by C genes.

globulin evolution, different groups of immunoglobulin chains having arisen by duplication and various chromosomal rearrangements of a precursor gene cluster. Presumably, gene duplication during evolution also led to the appearance of V region subgroups within each translocon.

The key problem of the generation of immunoglobulin diversity has been converted by the work on chains and subgroups to the problem of the origin of sequence variations within each V region subgroup. It is still not known whether there is a germ line gene for each V region within a subgroup or whether each subgroup contains only a few genes (see Figure 7) and intrasubgroup variation arises by somatic genetic rearrangements of translocons within precursors of antibody forming cells. At this time, therefore, we can conclude that only the basis but not the origin of diversity has been adequately explained by the work on structure. Although structural analysis of various immunoglobulin classes will continue to be important, it does not in itself seem likely to lead to an explanation of the origin of antibody diversity. What will probably be required are imaginative experiments on DNA, RNA and their associated enzymes obtained from lymphoid cells at the proper stage of development.

In this abbreviated and necessarily incomplete account, I have attempted to show how structural work on immunoglobulins has provided a molecular basis for a number of central features of the theory of clonal selection. The work on humoral antibodies is just a beginning, however, for two great problems of molecular and cellular immunology remain to be solved. The first problem, the origin of intrasubgroup diversity, will undoubtedly receive great attention in the next few years. The second problem is concerned with the triggering of the clonal expansion of lymphocytes after combination of their receptor antibodies with antigens and the quantitative description of the population dynamics of the responding cells. An adequate solution to this problem must also account for the phenomenon of specific immune tolerance as described by the original work of Medawar and his associates (53).

For the remainder of this lecture, I shall turn my attention to some recent attempts that my colleagues and I have made to see whether these problems can be profitably studied using molecular approaches.

Lymphocyte Stimulation by Means of Lectins

The mechanisms of the cellular events underlying immune responses and immune tolerance remain a major challenge to theoretical and practical immunology (53, 60). How does a given antigen induce clonal proliferation or immune tolerance in certain subpopulations of cells?

Cells reactive to a given antigen constitute a very small portion of the lymphocyte population and are difficult to study directly. Two means have been used to circumvent this difficulty: the application of molecules that can stimulate lymphocytes independent of their antigen binding specificity, and fractionation of specific antigen binding lymphocytes for studies of stimulation by antigens of known structure. Although the problem of lymphocyte stimulation is far from being solved, both of these approaches are valuable particularly when used together.

Antigens are not the only means by which lymphocytes may be stimulated. It has been found that certain plant proteins called lectins can bind to glycoprotein receptors on the lymphocyte surface and induce blast transformation, mitosis and immunoglobulin production (see reference 54 for a review). Different lectins have different specificities for cell surface glycoproteins and different molecular structures although their mitogenic properties can be quite similar. In addition, they have a variety of effects on cell metabolism and transport. Such effects are independent of the antigen binding specificity of the cell and they may therefore be studied prior to specific cell fractionation.

The fact that antigens and lectins of different specificity and structure may stimulate lymphocytes suggests that the induction of mitosis is a property of membrane-associated structures that can respond to a variety of receptors. Triggering appears to be independent of the specificity of these receptors for their various ligands. To understand mitogenesis, it is therefore necessary to solve two problems. The first is to determine in molecular detail how the lectin binds to the cell surface and to compare it to the binding of antigens. The second is to determine how the binding induces metabolic changes necessary for the initiation of cell division. These changes are likely to include the production or release of a messenger which is a final common pathway for the stimulation of the cell by a particular lectin or antigen.

One of the important requirements for solving these problems is to know the complete structure of several different mitogenic lectins. This structural information is particularly useful in trying to understand the molecular transformation at the lymphocyte surface required for stimulation. With the knowledge of the three-dimensional structure of a lectin, various amino acid side chains at the surface of the molecule may be modified by group reagents which also may be used to change the valence of the molecule. The activities of the modified lectin derivatives may then be observed in various assays of their effects on cell surfaces and cell functions.

My colleagues and I (55) have recently determined both the amino acid sequence and three-dimensional structure of the lectin, concanavalin A (Con A) (Figure 8). This lectin has specificity for glucopyranosides, mannopyranosides and fructofuranosides and binds to glycoproteins and possibly glycolipids at a variety of cell surfaces. The purpose of our studies was to know the exact size and shape of the molecule, its valence and the structure and distribution of its binding sites.

With this knowledge in hand, we have been attempting to modify the structure and determine the effects of that modification on various biological activities of the lymphocyte. So far, there are several findings suggesting that such alterations of the structure have distinct effects. Con A in free solution stimulates thymus-derived lymphocytes (T cells) but not bone marrow-derived lymphocytes (B cells), leading to increased uptake of thymidine and blast transformation. The curve of stimulation of T cells by native Con A shows a rising limb representing stimulation and a falling limb (Figure 9)

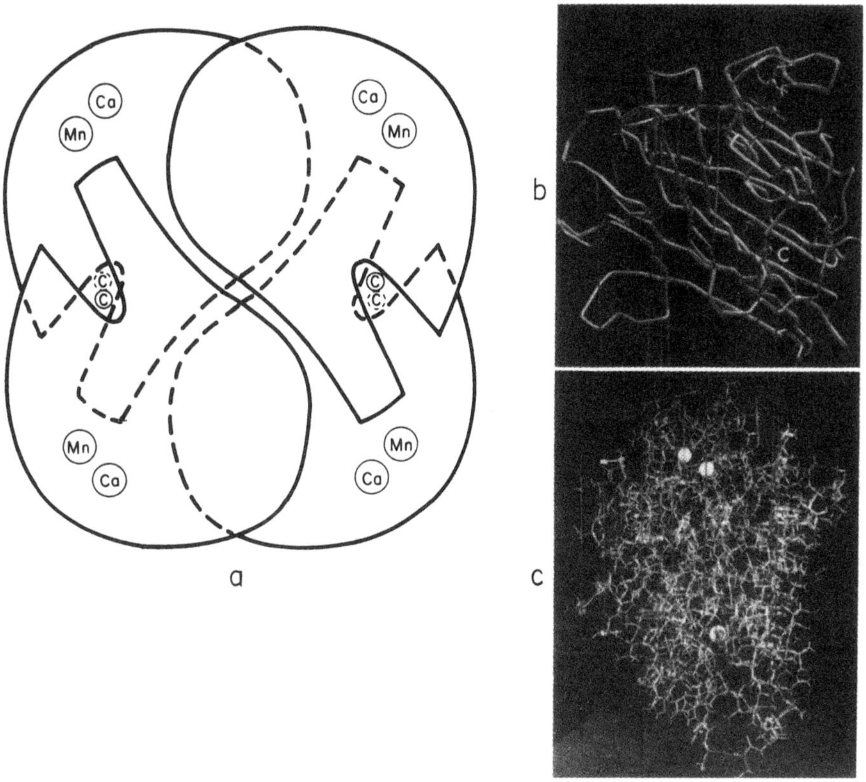

Fig. 8.
Three-dimensional structure of concanavalin A, a lectin mitogenic for lymphocytes. (a) Schematic representation of the tetrameric structure of Con A viewed down the z axis. The proposed binding sites for transition metals, calcium, and saccharides are indicated by Mn, Ca and C, respectively. The monomers on top (solid lines) are related by a twofold axis, as are those below. The two dimers are paired across an axis of D_2 symmetry to form the tetramer. (b) Wire model of the polypeptide backbone of the concanavalin A monomer oriented approximately to correspond to the monomer on the upper right of the diagram in (a). The two balls at the top represent the Ca and Mn atoms and the ball in the center is the position of an iodine atom in the sugar derivative, b-iodophenylglucoside, which is bound to the active site. Four such monomers are joined to form the tetramer as shown in (a). (c) A view of the Kendrew model of the Con A monomer rotated to show the deep pocket formed by the carbohydrate binding site. (White ball at the bottom of the figure is at the position of the iodine of b-iodophenylglucoside). The two white balls at the top represent the metal atoms.

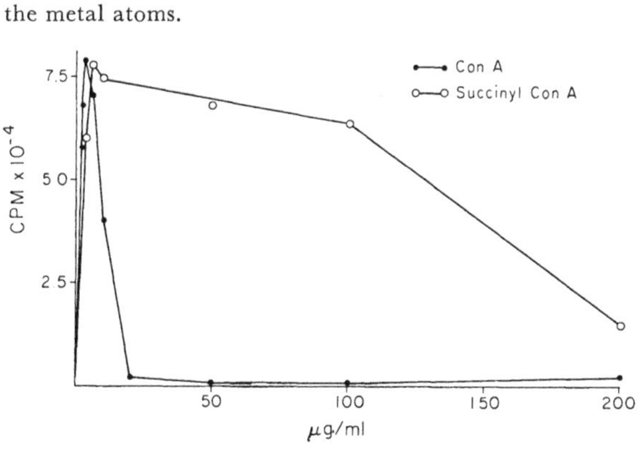

Fig. 9.
Stimulation of uptake of radioactive thymidine by mouse spleen cells after addition of concanavalin A and succinylated concanavalin A in increasing doses (μg/ml). CPM — counts per minute.

probably the result of cell death. The fact that the mitogenic effect and inhibition effect are dose dependent suggests an analogy to stimulation and tolerance induction by antigens. When Con A is succinylated, it dissociates from a tetramer to a dimer without alteration of its carbohydrate binding specificity. Although succinylated Con A is just as mitogenic as native Con A, the falling limb is not seen until much higher doses are reached.

Succinylation of Con A also alters another property of the lectin. It has been shown that, at certain concentrations, the binding of Con A to the cell surface restricts the movement of immunoglobulin receptors (56, 57). This suggests that it somehow changes the fluidity of the cell membrane resulting in reduction of the relative mobility of these receptors. In contrast, succinylated Con A has no such effect although it binds to lymphocytes to the same extent as the native molecule. Both the abolition of the killing effect in mitogenic assays and the failure to alter immunoglobulin receptor mobility in B cells after succinylation of Con A may be the result of change in valence or o alteration in the surface charge of the molecule. Examination of other derivatives and localization of the substituted side chains in the three-dimensional structure will help to establish which is the major factor. Recent experiments suggest that the valence is probably the major factor, for addition of divalent antibodies against Con A to cells that had bound succinylated Con A resulted again in restriction of immunoglobulin receptor mobility.

Con A may also be modified by cross-linking several molecules. A very striking effect is seen if the surface density of the Con A molecules presented to the lymphocyte is increased by cross-linking it at solid surfaces (58). Con A in free solution stimulates mouse T cells to an increased incorporation of radioactive thymidine but has no effect on B cells. When cross-linked at a solid surface, however, it stimulates mainly mouse B cells, although both T and B cells have approximately the same number of Con A receptors (58). Similar results have been obtained with other lectins (59). A reasonable interpretation of these phenomena (although not the only one) is that the lectin acts at the cell surface rather than inside the cell, that the presence of a high surface density of the mitogen is an important variable in exceeding the threshold for the lymphocyte stimulation, and that the threshold differs in the two kinds of lymphocytes.

Alteration of the structure and function of various lectins appears to be a promising means of analyzing the mechanism of lymphocyte stimulation. One intriguing hypothesis is that cross-linkage of the proper subsets of glycoprotein receptors by lectins is essentially equivalent in inducing cell transformation to cross-linkage of immunoglobulin receptors in the lymphocyte membrane by multivalent antigens. The central effector function of receptor antibodies, triggering of clonal proliferation, may turn out to be specifically related to the mode of anchorage of the antibody molecule to the cell membrane. The mode of attachment of antibody and lectin receptors to membrane-associated structures and their perturbation by crosslinkage at the cell surface may be similar and have similar effects despite the difference in their specificities and molecular structures.

Antibodies on the Surfaces of Antigen-Binding Cells

The most direct attack on the problem of lymphocyte stimulation is to explore the effects of antigens of known molecular geometry on specifically purified populations of lymphocytes. For this and other reasons, it is necessary to develop methods for the specific fractionation of antigenbinding cells.

In carrying out this task it is important both theoretically and operationally to discriminate between antigen-binding and antigen-reactive cells. In clonal selection, the phenotypic expression of the immunoglobulin genes is mediated in the animal by somatic division of precommitted cells (Figure 10). The pioneering work of Nossal and Mäkelä and later of Ada and Nossal (see reference 60) clearly showed that each cell makes antibodies of a single specificity and that there are different populations of specific antigen-binding cells. An animal is capable of responding specifically to an enormous number of antigens to which it is usually never exposed, and it therefore must contain genetic information for synthesizing a much larger number of different immunoglobulin molecules on cells than actually appear in detectable amounts in the bloodstream. In other words, the immunoglobulin molecules whose properties we can examine may represent only a minor fraction of those for which genetic information is available.

One may distinguish two levels of expression in the synthesis of immunoglobulins that I have termed for convenience the *primotype* and the *clonotype* (4). The primotype consists of the sum total of structurally different immunoglobulin molecules or receptor antibodies generated within an organism

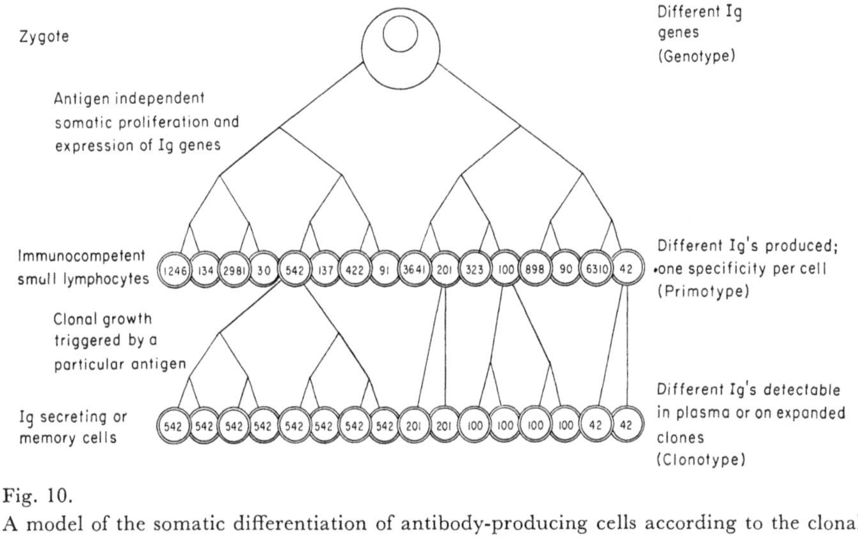

Fig. 10.
A model of the somatic differentiation of antibody-producing cells according to the clonal selection theory. The number of immunoglobulin genes may increase during somatic growth so that, in the immunologically mature animal, different lymphoid cells are formed each committed to the synthesis of a structurally distinct receptor antibody (indicated by an arabic number). A small proportion of these cells proliferate upon antigenic stimulation to form different clones of cells, each clone producing a different antibody. This model represents bone marrow-derived (B) cells but with minor modifications it is also applicable to thymus-derived (T) cells.

during its lifetime. The number of different molecules in the primotype is probably orders of magnitude greater than the number of different effective antigenic determinants to which the animal is ever exposed (Figure 10). The clonotype consists of those different immunoglobulin molecules synthesized as a result of antigenic stimulation and clonal expansion. These molecules can be detected and classified according to antigen-binding specificity, class, antigenic determinants, primary structure, allotype, or a variety of other experimentally measurable molecular properties. As a class, the clonotype is smaller than the primotype and is wholly contained within it (Figure 10).

Although a view of the clonotype is afforded by the analysis of humoral antibodies, we know very little about the primotype. It is therefore important

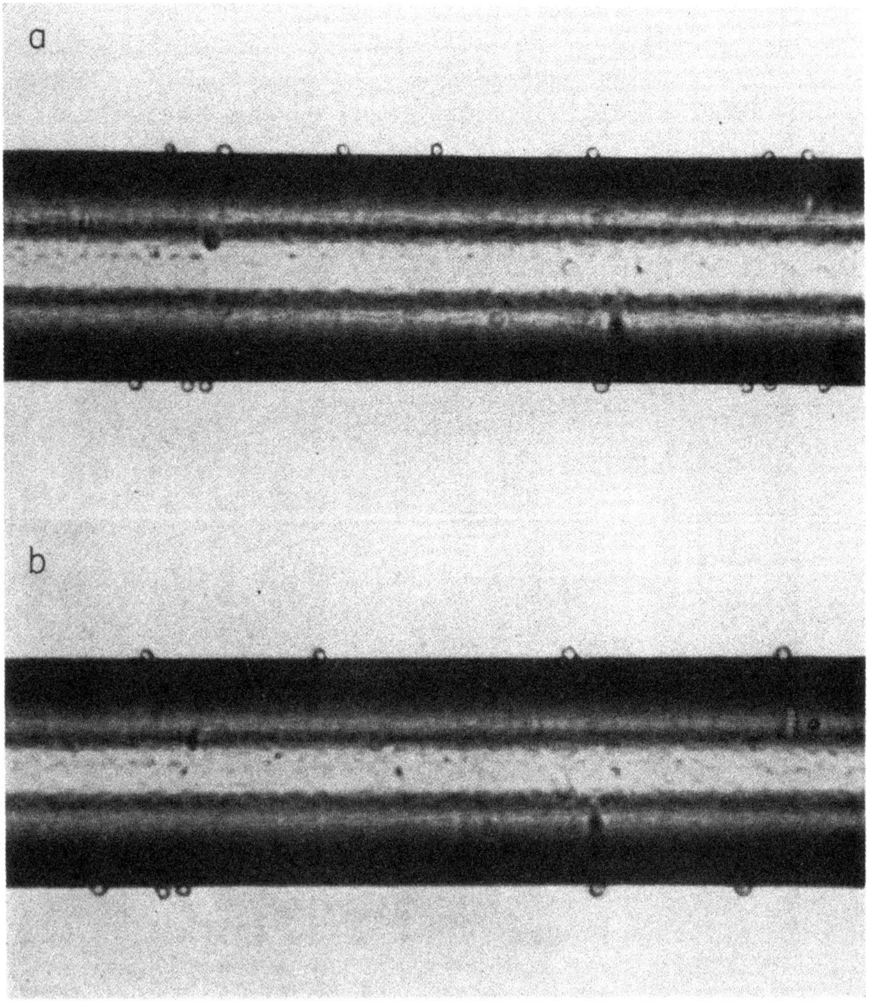

Fig. 11.
Lymphoid cells from mouse spleen bound by their antigen-specific receptors to a nylon fiber to which dinitrophenyl bovine serum albumin has been coupled. Treatment of bound cells in (a) with antiserum to the T cell surface antigen θ and with serum complement destroys the T cells leaving B cells still viable and attached (b). See Table 2. Magnification: ×235.

to attempt to fractionate the cells of the immune system according to the specificity of their antigen-binding receptors (61). We have been attempting to approach this problem of the specific fractionation of lymphocytes using nylon fibers to which antigens have been covalently coupled (62, 63). The derivatized fibers are strung tautly in a tissue culture dish so that cells in suspension may be shaken in such a way as to collide with them. Some of the cells colliding with the fibers are specifically bound to the covalently coupled antigens by means of their surface receptors. Bound cells may be counted microscopically *in situ* by focusing on the edge of the fiber (Figure 11). After washing away unbound cells, the specifically bound cells may be removed by plucking the fibers and shearing the cells quantitatively from their sites of attachment. The removed cells retain their viability provided that the tissue culture medium contains serum.

Derivatized nylon fibers have the ability to bind both thymus-derived lymphocytes (T cells) and bone marrow-derived (B cells) (64) according to the specificity of their receptors for a given antigen (65) (Figure 11, Table 2).

Table 2.
Characterization of mouse lymphoid cells fractionated according to their antigen-binding specificities. Nylon fibers were derivatized with hapten conjugates of bovine serum albumin and mice were immunized with each of the designated haptens coupled to hemocyanin. Inhibition of binding was achieved by addition of hapten-protein conjugates (250 μg/ml) or rabbit anti-mouse immunoglobulin (Ig) (250 μg/ml) to the cell suspension. High avidity cells are defined as those which are prevented from binding by concentrations of Dnp-bovine serum albumin of less than 4 μg/ml in the cell suspensions. Cells inhibited by higher concentrations are defined as low avidity cells. Virtually complete inhibition occurs at levels of homologous hapten greater than 100 μg/ml.

Antigen on Fiber	Dnp	Dnp	Tosyl	Tosyl
	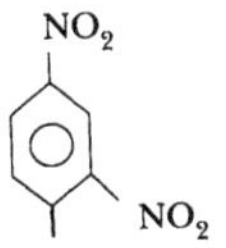		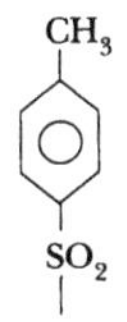	
Immunization	none	Dnp	none	Tosyl
Cells Bound to Fiber (per cm)	1200	4000	800	2000
% Inhibition of Binding by:				
Dnp	90	95	5	10
Tosyl	1	2	75	87
Anti-Ig	85	93	73	90
High Avidity Cells (per cm)	100	2800	—	—
Low Avidity Cells (per cm)	1200	1200	—	—
% T Cells	41	39	43	—
% B Cells	59	56	54	—

About 60 % of spleen cells specifically isolated are B cells and the remainder are T cells. By the use of appropriate antisera to cell surface receptors (Table 2), the cells of each type can be counted on the fibers and most of the cells of one type or the other may then be destroyed by the subsequent addition of serum complement. In this way, one can obtain populations of either T or B cells that are highly enriched in their capacity to bind a given antigen (Figure 11).

Cells of either kind may be further fractionated according to the relative affinity of their receptors. This can be accomplished by prior addition of a chosen concentration of free antigen, which serves to inhibit specific attachment of subpopulations of cells to the antigen-derivatized fibers by binding to their receptors. As defined by this technique, cells capable of binding specifically to a particular antigen constitute as much as 1 % of a mouse spleen cell population. Very few of these original antigen-binding cells appear to increase in number after immunization, however, and the cells that do respond are those having receptors of higher relative affinities (62) (Table 2).

Whether these populations correspond to the primotype and clonotype remains to be determined. It is significant, however, that fiber-binding cells do not include plaque forming (66) cells, and it is therefore possible to fractionate antigen-binding cells from cells that are already actively secreting antibodies. Recent experiments indicate that the antigen-binding cells isolated by this method may be transferred to irradiated animals to reconstitute a response to the antigen used to isolate them. This suggests that the antigen-specific population of cells removed from the fibers contains precursors of plaque-forming cells.

We have been rather encouraged by these findings, for the various methods of cell fractionation appear to have promise not only in determining the specificity and range of T and B cell receptors for antigens but also in analyzing the population dynamics of T and B cells in both adult and developing animals. Now that fractionated populations of lymphocytes specific for particular antigens are available, it should be possible to determine the connection between lectin-induced and antigen-induced changes by comparing responses to both agents on the same cells.

Although many experiments remain to be done in this area of the molecular immunology of the cell surface, continued analysis of the mitogenic mechanism should undoubtedly clarify the problems of immune induction and tolerance. The results obtained using lymphocytes may also have general significance, however, and bear upon the nature of cell division in normal and tumor cells as well as upon growth control and cell-cell interactions in developmental biology. Immunology can be expected to play a double role in these area of study, for it will be a tool as well as a model system of central importance.

Conclusion

Immunology has been and is a curiously reflexive science, generating its own tools for understanding, such as antibodies to antibody molecules themselves. While this approach is a powerful one, a fundamental understanding of immunological problems requires chemical analysis. The determination of the molecular structure of antibodies is a persuasive example and its virtual completion has allied immunology to molecular biology in a very satisfying way:

1) The heterogeneity of antibodies and complexity of immunoglobulin classes have been rationalized in a fashion consistent with selective theories of immunity.

2) The structural basis for differentiation of the biological activity of antibodies into antigen-binding and effector functions has been made clear.

3) The detailed analysis of antibody primary structure has provided a basis for studying the molecular genetics of the immune response, particularly the origin of diversity and the commitment of each cell to the synthesis of one kind of antibody.

4) A general framework has been provided for studying antibodies at the cell surface, opening several molecular approaches for analyzing stimulation and cell triggering.

5) Finally, it is perhaps not too extravagant to suggest that the extensions of the ideas and methods of molecular immunology to fields such as developmental biology has been facilitated. In this sense, immunology provides an essential tool as well as a model with distinct advantages: dissociable cells with unique gene products of known structure; the capacity to induce specific cloned cell lines for *in vitro* analysis; the means to fractionate cells according to their state of differentiation and binding specificity, allowing quantitative studies of their selection, interaction and population dynamics.

Whether or not the immune response turns out to be a uniquely useful model, we can expect that continued work by molecular and cellular immunologists will solve the major problems of the origin of diversity and the induction of antibody synthesis and tolerance. In view of the intimate connection of these problems with problems of gene expression and cellular regulation, their solution should bring valuable insights to other important areas of eukaryotic biology and again transform immunology both as a discipline and as an increasingly important branch of medicine.

ACKNOWLEDGEMENTS

By its very nature, science is a communal enterprise. I am deeply aware of the essential contributions to this work made by my many colleagues and friends throughout the last fifteen years. This occasion recalls the daily life we have shared with warmth and affection as well as the personal debt of gratitude that I owe them. I am equally cognizant of the fact that the knowledge of antibody structure was developed by many laboratories and researchers throughout the world. Not all of this work has been cited, for specific recognition here runs the risk of an unintentional omission; reference may be made to the reviews cited in the bibliography.

In addition to the fundamental support of the Rockefeller University, the work of my colleagues and myself was supported by grants from the National Institutes of Health and the National Science Foundation.

References

1. Cold Spring Harbor Symposia on Quantitative Biology, "Antibodies," *32*, 1967.
2. Nobel Symposium, 3, Gamma Globulins, Structure and Control of Biosynthesis (Killander, J. editor), Almqvist and Wiksell, Stockholm, 1967.
3. Edelman, G. M. and Gall, W. E., Ann. Rev. Biochem., *38*, 415, 1969.
4. Gally, J. A. and Edelman, G. M., Ann. Rev. Genet., *6*, 1, 1972.
5. Jerne, N. K., Proc. Natl. Acad. Sci. U.S., *41*, 849, 1955.
6. Burnet, F. M., The Clonal Selection Theory of Acquired Immunity, Vanderbilt University Press, Nashville, Tennessee, 1959.
7. Landsteiner, K., The Specificity of Serological Reactions, 2nd ed., Harvard University Press, Cambridge, Massachusetts, 1945.
8. Marrack, J. R., The Chemistry of Antigens and Antibodies, 2nd ed., (Medical Research Council Special Report Series, No. 230), London, His Majesty's Stationery Office, 1938.
9. Pedersen, K. O., Ultracentrifugal Studies on Serum and Serum Fractions, Uppsala, Almqvist and Wiksell, 1945.
10. Tiselius, A., Biochem. J., *31*, 313; 1464, 1937.
11. Karush, F., Advan. Immunol., *2*, 1, 1962.
12. Petermann, M. L., J. Biol. Chem., *144*, 607, 1942.
13. Porter, R. R., Biochem., *73*, 119, 1959.
14. Nisonoff, A., Wissler, F. C., Lipman, L. N. and Woernley, D. L. Arch. Biochem. Biophys., 89, 230, 1960.
15. Edelman, G. M., J. Am. Chem. Soc., *81*, 3155, 1959.
16. Edelman, G. M. and Poulik, M. D., J. Exp. Med., *113*, 861, 1961.
17. Edelman, G. M. and Gally, J. A., J. Exp, Med., *116*, 207, 1962.
18. Schwartz, J. and Edelman, G. M., J. Exp. Med., *118*, 41, 1963.
19. Berggård, I. and Edelman, G. M. Proc. Natl. Acad. Sci. U.S., *49*, 330, 1963.
20. Edelman, G. M., Benacerraf, B., Ovary, Z., and Poulik, M. D., Proc. Natl. Acad. Sci. U.S., *47*, 1751, 1961.
21. Koshland, M. E. and Englberger, F. M., Proc. Natl. Acad. Sci. U.S., *50*, 61, 1963.
22. Hilschmann, N. and Craig, L. C., Proc. Natl. Acad. Sci. U.S., *53*, 1403, 1965.
23. Titani, K., Whitley, E., Jr., Avogardo, L. and Putnam, F. W., Science, *149*, 1090, 1965.
24. Bull. World Health Org., *30*, 447, 1964.
25. Marchalonis, J. and Edelman, G. M., J. Exp. Med., *122*, 601, 1965; *124*, 901, 1966.
26. Fleishman, J. B., Pain, R. H., and Porter, R. R., Arch. Biochem. Biophys., Supplement 1, 174, 1962; Fleishman, J. B., Porter, R. R. and Press, E. M., Biochem. J., *88*, 220, 1963.
27. Fougereau, M. and Edelman, G. M., J. Exp. Med., *121*, 373, 1965.
28. Edelman, G. M., Gall, W. E., Waxdal, M. J. and Konigsberg, W. H., Biochemistry, *7*, 1950, 1968;
   Waxdal, M. J., Konigsberg, W. H., Henley, W. L. and Edelman, G. M., Biochemistry, *7*, 1959, 1968;
   Gall, W. E., Cunningham, B. A., Waxdal, M. J., Konigsberg, W. H. and Edelman, G. M., Biochemistry, *7*, 1973, 1968;
   Cunningham, B. A., Gottlieb, P. D., Konigsberg, W. H. and Edelman, G. M., Biochemistry, *7*, 1983, 1968;
   Edelman, G. M., Cunningham, B. A., Gall, W. E., Gottlieb, P. D., Rutishauser, U. and Waxdal, M. J., Proc. Natl. Acad. Sci. U.S., *63*, 78, 1969.
   Gottlieb, P. D., Cunningham, B. A., Rutishauser, U. and Edelman, G. M., Biochemistry, *9*, 3155, 1970;
   Cunningham, B. A., Rutishauser, U., Gall, W. E., Gottlieb, P. D., Waxdal, M. J. and Edelman, G. M., Biochemistry, *9*, 3161, 1970;
   Rutishauser, U., Cunningham, B. A., Bennett, C., Konigsberg, W. H. and Edelman, G. M., Biochemistry, *9*, 3171, 1970;

Bennett, C., Konigsberg, W. H., and Edelman, G. M., Biochemistry, *9*, 3181, 1970; Gall, W. E., and Edelman, G. M., Biochemistry, *9*, 3188, 1970; Edelman, G. M., Biochemistry, *9*, 3197, 1970.
29. Amiraian, K. and Leikhim, E. J., Proc. Soc. Exptl. Biol. Medl, *108*, 454, 1961; Taranta, A. and Franklin, E. C., Science, *134*, 1981, 1961.
30. Hill, R. L., Delaney, R., Fellows, R. R., Jr., and Lebovitz, H. E., Proc. Natl. Acad. Sci. U.S., *56*, 1762, 1966.
31. Singer, S. J. and Doolittle, R. E., Science, *153*, 13, 1966.
32. Press, E. M. and Hogg, N. M., Nature, *223*, 807. 1969.
33. Cunningham, B. A., Gottlieb, P. D., Pflumm, M. N. and Edelman, G. N. in *Progress in Immunology*, (B. Amos, ed.), Academic Press, Inc., N.Y., pp. 3–24, 1971; Cunningham, B. A., Pflumm, M. N., Rutishauser, U., and Edelman, G. M., Proc. Natl. Acad. Sci. U.S., *64*, 997, 1969.
34. Franěk, F. and Nezlin, R. S., Biokhimia, *28*, 193, 1963.
35. Edelman, G. M., Olins, D. E., Gally, J. A. and Zinder, N. D., Proc. Natl. Acad. Sci. U.S., *50*, 753, 1963.
36. Olins, D. E. and Edelman, G. M., J. Exp. Med., *119*, 799, 1964.
37. Haber, E., Proc. Natl. Acad. Sci. U.S., *52*, 1099, 1964.
38. Inbar, D., Hachman, J. and Givol, D. Proc. Natl. Acad. Sci. U.S., *69*, 2659, 1972.
39. Wu, T. T. and Kabat, E. A., J. Exp. Med., *132*, 211, 1970.
40. Edelman, G. M. Ann. N.Y. Acad. Sci., *190*, 5, 1971.
41. Kehoe, J. M. and Fougereau, M., Nature, *224*, 1212, 1970.
42. Yasmeen, D., Ellerson, J. R., Dorrington, K. J., and Pointer, R. H., J. Immunol., in press.
43. Berggård, I. and Bearn, A. G., J. Biol. Chem., *243*, 4095, 1968.
44. Smithies, O. and Poulik, M. D., Science, *175*, 187, 1972.
45. Peterson, P. A., Cunningham, B. A., Berggård, I. and Edelman, G. M. Proc. Natl. Acad. Sci. U.S., *69*, 1967, 1972.
46. Poljak, R. J., Amzel, L. M., Avey, H. P., Becka, L. N. and Nisonoff, A., Nature New Biol., *235*, 137, 1972.
47. Davies, D. R., Sarma, V. R., Labaw, L. W., Silverton, E. W. and Terry, W. D., in *Progress in Immunology*, (B. Amos, ed.), Academic Press, N.Y., pp. 25–32, 1971.
48. Gall, W. E. and D'Eustachio, P. G., Biochemistry. *11*, 4621, 1972.
49. Hood, L., Gray, W. R., Sanders, B. G. and Dreyer, W. J., Cold Spring Harbor Symp. Quant. Biol., *32*, 133, 1967.
50. Milstein, C., Nature, *216*, 330, 1967.
51. Grubb, R., *The Genetic Markers of Human Immunoglobulins*, Springer-Verlag Berlin-Heidelberg-New York, 1970; Acta Path. Microbiol. Scand., *39*, 195, 1956.
52. Oudin, J., Compt. Rend. Acad. Sci., *242*, 2489; 2606, 1956; J. Exp. Med., *112*, 125, 1960.
53. Medawar, P. B. in *Les Prix Nobel en 1960*, Imprimerie Royal, P. A. Norstedt and Söner, Stockholm.
54. Sharon, N. and Lis, H., Science, *177*, 949, 1972.
55. Edelman, G. M., Cunningham, B. A., Reeke, G. N., Jr., Becker, J. W., Waxdal, M. J., and Wang, J. L., Proc. Natl. Acad. Sci. U.S., *69*, 2580, 1972.
56. Yahara, I. and Edelman, G. M. Proc. Natl. Acad. Sci. U.S., *69*, 608, 1972.
57. Taylor, R. B., Duffus, P. H., Raff, M. C. and DePetris, S., Nature New Biol., *233*, 225, 1971; Loor, R., Forni, L. and Pernis, B., Eur. J. Immunol., *2*, 203, 1972.
58. Andersson, J., Edelman, G. M., Möller, G. and Sjöberg, O., Eur. J. Immunol., *2*, 233, 1972.
59. Greaves, M. F. and Bauminger, S., Nature New Biol., *235*, 67, 1972.
60. Nossal, G. J. V. and Ada, G. L., *Antigens, Lymphoid Cells and the Immune Response*, Academic Press, N.Y., 1971.
61. Wigzell, H. and Andersson, B., J. Exp. Med., *129*, 23, 1969.
62. Edelman, G. M., Rutishauser, U. and Millette, C. F. Proc. Natl. Acad. Sci. U.S., *68*, 2153, 1971.

63. Rutishauser, U., Millette, C. F. and Edelman, G. M. Proc. Natl. Acad. Sci. U.S., *69*, 1596, 1972.
64. Gowans, J. L., Humphrey, J. H. and Mitchison, N.A., A discussion on cooperation between lymphocytes in the immune response. Proc. Roy. Soc. London B, *176*, No. 1045, pp. 369–481, 1971.
65. Rutishauser, U. and Edelman, G. M. Proc. Natl. Acad. Sci. U.S., *69*, 3774, 1972.
66. Jerne, N. K., Nordin, A. A. and Henry, C. in "Cell Bound Antibodies," (B. Amos and H. Koprowski, eds.), Wistar Institute Press, pp. 109–125, 1963.

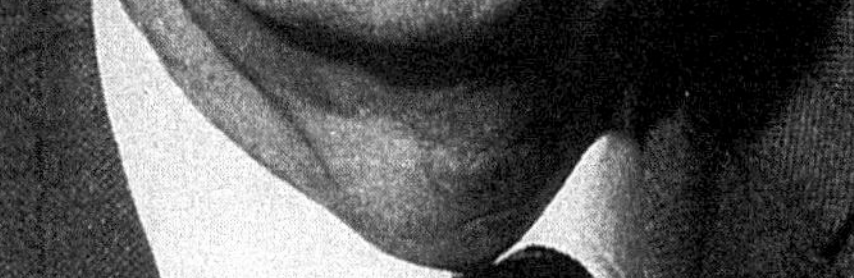

RODNEY R. PORTER[†]

Rodney Robert Porter born 8 October 1917 at Newton-le-Willows, Lancashire, England.

He was educated at the Ashton-in-Makerfield Grammar School taking his Hons.B.Sc. (Biochemistry) in 1939 at the University of Liverpool and his Ph.D. at the University of Cambridge in 1948.

After one year's postdoctoral work at Cambridge, Professor Porter joined the scientific staff of the National Institute of Medical Research in 1949 and was there until 1960 when he joined St. Mary's Hospital Medical School, London University as the first Pfizer Professor of Immunology.

In 1967, he was appointed Whitley Professor of Biochemistry in the University of Oxford and Fellow of Trinity College, Oxford.

Amongst his awards are those of:

Fellow of the Royal Society, 1964

Gairdner Foundation Award of Merit, 1966

Ciba Medal (Biochemical Society), 1967

Karl Landsteiner Memorial Award from the American Association of Blood Banks, 1968

National Academy of Science, U.S.A., Foreign Member 1972

He took his Ph.D. at Cambridge under the supervision of Dr. F. Sanger investigating protein chemistry. In 1948 Professor Porter started to investigate the structure of antibodies, but on moving to Mill Hill he worked on methods of protein fractionation collaborating with Dr. A. J. P. Martin. The particular interest was in chromatographic methods of fractionation.

He returned to the study of the chemical structure of antibodies leading to the finding of the three fragments produced by splitting with papain in 1958—59. He continued this work at St. Mary's Hospital Medical School and put forward the peptide chain structure of antibodies in 1962.

Since moving to Oxford he has been concerned with the structure of antibody combining site, of the genetic markers of immunoglobulins and recently in the chemical structure of some of the early complement components.

During the war years 1940—46 Professor Porter was in the army serving with the R.A., R.E., and R.A.S.C., finishing with the rank of Major. He was with the First Army in 1942 in the invasion of Algeria and with the 8th Army during the invasion of Silicy and then Italy. He remained with the Central Mediterranean Forces in Italy, Austria, Greece and Crete until January 1946.

[†] Dr Porter died in 1985.

STRUCTURAL STUDIES OF IMMUNOGLOBULINS

Nobel Lecture, December 12, 1972

by

R. R. PORTER

Department of Biochemistry, University of Oxford, England

In 1946, when I was starting work as a research student under the supervision of Dr. F. Sanger, the second edition of Karl Lansteiner's book 'The Specificity of Seriological Reactions' (1) reached England. In it was summarised the considerable body of information available on the range of antibody specificity and much of it was Landsteiner's own work or by others using his basic technique of preparing antibodies against haptenes and testing their ability to inhibit the precipitation of the antisera and the conjugated protein. Also described in this book was the work in Uppsala of Tiselius and Pederson in collaboration with Heidelberger and Kabat in which they showed that all rabbit antibodies were in the γ globulin fraction of serum proteins and that they had a molecular weight of 150,000. This combination of an apparently infinite range of antibody combining specificity associated with what appeared to be a nearly homogenous group of proteins astonished me and indeed still does.

ACTIVE FRAGMENTS OF ANTIBODIES

The preparation of antibodies by dissociation of specific precipitates with strong salt solutions or in acidic conditions had been described, as had the preparation in fair yield of γ globulin fractions by salting out techniques from whole serum, so an experimental approach to the structural basis of antibody combining specificity was possible. A start had indeed been made by showing that the whole molecule was not required for the combining specificity. Parventjev (2) had introduced pepsin treatment of serum as a commercial method of purification of horse antitoxins and Petermann and Pappenheimer (3) studied the reaction but using purified horse antidiphtheria toxin rather than whole serum for the peptic digestion. They showed that a product able to flocculate with the toxoid or neutralise toxin could be obtained and that it had a molecular weight of 113,000, i.e. substantially less than that of the original molecule. Petermann (4) showed later that human γ globulin could be split by papain to give what she estimated, by using the ultracentrifuge, to be quarter molecules. No antibody activities were, however, investigated in this study.

At about the same time Landsteiner (5) was extending his investigation of the antigenic specificity of protein antigens and had found that crude but apparently low molecular weight peptides from an acid digest of silk fibroin could inhibit the precipitation of soluble fibroin with its rabbit antiserum. This finding in conjunction with the haptene and other studies suggested that antigenic sites and presumably, therefore, antibody combining sites were small, certainly very much smaller than the antibody molecules and further attempts to obtain fragments of an antibody molecule, which retained the power to combine with the antigen seemed worthwhile. Testing for such active fragments was by their ability to inhibit the combination of the antigen and whole antibody. However, although a variety of conditions of hydrolysis by acids or enzymes were investigated (6) only papain gave an active product and this appeared to be from N terminal amino acid assay, the quarter molecules previously described by Petermann (4). There was no doubt that the combining site was in these smaller fragments and hence a substantial reduction in the magnitude of the structural problem had been achieved, but protein molecules of molecular weight 40,000 were still a formidable prospect. This work was carried out in Sanger's laborabory in Cambridge and with his guidance N terminal amino acid assay and the terminal sequences were attempted but proved unhelpful in that they suggested that the rabbit γ globulins and antibodies had a single open polypeptide chain and that the biologically active quarter molecules also had the same alanine N terminal

Fig. 1.
Crystals formed from a papain digest of rabbit IgG, the Fc fragment.

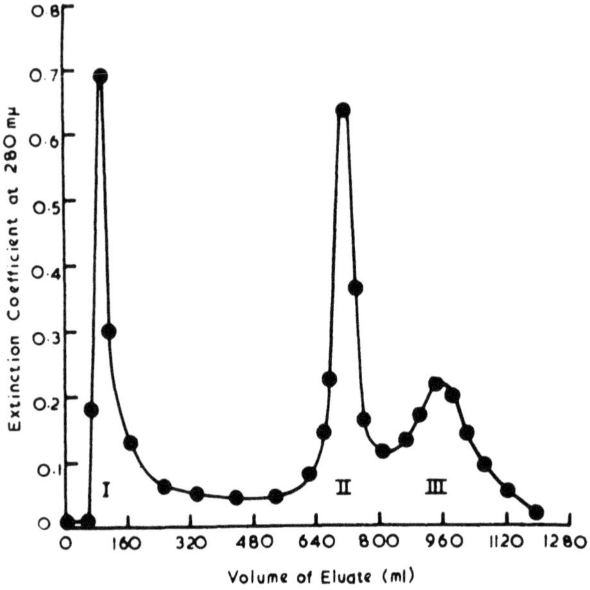

Fig. 2.
Fractionation of a papain digest of rabbit IgG on carboxylmethyl cellulose in sodium acetate buffer pH 5.5 with a gradient from 0.01M to 0.9M. Fractions I and II (Fab) carry the antibody combining sites and Fraction III (Fc) will crystallise easily.

acid. The possibility that there might be blocked N terminal residues was not considered.

A return to papain digestion of rabbit γ globulin was made seven years later, but in place of the crude enzyme preparation used earlier, a crystalline enzyme (7) was used in much lower concentrations, at 1/100 the weight of substrate (8). Under these conditions a number of points missed previously became apparent. First, there was a very high recovery of total protein after dialysis of the digest—very few small peptides were formed in spite of the wide specificity of the enzyme. The products, all of very similar size (sedimentation value 3.5S) were one third of the original rather than one quarter and most surprising one of the products of digestion crystallised very easily in diamond shaped plates (fig. 1) during the dialysis at neutrality in the cold room. This last observation suggesting that a protein which itself would never crystallise could give a fragment which presumably was more homogenous and hence able to crystallise was quite unexpected and indeed was unacceptable. The crystals were dismissed as coming from the less soluble amino acids and discarded without further consideration over several months. Fortunately, my neighbour in the adjoining laboratory at the National Institute of Medical Research in London was the X-ray crystallographer, Dr. Olga Kennard, and when I eventually asked her opinion she immediately gave the view that they were protein crystals. They were then identified with the third peak obtained by fractionation of the digest products on CM cellulose (fig. 2). They were named Fraction III and are now known as the Fc fragment. Fractions I and II were the components of the digest which retained the combining specificity of the orginal antibody. Fraction III had no such activity but did carry most of the antigenic specificity of the rabbit γ globulins when tested with antiserum from goats, rats and guinea pigs. More detailed studies (9) showed that fraction I and II were very similar, chemically and antigenically, and later Nisonoff and colleagues (10) showed that the distinction between these two fractions was artificial. If a basic fraction of γ globulin was treated with papain two molecules of II and one of III were obtained and an acidic fraction of γ globulin gave two molecules of I and one of III. The slight differences in charge between fractions I and II reflected the charge heterogeneity of the starting material.

Nisonoff *et al.* (11) had also returned to the peptic digestion of γ globulins but using the rabbit protein rather than horse antitoxin and showed that on reduction of the 100,000 molecular weight product comparable to that reported earlier would give half molecules very similar to the fractions I and II. The latter are now named Fab and the peptic digest product $(Fab')_2$.

The papain digest studies established that γ globulin, now named immunoglobulin gamma or IgG, was formed from three globular sections which were probably rather tightly folded as they were exceptionally resistant to further degradation by papain. The Fc fragment was apparently common to all molecules while the two identical Fab fragments each carried a combining site and with it the inherent variability associated with the whole antibody. An attempt was made (9) to relate this tripartite structure with the supposedly single open polypeptide structure deduced from end group analysis of rabbit IgG. Of course, it made no sense and progress depended upon the demonstration by Edelman (12) that in fact, human IgG and therefore presumably IgG of all species were multichain proteins. It followed that there must be blocked N terminal amino acids and that estimation of the free N terminal amino acid was of only limited significance.

The four peptide chain structure

The solution of the gross structure of immunoglobulins depended upon establishing the relationship between the peptide chains identified by Edelman and the products of papain digestion. This was achieved easily when the con-

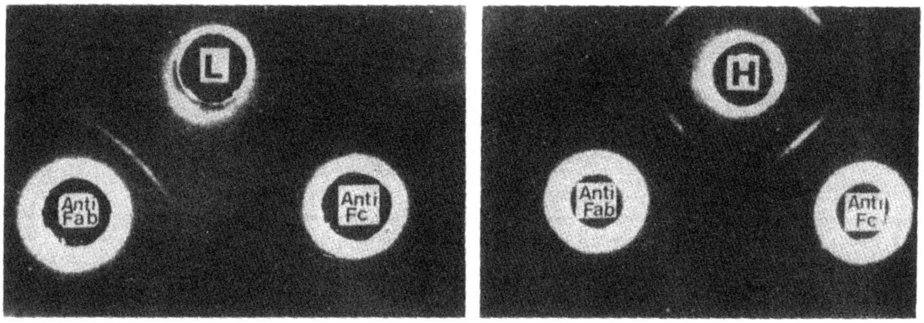

Fig. 3.
Double diffusion of the heavy and light chains of rabbit IgG against goat anti rabbit Fab and goat anti rabbit Fc. Note that light chain reacts only with anti Fab while heavy chain reacts with both anti Fab and anti Fc, i.e. Fab contains parts of heavy and light chains while Fc contains only parts of heavy chain.

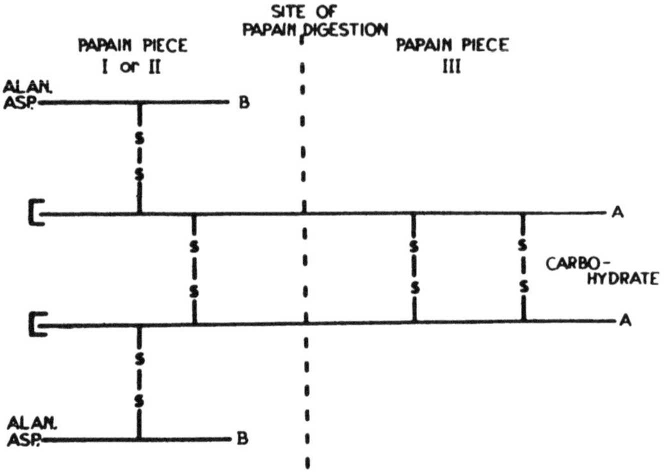

Fig. 4.
The four chain structure of rabbit IgG postulated on the basis of the double diffusion experiment of figure 3 and supporting chemical evidence.

ditions of isolation of the chains were modified using reduction in the absence of denaturing agents—conditions in which predominantly interchain disulphide bonds are broken. No fall in molecular weight followed such reduction but the chains were dissociated and separated when run on Sephadex columns in acetic or propionic acid to give heavy and light chains of molecular weights approximately 50,000 and 20,000 respectively. These chains now remained soluble at neutrality and retained antigenic specificities. A double diffusion plate using antisera specific to Fab or Fc showed that Fab contained antigenic sites common to both heavy and light chains but Fc those common to heavy chains only (fig. 3). This led to the postulated four chain structure (13) (fig. 4). More detailed studies were in agreement with this structure (14, 15 and 16)

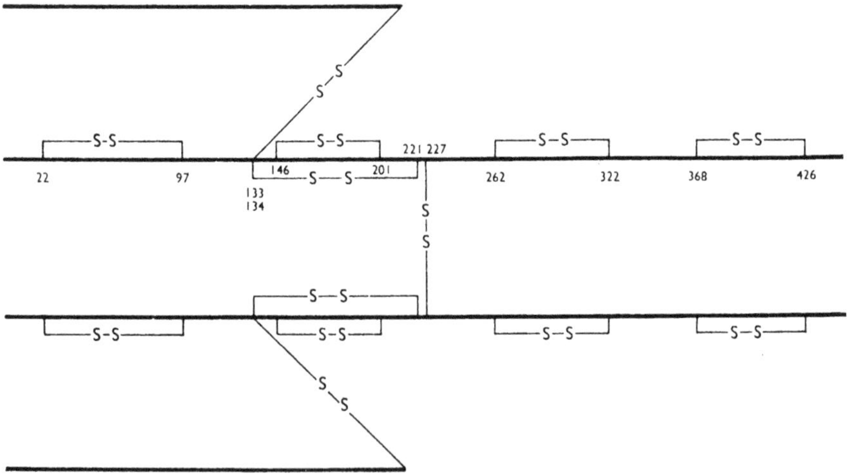

Fig. 5.
The structure of rabbit IgG showing the rather complex arrangement of inter and intra disulphide bonds in the heavy chain.

and confirmed the position of papain hydrolysis to be about the mid-point of the heavy chain.

It was some years before the rather complex arrangement of the inter and intra chain disulphide bonds of the rabbit heavy chains was resolved (17) (fig. 5). The mechanism of blocking of the α amino groups of the N terminal amino acids also proved more difficult to establish than expected as it was found to be due to the ringed residue pyrrolidone carboxylic acid (PCA) (18, 19) which was well known as an artefact arising from N terminal glutamine residues. All attempts to find N terminal glutamine in IgG were unsuccessful even when isolated under conditions in which conversion to PCA during handling appeared to be excluded. Evidence has been given that glutamine is the residue incorporated into the peptide chain during synthesis (20) but PCA appears to be the terminal residue present in the immunoglobulins in the blood, and it has been suggested that enzymatically catalysed cyclisation occurs intracellularly. It has been assumed that PCA is the only blocked N terminal residue in immunoglobulin molecules from all species, but there appears to have been little careful study.

Antibody combining sites

While the four peptide chain model clarified many aspects of the structure of antibodies, it made no contribution to our understanding of what features gave the possibility of forming antibodies of innumerable different specificities. It seemed at the time to increase the difficulties as the possibilities of variation were reduced. I made repeated attempts to obtain digest products of Fab which still bound the antigen but without success (21), but in fact progress has just been reported in isolating a peptic digest product of a mouse myeloma protein MOPC 315. This fragment named Fv appears to be formed from the N terminal half of the Fab molecules, i.e. of the light and Fd chains, and retains its full affinity for a dinitrophenyl hapten (22).

Understanding of the origin of the multiple binding sites came of course with the discovery of the phenomenon of the variable and constant parts in Bence Jones proteins (23, 24). Earlier it had been shown that the urinary Bence Jones proteins were the equivalent of the light chains of the myeloma protein in the blood of the same patient (25). This observation of the variability in amino acid sequence in the N terminal 107 residues of the human kappa Bence Jones proteins while the remainder were constant immediately made possible an understanding as to how millions of different combining sites could be formed within the same structural framework. That the phenomenon was common to the N terminal 110 residues or so of the heavy chains was shown (26, 27) and clearly the combining site was likely to be formed from the variable sections of both these chains.

Many lines of evidence have been brought to bear in an attempt to define more precisely in chemical terms just which residues in the variable regions of the heavy and light chains are likely to be directly concerned in determining the specificity of the binding site. They have been reviewed recently (28), and so the main conclusions could perhaps be just listed here:

1. The size of the antigenic site appears to be of the order of a hexapeptide or hexasaccharide.

2. This is comparable to the size of the substrate of a hydrolytic enzyme such as lysozyme. In this enzyme 15—20 amino acid residues have been identified as probable 'contact amino acids' i.e. residues forming a bond with the substrate. Hence this appears to be the likely number of residues to be expected to line the antibody combining site and to play a direct role in determining specificity. If any residue could occupy any of these 15—20 positions, the possible number of variants is indeed high.

3. There are at least three hypervariable sections in each of the heavy and light chain variable regions. They have been demonstrated rather clearly in the plots of Kabat (29, 30) of the frequence of occurrence of different residues in myeloma proteins along the 110 or so positions of the variable regions. This hypervariability is also apparent in sequence studies of the heavy chains of rabbit IgG, and both agree in suggesting that in most cases the hypervariability is confined to one or two positions but in the region 96—110 of the heavy chains four or five positions may be exceptionally variable. In the rabbit γ chain there is a section in this position across which no satisfactory sequence has been obtained, presumably because of the complexity of the sequences (31).

4. Several pieces of evidence suggest that these six hypervariable sections in the two variable regions may be brought together in the intact molecule to contribute to the structure of the combining site. The most direct evidence comes from affinity labelling studies in which an antibody is allowed to bind a haptene to which a reactive group has been attached. Covalent reaction will follow and after subsequent hydrolysis the labelled peptides can be identified and placed by comparison with the known amino acid sequence. A variety of affinity labelling techniques has been introduced and used both with natural antibodies and with a mouse myeloma protein showing high affinity for the dinitrophenyl group. Though the work is in some cases incomplete, all agree in that the labelled reagents are found attached to residues in or near one or other of the six hypervariable sections.

5. It is likely that there is a hydrophobic region adjacent to the combining site which may not contribute to specificity but could increase the affinity of binding for an appropriate antigenic site.

The precise details of the combining site must await the completion of the crystallographic studies on immunoglobulins and their fragments now being undertaken in several laboratories. There will be considerable interest in seeing how far the above prediction from chemical studies prove to be correct when the full structures become available.

The genetic origin of the multiple forms of antibodies

While the discovery of the variable region and particularly of the hypervariable sections in it, seemed likely to answer the basic question about the structural origin of multiple binding specificities it clearly raised very difficult problems as to the genetic origin of these many different amino acid sequences. This topic has formed the basis of many discussions and reviews and decisive evidence in favour of any of the many theories advanced is lacking. I only wish to make a brief comment here about the sequence work on the variable region of rabbit IgG heavy chains which developed as part of the other structural work discussed above.

There are perhaps two main points under immediate discussion. First, are both heavy and light chains each the products of two genes one coding for the variable region and one for the constant? Second, are the multiple genes which code for the variable regions, germ line genes or are they the product of somatic mutation of a much smaller number of germ line genes?

In each case it would be of obvious value if allelic variants able to act as genetic markers could be found in both the variable and constant regions. Markers in the constant regions have, in fact, been found in many of the different classes and subclasses of immunoglobulins of different species. The phenotypic character followed is the antigenic specificity of the immunoglobulin. This specificity has been correlated with amino acid changes but no direct identification of the specificity and the sequence change by demonstration of, say, the inhibitory power of a small peptide, has proved to be possible. Presumably a much larger section of molecule is necessary for the integrity of the antigenic site. However, in many cases there is no doubt that the genetic markers of the constant region have been found as they can be shown to be present in the Fc fragment. Allocation of genetic markers to the variable region is less clear though it is likely that this is where the rabbit 'a' locus allotypes orginate.

In 1963, Todd (32) made the very surprising observation that the 'a' locus allelic specificities of rabbit immunoglobulins were common to IgM and IgG. As the structural work progressed it became clear that these antigenic specificities were carried by the μ and γ chains respectively though these chains differed obviously in chemical structure and hence in their structural genes. It then seemed possible that the 'a' locus specificities could be determined by the variable regions which might be common to both chains. This observation of Todd's was the first to raise the possibility of two genes being concerned in the structure of the heavy chain, and now two examples of crossovers out of about 400 offspring between allelic specificities undoubtedly determined by structure of the constant parts of the γ chains and the 'a' locus specificities have been reported. If extended, these studies will strengthen the evidence that two genes code for the γ chains. However, the establishment of the structural basis of the 'a' locus specificities depends at present on correlation of amino acid sequences in the variable region with the specificity and lack of any similar correlation in the constant region sequences. This is rather indirect, but as far as it is acceptable, such correlation extending to about 16 positions in the γ chains has been found working with pooled rabbit IgG (33). Work with homogeneous rabbit anti polysaccharide antibodies has confirmed this correlation for some but not all these positions (34, 35).

These 'a' locus markers could clearly be of decisive importance in genetic studies, and indeed their existence in the variable region would be taken by some as strong evidence against the likelihood of many million copies of the variable regions being present in the germ line. It is, therefore, worthwhile to attempt to obtain unequivocal evidence that these apparent genetic markers of the variable region are indeed such. Alternatively, however, it should also be possible using the chemical evidence now obtained to follow the inheritance directly of a given allelic peptide rather than an antigenic specificity. One such peptide which occurs in two or three forms among different rabbits can be identified rather easily by autoradiography after reaction of the half cysteine residue at position 92 with ^{14}C iodoacetic acid (36). Early evidence suggests that it is indeed behaving as an heritable character and if confirmed will be direct evidence for a structural gene marker in the variable region. Further work is in progress and this should contribute to knowledge of the genetic origin of the variable region.

Some aspects of the structural studies of immunoglobulins have reached completion in that full chemical structures are now available for several human myeloma proteins and almost complete structures for rabbit immunoglobulins. The solution of the structural basis of the combining specificity of antibodies which seemed to me the central problem also appears to be nearing completion. There is still a role for structural work in the solution of the genetic origins of antibodies, and obviously there are many other applications not discussed during the lecture. Interaction of immunoglobulins with complement components and cell surfaces are two which are already arousing rapidly increasing interest.

References

1. Landsteiner, K., The Specificity of Serological Reaction. Harvard University Press, Cambridge U.S.A. (1946).
2. Parventjev, I. A., U.S. Patent 2,065,196 (1936).
3. Petermann, M. L., and Pappenheimer, A. M., J. Phys. Chem. *45*, 1 (1941).
4. Petermann, M. L., J. Am. Chem. Soc. *68*, 106 (1946).
5. Landsteiner, K., J. Exptl. Med. *75*, 269 (1942).
6. Porter, R. R., Biochem. J. *46*, 479 (1950).
7. Kimmel, J. R., and Smith, E. L., J. Biol. Chem. *207*, 515 (1954).
8. Porter, R. R., Nature *182*, 670 (1958).
9. Porter, R. R.. Biochem. J. *73*, 119 (1959).
10. Palmer, J. L., Mandy, W. J., and Nisonoff, A., Proc. Nat. Acad. Sci. US *48*, 49 (1962).
11. Nisonoff, A., Wissler, F. C., Lipman, L. N., and Woernley, D. L., Arch. Biochem. Biophys. *89*, 230 (1960).
12. Edelman, G. M., J. Am. Chem. Soc. *81*, 3155 o1959).
13. Porter, R. R., Basic Problems of Neoplastic Disease (A. Gellhorn and E. Hirschberg Eds) Columbia Univ. Press, New York (1962).
14. Fleishman, J. B., Porter, R. R., and Press, E. M., Biochem. J. *88*, 220 (1963).
15. Crumpton, M. J., and Wilkinson, J. M., Biochem. J. *88*, 228 (1963).
16. Pain, R. H., Biochem. J. *88*, 234 (1963).
17. O'Donnell, I. J., Frangione, B., and Porter, R. R., Biochem. J. *116*, 261 (1970).
18. Press, E. M., Piggot, P. J., and Porter, R. R., Biochem. J. *99*, 356 (1966).
19. Wilkinson, J. M., Press, E. M., and Porter, R. R., Biochem. J. *100*, 303 (1966).
20. Stott, D. I., and Munro, A. J., Biochem. J. *128*, 1221 (1972).
21. Porter, R. R., Brookhaven Symposium in Biology *13*, 203 (1960).
22. Inbar, D., Hochman, J., and Givol, D., Proc. Nat. Acad. Sci. US. *69*, (1972).
23. Hilschmann, N., and Craig, L. C., Proc. Nat. Acad. Sci. US *53*, 1403 (1965).
24. Titani, K., Whitley, E., Avogardo, L., and Putnam, F. W., Science *149*, 1090 (1965).

25. Edelman, G. M., and Gally, J. A., J. Exptl. Med. *116*, 207 (1962).
26. Press, E. M., and Hogg, N. M., Nature *223*, 807 (1969).
27. Edelman, G. M., Cunningham, B. A., Gall, W. E., Gottlieb, P. D., Rutishauser, U., and Waxdal, M. J., Proc. Nat. Acad. Sc. US *63*, 78 (1969).
28. Porter, R. R., Contemporary Topics in Immunochemistry (Ed. F. P. Inman) Plenum Press New York Vol *1*, 145 (1972).
29. Wu, T. T., and Kabat, E. A., J. Exptl. Med. *132*, 211 (1970).
30. Kabat, E. A., Wu, T. T., Ann. New York Acad. Sci. *190*, 382 (1971).
31. Fruchter, R. G., Jackson, S. A., Mole, L. E., and Porter, R. R., Biochem. J. *116*, 249 (1970).
32. Todd, C. W., Biochem. Biophys. Res. Commun. *11*, 170 (1963).
33. Mole. L. E., Jackson, S. A., Porter, R. R., and Wilkinson, J. M., Biochem. J. *124*, 301 (1971).
34. Fleischman, J. B., Biochemistry *10*, 2753 (1971).
35. Jaton, J. C., and Braun, D. G., Biochem. J. *130*, 539 (1972).
36. Mole, L. E., Unpublished Work (1972).

1973
Physiology or Medicine

KARL VON FRISCH, KONRAD LORENZ and NIKOLAAS TINBERGEN

"for their discoveries concerning organization and elicitation of individual and social behaviour patterns"

THE NOBEL PRIZE FOR PHYSIOLOGY OR MEDICINE

Speech by Professor BÖRJE CRONHOLM of the Karolinska Medico-Chirurgical Institute

Translation from the Swedish text

Your Majesty, Your Royal Highnesses, Ladies and Gentlemen,

Animal behavior has fascinated man since time immemorial as can be witnessed by the important role of animals in myths, fairy-tales and fables. However, for too long man has tried to understand it from his own experiences, from his own way of thinking, feeling and acting. Descriptions along these lines may be quite poetic, but they do not lead to any increase in knowledge. Various pre-scientific ideas have been especially tenacious in this field. Thus, it is not long ago that the vitalists maintained that the instincts bore witness of a wisdom that was inherent in the organism and could not be further analyzed. It was not until behavior problems were studied by means of scientific methods, by systematic observation and by experimentation, that real progress was made. Within that research field this year's Nobel prize laureates have been pioneers. They have collected numerous data about animal behavior both in natural settings and in experimental situations. Being biological scholars they have also studied the functions of behavior patterns, their role in the individual struggle for life and for the continuation of the species. Thus, behavior patterns have stood out as results of natural selection just as morphological characteristics and physiological functions.

It is of fundamental importance that some behavior patterns evidently are genetically programmed. The so-called fixed action patterns do not request any previous experience and they will be automatically elicited by definite key stimuli. They proceed in a mechanical, robot-like way, and when they have started they are no more influenced by external circumstances. In insects, fishes and birds, such important procedures as courtship, nesting and taking care of the brood, to a large extent consist in fixed action patterns. With development of the brain hemispheres, behavior has become increasingly modifiable and dependent on learning in mammals and especially in man, but fixed action patterns still play an important role.

For more than sixty years, Karl von Frisch has devoted himself to studies of the very complicated behavior of honeybees. Above all, he has elucidated what has rightly been called 'the language of bees'. When a bee has found flowers containing nectar, it performs a special dance when returning to the hive. The dance informs the bees in the hive of the existence of food, often also about the direction where the flowers will be found and about the distance to them. The foraging bee is able to indicate the direction of the food source in relation to the sun by means of analyzing polarized, ultraviolet light from the sky, light that is invisible to us. The honeybees do not learn, either to dance or to understand the message of the dance. Both the dancing and the appropriate reactions to it are genetically programmed behavior patterns.

Konrad Lorenz has studied among many other things the fixed action patterns of various birds. His experiments with inexperienced animals, *e.g.* young birds from an incubator, are of great importance in this context. In these young birds he observed behavior patterns that could not reasonably have been learnt but were to be interpreted as being genetically programmed. He also found that experiences of young animals during a critical period could be decisive for their future development. Newborn ducks and geese follow the first moving object that they catch sight of, and later on they will follow those particular objects only. Normally, they will follow their mother, but they may be seduced to follow almost any moving object or creature. This phenomenon has been called 'imprinting'.

While Konrad Lorenz has above all been a systematic observer of animal behavior, Nikolaas Tinbergen has to a large extent tested various hypotheses by means of comprehensive, careful, and quite often ingenious experiments. Among other things, he has used dummies to measure the strength of different key stimuli as regards their ability to elicit corresponding fixed action patterns. He made the important observation that 'supranormal' stimuli eliciting more intense behavior than those of natural conditions, may be produced by exaggerating certain characteristics.

The discoveries made by this year's Nobel prize laureates were based on studies of insects, fishes and birds and might thus seem to be of only minor importance for human physiology or medicine. However, their discoveries have been a prerequisite for the comprehensive research that is now pursued also on mammals. Studies are devoted to the existence of genetically programmed behavior patterns, their organization, maturation and their elicitation by key stimuli. There are also studies concerning the importance of specific experiences during critical periods for the normal development of the individual. Research into the behavior of monkeys have demonstrated that serious and to a large extent lasting behavior disturbances may be the result when a baby grows up in isolation without contact with its mother and siblings or with adequate substitutes. Another important research field concerns the effects of abnormal psychosocial situations on the individual. They may lead not only to abnormal behavior but also to serious somatic illness such as arterial hypertension and myocardial infarction. One important conclusion is that the psychosocial situation of an individual cannot be too adverse to its biological equipment without serious consequences. This holds true for all species, also for that which in shameless vanity has baptized itself '*Homo sapiens*'.

Karl von Frisch, Konrad Lorenz, Nikolaas Tinbergen,

Einer alten Fabel nach, von einem von Ihnen zitiert, wird behauptet, König Salomo hätte in seinem Besitz einen Ring gehabt, der die mystische Kraft besässe, ihm die Sprache der Tiere verständlich zu machen. Sie sind die Erben König Salomos gewesen insofern als Sie in der Lage gewesen sind, die Information zu entziffern, welche Tiere unter sich vermitteln und uns also den Sinn ihrer Verhaltensweisen zu deuten. Ihre Fähigkeit, massgebende Regeln zu finden, welche dem erstaunlichen Reichtum an Verhaltensweisen der Tiere zugrunde liegen, veranlässt uns manchmal zu glauben, dass Sie auch in den Besitz des Ringes König Salomos gekommen wären. Wir wissen aber, dass Sie nach empirischen Richtlinien gearbeitet haben, indem Sie Daten gesammelt haben und diese in Übereinstimmung mit harten und festen Regeln gedeutet haben.

Abgesehen von ihrer Bedeutung an sich haben ihre Entdeckungen einen weitreichenden Einfluss auf solche medizinischen Disziplinen gehabt wie Sozialmedizin, Psychiatrie und psychosomatische Medizin. Aus diesem Grunde geschah es sehr in Übereinstimmung mit dem Geiste in Alfred Nobels Testament, als die Medizinische Fakultät des Karolinska Institutet Ihnen den Nobelpreis dieses Jahres verlieh.

Wir sind stolz, heute zwei von Ihnen, Professor Konrad Lorenz und Professor Nikolaas Tinbergen, als Gäste begrüssen zu können und wir sind Professor Karl von Frisch dankbar dass er seinen Sohn, Professor Otto von Frisch, gesandt hat um ihn hier zu vertreten.

Im Auftrage des Karolinska Institutet möchte ich Ihnen die herzlichsten Glückwünsche aussprechen und ich darf Sie nun bitten, Ihren Preis aus den Händen Seiner Majestät des Königs zu empfangen.

Karl von Frisch, Konrad Lorenz, Nikolaas Tinbergen,

According to an old fable, cited by one of you, king Solomon is said to have been the owner of a ring that had the mystical power to give him the gift of understanding the language of animals. You have been the successors of king Solomon in the respect that you have been able to decode the information that animals pass to each other, and also to elucidate the meaning of their behavior to us. Your ability to find general rules underlying the confusing manifold of animal behavior makes us sometimes believe that king Solomon's ring has in fact been available also to you. But we know that you have been working in an empirical way, collecting data and interpreting it according to hard and fast scientific rules.

Aside from their value in themselves, your discoveries have had a far-reaching influence on such medical disciplines as social medicine, psychiatry and psychosomatic medicine. For that reason it was very much in agreement with the spirit of Alfred Nobel's will when the medical faculty of the Karolinska Institute awarded you this year's Nobel Prize.

We are proud to have two of you, professor Konrad Lorenz and professor Nikolaas Tinbergen, with us today, and we are also grateful to professor Karl von Frisch that he has sent his son, professor Otto von Frisch, to represent him here.

On behalf of the Karolinska Institute I wish to convey to you the warmest congratulations and I now ask you to receive your prize from the hands of His Majesty the King.

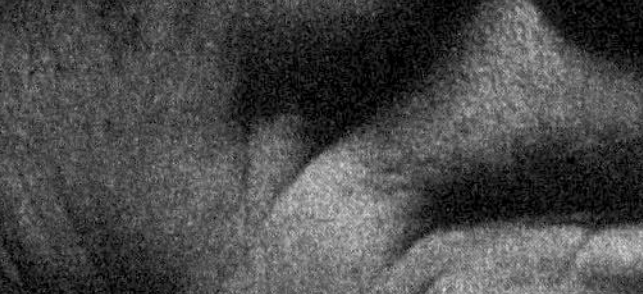

KARL VON FRISCH

I was born on 20 November 1886 in Vienna, the son of university professor Anton Ritter von Frisch and his wife Marie, *née* Exner. I studied at a grammar school and later at the University of Vienna in the Faculty of Medicine. After the first exams, I switched to the Faculty of Philosophy and studied Zoology in Munich and Vienna. I received my doctorate from the University of Vienna in 1910. In the same year I became assistant to Richard Hertwig at the Zoological Institute at the University of Munich. There I gained my University Teaching Certificate in Zoology and Comparative Anatomy.

In 1921 I went to the University of Rostock as Professor and Director at the Zoology Faculty; in 1923 I moved to Breslau and in 1925 I succeeded my former teacher Richard Hertwig in Munich. With support from the Rockefeller Foundation I oversaw the building of a new Zoological Institute with the best facilities available. After the destruction of the latter during the Second World War, I went to Graz in 1946, but returned to Munich in 1950 after the Institute had been reopened. I have been a Professor Emeritus since 1958, and have continued my scientific studies. Of my published papers the following are the most important:

Der Farben und Formensinn der Bienen: Zoologische Jarbücher (Physiologie) 35, 1–188, (1914–15). (*The bee's sense of colour and shape.*)

Über den Geruchssinn der Bienen und seine blütenbiologische Bedeutung: Zoologische Jahrbücher (Physiologie) 37, 1–238 (1919). (*The bee's sense of smell and its significance during blooming.*)

Über die "Sprache" der Bienen. Eine tierpsychologische Untersuchung: Zoologischer Jahrbücher (Physiologie) 40, 1–186 (1923). (*Bee's 'language' — an examination of animal psychology.*)

Untersuchung über den Sitz des Gehörsinnes bei der Elritze: Zeitschrift für vergleichende Physiologie 17, 686–801 (1932), with R. Stetter. (*Examination into the position of the sense of hearing in small insects.*)

Über den Geschmachsinn der Bienen: Zeitschrift für vergleichende Physiologie 21, 1–156 (1934). (*The bee's sense of taste.*)

Über einen Schreckstoff der Fischhaut und seine biologische Bedeutung: Zeitschrift für vergleichende Physiologie 29, 46–145 (1941). (*On the repellant substance on fish skin and its biological significance.*)

Die Tänze der Bienen: Österreichische Zooloische Zeitschrfit 1, 1–48 (1946). (*The bee's dances.*)

Die Polarisation des Himmelslichtes als orientierender Faktor bei den Tänzen der Bienen: Experientia (Basel) 5, 142–148 (1949). (*The polarisation of skylight as a means of orientation during the bee's dances.*)

Die Sonne als Kompaß im Leben der Bienen: Experientia (Basel) 6, 210–221 (1950). (*The sun as compass in the life of bees.*)

Tanzsprache und Orientierung der Bienen, Springer Verlag Berlin-Heidelberg-New York (1965). (*The Dance Language and Orientation of Bees,* Harvard University Press, 1967.)

DECODING THE LANGUAGE OF THE BEE

Nobel Lecture, December 12, 1973

by

KARL VON FRISCH[†]

University of Munich, Federal Republic of Germany

Some 60 years ago, many biologists thought that bees and other insects were totally color-blind animals. I was unable to believe it. For the bright colors of flowers can be understood only as an adaptation to color-sensitive visitors. This was the beginning of experiments on the color sense of the bee (*1*). On a table outdoors I placed a colored paper between papers of different shades of gray and on it I laid a small glass dish filled with sugar syrup. Bees from a nearby hive could be trained to recognize this color and demonstrated their ability to distinguish it from shades of gray. To prevent too great a gathering of bees, I instituted breaks between feedings. After these breaks, only sporadic scout bees came to the empty bowl and flew back home; the feeding table remained deserted. If a scout bee, however, found the bowl filled and returned home successfully, within a few minutes the entire forager group was back. Had she reported her findings to the hive? This question subsequently became the starting point for further investigations.

In order that the behavior of foragers could be seen after their return to the hive, a small colony was placed in an observation hive with glass windows, and a feeding bowl was placed next to it. The individual foragers were marked with colored dots, that is, numbered according to a certain system. Now an astonishing picture could be seen in the observation hive: Even before the returning bees turned over the contents of their honey sack to other bees, they ran over the comb in close circles, alternately to the right and the left. This round dance caused the numbered bees moving behind them to undertake a new excursion to the feeding place.

But foragers from one hive do not always fly to the same feeding source. Foraging groups form: One may collect from dandelions, another from clover, and a third from forget-me-nots. Even in flowering plants the food supply often becomes scarce, and a "feeding break" ensues. Were the bees in the experiment able to alert those very same foragers who were at the bowl with them? Did they know each other individually?

To settle the question, I installed two feeding places at which two groups from the same observation hive collected separately. During a feeding break, both groups stayed on the honey-comb and mingled with each other. Then one of the bowls was refilled. The bees coming from the filled bowl alerted by their dances not only their own group but also bees of the second group, which responded by flying to their customary feeding place where they investigated the empty bowl.

However, the natural stopping places of bees are not glass bowls but flowers. Therefore, the experiment was modified; one of two groups of bees collected food from linden blossoms, the other one from robinias. Now the picture

[†]Dr Karl von Frisch died in 1982.

changed. After the feeding break, the bees returning from the linden blossoms caused only the linden bees to fly out again; the robinia collectors paid no attention to their dances. On the other hand, when bees returned successfully from robinia blossoms, the linden bees showed no interest in their dances, while members of the robinia group immediately ran to a dancer in their vicinity, following along behind her and then flying out. Some clever bees also learned to use both sources of food, depending on the occasion. They would then send out the linden gatherers after returning from the linden source, and the robinia gatherers after visiting the robinias. Thus, the bees did not know each other individually. It appeared that the fragrance of the specific blossom attached to their bodies was decisive. This was confirmed when essential oils or synthetic scents at the feeding place produced the same effect.

When feeding was continuous, new recruits showed up at the food source next to the old foragers. They, too, were alerted by the dance. But how did they find their goal?

Peppermint oil was added to the feeding place next to the hive. In addition, bowls with sugar syrup were put on small cardboard sheets at various places in the nearby meadow; some of the sheets were scented with peppermint oil and the others with other essential oils. The result was unequivocal: A few minutes after the start of feeding, recruits from the observation hive appeared not only at the feeding place next to the foraging bees but also at the other peppermint bowls posted at some distance in the meadow. The other scented bowls, however, remained undisturbed. The smell of lavender, fennel, thyme oil, and so forth had no attraction. When the scent at the feeding place was replaced by a different one, the goal of the swarming recruits changed accordingly. They let themselves be guided by the scent on the dancers.

Scent is a very simple but effective means of communication. It attains full significance, however, only in combination with another condition. If the sugar syrup becomes scarce or is offered in weaker concentrations, after a certain point the dancing becomes slower and finally stops even though the collecting may continue. On the other hand, the sweeter the sugar syrup, the more lively and lengthier the various dances. The effect of advertising is thereby enhanced, and it is increased further by the scent gland in the forager's abdomen which is activated upon arrival at a good source of supply. Thus it signals "Come hither!" to recruits searching in the vicinity. Many female insects have scent glands to attract the male. In worker bees, which are mere workhorses devoid of any sexual interest, the scent organ is put to the service of the community.

Let us now imagine a meadow in the spring. Various types of plants blossom simultaneously, producing nectar of differing concentrations. The richer and sweeter its flow, the livelier the dance of the bees that discover and visit one type of flower. The flowers with the best nectar transmit a specific fragrance which ensures that they are most sought after. Thus, in this simple fashion, traffic is regulated according to the law of supply and demand not only to benefit the bees but also to promote pollination and seed yield of plant varieties rich in nectar. A new and hitherto unknown side of the biological significance of flower fragrance is thus revealed. Its great diversity and strict species specificity communicate a truly charming scent language.

This was how things stood in 1923 (*2*), and I believed I knew the language of the bees. On resuming the experiments 20 years later, I noticed that the most beautiful aspect had escaped me. Then, for the first time, I installed the feeding place several hundred meters away instead of next to the hive, and saw to my astonishment that the recruits immediately started foraging at that great distance while paying hardly any attention to bowls near the hive. The opposite occurred when the foragers located the sugar syrup, as before, near the hive. Could they possess a signal for distance?

Two foraging groups were formed from one observation hive. One feeding place was located 12 m from the hive, the other at a distance of 300 m. On opening the observation hive, I was astonished to see that all foragers from nearby performed round dances, while long-distance foragers did tail-wagging dances (Fig. 1). Moving the nearby feeding place step by step to greater distances resulted in the round dances changing to tail-wagging dances at a distance of about 50 m. The second feeding place was brought back step by step, past the first feeding place close to the hive. At the same critical distance of about 50 m, the tail-wagging dances became round dances (*3, 4*). I had been aware of the tail-wagging dance for a long time, but considered it to be typical of pollen collectors. My mistake was due to the fact that, at that time, bees with pollen baskets always arrived from a greater distance than my sugar syrup collectors.

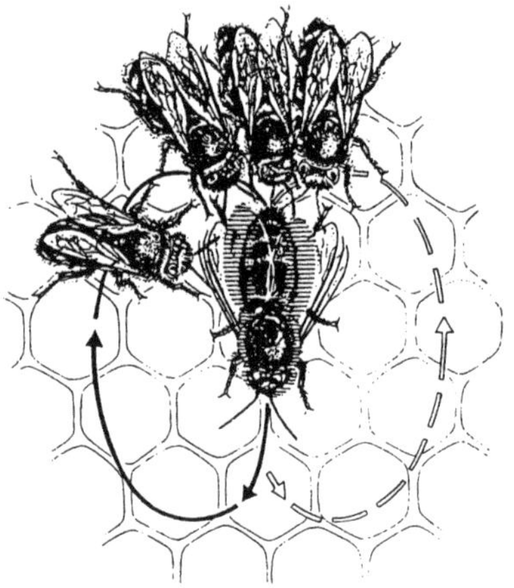

a. Round dance

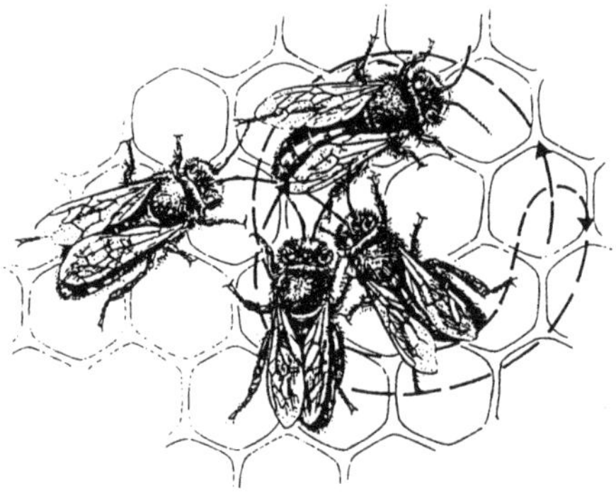

b. Tail-wagging dance

Fig. 1. Running curve of the bee (a) during round dance and (b) during tail-wagging dance. Bees that follow the dancer take in information.

Thus it became evident, and subsequent experiments confirmed (*5*), that the round dance is a signal that symbolically invites the hive members to search the immediate vicinity of the hive. The tail-wagging dance sends them to greater distances, not infrequently several kilometers. The signal "closer than 50 m" or "farther than 50 m" alone would not be of much help. In fact, however, the pace of the tail-wagging dance changes in a regular manner with increasing distance: its rhythm decreases. According to the present state of our knowledge, information on flight distance is given by the length of time required to go through the straight part of the figure - eight dance in each repeat. This straight stretch is sharply marked by tail-wagging dance movements and simultaneously toned (in the true meaning of the word) by a buzzing sound (*6, 7*). Longer distances are expressed symbolically by longer tail-wagging times. For distances of 200 to 4500 m, they increase from about 0.5 second to about 4 seconds (*6, 8*) (Fig. 2).

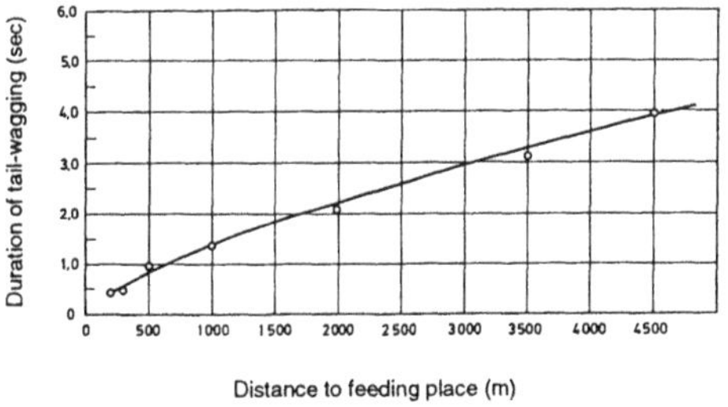

Fig. 2. Duration of the tail-wagging run for feeding places at various distances; based on film data.

The tail-wagging dance not only indicates distance but also gives the direction to the goal. In the observation hive, the bees that come from the same feeding place make their tail-wagging runs in the same direction, whereas these runs are oriented differently for bees coming from other directions. However, the direction of the tail-wagging runs of bees coming from one feeding place does not remain constant. As the day advances the direction changes by the same angle as that traversed by the sun in the meantime, but in the opposite rotation. Thus, the recruiting dancer shows the other bees the direction to the goal in relation to the position of the sun (*5, 6*). Those hours at the observation hive when the bees revealed this secret to me remain unforgettable. The fascinating thing is that the angle between the position of the sun and the dancer's path to the goal is expressed by the dancer in the darkness of the hive, on the vertical surface of the comb, as an angular deflection from the vertical. The bee thus transposes the angle to a different area of sense perception. Figure 3 shows the key to the transposition. If the

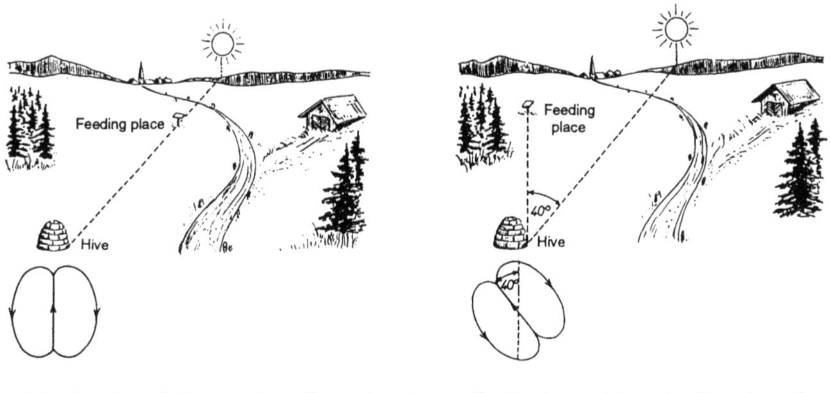

Fig. 3. Indication of direction by tail-wagging dance. (Left), the goal is in the direction of the sun; (right), the goal is 40° to the left of the sun's position. Dance figures, enlarged, are on the bottom left of the pictures.

goal lies in the direction of the sun, the tail-wagging dance points upward. If the goal is located 40° to the left of the sun's position, the dancer shifts the straight run 40° to the left of the vertical, and so forth (*5, 6*). On the comb, members of the hive move after the dancer and maintain close contact with her, especially during the tail-wagging runs, and take in the information offered. Can they follow it and with what accuracy?

The indication of direction was tested by us using the following method (*9*). At a certain distance from the hive, a feeding place was installed at which numbered bees were fed on an unscented platform with a sugar solution so dilute that they did not dance in the hive and therefore did not alert forager recruits. Only at the start of the experiment did they receive concentrated sugar solutions slightly scented with (for example) lavender oil. At 50 m closer to the hive, plates baited with the same scent but without food were placed in a fan-shaped arrangement. The number of forager recruits arriving at the plates was an indication of the intensity with which they searched in various directions. Figure 4 shows, as an example, the result of an experiment in which the feeding place was located 600 m from the hive.

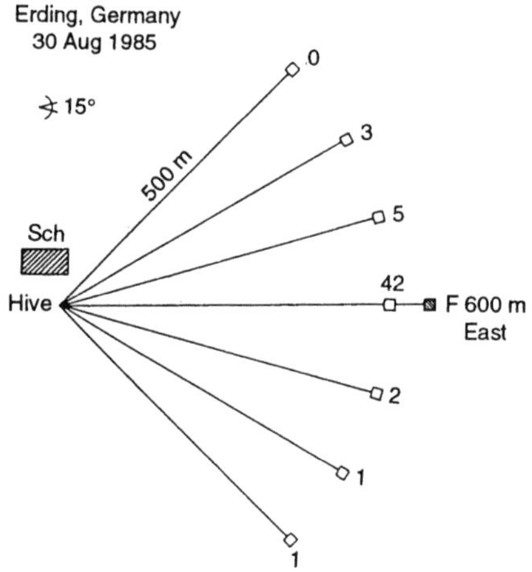

Fig. 4. Fan experiment. The feeding place (*F*) is 600 m from the observation hive. Scented plates without food are arranged in fan shape 550 m from the hive. The numbers indicate the number of forager recruits arriving during the first 50 minutes of the experiment; *Sch*, shed.

Since such fan experiments proved that indication of direction was successful, we made a step-by-step test of distance-indicating procedures. Here, all scented plates were located in the same direction as the feeding place, from the hive area to a distance well beyond the feeding place. Figure 5 gives an example of an experiment in which the feeding place was located 2 kilometers from the hive. Incoming flights of forager recruits to the feeding site itself were of course not evaluated because here an additional attractant was created by the food and the visiting bees (*6*).

To sum up, this and preceding experiments taught us that the information on the direction and distance of the goal was adhered to with astonishing accuracy—and not only in gathering nectar and pollen. The same dances are observed on a swarm. Here the scout bees indicate to the waiting bees the location of the domicile they have discovered. Of greatest interest here is that the intensity of the promotional message depends on the quality of the domicile discovered, that the various groups of scouting bees compete with each other, and that therefore the decision is finally made in favor of the best domicile (*10*).

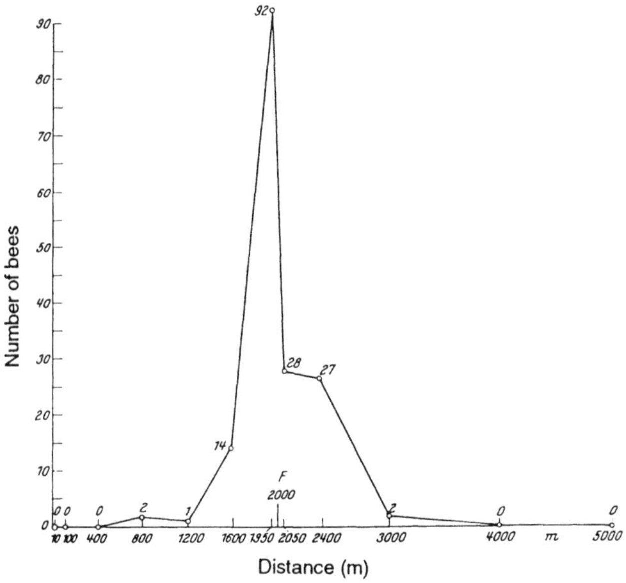

Fig. 5. Step-by-step experiment. The feeding place (*F*) is 2000 m from the observation hive. The numbers indicate the number of forager recruits that settled on the scented plates (without food) during the 3-hour observation period.

While not doubting that direction and distance of the goal can be discerned from the tail-wagging dances, a group of American biologists led by A. M. Wenner does not agree that the forager recruits make us of this information. According to them, these bees find the goal by using their olfactory sense only (*11*). This view is incompatible with many of our results (*6*, *12*). It is refuted by the following experiment, to cite only one.

Numbered bees from an observation hive collected at a feeding place 230 m from the hive. The hive was turned on its side so that the comb surface was horizontal; the sky was screened. Under these conditions, the dancers could orient themselves neither by gravity nor by the sky, and danced confusedly in all directions. Plates with the same scent as that at the feeding place were located at various distances in the direction of the feeding place and in three other directions. They were visited in all directions and in great numbers by forager recruits (Fig. 6), with no preferences being given to the direction of the feeding place. The observation hive was now turned back 90° to its normal position so that the dancers could indicate the direction of the goal on the vertical comb surface. Within a few minutes, the stream of newly alerted bees flew out in the direction of the feeding place; the scented plates in this direction were increasingly frequented, and in a short time no forager recruits at all appeared at the scented plates in the three other directions (Fig. 7). No change had occurred at the sources of scent in the open field or in the other external conditions. The change in the behavior of the forager recruits could be attributed only to the directional dances.

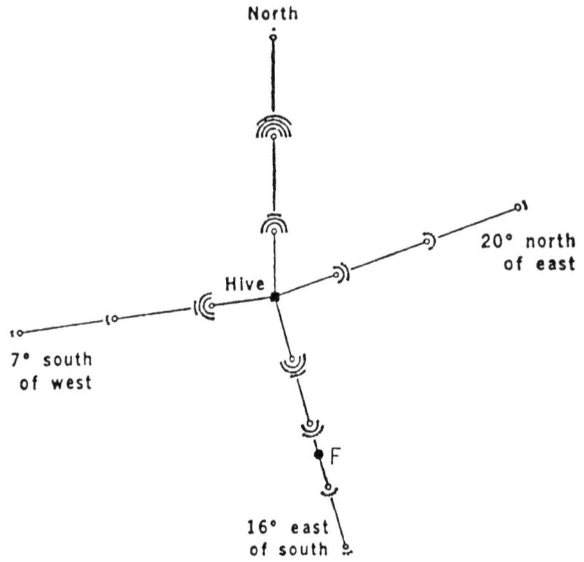

Fig. 6. Effect of placing observation hive horizontally. The dances are disoriented. Scented plates with the scent of the feeding place are visited by great numbers of forager recruits (small dots) in all four directions; *F*, feeding place.

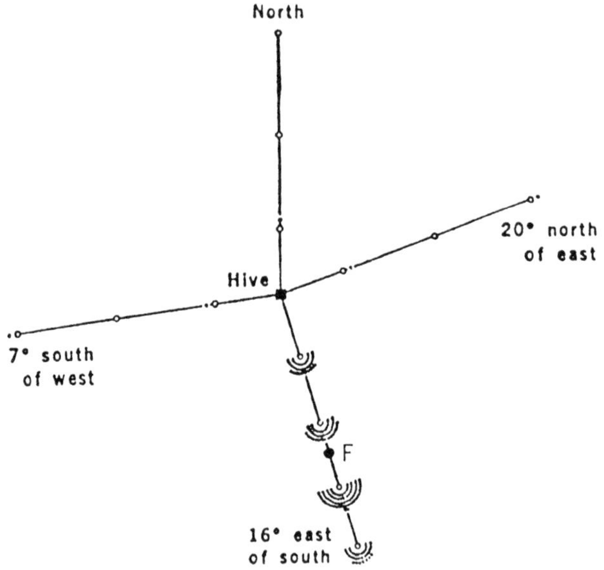

Fig. 7. Hive placed vertically after experiment in Fig. 6. The dances now indicate the direction of the feeding place. Within 10 minutes the stream of forager recruits turns in this direction. Flights no longer arrive in the three other directions.

It is conceivable that some people will not believe such a thing. Personally, I also harbored doubts in the beginning and desired to find out whether the intelligent bees of my observation hive had not perhaps manifested a special behavior. I opened an ordinary hive, lifted up one of the combs and watched the expected dances. Curious as to what would happen, I turned the comb in such a way that the dancing area became horizontal. Gravity as a means of orientation was thus eliminated. However, without any signs of perplexity, the bees continued to dance and by the direction of their tail-wagging runs pointed directly to the feeding place, just as we show the way by raising an arm. When the comb was turned like a record on a turntable, they continued to adjust themselves to their new direction, like the needle of a compass (*13*).

This behavior can be studied at leisure at a horizontal observation hive. It is basically very easy if we recall that the direction of the tail-wagging run relates to the sun's position. During the tail-wagging run on the comb, the bee has only to set itself at the same angle to the sun as it maintained during its flight to the feeding place (Fig. 8). Afterward, when the recruits set their line of flight at the same angle to the sun, they are flying in the direction of the goal.

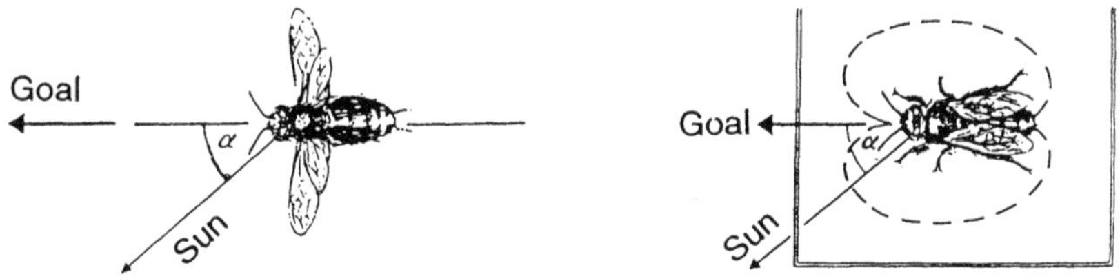

Fig. 8. The principle of direction indication during the dance on a horizontal plane. The bee (right) during the tail-wagging run positions itself in such a way that it views the sun from the same angle as earlier during its flight to the feeding place (left).

This type of discretional indication is nothing unusual. Incoming foragers not infrequently begin to dance facing the sun on the horizontal alighting board of the hive if they are met here by nonworking comrades. The transmission of information through horizontal dancing is easier to understand than that when the angle is transposed to the vertical comb surface. We also seem to have here the original, phylogenetically older type of directional indication. In India there still exist several strains of the species *Apis.* My student and co-worker, Martin Lindauer, went there to use them for "comparative language studies." The small honeybee, *Apis florea,* is on a more primitive level than our honeybee and other Indian strains. The colony builds a single comb out in the open on a branch; the comb has a horizontally extended top edge that serves exclusively as a dancing floor. When these bees are forced onto the vertical comb surface of the side, they cannot render the sun's angle by dancing and their tail-wagging dances become disoriented (*14*).

Let us now return to our own bees and the observation of dances on a horizontal hive. There can be no doubt that the sun's position is decisive for the direction of their dancing. The sun may be replaced by a lamp in a dark tent. By changing its position, the bees are made to dance in any desired direction. But there was one big puzzle. To prevent excessive heating during most of the experiments, a protective roof was installed over the observation hive. The dancers were unable to see the sun. Nevertheless their dance was usually correct. Orientation by heat rays, by penetrating radiation, as well as other explanations seemed possible and had to be discarded — until I noticed that a view of the blue sky is the same as a view of the sun. When clouds passed over the section of the sky visible to the bees, disoriented dances immediately resulted. Therefore they must have been able to read the sun's position from the blue sky. The direction of vibration of polarized blue light differs in relation to the sun's position across the entire vault of the sky. Thus, to one that is able to perceive the direction of vibration, even a spot of blue sky can disclose the sun's position by its polarization pattern. Are bees endowed with this capacity?

The following test furnished an answer. The observation hive was set horizontally in a dark tent from which the dancers had a lateral view of a small area of blue sky. They danced correctly toward the west where their feeding place was located 200 m away. When a round, rotatable polarizing foil was placed over the comb in a way as not to change the direction of the vibration of the polarized light from that part of the sky, they continued to dance correctly. If, however, I turned the foil right or left, the direction of the bees' dance changed to the right or the left by corresponding angle values.

Thus, bees are able to perceive polarized light. The sky, which to our eyes is a uniform blue, is distinctly patterned to them (*13, 15*). They use this extensively and, in their orientation, guide themselves not only by the sun's position but also by the resulting polarization patterns of the blue sky. They also continue to recognize the sun's position after it has set or when it is obscured by a mountain. Once again the bees appear to us miraculous. But it is now clear that ants and other insects, crayfish, spiders, and even octopuses perceive polarized light and use it for orientation, and that among all these animals the human being is the unendowed one, together with many other vertebrates. In one respect, however, bees remain singular: Only they use polarized light not only for their own orientation but also to communicate to their colonies the direction to a distant goal (*6*).

Thus the language of the bee, which was initially brought to our attention by the physiology of sense perception, has now led us back to it. It also had already led to general questions of orientation in time and space. When bees use the sun as a compass during their own flights as well as to inform their comrades, one difficulty arises: With the advancing hour of the day, the sun's position changes, and one would imagine that it can serve as a geographic marker for a short time only.

I had long contemplated an experiment whose execution was postponed from one year to the next by the feeling that it would not amount to much. However, in the early morning of a fall day in 1949, we sealed the entrance of our observation hive standing in Brunnwinkl on the shore of the Wolfgangsee, transported it across the lake, and placed it 5 km away in a completely different area unknown to the bees (*15*). Numbered bees from this colony had visited a feeding place 200 m to the west on previous days (Fig. 9). From the familiar

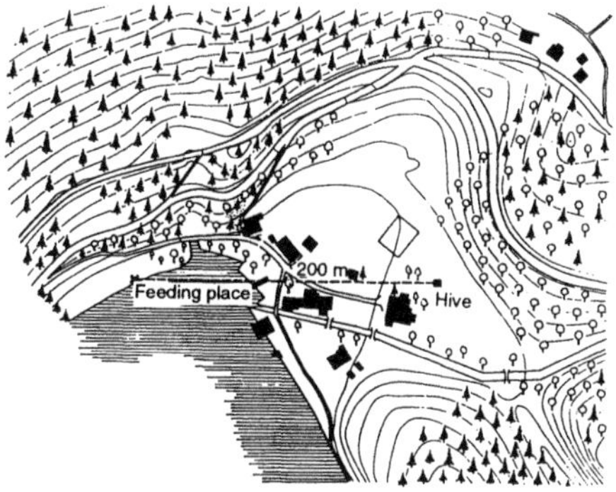

Fig. 9. Observation hive in Brunnwinkl on the Wolfgangsee and line of flight of a group of numbered bees to feeding place 200 m west.

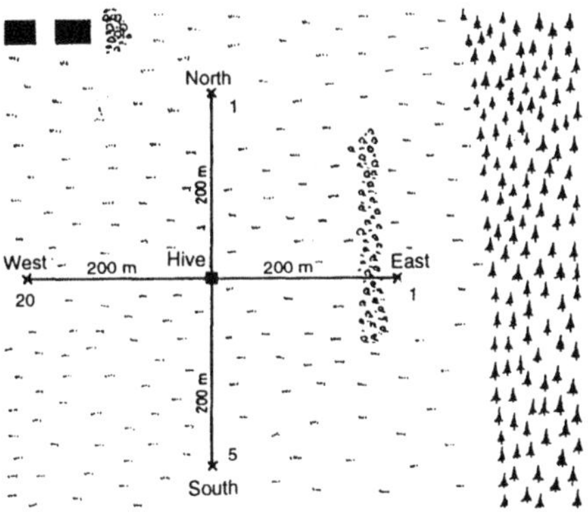

Fig. 10. The hive in Fig. 9 transported to a scene unfamiliar to the bees. Small feeding platforms with the familiar scent were placed 200 m from the hive in each of the four directions. The numbers indicate the numbers of arriving bees in the experimental group.

lakeshore and steep wooded hills they now found themselves in flat meadows; none of the known landmarks could be seen. Four feeding bowls with the same scent as at the former feeding place were placed 200 m from the hive toward the west, east, north, and south, and the entrance was then opened. Of the 29 marked bees that had visited in the west during the previous afternoon in Brunnwinkl, 27 found the bowls within 3 hours: 5 in the south, 1 each in the east and north, but 20 in the west (Fig. 10). Each was captured on arrival and was thus unable to send others out by dancing in the hive. Only the sun could have guided those who arrived. It, however, was southeast of the hive, while on the preceding day during the last foraging flights it had been close to the western horizon. Bees possess excellent timing, an inner clock, so to speak. During earlier experiments, by feeding at certain hours only they trained themselves to arrive promptly at the table at that time — even if the table was not set. The above trial, repeated in many modifications (*6*, *15*, *16*), has now taught us that they are also familiar with the sun's daily movement and can, by calculating the hour of the day, use this star as a true compass. The same discovery was simultaneously and independently made by Gustav Kramer using birds (*6*).

During the past few years, an old and persistent question has opened a new field of work for bee researchers. In discussing the direction indication, I initially kept something from you. The dancers did not always point correctly to the food sources. At certain hours they were markedly off to the left or the right. However, no inaccuracies or accidental deviation were involved; the errors were consistent and, when recorded under the same conditions, time and again gave the same curves for a typical daily routine. Thus they could correct, for example, for a different spatial position of the comb. Errors arose only with transposition of the dancing angle; in horizontal dances there is no "incorrect indication of direction." Observations over many years, made jointly with my co-worker Lindauer, finally led us to a conclusion which seemed acceptable (*6*). However, it was disproved by Lindauer, who persisted in his experiments together with his student H. Martin. They recognized the magnetic field of the earth as a cause for incorrect indication of direction. If this is artificially screened out, the error disappears; and by artificially altering the course of the lines of flux, the incorrect indication of direction was changed correspondingly (*17*). The idea that the magnetic field might play a role in the puzzling orientation performance of animals was rejected for a long time. During the past years it has been confirmed by new observations, especially in birds and insects (*18*). Nothing so far points to the possibility that bees, in their purposeful flights cross-country, are making use of the earth's magnetic field. Unexpectedly, however, it proved equally significant biologically but in a different way. When a swarm of bees builds its combs in a hive furnished to them by the beekeeper, their position in space is prescribed by the small suspended wooden frames. In the natural habitat of the bee, perhaps in the hollow of a tree, there are no wooden frames present. Nevertheless, thousands of bees labor together and in the course of one night achieve an orderly structure of parallel combs; the individual animal works here and there without getting instructions from a superintendent. They orient themselves by the earth's magnetic field and uniformly have in mind the comb position which they knew from the parent colony (*20*).

However, these are problems whose solution is fully underway, and we may expect quite a few surprises. By this I do not mean that problems such as the perception of polarized light have been conclusively solved. On the contrary: A question answered usually raises new problems, and it would be presumptuous to assume that an end is ever achieved.

It was not possible to present more than just a sketchy illustration in this lecture and to point out a few important steps in the development of our knowledge. To corroborate and extend them requires more time and work than the outsider can imagine. The effort of one individual is not sufficient for this. Helpers presented themselves, and I must express my appreciation to them at this time. If one is fortunate in finding capable students of whom many become permanent co-workers and friends, this is one of the most beautiful fruits of scientific work.

REFERENCES AND NOTES

1. K. von Frisch, *Zool. Jahrb. Abt. Allg. Zool. Physiol. Tiere* **35**, 1 (1914–1915).
2. _______, *ibid.* **40**, 1 (1923).
3. _______, *Experientia* **2**, No. 10 (1946).
4. The threshold of transition from the round dance to the tail-wagging dance varies with each race of honeybees; according to R. Boch [*Z. Vergl. Physiol.* **40**, 289 (1957)], it is about 50 m for *Apis mellifica carnica*, about 30 m for *A. mellifica mellifica* and *A. mellifica intermissa*, about 20 m for *A. mellifica caucasia* and *A. mellifica ligustica*, and 7 m for *A. mellifica fasciata*. The fact that the strain we used mostly in our experiments, the Carniolan bee, has the largest round dance circumference was of benefit in our experiments.
5. K. von Frisch, *Österreich. Zool.* Z. 1, 1 (1946).
6. _______, *Tanzsprache und Orientienung der Bienen* (Springer-Verlag, Berlin, 1965) (English translation: *The Dance Language and Orientation of Bees* (Belknap, Cambridge, Mass., 1967)]. Further references are found in this book.
7. H. Esch, Z. *Vergl. Physiol.* **45**, 1 (1961); A. M. Wenner, *Anim. Behav.* **10**, 79 (1962).
8. K. von Frisch and R. Jander, *Z. Vergl. Physiol.* **40**, 239 (1957).
9. I use the word "us," since the open-field experiments had assumed such proportions that they could no longer be carried out without trained assistants.
10. M. Lindauer, *Z. Vergl. Physiol.* **37**, 263 (1955).
11. A. M. Wenner, *The Bee Language Controversy: An Experience in Science* (Educational Programs Improvement Corp., Boulder, Colo., 1971).
12. K. von Frisch, *Anim. Behav.* **21**, 628 (1973).
13. _______, *Naturwissenschaften* **35**, 12 (1948): *ibid.*, p. 38.
14. M. Lindauer, *Z. Vergl. Physiol*, **38**, 521 (1956).
15. K. von Frisch, *Experientia* **6**, 210 (1950).
16. M. Renner, Z. *Vergl. Physiol.* **40**, 85 (1957); *ibid.* **42**, 449 (1959).
17. M. Lindauer and H. Martin, *ibid.* **60**, 219 (1968); M. Lindauer, *Rhein. Westjäl. Akad. Wiss. Rep. No. 218* (1971).
18. H. Martin and M. Lindauer, *Fortschr. Zool.* **21**, Nos. 2 and 3 (1973).

KONRAD LORENZ†

I consider early childhood events as most essential to a man's scientific and philosophical development. I grew up in the large house and the larger garden of my parents in Altenberg. They were supremely tolerant of my inordinate love for animals. My nurse, Resi Führinger, was the daughter of an old patrician peasant family. She possessed a "green thumb" for rearing animals. When my father brought me, from a walk in the Vienna Woods, a spotted salamander, with the injunction to liberate it after 5 days, my luck was in: the salamander gave birth to 44 larvae of which we, that is to say Resi, reared 12 to metamorphosis. This success alone might have sufficed to determine my further career; however, another important factor came in: Selma Lagerlöf's Nils Holgersson was read to me—I could not yet read at that time. From then on, I yearned to become a wild goose and, on realizing that this was impossible, I desperately wanted to *have* one and, when this also proved impossible, I settled for having domestic ducks. In the process of getting some, I discovered imprinting and was imprinted myself. From a neighbour, I got a one day old duckling and found, to my intense joy, that it transferred its following response to my person. At the same time my interest became irreversibly fixated on water fowl, and I became an expert on their behaviour even as a child.

When I was about ten, I discovered evolution by reading a book by Wilhelm Bölsche and seeing a picture of Archaeopteryx. Even before that I had struggled with the problem whether or not an earthworm was in insect. My father had explained that the word "insect" was derived from the notches, the "incisions" between the segments. The notches between the worm's metameres clearly were of the same nature. Was it, therefore, an insect? Evolution gave me the answer: if reptiles, via the Archaeopteryx, could become birds, annelid worms, so I deduced, could develop into insects. I then decided to become a paleontologist.

At school, I met one important teacher, Philip Heberdey, and one important friend, Bernhard Hellmann. Heberdey, a Benedictine monk, freely taught us Darwin's theory of evolution and natural selection. Freedom of thought was, and to a certain extent still is, characteristic of Austria. Bernhard and I were first drawn together by both being aquarists. Fishing for Daphnia and other "live food" for our fishes, we discovered the richness of all that lives in a pond. We both were attracted by Crustacea, particularly by Cladocera. We concentrated on this group during the ontogenetic phase of collecting through which apparently every true zoologist must pass, repeating the history of his

† Dr Lorenz died in 1989.

science. Later, studying the larval development of the brine shrimp, we discovered the ressemblance between the Euphyllopod larva and adult Cladocera, both in respect to movement and to structure. We concluded that this group was derived from Euphyllopod ancestors by becoming neotenic. At the time, this was not yet generally accepted by science. The most important discovery was made by Bernhard Hellmann while breeding the aggressive Cichlid Geophagus: a male that had been isolated for some time, would kill any conspecific at sight, irrespective of sex. However, after Bernhard had presented the fish with a mirror causing it to fight its image to exhaustion, the fish would, immediately afterwards, be ready to court a female. In other words, Bernhard discovered, at 17, that "action specific potentiality" can be "dammed up" as well as exhausted.

On finishing high school, I was still obsessed with evolution and wanted to study zoology and paleontology. However, I obeyed my father who wanted me to study medicine. It proved to be my good luck to do so. The teacher of anatomy, Ferdinand Hochstetter, was a brilliant comparative anatomist and embryologist. He also was a dedicated teacher of the comparative method. I was quick to realize not only that comparative anatomy and embryology offered a better access to the problems of evolution than paleontology did, but also that the comparative method was as applicable to behaviour patterns as it was to anatomical structure. Even before I got my medical doctor's degree, I became first instructor and later assistant at Hochstetter's department. Also, I had begun to study zoology at the zoological institute of Prof. Jan Versluys. At the same time I participated in the psychological seminars of Prof. Karl Bühler who took a lively interest in my attempt to apply comparative methods to the study of behaviour. He drew my attention to the fact that my findings contradicted, with equal violence, the opinions held by the vitalistic or "instinctivistic" school of MacDougall and those of the mechanistic or behavioristic school of Watson. Bühler made me read the most important books of both schools, thereby inflicting upon me a shattering disillusionment: none of these people *knew* animals, none of them was an expert. I felt crushed by the amount of work still undone and obviously devolving on a new branch of science which, I felt, was my responsibility.

Karl Bühler and his assistant Egon Brunswick made me realize that theory of knowledge was indispensable to the observer of living creatures, if he were to fulfill his task of scientific objectivation. My interest in the psychology of perception, which is so closely linked to epistemology, stems from the influence of these two men.

Working as an assistant at the anatomical institute, I continued keeping birds and animals in Altenberg. Among them the jackdaws soon became most important. At the very moment when I got my first jackdaw, Bernhard Hellmann gave me Oskar Heinroth's book "Die Vögel Mitteleuropas". I realized in a flash that this man knew everything about animal behaviour that both, MacDougall and Watson, ignored and that I had believed to be the only one to know. Here, at last, was a scientist who also was an expert! It is hard to assess the influence which Heinroth exerted on the development of my ideas. His classical comparative paper on Anatidae encouraged me to regard the comparative study of behaviour as my chief task in life. Hochstetter generously considered my ethological work as being comparative anatomy of sorts and permitted me to work on it while on duty in his department. Otherwise the papers I produced between 1927 and 1936 would never have been published.

During that period I came to know Wallace Craig. The American Ornitologist Margaret Morse Nice knew about his work and mine and energetically put us into contact. I owe her undying gratitude. Next to Hochstetter and Heinroth, Wallace Craig became my most influential teacher. He critisized my firmly-held opinion that instinctive activities were based on chain reflexes. I myself had demonstrated that long absence of releasing stimuli tends to lower their threshold, even to the point of the activity's eruption in vacuo. Craig pointed out that in the same situation the organism began actively to seek for the releasing stimulus situation. It is obviously nonsense, wrote Craig, to speak of a re-action to a stimulus not yet received. The reason why in spite of the obvious spontaneity of instictive behaviour, I still clung to the reflex theory, lay in my belief, that any deviation from Sherringtonian reflexology meant a concession to vitalism. So, in the lecture I gave in February 1936 in the Harnackhaus in Berlin, I still defended the reflex theory of instinct. It was the last time I did so.

During that lecture, my wife was sitting behind a young man who obviously agreed with what I said about spontaneity, murmuring all the time: "It all fits in, it all fits in." When, at the end of my lecture, I said that I regarded instinctive motor patterns as chain reflexes after all, he hid his face in his hands and moaned: "Idiot, idiot". That man was Erich von Holst. After the lecture, in the commons of the Harnackhaus, it took him but a few minutes to convince me of the untenability of the reflex theory. The lowering thresholds, the eruption of vacuum activities, the independence of motor patterns of external stimulation, in short all the phenomena I was struggling with, not only could be explained, but actually were to be postulated on the assumption that they were based not on chains of reflexes but on the processes of endogenous generation of stimuli and of central coordination which had been discovered and demonstrated by Erich von Holst. I regard as the most important break-through of all our attempts to understand animal and human behaviour the recognition of the following fact: the elemental neural organisation underlying behaviour does not consist of a receptor, an afferent neuron stimulating a motor cell and of an effector activated by the latter. Holst's hypothesis which we confidently can make our own, says that the basic central nervous organisation consists of a cell permanently producing endogenous stimulation, but prevented from activating its effector by another cell which, also producing endogenous stimulation, exerts an inhibiting effect. It is this inhibiting cell which is influenced by the receptor and ceases its inhibitory activity at the biologically "right" moment. This hypothesis appeared so promising that the Kaiser-Wilhelmsgesellschaft, now renamed Max-Planck-Gesellschaft, decided to found an institute for the physiology of behavior for Erich von Holst and myself. I am convinced that if he were still alive, he would be here in Stockholm now. At the time, the war interrupted our plans.

When, in autumn 1936, Prof. van der Klaauw convoked a symposium called "Instinctus" in Leiden in Holland, I read a paper on instinct built up on the theories of Erich von Holst. At this symposium I met Niko Tinbergen and this was certainly the event which, in the course of that meeting, brought the most important consequences to myself. Our views coincided to an amazing degree but I quickly realized that he was my superior in regard to analytical thought as well as to the faculty of devising simple and telling experiments. We discussed the relationship between spatially orienting responses (taxes in the sense of Alfred Kühn) and releasing mechanism on one hand, and the spontaneous endogenous motor patterns on the other. In these discussions some conceptualisations took form which later proved fruitful to ethological research. None of us knows who said what first, but it is highly probable that the conceptual separation of taxes, innate releasing mechanisms and fixed motor patterns was Tinbergen's contribution. He certainly was the driving force in a series of experiments which we conducted on the egg-rolling response of the Greylag goose when he stayed with us in Altenberg for several months in the summer of 1937.

The same individual geese on which we conducted these experiments, first aroused my interest in the process of domestication. They were F_1 hybrids of wild Greylags and domestic geese and they showed surprising deviations from the normal social and sexual behaviour of the wild birds. I realised that an overpowering increase in the drives of feeding as well as of copulation and a waning of more differentiated social instincts is characteristic of very many domestic animals. I was frightened—as I still am—by the thought that analogous genetical processes of deterioration may be at work with civilized humanity. Moved by this fear, I did a very ill-advised thing soon after the Germans had invaded Austria: I wrote about the dangers of domestication and, in order to be understood, I couched my writing in the worst of nazi-terminology. I do not want to extenuate this action. I did, indeed, believe that some good might come of the new rulers. The precedent narrow-minded catholic regime in Austria induced better and more intelligent men than I was to cherish this naive hope. Practically all my friends and teachers did so, including my own father who certainly was a kindly and humane man. None of us as much as suspected that the word "selection", when used by these rulers, meant murder. I regret those writings not so much for the undeniable discredit they reflect on my person as for their effect of hampering the future recognition of the dangers of domestication.

In 1939 I was appointed to the Chair of Psychology in Köningsberg and this appointment came about through the unlikely coincidence that Erich von Holst happened to play the viola in a quartette which met in Göttingen and in which Eduard Baumgarten played the first violin. Baumgarten had been professor of philosophy in Madison, Wisconsin. Being a pupil of John Dewey and hence a representative of the pragmatist school of philosophy, Baumgarten had some doubts about accepting the chair of philosophy in Köningsberg—

Immanuel Kant's chair—which had just been offered to him. As he knew that the chair of psychology was also vacant in Köningsberg, he casually asked Erich von Holst whether he knew a biologically oriented psychologist who was, at the same time, interested in theory of knowledge. Holst knew that I represented exactly this rather rare combination of interests and proposed me to Baumgarten who, together with the biologist Otto Koehler and the botanist Kurt Mothes—now president of the Academia Leopoldina in Halle —persuaded the philosophical faculty in Köningsberg of putting me, a zoologist, in the psychological chair. I doubt whether perhaps the faculty later regretted this choice, I myself, at any rate, gained enormously by the discussions at the meetings of the Kant-Gesellschaft which regularly extended late into the night. My most brillant and instructive opponents in my battle against idealism were the physiologist H. H. Weber, now of the Max-Planck-Gesellschaft, and Otto Koehler's late first wife Annemarie. It is to them that I really owe my understanding of Kantian philosophy—as far as it goes. The outcome of these discussions was my paper on Kant's theory of the à priori in the view of Darwinian biology. Max Planck himself wrote a letter to me in which he stated that he thoroughly shared my views on the relationship between the phenomenonal and the real world. Reading that letter gave me the same sort of feeling as hearing that the Nobel Prize had been awarded to me. Years later that paper appeared in the Systems Year Book translated into English by my friend Donald Campbell.

In autumn 1941 I was recruited into the German army as a medical man. I was lucky to find an appointment in the department of neurology and psychiatry of the hospital in Posen. Though I had never practised medicine, I knew enough about the anatomy of the nervous system and about psychiatry to fill my post. Again I was lucky in meeting with a good teacher, Dr. Herbert Weigel, one of the few psychiatrists of the time who took psychoanalysis seriously. I had the opportunity to get some first-hand knowledge about neurosis, particularly hysteria, and about psychosis, particularly schizophrenia.

In spring 1942 I was sent to the front near Witebsk and two months later taken prisioner by the Russians. At first I worked in a hospital in Chalturin where I was put in charge of a department with 600 beds, occupied almost exclusively by cases of so-called field polyneuritis, a form of general inflammation of nervous tissues caused by the combined effects of stress, overexertion, cold and lack of vitamins. Surprisingly, the Russian physicians did not know this syndrome and believed in the effects of diphteria—an illness which also causes a failing of all reflexes. When this hospital was broken up I became a camp doctor, first in Oritschi and later in a number of successive camps in Armenia. I became tolerably fluent in Russian and got quite friendly with some Russians, mostly doctors. I had the occasion to observe the striking parallels between the psychological effects of nazi and of marxist education. It was then that I began to realize the nature of indoctrination as such.

As a doctor in small camps in Armenia I had some time on my hand and I started to write a book on epistemology, since that was the only subject for which I needed no library. The manuscript was mainly written with potassium permanganate solution on cement sacking cut to pieces and ironed out. The Soviet authorities encouraged my writing, but, just when it was about finished, transferred me to a camp in Krasnogorsk near Moscow, with the injunction to type the manuscript and send a copy to the censor. They promised I should be permitted to take a copy home on being repatriated. The prospective date for repatriation of Austrians was approaching and I had cause to fear that I should be kept back because of my book. One day, however, the commander of the camp had me called to his office, asked me, on my word of honor, whether my manuscript really contained nothing but unpolitical science. When I assured him that this was indeed the case, he shook hands with me and forthwith wrote out a "propusk", an order, which said that I was allowed to take my manuscript and my tame starling home with me. By word of mouth he told the convoy officer to tell the next to tell the next and so on, that I should not be searched. So I arrived in Altenberg with manuscript and bird intact. I do not think that I ever experienced a comparable example of a man trusting another man's word. With a few additions and changes the book written in Russia was published under the title "Die Rückseite des Spiegels". This title had been suggested by a fellow prisoner of war in Erivan, by name of Zimmer.

On coming home to Austria in February 1948, I was out of a job and there was no promise of a chair becoming vacant. However, friends rallied from all sides. Otto Storch, professor of zoology, did his utmost and had done so for my wife even before I came back. Otto König and his "Biologische Station Wilhelminenberg", received me like a longlost brother and Wilhelm Marinelli, the second zoologist, gave me the opportunity to lecture at his "Institut für Wissenschaft und Kunst". The Austrian Academy of Sciences financed a small research station in Altenberg with the money donated for that purpose by the English poet and writer J. B. Priestley. We had money to support our animals, no salaries but plenty of enthusiasm and enough to eat, as my wife had given up her medical practice and was running her farm near Tulln. Some remarkable young people were ready to join forces with us under these circumstances. The first was Wolfgang Schleidt, now professor at Garden University near Washington. He built his first amplifier for supersonic utterances of rodents from radio-receivers found on refuse dumps and his first terrarium out of an old bedstead of the same provenance. I remember his carting it home on a wheel-barrow. Next came Ilse and Heinz Prechtl, now professor in Groningen, then Irenäus and Eleonore Eibl-Eibesfeldt, both lady doctors of zoology and good scientists in their own right.

Very soon the international contact of ethologists began to get re-established. In autumn 1948 we had the visit of Professor W. H. Thorpe of Cambridge who had demonstrated true imprinting in parasitic wasps and was interested in our work. He predicted, as Tinbergen did at that time, that I should find it impossible to get an appointment in Austria. He asked me in confidence whether I would consider taking on a lectureship in England. I said that I preferred, for the present, to stick in Austria. I changed my mind soon afterwards: Karl von Frisch who left his chair in Graz, Austria, to go back to

Munich, proposed me for his successor and the faculty of Graz unanimously concurred. When the Austrian Ministry of Education which was strictly Catholic again at this time, flatly refused Frisch's and the faculty's proposal, I wrote two letters to Tinbergen and to Thorpe, that I was now ready to leave home. Within an amazingly short time the University of Bristol asked me whether I would consider a lectureship there, with the additional task of doing ethological research on the water-fowl collection of the Severn Wild-fowl Trust at Slimbridge. So my friend Peter Scott also must have had a hand in this. I replied in the affirmative, but, before anything was settled, the Max-Planck-Gesellschaft intervened offering me a research station adjunct to Erich von Holst's department. It was a hard decision to take; finally I was swayed by the consideration that, with Max Planck, I could take Schleidt, Prechtl and Eibl with me. Soon afterwards, my research station in Buldern in Westfalia was officially joined to Erich von Holst's department in a newly-founded "Max-Planck-Institut für Verhaltensphysiologie". Erich von Holst con-voked the international meeting of ethologists in 1949. With the second of these symposia, Erich von Holst and I celebrated the coming-true of our dream in Buldern in autumn 1950.

Returning to my research work, I at first confined myself to pure observa-tion of waterfowl and of fish in order to get in touch again with real nature from which I had been separated so long. Gradually, I began to concentrate on the problems of aggressivity, of its survival function and on the mechanisms counteracting its dangerous effects. Fighting behaviour in fish and bonding behaviour in wild geese soon became the main objects of my research. Looking again at these things with a fresh eye, I realized how much more detailed a knowledge was necessary, just as my great co-laureate Karl von Frisch found new and interesting phenomena in his bees after knowing them for several decades, so, I felt, the observation of my animals should reveal new and in-teresting facts. I found good co-workers and we all are still busy with the same never-ending quest.

A major advance in ethological theory was triggered in 1953 by a violent critique by Daniel D. Lehrmann who impugned the validity of the ethological concept of the innate. As Tinbergen described it, the community of ethologists was humming like a disturbed bee-hive. At a discussion arranged by Profes-sor Grassé in Paris, I said that Lehrmann, in trying to avoid the assumption of innate knowledge, was inadvertently postulating the existence of an "innate school-marm". This was meant at a reduction to the absurd and shows my own error: it took me years to realize that this error was identical with that committed by Lehrmann and consisted in conceiving of the "innate" and of the "learned" as of disjunctive contradictory concepts. I came to realize that, of course, the problem why learning produces adaptive behaviour, rests exclusively with the "innate school-marm", in other words with the phylogenetically programmed teaching mechanism. Lehrmann came to realize the same and on this realisation we became friends. In 1961 I published a paper "Phylogenetische Anpassung und adaptive Modifikation des Verhaltens", which I later expanded into a book called "Evolution and

Modification of Behaviour" (Harvard University Press, 1961).

Until late in my life I was not interested in human behaviour and less in human culture. It was probably my medical background that aroused my awareness of the dangers threatening civilized humanity. It is sound strategy for the scientist not to talk about anything which one does not know with certainty. The medical man, however, is under the obligation to give warning whenever he sees a danger even if he only suspects its existence. Surprisingly late, I got involved with the danger of man's destruction of his natural environment and of the devastating vicious circle of commercial competition and economical growth. Regarding culture as a living system and considering its disturbances in the light of illnesses led me to the opinion that the main threat to humanity's further existence lies in that which may well be called mass neurosis. One might also say that the main problems with which humanity is faced, are moral and ethical problems.

Todate I have just retired from my directorship at the Max-Planck-Institut für Verhaltensphysiologie in Seewiesen, Germany, and am at work building up a department of animal sociology pertaining to the Institut für Vergleichende Verhaltensforschung of the Austrian Academy of Science.

ANALOGY AS A SOURCE OF KNOWLEDGE

Nobel Lecture, December 12, 1973

by

KONRAD Z. LORENZ

Austrian Academy of Sciences
Institute of Comparative Behaviour Research, Altenberg, Austria

1. THE CONCEPT OF ANALOGY

In the course of evolution it constantly happens that, independently of each other, two different forms of life take similar, parallel paths in adapting themselves to the same external circumstances. Practically all animals which move fast in a homogeneous medium have found means of giving their body a streamlined shape, thereby reducing friction to a minium. The "invention" of concentrating light on a tissue sensitive to it by means of a diaphanous lens has been made independently at least four times by different phyla of animals; and in two of these, in the cephalopods and in the vertebrates, this kind of "eye" has evolved into the true, image-projecting camera through which we ourselves are able to see the world.

Thanks to old discoveries by Charles Darwin and very recent ones by biochemists, we have a fairly sound knowledge of the processes which, in the course of evolution, achieve these marvellous structures. The student of evolution has good reason to assume that the abundance of different bodily structures which, by their wonderful expediency, make life possible for such amazingly different creatures under such amazingly different conditions, *all* owe their existence to these processes which we are wont to subsume under the concept of *adaptation*. This assumption, whose correctness I do not propose to discuss here, forms the basis of the reasoning which the evolutionist applies to the phenomenon of *analogy*.

2. DEDUCING COMPARABLE SURVIVAL VALUE FROM SIMILARITY OF FORM

Whenever we find, in two forms of life that are unrelated to each other, a similarity of form or of behaviour patterns which relates to more than a few minor details, we assume it to be caused by parallel adaptation to the same life-preserving function. The improbability of coincidental similarity is proportional to the number of independent traits of similarity, and is, for n such characters, equal to 2^{n-1}. If we find, in a swift and in an airplane, or in a shark, or a dolphin and in a torpedeo, the striking resemblances illustrated in Fig. 1, we can safely assume that in the organisms as well as in the man-made machines, the need to reduce friction has led to parallel adaptations. Though the independent points of similarity are, in these cases, not very many,

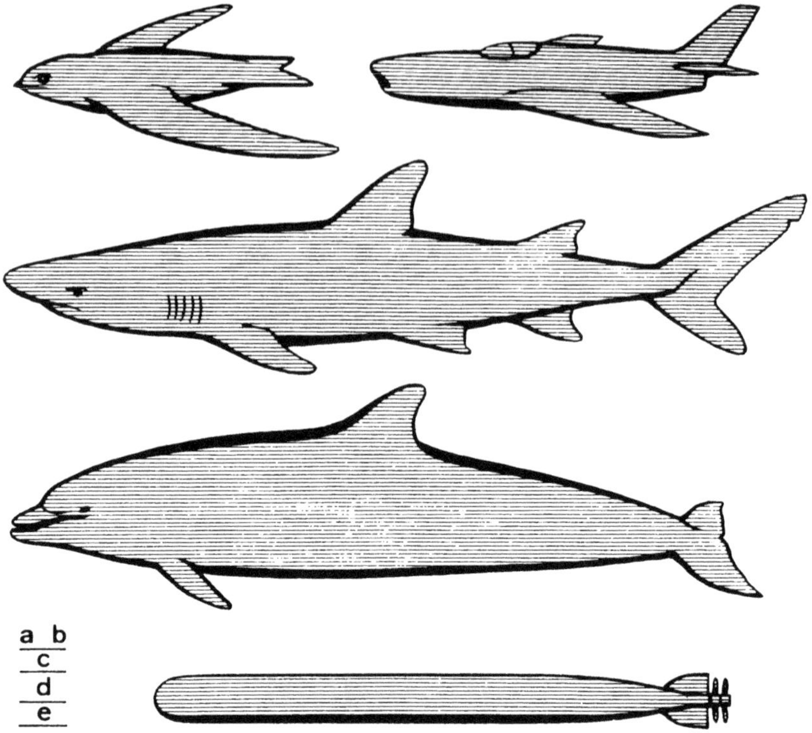

Fig. 1. Analogy of form due to adaptation to an identical function. Streamlining in a) a swift, b) a fighter plane, c) a shark, d) a dolphin, e) a torpedo.

it is still a safe guess that any organism or vehicle possessing them is adapted to fast motion.

There are conformities which concern an incomparably greater number of independent details. Fig. 2 shows cross sections through the eyes of a vertebrate and a cephalopod. In both cases there is a lens, a retina connected by nerves with the brain, a muscle moving the lens in order to focus, a contractile iris acting as a diaphragm, a diaphanous cornea in front of the camera and a layer of pigmented cells shielding it from behind—as well as many other matching details. If a zoologist who knew nothing whatever of the existence of cephalopods were examinining such an eye for the very first time, he would conclude without further ado that it was indeed a light-perceiving organ. He would not even need to observe a live octopus to know this much with certainty.

3. The allegation of "false analogy'

Ethologists are often accused of drawing *false* analogies between animal and human behaviour. However, no such thing as a *false* analogy exists: an analogy can be more or less detailed and hence more or less informative. Assiduously searching for a really false analogy, I found a couple of technological

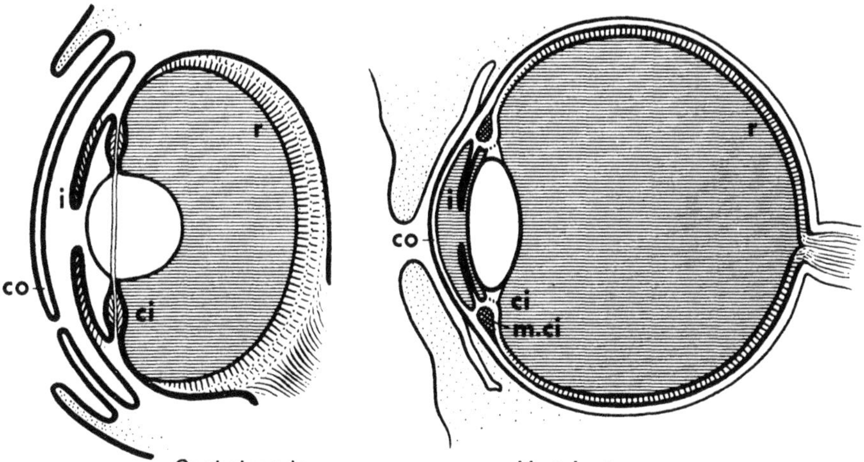

Fig. 2. Detailed analogy in two independently evolved light-perceiving organs, left the eye of an octopus, right of a man, co Cornea, ci Corpus ciliare, m.ci musculus ciliaris, i iris, r retina.

examples within my own experience. Once I mistook a mill for a sternwheel steamer. A vessel was anchored on the banks of the Danube near Budapest. It had a little smoking funnel and at its stern an enormous slowly-turning paddle-wheel. Another time, I mistook a small electric power plant, consisting of a two-stroke engine and a dynamo, for a compressor. The only biological example that I could find concerned a luminescent organ of a pelagic gastropod, which was mistaken for an eye, because it had an epidermal lens and, behind this, a high cylindrical epithelium connected with the brain by a nerve Even in these examples, the analogy was false only in respect of the direction in which energy was transmitted.

4. The concept of homology

There is, in my opinion, only one possibility of an error that might conceivably be described as the "drawing of a false analogy" and that is mistaking an *homology* for an analogy. An homology can be defined as any resemblance between two species that can be explained by their common descent from an ancestor possessing the character in which they are similar to each other. Strictly speaking, the term homologous can only be applied to characters and not to organs. Fig. 3 shows the forelimbs of a number of tetrapod vertebrates intentionally chosen to illustrate the extreme variety of uses to which a front leg can be put and the evolutional changes it can undergo in the service of these different functions. Notwithstanding the dissimilarities of these functions and of their respective requirements, all these members are built on the same basic plan and consist of comparable elements, such as bones, muscles, nerves. The very dissimilarity of their functions makes it extremely improbable that

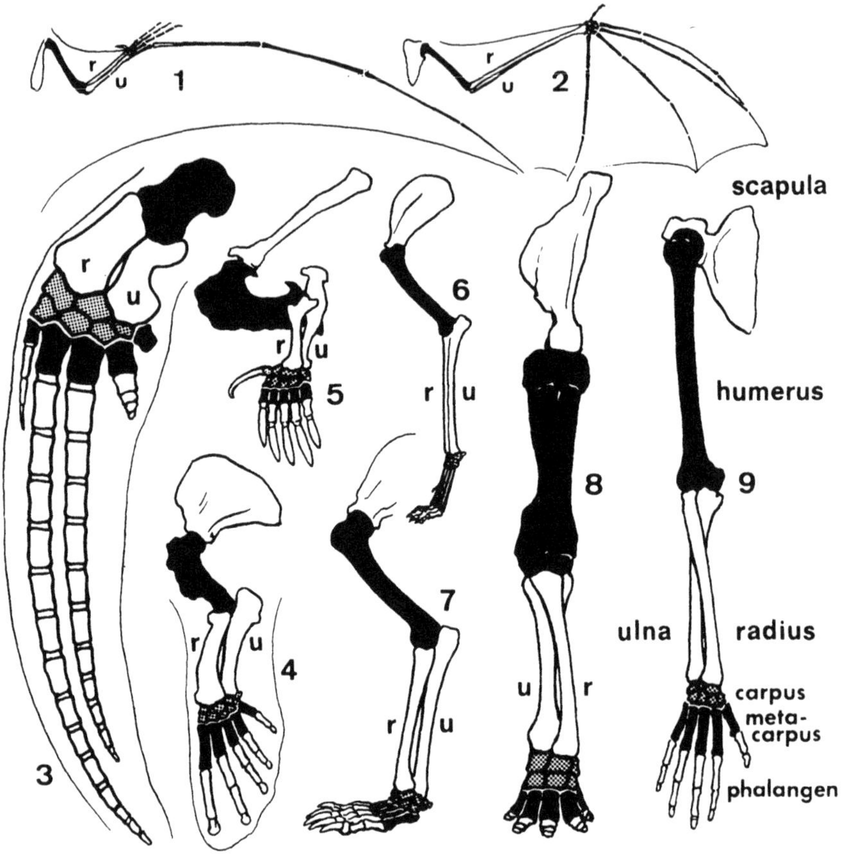

Fig. 3. Anterior limbs of vertebrates. 1. Jurassic flying reptile. 2. Bat. 3. Whale. 4. Sea Lion. 5. Mole. 6. Dog. 7. Bear. 8. Elephant. 9. Man. The humerus and the metacarpal bones are tinged in black, the carpal bone grey

the manifold resemblances of their forms could be due to parallel adaptation, in other words to analogy.

As a pupil of the comparative anatomist and embryologist Ferdinand Hochstetter, I had the benefit of a very thorough instruction in the methodological procedure of distinguishing similarities caused by common descent from those due to parallel adaptation. In fact, the making of this distinction forms a great part of the comparative evolutionist's daily work. Perhaps I should mention here that this procedure has led me to the discovery which I personally consider to be my own most important contribution to science. Knowing animal behaviour as I did, and being instructed in the methods of phylogenetic comparison as I was, I could not fail to discover that the very same methods of comparison, the same concepts of analogy and homology are as applicable to characters of behaviour as they are in those of morphology. This discovery is implicitly contained in the works of Charles Otis Whitman and of Oskar Heinroth; it is only its explicit formulation and the realization of its far-reaching inferences to which I can lay claim. A great part of my life's work has

Fig. 4. Change of function in a pieve of medieval armour which, losing its protective function, becomes a status symbol of officers.

consisted in tracing the phylogeny of behaviour by disentangling the effects of homology and of parallel evolution. Full recognition of the fact that behaviour patterns can be hereditary and species-specific to the point of being homologizable was impeded by resistance from certain schools of thought, and my extensive paper on homologous motor patterns in *Anatidae* was necessary to make my point.

5. Cultural homology

Much later in life I realized that, in the development of human cultures, the interaction between historically-induced similarities and resemblances caused by parallel evolution—in other words between homologies and analogies—was very much the same as in the phylogeny of species and that it posed very much the same problems. I shall have occasion to refer to these later on; here I want to illustrate the existence of cultural homology. Fig. 4 illustrates the cultural changes by which the piece of medieval armour that was originally designed to protect throat and chest was gradually turned, by a change of function, into a status symbol. Otto Koenig, in his book *Kulturethologie,* has adduced many other examples of persistent historically-induced similarity of characters to which the adjective "homologous" can legitimately be applied.

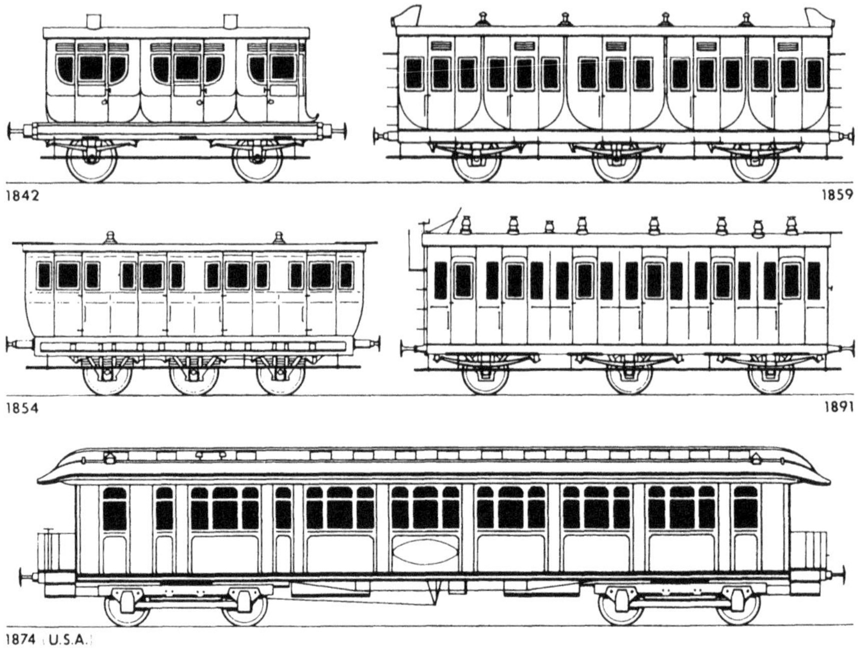

Fig. 5. Homology of technical products. Characters traceable to the ancestor, the horse-drawn coach persists, against the interests of technical progress, in railway carriages.

Ritualization and symbolisms play a large role in traditional clothing and particularly in military uniforms in their historical changes, so that the appearance of historically-retained similarities is, perhaps, not very surprising. It is, however, surprising that the same retention of historical features, not only independently of function, but in clear defiance of it, is observable even in that part of human culture which one would suppose to be free of symbolism, ritualization and sentimental conservativism, namely in technology. Fig. 5 illustrates the development of the railway carriage. The ancestral form of the horse-drawn coach stubbornly persists, despite the very considerable difficulties which it entails, such as the necessity of constructing a runningboard all along the train, on which the conductor had to climb along, from compartment to compartment, exposed to the inclemency of the weather and to the obvious danger of falling off. The advantages of the alternative solution of building a longitudinal corridor within the carriage are so obvious that they serve as a demonstration of the amazing power exerted by the factors tending to preserve historical features in defiance of expediency.

The existence of these cultural homologies is of high theoretical importance, as it proves that, in the passing-on of cultural information from one generation to the next, processes are at work which are entirely independent of rational considerations and which, in many respects, are functionally analogous to the factors maintaining invariance in genetical inheritance.

6. Deducing function from behavioural analogies

Let me now speak of the value of analogies in the study of behaviour. Not being vitalists, we hold that any regularly observable pattern of behaviour which, with equal regularity, achieves survival value is the function of a sensory and nervous mechanism evolved by the species in the service of that particular function. Necessarily, the structures underlying such a function must be very complicated, and the more complicated they are, the less likely it is, as we already know, that two unrelated forms of life should, by sheer coincidence, have happened to evolve behaviour patterns which resemble each other in a great many independent characters.

A striking example of two complicated sets of behaviour patterns evolving independently in unrelated species, yet in such a manner as to produce a great number of indubitable analogies, is furnished by the behaviour of human beings and of geese when they fall in love and when they are jealous. Time and again I have been accused of uncritical anthropomorphism when describing, in some detail, this behaviour of birds and people. Psychologists have protested that it is misleading to use terms like falling in love, marrying or being jealous when speaking of animals. I shall proceed to justify the use of these purely functional concepts. In order to assess correctly the vast improbability of two complicated behaviour patterns in two unrelated species being similar to each other in so many independent points, one must envisage the complication of the underlying physiological organization. Consider the minimum degree of complication which even a man-made electronic model would have to possess in order to simulate, in the simplest possible manner, the behaviour patterns here under discussion. Imagine an apparatus, A, which is in communication with another one, B, and keeps on continuously checking whether apparatus B gets into communication with a third apparatus C, and which furthermore, on finding that this is indeed the case, does its utmost to interrupt this communication. If one tries to build models simulating these activities, for example in the manner in which Grey-Walter's famous electronic tortoises are built, one soon realizes that the minimum complication of such a system far surpasses that of a mere eye.

The conclusion to be drawn from this reasoning is as simple as it is important. Since we know that the behaviour patterns of geese and men cannot possibly be homologous—the last common ancestors of birds and mammals were lowest reptiles with minute brains and certainly incapable of any complicated social behaviour—and since we know that the improbability of coincidental similarity can only be expressed in astronomical numbers, *we know for certain* that it was a more or less identical survival value which caused jealousy behaviour to evolve in birds as well as in man.

This, however, is *all* that the analogy is able to tell us. It does not tell us wherein this survival value lies—though we can hope to ascertain this by observations and experiments on geese. It does not tell us anything about the physiological mechanisms bringing about jealousy behaviour in the two species; they may well be quite different in each case. Streamlining is achieved in the shark by the shape of the musculature, in the dolphin by a thick layer of blubber, and in the torpedo by welded steel plates. By the same token, jealousy may be—and probably is—caused by an inherited and genetically fixed programme in geese, while it might be determined by cultural tradition in man—though I do not think it is, at least not entirely.

Limited though the knowledge derived from this kind of analogy may be, its importance is considerable. In the complicated interaction of human social behaviour, there is much that does not have any survival value and never had any. So it is of consequence to know that a certain recognizable pattern of behaviour does, or at least once did, possess a survival value for the species, in other words, that it is *not pathological.* Our chances of finding out wherein the survival value of the behaviour pattern lies are vastly increased by finding the pattern in an animal on which we can experiment.

When we speak of falling in love, of friendship, personal enmity or jealousy in these or other animals, we are *not* guilty of anthropomorphism. These terms refer to functionally-determined concepts, just as do the terms legs, wings, eyes and the names used for other bodily structures that have evolved independently in different phyla or animals. No one uses quotation marks when speaking or writing about the eyes or the legs of an insect or a crab, nor do we when discussing analogous behaviour patterns.

However, in using these different kinds of terms, we must be very clear as to whether the word we use at a given moment refers to a concept based on functional analogy or to one based on homology, e.g. on common phyletic origin. The word "leg" or "wing" may have the connotation of the first kind of concept in one case and of the second in another. Also, there is the third possibility of a word connoting the concept of physiological, causal identity. These three kinds of conceptualization may coincide or they may not. To make a clear distinction between them is particularly important when one is speaking of behaviour. A homologous behaviour pattern can retain its ancestral form *and* function in two descendants, and yet become physiologically different. The rhythmical beat of the umbrella is caused by ondogenous stimulus generation in many hydrozoa and in larva *(Ephyrae)* in other medusae. In adult *Scyphomedusae,* however, it is caused by reflexes released through the mechanism of the so-called marginal bodies. A homologous motor pattern may retain its original physiological causation as well as its external forms, yet undergo an entire change of function. The motor pattern of "inciting" common to the females of most *Anatidae* is derived from a threatening movement and has the primary function of causing the male to attack the adversary indicated by the female's threat. It has entirely lost this function in some species, for instance in the Golden-eyes, in which it has become a pure courtship movement of the female. Two non-homologous motor patterns of two related species may, by a change of function, be pressed into the service of the same survival value. The pre-flight movement of ducks is derived from an intention movement of flying, an upward thrust of head and neck, while the corresponding signal of geese is derived from a displacement shaking ot the head. When we speak of "pre-flight movements of *Anatidae*" we form a functional concept embracing both. These examples are sufficient to demonstrate the importance of keeping functional, phylogenetical and physiological conceptualizations clearly apart. Ethologists are not guilty of "reifications" or of illegitimate anticipations of physiological explanations when they form concepts that are only functionally defined—like, for instance, the concept of the IRM, the innate releasing mechanism. They are, in fact, deeply aware that this function may be performed by the sensory organ itself—as in the cricket—or by a complicated organization of the retina—as in the frog—or by the highest and most complicated processes within the central nervous system.

Deducing the existence of physiological mechanisms from known analogous functions

Recognizing analogies can become an important source of knowledge in quite another way. We can assume with certainty that, for instance, the functions of respiration, of food intake, of excretion, of propagation, etc., must somehow be performed by any living organism. In examining an unknown living system, we are, therefore, justified in *searching* for organs serving functions which we know to be indispensable. We are surprised if we miss some of them, for instance the respiratory tract in some small salamanders which breathe exclusively through their skin.

A human culture is a living system. Though it is one of the highest level of integration, its continuance is nevertheless dependent on all the indispensable functions mentioned above. The thought obtrudes itself that there is one of these necessary functions which is insufficient in our present culture, that of *excretion*. Human culture, after enveloping and filling the whole globe, is in danger of being killed by its own excretion, of dying from an illness closely analogous to uraemia. Humanity will be forced to invent some sort of planetary kidney—or it will die from its own waste products.

There are other functions that are equally indispensable to the survival of *all* living systems, ranging from bacteria to cultures. In any of these systems, adaptation has been achieved by the process, already mentioned, which hinges on the *gaining of information* by means of genetic change and natural selection, as well as on the storing of knowledge in the code of the chain molecules in the genome.

This storing, like *any* retention of information, of knowledge, is achieved by the formation of *structure*. Not only in the little double helix, but also in the programming of the human brain, in writing, or any other form of "memory bank", knowledge is laid down in structures.

The indispensable supporting and retaining function of structure always has to be paid for by a "stiffening", in other words, by the sacrifice of certain *degrees of freedom*. The structure of our skeleton provides an example; a worm can bend its body at any point, whereas we can flex our limbs only where joints are provided; but we can stand upright and the worm cannot.

All the adaptedness of living systems is based on knowledge laid down in structure; structure means *static* adapted*ness,* as opposed to the dynamic process of adap*tation*. Hence, new adaptation unconditionally presupposes a *dismantling* of some structures. The gaining of new information inexorably demands the breaking down of some previous knowledge which, up to that moment, had appeared to be final.

The dynamics of these two antagonistic functions are universally common to all living systems. Always, a harmonious equilibrium must be sustained between, on the one hand, the factors maintaining the necessary degree of *invariance* and, on the other, the factors which tend to break up firm structures and thereby create the degree of *variability* which is the prerequisite of all further gaining of information, in other words, of all new adaptation.

All this is obviously true of human culture as well as of any other living system whose lifespan exceeds that of the individual, e.g. of any species of bacteria, plants or animals. It is, therefore, legitimate to search for the mechanisms which, in their harmonious antagonism of preserving and dismantling structures, achieve the task of keeping a culture adapted to its ever-changing environment. In my latest book Die Rückseite des Spiegels, I have tried to demonstrate these two antagonistic sets of mechanisms in human culture.

The preservation of the necessary invariance is achieved by procedures curiously reminiscent of genetic inheritance. In much the same manner as the new nucleotids are arranged along the old half of a double helix, so as to produce a *copy* of it, the invariant structures of a culture are passed on, from one generation to the next, by a process *in which the young generation makes a copy* of the cultural knowledge possessed by the old. Sheer imitation, respect for a father-figure, identification with it, force of habit, love of old ritualized customs and, last not least, the conservativism of "magical thinking" and superstition—which as we have seen influences even the construction of railway carriages—contributes to invest cultural tradition with that degree of invariance which is necessary *to make it inheritable at all.*

Opposed to these invariance-preserving mechanisms, there is the specifically human urge to curiosity and freedom of thought which with some of us, persists until senescence puts a stop to it. However, the age of puberty is typically the phase in our ontogeny during which we tend to rebel against tradition, to doubt the wisdom of traditional knowledge and to cast about for new causes to embrace, for new ideals.

In a paper which I read a few years ago—at a Nobel symposium on "The Place of Value in a World of Facts"—I tried to analyse certain malfunctions of the antagonistic mechanisms and the dangers of an enmity between the generations arising from these disturbances. I tried to convince my audience that the question whether conservativism is "good" or "bad", or whether the rebellion of youth is "good" or "bad", is just as inane as the question whether some endocrine function, for instance that of the thyroid gland, is "good" or "bad". Excesses as well as deficiency of any such function cause illness. Excess of thyroid function causes Basedow's disease, deficiency myxoedema. Excess of conservativism produces living fossils which will not go on living for long, and excess of variability results in the appearance of monsters which are not viable at all.

Between the conservative representatives of the "establishment" on the one hand and rebelling youth on the other, there has arisen a certain enmity which makes it difficult for each of the antagonists to recognize the fact that the endeavours of *both* are *equally* indispensable for the survival of our culture. If and when this enmity escalates into actual *hate,* the antagonists cease to interact in the normal way and begin to treat each other as different, hostile cultures; in fact they begin to indulge in activities closely akin to tribal warfare. This represents a great danger to our culture, inasmuch as it may result in a complete disruption of its traditions.

Niko Tinbergen

NIKOLAAS TINBERGEN

I was born in The Hague, Netherlands, on 15th April 1907, the third of five children of Dirk C. Tinbergen and Jeannette van Eek. We were a happy and harmonious family. My mother was a warm, impulsive person; my father—a grammar school master in Dutch language and history—was devoted to his family, a very hard worker, and an intellectually stimulating man, full of fine, quiet humour and *joie de vivre*.

I was not much interested in school, and both at secondary school and at University, I only just scraped through, with as little effort as I judged possible without failing. Wise teachers, including my University Professors in Leiden, H. Boschma and the late C. J. van der Klaauw, allowed me plenty of freedom to engage in my hobbies of camping, bird watching, skating and games, of which playing left-wing in grass hockey teams gave me free rein for my almost boundless youthful energies.

Throughout my life, Fortune has smiled on me. Holland's then unparalleled natural riches—its vast sandy shores, its magnificent coastal dunes, the abundant wildlife in its ubiquitous inland waters, all within an hour's walk of our urban home—delighted me, and I was greatly privileged in having access to the numerous stimulating writings of the two quite exceptional Dutch naturalists, E. Heimans and Jac P. Thijsse—still household names in the Netherlands.

As a boy, I had two small aquaria in our backyard, in which I watched, each spring, the nest building and other fascinating behaviours of Sticklebacks. My natural history master at our High School, Dr. A. Schierbeek, put some of us in charge of the three seawater aquaria in the classroom, rightly arguing to the Head Master that I got plenty of fresh air, so that no one needed to worry about my spending the morning break indoors.

Having been frightened off by what I had been told of academic Biology as it was then taught in Leiden, I was at first disinclined to go to University. But a friend of the family, Professor Paul Ehrenfest, and Dr. Schierbeek urged my father to send me, in 1925, to Professor J. Thienemann, the founder of the famous 'Vogelwarte Rossitten', and the initiator of bird ringing. While Thienemann did not quite know what to do with this awkward youth, the photographer Rudy Steinert and his wife Lucy took me along on their walks along the uniquely rich shores and dunes of the *Kurische Nehrung*, where I saw the massive autumn migration of birds, the wild Moose, and the famous *Wanderdünen*. Upon my return to Holland, Christmas 1925, I had decided to read Biology at Leiden University after all. Here I had the good fortune to be befriended by Holland's most gifted naturalist Dr. Jan Verwey, who instilled in me, by his example, a professional interest in animal behaviour (he also beat me, much to my humiliation, in an *impromptu* running match along the deserted *Noordwijk* seashore—two exuberant Naked Apes!). I owe my interest in seagulls to early imprinting on a small protected Herring Gull colony not far from the Hague, and to the example of two fatherly friends, the late G. J. Tijmstra and Dr. h. c. A. F. J. Portielje.

Having scraped through my finals without much honour, I became engaged to Elisabeth Rutten, whose family I had often joined on skating trips on the Zuiderzee; this made me realise that some day I would have to earn a living. Influenced by the work of Karl von Frisch, and by J.-H. Fabre's writings on insects, I decided to use the chance discovery of a colony of Bee-wolves (*Philanthus*—a digger wasp) for a study of their remarkable homing abilities. This led to an admittedly skimpy but still quite interesting little thesis, which (as I was told later) the Leiden Faculty passed only after grave doubts; 32 pages of print were not impressive enough. But I was itching to get this milestone behind me, for, through the generosity of Sidney Van den Bergh, I had been offered the opportunity of joining the Netherlands' small contingent for the International Polar Year 1932—33, which was to have its base in Angmagssalik, the homeland of a small, isolated Eskimo tribe. My wife and I lived with these fascinating people for two summers and a winter just before they were westernised. Our first-hand experience of life among this primitive community of hunter-gatherers stood us in good stead forty years laters when I tried to reconstruct the most likely way of life of ancestral Man.

Upon our return to Holland, I was given a minor instructor's job at Leiden University, where in 1935 Professor C. J. van der Klaauw, who knew how to stretch his young staff members, told me to teach comparative anatomy and to organise a teaching course in animal behaviour for undergraduates. I was also allowed to take my first research graduates into the field and so could extend my official 12-day annual holiday to an annual two months' period of field work. This we used for further studies of the homing of Bee-wolves and behaviour studies of other insects and birds.

In 1936 Van der Klaauw invited Konrad Lorenz to Leiden for a small symposium on 'Instinct', and it was then that Konrad and I first met. We 'clicked' at once. The Lorenzes invited us, with our small son, for a four-months' stay in their parental home in Altenberg near Vienna, where I became Lorenz' second pupil (the first being Dr. Alfred Seitz, of the *Seitz's Reizsummenregel*). But from the start 'pupil' and 'master' influenced each other. Konrad's extraordinary vision and enthusiasm were supplemented and fertilised by my critical sense, my inclination to think his ideas through, and my irrepressible urge to check our 'hunches' by experimentation—a gift for which he had an almost childish admiration. Throughout this we often burst into bouts of hilarious fun—in Konrad's words, in *Lausbuberei*.

These months were decisive for our future collaboration and our lifelong friendship. On the way back to Holland, I shyly wrote to the great Von Frisch asking whether I could call at his already famous Rockefeller-built laboratory in Munich. My recollection of that visit is a mixture of delight with the man Von Frisch, and an anxiety on his behalf when I saw that he refused to reply to a student's aggressive *Heil Hitler* by anything but a quiet *Grüss Gott.*

In 1938 the Netherlands-America Foundation gave me free passage to and from New York, which I used for a four months' stay, eked out by fees for lectures given in halting English, by living for one dollar a day in YMCAs (40 c for a room, 50 c for a day's food, and 2 nickels for the subway), and travelling by Greyhound. During that visit I met Ernst Mayr, Frank A. Beach, Ted Schneirla, Robert M. Yerkes (who offered me hospitality both in Yale and Orange Park, Florida) and many others. I was frankly bewildered by what I saw of American Psychology. I sailed for home shortly after the Munich crisis, bracing myself for the dark years that we knew were lying ahead.

There followed a year of intense work, and of lively correspondence with Lorenz, which was broken off by the outbreak of war. Both of us saw this as a catastrophe. *Wir hätten soviel Gutes vor,* wrote Lorenz before the evil forces of nazism descended on Holland.

In the war I spent two years in a German hostage camp while my wife saw our family through the difficult times; Lorenz was conscripted as an Army doctor and disappeared during the battle of Witebsk; he did not emerge from Russian prison camps until 1947. Our reunion, in 1949, in the hospitable home of W. H. Thorpe in Cambridge, was to both of us a deeply moving occasion.

Soon after the war I was once again invited to the United States, and to Britain, to lecture on our work on animal behaviour. Lasting friendships with Ernst Mayr and David Lack proved decisive for my later interest in evolution and ecology. The lectures in the U.S. were worked out to a book 'The Study of Instinct' (1951); and my visit to Oxford, where David Lack had just taken over the newly founded Edward Grey Institute of Field Ornithology, ultimately led to our accepting the invitation of Sir Alister Hardy to settle in Oxford.

Apart from establishing, as Hardy had asked me, a centre of research and teaching in animal behaviour, I spent my Oxford years in seeing our newly founded journal Behaviour through its early years, in helping to develop contact with American psychology (of which we were perhaps excessively critical), and in fostering international cooperation. This work would not have been possible without the active help, behind the scenes, of Sir Peter Medawar (who urged the Nuffield Foundation to finance our little research group through its first ten years) and of E. M. Nicholson, who allocated generous funds from the Nature Conservancy which, with hardly any strings, was to last until my retirement. When Professor J. W. S. Pringle succeeded Alister Hardy as Head of the Department of Zoology in Oxford, he not only supported and encouraged our group, but also interested us in bridging the gap (so much wider than we had realised) between ethology and neuro-physiology. By founding the new inter-disciplinary Oxford School of Human Sciences he stimulated my still dormant desire to make ethology apply its methods to human behaviour.

Our research group was offered unique opportunities for ecologically oriented field work when Dr. h. c. J. S. Owen, the then Director of Tanzania's National Parks, asked me to help him in founding the Serengeti Research Institute. A number of my pupils have since helped to establish this Institute's world fame; and the scientific ties with it have remained strong ever since.

Our work received recognition by various proofs of acceptance by the scientific community, among which I value most my election as a Fellow of the Royal Society in 1962; as a Foreign Member of the *Koninklijke Nederlandse Akademie van Wetenschappen* in 1964; the conferment, in 1973, of the honorary degree of D. Sc. by Edinburgh University; and the awarding of the Jan Swammerdam medal of the *Genootschap voor Natuur-, Genees-, en Heelkunde* of Amsterdam in 1973.

In recent years I have, with my wife, concentrated my own research on the socially important question of Early Childhood Autism. This and other work on the development of children has recently brought us in contact with Professor Jerome S. Bruner, whose invigorating influence is already being felt throughout Britain. My only regret is that I am not ten years younger, so that I could more actively join him in developing his centre of child ethology in Oxford.

Among my publications the following are representative of my contributions to the growth of ethology:

1951 The Study of Instinct—Oxford, Clarendon Press
1953 The Herring Gull's World—London, Collins
1958 Curious Naturalists—London, Country Life
1972 The Animal in its World Vol. 1.—London, Allen & Unwin; Harvard University Press
1973 The Animal in its World Vol. 2.—London, Allen & Unwin; Harvard University Press
1972 (together with E. A. Tinbergen) Early Childhood Autism — an Ethological Approach—Berlin, Parey

ETHOLOGY AND STRESS DISEASES

Nobel Lecture, December 12, 1973

by

NIKOLAAS TINBERGEN

Department of Zoology, University of Oxford, England

Many of us have been surprised at the unconventional decision of the Nobel Foundation to award this year's prize 'for Physiology or Medicine' to three men who had until recently been regarded as 'mere animal watchers'. Since at least Konrad Lorenz and I could not really be described as physiologists, we must conclude that our *scientia amabilis* is now being acknowledged as an integral part of the eminently practical field of Medicine. It is for this reason that I have decided to discuss today two concrete examples of how the old method (1) of 'watching and wondering' about behaviour (which incidentally we revived rather than invented) can indeed contribute to the relief of human suffering—in particular of suffering caused by stress. It seems to me fitting to do this in a city already renowned for important work on psychosocial stress and psychosomatic diseases (2).

My first example concerns some new facts and views on the nature of what is now widely called Early Childhood Autism. This is a set of behavioural aberrations which Leo Kanner first described in 1943 (3). To us, i.e. my wife Elisabeth and me, it looks as if it is actually on the increase in a number of western and westernised societies. From the description of autistic behaviour—or Kanner's syndrome (4)—it is clear, even to those who have not themselves seen these unfortunate children, how crippling this affliction is. In various degrees of severity, it involves, among other things: a total withdrawal from the environment; a failure to acquire, or a regression of overt speech, and a serious lagging behind in the acquisition of numerous other skills; obsessive preoccupation with a limited number of objects; the performance of seemingly senseless and stereotyped movements; and an EEG pattern that indicates high overall arousal. A number of autists recover (some of them 'spontaneously') but many others end up in mental hospitals, where they are then often diagnosed, and treated, as schizophrenics.

In spite of a growing volume of research on the subject (5), opinions of medical experts on how to recognise autism, on its causation, and therefore on the best treatment vary widely. Let me consider this briefly, point by point.

1. There is disagreement already at the level of diagnosis and labeling. For instance, for 445 children Rimland compared the diagnosis given by the doctor who was consulted first, with a 'second opinion' (6). If the art of diagnosis had any objective basis, there should be a positive correlation between first and second opinions. In fact, as Rimland points out, there is not a trace of such a correlation—the diagnoses are practically random (Table 1). What

Table 1. Comparison of first and second opinions about 445 children showing severe behaviour disorders. From Rimland, 1971.

2nd Opinion → / 1st Opinion ↓	Autistic	Infantile autism or early infantile autism	Child-hood schizo-phrenia	Emotionally disturbed or mentally ill	Brain damaged, neuro-logically damaged	Retarded	Psychotic (symbiotic psychosis), etc.	Deaf or partly deaf	Total
Autistic	**33**	5	53	18	23	51	10	7	200
Infantile autism or early infantile autism	1	**10**	6	0	4	6	0	2	29
Childhood schizophrenia	17	3	**1**	2	8	1	0	0	32
Emotionally disturbed or mentally ill	12	2	4	**2**	9	13	3	0	45
Brain damaged or neurologically damaged	14	3	2	5	**4**	15	0	1	44
Retarded	21	2	6	18	16	**5**	2	2	72
Psychotic (symbiotic psychosis), etc.	4	0	1	1	2	2	**0**	0	10
Deaf or partly deaf	4	1	0	2	0	5	1	**0**	13
Total	106	26	73	48	66	98	16	12	445

these doctors have been saying to the parents is little more than: 'You are quite right; there is something wrong with your child'.

And yet, if we use the term autism in the descriptive sense of 'Kanner's syndrome', it does name a relatively well-defined cluster of aberrations.

2. The disagreement about the causation of autism is no less striking. It expresses itself at two levels. First, there is the usual 'nature-nurture' controversy. The majority of experts who have written on autism hold that it is due either to a genetic defect, or to equally irreparable 'organic' abnormalities—for instance brain damage such as can be incurred during a difficult delivery. Some of the specialists are certainly emphatic in their assertion that autism is 'not caused by the personalities of the parents, nor by their child-rearing practices' (7). If this were true, the outlook for a real cure for such children would of course be bleak, for the best one could hope for would be an amelioration of their suffering. But there are also a few experts who are inclined to ascribe at least some cases of autism to damaging environmental causes—either traumatising events in early childhood, or a sustained failure in the parent-infant interaction (8). If this were even partially correct, the prospect for a real cure would be brighter of course.

The confusion about causation is also evident in the disagreement about the question what is 'primary' in the overall syndrome—what is 'at the root of the trouble'—and what are mere symptoms. Some authors hold that autism is primarily either a cognitive, or (often mentioned in one breath) a speech defect (9). Others consider the hyperarousal as primary (10). Those who subscribe to the environmental hypothesis think either in terms of too much overall input (11), or in terms of failures in the processes of affiliation, and of subsequent socialisation (8).

3. In view of all this it is no wonder that therapies, which are often based on views concerning causation, also differ very widely. Nor is it easy to judge the success rates of any of these therapies, for the numbers of children treated by any individual therapist or institution are small; also, the descriptions of the treatments are inevitably incomplete and often vague. Unless one observes the therapist in action it is not really possible to judge what he has actually been doing.

In short, as O'Gorman put it not long ago (4. p. 124) '... our efforts in the past have been largely empirical, and largely ineffectual'.

In view of all this uncertainty any assistance from outside the field of Psychiatry could be of value. And it is such assistance that my wife and I have recently tried to offer (12). Very soon our work led us to conclusions which went against the majority opinion, and we formulated proposals about therapies which, with few exceptions, had not so far been tried out. And I can already say that, where these treatments have been applied, they are leading to highly promising results, and we feel that we begin to see a glimmer of hope.

Before giving my arguments for this optimistic prognosis, let me describe how and why we became involved.

Our interest in autistic children, aroused initially by what little we had seen of the work that was being done in the Park hospital in Oxford, remained dormant for a long time. But when, in 1970, we read the statement by Drs. John and Corinne Hutt that '... apart from gaze aversion of the face, all other components of the social encounters of these autistic children are those shown by normal non-autistic children'. (13, p. 147), we suddenly sat up, because we knew from many years of child watching that normal children quite often show *all* the elements of Kanner's syndrome.

Thinking this over we remembered the commonsense but sound warning of Medawar, namely that 'it is not informative to study variations of behaviour unless we know beforehand the norm from which the variants depart' (14, p. 109), and we realised that these words had not really been heeded by psychiatrists. In their literature we had found very little about normal children that could serve as a basis for comparison.

We also realised that, since so many autists do not speak (and are often quite wrongly considered not to understand speech either) a better insight into their illness would have to be based on the study of their non-verbal behaviour. And it is just in this sphere that we could apply some of the methods that had already proved their value in studies of animal behaviour (15).

Therefore we began to compare our knowledge of the non-verbal behaviour that normal children show only occasionally, with that of true autists, which we had not only found described in the literature but also began to observe more closely at first hand.

The types of behaviour to which we soon turned our attention included such things as: the child keeping its distance from a strange person or situation; details of its facial expressions; its bodily stance; its consistent avoid-

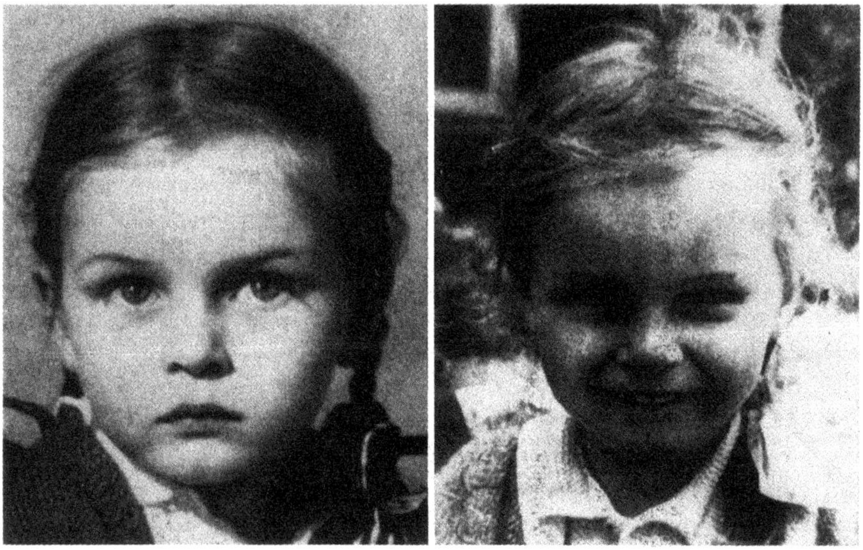

Fig. 1. Two photographs of a girl (aged 6 years) taken in the same spring. Left: taken by a school photographer. Right: taken by her elder sister. They illustrate some non-verbal expressions as used in motivational analysis. From Tinbergen and Tinbergen, 1972.

Fig. 2. An example of 'temporary' and permanent autistic behaviour. Left: Typical slight rejection by a twelve months old normal boy, photographed in his own home in the presence of his mother, who was smiling at him, and faced by him, from approximately four metres distance. The photographer, who was his (rarely met) grandfather, was approximately 1 1/2 metres away from the child. From Tinbergen and Tinbergen 1972. Right: 'Response of an autistic child to repeated attempts of adult to make eye-to-eye contact (drawn from 8 mm motion picture film) from Hutt and Ounsted 1966'. Reproduced from Hutt and Hutt 1970.

ance of making eye contact etc.—an extremely rich set of expressions that are all correlated with overt avoidance (Figs 1 and 2). The work of professional child ethologists is beginning to show us how immensely rich and subtle the repertoire is of such non-verbal expressions (16).

But, apart from observing these behaviours themselves, we also collected evidence about the circumstances which made normal children revert to bouts of autistic behaviour.

What emerged from this dual approach was quite clear: such passing attacks of autistic behaviour appear in a normal child when it finds itself in a situation that creates a conflict between two incompatible motivations. On the one hand, the situation evokes fear (a tendency to withdraw, physically and mentally) but on the other hand it also elicits social, and often exploratory behaviour—but the fear prevents the child from venturing out into the world. And, not unexpectedly, it is 'naturally' timid children (by nature or by nurture, or both) that show this conflict behaviour more readily than more resilient, confident children. But my point is that they all respond to the environment.

Once we had arrived at this interpretation, we tested it in some simple experiments. In fact, we realised that in our years of interaction with children we had already been experimenting a great deal. Such experiments had not been aiming at the elicitation of autistic behaviour, but rather at its opposite: its elimination. As we have written before, each of these experiments was in reality a subtly modulated series of experiments. For a description of what we actually did, I quote from our original publication. We wrote (12, pp. 29—30):

'What we invariably do when visiting, or being visited by a family with young children is, after a very brief friendly glance, ignoring the child(ren) completely, at the same time eliciting, during our early conversations, friendly responses from the parent(s). One can see a great deal of the behaviour of the child out of the corner of one's eye, and can monitor a surprising amount of the behaviour that reveals the child's state. Usually such a child will start by simply looking intently at the stranger, studying him guardedly. One may already at this stage judge it safe to now and then look briefly at the child and assess more accurately the state it is in. If, on doing so, one sees the child avert its glance, eye contact must at once be broken off. Very soon the child will stop studying one. It will approach gingerly, and it will soon reveal its strong bonding tendency by touching one—for instance by putting its hand tentatively on one's knee. This is often a crucial moment: one must *not* respond by looking at the child (which may set it back considerably) but by cautiously touching the child's hand with one's own. Again, playing this 'game' by if necessary stopping, or going one step back in the process, according to the child's response, one can soon give a mildly reassuring signal *by touch,* for instance by gently pressing its hand, or by touching it quickly, and withdrawing again. If, as is often the case, the child laughs at this, one can laugh oneself, but still without looking at the child. Soon it will become more daring, and the continuation of contact, by touch and by indirect vocalisation, will begin to cement a bond. One can then switch to the first, tentative eye contact. This again must be done with caution, and step by step; certainly with a smile, and for brief moments at first. We find that first covering one's face with one's hands, then turning towards the child (perhaps saying 'where's Andrew?' or whatever the child's name) and then briefly showing one's eyes and covering them up at once, is very likely to elicit a smile, or even a laugh. For this, incidentally, a child often takes the initiative (see, e.g. Stroh and Buick (11)). Very soon the child will then begin to solicit this; it will rapidly tolerate increasingly long periods of direct eye contact and join one. If this is played further, with continuous awareness of and adjustments to slight reverses to a more negative attitude, one will soon find the child literally clamouring for intense play contact. Throughout this process the vast variety of expressions of the child must be *understood* in order to monitor it correctly, and one must oneself *apply* an equally large repertoire in order to give, at any moment, the best signal. The 'bag of tricks' one has to have at one's disposal must be used to the full, and the 'trick' selected must whenever possible be adjusted to the child's individual tastes. Once established, the bond can be maintained by surprisingly slight signals; a child coming to show proudly a drawing it has made is often completely happy with just a 'how nice dear' and will then return to its own play. Even simpler vocal contacts can work; analogous to the vocal contact calls of birds (which the famous Swedish writer Selma Lagerloef correctly described in

'Nils Holgersson' as, 'I am here, where are you?') many children develop an individual contact call, to which one has merely to answer in the same language.

'The results of this procedure have been found to be surprisingly rapid, and also consistent *if one adjusts oneself to the monitoring results.* Different children may require different starting levels, and different tempos of stepping-up. One may even have to start by staying away from the child's favourite room. It is also of great significance how familiar to the child the physical environment is. Many children take more than one day; with such it is important to remenber that one has to start at a lower level in the morning than where one left off the previous evening. We have the impression that the process is on the whole completed sooner if one continually holds back until one senses the child longing for a more intense contact.'

With all these experiences with normal children in mind, we began to reconsider the evidence about permanently autistic children—again using our own observations as well as the reports we found in the literature. And two things became clear almost at once: neither for genetic abnormalities nor for gross brain damage was there any convincing, direct evidence; all we found were inferences, or arguments that do not hold water.

The main argument for a genetic abnormality is the statement (and one hears it time and again) 'these children have been odd from birth'. And we also found that, for various reasons, neither the specialists nor the parents are very willing to consider environmental influences. But in view of what we know about the effects of non-genetic agents that act in *utero*—of which the new indications about the effects of rubella contracted by pregnant women is only one (17)—the 'odd-from-birth' argument is of course irrelevant. And at least two cases are known of identical twins of whom only one developed Kanner's syndrome (18).

Equally unconvincing are the arguments in favour of gross brain damage—this idea too is based mainly on inference.

On the other hand, the body of positive evidence that points to environmental causes is growing. For instance, many workers report that the incidence of autism is not random. Relatively many autists are first-born children (19). There is also a pretty widespread conviction that the parents of autists are somehow different — for instance many of them are very serious people, or people who are themselves under some sort of strain. And to a trained observer it is also very obvious that autists respond to conditions, which to them are frightening or intrusive, by an intensification of all their symptoms. Conversely, we have tried out our 'taming procedure' as described for normal children on some severely autistic children, and succeeded in 'drawing them out of their shells', an making them snuggle up to us, and even in making them join us in, for instance, 'touch games'. I cannot possibly go into all the evidence, but there are several good indications, firstly, that many autists are potentially normal children, whose affiliation and subsequent socialisation processes have gone wrong in one way or another, and secondly: this can often be traced back to something in the early environment—on occasion a frightening accident, but most often something in the behaviour of the parents, in particular the mothers. Let me hasten to add that in saying this we are not *blaming* these unfortunate parents. Very often they seem to have been either simply inexperienced (hence perhaps the high incidence among first-borns); or over-apprehensive; or over-efficient and intrusive; or—perhaps most often —they are people who are themselves under stress. For this, and many other reasons, the parents of autists deserve as much compassion, and may be as much in need of help, as the autists themselves.

Now if we are only partially right in assuming that at least a large proportion of autists are victims of some kind of environmental stress, whose basic trouble is of an emotional nature, then one would expect that those therapies that aim at reducing anxiety—by *allowing* spontaneous socialisation and exploration whenever it occurs—would be more successful than those that aim at the teaching of specific skills. Unfortunately (as I have already said), it is hardly possible to judge from published reports what treatment has actually been applied. For instance one speech therapist may behave rather intrusively, and turn a child into a mere 'trained monkey', leaving all the other symptoms as they were, or even making them worse. Another speech therapist may have success simply by having proceeded in a very gentle, motherly way. One has to go by those instances where one has either been involved oneself or where one knows pretty precisely how the therapist has in fact proceeded. It is whith this in mind that I will now mention briefly three examples of treatments that seem to hold great promise.

Firstly, even before we published our first paper, the Australian therapist Helen Clancy had been treating autistic children and their families along lines that are very similar to, in fact are more sophisticated than those recommended by us in 1972.

The gist of Clancy's method is as follows (8): firstly, since she considers the restoration of affiliation as the first goal of treatment of autism, she treats both mother and child, and the family as well. She does this by provoking in the mother an increase in maternal, protective behaviour. Secondly, she uses a form of operant conditioning for speeding up the child's response to this change in the mother. In other words, she tries to elicit a mutual emotional bond between mother and child, and refrains, at least at first, from the piecemeal teaching of particular skills.

With those mothers who were willing to cooperate, Clancy has achieved highly encouraging success—although of course a few families (4 out of appr. 50 treated over a period of 14 years) have failed to benefit.

Secondly, after the first public discussion of our work, my wife received invitations to visit some schools for autists, and to observe what was being done. She found that in one of them, a small day school which already had an impressive record of recoveries, the treatment was likewise aimed at the restoration of emotional security, and teaching as such, including some gentle speech therapy, was never started until a child had reached a socially positive attitude. Much to our dismay, this school has since been incorporated into a school for maladjusted children—the experiment has been discontinued.

Thirdly, a regional psychiatrist invited us a year ago to act as advisors in a fascinating experiment which she too had begun well before she had heard of our work. Three boys, who are now 9, 9 1/2 and 11 1/2 years of age, and who had all been professionally diagnosed as severely autistic, are now being gently integrated into a normal primary school. This involves a part-time home-tutor for each boy, a sympathetic headmaster, and willingness of the parents to cooperate. The results are already little short of spectacular. In fact, a specialist on autistic children who visited the school recently said to us: 'Had the records not shown that these three children were still severely autistic a couple of years ago, I would not now believe it'. This experiment, which is also run along lines that are consistent with our ideas, is being carefully documented.

It is this type of evidence, together with that provided by a number of already published case histories (20), that has by now convinced us that many autists can attain full recovery, if only we act on the assumption that they have been traumatised rather than genetically or organically damaged.

I cannot go into further details here, but I can sum up in a few sentences the gist of what the ethological approach to Early Childhood Autism has produced so far:

1. There are strong indications that many autists suffer primarily from an emotional disturbance, from a form of anxiety neurosis, which prevents or retards normal affiliation and subsequent socialisation, and this in its turn hampers or suppresses the development of overt speech, of reading, of exploration, and of other learning processes based on these three behaviours.

2. More often than has so far been assumed these aberrations are not due to either genetic abnormalities or to gross brain damage, but to early environmental influences. The majority of autists—as well as their parents—seem to be genuine victims of environmental stress. And our work on normal children has convinced us not only that this type of stress disease is actually on the increase in western and westernised countries, but also that very many children must be regarded as semi-autistic, and even more as being seriously at risk.

3. Those therapies that aim at the reduction of anxiety and at a re-starting of proper socialisation seem to be far more effective than for instance speech therapy *per se* and enforced social instruction, which seem to be at best symptom treatments, and to have only limited success. Time and again treatment at the emotional level has produced an explosive emergence of speech and other skills.

If I now try to assess the implications of what I have said, I feel at the same time alarmed and hopeful.

We are alarmed because we found this corner of Psychiatry in a state of disarray, and because we discovered that many of the established experts—doctors, teachers and therapists—are so little open to new ideas and even facts. Another cause for alarm is our conviction that the officially recognised autists are only a fraction of a much larger number of children who obviously suffer to some degree from this form of social stress.

We feel hopeful because attempts at curing such children at the emotional level, while still in the experimental stage, are already leading to positive results. And another encouraging sign is that, among the young psychiatrists, we have found many who are sympathetic to our views, or even share them, and begin to act on them.

In the interest of these thousands of unfortunate children we appeal to all concerned to give the 'stress view' of autism at least the benefit of the doubt, and to try out the forms of therapy that I have mentioned.

My second example of the usefulness of an ethological approach to Medicine has quite a different history. It concerns the work of a very remarkable man, the late F. M. Alexander (21). His research started some fifty years before the revival of Ethology for which we are now being honoured, yet his procedure was very similar to modern observational methods, and we believe that his achievements and those of his pupils deserve close attention.

Alexander, who was born in 1869 in Tasmania, became at an early age a 'reciter of dramatic and humerous pieces'. Very soon he developed serious vocal trouble and he came very near to losing his voice altogether. When no doctor could help him, he took matters into his own hands. He began to observe himself in front of a mirror, and then he noticed that his voice was at its worst when he adopted the stances which to him felt appropriate and 'right' for what he was reciting. Without any outside help he worked out, during a series of agonising years, how to improve what is now called the 'use' of his body musculature in all his postures and movements. And, the remarkable outcome was that he regained control of his voice. This story, of perceptiveness, of intelligence, and of persistence, shown by a man without medical training, is one of the true epics of medical research and practice (22).

Once Alexander had become aware of the mis-use of his own body, he began to observe his fellow men, and he found that, at least in modern western society, the majority of people stand, sit and move in an equally defective manner.

Encouraged by a doctor in Sydney, he now became a kind of missionary. He set out to teach—first actors, then a variety of people—how to restore the proper use of their musculature. Gradually he discovered that he could in this way alleviate an astonishing variety of somatic and mental illnesses. He also wrote extensively on the subject. And finally he taught a number of his pupils to become teachers in their turn, and to achieve the same results with their patients. Whereas it had taken him years to work out the technique and to apply it to his own body, a successful course became a matter of months—with occasional 'refresher' sessions afterwards. Admittedly, the training of a good Alexander teacher takes a few years.

For scores of years a small but dedicated number of pupils have continued his work. Their combined successes have recently been described by Barlow (23). I must admit that his physiological explanations of how the treatment could be supposed to work (and also a touch of hero worship in his book) made me initially a little doubtful, and even sceptical. But the claims made, first by Alexander, and reiterated and extended by Barlow sounded so extraordinary that I felt I ought to give the method at least the benefit of the doubt. And so, arguing that medical practice often goes by the sound empirical principle of 'the proof of the pudding is in the eating', my wife, one of our daughters and I decided to undergo treatment ourselves, and also to use the opportunity for observing its effects as critically as we could. For obvious reasons, each of us went to a different Alexander teacher.

We discovered that the therapy is based on exceptionally sophisticated observation, not only by means of vision but also to a surprising extent by using the sense of touch. It consists in essence of no more than a very gentle, first exploratory, and then corrective manipulation of the entire muscular system. This starts with the head and neck, then very soon the shoulders and chest are involved, and finally the pelvis, legs and feet, until the whole body is under scrutiny and treatment. As in our own observations of children, the therapist is continuously monitoring the body, and adjusting his procedure all the time. What is actually done varies from one patient to another, depending on what kind of mis-use the diagnostic exploration reveals. And naturally, it affects different people in different ways. But between the three of us, we already notice, with growing amazement, very striking improvements in such diverse things as high blood pressure, breathing, depth of sleep, overall cheerfulness and mental alertness, resilience against outside pressures, and also in such a refined skill as playing a stringed instrument.

So from personal experience we can already confirm some of the seemingly fantastic claims made by Alexander and his followers, namely that many types of under-performance and even ailments, both mental and physical, can be alleviated, sometimes to a surprising extent, by teaching the body musculature to function differently. And although we have by no means finished our course, the evidence given and documented by Alexander and Barlow, of beneficial effects on a variety of vital functions no longer sounds so astonishing to us. Their long list includes first of all what Barlow calls the 'rag bag' of rheumatism, including various forms of arthritis; but also respiratory troubles, even potentially lethal asthma; following in their wake, circulation defects, which may lead to high blood pressure and also to some dangerous heart conditions; gastro-intestinal disorders of many types; various gynaecological conditions; sexual failures; migraines and depressive states that often lead to suicide — in short a very wide spectrum of diseases, both 'somatic' and 'mental', that are not caused by identifiable parasites.

Although no one would claim that the Alexander treatment is a cure-all in every case, there can be no doubt that it often does have profound and beneficial effects—and, I repeat once more, both in the 'mental' and 'somatic' sphere.

The importance of the treatment has been stressed by many prominent people, for instance John Dewey (24, Aldous Huxley (25), and—perhaps more convincing to us—by scientists of renown, such as Coghill (26), Raymond Dart (27), and the great neurophysiologist Sherrington (28). Yet, with few exceptions, the medical profession has largely ignored Alexander—per-

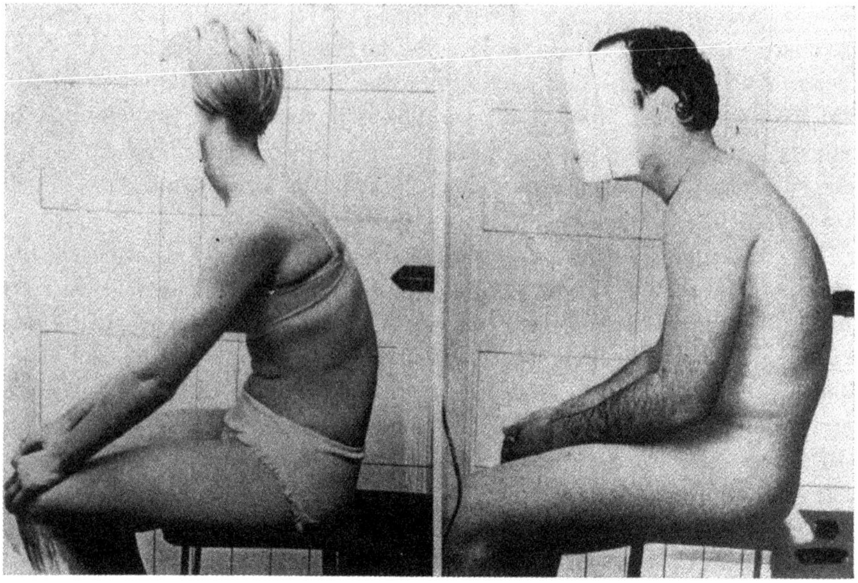

Fig. 3. Typical slumped sitting positions. From Barlow, 1973.

haps under the impression that he was the centre of some kind of 'cult', and also because the effects seemed difficult to explain. And this brings me to my next point.

Once one knows that an empirically developed therapy has demonstrable effects, one likes to know how it could work; what its physiological explanation could be. And here some recent discoveries in the borderline field between neurophysiology and ethology can make some aspects of the Alexander therapy more understandable and more plausible than they could have been in Sherrington's time.

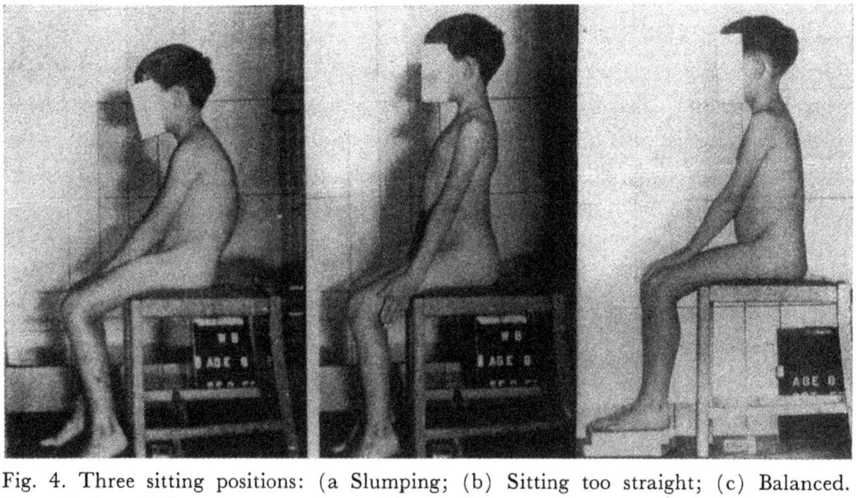

Fig. 4. Three sitting positions: (a Slumping; (b) Sitting too straight; (c) Balanced. From Barlow, 1973.

Fig. 5. Position of pelvis, back, neck and head in 'slumping' position. From Barlow, 1973.

One of these new discoveries concerns the key-concept of 're-afference' (29). There are many strong indiciations that, at various levels of integration, from single muscle units up to complex behaviour, the correct performance of many movements is continuously checked by the brain. It does this by comparing a *feedback report,* that says 'orders carried out', with the feedback *expectation* for which, with the initiation of each movement, the brain has been alerted. Only when the expected feedback and the actual feedback match does the brain stop sending out commands for corrective action. Already the discoverers of this principle, von Holst and Mittelstaedt, knew that the functioning of this complex mechanism could vary from moment to moment with the internal state of the subject—the 'target value' or *Sollwert* of the ex-

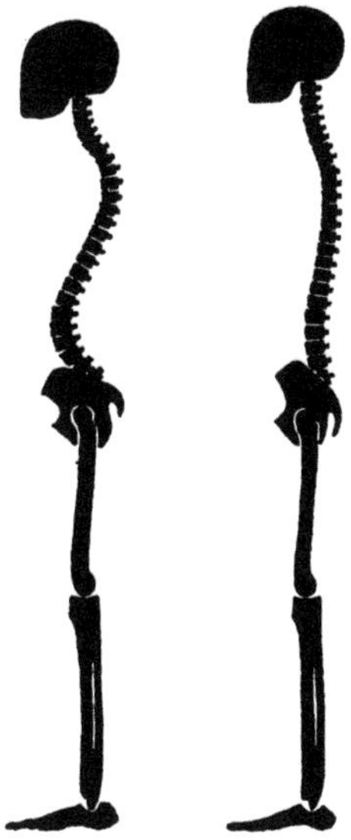

Fig. 6. Standing in 'hunched' position (left) and well balanced (right). From Barlow, 1973.

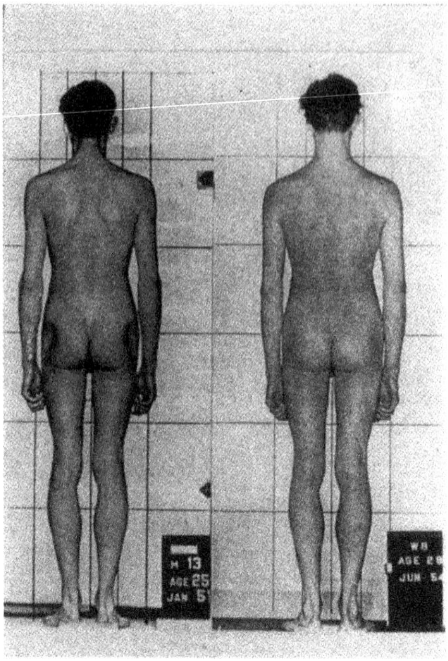

Fig. 7. Posture before (left) and after Alexander treatment. The photograph on the left shows muscle contractions at the back of the neck; raised shoulders and tightened buttocks. After treatment these tensions had disappeared and the patient was overall taller. From Barlow, 1973.

pected feedback changes with the motor commands that are given. But what Alexander has discovered beyond this is that a lifelong mis-use of the body-muscles (such as caused by, for instance, too much sitting and too little walking) can make the entire system go wrong. As a consequence, reports that 'all is correct' are received by the brain (or perhaps interpreted as correct) when in fact all is very wrong. A person can feel 'at ease' e.g. when slouching in front of a television set, when in fact he is grossly abusing his body. I can show you only a few examples, but they will be familiar to all of you. (Figs. 3—7).

It is still an open question exactly where in this complex mechanism the matching procedure goes wrong under the influence of consistent mis-use. But the modern ethologist feels inclined, with Alexander and Barlow, to blame phenotypic rather than genetic causes for mis-use. It is highly unlikely that in their very long evolutionary history of walking upright, the Hominids have not had time to evolve the correct mechanisms for bipedal locomotion. This conlusion receives support from the surprising, but indubitable fact that even after forty to fifty years of obvious mis-use one's body can (one might say) 'snap' back into proper, and in many respects more healthy use as a result of a short series of half-hourly sessions. Proper stance and movement are obviously genetically old, environment-resistant behaviours (30). Mis-use, with all its psychosomatic or rather: somato-psychic consequences, must therefore be considered a result of modern living conditions—of a culturally determined stress. I might add here that I am not merely thinking of too much sitting, but just as much of the 'cowed' posture that one assumes when one feels that one is not quite up to one's work—when one feels insecure.

Secondly, it need not cause surprise that a mere gentle handling of body muscles can have such profound effects on both body and mind. The more that is being discovered about psychosomatic diseases, and in general about the extremely complex two-way traffic between the brain and the rest of the body, the more obvious it has become that too rigid a distinction between 'mind' and 'body' is of only limited use to medical science—in fact can be a hindrance to its advance.

A third biologically interesting aspect of the Alexander therapy is that every session clearly demonstrates that the innumerable muscles of the body are continuously operating as an intricately linked web. Whenever a gentle pressure is used to make a slight change in leg posture, the neck muscles react immediately. Conversely, when the therapist helps one to 'release' the neck muscles, it is amazing to see quite pronounced movements for instance of the toes, even when one is lying on a couch.

In this short sketch, I can do no more than characterise, and recommend, the Alexander treatment as an extremely sophisticated form of rehabilition, or rather of re-deployment, of the entire muscular equipment, and through that of many other organs. Compared with this, many types of physiotherapy which are now in general use look surprisingly crude and restricted in their effect — and sometimes even harmful to the rest of the body.

What then is the upshot of these few brief remarks about Early Childhood Autism and about the Alexander treatment? What have these two examples in common? First of all they stress the importance for medical science of open-minded observation—of 'watching and wondering'. This basic scientific method is still too often looked down on by those blinded by the glamour of apparatus, by the prestige of 'tests' and by the temptation to turn to drugs. But it is by using this old method of observation that both autism and general mis-use of the body can be seen in a new light: to a much larger extent than is now realised both could very well be due to modern stressful conditions.

But beyond this I feel that my two excursions into the field of medical research have much wider implications. Medical science and practice meet with a growing sense of unease and of lack of confidence from the side of the general public. The causes of this are complex, but at least in one respect the situation could be improved: a little more open-mindedness (31), a little more collaboration with other biological sciences, a little more attention to the body as a whole, and to the unity of body and mind, could substantially enrich the field of medical research. I therefore appeal to our medical colleagues to recognise that the study of animals—in particular 'plain' observation—can make useful contributions to human biology not only in the field of somatic malfunctioning, but also in that of behavioural disturbances, and ultimately help us understand what psychosocial stress is doing to us. It is stress in the widest sense, the inadequacy of our adjustability, that will become perhaps the most important disruptive influence in our society.

If I have today stressed the applicability of animal behaviour research I do not want to be misunderstood. As in all sciences, applications come in the wake of research motivated by sheer intellectual curiosity. What this occasion enables me to emphasise is that biologically oriented research into animal behaviour, which has been done so far with very modest budgets, deserves encouragement—whatever the motivation and whatever the ultimate aims of the researcher. And we ethologists must be prepared to respond to the challenge if and when it comes.

Notes and references

1. I call the method old because it must already have been highly developed by our ancestral hunter-gatherers, as it still is in non-westernised hunting-gathering tribes such as the Bushmen, the Eskimo and the Australian Aborigines. As a scientific method applied to Man it could be said to have been revived first by Charles Darwin in 1872 in 'The Expression of the Emotions in Man and the Animals.' London, John Murray.
2. Levi, L. (ed.) 1971 Society, Stress, and Disease. Vol. 1.: The Psychosocial Environment and Psychosomatic Diseases. London, Oxford University Press.
3. Kanner, L. 1943 Austistic disturbances of affective contact. Nerv. Child *2,* 217—250. Recently, Kanner has published a selection of his papers in Childhood Psychosis 1973 Washington D.C., Winston & Son (distributed by John Wiley & Sons).
4. When I speak of Kanner's syndrome, I refer to the largely descriptive list of symptoms given by O'Gorman, G. 1970 The Nature of Childhood Autism. London, Butterworth. This is a slightly modified version of the description given in Creak, M. 1961 (Chairman) The schizophrenic syndrome in childhood. Progress report of a working party. Brit. Med. J. *2,* 889—890. Many other definitions of autism in its various forms are mixtures of observed behavioural deviations and interpretations. For a discussion of the confusion surrounding the word 'autism' see Tinbergen & Tinbergen (see below, note 12) pp. 45—46.
5. For the purpose of finding one's way in this literature we can refer to Rutter, M. 1965 Infantile Autism. London, Methuen; and to the quarterly, started in 1971: Journal of Autism and Childhood Schizophrenia, published by Winston & Son, Washington, D.C., which prints original articles as well as reviews.
   The most recent and most exhaustive review is:
   Ornitz, E. M. 1973 Childhood Autism. (Medical Progress). Calif. Med. *118,* 21—47.

   Throughout the literature (not only on autism but on many other psychiatric issues as well) one finds one fundamental error in scientific reasoning. Time and again we receive the comment that we overlook the 'hard' evidence of internal malfunctioning, in autists as well as in other categories of the mentally ill. I assure my readers that we do not *overlook* such evidence (such as that on blood platelets, on lead contents, on EEG patterns, etc.). The erroneous assumption underlying most of the arguments in which such facts are used for the purpose of throwing light on the causation of the behavioural deviation is almost invariably due to *the confusion between correlations and cause-effect relations.* With some exceptions (such as the deleterious effect of lead) the physiological or biochemical evidence is considered, without any ground whatsoever, to indicate causes, whereas the correlations found could just as well point to consequences or side-effects. *It is just as nonsensical to say that retarded bone growth, or abnormalities in the blood platelet picture (or for that matter speech defects, or high overall arousal) are 'causes' of autism as it is to say that a high temperature is the 'cause' of typhoid, or pneumonia.* Unless there is evidence, clinical and ultimately experimental, indicating what is cause and what is effect the opinions based on 'hard' evidence are in fact worthless. Our experimental evidence discussed on pp. 10 and 11 *is* hard, whereas evidence on correlations—however impressive the techniques might be by which

they are found—are scientifically useless until an attempt is made to place it into cause-effect context. This is what I mean in my final paragraphs by 'the glamour of apparatus'—the idolisation of techniques, coupled with the failure to think about the meaning of evidence, is a serious disease of medical research.

6. Rimland, B. 1971 The differentiation of childhood psychoses. J. Aut. Childh. Schizophr. *1,* 161—175.
7. Wing, L. 1970 The syndrome of early childhood autism. Brit. J. Hosp. Med. 381—392, (p. 381). See also Tinbergen and Tinbergen (see below, note 12) p. 51.
8. One of the most prominent exponents of this view is: Bettelheim, B. 1967 The Empty Fortress: Infantile Austism and the Birth of Self. London, Collier-Macmillan. See also: Clancy, H. and G. McBride 1969 The autistic process and its treatment. J. Child Psychol Psychiatry *10,* 233—244.
9. Rutter, M. et.al. 1971 Autism—a central disorder of cognition and language? In: M. Rutter (ed.) Infantile Autism: Concepts, Characteristics, and Treatment. London, Churchill.
10. Hutt, C., Hutt, S. J., Lee, D. and Ounsted, C., 1964 Arousal and childhood autism. Nature *204,* 908—909.
Hutt, S. J. and Hutt, C., (eds.) 1970 Behaviour Studies in Psychiatry. Oxford, Pergamon Press.
11. Stroh, G. Buick, D. 1970 The effect of relative sensory isolation on the behaviour of two autistic children. In: S. J. Hutt and C. Hutt (eds.) Behaviour Studies in Psychiatry. Oxford, Pergamon Press. pp. 161—174.
12. Tinbergen, E. A. and Tinbergen, N. 1972 Early Childhood Autism—an Ethological Approach. Advances in Ethology *10,* 1—53. Berlin, Paul Parey.
13. Hutt, S. J. and Hutt, C., 1970 Direct Observation and Measurement of Behaviour. Springfield, Ill., Charles C. Thomas.
14. Medawar, P. B. 1967 The Art of the Soluble. London, Methuen.
15. For a recent review about the analysis of non-verbal signs of mixed motivation, or motivational conflicts, see e.g. Manning, A. 1972 An Introduction to Animal Behaviour. London, Arnold. (Chapter 5), and Hinde, R. A. 1970 Animal Behavior. New York, McGraw-Hill. (Chapter 17). Both books give further references.
16. See e.g. Blurton Jones, N. G. (ed.) Ethological Studies of Infant Behaviour. London, Cambridge University Press.
17. Chess, S. 1971 Autism in children with congenital rubella. J. Aut. Childh. Schizophr. *1,* 33—47. The point I want to make with this brief reference is that, while one should call rubella an early environmental influence and therefore not 'congenital' in the sense of 'genetic', it might well be correct to call it 'organic', even though rubella could well create a state of anxiety already during pregnancy in mothers who have heard about other damaging effects of the disease. And this in itself could well cause a complex psychosomatic state.
18. Kamp, N. L. J. 1964 Autistic syndrome in one of a pair of monozygotic twins. Psychiatr., Neurol., Neurochir. *67,* 143—147; and Vaillant, G. E. 1963 Twins discordant for early infantile autism. Arch. Gen. Psychiatry *9,* 163—167. While I do not of course intend to underrate the possibility of genetic predisposition, the hypothesis of a *purely* genetic deviation conflicts with this type of observation. At the same time we know that even when twins grow up in the same family, their experiences can never be identical.
19. See, e.g. Wing, L. 1971 Autistic Children. London, Constable. p. 8.
20. Although not all the authors of the following books label their subject as 'autistic', I mention them because the descriptions of the initial behaviour conform in whole or in part to Kanner's syndrome; and, as I have said, I consider such descriptions the only acceptable starting points.
d'Ambrosio, R. 1971 No Language but a Cry. London, Cassell.
Axline, V. 1964 Dibs—in Search of Self. (since 1971 available as a Pelican Book,

Harmondsworth, Penguin Books Ltd.)
Copeland, J. and J. Hodges 1973 For the Love of Ann. London, Arrow Original.
Hundley, J. M. 1973 (first published in 1971) The Small Outsider. Sydney, Angus and Robertson.
Park, C. C. 1972 (first published in 1967) The Siege. Pelican Book, Harmondsworth, Penguin Books Ltd.
Wexler, S. S. 1971 (first published in 1955) The Story of Sandy. New York, A Signet Book.
Thieme, G. 1971 Leben mit unserem autistischen Kind. Lüdenscheid, Hilfe für das autistische Kind, e.V.
No two of these seven children received the same treatment, but on the whole one can say that those who were treated primarily at the emotional level rather than at the level of specific skills showed the most striking improvement.

21. The clearest introduction is: Alexander, F. M. 1932 The Use of Self. London, Chaterston; but a great deal of interest can also be found in Alexander, F. M. 1910 Man's Supreme Inheritance. London, Chaterston, and Alexander, F. M. 1942 The Universal Constant in Living. London, Chaterston.
22. The history of medical science is full of such examples of 'breakthroughs' due to a re-orientation of attention. Compare e.g. Jenner's discovery that milkmaids did not contract smallpox; Goldberger's observation that the staff of a 'lunatic asylum' did not develop pellagra; Fleming's wondering about empty areas round the *Pennicillium* in his cultures.
23. Barlow, W. 1973 The Alexander Principle. London, Gollancz.
24. Dewey, K. See e.g. 1932 Introduction to Alexander, F. M. The Use of Self. London, Chaterston.
25. Huxley, A. 1937 Ends and Means. London, Chatto and Windus. Huxley, A. 1965 End-gaining and means whereby. Alexander J. *4,* 19.
26. Coghill, G. E. 1941 Appreciation: the educational methods of F. Matthias Alexander. In: Alexander, F. M. 1941 The Universal Constant in Living. New York, Dutton.
27. Dart, R. A. 1947 The attainment of poise. South Afr. Med. J. *21,* 74—91. Dart, R. A. 1970 An anatomist's tribute to F. M. Alexander, London, the Shelldrake Press.
28. Sherrington, C. S. 1946 The Endeavour of Jean Fernel. London, Cambridge University Press.
Sherrington, C. S. 1951 Man on his Nature. London, Cambridge University Press.
29. von Holst, E. and Mittelstaedt, H. 1950 Das Reafferenzprinzip. Naturwiss. *37,* 464—476.
30. Tinbergen, N. 1973 Functional ethology and the human sciences. Proc. Roy. Soc. Lond. B. *182,* 385—410.
31. This plea is nowadays heard more often: see e.g. Kanner, L. 1971 Approaches: retrospect and prospect. J. Aut. Childh. Schizophr. *1,* 453—459 (see p. 457).

1974
Physiology or Medicine

ALBERT CLAUDE, CHRISTIAN DE DUVE and GEORGE E. PALADE

"for their discoveries concerning the structural and functional organization of the cell"

THE NOBEL PRIZE FOR PHYSIOLOGY OR MEDICINE

Speech by Professor JAN-ERIK EDSTRÖM of the Karolinska Medico-Chirurgical Institute
Translation from the Swedish text

Your Majesty, Your Royal Highnesses, Ladies and Gentlemen,
The 1974 Nobel Prize in Physiology or Medicine concerns the fine structure and the function of the cell, a subject designated Cell Biology. There are no earlier Prize Winners in this field, simply because it is one that has been newly created, largely by the Prize Winners themselves. It is necessary to go back to 1906 to find Prize Winners who are to some extent forerunners. In that year Golgi and Cajal were awarded the Prize for studies of cells with the light microscope. Although the light microscope certainly opened a door to a new world during the 19th century, it had obvious limitations. The components of the cell are so small that it was not possible to study their inner structure, their mutual relations or their different roles. To take a metaphor from an earlier Prize Winner, the cell was like a mother's work basket, in that it contained objects strewn about in no discernible order and evidently, for him, with no recognizable functions.

But, if the cell is a work basket, it is one on a very tiny scale indeed, having a volume corresponding to a millionth of that of a pinshead. The various components responsible for the functions of the cell correspond in their turn to a millionth of this millionth, and are far below the resolving powers of the light microscope. Nor would it have helped if researchers had used larger experimental animals: the cells of the elephant are not larger than those of the mouse.

Progress was quite simply at a standstill during the first few decades of this century, but then in 1938, the electron microscope became available, an innovation that held out great promise. The difference between this microscope and the ordinary light microscope is enormous, like being able to read a book instead of just the title. With such an instrument it should now be possible to see components almost down to the dimensions of single molecules. But the early hopes were succeeded by disappointment. It was found impossible to prepare the cells in such a way that they could be used. The book remained obstinately shut, even though it would have been possible to read it.

Albert Claude and coworkers were the first to get a glance inside the book. In the mid-forties they made a break-through and succeeded in preparing cells for electron microscopy. I say a glance, because much technical development still remained to be done, and *George Palade* should be mentioned foremost among those who developed electron microscopy further, to the highest degree of artistry.

In addition to form and structure it is necessary to know the chemical composition of the cell components in order to understand their functions.

It was hardly possible to analyse whole cells or tissues since these consist of so many different components, and so, one would get a confused picture. Each component has to be studied separately and obviously this is difficult when the components are so small. Here a new art was developed, and again Claude was the pioneer. He showed how one could first grind the cells into fragments and then sort out the different components on a large scale with the aid of the centrifuge. This was an important beginning. Palade made further contributions, but it was above all *Christian de Duve* who introduced brilliant developments within this field.

The functions of the cell could now be mapped with this armoury of methodology. Palade has taught us which components function when the cell grows and secretes. The Prize Winner of 1906, Camillo Golgi, discovered a cell component, the Golgi complex. Palade has demonstrated its role and he discovered the small bodies, ribosomes, in which cellular protein is produced.

Production of organic material must be balanced by scavenging and combustion of waste, even in the tiny world of the cell. de Duve discovered small components, lysosomes, which can engulf and dissolve, e.g., attacking bacteria or parts of the cell itself which are old and worn out. These are real acid baths, but the cell itself is normally protected by its surrounding membranes. Sometimes, however, the lysosomes are converted into veritable suicide pills for the cells. This occurs when the surrounding membranes are damaged, e.g. by ionizing radiation. The lysosomes play a role in many clinically important conditions and the foundations laid by de Duve are of the greatest significance for the interpretation of these states, and, thus, also for prophylactic and therapeutic measure.

To sum up, the 1974 Prize Winners have by their discoveries elucidated cellular functions that are of basic biological and clinical importance. Thus, they cover both aspects of the Prize, Physiology as well as Medicine.

Albert Claude, Christian de Duve and George Palade. During the last 30 years a new subject has been created, Cell Biology. You have been largely responsible for this development both by creating the basic methodology and by exploiting it to gain insight into the functional machinery of the cell. On behalf of the Karolinska Institute, I wish to convey to you our warmest congratulations, and I now ask you to receive the prize from the hands of his Majesty the King.

ALBERT CLAUDE[†]

I was born in Belgium, in 1899. Longlier, my birthplace, is located in a high point of the Belgian Ardennes, atop the rising spur of an eroded remnant of the foot of the Alps, next to a deep valley. In the Middle Ages, it had been a fortified place, of the Francs and Carolegian dynasties. Pepin le Bref, crowned King of the Francs in 751 spent in Longlier two winters, from October to Easter, in the years 750 and 763. His son Charlemagne who by then had become Emperor of the Occident called a High Court of Justice of the Empire to meet in Longlier: the diploma, still preserved, was signed by him there in 771. In the year 1050, the Charlemagne Villa became a Monastery, and renamed later "Ferme Charlemagne". In the 17th—18th century, it was adorned with a high sloping roof "à la mansarde", whereas the round towers, standing high at the wall corners, matched the roof with elegant, bell-shaped tops, a situation which remained unchanged until 1914.

The landscape of the Longlier region is covered with remnants of the primeval forest of oak trees, progressively invaded by evergreens. The blue-green color of the pines, which blends with the blue-grey color of the massif of slate rocks emerging through a meager soil gives the countryside an aspect, severe, but also of serene beauty, and even more, when the pure coat of the snow covers it during the long and cold winters.

The population was sparse, at least at the time I was a boy. Our agglomeration was made of scattered small farms, regrouped into hamlets which, with the village, amounted to about 800 inhabitants in all. Rarely, because the people were few, a funeral procession was climbing slowly from the valley, back of our house, and to the old church next to the Charlemagne farm, with the cemetery between them.

The unique school of the Longlier region was built at the outskirt, a kilometer from my home, and about equal distance from the surrounding hamlets, so that the children could leave their home, and reach the school at about the same time. Actually, this school was just a single room with high windows, and a central stove, fed with coal and wood, by the teacher himself. As I remember, there was a set of 5 benches at either side of the stove, with a common sitting board which could accomodate 5 children, in all 50 seats, for an average population, from year to year, of 40 pupils, at the most. The sexes and grades were mixed, and the ages, from 6 to 11 years old. All the courses were taught at the same time, in the same room, by the same and unique teacher. Under this highly pluralistic system, the school was running smoothly, and the results, as remembered over the years, turned out to be, in every respect, excellent.

As usual for the time, the roads were not lighted at night, and no water

[†] Dr Claude died in 1983.

distribution was available, nor in prospect. Due to the elevation of the site, we had to rely on rainwater, collected from the roofs, and on the clear water, filtering and running from the bare rocks, to the river and the streams below.

In the Ardennes, the washed soil is poor, and the configuration rugged. When the spring and summer came, the heat of the sun brought life and beauty to the land. The farmers, however, rose early and worked late, each on their farm, relatively far apart, without the occasion, or the need, to communicate between themselves. Even more than in the cold of the winter, there was a strange stillness, in the heat of the afternoon.

After supper, and when the daily work was over, we did not light the kerosene lamp, nor the makeshift carbide lamp we used, when the war came upon us, but sat outdoors, in the silence and the darkness of the night. As many have done before us, since the early rise of mankind, I reclined on the sloping back of a chair, and gazed intensely, and for hours, at the quivering milky way, and watched the coming of falling stars.

When I became old enough, I took my turn in getting up early, and ringing the church bells (there were two of them) calling for the daily mass, at six o'clock in the morning. The ropes of the bells were hanging freely down the hollow shaft of the church tower, so that we could seize them and pull them from the ground, with the bells seen overhead. When the bells were in full swing, we used to grasp the rope firmly and let ourselves be lifted, just when the hammer hit the roaring bronze. This little familiarity had created an affectionate and reciprocal understanding between us and the Bells. One night, during a heavy storm, we were awakened by a crash. The Pepin le Bref tower, as it was called, which had stood there for many centuries had collapsed, bringing down, with it, the church bells. A few years later, in 1914, the madness of war reached our peaceful shores; the Charlemagne Villa, and part of the village, next to our home, was burnt. I was 15 years old, and starting to become an adult. For us, and for the dying Europe, and the thousands years of its past, it was a new World, and the end of an Era.

My grandfather was born in 1830, just the year the Flemish and French speaking Catholics decided to secede from the Lutheran Dutch people of the low lands, governed by the House of Orange. His place of birth was not Longlier. For a number of generations tracing back to the 17th century, his ancestors had been active in maintaining a Relay, or Stagecoach stop, providing horses, food and lodging for travellers, and wagons for the conveyance of goods. The site of this undertaking was a small plateau, about the locality of Offaing, rising from the opposite side of the Longlier valley, away and higher up from the Charlemagne Villa. From this rather ancient time, I have a witness helping me to imagine and recreate the past. It is a chest of heavy oak with a secret lock, and a slit with a receptacle underneath, in which the hostess, my great-grandmother, would drop the coins she received from the customers, in payment for their expenditure at the inn. This chest, for the past twenty years, has been in my bedroom, next to my bed, supporting a lamp and a clock.

My great-grandfather, Godfroid, born on the heights of Offaing in 1800, or about, had five or six sons, including my grandfather, and a similar number of daughters, most of them promised to live well over ninety. In this healthy, no doubt dynamic, but crowded environment, my grandfather may have felt the pressure of competition, but most likely happened to the most adventurous and most farsighted: he decided to move and settle on his own.

Following the Belgian revolution of 1830 the new nation decided to give itself a King, the choice being Léopold, Prince of Saxe-Cobourg and recent widower of the heir of the throne of England, with the crowning in 1831. Léopold the First was a man of high character and wisdom. It is to his knowledge of the industry of England and to his own initiative that Belgium owed to have had the first railroad lines on the Continent, the first one connecting Bruxelles with Antwerp and its harbor. The next undertaking was much more ambitious. This second line was to be transcontinental, starting from Brussels, through Namur, Luxembourg, Vienna, and further on.

The Longlier valley gap, however, which happened to stand exactly across the projected direction of the new railroad line, would have to be bridged. In addition to this technical difficulty, it was found that the Devonian synclinal, which is the geologic substructure of the region was disturbed by a tectonic anomaly in the form of a narrow band, less than one kilometer in width, which had become deflected in front of the Longlier valley, passing just under the terminal point where the construction of the railroad had stopped. The problems were such that the construction of the line was postponed, for an undetermined length of time. My grandfather saw the opportunity and moved to Longlier. Apparently, he was not without means. Within a relatively short time he built a hotel, next to the freight depot of the railroad terminus. From the commissioned Agency handling the freight traffic for the line, the "Messageries Van Gent", he obtained some agreement whereby he would be responsible for the freight that landed at the Terminal, for its distribution outside the railroad areas. Very soon, he had horses and wagons distributing goods and wares in various directions, as far as the north of France, especially Sedan and Bazeilles, where we had some relatives. His business prospered rapidly, and he became relatively wealthy.

For me, this story of railroads and of a diligent grandfather, which I have recalled, has been more meaningful than the effect of a tectonic anomaly on a Devonian synclinal. Without the decision of my grandfather to move to Longlier, my mother would have been someone else, and there would have been no tales of ringing bells in a medieval church tower, and no ailing uncle to take care of. It was a question of being, or not being. Once the first step taken, what remains to deal with are the important but universal problems of the individual, versus his environment. My mother, Glaudice Watriquant, was 45 years old when I was born, and my father 43. I was the youngest of four, two brothers and one sister, with a gap of 9 years with the oldest. As it happened, most of my early years were spent in the company of old, or very old people, each having their problems and ailments, but never complaining. This created a pervading feeling of tolerance, kindness, and stoic strength which made me happy and feel secure.

For a while, my father worked for my grandfather. As a child eight years old he was already accompanying the driver, not much older than himself, returning by night bringing back fresh vegetables and labile goods from the renowned French market-garden of Fonds-de-Givonne. They took turn to rest, although the traffic was rare at night, especially in the long forest roads; moreover, the horses knew the way and kept on driving even if both drivers fell asleep, as occurred more than once. It was pleasant for youngsters to wake up at the songs of the birds, in a mellow summer night. I would have enjoyed it as they did. My father was gentle, and romantic, in tune with his century. He liked to memorize poetry, from Lamartine, and especially Victor Hugo, whom he admired the most. When he returned from his work and we were very young, we asked him to recite verses to us or sing a lieder, quite well, of the same vein. When he came of age, my father chose to become a baker and pastry maker, perhaps as a complement to the hotel, and for which he spent three years of training in Paris. He was there the year the poet Hugo died. On the Champs Elysées early, he found his way on the top of a gas lamp-post from where he watched pass the funeral procession of hundreds of thousands, for hours. It was in 1885, and my father was 29. It was also his last year of training. His first residence when married, two years later, was in the right wing of the Charlemagne Farm, next to the round tower, and my eldest brother was born there. The second residence, with the bakery and a store, where my second brother and my sister were born, was next to the railroad station. By the time of my birth, my father had taken over a former property of my grandfather, remodeled it and added a large building to serve as a kind of general store. During my time, the local work was already done by hired bakers, my father being away all day, taking care of orders and deliveries.

Two or 3 years after I was born, my mother developed a carcinoma of the breast which appeared shortly after she hurt herself in a fall. She died when I was 7. Too young to go to school, and my elder brothers away in the high school in the town nearby, I was with her most of the time. She suffered, but calmly. I was careful not to make demands on her, and tried to help her when I could. Neighbours and acquaintances came to visit her, sometimes two or three at a time. They didn't pay attention to me; on their way out I followed them to the door, and heard them describe, in their own way, the future course of the disease. I was sad but kept it to myself. Not to leave me alone at home, she took me with her when visiting some healer that had been recommended to her. For one of them, in Marbehan, we had to take the train. Living close to her and partaking of her pain, I felt more and more being as a little nurse at her side. But like the grown-ups of that time, I could not help.

The death of our mother made a big change in the family. After a few years of increasing difficulties (there was a pre-war depression going-on), the decision was made to move to Athus, a prosperous steel mills region bordering both France and the Grand Duchy of Luxembourg. A couple of years before we left, my eldest brother Léon, student at the high school was sitting, one day, at the kitchen table with a book flat beside him. Cautiously, I approached him and said, pointing the right hand page to him: what is this? I remember that, in order to see the page, I had to stand on the tip of my toes, and stretch my head forward. What I saw was the simple outline of a retord, drawn in a square, marginal indentation of the text. My brother did not turn me but began to explain, molding his words with his hands. I did not remember what he said, and could not understand their technical meaning, but as he was speaking I felt my chest, my heart, and the roots of my soul expand. It was a revelation, never to be forgotten. How beautiful this world within the book. I intensely wished to see and know more. In the innocence of my age, I did not doubt that I could. I was 8 ½ years old. The kitchen table of our youth followed in Athus, an is now in Brussels, in the kitchen of our home.

My attendance at the Longlier primary school was curtailed more than a year before the moving. When we arrived in Athus, we found ourselves in an essentially German speaking community. In the church, the hymns and prayers were said in German, and German was spoken in the school where I was received. Every day at 4 PM, each pupil in turn had to read aloud a chapter of the bible. The bible in use, and of which I had a copy, was printed in gothic characters of the oldest type. I may have practiced the sound of them at home, orduring the long, idle hours in the school: when my turn came, I succeeded in reading my part aloud without knowing the words. Again, as before in the world of the aged people I had lived in, I was made to observe my environment from without, in an abstract way, as visitors in an aquarium.

After a year or two, I was asked to return to Longlier to help in the care of an uncle who had suffered a major cerebral hemorrhage. His right side was paralyzed and he had lost the use of his speech. He was tall and heavy, and my aunt was in her 60th year, and ailing. Soon, I took over all the care of my uncle, day and night, and later, progressively the responsibility of the management and the routine work of the household. I was about 13 years old, and more duties and problems of other sorts were added when the war came. My only outside contacts then were the frequent visits of the doctor, who came regularly, or when we called for him in case of emergencies. He came driving himself his horse and cab which he used also when making the rounds of his patients in the country. To me, he looked old, but must have been less than 60. He had experience and common sense, and never seemed in a hurry. I reported to him about my uncle, and listened to his comments and advices. Finally, we conversed about other subjects and the news of the day. This familiarity with a respected physician and my appreciation of his work, or the tragedy I experienced with the long, tormented agony and death of my mother might have influenced me in wanting to study medicine. It was not the case. As far as I remember, even younger than eight, I have always been guided by reason. Not cold reason, but that which leads to the truth, to the real, and to sane Justice. When I went to the University, the medical school was the only place where one could hope to find the means to study life, its nature, its origins, and its ills.

Summarized Civic and Academic Status

Albert Claude was born in Longlier, Belgium, August 23, 1899, and obtained his medical degree from the Université de Liège, Belgium, in 1928.

He spent the winter of 1928—29 in Berlin, first at the Institute für Krebsforschung, and then at the Kaiser Wilhelm Institute, Dahlem, in the laboratory of tissues culture of Prof. Albert Fischer.

He joined the Rockefeller Institute (now the Rockefeller University), in the summer of 1929, and has been connected with this Institution, in different degrees, ever since.

He is Director emeritus of the Jules Bordet Institute for Cancer Researche and treatment, and Professor emeritus, the Faculty of Medicine, at the University of Brussels, Belgium.

He is now Professor, at the Rockefeller University, New York, N.Y., and Professor, at the Université Catholique de Louvain, Louvain, Belgium.

He is Director of the "Laboratoire de Biologie Cellulaire et Cancérologie", at the Université Catholique de Louvain, Louvain-la-Neuve, Belgium.

THE COMING OF AGE OF THE CELL

Inventory of living mechanisms by cell fractionation, biochemistry and electron microscopy, and a view of the impact of the findings on our status and thinking.

Nobel Lecture, December 12, 1974

by

ALBERT CLAUDE

The Rockefeller University, New York, N.Y., U.S.A. and the Université Catholique de Louvain, Louvain, Belgium.

Fifty years of cell research can hardly be summarized in the twenty to thirty minutes of a lecture; to expose only part of it might be unrepresentative, unfair, and altogether unnecessary, since by now you have already been informed of the essential facts and discoveries that have accumulated in the course of these years.

What I would like to do instead, is to discuss with you the impact of these discoveries on our daily life, and their significance for the present and the future. At the same time I will try to recall, first hand, what has been my own experience in this century's endeavor to uncover what were, not so long ago, the mysteries of life itself.

Until 1930 or thereabout biologists, in the situation of Astronomers and Astrophysicists, were permitted to see the objects of their interest, but not to touch them; the cell was as distant from us, as the stars and galaxies were from them. More dramatic and frustrating was that we knew that the instrument at our disposal, the microscope—so efficient in the 19th century—had ceased to be of any use, having reached, irremediably, the theoretical limits of its resolving power.

I remember vividly my student days, spending hours at the light microscope, turning endlessly the micrometric screw, and gazing at the blurred boundary which concealed the mysterious ground substance where the secret mechanisms of cell life might be found. Until I remembered an old saying, inherited from the Greeks—that the same causes, always produce the same effects. And I realized that I should stop that futile game, and should try something else. In the meantime, I had fallen in love with the shape and the color of the eosinophilic granules of leucocytes and attempted to isolate them. I failed—and consoled myself later on in thinking that this attempt was technically premature, especially for a pre-medical student, and that the eosinophilic granules were not pink, anyway. It was only postponed. That Friday, the 13th of September 1929, when I sailed from Antwerp on the fast liner "Arabic" for an eleven-day voyage to the United States, I knew exactly what I was going to do. I had mailed beforehand to Dr. Simon Flexner, Director of the Rockefeller Institute, my own research program, handwritten, in poor English, and it had been accepted. My proposition had been to isolate, and determine by chemical and biochemical means the constitution of the Rous, chicken Tumor I "Agent", at that time still controversial in its nature and not yet recognized as a bonafide Virus. This task occupied me for about five years. Two short years later the microsomes, basophilic components of the cell ground substance, had settled in one of my test tubes, still a structureless jelly, but now captive in our hands.

In the following ten years, the general method of cell fractionation by differential centrifugation was tested and improved, and the basic principles codified in two papers in 1946. This attempt to isolate cell constituents might have been a failure if they had been destroyed by the relative brutality of the technique employed. But this did not happen. The subcellular fragments obtained by rubbing cells in a mortar, and further subjected to the multiple cycles of sedimentations, washings and resuspensions in an appropriate fluid medium, continued to function in our test tubes, as they would in their original, cellular environment. The strict application of the balance sheet-quantitative analysis method permitted to trace their respective distribution among the various cellular compartments and thus, determine the specific role they performed in the life of the Cell.

Small bodies, about half a micron in diameter, and later referred to under the name of "mitochondria" were detected under the light microscope as early as 1894. Although they continued to be extensively investigated by microscopy in the course of the following 50 years, leaving behind an enormous and controversial litterature, no progress was achieved, and the chemical constitution and biochemical functions of mitochondria remained unknown, to the end of that period.

In the early 1940's, I began to make plans for an investigation on the distribution of respiratory pigments in cells. Considering the complexity of the problem, I realized that it should be a collaborative undertaking. A year or so before, I had collaborated with Dean Burk and Winsler in providing them a material of interest to them, Chicken Tumor N° 10, which they used in their studies of the respiratory function in tumor cells. We started experimenting, although they were but mildly impressed by the scientific value of my project, as they told me years later. Their laboratory was conveniently located at the corner of York and 68th, at street level with the Cornell University Department of Vincent du Vignaud. I remember turning across the street, handing them, through the window, each fraction as it was isolated, my share being the determination of the chemical constitution of the fractions, and their respective distribution within the Cell. One day, Rollin D. Hotchkiss appeared, returning from a one-yar fellowship spent in Cambridge, England, who was delighted to find on arrival, quote, "the golden fruits on my doorstep". We were soon rejoined by Hogeboom, and later by W. C. Schneider as regards the distribution of cytochrome c in the Cell, and its participation in respiratory processes. Together, the observations provided conclusive evidence to support the view that most, if not all, of cytochrome oxidase, succinoxidase and cytochrome c, three important members of the respiratory system responsible for most of the oxygen uptake, were segregated in mitochondria. In parallel with these biochemical studies, evidence was also obtained, by tests carried out with characteristic dyes, both under the microscope and in vitro, showing that the respiratory organelles and the mitochondria seen under the microscope were one and the same, a morphological information which would have remained meaningless, however, if we had not secured beforehand, the knowledge of their biochemical functions.

Altogether, these observations demonstrated that the power of respiration exists in a discrete state in the cytoplasm, a fact which led me to suggest, in my Harvey Lecture, that the mitochondria may be considered "as the real power plants of the Cell". At about the same time, with the help of electron microscopy, the microsomes became the endoplasmic reticulum.

Looking back 25 years later, what I may say is that the facts have been far better than the dreams. In the long course of cell life on this earth it remained for our age, for our generation, to receive the full ownership of our inheritance. We have entered the cell, the Mansion of our birth, and started the inventory of our acquired wealth.

For over two billion years, through the apparent fancy of her endless differentiations and metamorphosis the Cell, as regards its basic physiological mechanisms, has remained one and the same. It is life itself, and our true and distant ancestor.

It is hardly more than a century since we first learned of the existence of the cell: this autonomous and all-contained unit of living matter, which has acquired the knowledge and the power to reproduce; the capacity to store, transform and utilize energy, and the capacity to accomplish physical works and to manufacture practically unlimited kinds of products. We know that the cell has possessed these attributes and biological devices and has continued to use them for billions of cell generations and years.

In the course of the past 30 or 40 years, we have learned to appreciate the complexity and perfection of the cellular mechanisms, miniaturized to the utmost at the molecular level, which reveal within the cell an unparalleled knowledge of the laws of physics and chemistry. If we examine the accomplishments of man in his most advanced endeavors, in theory and in practice, we find that the cell has done all this long before him, with greater resourcefulness and much greater efficiency.

In addition, we also know that the cell has a memory of its past, certainly in the case of the egg cell, and foresight of the future, together with precise and detailed patterns for differentiations and growth, a knowledge which is materialized in the process of reproduction and the development of all beings from bacteria to plants, beasts, or men. It is this cell which plans and composes all organisms, and which transmits to them its defects and potentialities. Man, like other organisms, is so perfectly coordinated that he may easily forget, whether awake or asleep, that he is a colony of cells in action, and that it is the cells which achieve, through him, what he has the illusion of accomplishing himself. It is the cells which create and maintain in us, during the span of our lives, our will to live and survive, to search and experiment, and to struggle.

The cell, over the billions of years of her life, has covered the earth many times with her substance, found ways to control herself and her environment, and insure her survival. Man has now become an adjunct to perfect and carry forward these conquests. Is it absurd to imagine that our social behavior, from amoeba to man, is also planned and dictated, from stored information, by the cells? And that the time has come for men to be entrusted with the task, through heroic efforts, of bringing life to other worlds?

I am afraid that in this description of the cell, based on experimental facts, I may be accused of reintroducing a vitalistic and teleological concept which the rationalism and the scientific materialism of the 19th and early 20th centuries had banished from our literature and from our scientific thinking.

Of course, we know the laws of trial and error, of large numbers and probabilities. We know that these laws are part of the mathematical and mechanical fabric of the universe, and that they are also at play in biological processes. But, in the name of the experimental method and out of our poor knowledge, are we really entitled to claim that everything happens by chance, to the exclusion of all other possibilities?

About a year ago, I was invited to an official party by the Governor of a State. As the guests were beginning to leave, the Governor took me aside in a room nearby. He looked concerned and somewhat embarrassed. "Dr. Claude," he asked, "you seem to know much about life. Please tell me: what do you think about the existence of God." The question was unexpected, but I was not unprepared. I told him that for a modern scientist, practicing experimental research, the least that could be said, is that we do not know. But I felt that such a negative answer was only part of the truth. I told him that in this universe in which we live, unbounded in space, infinite in stored energy and, who knows, unlimited in time, the adequate and positive answer, according to my belief, is that this universe may, also, possess infinite potentialities. The wife of the Governor had joined us in the meantime. Hearing this, she seized her husband by the arm and said, "You see, I always told you so."

Life, this anti-entropy, ceaselessly reloaded with energy, is a climbing force, toward order amidst chaos, toward light, among the darkness of the indefinite, toward the mystic dream of Love, between the fire which devours itself and the silence of the Cold. Such a Nature does not accept abdication, nor skepticism.

No doubt, man will continue to weigh and to measure, watch himself grow, and his Universe around him and with him, according to the ever growing powers of his tools. For the resolving powers of our scientific instruments decide, at a given moment, of the size and the vision of our Universe, and of the image we then make of ourselves. Once Ptolemy and Plato, yesterday Newton, today Einstein, and tomorrow new faiths, new beliefs, and new dimensions.

As a result of the scientific revolution of the present century we are finding ourselves living in a magic world, unbelievable less than hundred years ago—magic our telephone, radio, television by multichannel satellites, magic our conversations with the moon, with Mars and Venus, with Jupiter—magic these means which transform our former solitude into a permanent simultaneity of presence, among the members of the Solar System.

And here, at home, thanks to these new media, and the ever increasing speed of transports, we are witnessing a vast mutation taking place, no longer local, but at the dimensions of the Globe: the birth of a new biological organism, in which all Continents, and all the human races participate.

For this equilibrium now in sight, let us trust that mankind, as it has occurred in the greatest periods of its past, will find for itself a new code of ethics, common to all, made of tolerance, of courage, and of faith in the Spirit of men.

CHRISTIAN DE DUVE

I was born on October 2nd 1917, in Thames-Ditton, near London. My parents, of Belgian-German extraction, were Belgian nationals who had taken refuge in England during the war. They returned to Belgium in 1920, and I grew up in the cosmopolitan harbour city of Antwerp, at a time when education in the Flemish part of the country was still half French and half Flemish. Due to these various circumstances, when I entered the Catholic University of Louvain in 1934, I had already travelled in a number of European countries and spoke four languages fairly fluently. This turned out to be a valuable asset in my subsequent career as a scientist.

That I would embrace such a career was, however, very far from my mind. My education, according to the tradition of the jesuit school which I attended, had been centered on the "ancient humanities", and I was strongly attracted to the more literary branches. I nevertheless decided to study medicine, largely because of the appeal of medical practice as an occupation. Medical studies left a fair amount of free time in those days, and there was a tradition at the university that the better students joined a research laboratory. So it was that I entered the physiology laboratory of Professor J. P. Bouckaert, whose rigorous analytical mind exerted a strong influence on my intellectual development. I was attached to a group investigating the effect of insulin on glucose uptake. By the time when I graduated as an MD in 1941, I had abandoned all thought of a medical career, and had only one ambition: to elucidate the mechanism of action of insulin.

In the meantime, war had broken out. After a brief interval in the army and a temporary stay in a prisoners' camp, from which I promptly escaped thanks to the general confusion which followed the disastrous defeat of the allies, I had returned to Louvain to complete my studies. I had become convinced that the problem of insulin action needed to be approached by means of biochemical methods. Since research activities were almost paralysed due to lack of essential supplies, I embarked an another four-year curriculum, to gain the degree of "Licencié en Sciences Chimiques". I combined these studies with a clinical internship in the Cancer Institute, with as much experimental work as war circumstances allowed, and with extensive reading of the earlier literature on insulin.

As a medical student, I had been rather relaxed, but I worked really hard during those four years. Still I could not have achieved what I did without the support of my clinical chief, Professor Joseph Maisin, who enthusiastically approved of my plans and gave me a great deal of free time. By 1945, I had presented a thesis on the mechanism of action of insulin, which earned me the degree of "Agrégé de l'Enseignement Supérieur", written a 400-page book entitled "Glucose, Insuline et Diabète", and prepared a number of research articles for publication.

By that time, the war had ended and I felt a great need of further training in biochemistry. In 1946—1947, I had the good fortune of spending 18 months at the Medical Nobel Institute in Stockholm, in the laboratory of Hugo Theorell, who was awarded the Nobel Prize in 1955. I then spent 6 months as a Rockefeller Foundation fellow at Washington University, under Carl and Gerty Cori who jointly received the Nobel Prize while I was there. In St. Louis, I collaborated with Earl Sutherland, Nobel laureate in 1971. Indeed, I have been very fortunate in the choice of my mentors, all sticklers for technical excellence and intellectual rigour, those prerequisites of good scientific work.

I returned to Louvain in March 1947 to take over the teaching of physiological chemistry at the medical faculty, becoming full professor in 1951. I started a small research laboratory, where I was joined by a young physician, Géry Hers, who had already worked with me during the war, and by an increasing number of first class students, including Jacques Berthet, Henri Beaufay, Robert Wattiaux, Pierre Jacques and Pierre Baudhuin. All have since carved distinguished careers for themselves.

Insulin, together with glucagon which I had helped rediscover, was still my main focus of interest, and our first investigations were accordingly directed on certain enzymatic aspects of carbohydrate metabolism in liver, which were expected to throw light on the broader problem of insulin action. But fate had a surprise in store for me, in the form of a chance observation, the so-called "latency" of acid phosphatase. It was essentially irrelevant to the object of our research but it was most intriguing. My curiosity got the better of me, and as a result I never elucidated the mechanism of action of insulin. I pursued my accidental finding instead, drawing most of my collaborators along with me.

Our investigations were very fruitful. They led to the discovery of a new cell part, the lysosome, which received its name in 1955, and later of yet another organelle, the peroxisome. At the same time, we were prompted to develop progressively improved instrumental, technical and conceptual tools in relation to the separation and analysis of cell components, and to apply them to an increasing variety of problems of biological and also medical interest.

In 1962, I was appointed a professor at the Rockefeller Institute in New York, now the Rockefeller University, the institution where Albert Claude had made his pioneering studies between 1929 and 1949, and where George Palade had been working since 1946. I retained my position in Louvain and have since shared my time more or less equally between the two universities. In New York, I was able to develop a second flourishing group, which follows the same general lines of research as the Belgian group, but with a program of its own. The two laboratories work closely together and complement each other in many respects.

Recently, with a number of colleagues, I have created a new institute, the International Institute of Cellular and Molecular Pathology, or ICP, located on the new site of the Louvain Medical School in Brussels. The aim of the ICP is to accelerate the translation of basic knowledge in cellular and molecular biology into useful practical applications.

In September 1943, I married the former Janine Herman, the daughter of a physician. We have four children, three of whom are married, and two grandchildren.

EXPLORING CELLS WITH A CENTRIFUGE

Nobel Lecture, December 12, 1974

by

CHRISTIAN DE DUVE

Université Catholique de Louvain, Belgium and The Rockefeller University, New York, N.Y., U.S.A.

INTRODUCTION

In one of her masterpieces, Nobel Laureate Selma Lagerlöf tells how the little boy Nils Holgersson visited the whole of Sweden, from Skåne to Lappland, on the wings of a friendly white gander.

I too have made a wonderful journey, using like Nils Holgersson an unconventional mode of travel. For the last 25 years, I have roamed through living cells, but with the help of a centrifuge rather than of a microscope.

On these trips I was never alone. I want to mention this at the outset, since I owe much to my travelling companions. Some of their names will come up as my tale unfolds; but there are so many of them that I will be quite unable to mention them all. My debt goes also to my early mentors in science: Joseph Bouckaert, Joseph Maisin, Hugo Theorell, Carl and Gerty Cori, Earl Sutherland. Four of them have preceded me on this podium. Three, unfortunately, are not with us any more.

THE DEVELOPMENT OF ANALYTICAL CELL FRACTIONATION

Thirty years ago, much of the living cell still remained virtually unexplored. The reasons for this are simple. Morphological examination was limited downward in the scale of dimensions by the resolving power of the light microscope, whereas chemical analysis stopped upward at the size of the smaller macromolecules. In between, covering almost two orders of magnitude, lay a vast "terra incognita", impenetrable with the means of the day. Invasion of this territory started almost simultaneously on its two frontiers, after electron microscopy became available to morphology and centrifugal fractionation to biochemistry.

When, in 1949, I decided to join the little band of early explorers who had followed Albert Claude in his pioneering expeditions, electron microscopy was still in its infancy. But centrifugal fractionation, the technique I wanted to use, was already well codified. It had been described in detail by Claude himself (1), and had been further refined by Hogeboom, Schneider and Palade (2) and by Schneider (3). According to the scheme developed by these workers, a tissue, generally rat or mouse liver, was first ground with a Potter-Elvehjem homogenizer, in the presence of either 0.88 M (2) or 0.25 M (3) sucrose. The homogenate was then fractionated quantitatively by means of three successive centrifugations and washings, under increasing centrifugal

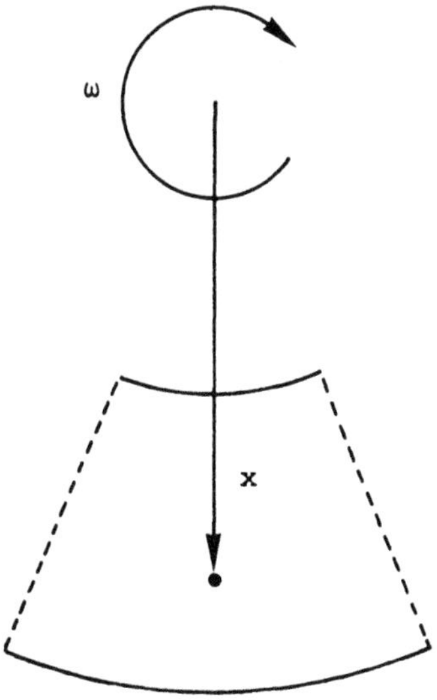

SEDIMENTATION VELOCITY

$$\frac{dx}{dt} = s.\omega^2.x$$

x = RADIAL DISTANCE (cm)

ω = ANGULAR VELOCITY $(rad.sec^{-1})$

s = SEDIMENTATION COEFFICIENT OF PARTICLE (sec)

FOR SPHERICAL PARTICLE OF RADIUS r (cm) AND OF DENSITY ρ_p $(g.cm^{-3})$

IN MEDIUM OF DENSITY ρ_m $(g.cm^{-3})$ AND OF VISCOSITY η (poises)

$$s = \frac{2\,r^2\,(\rho_p - \rho_m)}{9\,\eta}$$

Fig. 1. The Svedberg equation and its application to a spherical particle.

force x time integrals, to yield "nuclei", "mitochondria", "microsomes" and a final supernatant. The fractions, as well as the original homogenate, could then be analyzed for their chemical composition, enzyme content, and other properties.

All these details were available in the literature, and there seemed little more for us to do than to acquire the necessary equipment and follow instructions carefully, especially since our interest in cell fractionation itself was rather peripheral at that time. All we wanted was to know something about the localization of the enzyme glucose 6-phosphatase, which we thought might provide a possible clue to the mechanism of action, or lack of action, of insulin on the liver cell.

Fortunately, this is not exactly how things happened. Working with me on this project was Jacques Berthet, still a medical student at that time, but with an unusually mature and rigorous mind. He went about the job of setting up the technique in a careful and systematic fashion, paying special attention to all physical parameters. A few practical tips from Claude, who had just returned to Belgium, were also helpful.

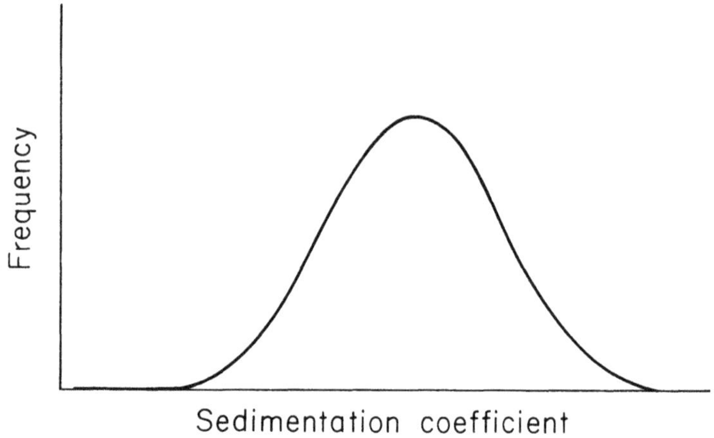

Fig. 2. Image of a polydisperse population of particles. Owing to individual differences in size and/or density, different members of the population do not have the same sedimentation coefficient. The centrifugal properties of the population as a whole are depicted by a *frequency distribution curve* of sedimentation coefficients. Size and/or density distribution can be similarly represented. Frequency is usually defined as $\frac{dn}{Ndx}$ (or in the case of histograms $\frac{\Delta n}{N\Delta x}$), in which $\frac{dn}{N}$ is the fraction of total particles having an abscissa value comprised between x and $x+dx$. Similar diagrams may be drawn in terms of relative mass, relative enzyme activity, etc . . ., instead of relative number.

Particularly important, I now realize in retrospect, was the fact that we took some time to study the theory of centrifugation, as beautifully exposed in the classical book by Svedberg and Pedersen (4).

Although separating mitochondria and microsomes might appear worlds apart from the determination of the molecular weight of macromolecules, certain concepts were common to the two operations and could be usefully transposed from the latter to the former.

One was that of *sedimentation coefficient* (Fig. 1), which obviously was applicable to any particle, irrespective of its size.

Another was that of *polydispersity* which, owing to biological variability, was likely to be a property of the populations made up by subcellular organelles. This meant that the centrifugal behavior of such populations could be described only by a frequency distribution curve of sedimentation coefficients (Fig. 2), not by a single s value as for most molecular populations.

A third important point related to the *resolving power* of differential sedimentation, which some elementary calculations revealed to be surprisingly low (Fig. 3).

There was much insistence in those days on the various artifacts that complicate centrifugal fractionation, such as, for instance, breakage or agglutination of particles, adsorption or leakage of soluble constituents. But these were only accidents, no doubt serious, but amenable to experimental correction. The problem, as it appeared to us, was a more fundamental one.

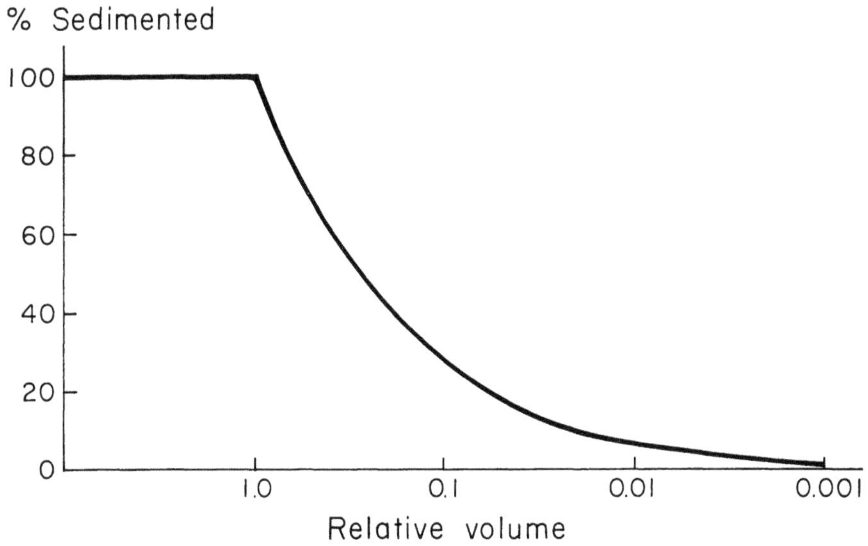

Fig. 3. Graph shows the percentage of particles recovered in a sediment as a function of relative particle volume. Particle density is assumed to be the same for all particles. The meniscus of fluid in the rotating centrifuge is assumed to be half-way between the axis and the bottom of the tube or cell.

What we were doing was trying to separate populations which, owing to overlapping polydispersities, might at best be only partly separable from each other. In addition, we were using a poorly discriminating method for this purpose.

I cannot claim that all this was immediately clear to us. But considerations of this sort undoubtedly colored our approach from the start (5). We fully expected centrifugally isolated fractions to be impure, while suspecting that populations of cell organelles might be difficult, if not impossible, to resolve quantitatively. Conscious also of the severe limitations of light microscopic examination of the fractions, we tried to extend the biochemical interpretation as far as possible. Instead of looking at each fraction separately and focusing on its enzyme content, as was usually done, we looked rather at each individual enzyme and contemplated its distribution between all the fractions.

In order to permit a comprehensive view of enzyme distribution patterns, I introduced a histogram form of representation, illustrated in Fig. 4. In this figure are shown the distribution patterns of three of the first enzymes we studied, on the left as determined by the classical 4-fraction scheme, and on the right as determined by the modified 5-fraction scheme that we worked out in an effort to elucidate the significance of the small difference in distribution observed between acid phosphatase and cytochrome oxidase (6). This difference, as can be seen, is very much magnified by the modification in fractionation scheme.

These histograms turned out to be very revealing, by more or less automatically conveying the notion of polydispersity, illustrated in Fig. 2. In fact,

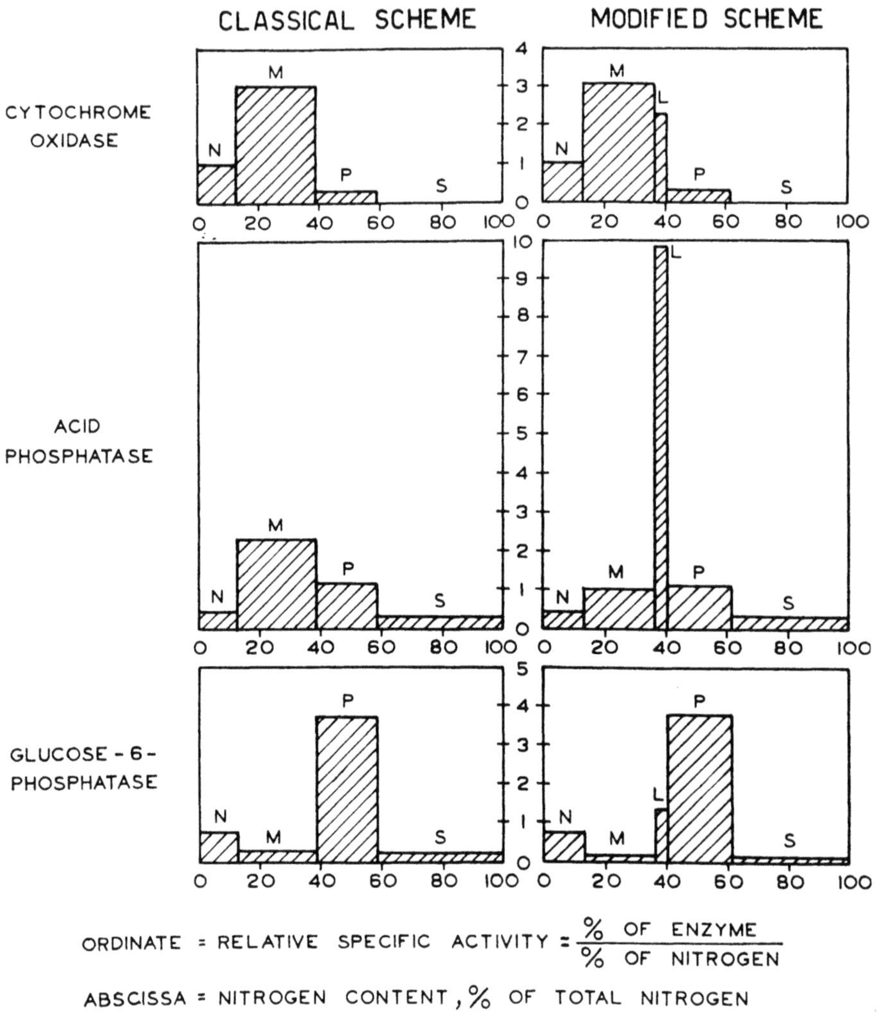

Fig. 4. Enzyme distributions represented in histogram form. The relative specific enzyme content (% activity/% protein) of the fractions is plotted against their relative protein content, inscribed cumulatively from left to right in their order of isolation (decreasing sedimentation coefficient): nuclear *N*, mitochondrial *M*, microsomal *P*, and supernatant *S*, in classical 4-fraction scheme; and nuclear *N*, heavy-mitochondrial *M*, light-mitochondrial *L*, microsomal *P* and supernatant *S*, in modified 5-fraction scheme (6). Although very crude, similarity with frequency distribution curves of polydisperse populations can be recognized. Distinction between three populations, now known to consist of mitochondria (cytochrome oxidase), lysosomes (acid phosphatase) and endoplasmic reticulum fragments (glucose 6-phosphatase), is enhanced by use of 5-fraction scheme. From reference 7.

since the fractions are aligned along the abscissa in order of decreasing sedimentation coefficient, one may, in a very crude fashion, look at the abscissa as a deformed scale of sedimentation coefficients, and at the histograms as correspondingly deformed frequency distribution histograms of sedimentation coefficients. The logical next step in this line of reasoning was to assimilate enzyme distributions to particle distributions, and therefore to interpret, at least tentatively, significant differences in the distribution patterns of two enzymes as reflecting association of the enzymes with distinct particle populations.

Extrapolation from enzymes to particles could, however, not be made without some sort of assumption concerning the relationship between relative enzyme activity, the numerator in the ordinate of Fig. 4, and relative particle number, the numerator in the ordinate of Fig. 2. The simplest, and at the same time most plausible, such assumption was that members of a given particle population have essentially the same biochemical composition, larger particles simply having more of everything than smaller particles. Within the limits of validity of this assumption, which I have called the *postulate of biochemical homogeneity,* the histograms of Fig. 4 could now be likened to distribution diagrams of total particle mass or protein (not of actual particle numbers, it should be noted, although further conversion to numerical distributions can be made with some additional information). We had to assume, of course, that the enzyme distributions were not grossly distorted by translocation artifacts, or to correct for such artifacts as much as possible.

Another postulate we made was that each enzyme is restricted to a single intracellular site. This *postulate of single location* is less essential than that of biochemical homogeneity, since bimodal or multimodal distributions are amenable to the same kind of interpretation. In practice, however, single location made a useful addition to biochemical homogeneity, supporting the use of enzymes as *markers* of their host-particles.

First used empirically as pure working hypotheses, the above considerations were progressively validated, as more enzymes were studied and a limited number of typical distribution patterns began to emerge. Actually, as shown by the results of Fig. 5, things were not quite as simple, and a number of complications of various sorts tended to blur the picture. But most of these could be dealt with satisfactorily by ancillary experiments (8).

In these studies, a second line of evidence proved very useful, based on *enzyme latency.* Owing to impermeability of particle membranes to one or more of the substrates used in the assay of enzymes, many particle-bound enzymes fail to display activity "in vitro" as long as the membrane surrounding them is intact. Various means, mechanical, physical or chemical, can be used to disrupt the membrane and to release the enzymes, as we first showed for rat-liver acid phosphatase (Fig. 6). If two or more enzymes are present together in the same particles, they will be released together in this kind of experiment; if in different particles, they may come out separately (Fig. 7). In our hands, such studies have been very useful, providing an independent verification of the significance of the similarities and differences revealed by centrifugation experiments.

By 1955, our results were sufficiently advanced to allow us to propose with a certain measure of assurance the existence of a new group of particles with lytic properties, the lysosomes, and to hint at the existence of another group of particles, the future peroxisomes (8). At the same time, we had, from the

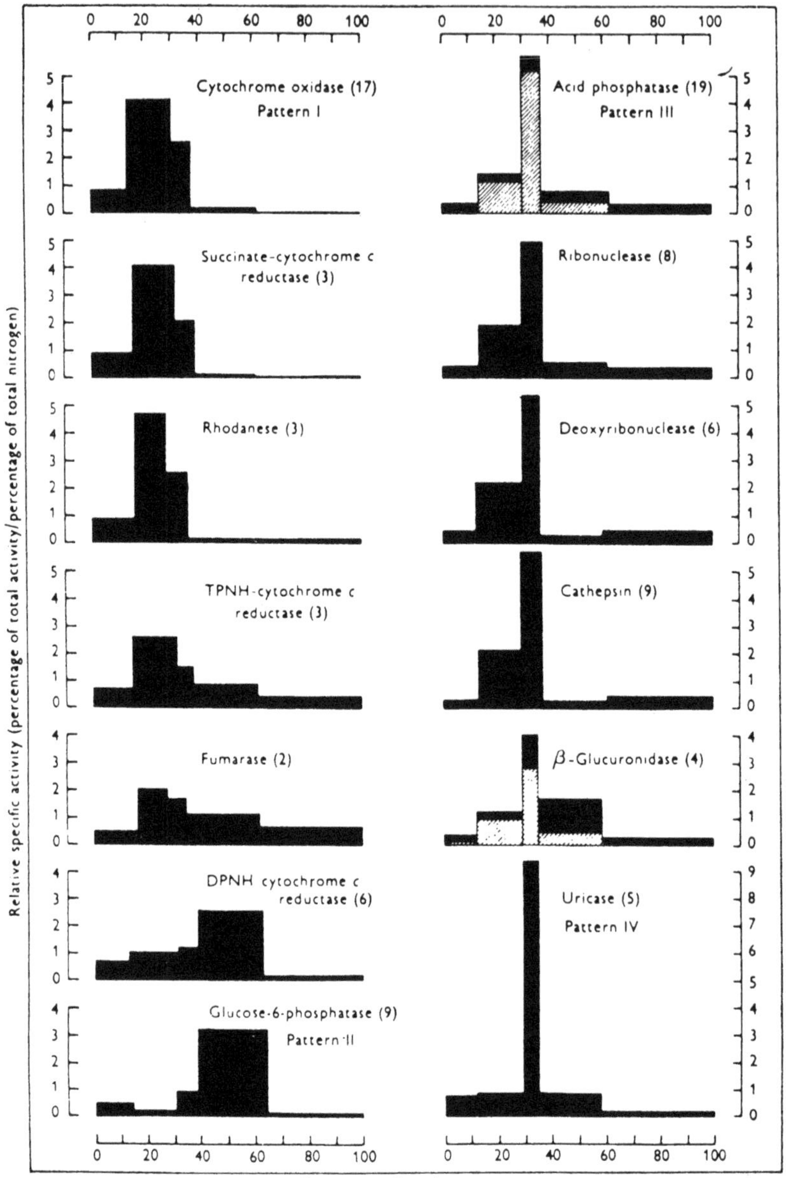

Fig. 5. Distribution patterns of enzymes in rat-liver fractions separated by 5-fraction procedure (see: Fig. 4). Pattern I, shared by 3 enzymes, represents the distribution of mitochondria; pattern II (glucose 6-phosphatase) that of microsomes. In between, in left-hand column, are complex combinations of patterns I and II. Pattern III is shared by 5 lysosomal acid hydrolases, except for β-glucuronidase which has an additional microsomal component. Pattern IV belongs to the peroxisomal urate oxidase. Details are given in original paper (8).

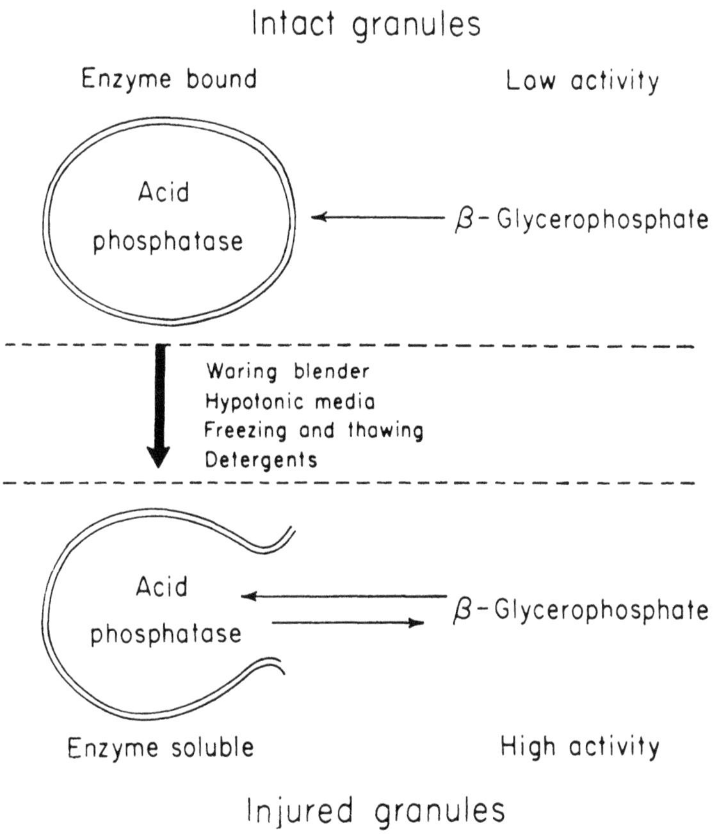

Fig. 6. Model of latency of rat-liver acid phosphatase, as proposed in 1951 (9). From reference 10.

mixture of theoretical considerations and experimental results that I have just briefly recalled, derived a certain "philosophy" of centrifugal fractionation, which I subsequently elaborated in greater detail in several publications (11). The key word here was *"analytical"*. Basically, we felt that our approach was no more than an extension of the classical Svedberg technique from the molecular to the submicroscopic and microscopic level.

A major difficulty at this stage, however, was that available techniques did not measure up to the kind of information we were hoping to extract. The answer to this problem was provided by density gradient centrifugation, which was introduced in the early 1950's. This new technique offered prospects of improved resolution; it allowed the use of density, as well as of sedimentation coefficient, as separation parameter; and, finally, its analytical character was unmistakable (Fig. 8). In fact, as shown as early as 1954 by Hogeboom and Kuff (13), it could even be used successfully for the determination of molecular weights.

Here again, we devoted some time to theoretical studies (12). In this, Berthet and I were joined by another young co-worker, Henri Beaufay, whose skills as a self-taught engineer proved particularly valuable for the design of

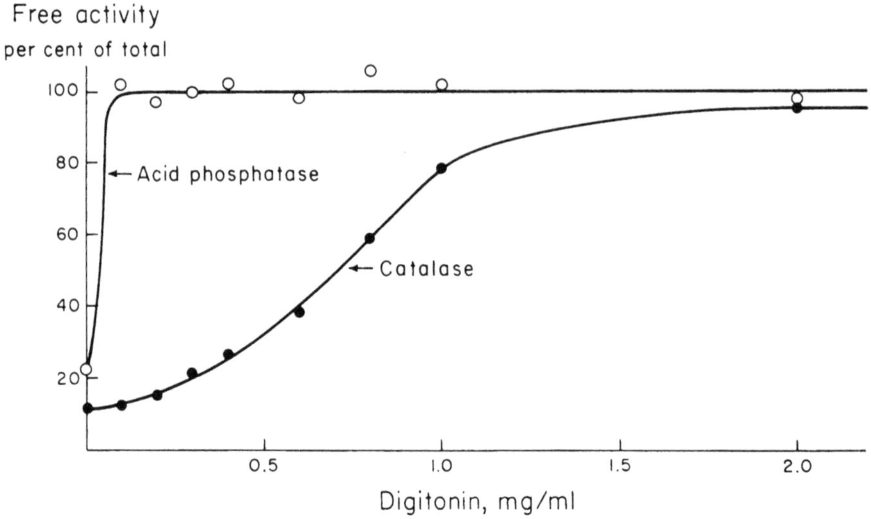

Fig. 7. Differential release of the lysosomal acid phosphatase and of the peroxisomal catalase by increasing concentrations of digitonin. From reference 10.

various accessories, culminating in the construction of a completely automatic rotor (14), different in principle from the zonal rotors built by Norman Anderson (15), and particularly adapted to rapid isopycnic separation at minimum hydrostatic pressure. The importance of the latter advantage has been emphasized by my former collaborator Robert Wattiaux (16).

Particles sedimenting through a density gradient are apt to undergo a progressive increase in density, due to inflow of solute or outflow of water or both, depending on the number and permeability properties of their membranes and on the nature of the solute(s) and solvent used to make the gradient. These factors we tried to incorporate in a theoretical model of particle behavior (12, 17), and at the same time to take into account in the design of our experiments. It appeared from our theoretical considerations that the sucrose concentration of the medium might be a particularly important variable, and that different types of particles might respond differently to changes in sucrose concentration. We therefore subfractionated large granule fractions from rat liver in iso-osmotic glycogen gradients prepared with sucrose solutions of different concentrations as solvent, as well as in sucrose gradients prepared with either H_2O or D_2O (18).

The results of these experiments confirmed and extended our earlier findings, establishing the existence of three distinct groups of enzymes, as defined by their centrifugal behavior. There was little doubt in our minds that these observations reflected the occurrence of three distinct populations of particles in the large granule fraction. By fitting our results to the theoretical equation, we were even able to evaluate a number of physical parameters for each putative particle population, and to construct, from purely biochemical data, a sort of "robot picture" of the particles themselves. This is shown in Table I. Due to heterogeneity within the population, the data given in this table for the

Table I. *Typical Physical Properties of Rat-Liver Particles*

From reference 10

Parameter	Mitochondria	Lysosomes		Peroxisomes		
Reference enzyme:	Cytochrome oxidase	Acid phosphatase	Acid DNase	Urate oxidase	Catalase	D-Amino acid oxidase
Dry weight (μg)	10^{-7}	2.7×10^{-8}	3.6×10^{-8}	2.4×10^{-8}	—	—
Dry density	1.315	1.300	1.331	1.322	1.319	1.315
Osmotically active solutes (milliosmoles/g dry weight)	0.157	0.128	0.334	0	0	0
Water compartments (cm^3/g dry weight)						
Hydration	0.430	0.256	0.212	0.214	0.295	0.296
Sucrose space	0.905	1.075	0.330	2.51	2.68	2.54
Osmotic space in 0.25 M sucrose	0.595	0.485	1.265	0	0	0
Total in 0.25 M sucrose	1.930	1.816	1.807	2.724	2.975	2.836
Sedimentation coefficient in 0.25 M sucrose (Svedberg units)	10^4	4.4×10^3	5×10^3	4.4×10^3	—	—
Diameter in 0.25 M sucrose (μm)	0.8	0.51	0.56	0.54	—	—
Density in 0.25 M sucrose	1.099	1.103	1.100	1.095	1.088	1.090

lysosomes are of questionable significance. On the other hand, those listed for mitochondria and peroxisomes agree very well with measurements made by other techniques.

Though analytically satisfactory, the results described so far still fell short from definitive proof, since they had unfortunately confirmed our fear that distinct populations of subcellular particles might prove intrinsically inseparable quantitatively due to overlapping of size and/or density distributions. It was possible to obtain pure samples by cutting off non-overlapping parts of the populations; but this introduced the danger of biased sampling. A means of almost complete separation, although under somewhat artificial conditions, was provided in 1962 by Wattiaux, Wibo and Baudhuin (19), when they discovered that pretreatment of the animals with Triton WR-1339 causes a selective decrease in the density of lysosomes, due to accumulation of the Triton within these particles (Fig. 9). Thanks to this finding and to the Beaufay rotor, large-scale separation of the three populations has now become possible, allowing a variety of biochemical and functional studies that were not feasible before (20).

While the biochemical approach I have outlined was being developed in our laboratory, electron microscopy was making great strides of its own, soon becoming available for the examination of subcellular fractions. For obvious

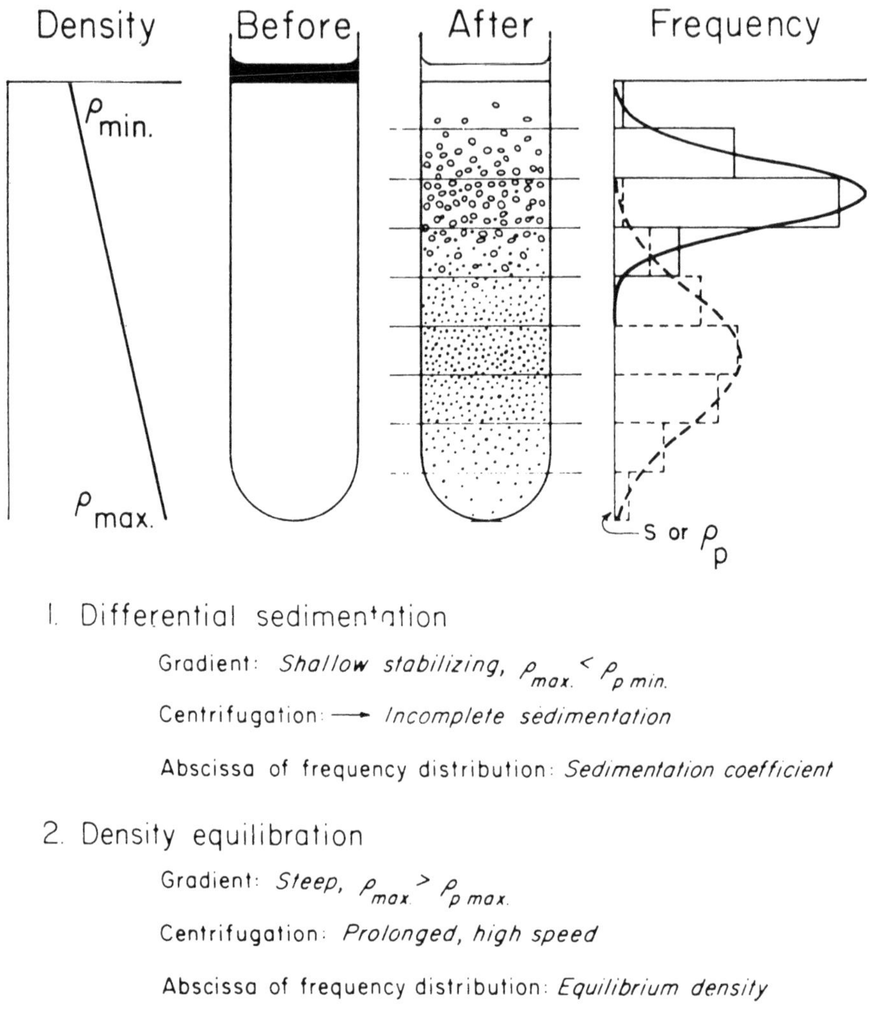

Fig. 8. Schematic representation of density gradient centrifugation, with initial top-layering of the sample. Two forms, based on differences in sedimentation coefficient and density respectively, are shown. Diagram at the right pictures frequency distribution of particles or markers as a function of tube height. Conversion to frequency distributions of sedimentation coefficients or densities generally requires readjustment of ordinate and abscissa values, leaving surface area of each block (% content in fraction) unchanged. For details of calculations, see reference 12. From reference 10.

reasons we were very anxious to take a look at our purest fractions, in order to test our conclusions and eventually identify our hypothetical particles. Already in 1955, thanks to the expert collaboration of Alex Novikoff and to the facilities of Albert Claude in Brussels and of Wilhelm Bernhardt in Paris, we were able to do this for lysosome-rich fractions, which were found to contain dense bodies, surrounded by a membrane and of about the size predicted for lysosomes (21). Later, we were able to acquire an instrument of our own, and Henri Beaufay taught himself another skill, which he later perfected under

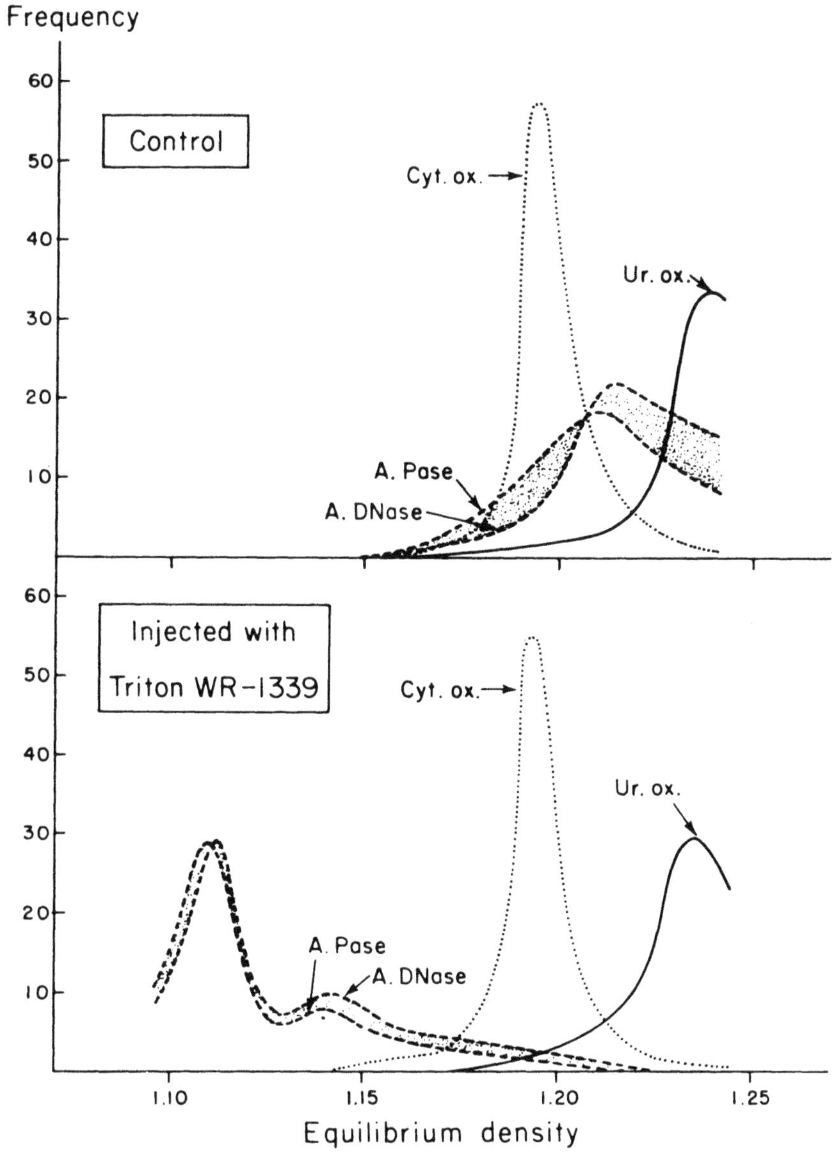

Fig. 9. Influence of a previous injection of Triton WR-1339 on the equilibrium density of rat-liver particles equilibrated in an aqueous sucrose gradient. Upper graph shows overlapping of lysosomes. (A.Pase = acid phosphatase; A.DNase = acid deoxyribonuclease) with mitochondria (Cyt.ox. = cytochrome oxidase) and peroxisomes (Ur.ox. = urate oxidase). Four days after intravenous injection of 170 mg of Triton WR-1339 to the animals, the density of the lysosomes has decreased drastically, that of mitochondria and of peroxisomes remains unchanged. Graph constructed from results of Wattiaux *et al.* (19), reproduced from reference 10.

the guidance of George Palade. With Pierre Baudhuin, he confirmed the identification of lysosomes as "pericanalicular dense bodies" and showed that the peroxisomes correspond to the particles known as "microbodies" (22).

Thus, the gap between biochemistry and morphology was finally bridged, after some 15 years of research.

More recently, Baudhuin has adapted quantitative morphometric methods to the examination of subcellular fractions, making it possible to compare measurements derived from biochemical data with those obtained by direct mensuration (23). In several instances, excellent agreement has been found between the two sets of data (20, 23, 24, 25).

Applications to Biology

I have chosen to dwell at some length on our theoretical and technical studies, because they were, I believe, the key to whatever achievements were made by our group. I know that others have accomplished important advances by the alternative process of first purifying a subcellular component and then analyzing it. For example, nuclei, secretion granules, plasma membranes and Golgi elements have been largely characterized in this fashion. But purification is generally a laborious procedure, it is difficult to control, and it is rarely quantitative. The advantage of the analytical approach is that it is widely applicable, and it can provide a considerable amount of quantitative information, even with a relatively poor resolving power. The important point is that with this kind of methodology, we derive the information, not from the properties of specific fractions believed to approximate a given intracellular component, but from the manner in which properties are distributed over a large number of fractions, which together represent the whole tissue.

In our laboratories, this general approach has been applied to a variety of biological materials and for the study of many different problems. In continuation of the work on liver, already described, it has supported a number of studies concerned with the functions of lysosomes, including those of Robert Wattiaux on intralysosomal storage (26), of Pierre Jacques on pinocytosis (27), of Russell Deter on autophagy (28), of Jack Coffey, Nick Aronson and Stanley Fowler on lysosomal digestion (29), and of André Trouet and Paul Tulkens on the effects of anti-lysosome antibodies (30). It has also allowed Brian Poole, Federico Leighton, Tokuhiko Higashi and Paul Lazarow to make a searching analysis of the biogenesis and turnover of peroxisomes (24, 31). In recent years, a large team grouped around Henri Beaufay and Jacques Berthet, and including Alain Amar-Costesec, Ernest Feytmans, Mariette Robbi, Denise Thinès-Sempoux and Maurice Wibo, has launched a major attack on microsomal and other membrane fractions, with the aim of characterizing physically, chemically, enzymically and immunologically the various types of cytomembranes occurring in these fractions (32).

In its applications to other mammalian tissues and cell types, analytical cell fractionation has allowed Pierre Baudhuin and Brian Poole to recognize peroxisomes in kidney (33); Gilbert Vaes to carry out a thorough study of bone lysosomes, leading to very revealing observations on the role of these particles in bone resorption (34); Bill Bowers to make a comprehensive biochemical dissection of lymphoid tissues and lymphocytes, as a preliminary to an analysis of cell-mediated immune cytotoxicity (35); Marco Baggiolini to characterize the two types of granules present in neutrophil polymorphonuclear leucocytes (36); Richard Schultz and Pierre Jacques to unravel some of the complexities of placental tissue (37); Tim Peters to fractionate aortic smooth muscle cells (38) and enterocytes (39); and Paul Tulkens to do the same for cultured fibroblasts (40), a system also used by Brian Poole and Maurice Wibo for investigations of protein turnover (41).

Under the leadership of Miklós Müller, a series of fascinating studies have been performed in New York on a number of different protozoa. In *Tetrahymena pyriformis,* Müller was able to identify two types of lysosomes, which discharge their enzymes, one in phagocytic vacuoles and the other in the outside medium (42). In collaboration with Pierre Baudhuin and later with Jim Hogg, he has shown the existence in the same organism of peroxisomes which, like plant glyoxysomes, contain enzymes of the glyoxylate cycle (33, 43). More recently, with Don Lindmark, he has characterized in *Trichomonads* a completely new type of cytoplasmic particle, with the capacity of converting pyruvate to acetate, CO_2 and molecular hydrogen, the hydrogenosome (44).

Other studies have dealt with the role of lysosomes in tissue regression, notably those of Denise Scheib-Pfleger and Robert Wattiaux on Müllerian ducts in chick embryos (45), and those of Yves Eeckhout on the tail of metamorphosing tadpoles (46).

It has been my good fortune to participate in most of these investigations, sometimes actively and sometimes simply in an advisory capacity, and to watch at the same time the growing interest of other laboratories in similar problems. After trying, with increasing difficulty, to review the field of lysosomes at regular intervals (7, 47), I welcomed with some relief the appearance in 1969, under the editorship of Professor John Dingle and Dame Honor Fell, of the multi-author treatise "Lysosomes in Biology and Pathology", of which volume 4 is now in press (48). The literature on peroxisomes and related particles has grown more slowly, but has now also reached an appreciable size (49).

It must be pointed out that many of these advances have been made by means of morphological rather than by biochemical methods, or by a combination of both. In this respect, the development of cytochemical staining reactions for enzymes previously identified biochemically as specific particle markers has been an invaluable aid, thanks to the pioneering work of Alex Novikoff, Stanley Holt, Werner Straus, Fritz Miller, Sidney Goldfischer, Marilyn Farquhar and many others.

Applications to Pathology and Therapeutics

In recent years, we have become increasingly concerned with the possible medical applications of our findings. The possibility that lysosomes might accidentally become ruptured under certain conditions, and kill or injure their host-cells as a result, was considered right after we got our first clues to the existence of these particles. We even made a number of attempts to test this hypothesis in ischemic tissue and in the livers of animals subjected to hepatotoxic treatments or to carcinogenic diets (50). But we became discouraged by problems of interpretation (47). Even today, clear-cut demonstration of the so-called "suicide bag" hypothesis remains very difficult, although there seem to be at least a few authenticated cases involving this mechanism of cell death. Much more clearly documented is the mechanism of tissue injury through extracellular release of lysosomal enzymes, a field which has been pioneered by Honor Fell and her co-workers.

The two mechanisms mentioned above rely on the very plausible instance of lysosomal enzymes exerting their lytic effect at abnormal sites. What we did not suspect in the beginning was that the failure of lysosomal enzymes to act at their normal site could also cause serious diseases. This fact was brought home to us in a rather surprising fashion through the work of my colleague Géry Hers, who in 1962 diagnosed glycogen storage disease type II as being due to a severe deficiency of a lysosomal enzyme (51). This finding initiated a series of fruitful investigations on other storage diseases, in which François Van Hoof played a major part (52). It also provided useful guidelines to the chemists and pathologists who, in various parts of the world, were trying to unravel the pathogeny of hereditary lipidoses and mucopolysaccharidoses. Today, with more than twenty distinct congenital lysosomal enzyme deficiencies identified, this mysterious chapter of pathology has been largely elucidated (53).

According to some results obtained over the last few years by Tim Peters, Miklós Müller, Tatsuya Takano, Bill Black and Helen Shio, with the collaboration of Marilyn Farquhar, lipid accumulation in arterial cells during the development of atherosclerosis could well be due to a mechanism similar to that involved in congenital lipidoses. At least in cholesterol-fed rabbits, there is strong evidence, both biochemical and morphological, that the lysosomes of the aortic smooth muscle cells are the main site of intracellular cholesterol ester accumulation, and there are indications that a relative deficiency of the lysosomal cholesteryl esterase may be responsible for this phenomenon (38, 54). Fig. 10 shows some of the biochemical evidence: after cholesterol feeding, lysosomes become considerably less dense due to lipid accumulation. This figure also illustrates the sensitivity of our present techniques. These fractionations were performed on a total of about 1 mg of cell protein. Similar experiments can be, and have been successfully, carried out on a needle biopsy.

Other interesting applications of the lysosome concept are in pharmacology and therapeutics. In line with the "suicide bag" hypothesis, early investigations in this area focused on "labilizers" and "stabilizers" of the lysosomal membrane (55). One outcome of this work has been the suggestion that certain anti-inflammatory agents, such as cortisone and hydrocortisone, might owe at least part of their pharmacological properties to their effect on the lysosomal membrane.

More recently, we have extended our interest to the various substances that are taken up selectively into lysosomes and owe some of their main pharma-

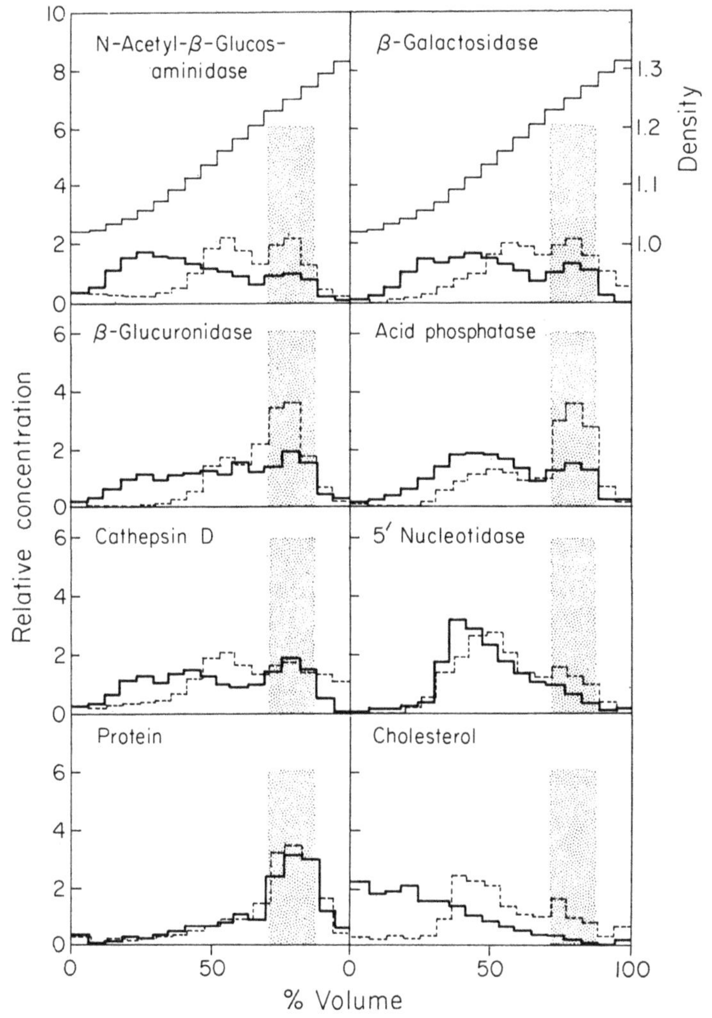

Fig. 10. Influence of cholesterol feeding on density of aortic smooth muscle cell lysosomes. Graphs show distribution patterns of enzymes after density equilibration (see: Fig. 8) in sucrose density gradient depicted by "staircase" on top. Starting material was a postnuclear supernatant of rabbit aortic cells brought to a density of 1.26 and layered initially at outer edge of gradient (dotted area). Broken lines give distributions in normal preparations, solid lines in preparation from a rabbit showing grade IV atheroma as a result of cholesterol feeding. Note extensive shift to the left of 5 acid hydrolases, indicating lowered density of lysosomes due to lipid accumulation. Distribution of protein, 5'-nucleotidase (plasma membranes) and mitochondrial cytochrome oxidase (not shown) was unchanged. From Peters and de Duve (54).

cological properties to this phenomenon. These "lysosomotropic" agents are surprisingly numerous, including such variegated compounds as neutral red, chloroquine, streptomycin, dextran, polyvinylpyrrolidone, Triton WR-1339 and trypan blue (56). Particularly interesting is the use of certain lysosomotropic agents as carriers for drugs. In Louvain, André Trouet has applied this principle to leukemia and cancer chemotherapy, by using DNA as carrier for the drugs daunorubicin and adriamycin. Experimentally, these DNA complexes proved less toxic and more effective on L1210 leukemia than the free drugs (57). Clinical trials under way over the last two years in several hospitals have given very encouraging results (58).

CONCLUSION

In the conclusion of his Nobel lecture delivered in 1955, Hugo Theorell asks the question: "What is the final goal of enzyme research?"

"The first stage", he answers, "is to investigate the entire steric constitution of all enzymes . . ."

"In the second stage," he continues, "it is a matter of deciding how the enzymes are arranged in the cell-structures. This implies, as a matter of fact, the filling of the yawning gulf between biochemistry and morphology".

The gulf still yawns today. But it is a particular pleasure for me to be able to tell my old friend Theo that it yawns a little less. In our efforts to narrow it, my co-workers and I have been privileged to contemplate many marvelous aspects of the structural and functional organization of living cells. In addition, we have the deep satisfaction of seeing that our findings do not simply enrich knowledge, but may also help to conquer disease.

REFERENCES

1. Claude, A., *J. Exp. Med.* **84**, 51—61 (1946).
2. Hogeboom, G. H., Schneider, W. C., and Palade, G. E., *J. Biol. Chem.* **172**, 619 (1948).
3. Schneider, W. C., *ibid.* **176**, 259 (1948).
4. Svedberg, T., and Pedersen, K. O., *The Ultracentrifuge* (Clarendon Press, Oxford, 1940).
5. de Duve, C., and Berthet, J., *Intern. Rev. Cytol.* **3**, 225 (1954).
6. Appelmans, F., Wattiaux, R., and de Duve, C., *Biochem. J.* **59**, 438 (1955).
7. de Duve, C., in *Subcellular Particles* (Ronald Press, New York, 1959), pp. 128—159.
8. —, Pressman, B. C., Gianetto, R., Wattiaux, R., and Appelmans, F., *Biochem. J.* **60**, 604 (1955).
9. Berthet, J., and de Duve, C., *Biochem. J.* **50**, 174 (1951); —, Berthet, L., Appelmans, F., and de Duve, C., *ibid.* p. 182.
10. de Duve, C., *The Harvey Lectures* **59**, 49 (1965).
11. —, *J. Theoret. Biol.* **6**, 33 (1964); —, in *Enzyme Cytology,* Roodyn, D. B., Ed. (Academic Press, New York, 1967), pp. 1—26; —, *J. Cell Biol.* **50**, 20D (1971).
12. —, Berthet, J., and Beaufay, H., in *Progr. Biophys. Biophys. Chem.* **9**, 325 (1959).
13. Hogeboom, G. H., and Kuff, E. L., *J. Biol. Chem.* **210**, 733 (1954).
14. Beaufay, H., *La Centrifugation en Gradient de Densité* (Ceuterick, Louvain, 1966), 132 pp.
15. Anderson, N. G., editor. *The Development of Zonal Centrifuges and Ancillary Systems for Tissue Fractionation and Analysis,* Nat. Cancer Inst. Monogr. 21 (1966).
16. Wattiaux, R., *Mol. Cell. Biochem.* **4**, 21 (1974).
17. Beaufay, H., and Berthet, J., *Biochem. Soc. Symp.* **23**, 66 (1963).
18. —, Jacques, P., Baudhuin, P., Sellinger, O. Z., Berthet, J., and de Duve, C., *Biochem. J.* **92**, 184 (1964).

19. Wattiaux, R., Wibo, M., and Baudhuin, P., in *Ciba Foundation Symposium on Lysosomes* (Churchill, London, 1963), pp. 176—196.
20. Leighton, F., Poole, B., Beaufay, H., Baudhuin, P., Coffey, J. W., Fowler, S., and de Duve, C., *J. Cell Biol.* **37**, 482 (1968).
21. Novikoff, A. B., Beaufay, H., and de Duve, C., *J. Biophys. Biochem. Cytol.* **2**, 179 (1956).
22. Baudhuin, P., Beaufay, H., and de Duve, C., *J. Cell Biol.* **26**, 219 (1965).
23. Baudhuin, P., *L'Analyse Morphologique Quantitative de Fractions Subcellulaires* (Ceuterick, Louvain, 1968) 183 pp.
24. Poole, B., Higashi, T., and de Duve, C., *J. Cell Biol.* **45**, 408 (1970).
25. Wibo, M., Amar-Costesec, A., Berthet, J., and Beaufay, H., *ibid*, **51**, 52 (1971).
26. Wattiaux, R., *Etude Expérimentale de la Surcharge des Lysosomes* (Imprimerie J. Duculot, Gembloux, 1966) 129 pp.
27. Jacques, P., *Épuration Plasmatique de Protéines Étrangères, Leur Capture et Leur Destinée dans l'Appareil Vacuolaire du Foie* (Librairie Universitaire, Louvain, 1968), 150 pp.
28. Deter, R. L., and de Duve, C., *J. Cell Biol.* **33**, 437 (1967); —, Baudhuin, P., and de Duve, C., *ibid.* **35**, C11 (1967).
29. Coffey, J. W., and de Duve C., *J. Biol. Chem.* **243**, 3255 (1968); Aronson, N. N. Jr and de Duve, C., *ibid.* p. 4564; Fowler, S., and de Duve, C., *ibid.* **244**, 471 (1969); —, *Biochim. Biophys. Acta* **191**, 481 (1969).
30. Trouet, A., *Caractéristiques et Propriétés Antigéniques des Lysosomes du Foie* (Vander, Louvain, 1969), 185 pp.; Tulkens, P., Trouet, A., and Van Hoof, F., *Nature* **228**, 1282 (1970).
31. Leighton, F., Poole, B., Lazarow, P. B., and de Duve, C., *J. Cell Biol.* **41**, 521 (1969); Poole, B., Leighton, F., and de Duve, C., *ibid.* p. 536; Lazarow, P. B., and de Duve, C., *Biochem. Biophys. Res. Commun.* **45**, 1198 (1971); — and de Duve, C., *J. Cell Biol.* **59** 491 (1973); — and de Duve, C., *ibid.* p. 507.
32. Thinès-Sempoux, D., Amar-Costesec, A., Beaufay, H., and Berthet, J., *J. Cell Biol.* **43**, 189 (1969); Beaufay, H., Amar-Costesec, A., Feytmans, E. Thinès-Sempoux, D., Wibo, M., Robbi, M., and Berthet, J., *ibid.* **61**, 188 (1974); Amar-Costesec, A., Beaufay, H., Wibo, M., Thinès-Sempoux, D., Feytmans, E., Robbi, M., and Berthet, J., *ibid.* p. 201; Beaufay, H., Amar-Costesec, A., Thinès-Sempoux, D., Wibo, M., Robbi, M., and Berthet, J., *ibid.* p. 213; Amar-Costesec, A., Wibo, M., Thinès-Sempoux, D., Beaufay, H., and Berthet, J., *ibid.* **62**, 717 (1974).
33. Baudhuin, P., Müller, M., Poole, B., and de Duve, C., *Biochem. Biophys. Res. Commun.* **20**, 53 (1965).
34. Vaes, G., and Jacques, P., *Biochem. J.* **97**, 380 (1965); —, and Jacques, P., *ibid.* p. 389; —, *ibid.* p. 393; —, *Exptl. Cell Res.* **39**, 470 (1965); —, *La Résorption Osseuse et l'Hormone Parathyroïdienne* (Imprimerie E. Warny, Louvain, 1966), 135 pp.; —, *J. Cell Biol.* **39**, 676 (1968).
35. Bowers, W. E., Finkenstædt, J. T., and de Duve, C., *J. Cell Biol.* **32**, 325 (1967); —, and de Duve, C., *ibid., p.* 339; — and de Duve, C., *ibid.*, p. 349; —, *J. Exp. Med.* **136**, 1394 (1972); —, *J. Cell Biol.* **59**, 177 (1973); —, *J. Immunol.* **113**, 1252 (1974).
36. Baggiolini, M., Hirsch, J. G., and de Duve, C., *J. Cell Biol.* **40**, 529 (1969); —, de Duve, C., Masson, P., and Heremans, J. F., *J. Exp. Med.* **131**, 559 (1970); —, Hirsch, J. G., and de Duve, C., *J. Cell Biol.* **45**, 586 (1970); Farquhar, M. G., Bainton, D. F., Baggiolini, M., and de Duve, C., *ibid.* **54**, 141 (1972).
37. Schultz, R. L., and Jacques, P. J., *Arch. Biochem. Biophys.* **144**, 292 (1971).
38. Peters, T. J., Müller, M., and de Duve, C., *J. Exp. Med.* **136**, 1117 (1972).
39. —, in *Peptide Transport in Bacteria and Mammalian Gut,* A Ciba Foundation Symposium (ASP, Amsterdam, 1971), pp. 107—122.
40. Tulkens, P., Beaufay, H., and Trouet, A., *J. Cell Biol.* **63**, 383 (1974).

41. Poole, B., and Wibo, M., in *Proceedings of Symposium on Intracellular Protein Catabolism* (Friedrichroda, GDR, May, 1973); in press; Wibo, M., and Poole, B., *J. Cell Biol.* **63,** 430 (1974).
42. Müller, M., Baudhuin, P., and de Duve, C., *J. Cell. Physiol.* **68,** 165 (1966); —, *Acta Biol. Acad. Sci. Hung.* **22,** 179 (1971); —, *J. Cell Biol.* **52,** 478 (1972).
43. Müller, M., Hogg, J. F., and de Duve, C., *J. Biol. Chem.* **243,** 5385 (1968).
44. —, *J. Cell Biol.* **57,** 453 (1973); Lindmark, D. G., and Müller, M., *J. Biol. Chem.* **248,** 7724 (1973); — and Müller, M., *J. Protozool.* **21,** 374 (1974); — and Müller, M., *J. Biol. Chem.* **249,** 4634 (1974).
45. Scheib-Pfleger, D., and Wattiaux, R., *Develop. Biol.* **5,** 205 (1962).
46. Eeckhout, Y., *Etude Biochimique de la Métamorphose Caudale des Amphibiens Anoures* (Académie Royale de Belgique, Mémoire, Classe des Sciences, 1969) **38,** No. 4, 113 pp.
47. de Duve, C., in *Ciba Foundation Symposium on Lysosomes* (Churchill, London, 1963), pp. 1—31; —, in *Injury, Inflammation and Immunity* (Williams & Wilkins Company, Baltimore, 1964), pp. 283—311; —, *Fed. Proc.* **23,** 1045 (1964); — and Wattiaux, R., *Ann. Rev. Physiol.* **28,** 435 (1966).
48. Dingle, J. T., and Fell, H. B., Eds., *Lysosomes in Biology and Pathology* (North-Holland, Amsterdam—London), Vol. 1 and 2 (1969), Vol. 3 (Dingle, J. T., ed., 1973).
49. de Duve, C., and Baudhuin, P., *Physiol. Rev.* **46,** 323 (1966); Hruban, Z., and Rechcigl, M., *Microbodies and Related Particles* (Academic Press, New York, 1968); de Duve, C., *Proc. Roy. Soc. Ser. B.* **173,** 71 (1969); Hogg, J. F., *Ann. N.Y. Acad. Sci.* **168,** 209 (1969); Tolbert, N. E., *Ann. Rev. Plant Physiol.* **22,** 45 (1971).
50. Deckers-Passau, L., Maisin, J., and de Duve, C., *Acta Unio Intern. Contra Cancrum* **13,** 822 (1957); de Duve, C., and Beaufay, H., *Biochem. J.* **73,** 610 (1959); Beaufay, H., Van Campenhout, E., and de Duve, C., *ibid,* p. 617.
51. Hers, H. G., *Biochem. J.* **86,** 11 (1963); Lejeune, N., Thinès-Sempoux, D., and Hers, H. G., *ibid.,* p. 16; Baudhuin, P., Hers, H. G., and Loeb, H., *Lab. Invest.* **13,** 1140 (1964).
52. Van Hoof, F., *Les Mucopolysaccharidoses en tant que thésaurismoses lysosomiales* (Vander, Louvain, 1972) 285 pp.
53. Hers, H. G., and Van Hoof, F., *Lysosomes and Storage Diseases* (Academic Press, New York, 1973).
54. Peters, T. J., and de Duve, C., *Exp. Mol. Pathol.* **20,** 228 (1974); Shio, H., Farquhar, M. G., and de Duve, C., *Am. J. Pathol.* **76,** 1 (1974); Takano, T., Black, W. J., Peters, T. J., and de Duve, C., *J. Biol. Chem.* **249,** 6732 (1974).
55. de Duve, C., Wattiaux, R., and Wibo, M., *Biochem. Pharmacol.* **9,** 97 (1962).
56. de Duve, C., de Barsy, T., Poole, B., Trouet, A., Tulkens, P., and Van Hoof, F., *Biochem. Pharmacol.* **23,** 2495 (1974).
57. Trouet, A., Deprez-De Campeneere, D., and de Duve, C., *Nature New Biology* **239,** 110 (1972); —, Deprez-De Campeneere, D., De Smedt-Malengreaux, M., and Atassi, G., *Europ. J. Cancer* **10,** 405 (1974).
58. Sokal, G., Trouet, A., Michaux, J. L. and Cornu, G., *Europ. J. Cancer,* **9,** 391 (1973); Cornu, G., Michaux, J. L., Sokal, G., and Trouet, A., *ibid.,* **10,** 695 (1974); Longueville, J., and Maisin, H., *in* "Adriamycin Review" (2d Intern. Symposium, Brussels), European Press, Ghent, p. 260 (1975).

GEORGE EMIL PALADE

I was born in November 1912 in Jassy (Iaşi), the old capital of Moldavia, the eastern province of Romania. My education was started in that city and was continued through a baccalaureate (continental style) at the "Al Hasdeu" Lyceum in Buzau. My father, Emil Palade, was professor of philosophy and my mother, Constanta Cantemir-Palade, was a teacher. The family environment explains why I acquired early in life great respect for books, scholars and education.

My father had hoped I was going to study philosophy at the University, like himself, but I preferred to deal with tangibles and specifics, and—influenced by relatives much closer to my age than he was—I entered the School of Medicine of the University of Bucharest (Romania) in 1930.

Early in my student years I developed a strong interest in basic biomedical sciences by listening to, and speaking with, Francisc Rainer and André Boivin, professors of Anatomy and Biochemistry, respectively. As a result, I started working in the Anatomy laboratory while still in medical school. I went, nonetheless, through six years of hospital training, mostly in internal medicine, but I did the work for my doctorate thesis in microscopic anatomy on a rather unusual topic (for an M.D.): the nephron of the cetacean *Delphinus delphi*. It was an attempt to understand its structure in terms of the functional adaptation of a mammal to marine life.

I graduated in 1940 and, after a short period as an assistant in internal medicine, I went back to Anatomy, since the discrepancy between knowledge possessed by, and expected from, the medical practitioners of that time made me rather uneasy.

During the second world war, I served in the medical corps of the Romanian Army, and after the war—encouraged by Grigore Popa, Rainer's successor—I came to the United States in 1946 for further studies. I worked for a few months in the Biology Laboratory of Robert Chambers at New York University and, while there, I met Albert Claude who had come to give a seminar on his work in electron microscopy. I was fascinated by the perspectives opened by his findings and extremely happy when, after a short discussion following his seminar, he asked me to come to work with him at The Rockefeller Institute for Medical Research in the fall of the same year. This was truly a timely development, since Chambers was retiring that summer.

At The Rockefeller Institute, Claude was working in the department of Pathology of James Murphy with George Hogeboom and Walter Schneider as direct collaborators; Keith Porter was in the same department but had developed his own line of research on the electron microscopy of cultured animal cells. At the beginning, I worked primarily on cell fractionation procedures, and I developed with Hogeboom and Schneider the "sucrose method" for the homogenization and fractionation of liver tissue. This first "Rockefeller group" had a rather short existence: Schneider returned to the University of Wisconsin, Hogeboom moved to the National Cancer Institute, and Claude went back to Belgium in 1949 to assume the directorship of the Jules Bordet Institute. Only Porter and I remained at The Rockefeller Institute; two years later, upon Murphy's retirement, we became "orphans" and were adopted by Herbert Gasser then the director of the Institute, since none of us had the rank required to head a laboratory.

Around that time, I started working in electron microscopy with the general aim of developing preparation procedures applicable to organized tissue. This line of research had been tackled before by a few investigators, Claude included, but there was still ample room for improvement. Taking advantage of whatever techniques were already available, Porter and I worked out enough improvements in microtomy and tissue fixation to obtain preparations which, at least for a while, appeared satisfactory and gratifying. A period of intense activity and great excitement followed since the new layer of biological structure revealed by electron microscopy proved to be unexpectedly rich and surprisingly uniform for practically all eukaryotic cells. Singly, or in collaboration with others, I did my share in exploring the newly open territory and, in the process, I defined the fine structure of mitochondria, and described the small particulate component of the cytoplasm (later called ribosomes); with Porter, I investigated the local differentiations of the endoplasmic reticulum and with Sanford Palay I worked out the fine structure of chemical synapses. With all this activity, our laboratory became reasonably well known and started functioning as a training center for biological electron microscopy. The circumstances that permitted this development were unusually favorable: we didn't have to worry about research funds (since we were well supported by Herbert Gasser), we had practically complete freedom in selecting our targets, strong competitors who kept us alert, and excellent collaborators who helped us in maintaining our advance.

In the middle 1950's, I felt that the time was ripe for going back to cell fractionation as a means of defining the chemical composition and the functional role of the newly discovered subcellular components. The intent was to use electron microscopy for monitoring cell fractionation. I was starting from structural findings and morphological criteria seemed appropriate for assessing the degree of homogeneity (or heterogeneity) of the cell fractions. Philip Siekevitz joined our laboratory in 1955 and together we showed that Claude's microsomes were fragments of the endoplasmic reticulum (as postulated by Claude in 1948) and that the ribosomes were ribonucleoprotein praticles. To find out more about the function of the endoplasmic reticulum and of the attached ribosomes, we started an integrated morphological and biochemical analysis of the secretory process in the guinea pig pancreas.

In 1961, Keith Porter who had been the head of our group since 1953 joined the Biological Laboratories of Harvard University and, with his departure, the history of the second "Rockefeller group" came to an end. It was during this period that cell biology became a recognized field of research in biological sciences and that the Journal of Cell Biology and the American Society for Cell Biology were founded. Our group participated actively in each of these developments.

In the 1960's, I continued the work on the secretory process using in parallel or in succession two different approaches. The first relied exclusively on cell fractionation, and was developed in collaboration with Philip Siekevitz, Lewis Greene, Colvin Redman, David Sabatini and Yutaka Tashiro; it led to the characterization of the zymogen granules and to the discovery of the segregation of secretory products in the cisternal space of the endoplasmic reticulum. The second approach relied primarly on radioautography, and involved experiments on intact animals or pancreatic slices which were carried out in collaboration with Lucien Caro and especially James Jamieson. This series of investigations produced a good part of our current ideas on the synthesis and intracellular processing of proteins for export. A critical review of this line of research is presented in the Nobel Lecture (page 175).

In parallel with the work on the secretory process in the pancreatic exocrine cell, I maintained an interest in the structural aspects of capillary permeability, that goes back to the early 1950's when I found a large population of plasmalemmal vesicles in the endothelial cells of blood capillaries. Along this line of research, Marilyn Farquhar and I investigated the capillaries of the renal glomeruli and recognized that, in their case, the basement membrane is the filtration barrier for molecules of 100A diameter or larger; a byproduct of this work was the definition of junctional complexes in a variety of epithelia. Visceral (fenestrated) capillaries were investigated with Francesco Clementi, and muscular capillaries with Romaine Bruns and Nicolae and Maia Simionescu.

The capillary work has relied primarily on the use of "probe" molecules of known dimensions detected individually or in mass (after cytochemical reactions) by electron microscopy. It led to the identification of the passageways followed by large water-soluble molecules in both types of capillaries and by small molecules in visceral capillaries. The pathway followed by small, water-soluble molecules in muscular capillaries is still under investigation.

In the middle of the 1960's our laboratory began a series of investigations on membrane biogenesis in eukaryotic cells using as model objects either the endoplasmic reticulum of mammalian hepatocytes (with P. Siekevitz, Gustav Dallner and Andrea Leskes), or the thylakoid membranes of a green alga (*Chlamydomonas reinhardtii*) (With P. Siekevitz, Kenneth Hoober and Itzhak Ohad). These studies showed that "new" membrane is produced by expansion of "old" preexisting membrane (there is no *de novo* membrane assembly), and that new molecules are asynchronously inserted, and randomly distributed throughout the expanding membrane. Asynchrony also applies to the turnover of membrane proteins in the endoplasmic reticulum as shown by work down with P. Siekevitz, Tsuneo Omura and Walter Bock.

In 1973, I left the Rockefeller University to join the Yale University

Medical School. The main reason for the move was my belief that the time had come for fruitful interactions between the new discipline of Cell Biology and the traditional fields of interest of medical schools, namely Pathology and Clinical Medicine. Besides, my work at the Rockefeller University was done: when I left there were at least five other laboratories working in different sectors of cell biology.

At present I am investigating, together with my collaborators, the interactions which occur among the membranes of the various compartments of the secertory pathway, namely the endoplasmic reticulum, the Golgi complex, the secretion granules, and the plasmalemma.

I have been a member of the National Academy of Sciences (U.S.A.) since 1961, and I have received in the past a number of awards and prizes for my scientific work, among them: the Lasker Award (1966), the Gairdner Special Award (1967), and the Hurwitz Prize—shared with Albert Claude and Keith Porter (1970).

Since my high school years I have been interested in history, especially in Roman history, a topic on which I have read rather extensively. The Latin that goes with this kind of interest proved useful when I had to generate a few terms and names for cell biology.

I have a daughter, Georgia Palade Van Duzen, and a son Philip Palade from a first marriage with Irina Malaxa, now deceased. In 1970 I married Marilyn Gist Farquhar who is a cell biologist like myself.

INTRACELLULAR ASPECTS OF THE PROCESS OF PROTEIN SECRETION

Nobel Lecture, December 12, 1974

by

GEORGE E. PALADE

Yale University Medical School, New Haven, Connecticut, U.S.A.

A SHORT HISTORY OF THE WORK

In the early 1950's, during the near avalanche of discoveries, rediscoveries, and redefinitions of subcellular components made possible by electrons microscopy, those prospecting in this newly opened field were faced with the problem of what to do with their newly acquired wealth. It could be increased by extending the inquiry on the horizontal to many other cell types prepared by many other techniques; it could be extended in further depth, instrumental resolution permitting ("ultra" was the preferred prefix of the period); or it could be used as a guide to monitor cell fractionation procedures of the type previously developed by Claude (1). The last alternative seemed particularly attractive since the small dimensions of many of the newly discovered structures suggested that they were relatively simple macromolecular assemblies. At their level, structure—as traditionally envisaged by the microscopist—was bound to merge into biochemistry, and biochemistry of mass-isolated subcellular components appeared to be the best way to get at the function of some of the newly discovered structures. The example provided by the work on isolated mitochondria was recent and still shining (2, 3).

At the time the structures of interest were the "small particulate component of the cytoplasm" (4) soon to become in succession "ribonucleoprotein particles" (5) and "ribosomes" (6), and the endoplasmic reticulum (ER) originally discovered by Porter, Claude and Fullam (7) and then studied by Porter (8) and by Porter and myself (9—11). Philip Siekevitz joined me in 1955 and together we started a long series of integrated morphological and biochemical studies on the pancreas of the guinea pig using primarily a combination of electron microscopy and cell fractionation procedures.

The choice of the pancreatic exocrine cell, a very efficient protein producer, as the object for our studies reflected in part our training, and in part our environment. I was coming from a medical school where I had acquired an interest in "microscopical anatomy" and "physiological chemistry" and great respect for the work of Claude Bernard, Rudolf Heidenhain and Charles Garnier. Philip Siekevitz was coming from a graduate school with a Ph.D. in Biochemistry and had recently worked out one of the first *in vitro* systems for protein synthesis (12). Our environment was the Rockefeller Institute for Medical Research where a substantial amount of work had been carried out on the isolation, crystallization and characterization of pancreatic secretory proteins (cf. 13). But perhaps the most important factor in this selection was the appeal of the amazing organization of the pancreatic acinar cell whose cytoplasm is packed with stacked endoplasmic reticulum cirsternae studded with ribosomes. Its pictures had for me the effect of the song of a mermaid: irresistible and half transparent. Its meaning seemed to be buried only under a few years of work, and reasonable working hypotheses were already suggested by the structural organization itself.

The general aim of the project was to define the role played by the ribosomes, endoplasmic reticulum and other subcellular components in the synthesis and subsequent processing of the proteins produced for export by the exocrine cells of the gland. The approach worked rather well for a while (14, 15), but after a few years we ran into the common limitations of the cell fractionation procedures then in use: imperfect separation, incomplete recovery, and incomplete representation of subcellular components in the fractionation scheme. To resume the advance of the inquiry, Lucien Caro and I shifted to radioautography adapted to electron microscopy and obtained, in experiments carried out *in vivo,* a reasonable approximation of the route and timetable followed by newly synthesized, radioactive proteins from their site of synthesis to their site of discharge from the cell (16). Radioautography has, however, its own limitations connected primarily with its low resolution, so that in subsequent experiments uncertain radioautographic findings had to be checked by going back to cell fractionation procedures—this time with an advised mind. The experimental protocols were also changed to obtain better time resolution of the events under study, the major changes being the use of an *in vitro* subcellular system (17) and the adaptation by James Jamieson of an *in vitro* slice system (18) which later on evolved into a lobule system (19, 20).

Analysis of the Secretory Process in the Pancreatic Exocrine Cell

Out of his combination of complementary techniques came a coherent representation of the secretory process, a "model" which has stood well the test of time. The current trend is to move from the subcellular to the molecular level in the analysis of the model, which means that its subcellular stage has been widely enough accepted.

The analysis of the secretory process of the pancreatic exocrine cell has not been the only research line pursued in our laboratory; membrane biogenesis, intercellular junctions and structural aspects of capillary permeability are other examples. But the corresponding bodies of information are either less fully developed or still under scrutiny by us and by others; besides none of them has affected the general thinking in our field to the same extent as the story of the secretory process. With these considerations in mind, I believe that this unique and solemn occasion would be put to good use if I were to depart from the apparent tradition, which favors a summary of past or current work, and assess instead the available evidence on the secretory process, pointing out its strengths as well as its weaknesses, and trying to figure out what can be done in the future to advance our knowledge still further.

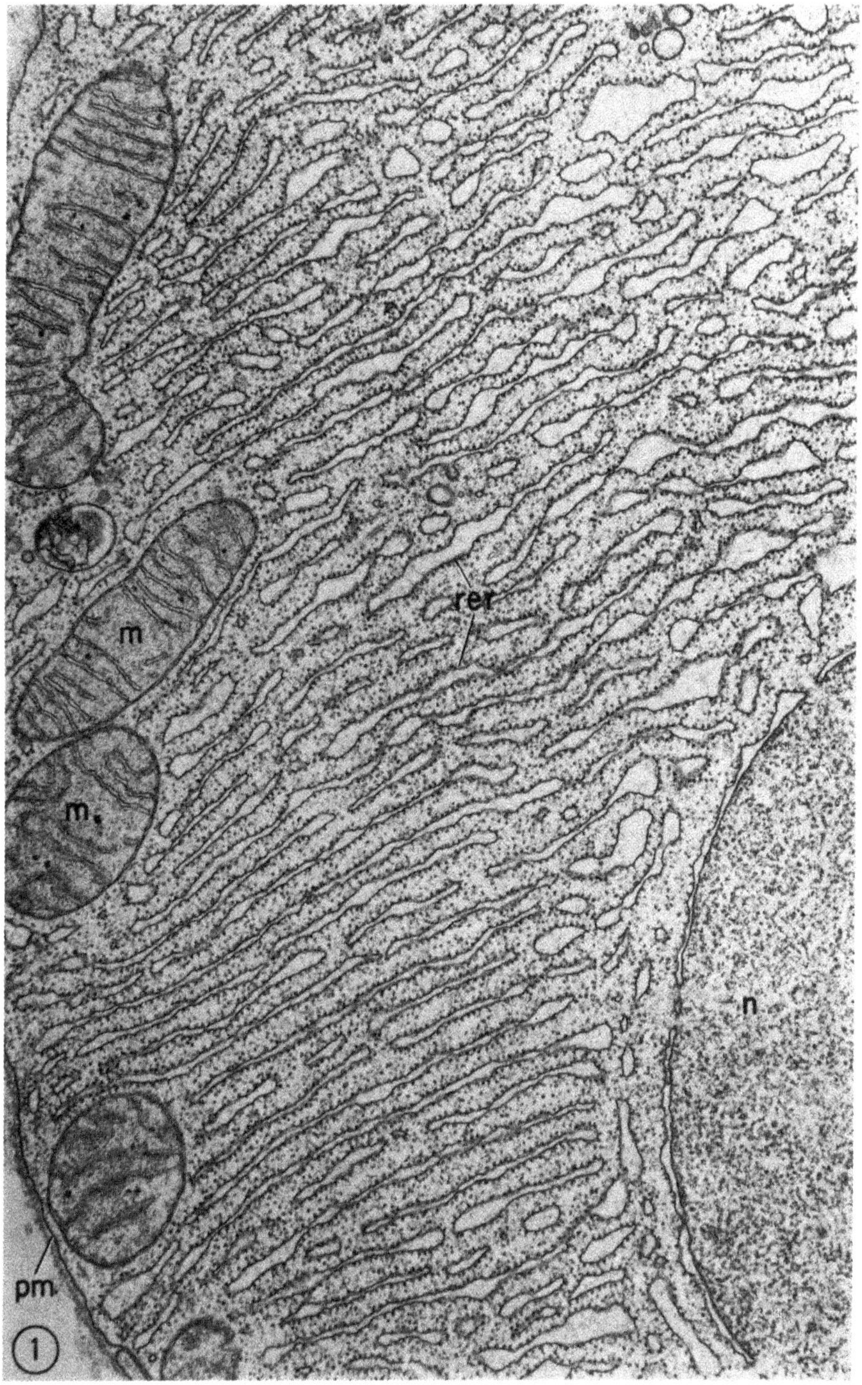

Fig. 1. Pancreatic exocrine cell. The basal region of the cell between the nucleus (*n*) and the plasmalemma (*pm*) is occupied by numerous cisternae of the rough endoplasmic reticulum (*rer*) and a few mitochondria (m).
× 12,000

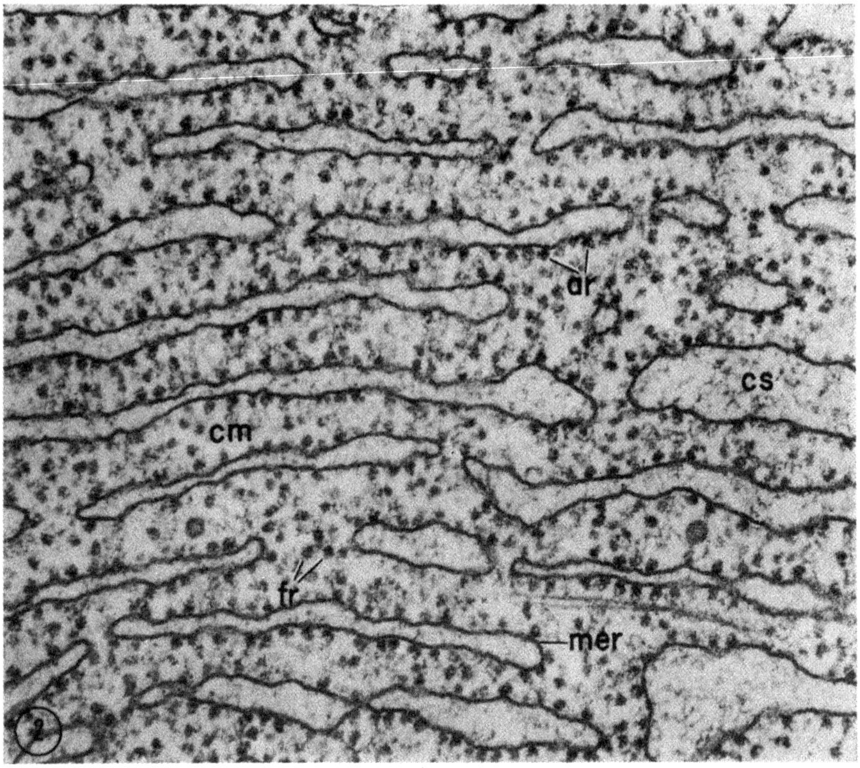

Fig. 2. Pancreatic exocrine cell. Array of cisternae of the rough surfaced endoplasmic reticulum.
cs, cisternal space; *cm,* cytoplasmic matrix (cell sol); *fr,* free ribosomes; *ar,* attached ribosomes; *mer,* membrane of the endoplasmic reticulum.
× 50,000

Our analysis recognizes in the secretory process[1] of the pancreatic exocrine cell 6 successive steps or operations of which the object is the secretory proteins. These steps are: *1) synthesis, 2) segregation, 3) intracellular transport, 4) concentration, 5) intracellular storage, and 6) discharge.* Each of them will be considered in some detail in what follows.

1. Synthesis

Proteins for export are synthesized on polysomes attached to the membrane of the rough endoplasmic reticulum (Figs. 1—2). The first clear indication that this is the case came from early work carried out with Philip Siekevitz. After a short *in vivo* exposure to [^{14}C]leucine, radioactive chymotrypsinogen appeared preferentially associated with attached polysomes isolated from the guinea pig pancreas (21) (Table I). The products of free polysomes were not investigated, but by analogy with the situation studied by others in the liver (22, 23) these polysomes probably synthesize proteins for intracellular

[1] For convenience, the term "secretory process" will be used in the rest of the text as a shorthand for "the process of protein secretion".

Table I. Specific radioactivity of chymotrypsinogen isolated from attached and free polysomes** after *in vivo* labeling with [^{14}C]leucine.

Fraction	Time after [^{14}C]leucine	
	1 min	3 min
Attached polysomes	22,100*	10,000*
Free polysomes	2,800	3,000

* cpm/mg chymotrypsinogen (estimated from enzyme activity).
** Guinea pig pancreas, from P. Siekevitz and G. E. Palade, J. Biophys. Biochem. Cytol. 7(1960)631.

use. Yet in all these cases, the results are—to some extent—ambiguous, since —as isolated—both polysome classes carry newly synthesized proteins irrespective of the latter's final destination. The differences are not qualitative as would be expected for strict specialization; they are definitely large, but only quantitative.

This finding could have a trivial explanation: e.g., leakage of newly synthesized proteins from cell compartments ruptured during tissue homogenization, followed by relocation by adsorption on the "wrong" class of polysomes. Available data indicate that artifactual relocation definitely occurs under these circumstances (24), but so far there is no reliable information concerning its extent. Alternatively the dual location may have functional significance since the position of the polysomes at the time of the initiation of translation is still unknown. Initiation in the free condition followed by enough elongation to expose either enzymic active sites or antigenic determinants before attachment seems unlikely but may occur, in principle. And the special sequence detected at the N-terminal of IgG light chains synthesized on detached polysomes (25) may function as a signal for attachment (cf. 26). To understand the situation, we need more information than we have at present on the relationship between free and attached ribosomes, on the position of polysomes at the time of initiation, and on the duration of polysomes attachment to the ER membrane.

Another aspect that should be considered at this point is the existence of two subclasses of attached polysomes: one synthesizing proteins for export and the other involved in the production of ER membrane proteins coupled with their insertion in this membrane (27). Much less is known about this second subclass, except that in its case the same uncertainties apply as to the location of the polysomes at the time of initiation. By analogy with a rather different system (chloroplast polysomes attached to thylakoid membranes during the synthesis of certain membrane proteins (28)), this type of attachment may be essentially transient, perhaps limited to a single round of translation for each site of attachment. It is generally assumed that all the soluble factors necessary for protein synthesis are present in molecular dispersion or in the form of soluble complexes in the cell sol or cytoplasmic matrix, but very few actual data are available in the case of the pancreatic exocrine cell—although this

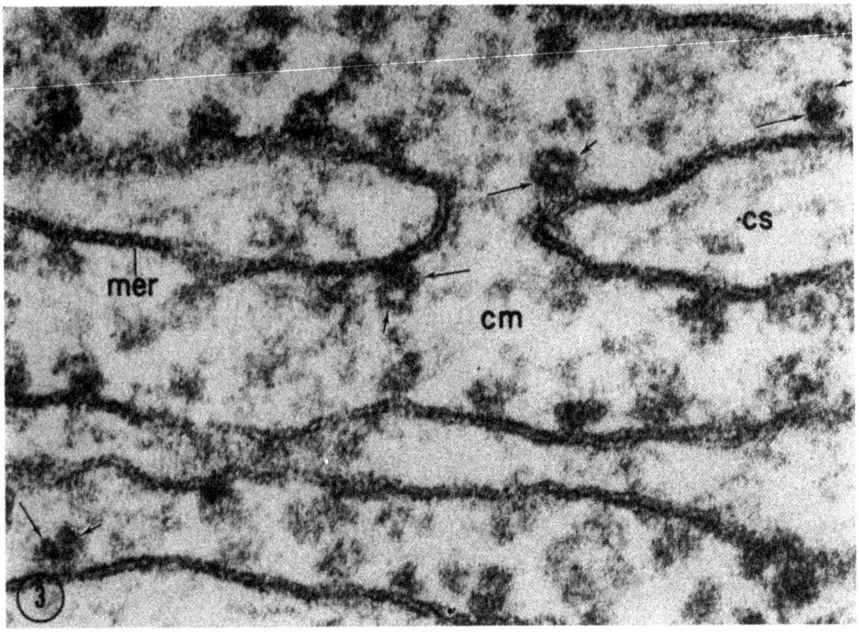

Fig. 3. Pancreatic exocrine cell. High magnification of a cytoplasmic region occupied by cisternal elements of the rough surfaced endoplasmic reticulum.
mer, membrane of the endoplasmic reticulum, *cs*, cisternal space, *cm,* cytoplasmic matrix (cell sol). The short arrows point to small subunits and the long arrows to large subunits of attached ribosomes.
× 275,000

cell is potentially a rich source of aminoacyl-t RNA synthetases, tRNAs and mRNAs. The presence of an active RNase among the secretory proteins produced by the cell has discouraged work along such lines, but this whole field may be open by using tissue taken from species known to have a very low pancreatic RNase content. Pancreatic proteolytic zymogens do not appear to constitute a problem, since their activation is either nil or controllable during cell fractionation.

2. Segregation

The newly synthesized secretory proteins are segregated in the cisternal space of the rough endoplasmic reticulum. The first evidence that this is the case came from work carried out by Redman *et al* (17) on pigeon pancreatic microsomes synthesizing *in vitro* [^{14}C] amylase. This radioactive secretory protein, initially associated with attached polysomes, preferentially appeared after ~ 3 min in the microsomal cavities. Experiments bearing on segregation were further refined in our laboratory by Redman and Sabatini (29) and Blobel and Sabatini (30). Their results indicate that the growing polypeptide chain is extruded through the microsomal membrane into the microsomal cavity which is the *in vitro* equivalent of the cisternal space of the rough endoplasmic reticulum. Upon natural or experimentally induced termination, the

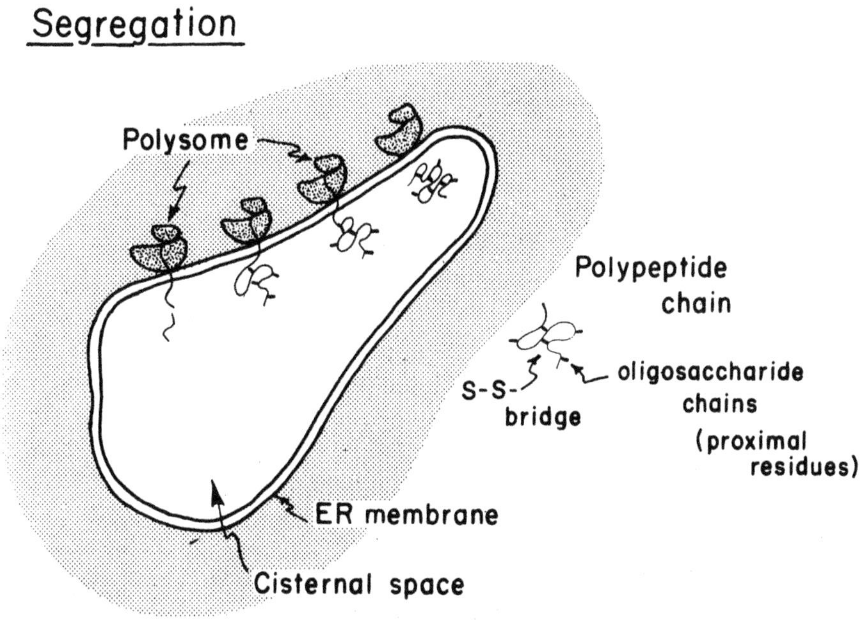

Fig. 4. Diagram of the segregation step.

newly synthesized chain separates with the microsomal vesicles and does not appear in the incubation medium, which topologically is the *in vitro* equivalent of the cell sol. Since it had already been established by Sabatini *et al* (31) that the ribosomes are attached to the ER membrane by their large subunits i.e., the bearers of nascent chains) (Fig. 3), it was concluded that segregation is the result of a vectorial transport of the newly synthesized polypeptide from the large ribosomal subunit through the ER membrane to the cisternal space.

This conclusion provides a satisfactory explanation for the basic structural features of the endoplasmic reticulum: a cavitary cell organ of complicated geometry which endows it with a large surface. All these features make sense if we assume that one of the main functions of the system is the trapping of proteins produced for export. With the exception of Ca^{2+} accumulation in the sarcoplasmic reticulum, i.e., the equivalent cell organ of muscle fibers, no other recognized function of the endoplasmic reticulum (e.g., phosphatide— and triacylglycerol synthesis, mixed function oxygenation, fatty acid desaturation) requires compellingly and directly a cavitary organ, at least according to our current knowledge. In detail, however, the forces and reactions involved in the trapping operation remain unknown. The interaction of the large ribosomal subunit with the ER membrane is understood only in very general terms (30), and precise information bearing on specific molecules involved in attachment is still lacking. Segregation appears to be an irreversible step: the nascent polypeptide is extruded in the cisternal space and, once inside, it can no longer get out (Fig. 4).

The membrane of isolated microsomes was found to be highly permeable to

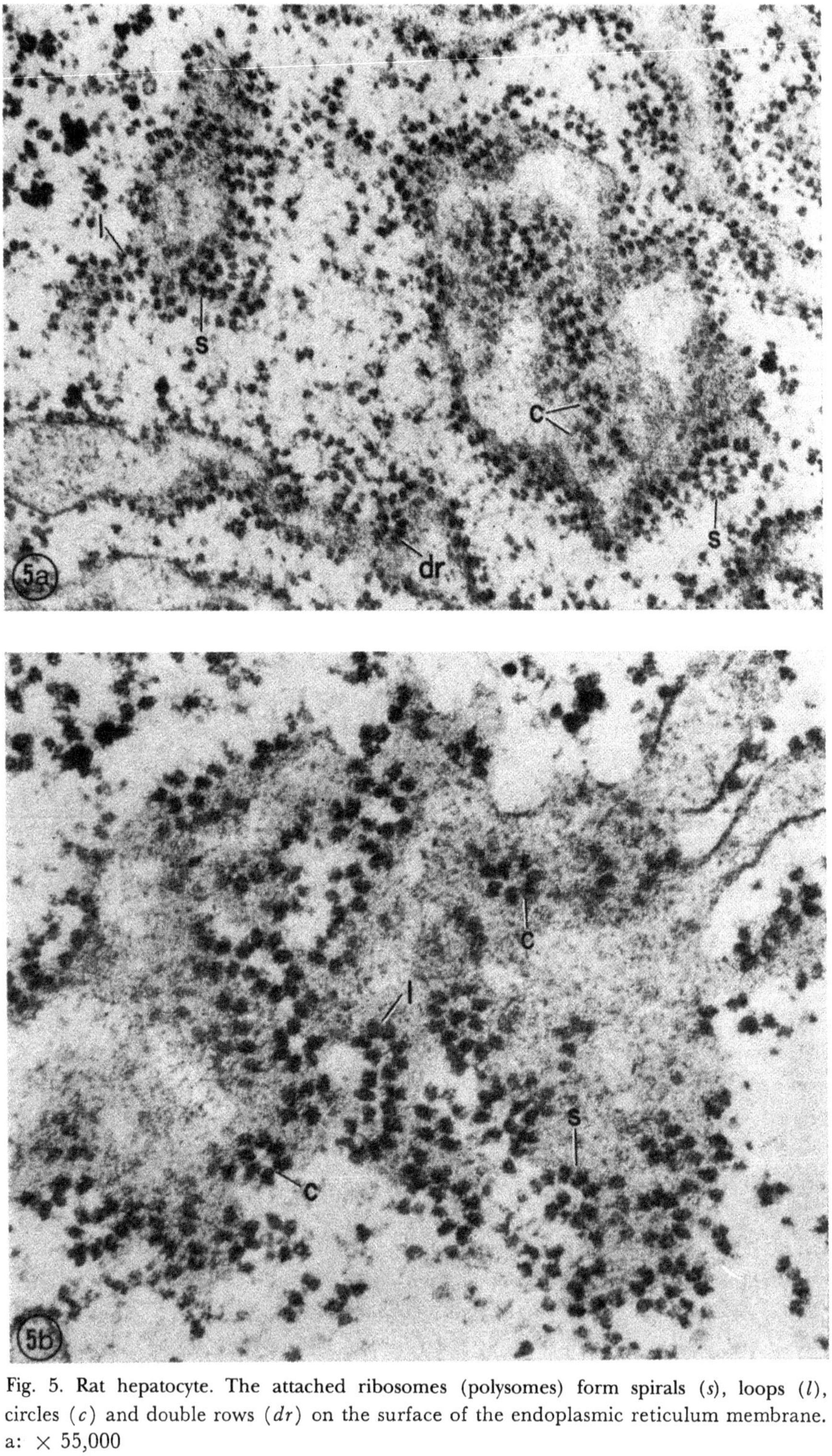

Fig. 5. Rat hepatocyte. The attached ribosomes (polysomes) form spirals (*s*), loops (*l*), circles (*c*) and double rows (*dr*) on the surface of the endoplasmic reticulum membrane.
a: × 55,000
b: × 90,000

molecules of ~ 10A diameter (32). Assuming that the same applies for the ER membrane *in situ,* it is reasonable to postulate that the imprisonment of the polypeptide is the consequence of its conversion into a globular protein too large (> 20A diameter) to permeate the membrane. This postulate is in keeping with a series of findings which show that enzymes associated with the ER membrane, or present in the cisternal space, are responsible for disulfide bridge formation (33), hydroxylation of proline and lysine residues (34), proximal glycosylation of polypeptide chains (35), and perhaps partial proteolysis (cf. 25). All these modifying operations are expected to affect directly or indirectly the tertiary structure of the secretory proteins which, once assumed, could render the proteins impermeant and their segregation irreversible (Fig. 4). Letting disulfide bridge formation aside, it would be of interest to know to what extent modifications of the type mentioned affect proteins produced for intracellular use. If the extent were nil or negligible, the differential modification of secretory proteins would provide an additional explanation for their segregation.

Available evidence either indicates or suggests that vectorial transport of secretory proteins to the cisternal space occurs in many other cell types (e.g., plasma cells (36), fibroblasts (37), granulocytes (38), parotid acinar cells (39) etc.) in addition to hepatocytes and pancreatic exocrine cells. Vectorial transport and its corollary—segregation—are most probably obligate functional features for all protein secreting cells, but further work is needed to check on the actual extent of their occurrence, as well as on possible exceptions (40).

Although the ER membrane is characterized by high fluidity (41), the polysomes attached to its cytoplasmic aspect maintain regular, characteristic patterns (Fig. 5) of rather constant geometry (4). One may wonder what prevents them from assuming a random coil conformation; or, in other words, how does the cell succeed in securing fixed attachment sites on a highly fluid membrane. This riddle must have an interesting answer.

3. Intracellular Transport

From the cisternal space of the rough endoplasmic reticulum, the secretory proteins are transported to the Golgi complex. In the case we have studied, i.e., the pancreatic exocrine cell of the guinea pig, the terminus of the transport operations is a set of large vacuoles on the trans side of the complex (16, 18) which, on account of their function (to be discussed later on), are called condensing vacuoles.

Intracellular transport was first recognized in radioautographic experiments carried out with Lucien Caro (16), but the details and requirements of this operation became evident only after James Jamieson and I shifted from intact animals to *in vitro* systems based on tissue slices (18). In such systems, short tissue exposure to radioactive amino acids ("labeling pulse") followed by effective removal of unincorporated label ("chase") became possible and, as a result, time-resolution in our experiments was considerably improved.

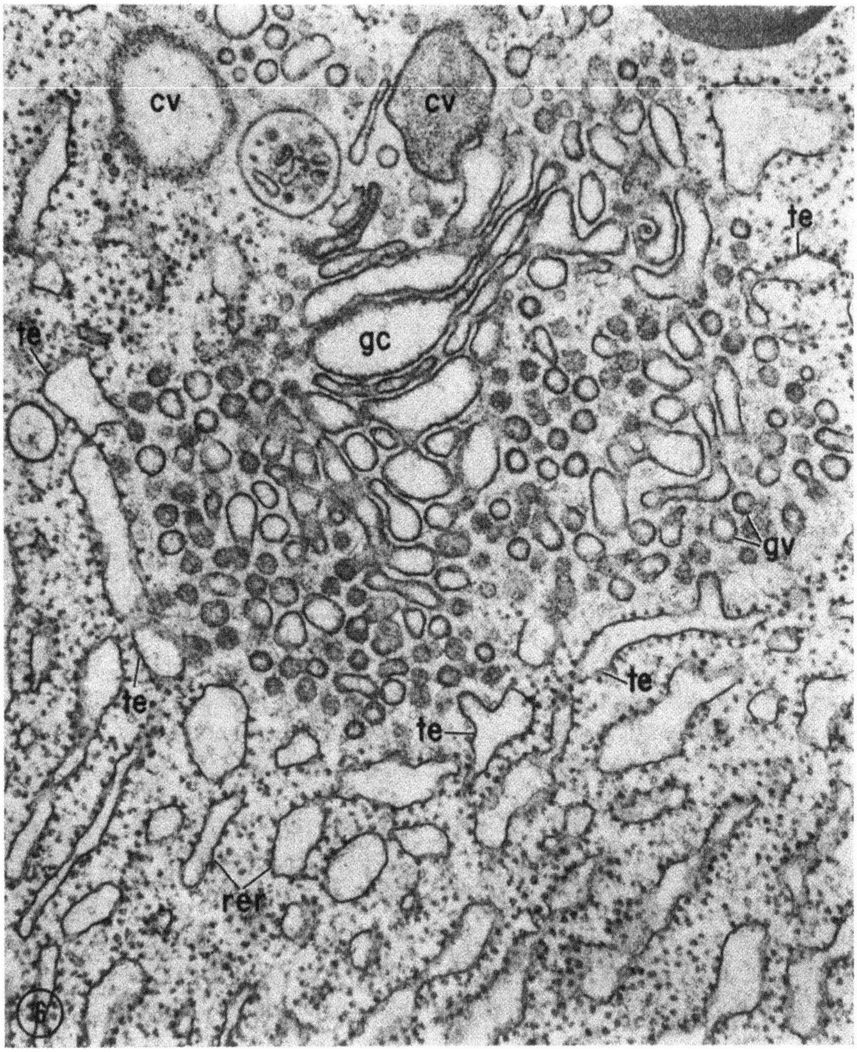

Fig. 6. Pancreatic exocrine cell. Golgi complex, partial view.
cv, condensing vacuoles; *gc,* Golgi cisternae; *gv,* Golgi vesicles; *te,* transitional elements; *rer,* rough endoplasmic reticulum.
× 26,000.

Results obtained in pulse-chase experiments showed that the pathway followed by the secretory proteins leads from the rough endoplasmic reticulum to the transitional elements of this system (Fig. 6), then to the small peripheral vesicles on the cis side of the Golgi complex (18) and finally, in about 30 min, to condensing vacuoles (42) (Table II, Fig. 7). An unexpected and intriguing finding was that intracellular transport requires energy (43) supplied (in the system investigated) by oxidative phosphorylation. In the absence of ATP synthesis, the secretory proteins remain in the rough endoplasmic reticulum, transport to condensing vacuoles being resumed upon resumption of ATP production. From these and other data, we concluded that the func-

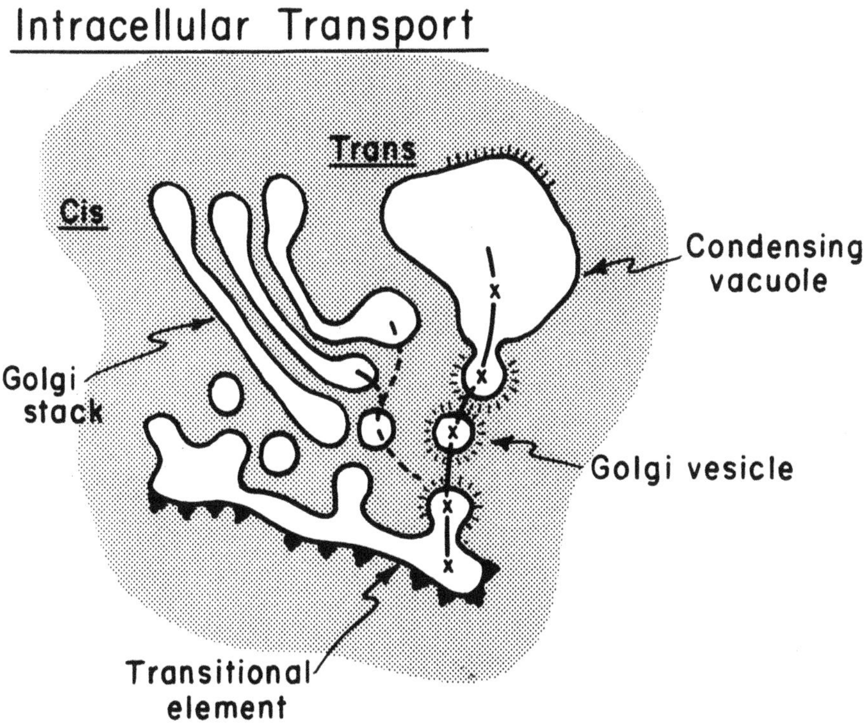

Fig. 7. Diagram of intracellular transport. X———X, pathway followed in the pancreatic exocrine cell of the guinea pig; - - - - -, pathway followed in other glandular cells.

Table II. Guinea pig pancreas. Slices incubated *in vitro**

Cell Compartments		% radioautographic grains			
		Pulse	chase minutes		
		3 min	+7	+37	+117
Rough endoplasmic reticulum		*86.3*	43.7	24.3	20.0
Golgi complex					
	peripheral vesicles	2.7	*43.0*	14.9	3.6
	condensing vacuoles	1.0	3.8	*48.5*	7.5
Zymogen granules		3.0	4.6	11.3	*58.6*
Acinar lumen		—	—	—	7.1
Other compartments**		7.0	4.6	1.1	3.2

* Simplified from J. D. Jamieson and G. E. Palade, J. Cell Biol. 34(1967)597.
pulse: 200 μCi/ml L-[^{3}H-4,5]leucine (40 μM).
chase: ^{1}H-leucine (2mM).
** Nuclei and mitochondria
For each compartment of the secretory pathway the maximal concentration figures are given in italics.

tional equivalent of a lock (or lock-gate) exists along the channels used for intracellular transport; that the lock is located at the level of the transitional elements of the endoplasmic reticulum, and that secretory proteins seem to flow vectorially to the Golgi complex, when the lock is opened.

The general pathway followed in intracellular transport appears to be the same in a variety of cell types (19, 44—48)), but direct evidence on the pre-Golgi lock has been obtained only in the case of the exocrine pancreatic cell. Extension to other systems of the inquiry dealing primarily with the lock-gate is clearly needed. In addition, many aspects of the transport operation remain either unknown or unsettled. The geometry of the connections between the endoplasmic reticulum and the Golgi complex is still a matter of debate: according to some investigators (49, 50), the two compartments are permanently connected by continuous tubules; according to us (18), the connection is intermittent and is probably established by shuttling vesicles. The energy-requiring reactions are unknown, and equally unknown are the forces involved in transport and the means by which macromolecules are moved from the endoplasmic reticulum to the condensing vacuoles against an apparent concentration gradient.

We have uncovered an interesting process, but we are only at the very beginning of its analysis. Every one of the points mentioned above remains to be elucidated by further work.

4. CONCENTRATION

The secretory proteins reach the condensing vacuoles in a dilute solution which is progressively concentrated at these sites to a level comparable to that eventually found in mature secretion granules. The exact concentration in each of the compartments involved in intracellular transport is unknown; but the increase in the density of the content in condensing vacuoles (as seen in electron micrographs), and the increase in number of radioautographic grains associated with the same vacuoles (42) (Fig. 8) suggest that the incoming solution is concentrated by a large factor. The final result of the concentration step is the conversion of the condensing vacuoles into mature secretion granules (16, 42), usually called zymogen granules in the case of the pancreatic exocrine cell.

Concentration is not dependent on a continuous supply of energy. *In situ,* neither condensing vacuoles nor zymogen granules swell when ATP production is blocked; and *in vitro,* isolated secretion granules are rather insensitive to the osmolality of the suspension medium at, or below, neutrality (51). They are instead highly sensitive to variations in pH and lyse promptly above pH 7.2 (52, 53). The findings rule out the hypothesis that concentration is achieved by ion pumps located in the membrane of the condensing vacuoles, and suggest that the cell uses for this step some other, energetically more economical mechanism. The synthesis of sulfate containing macrocolecules in Golgi elements and their presence in secretion granules in murine, pancreatic acinar cells (54) as well as in other murine glandular cells (55) have been

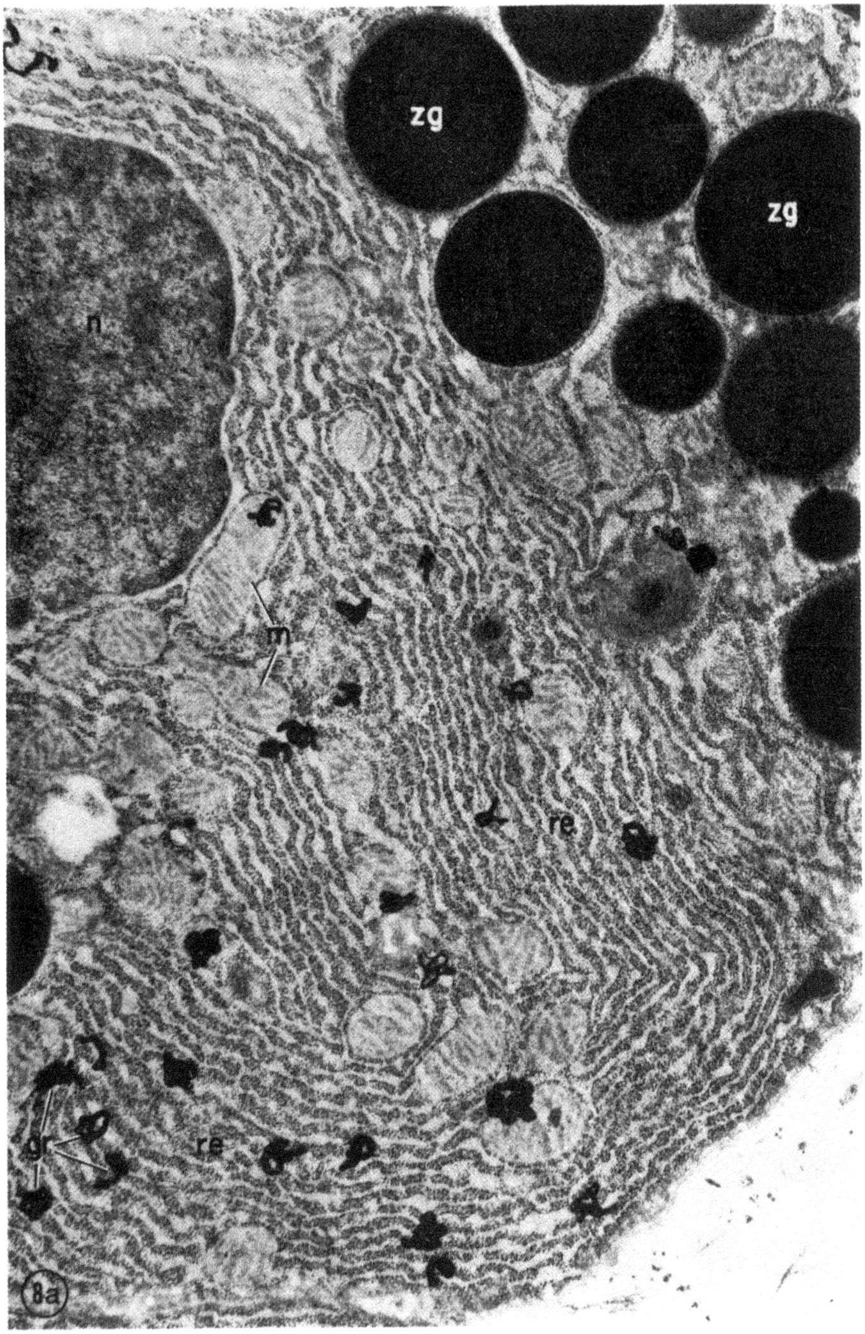

Fig. 8a. Pancreatic exocrine cell. (Guinea pig). Distribution of radioautographic grains in specimen fixed at the end of a 3 min. pulse with L-[^{3}H-4,5]leucine.
gr, radioautographic grains; *n*, nucleus; *m*, mitochondria; *zg*, zymogen granules; *re*, region of the cytoplasm occupied by the rough surfaced endoplasmic reticulum. At this time, ~ 85 % of the grains are found associated with such regions.
× 12,000

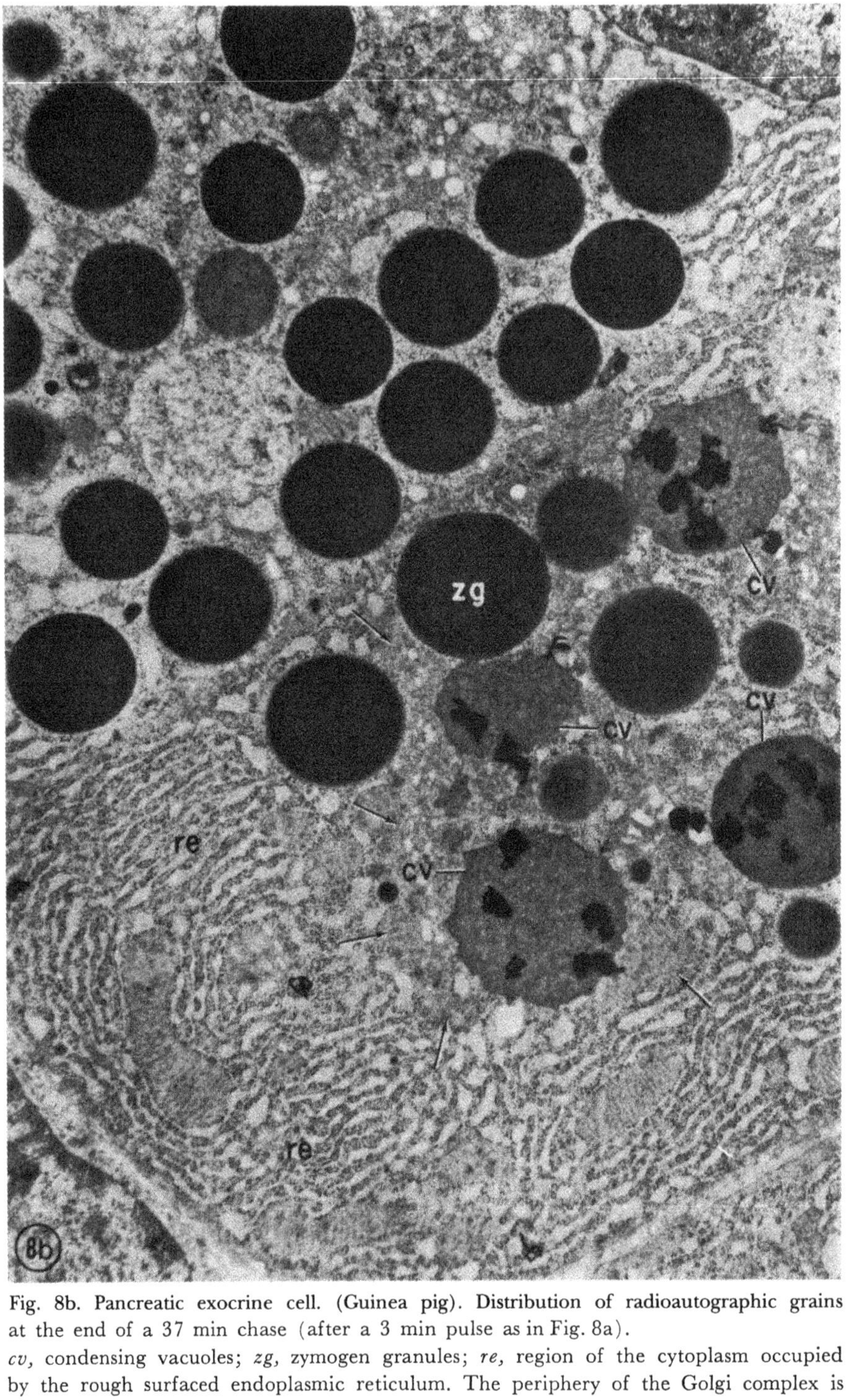

Fig. 8b. Pancreatic exocrine cell. (Guinea pig). Distribution of radioautographic grains at the end of a 37 min chase (after a 3 min pulse as in Fig. 8a).

cv, condensing vacuoles; *zg*, zymogen granules; *re*, region of the cytoplasm occupied by the rough surfaced endoplasmic reticulum. The periphery of the Golgi complex is marked by arrows. At this time, ~ 50 % of the radioautographic grains are associated with condensing vacuoles.

× 12,000

Figures 8a and 8b are taken from J. D. Jamieson and G. E. Palade, J. Cell Biol. *34*, (1967) 597.

established by radioautography. Moreover, Tartakoff *et al* (56) have recently detected a sulfated polyanion (pI $\simeq$ 3.4), presumably a sulfated peptidoglycan, in the content of zymogen granules and in discharged secretion in the guinea pig pancreas. The formation of large aggregates by ionic interactions between this polyanion and secretory proteins, which are known to be predominantly cationic (56), could cause a reduction in osmotic activity within condensing vacuoles with concomitant outflow of water. In this case, energy would no longer be required past the synthesis of the polyanion and concentration would depend primarily on the stability of the postulated aggregates.

This hypothesis remains to be validated by the isolation and characterization of the sulfated polyanion, and especially by the demonstration of relevant aggregate formation under conditions likely to prevail *in vivo* within condensing vacuoles. The hypothesis is particularly attractive because it could explain not only concentration *per se,* but also intracellular transport against an apparent chemical gradient. Such a gradient may not exist, or may be reversed, if the secretory proteins of every new batch were to be aggregated and thereby osmotically inactivated upon their entry into condensing vacuoles.

In the pancreatic exocrine cell of the guinea pig concentration is effected in trans Golgi condensing vacuoles, but in the same cell of other species (rat, for instance) the step under discussion takes places in the last cisterna on the trans side of each Golgi stack. Finally in many other glandular cells (cf. 57) the same operation is carried out in the dilated rims of the last 2—3 trans Golgi cisternae (Fig. 7). Moreover, in guinea pig pancreatic lobules hyperstimulated *in vitro,* the usual condensing vacuoles are no longer present, and concentration of secretory proteins begins already in the Golgi cisternae, preferentially in those located on the trans side of the stacks (58). There are, therefore, variations according to species, cell type, and physiological conditions in the location of concentration sites within the Golgi complex, and it would be of interest to find out whether these variations reflect changes in the distribution of the sulfated polyanion (or other functionally equivalent compounds) within the complex.

Radioautographic findings (45—47, 59) and cell fractionation data (60) obtained on a variety of tissues indicate that terminal glycosylation of secretory proteins occurs in the Golgi complex. This operation is expected to affect only a fraction, not the totality, of the proteins produced for export.

In addition, the Golgi complex appears to be the site of partial proteolysis of proinsulin (61) and perhaps other secretory proteins. It is also the site of synthesis of polysaccharides in plant cells (cf. 62). The Golgi apparatus has, therefore, a multiplicity of functions in the processing of secretory products, but—with the exception of concentration—the location of the other activities among its elements is either uncertain or still unknown.

On the one hand, there is a rather extensive literature dealing with differences in cytochemical reactions within the same cisterna (63, 64) or among the cisternae of the same stack (65, 66) without any obvious functional correlation. On the other hand, we begin to have biochemical data on Golgi subfractions, but so far they reveal no differences between Golgi cisternae and Golgi vacuoles (67).

Finally, at the level of the Golgi complex the secretory product is transferred from a high permeability membrane (i.e., the membrane of the endoplasmic reticulum), to a membrane whose lipid composition approaches that of the plasmalemma by its high content of cholesterol and sphingomyelin, and by the low degree of unsaturation of fatty acids in its phospholipids (68, 69)). Such a membrane is expected to have a low permeability, and therefore to be "exposable" without danger to the external medium at the time of discharge (see below).

In general, our knowledge of the functions of the complex is still rudimentary primarily because the isolation of Golgi fractions from tissue homogenates was achieved only recently (70—73) and is still limited to a few sources (liver, pancreas (68) and kidney (74)). The extent of compartmentation within the complex as well as the precise pathway followed by secretory products through it is still unknown. Finally, as a telling measure of our ignorance, it is worthwhile pointing out that we do not have any good idea about the functional meaning of the most prominent structural feature of the Golgi complex: the stacking of its cisternae.

5. Intracellular Storage

Secretory proteins are temporarily stored within the cell in secretion granules which, as already mentioned, are condensing vacuoles that have reached the end of the concentration step. Their membrane comes, therefore, from the Golgi complex and their content is the product of attached polysomes, modified at subsequent steps as already described in the previous sections.

In the cases so far investigated, i.e., the exocrine pancreas of the cow (53, 75), rat (76), and guinea pig (56), and the parotid of the rabbit (77), the content of the secretion granules (more precisely, the extract obtained from reasonably homogeneous secretion granule fractions) and the physiologically discharged secretion contain the same proteins in the same relative amounts (Fig. 9). Since no other intracellular sources has been revealed or suggested by our evidence, we have concluded that the content of these granules is the sole precursor of the proteins found in the juice secreted by the gland.

In the case of glands which, like the exocrine pancreas, consist of an apparently homogeneous population of secretory cells which produce a complex mixture of secretory proteins, the question of specialization at the cellular or subcellular level was asked repeatedly and answered only in part. So far all the proteins looked for in the bovine pancreas (trypsinogen (78), chymotrypsinogen, DNase (79) and RNase (80)) were detected by immunocytochemical procedures in all the secretion granules of all cells examined. Each granule probably contains a sample of the mixture discharged by the gland, but it is hard to believe that all these microsamples are quantitatively strictly identical. Specialization at the cellular level is well established in a

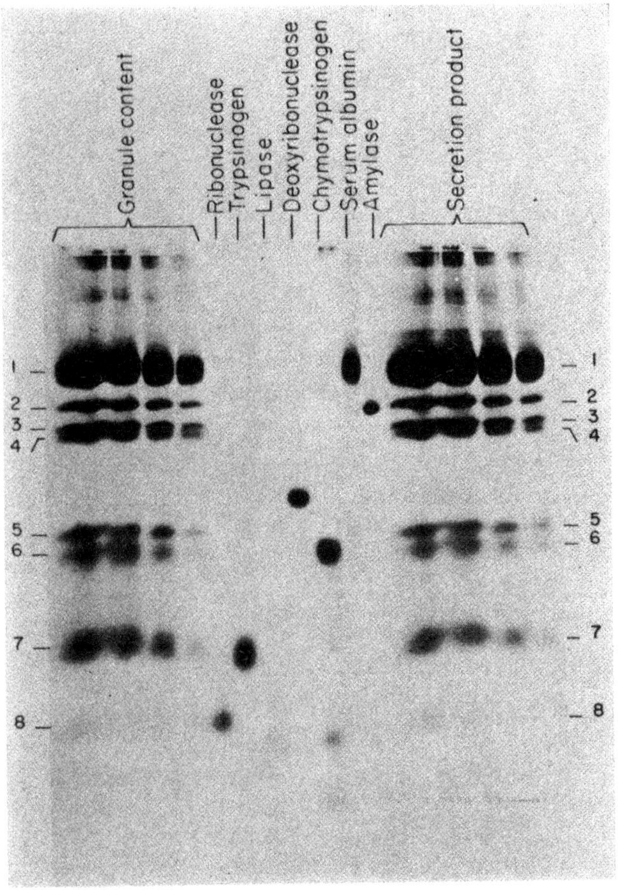

Fig. 9. Sodium dodecyl sulfate—polyacrylamide gel electropherograms of (left to right) zymogen granule content, standards, and secretion discharged by pancreatic lobules incubated and stimulated in vitro. Identification of bands: 1, unknown secretory protein and carrier bovine plasma albumin; 2, amylase; 3—4, procarboxypeptidases A and B and unknown secretory proteins; 5, unknown protein; 6, chymotrypsinogen; 7, trypsinogen; 8, ribonuclease.
From A. M. Tartakoff, L. J. Greene and G. E. Palade, J. Biol. Chem., *249,* (1974) 7420.

number of endocrine glands which are characterized by a morphologically heterogeneous cell population (cf. 57). Specialization at the subcellular level exists in polymorphonuclear neutrophil granulocytes (35). The formula used in the pancreas, i.e., intracellular storage of a complex mixture in apparently equivalent quanta, probably explains the lack of short term qualitative modulation of the secretory output (see (20, 81) for a more detailed discussion of this point). It can be assumed that this type of modulation is rendered unnecessary by the specialized nutritional habits of each species.

In the exocrine cells of the pancreas, secretion granules usually occupy the apical region between the Golgi complex and the acinar lumen. There are few microtubules in this region and few microfilaments, and there is no consistent pattern in their organization and distribution (except for the micro-

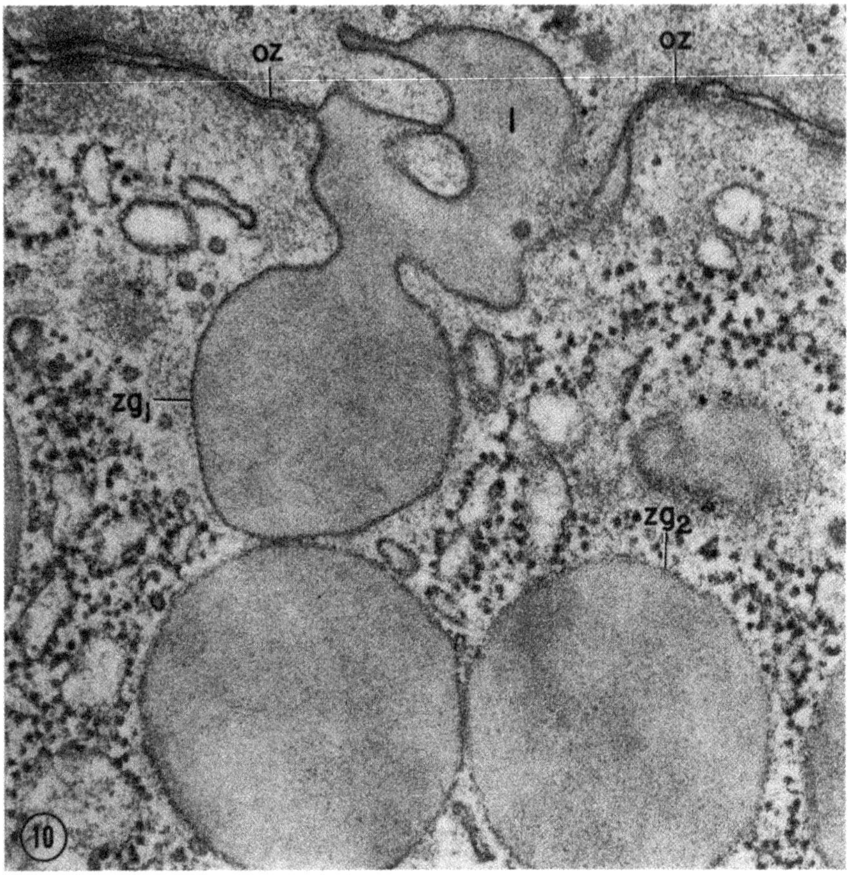

Fig. 10. Pancreatic exocrine cell. Apical region. *l,* lumen; *oz,* occluding zonules; zg_1, discharging zymogen granule; zg_2, zymogen granule still in storage.
× 110,000

filaments associated with junctional elements and microvilli). In other cell types, it has been postulated that microtubules and microfilaments play a role in effecting secretory discharge (se below), as well as in directing or moving secretory granules to their sites of discharge. In pancreatic acinar cells, radioautographic findings show that newly formed, i.e., labeled, granules are distributed at random within the preexisting granule population (42), and biochemical data indicate that newly synthesized and preexisting proteins are discharged at random from the total zymogen granule population (20, 81). With the evidence at hand, these results can be ascribed to slow diffusion leading to thorough mixing of old and new granules within the apical region. In other cell types, the situation may be different on account of incomplete mixing within the granule population and uneven distribution of discharge sites (see below).

6. Discharge

Relatively early in the investigation of the secretory process it was found that

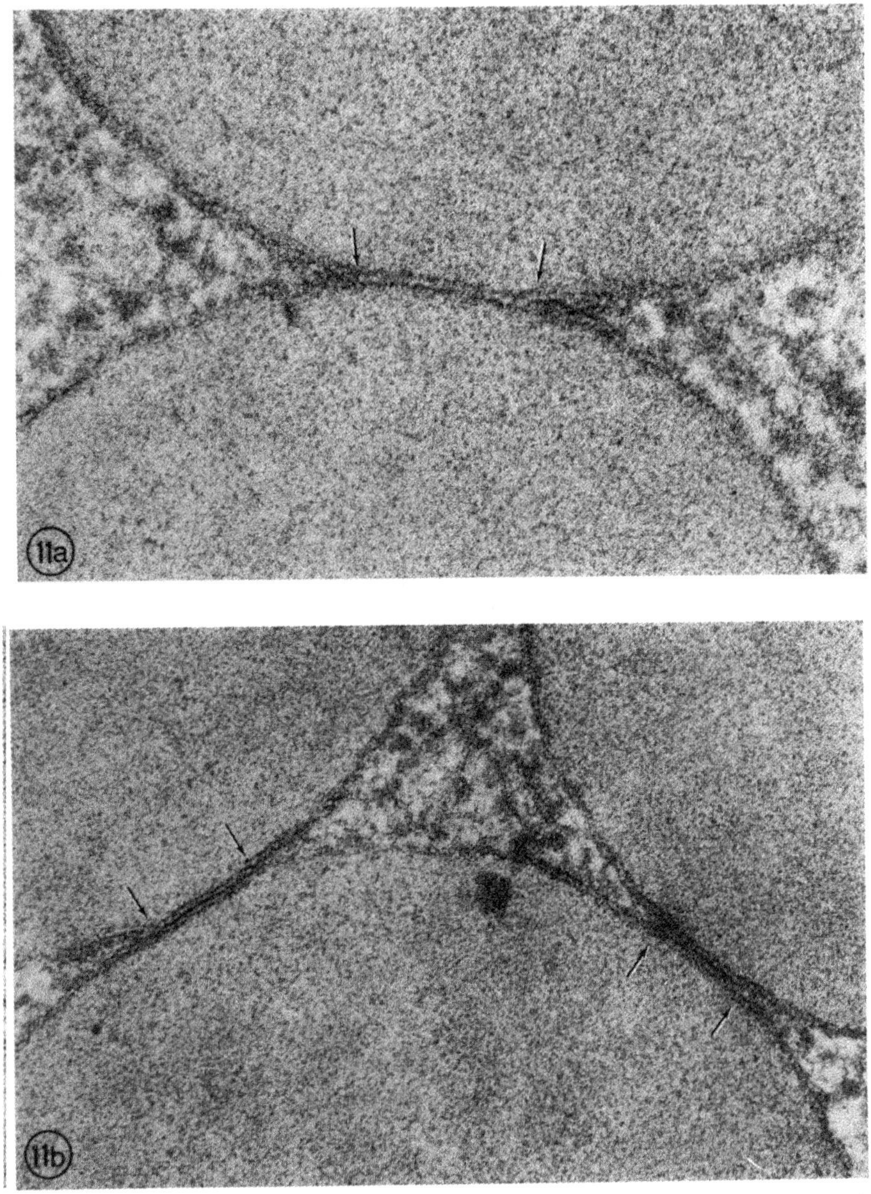

Fig. 11a, b. Pancreatic exocrine cells, Apical region. a. fusion of zymogen granule membranes followed by partial elimination of membrane layers (arrows), b. fusion of zymogen granule membranes (arrows);
a: × 220,000
b: × 160,000

secretion granules discharge their content into glandular lumina (Fig. 10) by a process originally called "membrane fusion" (82) and later on exocytosis (83). Morphological findings suggest that in preparation for discharge the membrane of the secretion granule fuses with the plasmalemma and that subsequent reorganization (i.e., progressive elimination of layers (Figs. 11, 12).

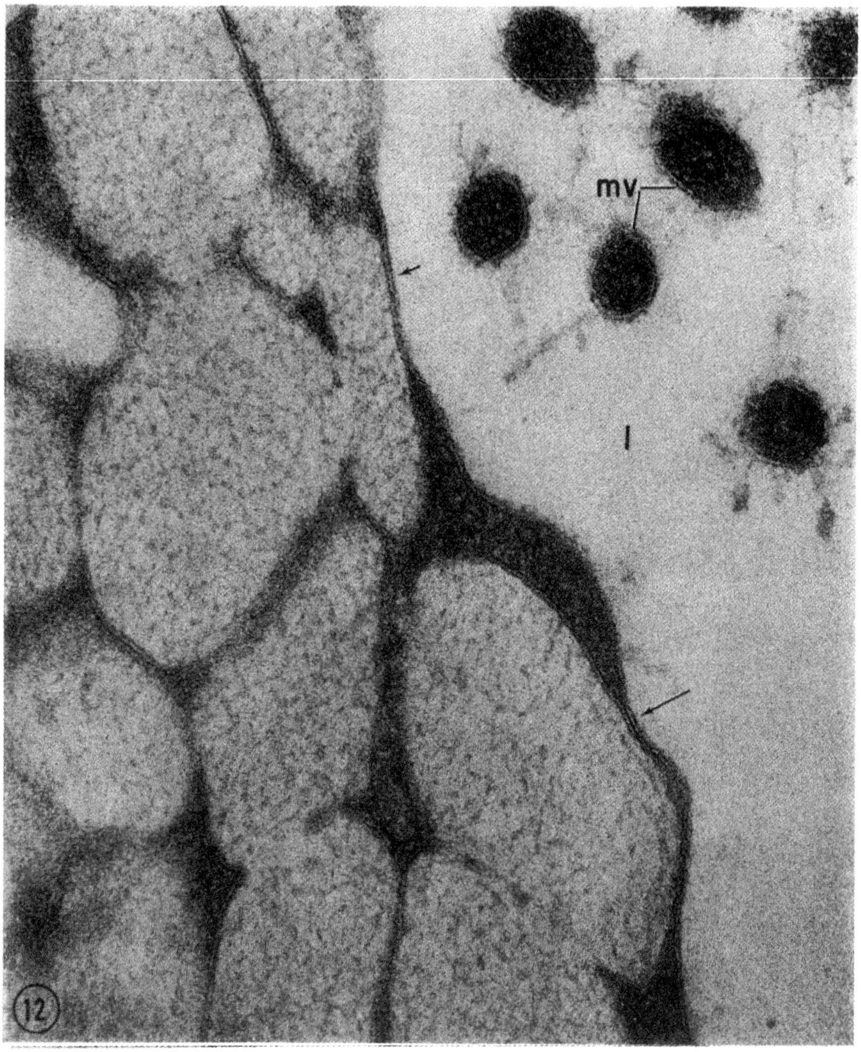

Fig. 12. Intestinal epithelium, Goblet cell. (Rat). Fusion of secretion granule membranes with the plasmalemma. Long arrows: simple fusion; short arrow: fusion with partial elimination of membrane layers.
l, lumen; *mv*, microvilli.
× 140,000

leads to fission of the fused membranes within the area of fusion. The final result is continuity established between the granule compartment and the extracellular medium (lumen), concomitantly with continuity of the granule membrane with the plasmalemma all around the orifice through which the granule content reaches the lumen (Fig. 13). This operation allows the discharge of the secretory product while insuring the maintenance of a continuous diffusion barrier between the cell sol and the extracellular medium. At the beginning, a few alternatives were considered, but by now exocytosis is recognized as a widely occurring, probably general mechanism for the dis-

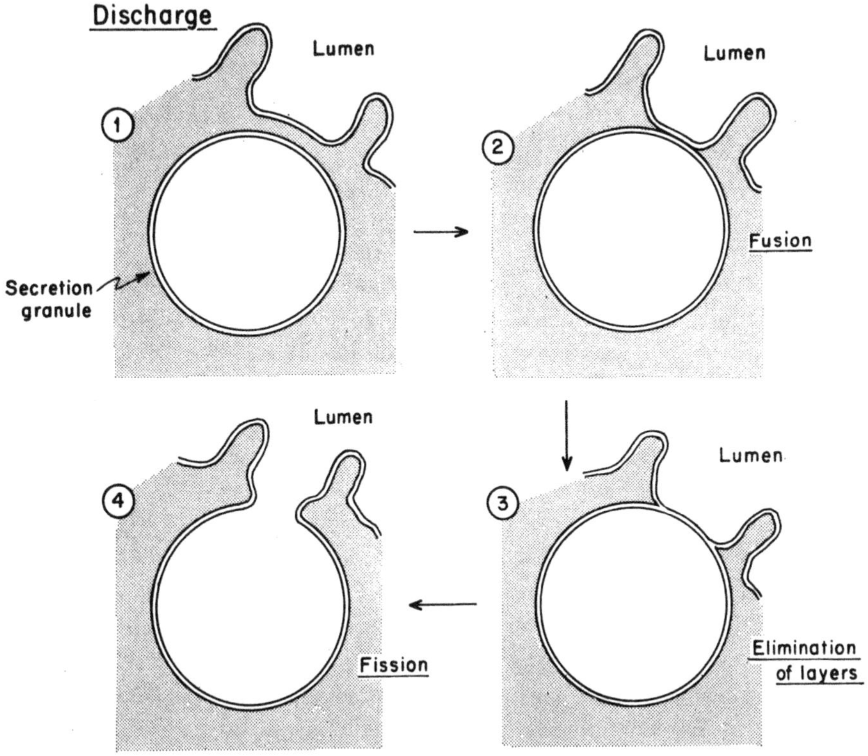

Fig. 13. Diagram of membrane interactions during secretory discharge.

charge of macromolecular secretory products.

The membrane fusion involved in secretory discharge has a high degree of specificity. The membrane of secretion granules fuses only with the plasmalemma, although there are at the time of this event and at comparable distances around the interacting pair many other types of cellular membranes. In the exocrine cells the specificity is even more stringent since ability to fuse is limited to the apical or luminal domain of the plasmalemma. The only permanently operative alternative is preliminary fusion of granule membrane to granule membrane leading eventually to discharge of two or more secretion granules in tandem (84). This type of specificity suggests the existence of complementary recognition sites in each interacting membrane which may be involved in binding preliminary to fusion. In some respects the postulated situation is reminiscent of the interaction between a hormone and its membrane receptor (85), except that in this case the events are intracellular and receptors as well as agonists are assumed to be membrane-bound.

Exocytosis has been extensively studied in a variety of secretory cells and by now its basic requirements for Ca^{2+} and energy are well established (86—88). Our own data demonstrated a stringent energy requirement for secretory discharge in the exocrine pancreatic cell and, hence, the existence of a second energy-dependent lock that controls the flow of secretory products from secretion granules in to the acinar lumina (58). Our data also showed that discharge can proceed in the absence of continuous protein synthesis (58).

In certain glandular cells, pancreatic exocrine cells included, discharge is intermittent and well integrated with other activities of the organism. In such cases, the cell which without stimulation discharges at a slow, liminal rate, responds to stimulation by either hormones or neurotransmitters by a dramatic step-up in the rate of exocytosis. The stimulus-secretion coupling (87) often involves of cyclic nucleotide generating system (adenylate cyclase in most cases) and one or more protein kinases (89). But this coupling also involves a depolarization of the plasmalemma. In the case of the pancreatic exocrine cell stimulation definitely leads to membrane depolarization (90), while the activation of a cyclic nucleotide system is still uncertain (91 vs. 89). The final target of the protein kinases is unknown in secretory cells. A hypothesis advanced a few years ago ascribes this role to tubulin (92), but the evidence in case is open to question. Results obtained on other systems (93, 94) suggest that the target might be a membrane protein.

In recent years, a number of agents activating or inhibiting exocytosis have been described and among the latter colchicine and the vinca alkaloids have received considerably attention (95, 96), the general assumption being that their inhibitory effect implies the involvement of microtubules in exocytosis. At present the situation is rather confused and a reasonable interpretation of the numerous and in part contradictory data is hardly possible. A distinction should be made between agents affecting directly membrane fusion-fission, and agents affecting the superimposed regulatory systems which activate and inactivate the coupling between stimulation and secretion. Colchicine appears to affect the basic mechanism, rather than its controls, since it inhibits discharge in hepatocytes, (97, 99), i.e., in cells that appear to lack a stimulus—secretion coupling. In these cells the effect has been localized at discharge, the last step in the secretory process, all previous steps being unaffected (99). But the involvement of microtubules remains open to question since, at least in hepatocytes, the inhibitory effect is prompt and reaches its maximum long before the depolymerization of the microtubules becomes morphologically detectable. Hence, alternative targets should be considered, especially because colchicine binds to membranes (100) and inhibits a number of transport mechanisms in the plasmalemma (101).

As already mentioned, there is no elaborate organization involving microtubules and microfilaments in the apical region of the pancreatic exocrine cells. A rather modest fibrillar feltwork (terminal web) is found under the luminal plasmalemma, but there is no fibrillar lining on the cytoplasmic aspect of the membrane of the zymogen granules while still in storage. However, a fibrillar shell[2] often appears around discharging zymogen granules when their membrane is already in continuity with the plasmalemma. It is continuous with the terminal web, it may consist of contractile proteins (actin?

[2] A fibrillar feltwork exists also at the periphery of the Golgi complex in association with the transitional elements of the ER. Its function, and the function of fibrillar coats or layers occasionally found around Golgi vesicles and vacuoles are unknown.

myosin?), and it may promote the expulsion of the secretion granule content.

Effects of Exocytosis and Intracellular Transport on Membrane Distribution

The end result of exocytosis is—on the one hand—discharge of a secretory product, and—on the other hand—relocation of secretory granule membranes in the plasmalemma. Under normal steady state conditions, excess membrane must be removed from the receiving compartment (lumen) and membrane added to the donor compartment (secretion granules, or Golgi complex), since the distribution of membrane amounts among these compartments remains relatively constant with time.

The procedures used by the cell to recover and redistribute membrane after exocytosis are unknown. Morphological findings suggest coupled endocytosis and in a few cases, namely in nerve endings (102, 103) and in anterior pituitary cells (104, 105), recovery of organized membrane in the form of endocytic vesicles has convincingly been demonstrated with the help of cytochemical tracers. Moreover, in the case of pituitary cells the recovered membrane was eventually traced to trans Golgi vacuoles and cisternae (104, 105). But the exact nature of this membrane and its ultimate fate remain a matter of speculation.

In the case of discharge, the membranes of the secretory granules can be viewed as a set of individual vesicular containers which move forward from the Golgi complex to the surface during exocytosis and presumably back to the Golgi during coupled endocytosis. In the pancreas (106) as well as in the parotid (107), the rate of synthesis of the proteins of the granule membranes is generally slower than the rate of synthesis of the secretory proteins contained in the granules. Hence, reutilization or recycling of the membrane containers is possible, in principle, but so far it has not been proven.

Assuming that a similar shuttling system of membrane containers operates between the rough endoplasmic reticulum and the Golgi complex, recently obtained evidence indicates that there is no mixing among either the lipid (68, 69) or the protein (67, 108) components of the membranes of the two compartments in the pancreas (guinea pig) and in the liver (rat). These findings impose stringent limitations on membrane interactions since they suggest that lateral diffusion of components is prevented at the time the membranes of the two compartments establish continuity, and that incoming membrane is removed from the receiving compartment according to a non-random formula (67).

The situation may appear unexpectedly complicated, even confusing, but in fact it makes sense since the final result of the restrictions mentioned is the preservation of functional specificity for the membrane of each compartment. This specificity is implied in both the old concept of "marker enzyme," and the newer ideas on sequential modification of secretory proteins as they move along the secretory pathway. The most convincing example is that of the successive glycosylation of glycoproteins (45—47, 60). The main difficulty is that we do not have at present any clear idea about the means used by the cell to carry through the various steps of the secretory process while imposing and maintaining the restrictions mentioned.

These are intriguing and challenging problems which stress the need for extending the inquiry from the processed product to the processing apparatus, expecially to the membranes that outline the compartments which form the processing apparatus. Further understanding of the secretory process is now becoming dependent on adequate information on the chemistry of these membranes and on the reactions involved in their interactions.

Variations on a Common Theme

The functional analysis of the pancreatic exocrine cell gave us a reasonably good representation of the steps generally involved in the secretory process. In addition, it helped us understand a series of special cases in other cell types which now appear to be recognizable variations on the theme already described. (Table III).

Table III. Secretory Process. Variations on a common theme.

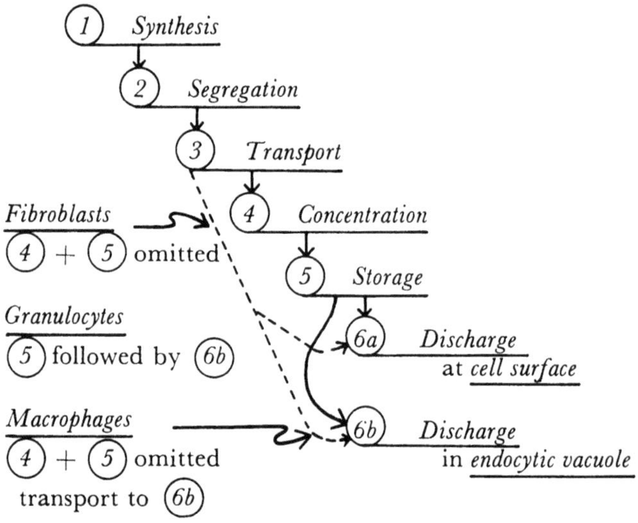

Endocrine cells producing peptide or protein hormones follow the same sequence of operations but apparently discharge their secretory product along the entire plasmalemma (57), instead of discharging within restricted plasmalemmal domains as exocrine cells do. In many secretory cells (e.g., fibroblasts, chondrocytes, plasma cells), the concentration step is omitted, secretion granules of usual appearance are absent, intracellular storage is reduced in duration or eliminated, and discharge seems to take place continuously. In such cells, the applicability of the last 3 steps of the general scheme was in doubt and the possibility of direct discharge from the cisternal space of the endoplasmic reticulum was considered (109). But recently, equivalents of secretion granules were recognized in special fibroblasts, i.e., odontoblasts (110), as well as in ordinary fibroblasts after treatment with colchicine (111). Their secretory process now appears as a variation on the common theme with the variant step resulting from lack of extensive concentration in the Golgi complex. In plasma cells the equivalent of secretion granules is still not yet identified (47).

In polymophonuclear neutrophil and eosinophil granulocytes, secretion granules are preferentially discharged into endocytic vacuoles (112, 113), discharge at the cell surface occurring only under special conditions (114). In eosinophils, the entire population of secretion granule consists of primary lysosomes, while in neutrophils the population includes "specific granules" in addition to primary lysosomes. In both cell types, all secretory proteins—irrespective of their nature—appear to be produced and processed according to the general scheme worked out for the pancreatic exocrine cell, except for the variant already mentioned at the discharge step.[3]

In macrophages, discharge of secretory proteins is also preferentially effected into endocytic vacuoles, but in addition the concentration step is apparently omitted. A dilute solution of acid hydrolases is carried probably by small vesicles (the local equivalent of primary lysosomes) from the Golgi complex to endocytic vacuoles. The latter are also able to fuse with secondary lysosomes which provide a second hand source of hydrolases (115). The variation on the common theme used by macrophages seems to be applied in all cells capable of autophagy and low efficiency heterophagy including cells specialized in protein production for export, like the hepatocytes, exocrine cells of the pancreas and cells of the anterior pituitary. A special problem arises in this case in connection with the separation of regular exportable proteins from lysosomal hydrolases. The separation seems to be reasonably efficient, though not perfect, since acid phosphatase activity has been repeatedly detected by histochemical procedures within regular secretion granules—mature and immature—and within trans Golgi cisternae (65, 116). In addition, it has been postulated that in a number of cell types lysosome formation takes place in a special compartment, called GERL (117), intercalated between the endoplasmic reticulum and trans Golgi elements. It is evident that all these cells are capable of handling concomitantly, and probably in the same production apparatus two "incompatible" lines of secretory proteins, but the means by which the products are separated or their inactivation prevented (in case of mixing) remain unknown. This riddle must also have an interesting answer.

Finally, another variation on the common theme has been found in glandular cells which produce protein or glycoprotein hormones, and are faced with an excess of stored product (116, 57). In this case the secretion granules are discharged directly into secondary lysosomes by simple membrane fusion. The process, called crinophagy was originally discovered in pituitary mammotrophs (116), but further work has shown that it probably occurs in all the cells of

[3] And except also for the fact that specific granules and primary lysosomes are formed on opposite sides of the Golgi complex of the neutrophil granulocytes (38).

the anterior pituitary (57) and probably in those of many other glands. The use of lysosomes for degrading excess secretory proteins stresses once more the need for understanding protective means against lysosomal hydrolases which must be at work along the entire secretory pathway beginning with the endoplasmic reticulum.

On the Generality of the Secretory Process

The evidence already discussed stresses the role played by the endoplasmic reticulum and the Golgi complex in the production and processing of secretory proteins. The stress put on secretion leads, however, to an apparent impasse. Since every eukaryotic cell possesses both an endoplasmic reticulum and a Golgi complex, it follows that all eukaryotic cells secrete proteins or that the organs of the secretory pathway have additional, perhaps more general and more important functions than secretion, which have been ignored or are still unknown.

This problem actually concerns fewer cell types than generally assumed since secretion of macromolecules has been recognized in recent years as an important activity in a wide variety of cells. Interestingly enough, all plant eukaryotes are secretory cells since they produce and discharge the polysaccharides and proteins of their cell walls (118). Among animal eukaryotes, male (119) and female (120, 121) gametes produce protein for extracellular use[4] and so do secretory nerve cells (122) including adrenergic (87) and presumably cholinergic (123) neurons. Smooth muscle cells have been recently recognized as producers of collagen, elastin and other proteins of the intracellular matrices (124), and the same probably applies for a variety of epithelia (including the vascular endothelium) in relation to the production of the corresponding basement membranes (125, 126).

For those animal cells for which a protein product for extracellular use has not been identified, an acceptable answer is provided by the production of lysosomal enzymes. As already mentioned, the production of these enzymes involves the same secretory apparatus (i.e., the endoplasmic reticulum and the Golgi complex) and the same sequence of steps (except for extracellular discharge) as in bona fide glandular or secretory cells. It appears, therefore, that—for the moment and with the evidence at hand—the problem can be solved in favor of the first alternative, i.e., all eukaryotic cells produce secretory proteins, the basic general secretory functions being the production of cell wall components in plant cells and the production of lysosomal enzymes in animal cells. To some extent, each type of basic production must be represented in the other kingdom. On top of these lowly but ubiquitous secretory activities, appears to be superimposed the production of highly specialized proteins exported by a variety of differentiated cell types. Our attention has been focused on the latter long enough to lose proper perspective and to

[4] In many species, female gametes produce vitellus proteins by using in part or in toto the secretory pathway (127).

assume (as we did until recently) that the secretion of proteins is a specialized function restricted to a few, highly differentiated, glandular cells.

Notwithstanding the conclusion reached in the preceding paragraph, the second alternative, i.e., the involvement of the secretory pathway in another general, but still unrecognized function, is not excluded. Among the non-secretory functions postulated for the endoplasmic reticulum and the Golgi complex is the production of cellular membranes, plasmalemma included (cf. 62). At present this postulate rests only on suggestive evidence, most of it morphological. This situation brings us back to the necessity of obtaining detailed and—if possible—comprehensive data on the chemistry and function of the different membranes of the secretory pathway and on their interactions. With this type of information, the second alternative could be put to test, and in the same time our understanding of the secretory process and of the basic organization of eukaryotic cells could be further advanced.

References

1. Claude, A., J. Exper. Med., *84* (1946) 51, 61.
2. Hogeboom, G. H., Schneider, W. C., and Palade, G. E., J. Biol. Chem., *172* (1948) 619.
3. Kennedy, E. P., and Lehninger, A. L., J. Biol. Chem., *179* (1949) 957.
4. Palade, G. E., J. Biophys. Biochem. Cytol., *1* (1955) 59.
5. Palade, G. E., in Microsomal particles and protein synthesis, Roberts, R. B., editor, Pergamon Press, 1958.
6. Roberts, R. B. in Introduction to Microsomal particles and protein synthesis, Roberts, R. B., editor, Pergamon Press, 1958.
7. Porter, K. R., Claude, A. and Fullam, E., J. Exper. Med. *81* (1945) 233.
8. Porter, K. R., J. Exper. Med., *97* (1953) 727.
9. Palade, G. E., and Porter, K. R., J. Exper. Med., *100* (1954) 641.
10. Porter, K. R., and Palade, G. E., Biophys. Biochem. Cytol., *3* (1957) 269.
11. Palade, G. E., J. Biophys. Biochem. Cytol., *2* (suppl.) (1956) 85.
12. Siekevitz, P., J. Biol. Chem., *195* (1952) 549.
13. Northrop, J. H., Kunitz, M., and Herriott, R. M., Crystalline Enzymes, Columbia University Press (1948).
14. Palade, G. E., and Siekevitz, P., J. Biophys. Biochem. Cytol., *2* (1956) 171, 671.
15. Siekevitz, P., and Palade, G. E., J. Biophys. Biochem. Cytol., *4* (1958) 203, 309, 557; *5* (1959) 1.
16. Caro, L. G., and Palade, G. E., J. Cell Biol., *20* (1964) 473.
17. Redman, C. M., Siekevitz, P., and Palade, G. E., J. Biol. Chem., *241* (1966) 1150.
18. Jamieson, J. D., and Palade, G. E., J. Cell Biol., *34* (1967) 577.
19. Castle, J. D., Jamieson, J. D., and Palade, G. E., J. Cell Biol., *53* (1972) 290.
20. Scheele, G. A., and Palade, G. E., J. Biol Chem., *250* (1975) 2660.
21. Siekevitz, P., and Palade, G. E., J. Biophys. Biochem. Cytol., *7* (1960) 619, 631.
22. Redman, C. M., J. Biol. Chem., *244* (1969) 4308.
23. Hicks, S. J., Drisdale, J. W., and Munro, H. N., Science (Washington) *164* (1969) 584.
24. Tartakoff, A., and Palade, G., unpublished observations.
25. Milstein, C., Brownlee, G .G., Harrison, T. M., and Mathews, M. B., Nature, *239* (1972) 117.
26. Blobel, G., and Sabatini, D. D., in Biomembranes 2 (1971) 193; L. A. Menton

editor, Plenum Publish. Co., New York.
27. Dallner, G., Siekevitz, P., and Palade, G. E., J. Cell Biol., *30* (1966) 73, 97.
28. Chua, N. H., Blobel, G., Siekevitz, P., and Palade, G. E., Proc. Nat. Acad. Sci., U.S.A., *70* (1973) 1554.
29. Redman, C. M., and Sabatini, D. D., Proc. Nat. Acad. Sci., U.S.A. *56* (1966) 608.
30. Blobel, G., and Sabatini, D. D., J. Cell Biol., *45* (1970) 146.
31. Sabatini, D. D., Tashiro, Y., and Palade, G. E., J. Mol. Biol., *19* (1966) 503.
32. Tedeschi, H., James, J. M., and Anthony, W., J. Cell Biol., *18* (1963) 503.
33. Anfinson, C. B., Harvey Lectures *61* (1966) 95.
34. Olsen, B. R., Berg, R. A., Kishida, Y., and Prockop, D. J., Science (Washington) *182* (1973) 825.
35. Molnar, J., Robinson, G. B., and Winzler, R. J., J. Biol. Chem., *240* (1965) 1882.
36. Mach, B., Koblet, H., and Gros, D., Proc. Nat. Acad. Sci., U.S.A. *59* (1968) 445.
37. Grant, M. G., and Prockop, D. J., New England J. Med., *286* (1972) 194.
38. Bainton, D. F., and Farquhar, M. G., J. Cell Biol., *39* (1968) 299 and *45* (1970) 54.
39. Herzog, V., and Miller, F., Z. Zellforsch. Mikrosk. Anat., *107* (1970) 403.
40. Lisowska-Berstein, B., Lamm, M. E., and Vassali, P., Proc. Nat. Acad. Sci., U.S.A. *66* (1970) 425.
41. Rogers, M. J., and Strittmatter, P., J. Biol. Chem. *249* (1974) 895, 5565.
42. Jamieson, J. D., and Palade, G. E., J. Cell Biol., *34* (1967) 597.
43. Jamieson, J. D., and Palade, G. E., J. Cell Biol., *39* (1968) 589.
44. Swenson, R. M., and Kern, M., Proc. Nat. Acad. Sci., U.S.A. *57* (1967) 417.
45. Wuhr, P., Herscovics, A., and Leblond, C. P., J. Cell Biol., *43* (1969) 289.
46. Haddad, A., Smith, M. D., Herscovics, A., Nadler, N. J., and Leblond, C. P., J. Cell Biol., *49* (1971) 856.
47. Zagury, D. Uhr, J. W., Jamieson, J. D., and Palade, G. E., J. Cell Biol., *46* (1970) 52.
48. Hopkins, C. R., and Farquhar, M. G., J. Cell Biol., *59* (1973) 276.
49. Morré, D. J., Keenan, T. W., and Mollenhauer, H. H., in Advances in Cytopharmacology, Clementi, F., and Ceccarelli, B., editors, Raven Press, New York 1971.
50. Claude, A., J. Cell Biol., *47* (1970) 745.
51. Jamieson, J. D., and Palade, G. E., J. Cell Biol., *48* (1971) 503.
52. Hokin, L. E., Biochim. et Biophys. Acta *18* (1955) 379.
53. Greene, L. J., Hirs, C. H. W., and Palade, G. E., J. Biol. Chem. *238* (1963) 2054.
54. Berg, N. B., and Young, R. W., J. Cell Biol., *50* (1971) 469.
55. Young, R. W., J. Cell Biol., *57* (1973) 175.
56. Tartakoff, A. M., Greene, L. J., and Palade, G. E., J. Biol. Chem., *249* (1974) 7420.
57. Farquhar, M. G., Memoirs Soc. for Endocrinology, *19* (1971) 79.
58. Jamieson, J. D., and Palade, G. E., J. Cell Biol., *50* (1971) 135.
59. Neutra, M., and Leblond, C. P., J. Cell Biol., *30* (1966) 137.
60. Schachter, H., Jabbal, I., Hudgin, R. L., Pinteric, L., McGuire, J., and Roseman, S., J. Biol. Chem., *245* (1970) 1090.
61. Steiner, D. F., Clark, J. L., Nolan, C., Rubenstein, A. H., Margoliash, E., Melani, F., and Oyer, P. E., Proc. 13th Nobel Symposium, (1970) 123.
62. Dauwalder, M., Whaley, W. G., and Kephart, J. E., Subcell. Biochem., *1* (1972) 225.
63. Farquhar, M. G., Bergeron, J. J. M., and Palade, G. E., J. Cell Biol., *60* (1974) 8.
64. Ovtracht, L., and Thiéry, J. P., J. Microscopie, *15* (1972) 135.

65. Novikoff, A. B., Essner, E., and Goldfischer, S., in The Interpretation of Ultrastructure, Harris, R. J. C., editor, Acad. Press, New York. (1962).
66. Friend, D. S., J. Cell Biol., *41* (1969) 269.
67. Bergeron, J. J. M., Ehrenreich, J. H., Siekevitz P., and Palade, G. E., J. Cell Biol., *59* (1973) 73.
68. Meldolesi, J., Jamieson, J. D., and Palade, G. E., J. Cell Biol., *49* (1971) 109, 130.
69. Keenan, T. W., and Morré, D. J., Biochemistry, *9* (1970) 19.
70. Fleischer, B., Fleischer, S., and Ozawa, H., J. Cell Biol., *43* (1969) 59.
71. Fleischer, B., and Fleischer, S., Biochim. Biophys. Acta, *219* (1970) 301.
72. Morré, D. J., Hamilton, R. L., Mollenhauer, H. H., Mahley, R. W., Cunningham, W. P., Cheetham, R. D., and LeQuire, V. S., J. Cell Biol., *44* (1970) 484.
73. Ehrenreich, J. H., Bergeron, J. J. M., Siekevitz, P., and Palade, G. E., J. Cell Biol., *59* (1973) 45.
74. Fleischer, B., and Zambrano, F., J. Biol. Chem., *249* (1974) 5995.
75. Keller, P. J., and Cohen, E., J. Biol. Chem., *236* (1961) 1407.
76. Palla, J. C., Thèse de Doctorat-ès-Sciences, Marseilles, 1970.
77. Castle, J. D., Jamieson, J. D., and Palade, G. E., J. Cell Biol., *64* (1975) 182.
78. Kraehenbuhl, J. P., and Jamieson, J. D., Proc. Nat. Acad. Sci., U.S.A. *69* (1972) 1771.
79. Kraehenbuhl, J. P., and Jamieson, J. D., unpublished observations.
80. Painter, R. G., Tokuyashu, K. T., and Singer, S. J., Proc. Nat. Acad. Sci., U.S.A., *70* (1973) 1649.
81. Tartakoff, A. M., Jamieson, J. D., Scheele, G. A., and Palade, G. E., J. Biol. Chem., *250* (1974) 2671.
82. Palade, G. E., in Subcellular Particles, Hayashi, T., editor, Ronald Press, New York, 1959.
83. de Duve, C., in Lysosomes, Ciba Foundation Symposium, (1963) 126.
84. Ishikawa, A., J. Cell Biol., *24* (1965) 369.
85. Cuatrecasas, P., Proc. Nat. Acad. Sci., U.S.A. *68* (1971) 1264.
86. Douglas, W. W., and Rubin, R. P., J. Physiol., *159* (1961) 40 and *167* (1963) 288.
87. Douglas, W. W., Br. J. Pharmacol., *34* (1968) 451.
88. Schramm, M., Annu. Rev. Biochem., *36* (1967) 307.
89. Rasmussen, H., Science *170* (1970) 404.
90. Mathews, E. K., and Petersen, O. H., J. Physiol., *231* (1973) 283.
91. Kulka, R. G., and Sternlicht, E., Proc. Nat. Acad. Sci., U.S.A. *61* (1968) 1123
92. Goodman, D. P. B., Rasmussen, H., DiBella, F., and Guthrow, C. E., Jr., Proc. Nat. Acad. Sci., U.S.A., *67* (1970) 652.
93. Dilorenzo, R. J., Walton, K. G., Curran, P. F., and Greengard, P., Proc. Nat. Acad. Sic., U.S.A., *70* (1973) 880.
94. Johnson, E. M., Ueda, T., Moeno, H., and Greengard, P., J. Biol. Chem., *247* (1972) 5650.
95. Lacy, P. E., Howell, S. L., Young, D. A., and Fink, C. J., Nature, *219* (1968) 1177.
96. Williams, J. A., and Wolff, J., J. Cell Biol., *54* (1972) 158.
97. LeMarchand, Y., Single, A., Assimacopoulos-Jeannet, F., Orci, L., Rouillier, C., and Jeanrenard, B., J. Biol. Chem., *248* (1973) 6862.
98. Stein, O., and Stein, Y., Biochim. Biophys. Acta *306* (1973) 142.
99. Redman, C. M., Banerjee, D., Howell, K., and Palade, G. E., J. Cell Biol., *64* (1975) in press.
100. Stadler, J., and Franke, W. W., J. Cell Biol., *60* (1974) 297.
101. Wilson, L., Bamberg, J. R., Mizel, S. B., Grisham, L. M., and Creswell, K. M.,

Fed. Proc., *33* (1973) 158.
102. Heuser, J. E., and Reese, T. S., J. Cell Biol., *57* (1973) 315.
103. Ceccarelli, B., Hurlbut, W. P., and Mauro, A., J. Cell Biol., *57* (1973) 449.
104. Pelletier, G., J. Ultrastructure. Res. *43* (1973) 445.
105. Farquhar, M. G., Skutelsky, E., andHopkins, C. R., in The Anterior Pituitary, Tixier-Vidal, A., and Farquhar, M. G., eds. Academic Press, New York 1975, p. 83.
106. Meldolesi, J., J. Cell Biol., *61* (1974) *1*.
107. Castle, J. D., Thesis, Rockefeller University, 1974.
108. van Golde, L. M. G., Fleischer, B., and Fleischer, S., Biochim. Biophys. Acta *249* (1971) 318.
109. Ross, R., and Benditt, E., J. Cell Biol., *27* (1965) 83.
110. Weinstock, M., and Leblond, C. P., J. Cell Biol., *60*(1974) 92.
111. Olsen, B. R., and Prockop, D. J., Proc. Nat. Acad. Sci., U.S.A. *71* (1974) 2033.
112. Zucker-Franklin, D., and Hirsch, J. G., J. Exper. Med., *120* (1964) 569.
113. Bainton, D. F., in Phagocytic mechanisms in health and disease, Williams, R. C., and Fudenberg, H. H., eds. Intercontinental Medical Book Corp., New York, (1972).
114. Henson, P. M., J. Immunol. *107* (1971) 1547.
115. Cohn, Z. A., Fedorko, M. E., and Hirsch, J. G., J. Exp. Med., *123* (1966) 157.
116. Smith, R. E., and Farquhar, M. G., J. Cell Biol. *31* (1966) 319.
117. Novikoff, P. M., Novikoff, A. B., Quintana, N., and Hauw, J., J. Cell Biol., *50* (1971) 859.
118. Albersheim, P., Bauer, W. D., Keestra, K., and Talmadge, K. W., in Biogenesis of plant cell wall polysaccharides, Loewus, F., ed. Academic Press, New York (1973).
119. Fawcett, D. W., Biology of Reproduction, *2* (1970) 90.
120. Anderson, E., J. Cell Biol., *37* (1968) 514.
121. Szollosi, D., Anat. Record, *159* (1967) 431.
122. Douglas, W. W., and Poisner, A. M., J. Physiol. *172* (1964) 1.
123. Whittacker, V. P., in Advances in Cytopharmacology, *2* (1973) 311, Raven press, New York.
124. Ross, R., J. Cell Biol., *50* (1971) 172.
125. Hay, E. D., in The Epidermis, Montagna, W., and Lobitz, W. C., eds., Academic Press, New York, 1964, p. 97.
126. Hay, E. D., and Dodson, Y. W., J. Cell Biol., *57* (1973) 190.
127. Kessel, R. G., Zeitschr. Zellforschung, *89* (1968) 17.

Dedication

This lecture is dedicated with affection and gratitude to Keith Porter, Philip Siekevitz, James Jamieson, Lucien Caro, Lewis Greene, Lars Ernster, David Sabatini, Colvin Redman, Jacopo Meldolesi, Gustav Dallner, Yutaka Tashiro, Tsuneo Omura, Gunter Blobel, Alan Tartakoff, David Castle and George Scheele, my good colleagues and companions in the work carried out on the endoplasmic reticulum and secretory process.

1975
Physiology or Medicine

DAVID BALTIMORE, RENATO DULBECCO and HOWARD M. TEMIN

"for their discoveries concerning the interaction between tumour viruses and genetic material of the cell"

THE NOBEL PRIZE FOR PHYSIOLOGY OR MEDICINE

Speech by Professor PETER REICHARD of the Karolinska Medico-Chirurgical Institute
Translation from the Swedish text

Your Majesties, Your Royal Highnesses, Ladies and Gentlemen,
How does a cancercell arise? What distinguishes it from a normal cell? Cancercells are social misfits, outside the control of the organism. The capacity for unlimited growth is inherited from generation to generation. The difference between a normal cell and a cancercell resides in their genes.

Transformation of a normal cell to a cancercell thus requires a change in the genetic material. This may be caused by radiation, by treatment with different chemicals or by infection with tumor viruses. This year's Nobel prize winners have clarified what happens on infection of cells with tumor viruses.

The genetic material of the cell—or with a chemical term, its DNA—is locked up in the cell nucleus. The cell is constructed in accordance with the master plan laid down in the chemical structure of the genetic material. The building job is carried out by proteins, but the instructions are given by DNA. But the master plan is locked up in the nucleus, and the instructions reach the building place in the form of blue prints which we call RNA with a chemical term. The flow of information in the cell thus occurs first from DNA or RNA and then from RNA to proteins.

How does a tumor virus interfere with this process? As all other viruses, tumor viruses contain either DNA or RNA. Renato Dulbecco clarified the process of infection with DNA-tumor viruses. When a cell becomes infected this means that viral DNA enters it. Dulbecco made the basic observation that viral DNA is incorporated in the infected cell nucleus and is there joined to the DNA of the cell. Tumor transformation involves the incorporation of viral into cellular genetic material and results in the appearance of hereditary properties resulting in unlimited cell growth. At each cell division the new "cancer" DNA is then carried to the next generation.

What happens after infection with a tumor virus which contains RNA and not DNA? Also in this case transformation to a cancer cell is the result of an influence on the genetic material, but this now occurs in a more indirect manner. Already in the early 1960's Howard Temin proposed that a RNA tumor virus can give rise to a DNA copy which is incorporated into the genetic material of the cell. This requires a reversal of the normal flow of information from DNA to RNA, a rather unorthodox idea at that time, which was not well received by the scientific establishment. But in 1970 Howard Temin and David Baltimore independently discovered that RNA tumor viruses contain a special protein which can make a DNA copy from RNA. This protein carries out precisely the chemical reaction postulated by Temin. This discovery marked not only a new chapter in cancer research but also had more general and far reaching biological consequences. Through the work of David Baltimore and others after 1970 it is now clear that infection of cells with RNA tumor viruses indeed leads to the formation of a DNA copy which is incorporated into the DNA of the cell. Thus, the mechanism of transformation discovered by Renato Dulbecco now applies to both DNA and RNA tumor viruses. Furthermore, other scientists have also found that many normal cells in their DNA contain copies of viral RNA, closely related to the RNA of tumor viruses. In some cases these DNA copies have existed and been part of the genetic material for millions of years. Even though their function as yet has not been clarified it is known that they can be released and transformed into RNA viruses if the cell is treated with certain chemicals. Part of the genetic material of the cell is then suddenly set free and is reborn as a virus particle.

What is the medical relevance of these discoveries? Is human cancer a virus disease? Certain forms of cancer of animals, such as leukemia, mammary carcinoma and cancer of connective tissues can be caused by viruses. This does not of course mean that all animal cancer is caused by viruses. I do not believe that man is an exception in this connection, even though we still lack conclusive evidence for viral involvement in any form of human cancer. In the domestic cat leukemia is today a quite common cause of death. In all probability this will not be the case in the future since one already is beginning to find a vaccine against the cat virus which causes this form of cancer. We have now the tools which should make it possible to identify human cancers caused by tumor viruses. There is thus good hope that this will lead to methods which help us to prevent the disease.

David Baltimore, Renato Dulbecco, Howard Temin: 65 years ago Peyton Rous established that cancer can be caused by viruses. Through your experiments you have shown how the virus does its job. However, the ramifications of your discoveries go further than that. The borderline between viruses and genes is disappearing. Specific genetic information which for millions of years has been an integral part of chromosomes of higher organisms may suddenly be released and be reborn as a virus particle. On behalf of Karolinska Institutet I wish to convey to you our warmest congratulations and now ask you to receive your Nobel prizes from the hands of His Majesty the King.

DAVID BALTIMORE

My interest in Biology began when I was a high school student and spent a summer at the Jackson Memorial Laboratory in Bar Harbor, Maine. There I first experienced research biology and saw research biologists at work; this experience led me to become a biology major in college.

I went on to Swarthmore College where I began as a major in biology but switched to chemistry later so that I could carry out a research thesis. Between my last two years at Swarthmore I spent a summer at the Cold Spring Harbor Laboratories working with Dr. George Streisinger, and the experience of working with and watching that great teacher led me to molecular biology.

I started graduate school at Massachusetts Institute of Technology in biophysics, but when I decided to work on animal viruses I left M.I.T. to study for a summer with Dr. Philip Marcus at the Albert Einstein Medical College and to take the animal virus course at Cold Spring Harbor, then taught by Dr. Richard Franklin and Dr. Edward Simon. I joined Dr. Franklin at the Rockefeller Institute to do my thesis work and then continued in animal virology as a postdoctoral fellow with Dr. James Darnell. I had already found that much could be learned by studying virus-specific enzymes, so I studied for a while with Dr. Jerard Hurwitz at the Albert Einstein College of Medicine to learn from someone who knew enzymology as a professional.

My first independent position was at the Salk Institute in La Jolla, California where I had the rare opportunity to work in association with Dr. Renato Dulbecco. After 2 1/2 years away from a university setting, I returned to M.I.T. in 1968 and have remained there. In 1974, I joined the staff of the M.I.T. Center for Cancer Research under the directorship of Dr. Salvador Luria because I had found that my research interests, that previously had involved mainly the non-oncogenic RNA viruses, were more and more focused on the problems of cancer.

DATE AND PLACE OF BIRTH:
March 7, 1938 in New York, New York

EDUCATION:

1956—1960	Swarthmore College, Swarthmore, Pennsylvania B.A. with high honors in Chemistry, 1960
1960—1961	Massachusetts Institute of Technology, Cambridge Massachusetts, graduate courses toward Ph.D.
1961—1964	Rockefeller University, New York, New York. Ph.D. received in 1964

POSITIONS HELD:

1963—1964	Postdoctoral Fellow, Massachusetts Institute of Technology, Cambridge, Mass.
1964—1965	Postdoctoral Fellow, Albert Einstein College of Medicine, Bronx, New York
1965—1968	Research Associate, Salk Institute for Biological Studies, La Jolla, California
1968—1972	Associate Professor of Microbiology, Massachusetts Institute of Technology, Cambridge, Mass.
1972—present	Professor of Biology, Massachusetts Institute of Technology, Cambridge, Mass.
1973—present	American Cancer Society Professor of Microbiology

VIRUSES, POLYMERASES AND CANCER

Nobel Lecture, December 12, 1975

by

DAVID BALTIMORE

Massachusetts Institute of Technology, Cambridge, Mass., U.S.A.

The study of biology is partly an exercise in natural esthetics. We derive much of our pleasure as biologists from the continuing realization of how economical, elegant and intelligent are the accidents of evolution that have been maintained by selection. A virologist is among the luckiest of biologists because he can see into his chosen pet down to the details of all of its molecules. The virologist sees how an extreme parasite functions using just the most fundamental aspects of biological behavior.

A virus is a form of life with very simple requirements (1). The basic needs of a virus are a nucleic acid to be transmitted from generation to generation (the genome) and a messenger RNA to direct the synthesis of viral proteins. The critical viral proteins that the messenger RNA must encode are those that coat the genome and those that help replicate the genome. One of the great surprises of modern virology has been the discovery of the variety of genetic systems that viruses have evolved to satisfy their needs. Among the animal viruses, at least 6 totally different solutions to the basic requirements of a virus have been found (2).

If we look back to virology books of 15 years ago, we find no appreciation yet for the variety of viral genetic systems used by RNA viruses (3). Since then, the various systems have come into focus, the last to be recognized being that of the retroviruses ("RNA tumor viruses"). As each new genetic system was discovered, it was often the identification of an RNA or a DNA polymerase that could be responsible for the synthesis of virus-specific nuclei acids that gave the most convincing evidence for the existence of the new system.

Now that the life-styles of different types of viruses have been delineated we can ask what relation there is between a virus' multiplication cycle and the disease it causes. In general, this question has no simple answer because disease symptoms do not correlate with the biochemical class of the virus. For instance, both poliovirus and rhinovirus are picornaviruses but one causes an intestinal infection with paralysis while the other causes the common cold. One class of RNA viruses, however, does have a unique symptom associated with it: the biochemically-defined group of viruses called the retroviruses are the only RNA viruses known to cause cancer. For a virologist interested in cancer, the problem is to first understand the molecular biology of retroviruses and then to understand how they cause the disease.

In what follows, I will review my personal involvement in uncovering the different genetic systems of RNA viruses, a story which leads to the recognition of the unique style of retroviruses, I will then consider what is known about the relationship between the biochemistry of retroviruses and their ability to be oncogenic.

As I tell my story I will mention a few of the many co-workers, teachers and students who have influenced my thinking or contributed their labors and ideas to the products of my laboratory. To mention all of the people to whom I am indebted would make too long a list; I can only say that the honors I receive are in large measure testaments to their accomplishments.

PICORNAVIRUSES

My work on the genetic systems of RNA viruses dates back to my graduate school days. As part of my introduction to animal virology during a Cold Spring Harbor course, I heard Richard Franklin describe his then-recent experiments using autoradiography to show that Mengovirus, a picornavirus and a close relative of poliovirus, could shut off the nuclear synthesis of cellular RNA early after infection and could later induce new RNA synthesis in the cytoplasm which appeared to represent synthesis of viral RNA (4). I decided to go to the Rockefeller University as a graduate student with Richard Franklin in order to work on the system he had developed.

Before I began to study how picornavirus RNA was made, it was already known from the work of Simon that picornavirus RNA synthesis was independent of DNA synthesis (5). Furthermore, studies with actinomycin D had shown that neither synthesis nor expression of cellular DNA was involved in viral RNA synthesis (6). These results suggested that Mengovirus might make a cytoplasmic RNA-dependent RNA synthesis system. The concept that viruses induce synthesis of their own enzymes had strong precedents in bacteriophage systems—Seymour Cohen's work had shown decisively that new virus-specified enzymes were found in infected bacteria (7).

We approached the problem of the virus' effect on intracellular RNA synthesis as a question in enzymology. We first showed that the nuclei from Mengovirus-infected cells were greatly reduced in their ability to carry out cell-free DNA-dependent RNA synthesis compared to nuclei of uninfected cells (8). Later, we showed that cytoplasmic extracts of Mengovirus- or poliovirus-infected cells contained an RNA synthesis activity not evident in uninfected cells and not inhibited by actinomycin D (9). When we learned that the system made RNA of the size and structure of virion RNA (10), it became clear that it represented the postulated viral RNA-dependent RNA synthesis system.

While there has been extensive further analysis of crude cytoplasmic systems (11, 12) and impressive enrichment of the RNA synthesis system has been achieved (13), no pure enzyme able to make picornavirus RNA has ever been isolated so the detailed mechanism of viral RNA synthesis still remains obscure. Most of our knowledge of how picornavirus RNA is made has come from studies on the virus-specific RNA molecules in infected cells and their kinetics of labeling by radioactive precursors. Such research has been carried out in many laboratories (11, 12); my work in this area was done in association with James Darnell and Marc Girard. Together we found and studied the relations between the poliovirus double-stranded RNA, the poliovirus replicative intermediate and the poliovirus replication complex (14). Marc Girard's precise *in vitro* analysis of RNA synthesis capped this whole series of experiments (15).

A crucial part of the viral genetic system is the manner of translation of the viral messenger RNA. While working on viral maturation, my first graduate student, Michael Jacobson, and I began to realize that proteolytic cleavage was an important part of the process (16). Our work led us to suggest that the whole 7500-nucleotide viral genome might be translated into a single continuous polypeptide that we have called a polyprotein (17, 18). Recently, this work culminated in the demonstration that poliovirus RNA can be translated into this continuous polyprotein in a cell-free system and that some of the cleavages that make the polyprotein into the functional proteins appear to occur in extracts of uninfected cells (19).

The demonstration that the poliovirus genome RNA is the messenger RNA for the synthesis of viral proteins, coupled with the demonstration of the infectivity of viral RNA (20), implies that the poliovirus genetic system is very simple. The existing evidence confirms this simplicity—as seen diagrammatically in Figure 1, it appears that the incoming "plus" strand of

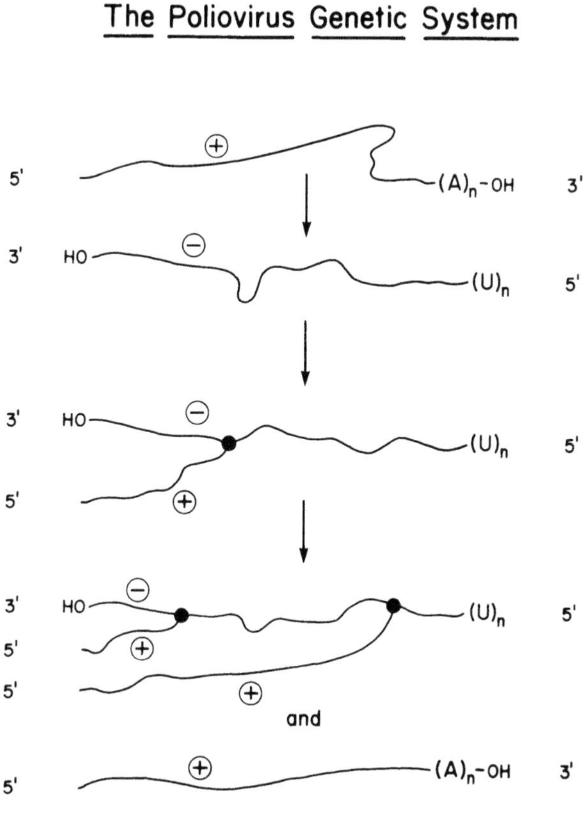

Figure 1. Schematic representation of poliovirus-specific RNA synthesis in the cytoplasm of infected cells.

RNA synthesizes a "minus" strand which in turn synthesizes a series of plus strands. This diagrammatic simplicity of poliovirus replication hides a fair amount of as yet undeciphered complexity as shown by the work of Eckard Wimmer and his colleagues as well as by work in my laboratory. For instance, the 3′-ends of the virion RNA and messenger RNA are both poly(A), the 5′-end of the minus strand is poly(U), so we assume that they are templates for each other (21). But these homopolymer stretches have very variable lengths even in the progeny of a cloned virus; what then determines their length? The poly(A) serves some obscure but necessary function in the life-cycle of the virus (22); what is this function? There is no triphosphate 5′-terminus, either free or capped, on the virion RNA or messenger RNA (23, 24); how then is the synthesis of these molecules initiated? The 5′-end of the virion RNA and messenger RNA are different (24); what does this mean?

VESICULAR STOMATITIS VIRUS

Most of the work in my laboratory until 1969 centered on poliovirus. We had assumed that all RNA viruses would be similar in their basic molecular biology but during the 1960's results emerging from various laboratories implied that poliovirus, rather than being a model for all RNA viruses, used one out of a collection of different viral genetic systems. Probably the first dramatic demonstration of the variety in RNA viruses came from next door to Richard Franklin's laboratory at the Rockefeller Institute where Peter Gomatos and Igor Tamm found that reovirus has double-stranded RNA as its genome (25). The peculiarity of reovirus was underscored by the demonstration later that an RNA polymerase in the virion of reovirus is able to assymetrically transcribe the double-stranded RNA (26). This was the first virion-bound RNA-dependent RNA polymerase ever found and followed after the finding of the first nucleic acid polymerase in a virion—the DNA-dependent RNA polymerase found by Kates and McAuslan and Munyon *et al* in virions of vaccinia virus (27).

Another observation that suggested there were profound differences among the RNA viruses was the finding that in cells infected by the paramyxoviruses, Newcastle disease virus or Sendai virus, much of the virus-specific RNA was complementary to the virion RNA (28). This result was in sharp contrast to what was found in poliovirus-infected cells where most of the virus-specific RNA was of the same polarity as virion RNA (11).

We branched away from concentration only on poliovirus to include the study of vesicular stomatitis virus (VSV) because of the lucky circumstances that brought Alice Huang to my laboratory. She joined me first at the Salk Institute and then we both came to M.I.T. in 1968. Alice had studied VSV as her graduate work with Robert R. Wagner at Johns Hopkins. We decided that the techniques developed for studying poliovirus should be applied to VSV, hoping that the peculiar ability of VSV to spawn and then be inhibited by short, defective particles could be understood at the molecular level. A graduate student, Martha Stampfer, joined in this work and together we found that we had bitten off an enormous problem because VSV induced synthesis of so many species of RNA. In poliovirus-infected cells, only three species of RNA are seen but we found at least 9 RNA's in VSV-infected cells and one of these RNA's was clearly heterogeneous (29)—later work showed it had 4 components (30, 31). In our description of this work we said that 9 RNA species "seems exorbitant" (29) but we soon realized that each RNA had its place in the cycle of growth and growth inhibition of VSV.

As we were beginning to unravel the multiple VSV RNA's, Schaeffer *et al* (32) published a paper showing that the major VSV-induced RNA's in infected cells, like those induced by Sendai and Newcastle disease viruses, were complementary in base sequence to the virion RNA. We confirmed and extended that observation, showing that the virus-specific RNA recovered from the polyribosomes of infected cells (the viral messenger RNA) was all complementary to virion RNA (33). This finding presented a pregnant paradox: if all viruses were like poliovirus and induced a new polymerase in the infected cell how could a virus that carried as its genome the strand of RNA complementary to messenger RNA ever start an infection? There seemed two possible solutions: the RNA came into the cells and was copied by a cellular enzyme to make the messenger RNA to initiate the infection cycle or the RNA came into the cell carrying an RNA polymerase with it.

Because no convincing evidence for RNA to RNA transcription existed (or exists) for any uninfected cell, the possibility of a polymerase with the incoming RNA seemed attractive. This possibility implied that there might be polymerase activity demonstrable in disrupted virions of VSV. Almost as soon as the power of this reasoning was clear to us, we had shown the existence of the virion RNA polymerase (34). The demonstration of this enzyme was the piece of evidence that led to the realization that there is a huge class of viruses, now called negative strand viruses (35), that all carry the strand of RNA complementary to the messenger RNA as their genome and that carry an RNA polymerase able to copy the genome RNA to form multiple messenger RNA's.

RETROVIRUSES

The discovery of a virion polymerase in VSV led us to search for such enzymes in other viruses. Because Newcastle disease virus made a lot of complementary RNA after infection it seemed a logical candidate and after an initial failure (34), we found activity in virions of that virus (36). But a more exciting possibility occurred to me; maybe by looking for a virion polymerase, light could be shed on the puzzle of how RNA tumor viruses multiply.

In his Nobel lecture, Howard Temin has outlined how he was led to postulate a DNA intermediate in the growth of RNA tumor viruses (37). Although his logic was persuasive, and seems in retrospect to have been flawless, in 1970 there were few advocates and many skeptics. Luckily, I had no experience in the field and so no axe to grind—I also had enormous respect for Howard dating back to my high school days when he had been the guru of the Summer School I attended at the Jackson Laboratory in Maine. So I decided to hedge my bets—I would look for either an RNA or a DNA polymerase in virions of RNA tumor viruses.

To make the foray into tumor virology, I needed some virus. Peter Vogt first sent me some Rous sarcoma virus and, although I later used it to good advantage, I initially assayed for an RNA polymerase in this viral preparation and failed to find anything. Then George Todaro helped me utilize the resources of the Special Virus Cancer Program of the National Cancer Institute to get some Rauscher mouse leukemia virus. Using that virus preparation I set out to look for a DNA polymerase activity. With little difficulty, I was able to demonstrate that virions of Rauscher virus contained a ribonuclease-sensitive DNA polymerase activity and, after confirming the results with Rous sarcoma virus, I knew that the machinery for making a DNA copy of the RNA genome was wrapped up inside the virions of RNA tumor viruses (38). Simultaneously with my work, Temin and Mizutani discovered the DNA synthesis activity in Rous sarcoma virus (39).

BIOCHEMISTRY OF REVERSE TRANSCRIPTASE

Once the DNA polymerase activity had been demonstrated in the virions of what we now call retroviruses, many laboratories began to study the enzyme. A correspondent of *Nature* dubbed the enzyme "reverse transcriptase" (40) and this romantic name has become common parlance. About 2 years after its discovery, Howard Temin and I reviewed the literature on the enzyme (41). That review and later compendia (42) make a detailed rehash of the biochemistry of retroviruses superfluous. So, I will only present a brief sketch of the picture we have today of how a retrovirus multiplies and how the reverse transcriptase functions. I will not attempt to credit all of the people who have helped to clarify this picture.

There are two separate time-periods that can be distinguished after infection of a cell by a retrovirus (Fig. 2). The first period, which lasts a few hours, involves reverse transcription and establishment of the DNA provirus as an integrated part of the cellular DNA; the second period involves the expression of the integrated genome and synthesis of progeny virions.

Analysis of the first period of retrovirus growth has focused on the types of virus-specific DNA molecules that are produced. One important type of DNA that has been found is a closed circular duplex DNA of about 5.7×10^6 daltons (43). This DNA can transfect cells with one-hit kinetics (44) and therefore contains the total viral genetic information. Other DNA's that may be on the way to becoming the closed circular form are also evident (45). It has been hard to get definitive evidence as to what DNA form integrates but presumably it is the circular duplex DNA. Whatever the form that integrates, the evidence is quite good for acquisition of proviral DNA by the chromosomes of infected cells (46, 47).

The second period in a productive retrovirus growth cycle starts when the integrated genome begins making viral RNA (48). Synthesis of viral proteins and progeny virions ensues and the cell ever-after continues to make viral products except for variations imposed by the cell cycle (49). Among the

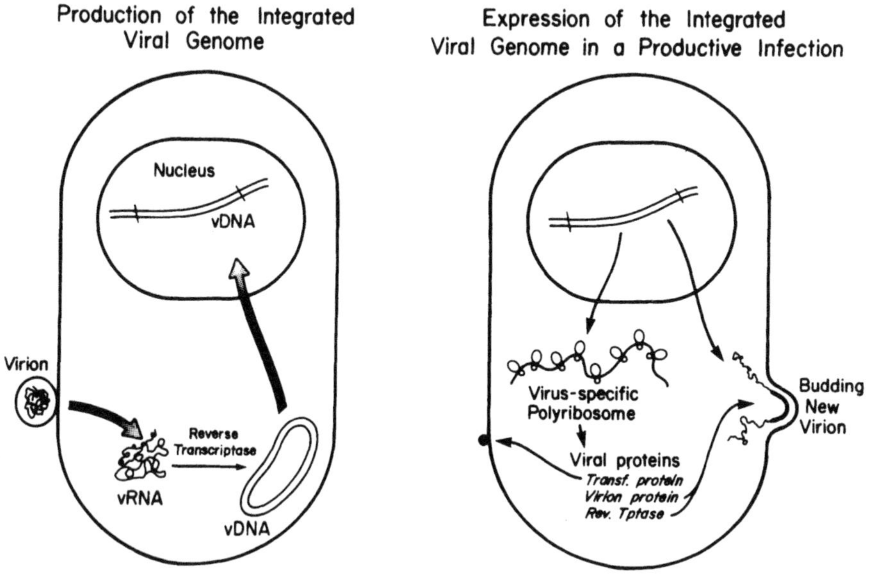

Figure 2. The life cycle of an RNA tumor virus.
Based on present knowledge (42), the life cycle of an RNA tumor virus can be separated into two parts. In the first part the virion attaches to the cell and somehow allows its RNA along with reverse transcriptase to get into the cell's cytoplasm. There the reverse transcriptase causes the synthesis of a DNA copy of the viral RNA. A fraction of the DNA can be recovered as closed, circular DNA (43) and it is presumably that form which integrates into the cellular DNA. Once the proviral DNA is integrated into cellular DNA it can then be expressed by the normal process of transcription. The two types of product which have been characterized are new virion RNA and messenger RNA. Much of the messenger RNA which specifies the sequence of viral protein is of the same length as the virion RNA but there may also be shorter messenger RNA's (48). The virus-specific proteins have 2 known functions: one is the transformation of cells which occurs when, for instance, a sarcoma virus infects a fibroblast, the second is to provide the protein for new virion production. The transforming protein is shown here as acting at the cell surface but that is only a hypothesis.

viral proteins made in the infected cell may be a product that changes the growth properties of the cell (50); in such a case the retrovirus becomes a tumor virus.

The second period of the infection cycle can be dissociated from the first in a number of experimental systems. For instance, mammalian cells infected by avian viruses can gain viral DNA but not express it (46). Also, cells can have viral genomes that they inherited from their ancestors and such genomes are generally not being transcribed. Nonexpressed genomes can be activated: bromodeoxyuridine and iododeoxyuridine, for instance, stimulate the expression of inherited, silent viral genomes (51). The exact mechanism of activation of the genome for transcription, initiation of the transcript and termination of transcription are obscure, as are any processing events of the initial transcript which may occur.

It is evident that the key piece of machinery provided by the virus for this unique life cycle is the reverse transcriptase. Purified reverse transcriptase has the properties of most DNA polymerases: it is a primer-dependent enzyme that makes DNA in a $5' \rightarrow 3'$ direction using deoxyribonucleoside triphosphates as substrates and taking the direction of a template for determining the base sequence of the product. The enzyme differs from normal cellular DNA polymerases by having a unique polypeptide structure, having an associated ribonuclease H activity and being able to make copies of RNA templates as readily as DNA templates (41). Genetics has shown us that the avian leukosis viral enzyme, at least, is encoded by viral RNA and needed only in the first period of the infection cycle (52).

The primer-dependence of the reverse transcriptase means that the enzyme can only elongate nucleic acid molecules, it cannot initiate DNA synthesis *de novo*. How then does the enzyme initiate the copying of viral RNA? The answer is that the genome RNA has attached to it a primer RNA molecule which is, in the case of avian leukosis viruses, cellular tryptophan transfer RNA (53). The avian leukosis virus reverse transcriptase has a high-affinity binding site for that transfer RNA which the enzyme presumably uses for precise initiation of reverse transcription (54).

RETROVIRUSES AND CANCER

The last 15 years of research in animal virology has allowed us to see the diversity of genetic systems used by the various types of RNA viruses and has most recently shown us how distinct the retroviruses are from the others. Rather than using an entirely RNA-dependent replication and transcription machinery, the retroviruses have included the DNA provirus in their life-cycle. Having a DNA intermediate does not make their mode of growth especially complicated—the DNA formally takes the place of the "minus" strand in the picornavirus genetic system—but the DNA is probably the clue to why retroviruses are the only ones able to cause cancer. The DNA provides the necessary stability to the virus-cell interaction so that a viral gene product can permanently change the growth properties of an infected cell. Equally significant, the DNA stage is probably important to the ease with which retroviruses carry out genetic recombination (55); it is quite possible that the recombination system can bring together host cell genetic information with viral information and that in this way non-oncogenic retroviruses become oncogenic (56).

So the inclusion of a proviral stage in the retrovirus life-cycle may provide critical capabilities towards the development of an oncogenic potential. But the actual transformation of cells by retroviruses is a highly selective process; each type of oncogenic virus transforms a very limited range of cell types (57). If we assume that all transforming genes of viruses are like those of Rous sarcoma virus, genes that are not necessary to the growth of the virus (50, 58), then we can postulate that each type of transforming virus makes a specific type of transforming protein. Such a protein, by this model, would not be critically involved in the multiplication cycle of the virus. Isolation of such transforming proteins and elucidation of their mechanism of action is the present challenge of cancer virology. Not only will such work help us to understand carcinogenesis, it may also be important to the study of developmental biology because of the intimate relationship between the differentiated state of cells and the type of virus able to transform them.

Another implication of the occurrence of a proviral stage in the life-cycle of retroviruses is that cells can harbor such viruses as genetically silent DNA molecules. In fact, in most, if not all, animal species, the normal cells of the body have DNA related to one or more types of retroviruses (59). They receive that DNA by inheritance, not infection, and in favorable cases it can be as precisely located in the chromosomes as any gene (60). What is the significance of these genes that look like viruses?

There have been three types of explanations for virus-related genes that are inherited in the germ line of so many animal species:

1) They are an aspect of the normal genetic complement of the animal and they are virus-related because they are the progenitors of retroviruses. These genes play some important role in the life of the animal and so are not dispensable. This explanation is basically the protovirus hypothesis put forward by Howard Temin (37).

2) They are genes inserted into the chromosome of some ancestral animal by a retrovirus infection of the germ cells of that animal. Because once the provirus is integrated it remains stably associated with the chromosome, the viral genes are inherited by progeny of the original infected animal. There is one force that can eliminate such genes from a species, the slow but inexorable process of mutation. As part of this explanation of inherited viral genomes, therefore, it has been suggested that the viral genes have some positive influence on the life of the animal and so are maintained by positive selection. This explanation is closely related to the virogene-oncogene hypothesis (61).

3) The previous explanation can be modified by the exclusion of any positive role for viral genes in the life of the infected animal. There are a number of reasons why positive selection may be an unnecessary attribute to postulate. For one thing, mutations are rare events and totally inactivating mutations are much rarer. Also, the virus can be genetically invigorated by becoming a replicating virus in the body of the animal and then reinfecting the germ line. When the virus starts multiplying as an independent entity, the burden of debilitating mutations it might have accumulated could be purged if a sufficient number of generations intervened between the activation of the latent provirus and its reintegration into the germ line. The reintegration might even replace the original provirus (62). If the viral genes were not transcribed in most cells that have the viral genome, as appears to be the case, the negative effect of having one or a few integrated genomes would be slight and probably insufficient to cause a serious negative selection against animals with inherited proviruses.

I would argue that the third explanation above is the one most likely to be correct. It is an explanation that maintains the separation of viral activities and cellular activities and does not require the *ad hoc* postulation of beneficial properties of viral products. It treats retroviruses like any other virus, as an entity with its own life-style and its own accomodation with evolution.

In summary, I have tried here to develop the view of retroviruses as one of a number of solutions to the problem of creating a virus. Each virus directs synthesis of two critical classes of proteins: proteins for replication and proteins for constructing the virus particle. By encoding the reverse transcriptase, retroviruses have evolved the ability to integrate themselves into the cell chromosome as a provirus. This is a very sheltered environment in which to live, only mutation interferes with the continual transmission of the virus to the progeny of an animal that is infected in its germ cells. In this context, the ability of some retroviruses to cause cancer is a gratuitous one. But it is today the most challenging and important attribute of these retroviruses and the one that will dominate future research efforts in this area.

REFERENCES:

1. Diener, T., in *Advances in Virus Research,* vol. 17, Smith, K. M., Lauffer, M. A. and Bang, F. B. Eds. (Academic Press, New York and London, 1972), pp. 295—313.
2. Baltimore, D. *Bacteriological Reviews* 35, 235 (1971).
3. Luria, S. E., in *The Viruses,* vol. I, Burnet, F. M. and Stanley, W. M., Eds. (Academic Press, New York, 1959), pp. 549—568.
4. Franklin, R. M. and Rosner, J. *Biochem. Biophys. Acta* 55, 240 (1962).
5. Simon, E., *Virology* 13, 105 (1961).
6. Reich, E., Franklin, R. M., Shatkin, A. J. and Tatum, E. L., *Science* 134, 556 (1961).
7. Cohen, S. S., *Fed. Proc.* 20, 641 (1961).
8. Baltimore, D. and Franklin, R. M., *Proc. Nat. Acad. Sci. U.S.A.* 48. 1383 (1962). Franklin, R. M. and Baltimore, D., *Cold Spring Harbor Symp. Quant. Biol.* 27, 175 (1962).
9. Baltimore, D. and Franklin, R. M. *Biochem. Biophys. Res. Commun.* 9, 388, (1962). Baltimore, D. and Franklin, R. M. *J. Biol. Chem.* 238, 3395 (1963). Baltimore, D., Franklin, R. M., Eggers, H. J. and Tamm, I. *Proc. Nat. Acad. Sci. U.S.A.* 49, 843 (1963). Baltimore, D. and Franklin, R. M., *Cold Spring Harbor Symp. Quant. Biol.* 28, 105 (1963).
10. Baltimore, D., *Proc. Nat. Acad. Sci. U.S.A.* 51, 450 (1964).
11. Baltimore, D., in *The Biochemistry of Viruses,* Levy, H. B., Ed., (Marcel Dekker Inc., New York, 1969), pp. 101—176.
12. Levintow, L., in *Comprehensive Virology,* vol. 2. Fraenkel-Conrat, H. and Wagner, R. R., Eds. (Plenum Press, New York, 1974), pp. 109—169.
13. Lundquist, R. E., Ehrenfeld, E. and Maizel, J. E., *Proc. Nat. Acad. Sci. U.S.A.* 71, 4773 (1974).
14. Baltimore, D., Becker, Y. and Darnell, J. E., *Science* 143, 1034 (1964). Baltimore, D., *J. Mol. Biol.* 18, 421 (1966). Baltimore, D., Girard, M. and Darnell, J., *Virology* 29, 179 (1966). Baltimore, D. and Girard, M., *Proc. Nat. Acad. Sci. U.S.A.* 56, 741 (1966). Girard, M., Baltimore, D. and Darnell, J. E., *J. Mol. Biol.* 24, 59 (1967). Baltimore, D., *J, Mol. Biol.* 32, 359 (1968).
15. Girard, M., *J. Virol.* 3, 376 (1969).
16. Jacobson, M. F. and Baltimore, D., *J. Mol. Biol.* 33, 369 (1968).
17. Jacobson, M. F. and Baltimore, D., *Proc. Nat. Acad. Sci. U.S.A.* 61, 77 (1968). Jacobson, M. F., Asso, J. and Baltimore, D., *J. Mol. Biol.* 49, 657 (1970).

18. Baltimore, D., Harvey Lectures (Academic Press, New York, 1975), in press.
19. Villa-Komaroff, L., Guttman, N., Baltimore, D. and Lodish, H. F., *Proc. Nat. Acad. Sci. U.S.A.* 72, 4157 (1975).
20. Colter, J. S., Birel, H. H., Mayer, A. W. and Brown, R. A., *Virology* 4, 522 (1957). Alexander, H. E., Koch, G., Mountain, I. M., Sprunt, K. and Van Damme, O., *Virology* 5, 172 (1958).
21. Yogo, Y. and Wimmer, E., *Proc. Nat. Acad. Sci. U.S.A.* 69, 1877 (1972). Yogo, Y., Teng, M. and Wimmer, E., *Biochem. Biophys. Res. Commun.* 61, 1101 (1974). Spector, D. H. and Baltimore, D., *J. Virol.* 15, 1418 (1975). Spector, D. H. and Baltimore, D., *Virology* 67, 498 (1975).
22. Spector, D. H. and Baltimore, D., *Proc. Nat. Acad. Sci. U.S.A.* 71, 2983 (1974).
23. Wimmer, E., *J. Mol. Biol.* 68, 537 (1972).
24. Nomoto, A., Lee, Y. F. and Wimmer, E., *Proc. Nat. Acad, Sci. U.S.A.*, 73, 375 (1976). Hewlett, M. J., Rose, J. K. and Baltimore, D., *Proc. Nat. Acad. Sci. U.S.A.*, 73, 327 (1976).
25. Gomatos, P. J. and Tamm, I., *Proc. Nat. Acad. Sci. U.S.A.* 49, 707 (1963).
26. Shatkin, A. J. and Sipe, J. D., *Proc. Nat. Acad. Sci. U.S.A.* 61, 1462 (1968). Borsa, J. and Graham, A. F., *Biochem. Biophys. Res. Commun.* 33, 895 (1968).
27. Kates, J. R. and McAuslan, B. R., *Proc. Nat. Acad. Sci. U.S.A.* 58, 1134 (1967). Munyon, W., Paoletti, E. and Grace, J. T., *Prot. Nat. Acad. Sci. U.S.A.* 58, 2280 (1967).
28. Kingsbury, D., *J. Mol. Biol.* 18, 204 (1966). Bratt, M. A. and Robinson, W. S., *J. Mol. Biol.* 23, 1 (1967).
29. Stampfer, M., Baltimore, D. and Huang, A. S., *J. Virol.* 4, 154 (1969).
30. Morrison, T. G., Stampfer, M., Lodish, H. F. and Baltimore, D., in *Negative Strand Viruses,* vol. 1, Mahy, B. W. J. and Barry, R. D., Eds. (Academic Press, London, New York, and San Francisco, 1975), pp. 293—300.
31. Rose, J. K. and Knipe, D., *J. Virol.* 15, 994 (1975).
32. Schaffer, F. L., Hackett, A. J. and Soergel, M. E., *Biochem, Biophys. Res. Commun.* 31, 685 (1965).
33. Huang, A. S., Baltimore, D. and Stampfer, M., *Virology* 42, 946 (1970).
34. Baltimore, D., Huang, A. S. and Stampfer, M., *Proc. Nat. Acad. Sci U.S.A.* 66, 572 (1970).
35. Mahy, B. W. J. and Barry, R.D., Eds., *The Negative Strand Viruses,* Vol. 1 & 2, (Academic Press, London, New York and San Francisco, 1975).
36. Huang, A. S., Baltimore, D. and Bratt, M. A., *J. Virol.* 7, 389 (1971).
37. Temin, H., Nobel address, 1975.
38. Baltimore, D., *Nature* 226, 1209 (1970).
39. Temin, H. and Mizutani, S., *Nature* 226, 1211 (1970).
40. Anonymous, *Nature* 228, 609 (1970).
41. Temin, H. and Baltimore, D., in *Advances in Virus Research,* vol. 17 Smith, K. M., Lauffer, M. A. and Bang, F. B. (eds), (Academic Press, New York and London, 1972), pp. 129—186.
42. Cold Spring Harbor Symp. Quant. Biol., vol. 39 (Cold Spring Harbor Laboratory, New York, 1975). Tooze, J., Ed., *The Molecular Biology of Tumor Viruses,* (Cold Spring Harbor Laboratory, New York, 1973). Bishop, J. M. and Varmus, H. E., in *Cancer,* vol. 2, Becker, F. F., Ed., (Plenum Press, New York, 1975), pp. 3—48.
43. Varmus, H. E., Guntaka, R. V., Fan, W. J. W., Heasley, S. and Bishop, J. M., *Proc. Nat. Acad. Sci. U.S.A.* 71, 3874 (1974). Gianni, A. M., Smotkin, D. and Weinberg, R. A., *Proc. Nat. Acad. Sci. U.S.A.* 72, 447 (1975).
44. Smotkin, D., Gianni, A. M., Rozenblatt, S. and Weinberg, R. A., *Proc. Nat. Acad. Sci. U.S.A.* 72, 4910 (1975).
45. Gianni, A. M. Weinberg, R. A., Nature 255, 646 (1975).
46. Varmus, H. E., Vogt, P. K. and Bishop, J. M., *Proc. Nat. Acad. Sci. U.S.A.* 70, 3067 (1973).

47. Markham, P. D. and Baluda, M. A., *J. Virol.* 12, 721 (1973).
48. Fan, H. and Baltimore, D., *J. Mol. Biol.* 80, 93 (1973).
49. Leong, J. A., Levinson, W. and Bishop, J. M., *Virology* 47, 133 (1972). Paskind, M. P., Weinberg, R. A. and Baltimore, D., *Virology* 67, 242 (1975).
50. Martin, G. S., *Nature* 227, 1021 (1970).
51. Lowy, D. R., Rowe, W. P., Teich, N. and Hartley, J. W., *Science,* 174, 155 (1971).
52. Verma, I. M., Mason, W. S., Drost, S. D. and Baltimore, D., *Nature* 251, 27 (1974).
53. Dahlberg, J. E., Sawyer, R. C., Taylor, J. M., Faras, A. J., Levinson, W. E., Goodman, H. M. and Bishop, J. M., *J. Virol.* 13, 1126 (1974).
54. Panet, A., Haseltine, W. A., Baltimore, D., Peters, G., Harada, F. and Dahlberg, J. E., *Proc. Nat. Acad. Sci. U.S.A.* 72, 2535 (1975).
55. Wyke, J., Bell, J. G. and Beamund, J. A., *Cold Spring Harbor Symp. Quant. Biol.* 39, 897 (1975). Wyke, J., *Reviews on Cancer* 417, 91 (1975).
56. Weiss, R. A., in *Possible Episomes in Eukaryotes, 4th Lepetit Symposium* Sylvestri, L. G., Ed., (North-Holland, Amsterdam, 1973), pp. 130—141. Stehelin, D., Varmus, H. E., and Bishop, J. M., *J. Mol. Biol.,* in press (1975).
57. Gross, L., *Oncogenic Viruses,* 2nd edition (Pergamon Press, New York, 1972).
58. Hanafusa, H., in *Cancer,* vol. 2, Becker, F. F., Ed. (Plenum Press, New York, 1975), pp. 49—90.
59. Todaro, G. J., Beneviste, R. E., Callahan, R., Lieber, R. R. and Sherr, C., *Cold Spring Harbor Symp. Quant. Biol.* 39, 1159 (1975). Beneviste, R. E. and Todaro, G. J., *Proc. Nat. Acad. Sci. U.S.A.* 72, 4090 (1975).
60. Chattopadhyay, S. K., Lowy, D. R., Teich, N. M., Levine, A. S. and Rowe, W. P., *Cold Spring Harbor Symp. Quant. Biol.* 39, 1085 (1975). Chattopadhyay, S. K., Rowe, W. P., Teich, N. M. and Lowy, D. R., *Proc. Nat. Acad. Sci. U.S.A.* 72, 906 (1975).
61. Huebner, R. J. and Todaro, G. J., *Proc. Nat. Acad. Sci. U.S.A.* 64, 1087 (1969). Todaro, G. J. and Huebner, R. J., *Proc. Nat. Acad. Sci. U.S.A.* 60, 1009 (1972).
62. Temin, H. M., *Virology,* 13, 159 (1961).

Renato Dulbecco

RENATO DULBECCO

I was born in Catanzaro, Italy, from a Calabrese mother and a Ligurian father. I stayed in that city for a short time; my father was called into the army (World War I) and we moved to the north, Cuneo and Torino. At the end of the war my father, who was in the "Genio Civile", was sent to Imperia, Liguria, where we stayed for many years. The life I remember begins at Imperia, where I went to school, including the Ginnasio-Liceo "De Amicis". What I remember most of that period, besides my family and the few friends, was the rocky beach where I spent most of my time during the summer holiday, and a small meterological observatory, where I used to spend lots of my free time throughout the year. There I developed a strong liking for physics, which I put to good use by building an electronic seismograph, probably one of the first of its kind, which actually worked.

I graduated from high school at 16 (1930) and went to the University in Torino. Although I liked especially physics and mathematics for which I had considerable talent, I decided to study medicine. This profession had for me a strong emotional appeal, which was reinforced by having an uncle who was an excellent surgeon.

In Torino I was a very successful student, but I soon realized that I was interested in biology more than in applied medicine. So I went to work with Giuseppe Levi, the professor of Anatomy, where I learned Histology and the rudiments of cell culture. For my degree, however, I went to morbid anatomy and pathology. In Levi's laboratory I met two students who later had a strong influence on my life: Salvador Luria and Rita Levi-Montalcini.

All through the student years I was at the top of my class although I was two years younger than everbody else.

After taking my MD degree in 1936 I was called up for military service as a medical officer. In 1938 I was disharged and returned to pathology. A year later, however, I was called up again because of the Second World War. I was sent briefly to the French front, and a year later to Russia. There I had a narrow escape on the front of the Don during a major Russian offensive in 1942: I was hospitalized for several months and sent home. When Mussolini's government collapsed and Italy was taken over by the German army I hid in a small village in Piemonte and joined the Resistance, as physician of the local partisan units. I continued to visit the Institute of Morbid Anatomy in Torino where I joined in underground political activities together with Giacomo Mottura, a senior collegue. I was part of the "Committee for National Liberation" of the city of Torino, and became a councillor of that city in the first postwar city council. However, the life of routine politics was not for me and within months I left that position to return to the laboratory. I also went back to school, enrolling in regular courses in physics, which I pursued for the next two years.

I moved back to Levi's Institute and worked together with Levi-Montalcini, who encouraged me to go to the USA to work in modern biology. My dream was to work in genetics of some very simple organism, possibly using radiations. This dream became a reality after Luria, who had been in the USA since the beginning of the war, and was working in this very field, came in the summer of 1946 to Torino. He encouraged me and offered me a small salary for working in his group. I was urged in this direction by Rita Levi-Montalcini, who was herself preparing to go to another laboratory in USA. So in the autumn 1947 we both embarked for the US.

I went to work with Luria in Bloomington, Indiana, where I shared with him a small laboratory under the roof, to be soon joined by Jim Watson. Within a year I had made two good pieces of work, using my mathematical knowledge, and discovered photoreactivation of phage inactivated by ultraviolet light. This attracted the interest of Max Delbrück, who offered me a job in his group at Caltech.

I moved to Caltech in the summer 1949. I remember that memorable trip from Indiana to California with my family in an old car, with our limited possessions in a small trailer behind. I was fascinated by the beauty and immensity of the USA and the kindness of its people. Reaching the Pacific Ocean in Oregon was like arriving at a new world, an impression that continued and increased as we made our way south to Pasadena. I resolved at that time that I would not like to live anywhere else in the world — a resolution that I changed only some twenty-three years later.

At Caltech I continued to work with phages for a few years. One day I was told by Delbrück that a rich citizien had given Caltech a fund for work in the animal virus field. He asked me whether I was interested. My medical background and the experience gained in Levi's laboratory came back to me and I accepted. After visiting the major centers of animal virus work in the US I set out to discover the way to assay animal viruses by a plaque technique, similar to that used for phages, using cell cultures. Within less than a year, I worked out such a method, which opened up animal virology to quantitative work. I used the technique for studying the biological properties of poliovirus. These successes brought me an appointment first to associate professor, then to full professor at Caltech.

In the late fifties I had as a student Howard Temin, who, together with Harry Rubin, then a postdoctoral fellow in my laboratory, worked on the Rous Sarcoma Virus. Their work started my interest in the tumor virus fields. I myself started working on an oncogenic virus, polyoma virus, in 1958, and continued until now. This work has led to discovering many aspects of the interaction of this virus (and of SV_{40}) with the host cells in lytic infection and transformation.

I moved from Caltech to the Salk Institute in 1962, and in 1972 to the Imperial Cancer Research Fund Laboratories in London. One of the reasons for the latter move was the opportunity to work in the field of human cancer.

My work throughout the years has been strongly influenced by my associates. Giuseppe Levi taught me the essential value of criticism in scientific work, Rita Levi-Montalcini helped me to determine my goals at an early stage; Salvador Luria introduced me to viruses; Herman Muller, at the University of Indiana taught me the significance of Genetics; Max Delbrück helped me understand the scientific method and the goals of biology, and Marguerite Vogt contributed to my knowledge of animal cell cultures. Perhaps more important than all this, the daily interaction through the years with a continuously changing group of young investigators shaped my work. For although I had general goals, the actual path followed by my research was pragmatically determined by what could be done at any given time, and my young collaborators were an essential part of this process. I always did as much as possible of the experimental work with my own hands, but in the later part of my research career this became progressively less feasible, both because the demand on my time increased and because the increasing technical sophistication and complexities of the experiments demanded a great deal of specialized skills.

Since 1962 my scientific life has had the support of my second wife, Maureen, who for some years helped in my experiments. Without her affectionate encouragement and sound advice I doubt whether I would have been able to accomplish what I have done.

FROM THE MOLECULAR BIOLOGY OF ONCOGENIC DNA VIRUSES TO CANCER

Nobel Lecture, December 12, 1975
by RENATO DULBECCO
Imperial Cancer Research Fund, London, England.
Also a Fellow of the Salk Institute.

Oncogenic viruses, able to elicit tumour formation in animals, have been on the scientific scene for many years. After the early discovery of Ellerman and Bang at the beginning of this century, Peyton Rous opened up the field in its second decade and in prophetic words gave a good hint of things to come. However, these discoveries were soon forgotten and only after a long eclipse was interest in oncogenic viruses revived in the fifties. My involvement in this field began at that time when Rubin and Temin worked in my laboratory with the Rous Sarcoma Virus. When in 1958 polyoma virus, a new oncogenic virus with different properties, was isolated, I jumped at the new opportunity and started working with it. Within a short time polyoma virus became the main interest of my laboratory, to be joined, a few years later by SV_{40}, another papovavirus. It became clear fairly soon that the molecular biology of these viruses could be worked out, and I set out to find the molecular basis of cancer induction. The results that I and a number of brilliant young collaborators have obtained during the following fifteen years have brought us close to that goal. I will review the most interesting steps of our work and will then ask some questions concerning the nature of cancer and about perspectives for prevention and treatment. I stress the relevance of my work for cancer research because I believe that science must be useful to man.

INTEGRATION: THE PROVIRUS

Let me start with a brief review of our work in the molecular events in transformation. The first results, crucial for future developments, showed that polyoma virus could be assayed in certain cell cultures (1), which we call permissive, and could induce a cancer-like state in other cultures (2, 3) in which the virus does not grow, which we call non-permissive. The induction of the cancer-like state in vitro was called transformation. We were able to show that the virus containes DNA (4), and within a few years we gave the first evidence of its cyclic, or circular, shape (5), which is important for two critical biological events: DNA replication and integration. In integration, which we discovered a few years later with SV_{40} (6), the viral DNA becomes a provirus, i. e. it establishes permanent, covalent bonds with the cellular DNA. The cyclic configuration explains how a complete molecule of the SV_{40} DNA can be integrated without losses.

Integration is one of the key events in virus induced cell transformation. It explained the persistence of the transformed state in the cell clone deriving from a transformed cell, since the provirus replicates with the cellular DNA. It also permitted us to resolve one of the main questions about the role of viruses in transformation. It was known at the time that papova viruses leave their footprints in the cells of the cancers they induce and those they transform *in vitro*, in the form of characteristic antigens. However, it was not known whether the antigens were expressed by viral genes or by derepressed cellular genes. Hence, it was uncertain whether cells were transformed by the expression of viral genes persisting in the cells or alternatively if the virus altered the cells by a hit-and-run mechanism, changing the expression of cellular genes and then leaving. The demonstration that viral DNA is integrated in the cells in conjunction with the finding that the provirus is transcribed into messenger RNA (7) hundreds of generations after the establishment of a transformed clone, made the hit-and-run hypothesis unlikely and supported a continuing role of viral gene functions in determining transformation. This possibility was later supported by observations with abortively transformed cells, which behave as transformed only for several generations after infection, but then return to normality (8). When they are back to normal these cells no longer contain the viral DNA (9).

The viral genes that remain unexpressed in the transformed cells, such as those for capsid protein in SV_{40}-transformed cells, were also interesting, although in a different way. In fact their expression could be renewed in heterokaryons formed by fusing transformed cells with permissive cells (10, 11), a result that gave the first evidence that the viral functions are under the control of cellular functions. The provirus thus became a tool for studying regulation of DNA transcription in animal cells. Subsequently, the presence of giant RNAs containing viral sequences in the nucleus of transformed or lytically infected cells (12) raised the question of the initiation and termination signals for transcription in animal cells, as well as the question of processing of nuclear RNA precursors of messenger RNA, questions that are still largely unresolved.

VIRAL FUNCTIONS IN TRANSFORMATION

Efforts were in the meantime directed at identifying the viral genes transcribed in the transformed cells. It was established that in lytic infection with SV_{40} the whole viral DNA is transcribed in two nearly equal parts, one early, before the inception of replication of the viral DNA, the other late after DNA replication has begun and that the early RNA is also present in transformed cells (7). Subsequently, the early and the late messengers were found to be transcribed from different DNA strands (13), an observation that facilitated the further characterization of the viral transcripts. Later work in other laboratories using specific fragments produced by restriction endonucleases confirmed and refined these findings; and the results were extended to adenoviruses by showing that a segment of the early part of that DNA is always present and transcribed in transformed cells (14, 15).

These facts suggested that some early viral function is essential for maintaining the transformed state but they could also be interpreted differently: for instance, transformation might be caused by the mere presence of the viral DNA in the cellular DNA, the persistent viral functions being perhaps required for establishing and maintaining integration.

Attempts were made to solve the dilemma by isolating temperature sensitive mutants affecting either initiation or maintenance of transformation. Many transformation mutants were found, all clustered in a segment of the early region of the viral DNA, designated as the A gene, but they were all initiation mutants (16, 17, 18). These mutations prevent the onset of transformation at high but not at low temperature, and cells transformed at low temperature remain transformed at high temperature. It was not possible to find clear cut maintenance mutants, i.e. mutations capable of causing a complete reversion of the phenotype when cells transformed at low temperature were shifted to high temperature: however, careful observation later showed that the initiation mutants are also partial maintenance mutants, since the cells they transform undergo a partial reversion of the phenotype at high temperature (19, 20, 21). This result shows that the viral genes play a continuing role in transformation; however, the failure to obtain complete maintenance mutants suggests that the relation between viral gene expression and cell phenotype is complex.

SEARCH FOR THE VIRAL TRANSFORMING PROTEIN

Further progress in this subject has been achieved by studying the proteins specified by the early region of the viral DNA. This work has centred around the so-called T antigen (22) present in the nucleus of cells infected or transformed by SV_{40} and whose synthesis and properties are affected by mutations of the A gene (23, 24, 25). In non-permissive transformed cells the antigen is a protein of molecular weight of about 94,000 daltons (26), which binds firmly to double-stranded DNA, but without much specificity (27, 28, 29). That the T antigen is specified by the viral DNA is strongly suggested by its *in vitro* synthesis using a wheat germ extract primed with various messengers (30), especially since the size of the product depends on the nature of the messenger. Thus, when the messenger was viral RNA made *in vitro* by transcribing SV_{40} DNA with *E. coli* RNA polymerase, an antigenic protein of about 62,000 daltons was synthesized; but when mRNA extracted from infected cells was used the protein synthesized was, like the T antigen of transformed cells, of about 94,000 daltons. The discrepancy of the two molecular weights makes it very unlikely that the T antigen is a cellular protein modified by a viral function, because two different proteins would have to be modified in the same extract depending on the messenger used. In contrast, the synthesis of a shorter polypeptide chain with the artificial messenger may be justified by the absence of accessory signals, such as cap, poly A, and possibly other modifications. Further definition of these findings awaits peptide maps of the various products.

Since the early, transforming part of the SV_{40} genome can specify proteins of a molecular weight of about 100,000 daltons altogether, the T antigen is likely to be its sole product and, therefore, to be the transforming protein. However, the same protein must also initiate viral DNA replication, which cannot initiate at high temperature in cells infected by mutants of the A gene. The different functions in transformation and lytic infections could be performed by different domains of the same protein, or could result from modifications (such as phosphorylation and glycosylation) or from processing. Processing of SV_{40} T antigen seems to occur in lytically infected cells which contain a smaller T antigen of about 84,000 daltons; this smaller size contrasts with the regular size (94,000 daltons) of the antigen specified *in vitro* by mRNA extracted from the same cells (26). Whether the two forms of the antigen have different roles in transformation and DNA replication still remains to be established.

Since the transforming protein should control both initiation and maintenance of transformation the partial reversion of the phenotype of cells transformed by A mutants when shifted to high temperature may be explained by a decreased requirement for the transforming protein once transformation has taken place, which in turn could result from a positive feedback stabilizing the transformed state. For instance, unstable protein monomers specified by the mutated gene might form self-stabilizing oligomers (31), or the transforming protein might generate changes that tend to favour the transformed state. An example of the latter model is the β galactosidase induction in *E. coli* which is maintained by inducer concentrations much smaller than that required for initiating induction, because inducer is pumped into the cells by the induced permease (32). I wonder whether a certain degree of self-stabilization of the state of gene expression is a general property of animal cells, which has developed for maintaining differentiation.

CELLULAR EVENTS IN TRANSFORMATION

I now wish to turn to cellular events participating in transformation, which will be the main problem after the remaining questions on the role of the virus have been answered. Among the cellular events are functional changes and mutations. Some functional changes, which affect many cellular properties, are associated with the shift of resting cells to growing state after infection with polyoma virus or SV_{40} (33, 34); other changes observed in transformed and in cancer cells in general consist of the re-expression of cellular genes normally expressed in a preceding state of differentiation, in foetal life (35). These functional changes might be caused by the binding of transforming proteins to DNA; if so they may be mediated by an alteration of transcription of the cellular DNA. However, we do not know whether the transcription pattern changes because experiments based on competition hybridization have given ambiguous results. Perhaps the methodology is not good enough. Cloning of cellular DNA fragments in phages or plasmids may afford the necessary probes for carrying out significant experiments.

In order to understand further how the virus deregulates cellular growth we would need a detailed knowledge of the mechanisms of growth regulation in animal cells, which is now lacking. However, certain useful ideas about growth regulation are now available and can be used to draw inferences on the action of the virus. Thus, it seems clear that with a given cell type growth regulation involves a complex chain of events, beginning with extracellular regulators of many kinds, probably interacting with the cell plasma membrane. Cytoplasmic mediators then appear to transmit regulatory signals from the plasma membrane to the nucleus, where they perhaps control DNA-binding proteins similar to the transforming protein of papova viruses. The complexity of growth regulation increases markedly when different cell types are considered, since they seem to recognize different sets of extracellular regulators and may have different mediators and DNA binding proteins.

Proceeding from this general picture it would be tempting to propose that the viral transforming protein replaces one of the normal nuclear regulatory proteins of the cell, and being unaffected by the mediators that control the normal protein, keeps growth related transcription going, bypassing the signals of the plasma membrane. If so, however, the transformed state should be dominant over the normal state in cell hybrids, whereas the contrary is usually true (36). On the other hand, the dominance of the normal state could be explained if the transformed cells had a changed surface, unable to respond to regulatory signals. Such a change could result from the re-expression of foetal functions to make the transformed cells anachronistic, i. e. belonging to a stage of differentiation inappropriate to that of the organism which contains them. The cells with an anachronistic surface being insensitive to the growth regulators present in the adult organisms, which operate on adult cells, would grow without control. A striking support of the role of cell anachronism in cancer has been obtained with teratoma, a tumour originating when cells from an early embryo are transplanted to an adult environment. When, after many transplants cells of this tumour are introduced back into a blastocyst, i. e. an early embryo, they return to normality (37, 38), presumably because the internal growth control of the cells becomes again matched by the environmental regulators of the recipient embryo. In this model a hybrid cell formed by fusing a transformed and a normal cell may be untransformed, if the normal partner contributes normal surface components which respond to the normal extracellular regulators. For this result to be possible, anachronistic transcription should not be initiated after cell fusion on the DNA deriving from the normal parent. The virological studies suggest that this may well be the case, since the initiation of transformation seems to require much more transforming protein than its maintenance.

It would be important to recognize the developmental period during which the anachronistic genes of transformed or cancer cells are normally expressed, not only for understanding but possibly also for controlling cancer. In fact, if the growth regulators specific for the periods expressed in cancer cells could be identified they could be used for halting the growth of the cancer cells.

ROLE OF CELLULAR MUTATIONS

I will now consider the other cellular events important in viral transformation, i. e. cellular mutations. Several results suggest that cellular mutations may be needed for obtaining the full state of transformation with papova viruses. Thus, after infection primary cultures generate clones with various degree of transformation, some of which appear to undergo full transformation in steps (39) which may correspond to the occurrence of cellular mutations. Cells that achieve full transformation inmediately, as are common with permanent lines, may have already undergone similar mutations before infection. Some cellular mutations occuring in transformed cells may even be virus-induced, because in the early stages of transformation by papova viruses cells of primary cultures have frequent chromatid breaks (39). Conversely, cells fully transformed by SV_{40} can revert to a relatively normal phenotype although they still contain normal viral DNA and T antigen (40, 41). It is conceivable that these mutations are reversions of mutations of the former kind, which enhance the transformed state of the cells. Stepwise transformation may not only occur with viruses. Thus, I have observed it in primary cultures exposed to a chemical carcinogen. In this experiment fully transformed cells evolved from the normal cells, which have limited life, generating first cells with unlimited life but unable to form colonies in agar, then cells with progressively increasing colony forming efficiency in agar, to finally reach 100 % efficiency.

All these observations show the important role of cellular mutations in cell transformation induced by different agents. This conclusion is reinforced and generalised by additional findings, such as 1) the experimental enhancement of the transforming activity of viruses by mutagenic agents (42), 2) by the elevated cancer frequency in some genetic diseases, and 3) by the evidence that most carcinogens are pro-mutagens, i. e. generate mutagenic substances when acted upon by normal metabolism (43). Most of the carcinogens themselves must be activated by metabolism in a similar way in order to induce cancer.

PROSPECTS FOR CANCER PREVENTION

I will now turn to some general deductions concerning the etiology and possible prevention of human cancer, which derive from the various points I have discussed so far. One deduction, deriving from the persistence of the viral DNA in the cells is that we can test whether a given DNA virus is a possible agent of human cancer by looking for its DNA in the cancer cells. I think that much more extensive surveys than those carried out so far are warranted, but they should have a sensitivity sufficient to detect fragments of viral DNA of about one million daltons, which is within the reach of modern technology, even with the most difficult viruses. A positive finding would be significant because DNA viruses do not appear to exist in widespread endogenous forms.

Another deduction is that somatic mutations are one of the fundamental ingredients of cancer although they appear to require the occurrence of several other events not yet understood. The role of mutations in turn suggest that the incidence of cancer in man could be reduced by identifying as many promutagens as possible, and by eliminating them from the environment. One important feature of this approach to cancer prevention is that it can be started now, since these substances can be identified with simple bacterial tests suitable for mass screening (44). The feasibility of prevention is shown by the fact that the promutagens already identified in a preliminary screening, such as tobacco or some hair dyes, are inessential for human life (45, 46).

However, it is practically difficult to achieve a substantial reduction of the use of these substances, as shown by the example of tobacco. According to epidemiological evidence tobacco smoke is the agent of lung cancer in man, which in Britain is responsible for one in eight of all male deaths (47). Yet only mild sanctions have been imposed on tobacco products, such as a vague health warning on cigarette packets, which sounds rather like an official endorsement. Any limitation on the use of tobacco is left to the individual, although it is clear that the individual cannot easily exercise voluntary restraint in the face of very effective advertisements, especially as he does not usually appreciate the danger of a cumulative action over a long period of time.

The lax attitude of governments towards tobacco probably also derives from the difficulty of appreciating epidemiological evidence, especially since this evidence is contradicted from time to time by single-minded individuals who use incomplete or even erroneous analyses of the data, and whose views are magnified out of all proportion by the media. However, the recent recognition that tobacco smoke contains promutagens contributes direct experimental evidence on the dangers of tobacco smoke, on which there cannot be any equivocation. I, therefore, call on governments to act towards severely discouraging tobacco consumption; and to act now because it will be at least thirty years before their action has its full effect.

Although tobacco smoke is a striking example of an environmental carcinogen, many others are known and probably many more remain to be identified. Identification by conventional tests is difficult because they are costly and laborious; but they can now be replaced by the bacterial tests for promutagens. Since the tests are easy and inexpensive it should be possible to investigate many normal constituents of the environment, and every new compound before it is offered to the public. The feasibility of such a programme is borne out by the finding that most of the commonly available substances are not promutagens (45, 46). Given the strong correlation between mutagenicity and carcinogenicity (43), any promutagen is suspect and, if all possible, should be withdrawn.

In fact, this is precisely the attitude that scientists have taken for themselves concerning the experiments in genetic engineering, which carry the theoretical possibility of creating new virus-like molecules, endowed with carcinogenic activity. Although the danger is only hypothetical, experiments which might be very useful for science and society have been postponed until they can be carried out under the strictest safeguards (48). Governments have accepted this position and are eager to impose severe restrictions on the performance of these experiments. While I fully approve of their concern, I cannot help noticing that they follow a double standard; if there is any doubt—

you must discourage experiments, but if there is any doubt—you cannot discourage cigarettes.

BIOLOGISTS AND SOCIETY

This discussion about cancer prevention is a development of the experimental results obtained in the field of oncogenic viruses, but it is also strongly influenced by the new social conscience of many scientists. Historically, science and society have gone separate ways, although society has provided the funds for science to grow and in return science has given society all the material things it enjoys. In recent years, however, the separation between science and society has become excessive, and the consequences are felt especially by biologists. Thus, while we spend our life asking questions about the nature of cancer and ways to prevent or cure it, society merrily produces oncogenic substances and permeates the environment with them. Society does not seem prepared to accept the sacrifices required for an effective prevention of cancer. The situation is clearly unacceptable, and we biologists would like to see it corrected. We have ourselves begun to put our house in order, by banning some experiments that may contain a risk for mankind. We would like to see society take a similar attitude, abandoning selfish practices that are dangerous for society itself. We would also like to see a new co-operation of science and society for the benefit of all mankind and hope that the dominant forces in society will recognize that this is a necessity.

REFERENCES:

1. Dulbecco, R., and Freeman, G., *Virology, 8*, 396 (1959).
2. Vogt, M., and Dulbecco, R., *Proc. Nat. Acad. Sci. USA., 46*, 365 (1960).
3. Dulbecco, R., and Vogt, M., *Proc. Nat. Acad. Sci. USA., 46*, 1617 (1960).
4. Smith, J. D., Freeman, G., Vogt, M., and Dulbecco, R., *Virology, 12, 185* (1960).
5. Dulbecco, R., and Vogt, M., *Proc. Nat. Acad. Sci. USA., 50*, 236, (1963)
6. Sambrook, J., Westphal, H., Srinivasan, P. R., and Dulbecco, R., *Proc. Nat. Acad. Sci. USA., 60*, 1288 (1968).
7. Oda, K., and Dulbecco, R., *Proc. Nat. Acad. Sci. USA., 60*, 525 (1968).
8. Stoker, M., *Nature (London), 218*, 234 (1968).
9. Berg, P., and Stoker, M., Personal communication.
10. Koprowski, H., Jensen, F. C., and Steplewski, Z., *Proc. Nat. Acad. Sci. USA., 58*, 127 (1967).
11. Watkins, J. F., and Dulbecco, R., *Proc. Nat Acad. Sci. USA., 58*, 1396 (1967).
12. Tonegawa, S., Walter, G., Bernardini, A., and Dulbecco, R., *Cold Spring Harbor Symp. Quant. Biol., 35*, 823 (1970).
13. Lindstrom, D. M., and Dulbecco, R., *Proc. Nat. Acad. Sci. USA., 69*, 1517, (1969).
14. Sharp. P. A., Petterson, U., and Sambrook, J., *J. Mol. Biol., 86*, 709 (1974).
15. Sharp, P. A., Gallimore, P. H., and Flint, S. J., *Cold Spring Harbor Symp. Quant. Biol., 39*, 457 (1974).
16. Fried, M., *Proc. Nat. Acad. Sci. USA., 53*, 486 (1965).
17. Eckhart, W., *Virology, 38*, 120 (1969).
18. Tegtmeyer, P., and Ozer, H. L., *J. Virol., 8*, 516 (1971).
19. Martin, R. G., Chou, J. Y., Avila, J., and Saral, R., *Cold Spring Harbor Symp. Quant. Biol., 39*, 17 (1974).

20. Butel, J. S., Brugge, J. S., and Noonan, C. A., *Cold Spring Harbor Symp. Quant. Biol., 39,* 25 (1974).
21. Kimura, G., and Itagaki, A., *Proc. Nat. Acad. Sci. USA., 72,* 673 (1974).
22. Black. P. H., Rowe, P. W., Turner, H. C., and Hubner, R. J., *Proc. Nat. Acad. Sci. USA., 50,* 1148 (1963).
23. Tegtmeyer, P., *Cold Spring Harbor Symp. Quant. Biol., 39,* 9 (1974).
24. Oxman, M., Takemoto, K. K., and Eckhart, W., *Virology, 49,* 675 (1972).
25. Paulin, D., and Cuzin, F., *J. Virol., 15,* 393 (1975).
26. Carroll, R. B., Personal communication.
27. Carroll, R. B., Hager, L., and Dulbecco, R., *Proc. Nat. Acad. Sci. USA., 71,* 3754 (1974).
28. Jessel, D., Hudson, J., Landau, T., Tenen, D., and Livingston, D. M., *Proc. Nat. Acad. Sci. USA., 72,* 1960 (1975).
29. Reed, S. I., Ferguson, J., Davis, R. W., Stark, G. R., *Proc. Nat. Acad. Sci. USA., 72,* 1605 (1975)
30. Smith, A. E., Bayley, S. T., Wheeler, T., and Mangel, W. F., In "*In vivo Transcription and Translation of Viral Genomes*"., Eds., Haenni, A., and Beaud, J., *INSERM,* Paris, *47,* 331 (1975).
31. Dulbecco, R., *Proc. Royal Society, Lond. B., 189,* 1—14 (1975).
32. Novick, A., and Weiner, M., *Proc. Nat. Acad. Sci. USA., 43,* 553 (1957).
33. Dulbecco, R., Hartwell, L. H., and Vogt, M., *Proc. Nat. Acad. Sci. USA., 53,* 403 (1965).
34. Hartwell, L. H., Vogt, M., and Dulbecco, R., *Virology, 27,* 262 (1965).
35. Coggin, J. H., *J. Immunol., 105,* 524 (1970).
36. Wiener, F., Klein, G., and Harris, H., *J. Cell Sci, 8,* 681 (1971).
37. Mintz, B., and Illmensee, K., *Proc. Nat. Acad. Sci. USA., 72,* 3585 (1975).
38. Papaionnou, V. E., McBurney, M. W., Gardner, R. L., and Evans, M. J., *Nature (Lond.), 258,* 70 (1975).
39. Vogt, M., and Dulbecco, R., *Cold Spring Harbor Symp. Quant. Biol., 27,* 367 (1962).
40. Pollack, R. E., Green, H., and Todaro, G. J., *Proc. Nat. Acad. Sci. USA., 60,* 126 (1968).
41. Renger, H. C., and Basilico, C., *Proc. Nat. Acad. Sci. USA., 69,* 109 (1972).
42. Stich, H. F., San, R. H. C., and Kawazoe, Y., *Nature (Lond)., 229,* 416, (1971).
43. McCann, J., Choi, E., Yamasaki, E., and Ames, B. N., *Proc. Nat. Acad. Sci. USA.,* In press.
44. Ames, B. N., Lee, F. D., and Durston, W. E., *Proc. Nat. Acad, Sci. USA.,* 70, 782 (1973).
45. Kier, L. D., Yamasaki, E., and Ames, B. N., *Proc. Nat. acad. Sci. USA., 71,* 4159 (1974).
46. Ames, B. N., Kammen, H. O., and Yamasaki, E., *Proc. Nat. Acad. Sci.·USA., 72,* 2423 (1975).
47. Doll, R., *J. Roy Stat. Soc., Series A, 134,* 133 (1971)
48. Berg, P., Baltimore, D., Brenner, S., Roblin, R. O., and Singer, M. F., *Proc. Nat. Acad. Sci. USA., 72,* 1981 (1975).

Howard M. Temin

HOWARD M. TEMIN

I was born on December 10, 1934 in Philadelphia, Pennsylvania, United States of America, the second of three sons of Annette and Henry Temin. My father was an attorney, and my mother has been continually active in civic affairs, especially educational ones. My older brother, Michael, is also an attorney in Philadelphia, and my younger brother, Peter, is a Professor of Economics at the Massachusetts Institute of Technology, Cambridge, Mass.

I received my elementary and high school education in the public schools of Philadelphia. My specific interest in biological research was focused by summers (1949—1952) spent in a program for high school students at the Jackson Laboratory in Bar Harbor, Maine, and a summer (1953) spent at the Institute for Cancer Research in Philadelphia. I attended Swarthmore College from 1951 to 1955, majoring and minoring in biology in the honors program. After another summer (1955) at the Jackson Laboratory, I became a graduate student in biology at the California Institute of Technology in Pasadena, California, majoring in experimental embryology. After a year and a half, I changed my major to animal virology, becoming a graduate student in the laboratory of Professor Renato Dulbecco. My doctoral thesis was on Rous sarcoma virus. Much of my early work on this virus was carried out with the close collaboration of Dr. Harry Rubin, then a postdoctoral fellow in Professor Dulbecco's laboratory. At Cal Tech, I was also greatly influenced by Professor Max Delbrück and by Dr. Matthew Meselson. After finishing my Ph.D. degree in 1959, I remained for an additional year in Professor Dulbecco's laboratory as a postdoctoral fellow. In that year, I performed the experiments that led to the formulation in the same year of the provirus hypothesis for Rous sarcoma virus.

In 1960, I moved to Madison as an Assistant Professor in the McArdle Laboratory for Cancer Research, which is also the Department of Oncology, in the Medical School, The University of Wisconsin-Madison. My first laboratory was in the basement, with a sump in my tissue culture lab and with steam pipes for the entire building in my biochemistry lab. Here I performed the experiments that led in 1964 to my formulating the DNA provirus hypothesis. In the fall of 1964, the entire department moved to a new building. I became successively Associate Professor, Full Professor, Wisconsin Alumni Research Foundation Professor of Cancer Research, and, in 1974, American Cancer Society Professor of Viral Oncology and Cell Biology. From 1964 to 1974, I also held a Research Career Development Award from the National Cancer Institute.

During my first years at Wisconsin, I worked with only two technicians.

My first postdoctoral fellow joined me in 1963, and my first graduate student, in 1965. I had no more than two or three postdoctoral fellows and graduate students at one time until about 1968.

During the late 1960's, about half of my time was spent in studying the control of multiplication of uninfected and Rous sarcoma virus-infected cells in culture. This work led to my appreciation of the role of specific serum factors in the control of cell multiplication and the demonstration that a multiplication-stimulating factor in calf serum for chicken fibroblasts was the same as somatomedin.

I serve on the editorial boards of several journals, including the *Journal of Cellular Physiology,* the *Journal of Virology,* and the *Proceedings of the National Academy of Sciences U.S.A.* I have also been a member of the Virology Study Section of the National Institutes of Health. In addition, I do much other paper and grant reviewing.

Since the general acceptance of the DNA provirus hypothesis in 1970, I have received many honors, including the Warren Triennial Prize (with David Baltimore); the Pap Award of the Papanicolaou Institute, Miami, Florida; the Bertner Award, M. D. Anderson Hospital and Tumor Institute, Houston, Texas; the U. S. Steel Foundation Award in Molecular Biology, National Academy of Sciences U.S.A.; the American Chemical Society Award in Enzyme Chemistry; the Griffuel Prize, Association Developpment Recherche Cancer, Villejuif, France; the G.H.A. Clowes Award, American Association for Cancer Research; the Gairdner International Award (with David Baltimore); the Albert Lasker Award in Basic Medical Research; and honorary degrees from Swarthmore College and New York Medical College. I have also presented several honorary lectures. I am a fellow of the American Academy of Arts and Sciences and a member of the National Academy of Sciences, U.S.A.

In 1962 I married Rayla Greenberg of Brooklyn, New York, a population geneticist. She has been a constant source of support and warmth. We have two daughters, Sarah Beth and Miriam.

THE DNA PROVIRUS HYPOTHESIS

The Establishment and Implications of RNA-directed DNA Synthesis

Nobel Lecture, December 12, 1975

by

HOWARD M. TEMIN

University of Wisconsin, Madison, Wisconsin, U.S.A.

I. INTRODUCTION

Your Majesty, fellow scientists, ladies and gentlemen: It is a great honor for me to be here today to discuss the DNA provirus and RNA-directed DNA synthesis, and it has been a pleasure for my family and me to be here in Stockholm this week. The Nobel Prize is an honor not only for me but also for all those who have been working with avian RNA tumor viruses. The Nobel Prize is also an honor for the American people, whose tax dollars and private contributions have supported my work.

The genetic information in RNA is transferred to DNA during the replication of some viruses, including some that cause cancer. This transfer of information from the messenger molecule, RNA, to the genome molecule, DNA, apparently contradicted the "central dogma of molecular biology", formulated in the late 1950's. This mode of information transfer was first postulated and established for the replication of Rous sarcoma virus, a strongly transforming avian C-type ribodeoxyvirus. (Ribodeoxyviruses are RNA viruses that replicate through a DNA intermediate.)

In this lecture, I shall discuss the experiments that led to the formulation of the DNA provirus hypothesis; the experiments that established the DNA provirus hypothesis and, therefore, the existence of RNA-directed DNA synthesis; some aspects of the present status of our knowledge of the mechanism of formation of the DNA provirus; and, finally, some implications of this work for the questions of the origin of animal viruses, how cancers may be caused by viruses, and how the majority of cancers, which do not involve infectious viruses, are caused.

The majority of the ideas I shall discuss today came from experiments with Rous sarcoma virus (RSV), the prototype RNA tumor virus. Rous sarcoma virus was originally described by Peyton Rous in 1911. He stated, "A transmissible sarcoma of the chicken has been under observation in this laboratory for the past fourteen months, and it has assumed of late a special interest because of its extreme malignancy and a tendency to wide-spread metastasis. In a careful study of the growth, tests have been made to determine whether it can be transmitted by a filtrate free of the tumor cells . . . Small quantities of a cell-free filtrate have sufficed to transmit the growth to susceptible fowl." (Rous, 1911).

Although Rous and his associates carried out many experiments with RSV, as the virus is now called, and had many prophetic insights into its behavior, they and other biologists of that time did not have the scientific concepts or the technical tools to exploit his discovery. And in about 1915 Rous himself stopped work with RSV.

The major scientific concepts required to understand the behavior of RSV were that genetic information was contained in and transferred from nucleic acids, developed especially by Avery, MacLeod and McCarthy (1944), and by Watson and Crick (1953), as well as the concept that viral genomes could become part of cell genomes, developed especially by Lwoff (1965). The major technical tools required were those of quantitative virology and of the study of animal viruses in cell culture, developed especially by Delbrück (Cairns, Stent, and Watson, 1966), Enders, Robbins, and Weller (1955), and Dulbecco (1966).

My first contact with RSV was in 1956 when, as a graduate student at the California Institute of Techonology, I was asked by Dr. Harry Rubin, a postdoctoral fellow in Professor Dulbecco's laboratory, to try and make more quantitative the observations of Manaker and Groupé (1956) that discrete foci of altered chicken embryo cells were associated with Rous sarcoma virus in tissue culture (see also Rubin, 1966; Temin, 1971c).

II. ASSAY FOR ROUS SARCOMA VIRUS

I soon found that addition of RSV to cultures of chicken embryo cells in a sparse layer, rather than in a crowded monolayer as then used for the assay of other animal viruses, led to the appearance of foci of transformed cells (Figure 1). The number of these foci was proportional to the concentration of virus, and the foci resulted from altered morphology and altered control of multiplication of the infected cells (Temin and Rubin, 1958). The foci were cell culture analogs of tumors in chickens.

This assay allowed RSV to be studied like other viruses, leading to the demonstration that RSV-infected cells could produce virus and divide (Temin and Rubin, 1959) and to the demonstration by Crawford and Crawford (1961) that the genome of RSV was RNA. The assay for RSV was also a model for the assay of other transforming viruses, such as polyoma virus, as discussed by Dr. Dulbecco (1976).

Further observation of RSV-induced foci revealed that some of the foci contained long fusiform cells rather than the rounded cells seen in the focus in Figure 1 (Temin, 1960). Virus produced by these fusiform foci caused the formation of further foci of long fusiform cells, that is, the virus from these foci was a genetic variant.

These and other observations indicated that viral genes controlled the morphology of transformed cells and led to the hypothesis that transformation is the result of the action of viral genes, that is, transformation is a conversion analogous to lysogenic conversion. This hypothesis has been amply confirmed for RSV by the isolation of variant viruses temperature-sensitive for transformation or defective for transformation (Martin, 1970; Vogt, 1971; Kawai and Hanafusa, 1972).

These observations also led to the study of differences between transformed and normal cells. At least two important results came from these stud-

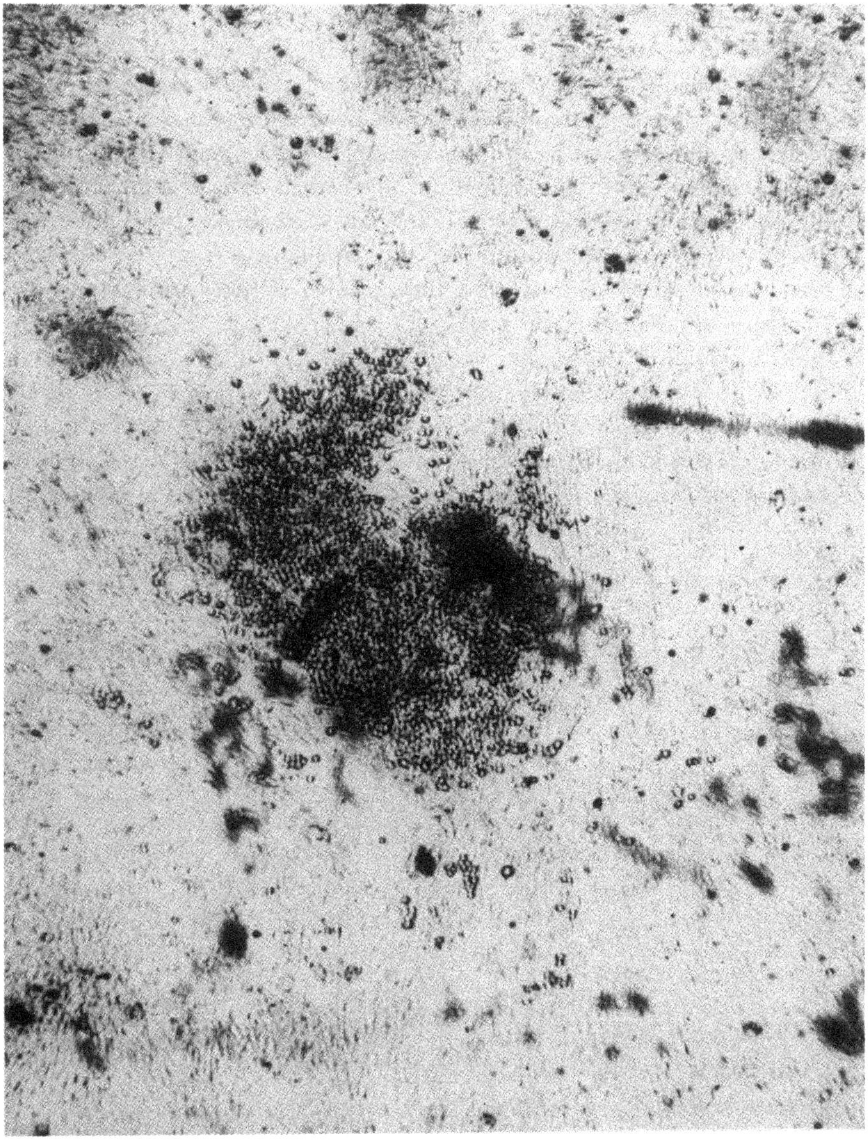

Figure 1. Focus induced by Rous sarcoma virus in chicken cells. A sparse monolayer of chicken embryo fibroblasts was exposed to Bryan standard Rous sarcoma virus. The cells were overlaid with tissue culture medium and incubated at 38° C for ten days. This photograph of one focus was taken with an inverted microscope at a magnification of 25.

ies: 1. the concept of an altered requirement of transformed cells for specific multiplication-stimulating factors in serum (Temin, 1967b; Pierson and Temin, 1972; Dulak and Temin, 1973); and 2. the discovery by Reich and coworkers of increased production by transformed cells of an activator of a serum protease (Reich, 1975).

III. THE PROVIRUS HYPOTHESIS

In 1960 I studied the kinetics of mutation of the viral genes controlling cell and focus morphology, the effects of mutation in these viral genes on the morphology of infected cells, and the inheritance of these genes in cells infected with two different Rous sarcoma viruses (Temin, 1961). These studies demonstrated that these viral genes mutated at a high rate, that mutation in a viral gene present in an infected cell often led to change in the morphology of that infected cell, that two different viruses infecting one cell were stably inherited, and that the intracellular viral genomes were probably located at only one or two sites in the cell genome.

These observations led to the provirus hypothesis (Figure 2) — infection of chicken cells by RSV leads to the formation of one or two copies of a regularly inherited structure with the information for progeny virus and for cell morphology. (Svoboda *et al.* (1963) from studies of RSV-infected rat cells independently postulated the existence of a provirus in RSV-infected cells.)

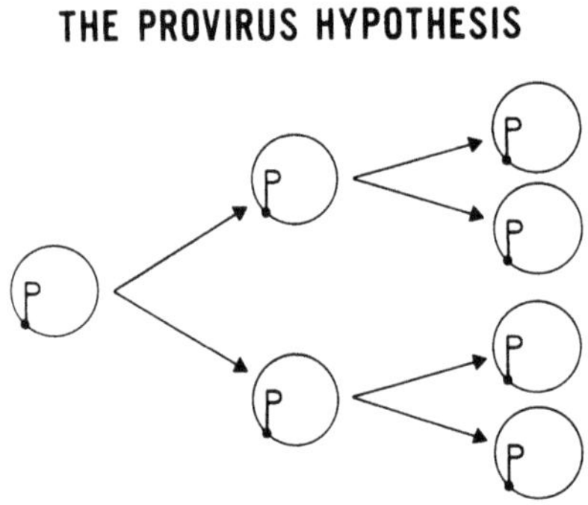

Figure 2. The provirus hypothesis. Virus information (P) is contained in infected cells in one or two copies of a regularly inherited structure with the information for progeny virus and for cell morphology.

The provirus hypothesis was a genetic hypothesis and contained no statement about the molecular nature of the provirus. However, the regular inheritance of the provirus led me to postulate that the provirus was integrated with the cell genome.

The provirus hypothesis was further supported by the behavior of converted RSV-infected chicken cells that were not producing infectious virus (Temin, 1962). (Analysis of similar cells by others led to the concept of defectiveness of some strongly transforming RNA tumor viruses (Hanafusa, Hanafusa, and Rubin, 1963; Hartley and Rowe, 1966).)

IV. THE DNA PROVIRUS HYPOTHESIS

At the time of my formulation of the provirus hypothesis in 1960, the general rules for information transfer in living systems were being clearly established in what was called "the central dogma of molecular biology", that is, genetic information is transferred from DNA to RNA to protein. RNA viruses were an apparent exception to this "dogma". Studies with the newly discovered RNA bacteriophage and with animal RNA viruses, especially using the antibiotic actinomycin D, indicated that RNA viruses transferred their information from RNA to RNA and from RNA to protein and that DNA was not directly involved in the replication of these RNA viruses (Reich *et al.*, 1962).

Although I was unable to reconcile the regular inheritance of the provirus with its being RNA, I still tried in 1962, after I had arrived at the University of Wisconsin—Madison, to use actinomycin D to isolate the provirus of Rous sarcoma virus, just as David Baltimore and others were using actinomycin to study the intermediates in the replication of other animal RNA viruses (Franklin and Baltimore, 1962).

However, when actinomycin D was added to Rous sarcoma virus-producing cells, it inhibited virus production (Figure 3). Control experiments demonstrated that this inhibition was neither of early events in infection (as was found by Barry, Ives, and Cruickshank (1962) with influenza virus) nor of the ability of the treated cells to support replication of other animal RNA viruses. These results indicated to me that the provirus was DNA.

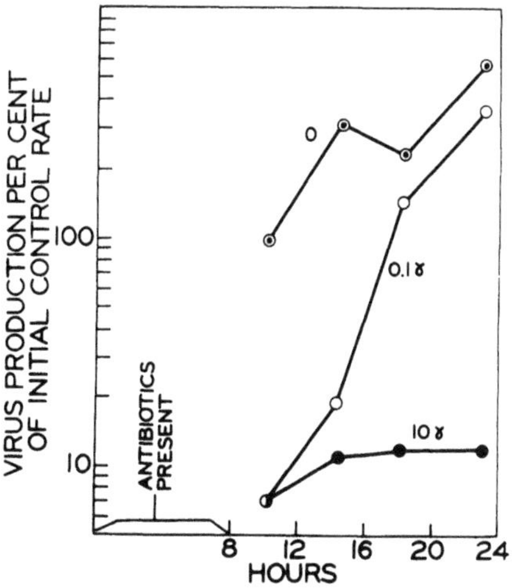

Figure 3. Effects of actinomycin D on production of Rous sarcoma virus. Chicken cells producing RSV were exposed to 0, 0.1 or 10 μg (γ)/ml of actinomycin D. After 8 hours, the medium was removed, the cells were washed, and fresh medium was added. At the indicated times, the medium was harvested and assayed for focus forming units of RSV. (Taken from Temin, 1963.)

I carried out further experiments that indicated that new DNA synthesis was required for RSV infection and that new RSV-specific DNA was found in infected chicken cells (Temin, 1964a,b; see also Bader, 1965).

Based on the results of these experiments, I proposed the DNA provirus hypothesis at a meeting in the Spring of 1964 (Temin, 1964c)—the RNA of infecting RSV acts as a template for the synthesis of viral DNA, the provirus, which acts as a template for the synthesis of progeny RSV RNA (Figure 4).

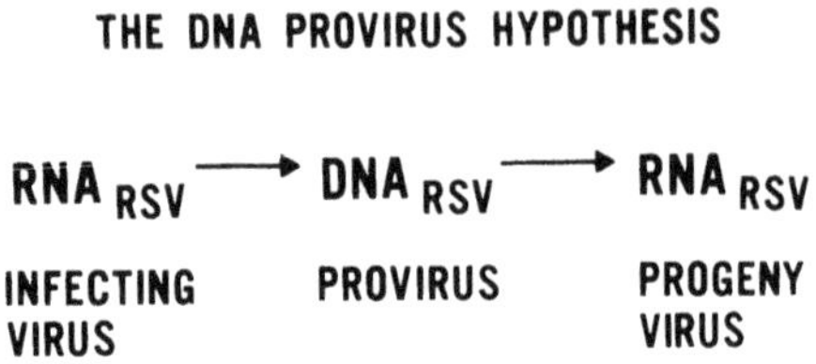

Figure 4. The DNA provirus hypothesis.

At this meeting and for the next 6 years this hypothesis was essentially ignored.

My co-workers and I tried in 1964 and 1965 to obtain direct molecular evidence for the DNA provirus by looking for RNA-directed DNA polymerase activity in cells soon after infection, for infectious DNA in infected cells, and for better systems of nucleic acid hybridization. These initial efforts were unsuccessful (Temin, 1966).

I then developed systems with better controlled cells to study RSV infection—at first, synchronized cells, and later stationary cells (Temin, 1967a, 1968a). Experiments with these cells indicated that a normal replicative cell cycle was needed for initiation of RSV production.

With this knowledge, I performed experiments that demonstrated more clearly a requirement for new non-S phase DNA synthesis for RSV infection (Temin, 1968b; see also Murray and Temin, 1970), and I demonstrated that this new DNA synthesis was virus-specific (Temin, 1970a). Finally, using infection of stationary cells, we demonstrated that the newly synthesized viral DNA could be labeled with 5-bromodeoxyuridine and inactivated by light (Figure 5) (Boettiger and Temin, 1970). However, our attempts at this time

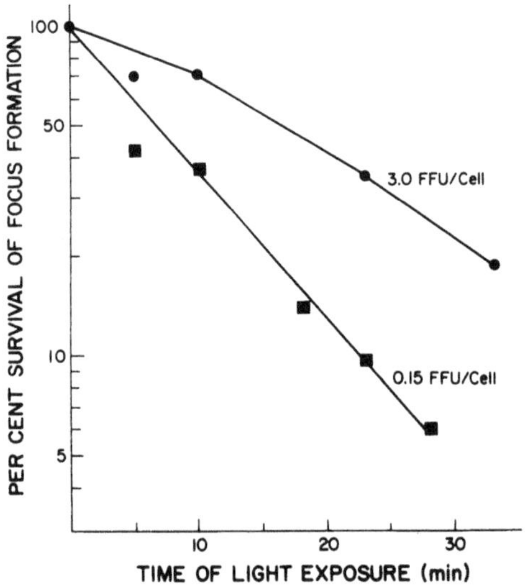

Figure 5. Light inactivation of focus formation by chicken cells infected with RSV in the presence of 5-bromodeoxyuridine. Stationary chicken cells were exposed to RSV at two multiplicities of infection (0.15 or 3.0 focus forming units per cell), incubated in medium containing 5-bromodeoxyuridine, exposed to light, and plated on rat cells to determine the number of focus forming cells surviving. (Taken from Boettiger and Temin, 1970.)

to isolate the bromodeoxyuridine-labeled viral DNA were unsuccessful (Boettiger, 1972).

V. RSV VIRION DNA POLYMERASE

In 1969 Dr. Satoshi Mizutani came to my laboratory. He demonstrated that no new protein synthesis was required for the synthesis of viral DNA during RSV infection of stationary chicken cells (quoted in Temin, 1971a), and, therefore, that the DNA polymerase that synthesized viral DNA existed before the infection of the chicken cells. This work was never published completely for, in December, 1969, we decided that the experiments indicated that RSV virions contain a DNA polymerase, and we decided to look for the virion polymerase first.

There were precedents for virion polymerases. In 1967 Kates and McAuslan, and Munyon, Paoletti, and Grace had found a DNA-directed RNA polymerase in poxvirus virions, and in 1968 Borsa and Graham, and Shatkin and Sipe had found an RNA-directed RNA polymerase in virions of reovirus. (The conclusion that RSV virions contain a DNA polymerase could have been deduced in 1967 or 1968 from the DNA provirus hypothesis and the existence of these virion polymerases, but it was not (but see Baltimore, 1976).)

RSV virions contain an endogenous DNA polymerase activity with the following characteristics (Figure 6). The virion polymerase activity incor-

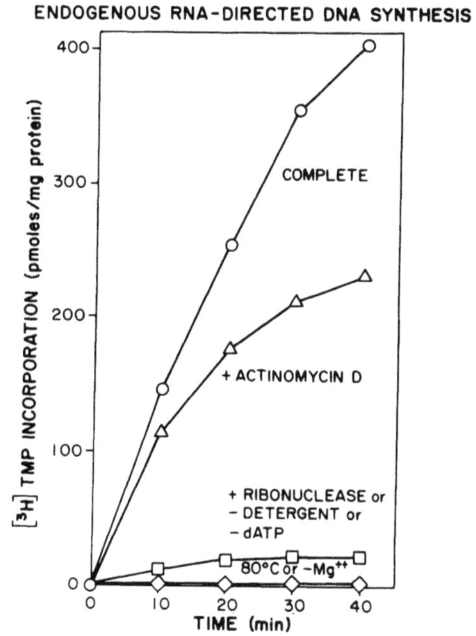

Figure 6. Endogenous RNA-directed DNA synthesis by avian leukosis virus virions. Purified virions (2 μg protein) of an avian leukosis virus were incubated in a complete system (Mizutani, Kang, and Temin, 1973) with the indicated pretreatments, additions, or subtractions, and the incorporation of label was measured.

porates deoxyribonucleoside monophosphates into DNA and requires all four deoxyribonucleoside triphosphates, a divalent cation, and a detergent to disrupt the virion envelope. Furthermore, the polymerase activity is inactivated by heat, which denatures the polymerase, and by ribonuclease, which destroys the template, and it is partially resistant to actinomycin D. (All but one of these characteristics, actinomycin D resistance (McDonnell *et al.*, 1970), were presented in our original paper (Temin and Mizutani, 1970), which was published together with the paper of Dr. Baltimore (Baltimore, 1970).) We call this virion enzyme activity "endogenous RNA-directed DNA polymerase activity".

The avian RNA tumor virus DNA polymerases are stable and easy to solubilize and study (see Temin and Baltimore, 1972). Numerous workers have purified these enzymes, especially from avian myeloblastosis virus, and this DNA polymerase has become a standard reagent for molecular biologists. It is especially useful because it has no deoxyribonuclease activity, but it does have ribonuclease H activity. (Ribonuclease H activity degrades the RNA strand of an RNA · DNA hybrid molecule, but not single-stranded RNA.)

VI. THE ESTABLISHMENT OF THE DNA PROVIRUS HYPOTHESIS

Although the discovery of the RSV virion DNA polymerase immediately provided convincing evidence for the DNA provirus hypothesis, actual proof of the existence of a DNA provirus depended upon later work involving nucleic acid hybridization and infectious DNA experiments.

Neiman (1972) was the first to demonstrate convincingly increased hybridization of labeled RSV RNA to DNA of infected chicken cells. We have confirmed his results with another avian RNA virus that replicates through a DNA intermediate, spleen necrosis virus, which gives a clearer and cleaner result (Figure 7). (Others have also confirmed Neiman's results (for example

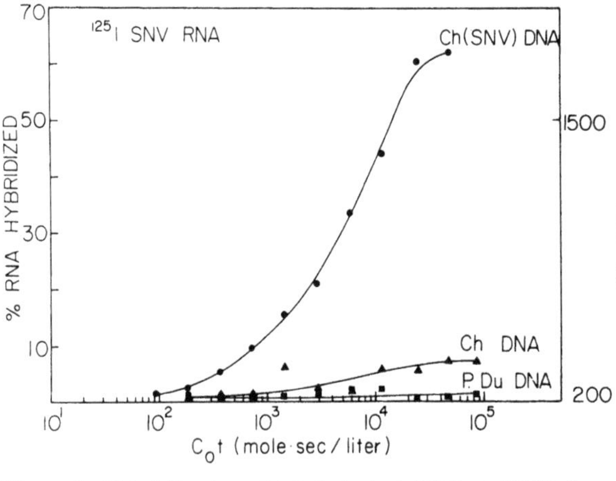

Figure 7. Hybridization of labeled viral RNA to DNA from infected and uninfected cells. ^{125}I-labeled RNA from spleen necrosis virus (SNV) was incubated for different times with a large excess of DNA from uninfected chicken (Ch) or Peking duck (P. Du) cells or from spleen necrosis virus-infected chicken (Ch(SNV)) cells, and the percentage of RNA that was ribonuclease-resistant was determined (Kang and Temin, 1974).

Varmus, Heasley, and Bishop, 1974; Shoyab, Baluda, and Evans, 1974).) Therefore, the DNA of ribodeoxyvirus-infected cells contains new nucleotide sequences complementary to the RNA of the infecting ribodeoxyvirus.

To a virologist an even more satisfying proof for the existence of the DNA provirus was the demonstration, first by Hill and Hillova (1972), of infectious DNA for RSV. We, as well as others, have repeated and extended their work, making it more quantitative (Table 1). Rous sarcoma virus-infected cells, but not uninfected cells, contain a nucleic acid with the information for RSV (the provirus). This information is contained in DNA as shown by its inactivation by deoxyribonuclease, its resistance to alkali, ribonuclease, and Pronase, and its density in equilibrium cesium chloride density gradient centrifugation. A single molecule of about 6×10^6 daltons of double-stranded DNA is sufficient to cause infection, and the efficiency of infection is similar to that of the DNA isolated from animal small DNA viruses (Cooper and Temin, 1974).

Table 1. Infectious Rous sarcoma virus DNA.[a]

DNA	Infectious dose 50 (ID_{50}) (μg)
RSV-infected chicken cell	0.1
RSV-infected chicken cell, deoxyribonuclease	> 10
RSV-infected chicken cell, alkali	1.0
RSV-infected chicken cell, ribonuclease	0.1
RSV-infected chicken cell, Pronase	0.1
RSV-infected chicken cell, cesium chloride density gradient centrifugation	0.1
RSV-infected rat cell	0.1

[a] DNA was isolated from RSV-infected chicken or rat cells, treated as indicated, and assayed for infectivity in chicken fibroblasts. Infectivity is presented as the amount of DNA required to infect half of the assay cultures. (Taken from Cooper and Temin, 1974.) The lower the amount of DNA required for infection, the more infectious the DNA was.

VII. STATUS OF KNOWLEDGE OF THE MECHANISM OF FORMATION OF THE DNA PROVIRUS AT THE PRESENT TIME (NOVEMBER, 1975)

The existence of a DNA provirus for RSV has been established. In addition, some knowledge has been gained of the details of the molecular mechanisms for the formation of the RSV provirus. Especially notable has been the work of Bishop and Varmus and their colleagues at the University of California—San Franscisco Medical School (Bishop and Varmus, 1975).

After infection of susceptible cells by RSV, the virion DNA polymerase synthesizes a DNA copy of the viral RNA, probably using a cellular transfer RNA molecule associated with the viral RNA as a primer for the DNA synthesis. After the formation of the RNA · DNA hybrid molecule, there is synthesis of a second strand of DNA, perhaps after degradation of the viral RNA by the ribonuclease H activity of the virion DNA polymearse. Double-stranded closed circular viral DNA appears. Viral DNA becomes integrated with host DNA. However, neither the mechanism for integration nor whether virion-associated enzymes (Mizutani *et al.*, 1971) are involved in integration is known.

We have been studying the formation of the provirus of spleen necrosis virus (SNV), a cytopathic member of a species of avian ribodeoxyviruses distinct from the avian leukosis viruses like RSV. Some interesting contrasts, as well as similarities, have been found.

Instead of using only a pre-formed primer for DNA synthesis, spleen necrosis virus may at times synthesize an RNA primer *de novo* (Mizutani and Temin, 1975). The virions of spleen necrosis virus contain an RNA polymerase activity as well as a DNA polymerase activity (Mizutani and Temin, 1976) (Figure 8). This RNA polymerase activity can initiate synthesis of new RNA chains, and its product RNA, a small molecule, is hydrogen-bonded to viral RNA. Thus, SNV virions contain both DNA polymerase and RNA polymerase activities—the only virions so far reported to contain both of these enzyme activities.

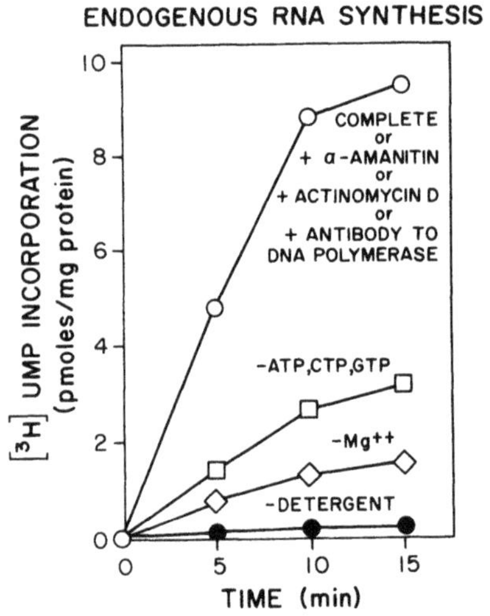

Figure 8. Endogenous RNA synthesis by reticuloendotheliosis virus virions. Purified virions (2 μg protein) of SNV were incubated in a complete system with the indicated additions, subtractions, or pretreatments, and the incorporation of label was measured. (Taken from Mizutani and Temin, 1976.)

We have also studied the kinetics of formation of infectious SNV DNA (Fritsch and Temin, 1976) (Figure 9). After infection of chicken cells by SNV, infectious viral DNA appeared in an unintegrated form, found in the supernatant of a Hirt extract (Hirt, 1967), shortly before it appeared in an integrated form, found in the pellet of a Hirt extract. Surprisingly there were large further increases in the amounts of both unintegrated and integrated viral DNAs, and some unintegrated viral DNA persisted for over a week after infection. In contrast to these results with dividing cells, little infectious vi-

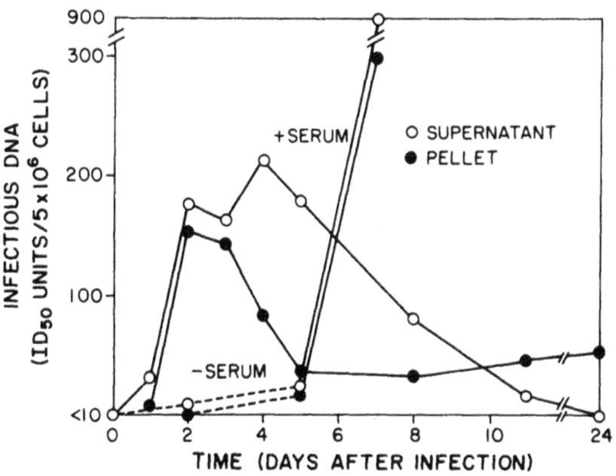

Figure 9. Kinetics of formation of infectious DNA in SNV-infected multiplying and stationary chicken cells. Chicken cells were exposed to SNV at a multiplicity of infection of 5 plaque forming units per cell, and medium with or without serum was added. At different times, the cells were fractionated by Hirt extraction (Hirt, 1967), and the DNAs in the supernatant and pellet fractions were assayed for infectivity. (Taken from Fritsch and Temin, 1976.)

ral DNA was formed in stationary cells exposed to SNV. This result indicates that a normal replicative cell cycle is required for formation of infectious viral DNA (also see Humphries and Temin, 1974).

The forms of unintegrated infectious viral DNA were analyzed by agarose gel electrophoresis (Figure 10). Three forms were found, reminiscent of the three forms of DNA in papovavirus virions (see Tooze, 1973). The majority of the infectious viral DNA was in linear molecules, but there were minor components of closed circular and nicked infectious SNV DNA.

Thus, the early events in ribodeoxyvirus infection are complex, and much remains to be learned before we can describe the formation of the provirus in molecular detail.

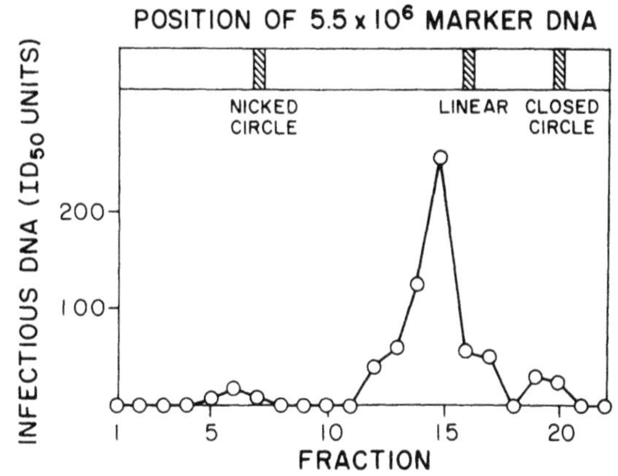

Figure 10. Electrophoresis of unintegrated infectious SNV DNA. The supernatant fraction from Hirt extraction of cells 65 hours after infection by SNV was electrophoresed in a 0.7 % agarose gel in the presence of ethidium bromide with DNA from plasmid RSF 1010 as a marker. The positions of the marker DNAs were established visually, and each fraction was assayed for infectivity. (Taken from Fritsch and Temin, 1976.)

VIII. ORIGIN OF RIBODEOXYVIRUSES

Avian RNA tumor viruses undergo a great amount of genetic variation (see Temin, 1971b; 1974a). This variation is the result of both mutation and recombination. Recombination takes place not only between viruses, but also between viruses and cells.

The recombination between viruses and cells does not appear to be random, but is primarily with specific cellular genes. These genes are called endogenous ribodeoxyvirus-related genes and are, of course, part of the normal cellular DNA.

Endogenous avian leukosis virus-related genes were first recognized about 10 years ago by the presence and Mendelian inheritance of a Rous sarcoma virus virion antigen in some uninfected chicken cells (Dougherty and DiStefano, 1966; Payne and Chubb, 1968). Later an avian leukosis virus virion envelope protein was found in some uninfected chicken cells, and, finally, nucleotide sequences of avian leukosis virus RNA were found in the DNA of all uninfected chicken cells (see Tooze, 1973; Temin, 1974a). (Similar results have been found with mammalian leukemia viruses and cells.)

Study of the phylogenetic distribution of the endogenous avian leukosis virus-related nucleotide sequences revealed (Table 2) a relationship between the amount of these sequences in cell DNA from a particular species of bird and the closeness of the relationship of that species to chickens; for example, more avian leukosis virus nucleotide sequences were found in pheasant DNA than in duck DNA (Kang and Temin, 1974; see also Benveniste and Todaro, 1974).

Table 2. Endogenous avian ribodeoxyvirus-related nucleotide sequences in avian cell DNAs.[a]

Virus	DNA				
	Chicken	Pheasant	Quail	Turkey	Duck
RAV-O	55	20	15	10	< 1
SNV	10	10	10	10	< 2

[a] ^{125}I-labeled RNAs of Rous-associated virus-0, an avian leukosis virus, and of spleen necrosis virus, a reticuloendotheliosis virus, were incubated with an excess of DNA from uninfected cells as described in the legend of Figure 7. The maximum amounts of hybridization from curves like those in Figure 7 are listed. (Taken from Kang and Temin, 1974.) In contrast to RAV-O RNA, SNV RNA hybridized equally to DNA of all the gallinaceous birds. This difference reflects the horizontal transmission of SNV and the vertical transmission of RAV-O.

This distribution is consistent with an hypothesis (the protovirus hypothesis) I originally proposed in 1970 to explain the origin of ribodeoxyviruses—ribodeoxyviruses evolved from normal cellular components (Temin, 1970b, 1974d). The normal cellular components are the endogenous ribodeoxyvirus-related genes. These genes are involved in normal DNA to RNA to DNA information transfer. This normal process of information transfer in cells could not exist only for its ability to give rise to viruses. It must exist as a result of its role in normal cellular processes, for example, cell differentiation, antibody formation, and memory (Temin, 1971d).

One prediction of this protovirus hypothesis is that there are relationships between ribodeoxyvirus and cell DNA polymerases. We have demonstrated such relationships by an antibody blocking test (Mizutani and Temin, 1974). In this test, for example, the activity of an antibody against avian leukosis virus DNA polymerases was blocked by incubation with chicken cell DNA polymerases or a DNA polymerase from an otherwise unrelated avian ribodeoxyvirus.

Therefore, certain predictions of the protovirus hypothesis for the origin of ribodeoxyviruses have been verified. But, obviously, much further work must be done to establish or disprove this hypothesis.

IX. FURTHER IMPLICATIONS OF THESE STUDIES

The protovirus hypothesis can explain the origin of ribodeoxyviruses, but it does not help in understanding the origin of other animal viruses. The presence of an RNA polymerase activity in virions of spleen necrosis virus might, however, present a clue to the origin of the other animal enveloped RNA viruses. As Dr. Baltimore has described, many animal enveloped RNA viruses contain an RNA polymerase activity (Baltimore, 1976). If there were genetic changes so that the SNV RNA polymerase activity synthesized a complete copy of SNV RNA rather than only a small molecule, the first step in the synthesis of a viral RNA intermediate would occur (Figure 11). Further genetic changes leading to copying of the newly synthesized RNA strand would complete the replication of the viral RNA. Therefore, I propose that other animal enveloped RNA viruses evolved from ribodeoxyviruses. (The recent reports of DNA intermediates in carrier cultures of some animal enveloped RNA viruses (Zhdanov, 1975; Simpson and Iinuma, 1975) could indicate a vestige of the origin of these viruses from ribodeoxyviruses.)

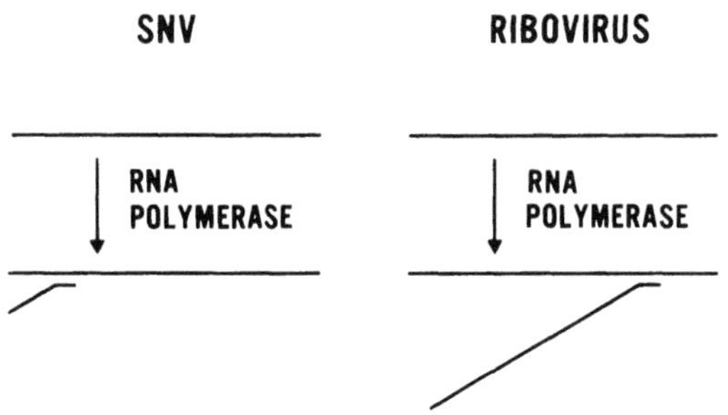

Figure 11. Initial RNA synthesis by SNV and by other RNA viruses.

Animal small DNA viruses might also have originated from ribodeoxyviruses. As discussed above, the unintegrated infectious DNA in SNV-infected cells exists in several forms, and the amount of the unintegrated DNA increases for several days after infection. This unintegrated ribodeoxyvirus DNA could represent a precursor of animal small DNA viruses. Continued replication of unintegrated viral DNA and encapsidation in viral proteins would also be required. Therefore, I propose that animal small DNA viruses also evolved from ribodeoxyviruses.

In most of this discussion of virus replication and virus origins, I have not mentioned cancer. In fact, the absence of such discussion makes an important point: RNA tumor virus replication is not sufficient for cancer formation by RNA tumor viruses. Strongly transforming RNA tumor viruses like RSV cause cancer by introducing genes for cancer into cells. But there are viruses that replicate in much the same way as RSV, for example, SNV or Rous-associated virus-O, that do not cause cancer because they do not contain genes for cancer (Temin, 1974c).

In addition, the majority of human cancers are not caused primarily by infectious viruses like RSV (Temin, 1974b), but by other types of carcinogens, for example, the chemicals in cigarette smoke (Hammond, 1964). These non-viral carcinogens probably act to mutate a special target in the cell DNA to genes for cancer (Figure 12).

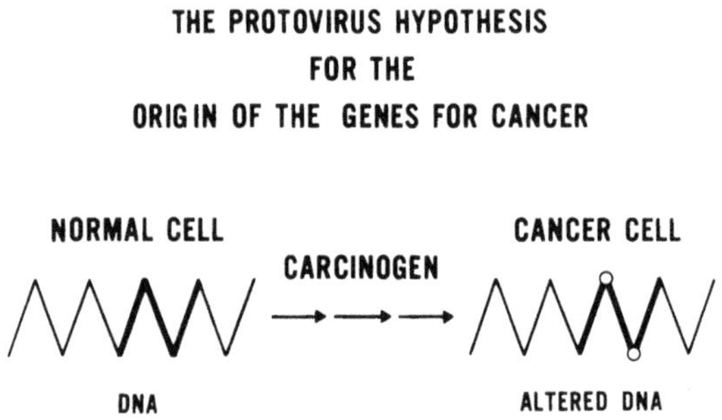

Figure 12. The protovirus hypothesis for the origin of the genes for cancer. The heavy lines indicate DNA involved in DNA to RNA to DNA information transfer.

To relate this hypothesis to the existence of animal RNA viruses like RSV, which do cause cancer efficiently, I have suggested that the targets for the non-viral carcinogens are the genes involved in information transfer from DNA to RNA to DNA (Temin, 1974b,c). Under this hypothesis, genes for cancer would be formed in a process involving RNA-directed DNA synthesis in both RNA virus-induced and non-viral carcinogen-induced cancers.

Finally, to end this lecture where it began with Peyton Rous and RSV, we can speculate on the origin of RSV. As I quoted at the beginning of my lecture, Rous noted a change with transplantation in the behavior of the chicken tumor. This change, I propose, was the result of the formation of RSV, that is, the Rous sarcoma appeared before the Rous sarcoma virus. More specifically, other events not involving a virus led to the formation of genes for cancer and the chicken sarcoma. This sarcoma was infected by an avian leukosis virus, and RSV was formed by a rare recombination (Figure 13).

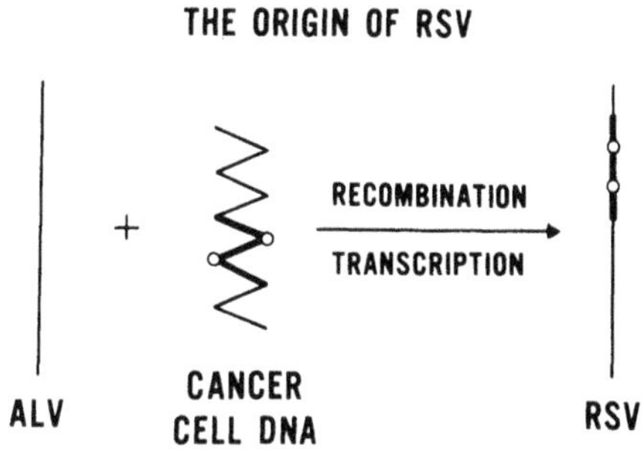

Figure 13. A hypothesis for the origin of Rous sarcoma virus (RSV). A straight line represents RNA, and a zig-zag line represents DNA. ALV is avian leukosis virus.

X. SUMMARY

I have discussed the observations and experiments that led to the formulation and establishment of the provirus hypothesis and the DNA provirus hypothesis, which includes RNA-directed DNA synthesis for the formation of the provirus.

I have also discussed some aspects of the present status of our knowledge of the mechanism of formation of the DNA provirus both to point out the work remaining to be done and to illustrate hypotheses for the origins of ribodeoxyviruses and the origins of other animal enveloped RNA viruses and of animal small DNA viruses.

Finally, I have indicated that I do not believe that infectious viruses cause most human cancers, but I do believe that viruses provide models of the processes involved in the etiology of human cancer.

ACKNOWLEDGEMENTS

I should like to acknowledge three types of support: financial, intellectual, and personal.

The work from my laboratory has been supported by grants from the National Cancer Institute and the American Cancer Society. I held a Research Career Development Award from the National Cancer Institute and am now an American Cancer Society Research Professor.

My work has been supported intellectually by colleagues in my laboratory, especially Satoshi Mizutani, by colleagues at McArdle Laboratory and the University of Wisconsin—Madison, and by co-workers in the field of avian RNA tumor viruses, especially Peter Vogt, Hidesaburo Hanafusa, Marcel Baluda, Jan Svoboda, Peter Duesberg, Robin Weiss, J. Michael Bishop, and Harold Varmus.

I have been supported personally by my family, especially my wife, Rayla.

I thank all these and the others who have supported me.

REFERENCES

Avery, O. T., C. M. MacLeod, and M. McCarty 1944. Studies on the chemical nature of the substance inducing transformation of pneumococcal types. Induction of transformation by a desoxyribonucleic acid fraction isolated from Pneumococcus Type III. J. Exp. Med. *79*: 137—158.

Bader, J. P. 1965. The requirement for DNA synthesis in the growth of Rous sarcoma and Rous-associated viruses. Virology 26: 253—261.

Baltimore, D. 1970. RNA-dependent DNA polymerase in virions of RNA tumor viruses. Nature *226:* 1209-1211.

Baltimore, D. 1976. Viruses, polymerases and cancer. *In:* Le Prix Nobel 1975, this volume.

Barry, R. D., D. R. Ives, and J. G. Cruickshank. 1962. Participation of deoxyribonucleic acid in the multiplication of influenza virus. Nature *194*: 1139—1140.

Bishop, J. M., and H. E. Varmus. 1975. The molecular biology of RNA tumor viruses. *In:* Cancer, Vol. 2. F. F. Becker (Ed.), pp. 3—48, Plenum Press, N.Y.

Benveniste, R. E., and G. J. Todaro. 1974. Evolution of type C viral genes: I. Nucleic acid from baboon type C virus as a measure of divergence among primate species. Proc. Nat. Acad. Sci. U.S.A., *71:* 4513—4518.

Boettiger, D. 1972. A probable early DNA intermediate in Rous sarcoma virus replication. Ph. D. Dissertation, University of Wisconsin-Madison.

Boettiger, D., and H. M. Temin. 1970. Light inactivation of focus formation by chicken embryo fibroblasts infected with avian sarcoma virus in the presence of 5-bromodeoxyuridine. Nature *228:* 622—624.

Borsa, J., and A. F. Graham. 1968. Reovirus: RNA polymerase activity in purified virions. Biochem. Biophys. Res. Commun. *33*: 895—901.

Cairns, J., G. S. Stent, and J. D. Watson (Eds). 1966. Phage and The Origins of Molecular Biology. Cold Spring Harbor Laboratory.

Cooper, G. M., and H. M. Temin. 1974. Infectious Rous sarcoma and reticuloendotheliosis virus DNAs. J. Virol. *14:* 1132—1141.

Crawford, L. V., and E. M. Crawford. 1961. The properties of Rous sarcoma virus purified by density gradient centrifugation. Virology *13:* 227—232.

Dougherty, R. M., and H. S. Di Stefano. 1966. Lack of relationship between infection with avian leukosis virus and the presence of COFAL antigen in chick embryos. Virology *29:* 586—595.

Dulak, N. C., and H. M. Temin. 1973. A partially purified polypeptide fraction from rat liver cell conditioned medium with multiplication-stimulating activity for embryo fibroblasts. J. Cell. Physiol. *81:* 153—160.

Dulbecco, R. 1966. The plaque technique and the development of quantitative animal virology. *In:* Phage and The Origins of Molecular Biology. J. Cairns, G. S. Stent, and J. D. Watson (Eds.), pp. 287—291. Cold Spring Harbor Laboratory.

Dulbecco, R. 1976. From the molecular biology of oncogenic DNA viruses to cancer. *In:* Le Prix Nobel 1975, this volume.

Enders, J. F., F. C. Robbins, and T. H. Weller. 1954. The cultivation of the poliomyelitis viruses in tissue culture. *In:* Nobel Lectures in Physiology or Medicine, 1942—1962, pp. 448—467. Elsevier Publ. Co., 1964.

Franklin, R. M., and D. Baltimore. 1962. Patterns of macromolecular synthesis in normal and virus-infected mammalian cells. Cold Spring Harbor Symp. Quant. Biol. XXVII: 175—198.

Fritsch, E., and H. M. Temin. 1976. Formation and structure of infectious DNA of spleen necrosis virus. Submitted for publication.

Hammond, E. C. 1964. Smoking in relation to mortality and morbidity. Findings in first thirty-four months of follow-up in a prospective study started in 1959. J. Nat. Cancer Inst. *32:* 1161—1188.

Hanafusa, H., T. Hanafusa, and H. Rubin. 1963. The defectiveness of Rous sarcoma virus. Proc. Nat. Acad. Sci. U.S.A. *49:* 572—580.

Hartley, J. W., and W. P. Rowe. 1966. Production of altered cell foci in tissue culture by defective Moloney sarcoma virus particles. Proc. Nat. Acad. Sci. U.S.A. *55:* 780—786.

Hill, M., and J. Hillova. 1972. Virus recovery in chicken cells tested with Rous sarcoma cell DNA. Nature New Biol. 237: 35—39.

Hirt, B. 1967. Selective extraction of polyoma DNA from infected mouse cell cultures. J. Mol. Biol. *26*: 365—369.

Humphries, E. H., and H. M. Temin. 1974. Requirement for cell division for initiation of transcription of Rous sarcoma virus RNA. J. Virol. *14*: 531—546.

Kang, C.-Y., and H. M. Temin. 1974. Reticuloendotheliosis virus nucleic acid sequences in cellular DNA. J. Virol. *14:* 1179—1188.

Kates, J. R., and B. R. McAuslan. 1967. Poxvirus DNA-dependent RNA polymerase. Proc. Nat. Acad. Sci. U.S.A. *58*: 134—141.

Kawai, S., and H. Hanafusa. 1972. Genetic recombination with avian tumor virus. Virology *49*: 37—44.

Lwoff, A. 1965. Interaction among virus, cell, and organism. *In:* Nobel Lectures in Physiology or Medicine, 1963—1970, pp. 174—185. Elsevier Publ. Co., 1972.

Manaker, R., and V. Groupé. 1956. Discrete foci of altered chicken embryo cells associated with Rous sarcoma virus in tissue culture. Virology *2*: 838—840.

Martin, G. S. 1970. Rous sarcoma virus: a function required for the maintenance of the transformed state. Nature *227:* 1021—1023.

McDonnell, J. P., A.-C. Garapin, W. E. Levinson, N. Quintrell, L. Fanshier, and J. M. Bishop. 1970. DNA polymerase of Rous sarcoma virus: delineation of two reactions with actinomycin. Nature *228*: 433—435.

Mizutani, S., H. M. Temin, M. Kodama, and R. D. Wells. 1971. DNA ligase and exonuclease activities in virions of Rous sarcoma virus. Nature New Biol. *230*: 232—235.

Mizutani, S., and H. M. Temin. 1974. Specific serological relationship among partially purified DNA poymerases of avian leukosis-sarcoma viruses, reticuloendotheliosis viruses, and avian cells. J. Virol. *13:* 1020—1029.

Mizutani, S., C.-Y. Kang, and H. M. Temin. 1974. Endogenous RNA-directed DNA polymerase activity in virions of RNA tumor viruses and in a fraction from normal chicken cells. *In:* Methods of Enzymology. Vol. XXIX. Nucleic Acids and Protein Synthesis, Part E. L. Grossman, and K. Moldave (Eds.), pp. 119—124. Academic Press, N.Y.

Mizutani, S., and H. M. Temin. 1975. Endogenous RNA synthesis is required for endogenous DNA synthesis by reticuloendotheliosis virus virions. *In:* Fundamental Aspects of Neoplasia. A. A. Gottlieb, O. J. Plescia, and D. H. L. Bishop (Eds.), pp. 235—242. Springer-Verlag. N.Y.

Mizutani, S., and H. M. Temin. 1976. An RNA polymerase activity in purified virions of avian reticuloendotheliosis viruses. J. Virol. 19: 610—619.

Munyon, W., E. Paoletti, and J. T. Grace, Jr. 1967. RNA polymerase activity in purified infectious vaccinia virus. Proc. Nat. Acad. Sci. U.S.A. *58:* 2280—2287.

Murray, R. K., and H. M. Temin. 1970. Carcinogenesis by RNA sarcoma virus. XIV. Infection of stationary cultures with murine sarcoma virus (Harvey). Int. J. Cancer *5:* 320—326.

Neiman, P. E. 1972. Rous sarcoma virus nucleotide sequences in cellular DNA: Measurement of RNA-DNA hybridization. Science *178:* 750—753.

Payne, L. N., and R. C. Chubb. 1968. Studies on the nature and genetic control of an antigen in normal chick embryos which reacts in the COFAL test. J. Gen. Virol. *3*: 379—391.

Pierson, R. W. Jr., and H. M. Temin. 1972. The partial purification from calf serum of a fraction with multiplication-stimulating activity for chicken fibroblasts in cell culture and with nonsuppressible, insulin-like activity. J. Cell. Physiol. *79:* 319—330.

Reich, E. 1975. Plasminogen secretion by neoplastic cells and macrophages. *In:* Proteases and Biological Control. E. Reich, D. B. Rifkin, and E. Shaw (Eds.), pp. 333—341. Cold Spring Harbor Laboratory.

Reich, E., R. M. Franklin, A. J. Shatkin, and E. L. Tatum. 1962. Action of actinomycin D on animal cells and viruses. Proc. Nat. Acad. Sci. U.S.A. *48*: 1238—1245.

Rous, P. 1911. A sarcoma of the fowl transmissible by an agent separable from the tumor cells. J. Exp. Med. *13:* 397—411.

Rubin, H. 1966. Quantitative tumor virology. *In:* Phage and the Origin of Molecular Biology. J. Cairns, G. S. Stent, and J. D. Watson (Eds.), pp. 292—300. Cold Spring Harbor Laboratory.

Shatkin, A. J., and J. D. Sipe. 1968. RNA polymerase activity in purified reoviruses. Proc. Nat. Acad. Sci. U.S.A. *61*: 1462—1469.

Shoyab, M., M. A. Baluda, and R. Evans. 1974. Acquisition of new DNA sequences after infection of chicken cells with avian myeloblastosis virus. J. Virol. *13:* 331—339.

Simpson, R. W., and M. Iinuma. 1975. Recovery of infectious proviral DNA from mammalian cells infected with respiratory syncytial virus. Proc. Nat. Acad. Sci. U.S.A. *72:* 3230—3234.

Svoboda, J., P. Chyle, D. Simkovic, and I. Hilbert. 1963. Demonstration of the absence of infectious Rous virus in rat tumor XC, whose structurally intact cells produce Rous sarcoma when transferred to chick. Folia Biol. (Praha) *9*: 77—81.

Temin, H. M. 1960. The control of cellular morphology in embryonic cells infected with Rous sarcoma virus *in vitro*. Virology *10*: 182—197.

Temin, H. M. 1961. Mixed infection with two types of Rous sarcoma virus. Virology *13:* 158—163.

Temin, H. M. 1962. Separation of morphological conversion and virus production in Rous sarcoma virus infection. Cold Spring Harbor Symp. Quant. Biol. *XXVII*: 407—414.

Temin, H. M. 1963. The effects of actinomycin D on growth of Rous sarcoma virus *in vitro*. Virology *20*: 577—582.

Temin, H. M. 1964a. The participation of DNA in Rous sarcoma virus production. Virology *23*: 486—494.

Temin, H. M. 1964b. Homology between RNA from Rous sarcoma virus and DNA from Rous sarcoma virus-infected cells. Proc. Nat. Acad. Sci. U.S.A. *52:* 323—329.

Temin, H. M. 1964c. Nature of the provirus of Rous sarcoma. Nat. Cancer Inst. Monograph *17*: 557—570.

Temin, H. M. 1966. Genetic and possible biochemical mechanisms in viral carcinogenesis. Cancer Res. *26*: 212—216.

Temin, H. M. 1967a. Studies on carcinogenesis by avian sarcoma viruses. V. Requirement for new DNA synthesis and for cell division. J. Cell. Physiol. *69:* 53—64.

Temin, H. M. 1967b. Studies on carcinogenesis by avian sarcoma viruses. VI. Differential multiplication of uninfected and of converted cells in response to insulin. J. Cell. Physiol. *69:* 377—384.

Temin, H. M. 1968a. Carcinogenesis by avian sarcoma viruses. X. The decreased requirement for insulin-replaceable activity in serum for cell multiplication. Int. J. Cancer *3:* 771—787.

Temin, H. M. 1968b. Carcinogenesis by avian sarcoma viruses. Cancer Res. *28:* 1835—1838.

Temin, H. M. 1970a. Formation and activation of the provirus of RNA sarcoma viruses. *In:* The Biology of Large RNA viruses. R. D. Barry and B. W. J. Mahy (Eds.), pp. 233—249. Academic Press, N.Y.

Temin, H. M. 1970b. Malignant transformation of cells by viruses. Perspectives Biol. Med. *14:* 11—26.

Temin, H. M. 1971a. The role of the DNA provirus in carcinogenesis by RNA tumor viruses. *In*: The Biology of Oncogenic Viruses. Lepetit Colloquia in Biology and

Medicine, Vol. 2. L. G. Silvestri (Ed.), pp. 176—187. North Holland Publ. Co., Amsterdam.

Temin, H. M. 1971b. Mechanism of cell transformation by RNA tumor viruses. Ann. Rev. Microbiol. *25*: 610—648.

Temin, H. M. 1971c. RNA tumor viruses and cancer. *In*: Virus Y Cancer—Homenaje a F. Duran-Reynals. V. Congress National Society Esp. Bioq. W. M. Stanley, J. Casals, J. Oro, and R. Segura (Eds.), pp. 331—357, Barcelona, Spain, Imprenta Socitra.

Temin, H. M. 1971d. Guest Editorial. The protovirus hypothesis: Speculations on the significance of RNA-directed DNA synthesis for normal development and carcinogenesis. J. Nat. Cancer Inst. *46:* III—VIII.

Temin, H. M. 1974a. The cellular and molecular biology of RNA tumor viruses, especially avian leukosis-sarcoma viruses, and their relatives. Advan. Cancer Res. *19*: 47—104.

Temin, H. M. 1974b. Introduction to virus-caused cancers. Cancer *34*: 1347—1352.

Temin, H. M. 1974c. On the origin of the genes for neoplasia: G. H. A. Clowes Memorial lecture. Cancer Res. *34:* 2835—3841.

Temin, H. M. 1974d. On the origin of RNA tumor viruses. Ann. Rev. Genetics *8:* 155—177.

Temin, H. M., and H. Rubin. 1958. Characteristics of an assay for Rous sarcoma virus and Rous sarcoma cells in tissue culture. Virology *6*: 669—688.

Temin, H. M., and H. Rubin. 1959. A kinetic study of infection of chick embryo cells *in vitro* by Rous sarcoma virus. Virology *8:* 209—222.

Temin, H. M., and S. Mizutani. 1970. RNA-dependent DNA polymerase in virions of Rous sarcoma virus. Nature *226:* 1211—1213.

Temin, H. M., and D. Baltimore. 1972. RNA-directed DNA synthesis and RNA tumor viruses. Advan. Virus. Res. *17*: 129—186.

Tooze, J. (Ed.). 1973. The Molecular Biology of Tumor Viruses. Cold Spring Harbor Laboratory.

Varmus, H. E., S. Heasley, and J. M. Bishop. 1974. Use of DNA-DNA annealing to detect new virus-specific DNA sequences in chicken embryo fibroblasts after infection by avian sarcoma virus. J. Virol. *14:* 895—903.

Vogt, P. K. 1971. Spontaneous segregation of nontransforming viruses from cloned sarcoma viruses. Virology *46*: 939—946.

Watson, J. D., and F. H. C. Crick. 1953. Genetical implications of the structure of deoxyribonucleic acid. Nature *171:* 964—967.

Zhdanov, V. M. 1975. Integration of viral genomes. Nature *256:* 471—473.

1976
Physiology or Medicine

BARUCH S. BLUMBERG and D. CARLETON GAJDUSEK

"for their discoveries concerning new mechanisms for the origin and dissemination of infectious diseases"

THE NOBEL PRIZE FOR PHYSIOLOGY OR MEDICINE

Speech by Professor ERLING NORRBY of the Karolinska Medico-Chirurgical Institute
Translation from the Swedish text

Your Majesties, Your Royal Highnesses, Ladies and Gentlemen,
An occasional encounter with infectious agents is part of our daily life. The smallest among these infectious agents are called viruses. In spite of their small size viruses may cause many different types of infections. When the virus of common cold comes into contact with our respiratory tract, certain well known symptoms appear after a few days. However, the body can defend itself against the attack by the virus. Under normal conditions we recover after a few days of illness.

Occasionally infections occur which take a completely different course. The Nobel prize winners of this year have described mechanisms involved in such infections. They have studied diseases of two different kinds.

Baruch Blumberg investigated, in the beginning of 1960, the inheritance of specific blood proteins. In connection with this, he discovered a new protein. Eventually it was shown that Blumberg, like the princes of Serendip, had found something completely different from the types of substances he was looking for. The protein he had discovered was not a part of normal body constituents but instead a virus causing jaundice.

It has been known since 1940 that there are two different forms of virus-induced jaundice. One form of the disease is transmitted as an intestinal infection whereas the other form is propagated primarily by blood transmission. The virus discovered by Blumberg caused the latter form of the disease. After exposure to this virus, disease of the liver may appear after 3–4 months. Normally symptoms of the disease wane after a few weeks. However, in certain individuals the body lacks the capacity to eliminate the virus infection, which therefore persists throughout life. Such a persistent infection occurs in about one out of 1000 persons in an industrialized society and altogether in more than 100 million individuals around the world. Individuals who carry this kind of persistent infection represent the source of further virus transmission. Due to Blumberg's discovery it is possible today to identify these individuals. Such a person, for example, should not become a blood donor. New possibilities for prevention of this type of jaundice have also become available. A vaccine is currently beeing tested.

Carleton Gajdusek studied in the end of 1950 a remarkable disease in a neolithic people living in the highlands of New Guinea. The disease, named kuru, involved a progressive destruction of the brain, which eventually resulted in death. Kuru lacks the regular signs of an infectious disease, e.g. fever and inflammation. In spite of this Gajdusek showed that the disease was caused by an infectious agent which in chimpanzees gave a disease identical with kuru in man. It took 1 1/2—3 years before the first symptoms appeared in infected animals. By this discovery it was made possible to clarify the origin of the disease kuru.

Amongst the people suffering from this infection about 3000 of 35,000 persons have died of the disease during a study period of 20 years. Transmission of the disease occurred in connection with a mourning ceremony at which dead relatives were cannibalized. This form of funeral ceremony ceased in 1959 and as a consequence kuru no longer occurs among children born after that year. However, cases of kuru still appear among adults. This implies that the infectious agent may remain in a dormant stage in the organism for many decades prior to appearance of disease.

However, the Karolinska Institute has not awarded the Nobel prize for this year to Gajdusek for his demonstration of the danger of cannibalism. The importance of his discovery of the origin of the kuru disease lies in the identification of a new class of human diseases caused by unique infectious agents. The fact that kuru, which lacks the classic signs of infections, still is caused by a contagious agent implies that we must investigate whether certain other diseases may arise in a similar way. An unusual form of presenile dementia of wide dissemination has also been shown by Gajdusek to be caused by an infectious agent.

Our normal defence mechanisms appear not to protect us against infectious agents of this kind. Furthermore, these infectious agents display a much greater resistance against destruction by e.g. boiling and irradiation than regular viruses. Thus, we are dealing with a completely different type of infectious agent the exact nature of which still remains to be demonstrated.

Baruch Blumberg and Carleton Gajdusek. You have made discoveries giving us new views on mechanisms of infectious diseases. The impact of your conceptual reformulations is wide. New directions have been given for future research. On behalf of the Karolinska Institute, I wish to convey to you our warmest congratulations, and I now ask you to receive your Nobel prizes from the hands of His Majesty the King.

Baruch S. Blumberg

BARUCH S. BLUMBERG

I was born in 1925, in New York City, the second of three children of Meyer and Ida Blumberg. My grandparents came to the United States from Europe at the end of the 19th century. They were members of an immigrant group who had enormous confidence in the possibilities of their adopted country. I received my elementary education at the Yeshiva of Flatbush, a Hebrew parochial school, and, at an early age, in addition to a rigorous secular education, learned the Hebrew Testament in the original language. We spent many hours on the rabbinic commentaries on the Bible and were immersed in the existential reasoning of the Talmud at an age when we could hardly have realized its impact.

After attending Far Rockaway High School I joined the U.S. Navy in 1943 and finished college under military auspices. I was commissioned as a Deck Officer, served on landing ships, and was the commanding officer of one of these when I left active duty in 1946. My interest in the sea remained. In later years I made several trips as a merchant seaman, held a ticket as a Ships Surgeon, and, while in medical school, occasionally served as a semi-professional hand on sailing ships. Sea experience placed a great emphasis on detailed problem solving, on extensive planning before action, and on the arrangement of alternate methods to effect an end. These techniques have application in certain kinds of research, particularly in the execution of field studies.

My undergraduate degree in Physics was taken at Union College in upstate New York, and in 1946 I began graduate work in mathematics at Columbia University. My father, who was a lawyer, suggested that I should go to medical school, and I entered The College of Physicians and Surgeons of Columbia University in 1947. I enjoyed my four years at the College immensely. Robert Loeb was the chairman of the Department of Medicine and exerted a marked influence on the entire college. There was a strong emphasis on basic science and research in the first two years (we hardly saw a patient till our third year), and we learned practical applications only in our last years.

Between my third and fourth years, Harold Brown, our professor of parasitology, arranged for me to spend several months at Moengo, an isolated mining town, accessible only by river, in the swamp and high bush country of northern Surinam. While there we delivered babies, performed clinical services, and undertook several public health surveys, including the first malaria survey done in that region. Different people had been imported into the country to serve as laborers in the sugar plantations, and they, along with the indigenous American Indians, provided a richly heterogeneous population.

Hindus from India, Javanese, Africans (including the Djukas, descendents of rebelled slaves who resided in autonomous kingdoms in the interior), Chinese, and a smattering of Jews descended from 17th century migrants to the country from Brazil, lived side by side. Their responses to the many infectious agents in the environment were very different. We were particularly impressed with the enormous variation in the response to infection with *Wuchereria bancroftia* (the filariad which causes elephantiasis), and my first published research paper was on this topic. This experience was recalled in later years when I became interested in the study of inherited variation in susceptibility to disease. Nature operates in a bold and dramatic manner in the tropics. Biological effects are profound and tragic. The manifestations of important variables may often be readily seen and measured, and the rewards to health in terms of prevention or treatment of disease can be great. As a consequence, much of our field work has been done in tropical countries.

I was an intern and assistant resident on the First (Columbia) Division at Bellevue Hospital in lower New York from 1951 to 1953. It is difficult to explain the fascination of Bellevue. In the days before widespread health insurance, many of the city's poor were hospitalized at Bellevue, including many formerly middle class people impoverished by the expenses of chronic illness. The wards were crowded, often with beds in the halls. Scenes on the wards were sometimes reminiscent of Hogarth's woodcuts of the public institutions of 18th century London. Despite this, morale was high. We took great pride that the hospital was never closed; any sick person whose illness warranted hospitalization was admitted, even though all the regular bed spaces were filled. A high scientific and academic standard was maintained. Our director, Dickinson W. Richards, and his colleague, Andre F. Cournand, received the Nobel Prize for their work on cardio-pulmonary physiology. Anyone who has been immersed in the world of a busy city hospital, a world of wretched lives, of hope destroyed by devastating illness, cannot easily forget that an objective of bio-medical research is, in the end, the prevention and cure of disease.

I spent the following two years as a Clinical Fellow in Medicine at Columbia Presbyterian Medical Center working in the Arthritis Division under Dr. Charles A. Ragan. I also did experimental work on the physical biochemistry of hyaluronic acid with Dr. Karl Meyer. From 1955 to 1957, I was a graduate student at the Department of Biochemistry at Oxford University, England, and a member of Balliol College. I did my Ph.D. thesis with Alexander G. Ogston on the physical and biochemical characteristics of hyaluronic acid. Professor Ogston's remarkable combination of theory and experiment guided the scientific activity in his laboratory. He has served as a model to me on how to train students; I hope I have measured up to his standard. Sir Hans Krebs was the chairman of the Department of Biochemistry. I have profited by conversations with him, particularly when (in 1972) I was a visiting fellow at Trinity College and we had opportunities to discuss our mutual interests in the history of science.

Oxford science at that time was influenced by the 19th and 20th century

British and European naturalists, scientists and explorers who went to the world of nature—often to distant parts of it—to make the observations which generated their hypotheses. Anthony C. Allison was then working in the Department of Biochemistry and introduced me to the concept of polymorphism, a term introduced by the lepidopterist E. B. Ford of the Department of Zoology. In 1957 I took my first West African trip (to Nigeria) and was introduced to the special excitement of that part of the world. I found the Nigerians warmhearted and friendly with a spontaneous approach to life. We collected blood specimens from several populations (including the nomadic pastoral Fulani and their domestic animals) and studied inherited polymorphisms of the serum proteins of milk and of hemoglobin. This approach was continued in many subsequent field trips, and it eventually led to the discovery of several new polymorphisms and, in due course, the hepatitis B virus.

I worked at the National Institutes of Health from 1957 until 1964. This was during a period of rapid growth for the NIH, and I continued to develop my research on polymorphisms and their relation to disease. This led to the formation of the Section on Geographic Medicine and Genetics, which was eventually assigned to an epidemiology branch directed by Thomas Dublin, from whom I learned the methods of epidemiology. The NIH was a very exciting place, with stimulating colleagues including J. Edward Rall, Jacob Robbins, J. Carl Robinson, Kenneth Warren, Seymour Geisser, and many others. The most important connection I made, however, was with W. Thomas London (who later came to The Institute for Cancer Research), who has become a colleague, collaborator, and good friend with whom I have worked closely for fifteen years. Tom was an essential contributor to the work on Australia antigen and hepatitis B, and without him it could not have been done.

I came to The Institute for Cancer Research in 1964 to start a program in clinical research. The Institute was, and is, a remarkable research organization. Our director, Timothy R. Talbot, Jr., has a deep respect for basic research and a commitment to the independence of the investigators. Above all, people are considered an end in themselves, and the misuse of staff to serve some abstract goal is not tolerated. Jack Schultz was a leading intellectual force in the Institute, and his foresighted, humane view of science, his honesty and his good sense influenced the activities of all of us. Another important characteristic is the dedication and intelligence of our administrative and maintenance staffs, which contributes to the strong sense of community which pervades our Institute.

Over the course of the next few years we built up a group of investigators from various disciplines and from many countries (Finland, France, Italy, Poland, Venezuela, England, India, Korea, China, Thailand, Singapore) who, taken together, did the work on Australia antigen. Alton I. Sutnick (now Dean of the Medical College of Pennsylvania) was responsible for much of the clinical work at Jeanes Hospital. Some of the early workers included Irving Millman, Betty Jane Gerstley, Liisa Prehn, Alberto Vierucci, Scott Mazzur, Barbara Werner, Cyril Levene, Veronica Coyne, Anna O'Connell,

Edward Lustbader, and others. There were many field trips during this period to the Philippines, India, Japan, Canada, Scandinavia, Australia, and Africa. It has been an exciting and pleasant experience surrounded by stimulating and friendly colleagues.

At present, we are conducting field work in Senegal and Mali, West Africa, in collaboration with Professor Payet of Paris, formerly the Dean of the Medical School of Dakar, with Professor Sankalé, his successor in Dakar, and a group of other French and Sengalese colleagues, including Drs. Larouzé and Saimot.

I am Professor of Medicine at the University of Pennsylvania and attend ward rounds with house staff and medical students. I am also a Professor of Anthropology and have taught Medical Anthropology for eight years. I have learned a great deal from my students.

My non-scientific interests are primarily in the out-of-doors. I have been a middle distance runner (very non-competitive) for many years and also play squash. We canoe on the many nearby lakes and rivers of Pennsylvania and New Jersey. I enjoy mountain walking and have hiked in many parts of the world on field trips. With several friends we own a farm in western Maryland which supplies beef for the local market. Shoveling manure for a day is an excellent counterbalance to intellectual work.

My wife, Jean, is an artist who has recently become interested in print making. We have four children of whom I am very proud: Anne, George, Jane, and Noah. They are all individualists, which makes for a turbulent and noisy household, still we miss the two oldest who are now away at college. We live in the center of old Philadelphia, a few blocks from Independence Hall. The city has appreciated its recognition by the Nobel Award in our Bicentennial Year.

AUSTRALIA ANTIGEN AND THE BIOLOGY OF HEPATITIS B.

Nobel Lecture, December 13, 1976

by

BARUCH S. BLUMBERG
The Institute for Cancer Research, The Fox Chase Cancer Center, Philadelphia, Pennsylvania, U.S.A.

The discovery of the infectious agent associated with hepatitis B and the elucidation of new mechanisms for its dissemination are the consequences of a series of studies involving many investigators in our laboratory in Philadelphia. The particular directions the work has followed have been a product of the interests and personalities of the investigators, physicians, technicians, students and others who have come to our laboratory. It has resulted in a complex body of data which crosses the boundaries of several disciplines. I have been fortunate in having as co-workers dedicated, and highly motivated scientists. We have had a warm, friendly and congenial atmosphere and I am grateful to my colleagues for bringing these qualities to their work.

POLYMORPHISM AND INHERITED VARIATION

E. B. Ford, the Oxford zoologist, lepidoptorist and geneticist, defined polymorphims as "the occurrence together in the same habitat of two or more (inherited) discontinuous forms of a species, in such proportions, that the rarest of them cannot be maintained merely by recurrent mutation" (1). Examples of polymorphism are the red blood cell groups in which the different phenotypes of a system may occur in high frequencies in many populations. This, according to Ford's view, would be unlikely to occur as a consequence of recurrent mutation operating alone to replace a phenotype lost by selection. Another example is the sickle cell homoglobin system. In this, Hb^{S} genes may be lost from the population each time a homozygote (who has sickle cell disease) fails to contribute to the next generation because of death before the reproductive age. The heterozygotes (Hb^{S}/Hb^{A}) are, however, thought to be differentially maintained in the population because individuals with this phenotype are less likely to succumb to falciparum malaria and consequently survive to contribute genes to the next generation. The theory implies that there are different selective values to the several forms of polymorphisms. This notion has been questioned recently since it has been difficult to demonstrate selective differences for most polymorphisms. Independent of the biological causes for the generation and maintenance of polymorphisms, the concept unifies a large number of interesting biological data. No two people are alike and polymorphisms probably account for a great deal of variation in humans. There are other interesting implications of polymorphisms. In some instances the presence of a small amount of a material may be associated with one effect and the presence of larger amounts of the same material with a very different effect. One gene for hemoglobin S protects against malaria, while two genes result in the (often) fatal sickle cell disease. Polymorphisms may produce antigenic differences. Antigenic variants of ABO and other red blood cell groups may result in transfusion reactions. Differences in Rh red blood cell groups may cause life threatening antigenic reactions between a mother and her child late in pregnancy and at the time of birth. Polymorphic antigens may have an effect when one human's tissues interacts with another's in blood transfusion, transplantation, pregnancy, intercourse, and possibly, as we shall see, when human antigens are carried by infectious agents.

Oliver Smithies (who had been a graduate student of A. G. Ogston, my mentor at Oxford) developed the ingenious starch gel electrophoresis method which allowed the seperation of serum proteins on the basis of complex characteristics of their size and shape. With this, he distinguished several electrophectically different polymorphic serum proteins (haptoglobins, transferrins, etc.). In 1957 and for several years after, in collaboration with Anthony C. Allison who was then in the Department of Biochemistry in Oxford, we studied these variants in Basque, European, Nigerian, and Alaskan populations and found striking variations in gene frequencies (2, 3). At the same time, I acquired experience and some skill in mounting field studies. Using this and similar techniques in the following years I studied inherited variants in other populations and regions. These included red blood cell and serum group in Spanish Basque, Alaskan and Canadian Indians, and Eskimos; β-amino isobutyric acid excretion in Eskimos, Indians, and Micronesians; protein and red blood cell antigens in Greece, and various variants in North and South American Indians and in U.S. blacks and whites (4—7). We identified several "new" polymorphisms in animals. With Michael Tombs, another of Ogstons' pupils, we discovered a polymorphism of alpha lactalbumin in the "Zebu" cattle of the pastoral Fulani of northern Nigeria (8). Later, Jacob Robbins and I found a polymorphism of the thyroxine binding pre-albumin of *Macaca mulatta* (9). From these studies, and those of other investigators, the richness and variety of biochemical and antigenic variation in serum became strikingly apparent.

In the summer of 1960 Allison came to my laboratory at the National Institutes of Health. We decided to test the hypothesis that patients who received large numbers of transfusions might develop antibodies against one or more of the polymorphic serum proteins (either known or unknown) which they themselves had not inherited, but which the blood donors had. We used the technique of double diffusion in agar gel (as developed by Professor Ouchterlony of Goteborg) to see whether precipitating antibodies had formed in the transfused patients which might react with constituents present in the sera of normal persons.

After testing sera from 13 multiply transfused patients (defined as a person who had received 25 units of blood or more), we found a serum which contained a precipitating antibody (10). It was a very exciting experience to see these precipitin bands and realize that our prediction had been fulfilled. The antibody developed in the blood of a patient (Mr. C. de B.), who had received many transfusions for the treatment of an obscure anemia. He was extremely cooperative and interested in our research and on several occasions came to Maryland from his home in Wisconsin for medical studies and to donate blood.

During the course of the next few months we found that the antibody in Mr. C. de B.'s blood reacted with inherited antigenic specificities on the low density lipoproteins. We termed this the Ag system; and it has subsequently been the subject of genetic, clinical and forensic studies (11).

We continued to search for other precipitating systems in the sera of transfused patients on the principle that this approach had resulted in one significant discovery and that a further search would lead to other interesting findings. During my last year at Bethseda, Dr. Harvey Alter, a hematologist, came to work with us. We also had been joined by Mr. Sam Visnich, a former Navy jet fighter and commercial airline pilot, who, during a slack period in aviation, came to work in our laboratory as a technician.

In 1963 we had been studying the sera of a group of hemophilia patients from Mt. Sinai Hospital in New York City which had been sent to us by Dr. Richard Rosenfield, the director of the blood bank. Antibodies against the Ag proteins were not common in this group of sera, but one day we saw a precipitin band which was unlike any of the Ag precipitins. It had a different configuration, it did not stain readily with Sudan black (suggesting a low lipid content compared to the Ag precipitin) but did stain red with Azocarmine, indicating that protein was a major component. There was a major difference in the distribution of the sera with which the transfused hemophilia patient reacted. Most of the anti-Ag antisera reacted with a large number (usually about 50—90%) of the panel sera, but the serum from the hemophilia patient reacted with only one of 24 sera in the panel, and that specimen was from an Australian aborigine (12, 13). We referred to the reactant as Australia antigen, abbreviated Au. The original Australian sera had been sent to us by Dr. Robert Kirk. We subsequently went to Western Australia to collect and test a large number of additional sera.

We now set out to find out why a precipitin band had developed between the sera of a hemophilia patient from New York and an aborigine from Australia. At the outset we had no set views on where this path might lead, although our investigation was guided by our prior experience with the Ag polymorphism. In preparing this "history" of the discovery of Au, I constructed an outline based on a hypothetico-deductive structure, showing the actual events which led to the discovery of the association of Au with hepatitis. From this it is clear that I could not have planned the investigation at its beginning to find the cause of hepatitis B. This experience does not encourage an approach to basic research which is based exclusively on specific-goal-directed programs for the solution of biological problems.

The next step was to collect information on the distribution of Au and anti-Au in different human populations and disease groups. We had established a collection of serum and plasma samples, later to develop into the Blood Collection of the Division of Clinical Research of The Institute for Cancer

Research which now numbers more than 200,000 specimens. The antigen was very stable; blood which had been frozen and stored for 10 years or more still gave strong reactions for Au. There were some instances in which blood had been collected from the same individual for 6 or more successive years. If the sera were positive on one occasion, they were in general positive on subsequent testings; if negative initially, they were consistently negative. Presence or absence of Au appeared, at least in the early experiments, to be an inherent characteristic of an individual.

We were able to use our stored sera for epidemiological surveys and in a short time accumulated a considerable amount of information on the world-wide distribution of Au. It was very rare in apparently normal populations from the United States; only 1 of 1000 sera tested was positive. However, it was quite common in some tropical and Asian populations (for example, 6% in Filipinos from Cebu, 1% in Japanese, 5—15 % in certain Pacific Ocean populations). I will come back to a consideration of the hypothesis which was generated from this set of epidemiologic observations after consideration of an interesting disease association discovered at about the same time.

Sam Visnich had been asked to select from our collection the sera of patients who had received transfusions, in order to search for more anti-Au antisera. He decided, however, to use them both as potential sources of antibody and also in the panels against which anti-Au sera were tested. Included among the transfused sera were specimens from patients with leukemia who had received transfusions. A high frequency of Au, rather than anti-Au, was found in this group. We subsequently tested patients with other diseases and found Au only in transfused patients.

On the basis of these observations we made several hypotheses. Although they sound like alternative ones, they in fact are not, and over the course of subsequent years, in a sense, all of them have been supported and are still being tested.

One hypothesis stated that, although Au may be rare in normal populations, individuals who have Au are more likely to develop leukemia than are individuals who do not have the antigen. That is, there is a common susceptibility factor which makes it more likely for certain people both to have Au and to develop leukemia. We also suggested that Au might be related to the infectious agent (virus) which is said to be the cause of leukemia.

A corollary of the susceptibility hypothesis is that individuals who have a high likelihood of developing leukemia would be more likely to have Au. Young Down's syndrome (mongolism) patients are more likely to develop leukemia than are other children; estimations of the increased risk vary from 20 to 2000 times that of children without Down's syndrome. I had in 1964 moved to The Institute for Cancer Research in Philadelphia to start its Division of Clinical Research. While there we tested the sera of Down's syndrome patients resident in a large institution and found that Au was very common in this group (~ 30 % were Au positive); the prediction generated by our hypothesis was fulfilled by these observations, a very encouraging finding (14). The presence of the antigen in people living closer to Philadelphia also made it possible to more readily study persons with Au. Until this time all the individuals with Au who had been identified either lived in Australia, or some other distant place, or were sick with leukemia.

Down's syndrome patients were admitted to the Clinical Research Unit (located in our sister institution, Jeanes Hospital) for clinical study. We found again that the presence or absence of Au seemed to be a consistent feature of an individual. If Au was present on initial testing, then it was present on subsequent testing; if absent initially, it was not found later. In early 1966 one of our Down's syndrome patients, James Bair, who had originally been negative, was found to have Au on a second test. Since this was an aberrant finding we admitted James to the Unit. There was no obvious change in his clinical status. Because he apparently had developed a "new" protein, and since many proteins are produced in the liver we did a series of "liver chemistry" tests. These showed that between the first testing (negative for Au) and the subsequent testing (positive for Au) James had developed a form of chronic anicteric hepatitis.

On 6/28/66, the day of his admission to the Clinical Research Unit, my colleague, Alton I. Sutnick, wrote the following dramatic note in the patient's chart.

"SGOT slightly elevated! Prothrombin time low! We may have an indication of [the reason for] his conversion to Au+."

His prediction proved correct. The diagnosis of hepatitis was clinically confirmed by liver biopsy on 7/20/66, and we now began to test the hypothesis that Au was associated with hepatitis (15). First, we compared the transaminase (SGPT) levels in males with Down's syndrome who had Au and those who did not. The SGPT levels were slightly but significantly higher in the Au(+) individuals. Secondly, we asked clinicians in Pennsylvania to send us blood samples from patients with acute hepatitis. W. Thomas London, and others in our laboratory soon found that many hepatitis patients had Au in their blood early in their disease, but the antigen usually disappeared from their blood, after a few days or weeks. Another dramatic incident occurred which added to our urgency in determining the nature of the relation of Au to hepatitis. Miss (now Dr.) Barbara Werner was the first technician in our laboratory in Philadelphia. She had been working on the isolation of Au by extensions of the methods developed by Alter and Blumberg during the earlier work in Bethesda. Early in April of 1967 she noticed that she was not in her usual state of good health. She was well aware of our observations that Au was related to hepatitis and, one evening, tested her own serum for the presence of Au. The following morning a faint but distinct line appeared, the first case of viral hepatitis diagnosed by the Au test. She subsequently developed icteric hepatitis and, fortunately, went on to a complete recovery.

By the end of 1966 we had found that Au was associated with acute viral hepatitis. In our published report (14) we said:

"Most of the disease associations could be explained by the association of Au(1) with a virus, as suggested in our previous publications. The discovery of the frequent occurrence of Au(1) in patients with virus hepatitis raises the possibility that the agent present in some cases of this disease may be Australia antigen or be responsible for its presence. The presence of Australia antigen in the thalassemia and hemophilia patients could be due to virus introduced by transfusions."

That is, we made the hypothesis that Au was (or was closely related to) the etiologic agent of "viral" hepatitis and we immediately set about to test it. Our original publication did not elicit wide acceptance; there had been many previous reports of the identification of the causative agent of hepatitis and our claims were naturally greeted with caution. Indeed, an additional paper on Australia antigen and acute viral hepatitis (15) which extended our findings published in 1967 was initially rejected for publication on the grounds that we were proposing another "candidate virus" and there were already many of these.

Confirmation of our findings and the first definitive evidence on the relation of Au to post-transfusion hepatitis came soon. Dr. (now Professor) Kazuo Okochi, then at the University of Tokyo, had followed a line of inquiry very similar to ours. He had started with the investigation of anti-Ag (lipoprotein) antisera, and we had corresponded on this subject. Professor Okochi then found an antiserum in a patient with chronic myelogenous leukemia which was different from the anti-Ag precipitins. He also found that it was associated with liver damage. During my several field trips to Japan I had lectured on Australia antigen. Professor Okochi sent the unusual antiserum to us to compare with anti-Australia antigen; we found that they were identical. He confirmed our finding of the association of Au with hepatitis and then proceeded to do the first definitive transfusion study. He found that Au could be transmitted by transfusion and that it led to the development of hepatitis in some of the people who received it, and that some transfused patients developed anti-Au (16, 17). The Au-hepatitis association was also confirmed in 1968 by Dr. Alberto Vierucci (18) who had worked in our laboratory and Dr. Alfred M. Prince (19).

We had made some preliminary observations in Philadelphia in collaboration with Dr. John Senior of the University of Pennsylvania on the transfusion of donor blood which was found to contain Au. We then developed a protocol for a controlled, long term study to determine whether donor bloods which had Au were more likely to transmit hepatitis than those which did not. In 1969 we heard from Prof. Okochi that he had already embarked on a similar transfusion studies. In June of that year he visited our laboratory in Philadelphia and showed us his data. These, in his (and our) opinion demonstrated with a high probability, that donor blood containing Australia antigen was much more likely to transmit hepatitis than donor blood which did not contain the antigen. (Similar studies were later done by Dr. David Gocke (20) in the United States and the same conclusions were reached.) We immediately stopped the experimental study and established the practice of excluding donor bloods with Australia antigens in the hospitals where we were testing donor units. This was a dramatic example of how technical information may completely change an ethical problem. Before Okochi's data had become available it was a moral necessity to determine the consequences of tranfusing blood containing Australia antigen; and it had to be done in a controlled and convincing manner since major changes in blood transfusion practice were consequent on the findings. As soon as the conclusion of Okochi's well controlled studies were known to us, it became untenable to administer donor blood containing Australia antigen. *Autre temps, autre moeurs.*

It was, however, possible to do a study to evaluate the efficacy of Au screening on post-transfusion hepatitis using historical controls, which appears to be valid. Senior and his colleagues had completed an analysis of post-transfusion hepatitis in Philadelphia General Hospital before the advent of screening and found an 18 % frequency of post-transfusion hepatitis. In the fall of 1969 we started testing all donor blood and excluding Au positive donors. Senior and others undertook a similar study one year after the screening program was in progress. They found that the frequency of post-transfusion hepatitis had been reduced to 6 %, a striking improvement (21).

The practical application of an initially esoteric finding had come about only two years after the publication of our paper on the association between Au and hepatitis (14). In retrospect, one of the major factors contributing to the rapid application of the findings was the simplicity of the immunodiffusion test. Another was our program of distributing reagents containing antigen and antibody to all investigators who requested them. We did this until this function was assumed by the National Institutes of Health.

Following the confirmation of the association of hepatitis with Australia antigen, a large number of studies were published, and, in a relatively short time, the routine use of the test in blood banks became essentially universal in the United States and many other countries. It has been estimated that the annual saving resulting from the prevention of post-transfusion hepatitis amounts to about half a billion dollars in the United States.

VIROLOGY

Virological methods (i.e. tissue culture, animal inoculation, etc.) had been used for many years prior to our work to search for hepatitis virus but had not been very productive. Our initial discoveries were based primarily on epidemiologic, clinical, and serological observations. Here, I will try to review the early virology work from our laboratory (Robinson has reviewed much of the recent work (22)).

Bayer *et al.* (23), using the isolation techniques initially introduced by Alter and Blumberg (24), examined isolated Au with the electron microscope. They found particles about 20 nm in diameter which were aggregated by anti-Au antiserum. There were also sausage like particles of the same diameter, but much elongated (Figs. 1—2). Subsequently Dane, Cameron and Briggs identified a larger particle about 42 nm in diameter with an electron dense core of about 27 nm (25). It is probable that this represents the whole virus particle. Both the 20 nm and 42 nm particles contain Australia antigen on their surfaces and this is now termed hepatitis B surface antigen (HBsAg). The surface antigen can be removed from Dane particles by the action of detergents to reveal the core which has its own antigen, hepatitis B core

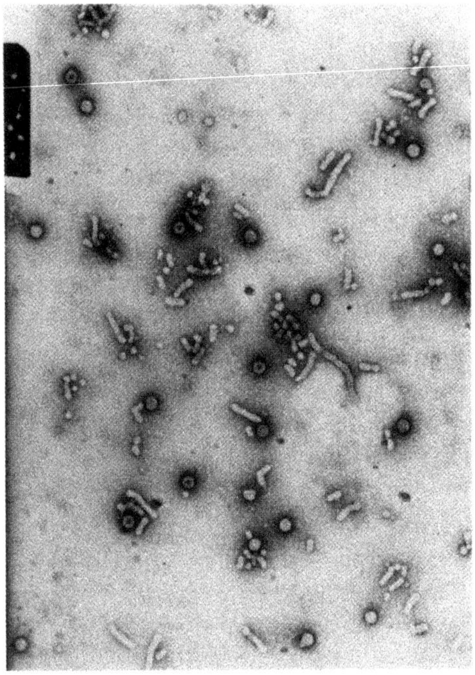

Fig. 1. Electron micrograph showing the several kinds of particles associated with hepatitis B virus (see Figure 2). Magnification = 90,000X. Electron micrograph prepared by E. Halpern and L. K. Weng.

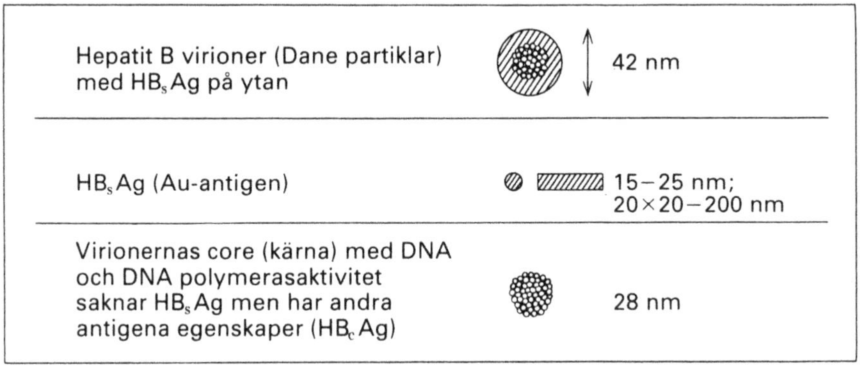

Fig. 2. Diagram showing appearance of particles associated with hepatitis B virus, the large or Dane particle (top), small surface antigen particle and the sausage shaped particle (middle), and the core of the Dane particle (bottom). (Adapted from E. Lycke, Läkartidningen *73*, 1976.)

antigen (HBcAg). Antibodies to both these antigens (anti-HBs, anti-HBc) can be detected in human blood. The surface antigen can be detected in the peripheral blood by the methods we initially introduced and by more sensitive methods which have since been developed. Anti-Hbs is often found in the peripheral blood after infection and may persist for many years. It may also be detected in people who have not had clinical hepatitis. Anti-HBc is usually associated with the carrier state (i.e. persistent HBsAg in the blood), but may occur without it. HBcAg itself has not been identified in the peripheral blood. Anti-HBc is also found commonly during the active phase of acute hepatitis, before the development of anti-HBs but in general does not persist as long as anti-Hbs.

DNA has been isolated from the cores of Dane particles and is associated with a specific DNA polymerase. Robinson has shown that the DNA is in the form of double stranded rings (22). Jesse Summers, Anna O'Connell and Irving Millman of our Institute have confirmed these findings and provided a model for the molecule which appears to have double and single stranded regions (Fig. 3) (26).

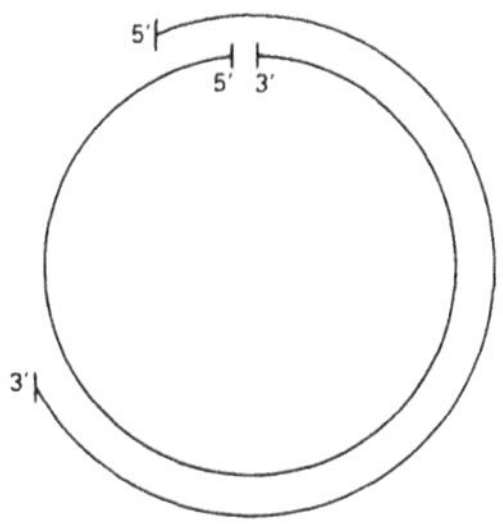

Fig. 3. Structure of the DNA extracted from Dane particles proposed by Summers et al. (26). The position of the gaps in the single strands, and the location of the 5′ and 3′ ends are shown.

Using immunofluorescent and electron microscope studies, hepatitis B core particles have been identified in the nuclei of liver cells of infected patients. HBsAg is found in the cytoplasm. It is thought that assembly of the large particles occurs in the cytoplasm and that large and small particles (surface antigen only) emerge from the cells and eventually find their way to the peripheral blood.

VACCINE AGAINST HEPATITIS B

In 1968 we were informed by the Federal government, who provided most of the funds for our work, that they would like to see applications of the basic research they had funded for many years. It occurred to us that the existence of the carrier state provided an unusual method for the production of a vaccine. We presumed that the very large amounts of HBsAg present in the blood could be separated from any infectious particles and used as an antigen for eliciting the production of antibodies. The antibodies in turn would protect against infection with the virus. Irving Millman and I applied separation techniques for isolating and purifying the surface antigen and proposed using this material as a vaccine. To our knowledge, this was a unique approach to the production of a vaccine; that is, obtaining the immunizing antigen directly from the blood of human carriers of the virus. In October, 1969, acting on behalf of the ICR we filed an application for a patent for the production of a vaccine. This patent was subsequently (January, 1972) granted in the United States and other countries (27).

There are observations in nature which indicate that antibody against the surface antigen is protective. In their early studies, Okochi and Murakami observed that transfused patients with antibody were much less likely to develop hepatitis than those without it (17). In a long term study, London (36) has shown that patients on a renal dialysis unit, and the staff who served them, were much less likely to develop hepatitis if they had anitbody then if they did not (Fig. 4). Edward Lustbader has used this data to develop a statistical method for rapidly evaluating the vaccine (28).

There have now been several animal and human studies of the vaccine and the results are promising (29—32, 54, 55). It should be possible to determine the value of the vaccine within the next few years.

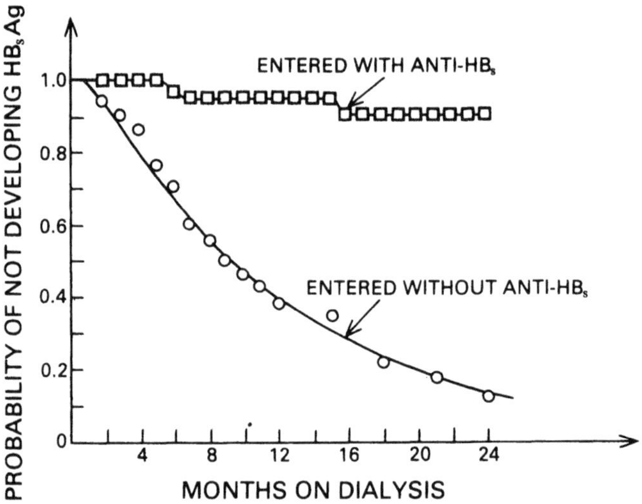

Fig. 4. Probability of not developing HBsAg for patients admitted to a renal dialysis unit with and without anti-HBs. The patients with anti-HBs are relatively well protected while those without antibody are very likely to develop infection (adapted from Lustbader et al. (28)).

VARIATION IN RESPONSE TO INFECTION WITH HEPATITIS B

A physician is primarily interested in how a virus interacts with humans to cause disease. But this is only part of the world of the virus. Our introduction to studies on hepatitis B was not through patients with the disease, but rather through asymptomatic carriers and infected individuals who developed antibody. Therefore, many of our investigations have been of infected but apparently healthy people. There are a variety of responses to infection:

1) Development of acute hepatitis proceeding to complete recovery. Transient appearance of HBsAg and anti-HBc. Subsequent appearance of anti-HBs which may be persistent.

2) Development of acute hepatitis proceeding to chronic hepatitis. HBsAg and associated anti-HBc are usually persistent.

3) Chronic hepatitis with symptoms and findings of chronic liver disease not preceeded by an episode of acute hepatitis. HBsAg and anti-HBc are persistent.

4) Carrier state. Persistent HBsAg and anti-HBc. Carrier is asymptomatic but may have slight biochemical abnormalities of the liver.

5) Development of persistent anti-HBs without detectable HBsAg or symptoms.

6) Persistent HBsAg in patients with an underlying disease often associated with immune abnormalities, i.e. Down's syndrome, lepromatous leprosy,

chronic renal disease, leukemia, primary hepatic carcinoma. Usually associated with anicteric hepatitis.

7) Formation of complexes of antigen and antibody. These may be associated with certain "immune" diseases such as periarteritis nodosa.

FAMILY STUDIES

In our first major paper on Australia antigen (13) we described family clustering of Au in a Samaritan family from Israel which had been studied by the anthropologist Batsheva Bonné. From it we inferred the hypothesis that the persistent presence of Au was inherited as a simple autosomal recessive trait. The genetic hypothesis has proven to be very useful not in the sense that it is necessarily "true" (exceptions to the simple hypothesis were noted by us and others very soon (33)), but because it has generated many interesting studies on the family distributions of responses to infection with hepatitis B. We suggested that hepatitis virus may have several modes of transmission. It can be transmitted horizontally from person to person similar to "conventional" infectious agents. This is seen in the transmission of hepatitis B virus (HBV) by transfusion. Other forms of direct and indirect horizontal transmission exist; for example, sputum, the fecal-oral route, and perhaps, by hematophagous insects (see below). It has even been reported to have been spread by computer cards (34), an extraordinary example of adaptation by this ingenious agent! HBV may also be transmitted vertically. If the genetic hypothesis were sustained, then it would imply that the capacity to become persistently infected is controlled (at least in part) as a Mendelian trait. The data are also consistent with the notion that the agent could be transmitted with the genetic material; that the virus could enter the nucleus of its host and in subsequent generations act as a Mendelian trait. The data also suggest a maternal effect. A re-analysis of our family data showed that in many populations more of the offspring were persistent carriers when the mother was a carrier than when the father was a carrier. Many investigators have now shown that women who have acute type B hepatitis just before and/or during delivery or women who are carriers can transmit HBV to their offspring, who then also become carriers. This may be a major method for the development of carriers in some regions, for example, Japan. Interestingly, this mechanism does not appear to operate in all populations. This suggests that some aspects of delivery and parent child interaction, differing in different cultures, as well as biological characteristics may affect transmission.

The family is an essential human social unit. It is also of major importance in the dissemination of disease. A large part of our current work is directed to an understanding of how the social and genetic relations within a family affect the spread of hepatitis virus.

HOST RESPONSES TO HUMAN ANTIGENS AND HEPATITIS B VIRUS. KIDNEY TRANSPLANTATIONS

W. Thomas London, Jean Drew, Edward Lustbader and others in our laboratory have undertaken an extensive study of the patients in a large renal dialysis unit in Philadelphia (36, 37). The renal patients can be characterized on the basis of their response to infection with hepatitis B. Patients who develop antibody to HBsAg are significantly more likely to reject transplanted kidneys that are not completely matched for HLA antigens than patients who become carriers of HBsAg (Fig. 5) (37). Since many of the patients became exposed to hepatitis B while on renal dialysis, their response to infection can be determined prior to transplantation. In this patient population there is a correlation between development of anti-HBs and the subsequent development of anti-HLA antibodies after transplantation. We have also found a correlation between the development of anti-HLA and anti-HBs in transfused hemophilia patients and in pregnant women. Hence, there appears to be a correlation between the response to infection with HBV and the immunologic response to polymorphic human antigens in tissue transplants. Further, from preliminary studies it appears that donor kidneys from males are much more likely to be rejected by patients with anti-HBs than by patients without anti-HBs. These differences were not observed when the kidneys were from female donors. Dr. London is now extending his observations to other transplants, in particular, bone marrow, to determine whether a similar relation exists.

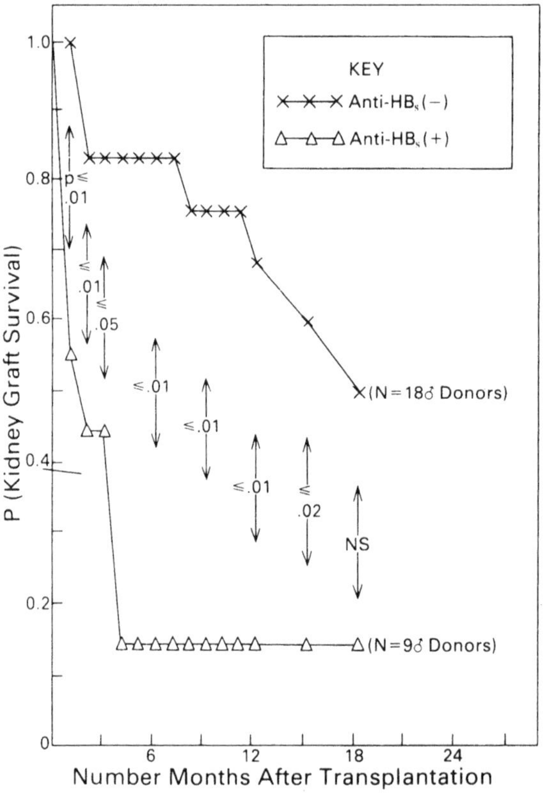

Fig. 5a. Probability of rejecting a kidney graft by renal dialysis patients who received kidneys from male donors. There is a significant difference in rejection rate between patients who were carriers and those who developed anti-HBs (37).

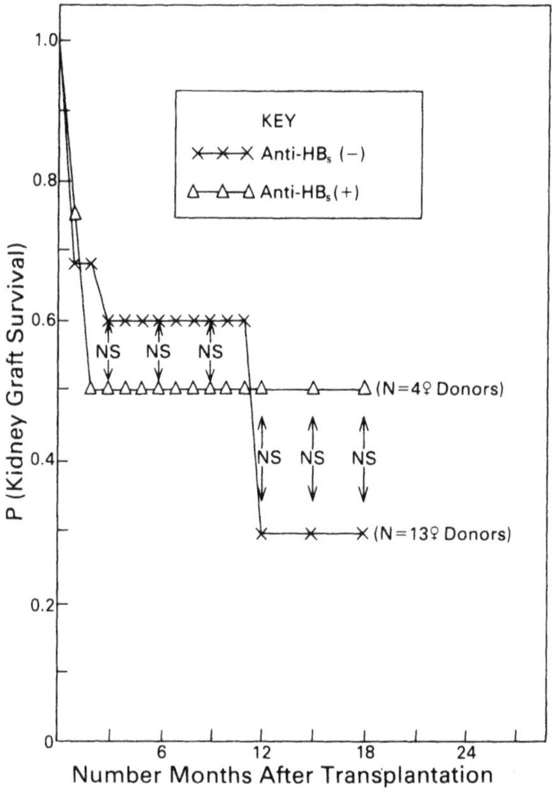

Fig. 5b. Probability of rejecting a kidney graft by renal dialysis patients who received kidneys from female donors. There is no difference in the rejection rates between the two groups of patients (37).

SEX OF OFFSPRING AND FERTILITY OF INFECTED PARENTS

In many areas of the world, including many tropical regions (i.e. the Mediterranean, Africa, southeast Asia, and Oceania) the frequency of HBsAg carriers is very high. In these regions most of the inhabitants will eventually become infected with HBV and respond in one of the several ways already described. Our family studies and the mother-child studies show that there is a maternal effect. Jana Hesser (then a graduate student in Anthropology working in our laboratory) and Ioanna Economidou, Stephanos Hadziyannis, and our other Greek colleagues collected information on the sex of the offspring of parents in a Greek town in southern Macedonia. In this community the probability of infection with HBV is very high and a majority of the parents had evidence of infection, i.e. detectable HBsAg and/or anti-HBs in their blood. It was found that if either parent was a carrier of HBsAg there were significantly more male offspring than in other matings (38). In subsequent studies using the Greek data and additional data from Mali in West Africa, London, Drew, and Veronique Barrois (a post-doctoral trainee from Paris) have found that there is a deficiency of male offspring when parents have anti-HBs (39). This had led London and his colleagues to test the hypothesis that anti-HBs has specificities in common with H-Y or other antigens determined by genes on the Y chromosome. If these observations are supported by additional studies, then HBV may have a significant effect on the composition of populations in places where it is common, which includes the most populous regions of the world. The ratio of males and females in a population has a profound effect on population size as well as on the sociology of the population. This connection of HBV with sex selection may also explain why there is a greater likelihood of rejection of male kidneys by renal patients with anti-HBs, and indicate how kidneys can be better selected for transplantation. Transplantation of organs and pregnancy have certain immunologic features in common. Rejection of male kidneys and "rejection" of the male fetus may be mediated by similar biological effects.

PRIMARY HEPATIC CARCINOMA

The project with which we are most concerned at present is the relation of hepatitis B to primary hepatic carcinoma (PHC), and methods for the prevention of the disease. PHC is the most common cancer in men in many parts of Africa and Asia. For many years investigators in Africa including Payet (40), Davies (41), and Steiner (42) have suggested that hepatitis could be the cause of PHC. With the availability of sensitive tests for Australia antigen it became possible to test this hypothesis; it has now been established that there is a striking association of hepatitis B with PHC (43, 45) (Table 1). In our studies in Senegal and Mali, Bernard Larouzé and others found that essentially all the patients had been infected with HBV and that most had

Table 1. Frequency of HBsAg, anti-HBc and anti-HBs in primary hepatic carcinoma (PHC) and controls in Senegal and Mali, West Africa. RIA = radioimmunoassay; "p" is the two tailed probability obtained from Fisher's Exact Test (adopted from Larouzé et al. (45).

		PATIENTS				CONTROLS				
	Test	No. Tested	+	−	% +	No. Tested	+	−	% +	p
Senegal	HBsAg									
PHC	RIA	39	31	8	79.4	53	6	47	11.3	4×10^{-11}
	Anti-HBc	39	35	4	89.7	58	16	42	27.6	1×10^{-9}
	Anti-HBs	39	8	31	20.5	58	26	32	44.8	0.02
	Total Exposed	39	37	2	94.8	58	38	20	65.1	8×10^{-4}
Mali	HBsAg									
PHC	RIA	21	10	11	47.6	38	2	36	5.2	4×10^{-4}
	Anti-HBc	20	15	5	75.0	40	10	30	25.0	5×10^{-4}
	Anti-HBs	21	8	13	38.0	40	17	23	42.5	0.95
	Total Exposed	21	19	2	90.4	40	25	15	62.0	0.02

evidence of current infection (presence of HBsAg and/or anti-HBc). Ohbayashi and his colleagues (44) had reported several families of patients with PHC in which the mothers were carriers. In our study in Senegal (56) we found that a significantly larger number of mothers of PHC patients were carriers of HBsAg compared with controls, and that none of the fathers of the cases had anti-HBs. In control families, on the other hand, 48 % of the fathers developed antibody (Table 2). The hypothesis we have made is that, in some families, children will be infected by their mothers, *in utero*, at the time of birth, and/or shortly afterwards during the period when there is intimate contact between mother and children. In some cases, the infected child will proceed through several stages to the development of PHC. At each stage, only a fraction of the infected individuals will proceed to the next stage, and this will depend on other factors in the host and in the environment. The stages include, retention of the antigen (carrier state), development of chronic hepatitis, development of cirrhosis and finally,

Table 2. Frequency of HBsAg, anti-HBc and anti-HBs in patients with primary hepatic carcinoma (PHC) and controls, and in the parents of patients and controls. The studies were conducted in Dakar, Senegal, West Africa. I.D. = HBsAg by immunodiffusion; RIA = HBsAg by radioimmunoassay. (Adopted from Larouzé et al. (56).)

	Primary Hepatic Carcinoma (PHC)			Controls			p
	N	+	% +	N	+	% +	
$HB_sAg(+)$ ID	28	9	32.1 %	28	5	17.9 %	0.35
$HB_sAg(+)$ RIA	28	22	78.6 %	28	16	57.1 %	0.15
anti-$HB_c(+)$	28	25	89.2 %	28	18	64.3 %	0.05
anti-$HB_s(+)$	28	7	25.0 %	28	18	64.3 %	6×10^{-3}
$HB_sAg(+)$, anti-$HB_c(+)$ and/or anti-$HB_s(+)$*	28	27	96.4 %	28	26	92.9 %	0.99
	Mothers of PHC			Mothers of Controls			
$HB_sAg(+)$ ID	28	15	53.6 %	28	3	10.7 %	1×10^{-3}
$HB_sAg(+)$ RIA	28	20	71.4 %	28	4	14.3 %	3×10^{-5}
anti-$HB_c(+)$	28	20	71.4 %	28	9	32.1 %	6.9×10^{-3}
anti-$HB_s(+)$	28	3	10.8 %	28	15	53.6 %	1×10^{-3}
$HB_sAg(+)$, anti-$HB_c(+)$ and/or anti-$HB_s(+)$*	28	21	75.0 %	28	19	67.9 %	0.76
	Fathers of PHC			Fathers of Controls			
$HB_sAg(+)$ ID	27	2	7.4 %	27	3	11.1 %	0.99
$HB_sAg(+)$ RIA	27	5	18.5 %	27	5	18.5 %	1.00
anti-$HB_c(+)$	27	5	18.5 %	27	8	29.6%	0.52
anti-$HB_s(+)$	27	0	0	27	13	48.1 %	3×10^{-5}
$HB_sAg(+)$, anti-$HB_c(+)$ and/or anti-$HB_s(+)$*	27	5	18.5 %	27	18	66.6 %	7×10^{-6}

* Add evidence of infection with HBV.

development of PHC (Fig. 6). We are currently testing this hypothesis in prospective studies in West Africa (45). If it is true, then prevention of PHC could be achieved by preventing infection with HBV, and the vaccine we have introduced, in association with appropriate public health measures, could reduce the amount of infection. This might also involve the use of gamma-globulin in the newborn children of carrier mothers and such studies are now being conducted by Beasley and his colleagues in Taipei. We are now considering the appropriate strategies that might be used to control hepatitis infection and, hopefully, cancer of the liver.

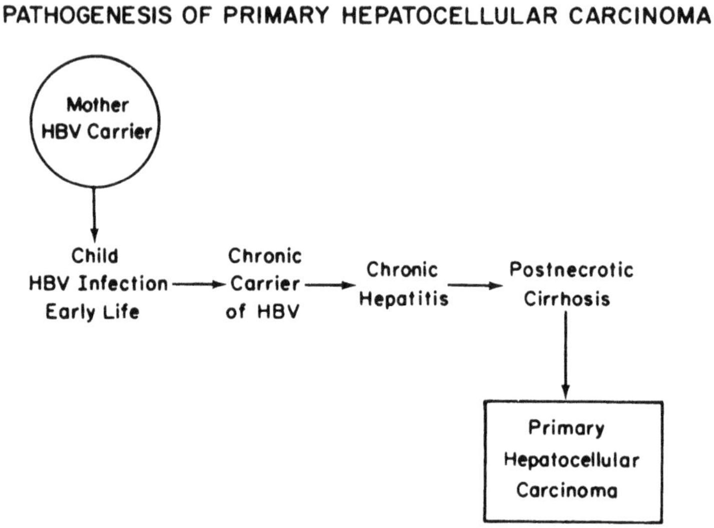

Fig. 6. Scheme for the pathogenesis of primary hepatic carcinoma showing the sequence of stages leading to PHC.

TRANSMISSION BY INSECTS

HBsAg has been detected by several investigators (including Prince, Smith, Muniz and others (46, 47, 48)) in mosquitos collected in the field in areas where HBsAg is common in the human population. In 1971 we collected mosquitoes in Uganda and Ethiopia and found Au antigen in individual mosquitoes (57). In more extensive studies in Senegal we found a field infection rate of about 1 in 100 for *Anopheles gambiae* and also identified the antigen in several other species of mosquitoes (58). It is not known if HBV replicates in mosquitoes, but it has been reported that it can be detected in mosquitoes many weeks after feeding and it has been found in a mosquito egg. Feeding experiments have been conducted with the North American bedbug (*Cimex lectularius*) which show that this insect can also carry the antigen (49). Wills, in our laboratory, has found a very high infection rate ($\sim 60\%$) in the tropical bedbug *Cimex hemipterus* collected from the beds of individuals known to be carriers of hepatitis B (61). Bedbugs could transfer blood (and the virus) from one occupant of a bed to another. If it is in fact a vector of hepatitis, then it could provide a frequent non-venereal (and unromantic) form of connubial spread. It may also provide a means for transmission from mother (or father) to young children who may share the parents' bed in early life; and this would be related to the child rearing practices of a community.

Insect transmission may be important in the program for the control of hepatitis B infection and for the prevention of chronic liver disease and primary hepatic carcinoma. An understanding of the role of insects in the spread of infection, particularly its transmission from mother to children, would help in designing effective strategies for control.

HEPATITIS B AS A POLYMORPHISM

The original discovery of hepatitis B resulted from the study of serum antigen polymorphisms. Its identification as an infectious agent does not diminish the value of this concept. It is useful to view infection with HBV not only as a "conventional" infection but also as a transfusion or transplantation reaction; and our studies on renal transplantation are an example of this.

HBV appears to have only a small amount of nucleic acid, probably only sufficient to code for a few proteins. Much of the coat (and possibly other portions of the virus) could be produced by the genes of the host. Millman and his colleagues found that the surface antigen contains material with antigenic specificities in common with serum proteins including IgG, transferrin, albumin, beta-lipoprotein and others (50). If this is true then the antigenic makeup of the virus would be, at least in part, a consequence of the antigenic characteristics of the host from whence it came; and this, as suggested by our sex studies, may include male antigens. In our discussion of the "Icron" concept (a name we introduced which is an acronym on the Institute for Cancer Research) (51), we pointed out that the responses of the putative host to HBV may be dictated in part by the nature of the "match" between the antigens of the host and virus (i.e., the virus acts as if it were a polymorphic human antigen). London and Werner have described this in a review of these concepts (52, 60). If person A is infected with HBV particles which contain proteins that are antigenically very similar to his own, then he will have little immunologic response and will tend to develop a persistent infection with the virus. On the other hand, if the proteins of the agent are antigenically different from his, he will develop an immune response to the virus (i.e. anti-HBs) and will have a transient infection. During the course of infection in person A, new particles will be synthesized which contain antigenic characteristics of A. Person A, in turn can infect person B and the same alternatives present themselves. If the relevant proteins of B are antigenically similar to the antigens of A and the antigens of the HBV produced by A, then B could develop a persistent infection. If they are different, then antibody can form, as described above. (A derivative of this hypothesis is that inflammatory disease of the liver is associated with the immune response to the infectious agent rather than solely with replication of the agent.) A further possibility is that the virus has complex antigens; that some may match the host and some may not, and that both persistent infection and development of antibodies may occur. The persistent antigens and the antibody in the same individual would have different specificities, and this occurrence has been described (59).

This view of the agent as an Icron introduces an interesting element into the epidemiology of infectious agents in which not only the host and virus are factors, but also the previous host or hosts of the agent. If, in fact, the agent does replicate in insects (see above), the antigenic characteristics of previous human hosts may be affected by transmission through another species. This in turn might have an effect on the response of the next host.

BIOETHICS AND THE CARRIER STATE

During the course of our work a number of bioethical questions have arisen (53). Experience has shown that these bioethical considerations cannot be separated from "science," that answers cannot be provided on a "purely scientific" basis, and that our technical knowledge is inseparably intertwined with bioethical concerns.

It has been recognized that hepatitis B may be transmitted by means other than transfusion, i.e. by contact, fecal-oral spread, insects, etc. With the introduction of the screening test many carriers were identified. It is estimated that there are one million such carriers in the United States and more than 100 million in the world. This has led to a situation which may be unique in medicine. Although some carriers may be able to transmit hepatitis by means other than blood transfusion, this is probably not true for many (or most) carriers. There are studies which show that spread of infection from carriers in health care occupations to patients may not be common. At present there is no satisfactory method of identifying the infectious carriers although it appears that carriers with "e" antigen, an unusual antigen originally described by Magnius (62), are much more likely to transmit disease. Despite this, carriers have had professional and social difficulties. Health care personnel who are carriers have been told that they must leave their jobs. In some cases carriers have changed their pattern of social beavior because of the fear they might spread disease to people with whom they come in contact. What appeared to be happening was the development of a class of individuals stigmatized by the knowledge that some member of the "class" could transmit hepatitis.

The bioethical problems that are raised from the studies of hepatitis carriers can be viewed as a conflict between public health interests and individual liberty. When the risk to the public is clear, and the restrictions on personal liberties are small, there is little problem in arriving at appropriate regulations. For example, the transfusion of blood containing hepatitis B antigen is a disadvantage to the patient recipient and it has been stopped. The denial of the right to donate blood is not a great infringement of personal activity and the individuals concerned and society have agreed to accept this moderate restriction. The problems raised by person-to-person transmission are more difficult. The extent of the hazard to the public is not clear, since it is not (now) possible to distinguish carriers who transmit disease from those who do not. On the other hand, if all carriers are treated as infectious, the hazards imposed on the carrier may be enormous, i.e. loss of job and ability to continue in the same profession, restriction of social and family contacts, etc. What is clear is that for a very large number of carriers, the risk of transmitting hepatitis by person-to-person contact must be very small. All members of the carrier class should not be stigmatized because some can transmit hepatitis.

On a broader level, the ethical issue is raised as to the extent to which biological knowledge about individuals should impinge on daily lives. Is it appropriate to regulate the risks inherent in people living together and interacting with each other? An issue has been raised with respect to hepatitis because the test can be easily done and because millions of people are tested as part of blood donor programs. As a consequence of these tets, this particular group of carriers has been identified. There are carriers of other agents, some of them potentially more hazardous (i.e. staphylococcus, typhoid) who are not routinely tested and therefore not placed at a disadvantage.

It is hoped that many of these problems can be resolved by continued research into the nature of the hepatitis carrier state, and that carriers who have already been identified will not be jeopardized during this period when necessary information is not available.

A characteristic of many large scale public health control programs is the emergence of problems that were not anticipated prior to the institution of the program. For example, the control of malaria has in many areas resulted in a markedly decreased infant mortality with a large increase in population. When this has not been accompanied by a concomitant increase in food production, the nourishment and well-being of the population have actually decreased.

With the availability of the serologic and environmental tests for hepatitis B, it is now possible to begin the design of control measures for this disease. If the heptatitis B vaccine is found to be effective, then it may also be of value in preventing the development of the carrier state. We are now attempting to investigate the biology of the hepatitis B agent to learn whether some of the consequences of control can be known before the program begins. An example already discussed is the possible effect of HBV infection on sex ratio. The role of the virus in the life of the insects in which it is found is not known, but may be profound; and there may be other effects on the ecology that are not now obvious.

We hope to continue the study of these broad problems to be as well prepared as possible when and if attempts are made to eliminate or decrease the frequency of the hepatitis B virus.

REFERENCES

1. Ford, E. B. Genetics for medical students, Metheun, London, 1956, p. 202.
2. Allison, A. C., Blumberg, B. S., and ap Rees, W. Haptoglobin types in British, Spanish Basque and Nigerian African populations, Nature, *181:* 824—825, 1958.
3. Blumberg, B. S., Allison, A. C., and Garry, W. The haptoglobins and hemoglobins of Alaskan Eskimos and Indians. Ann. Human Genet. *23:* 349—356, 1959.
4. Alberdi, F., Allison, A. C., Blumberg, B. S., Ikin, E. W., and Mourant, A. E. The blood groups of the Spanish Basques. J. Roy Anthrop. Inst. *87:* 217—221, 1957.
5. Corcoran, P. A., Allen, F. H., Jr., Allison, A. C., Blumberg, B. S. Blood groups of Alaskan Eskimos and Indians. Am. J. Phys. Anthropol. *17:* 187—193, 1959.
6. Allison, A. C., Blumberg, B. S., and Gartler, S. M. Urinary excretion of beta-amino-isobutyric acid in Eskimo and Indian populations of Alaska. Nature *183:* 118—119, 1959.
7. Blumberg, B. S., and Gartler, S. M. High prevalence of high level beta-amino-iso-butyric acid excretors in Micronesians. Nature *184:* 1990—1992, 1959.
8. Blumberg, B. S. and Tombs, M. T. Possible polymorphism of bovine alpha-lactalbumin. Nature *181:* 683—684, 1958.
9. Blumberg, B. S., and Robbins, J. Thryoxine-serum protein complexes. Advan. Thyroid Res. 461—465, 1961.
10. Allison, A. C., and Blumberg, B. S. An isoprecipitation reaction distinguishing human serum protein types. Lancet *1:* 634—637, 1961.
11. Blumberg, B. S., Dray, S., and Robinson, J. C. Antigen polymorphism of a low-density beta-lipoprotein. Allotypy in human serum. Nature *194:* 656—658, 1962.
12. Blumberg, B. S. Polymorphisms of serum proteins and the development of isoprecipitins in transfused patients. Bull. N. Y. Acad. Med. *40:* 377—386, 1964.
13. Blumberg, B. S., Alter, H. J. and Visnich, S. A "new" antigen in leukemia sera. J. Am. Med. Assoc. *191:* 541—546, 1965.
14. Blumberg, B. S., Gerstley, B. J. S., Hungerford, D. A., London, W. T., and Sutnick, A. I. A serum antigen (Australia antigen) in Down's syndrome leukemia and hepatitis. Ann. Int. Med. *66:* 924—931, 1967.
15. London, W. T., Sutnick, A. I. and Blumberg, B. S. Australia antigen and acute viral hepatitis. Ann. Int. Med. *70:* 55—59, 1969.
16. Okochi, K. and Murakami, S. Observations on Australia antigen in Japanese. Vox Sang. *15:* 374—385, 1968.
17. Okochi, K., Murakami, S., Ninomiya, K. and Kaneko, M. Australia antigen, transfusion and hepatitis. Vox Sang. *18:* 289—300, 1970.
18. Vierucci, A., Bianchini, A. M., Morgese, G., Bagnoli, F. and Messina, G. L'antigen Australia. 1. Rapporti con l'epatite infettiva e da siero. Una ricerca in pazienti pediatrici. Pediatria Internazione XVIII, No. 4, 1968.
19. Prince, A. M. An antigen detected in the blood during the incubation period of serum hepatitis. Proc. Nat. Acad. Sci. *60:* 814, 1968.
20. Gocke, D. J. and Kavey, N. B. Correlation with disease and infectivity of blood donors. Lancet *1:* 1055—1059, 1969.
21. Senior, J. R., Sutnick, A. I., Goeser, E., London, W. T., Dahlke, M. D. and Blumberg B. S. Reduction of post-transfusion hepatitis by exclusion of Australia antigen from donor blood in an urban public hospital. Am. J. Med. Sci. *267:* 171—177, 1974.
22. Robinson, W. S. and Lutwick, L. I. The virus of hepatitis, type B. N. Engl. J. Med. *295:* 1168—1175, 1976.
23. Bayer, M. E., Blumberg, B. S. and Werner, B. Particles associated with Australia antigen in the sera of patients with leukemia, Down's syndrome and hepatitis. Nature *218:* 1057—1059, 1968.
24. Alter, H. J. and Blumberg, B. S. Studies on a "new" human isoprecipitin system (Australia antigen). Blood *27*(3): 297—309, 1966.

25. Dane, D. S., Cameron, C. H. and Briggs, M. Virus-like particles in serum of patients with Australia antigen-associated hepatitis. Lancet *1:* 695—698, 1970.
26. Summers, J., O'Connell, A. and Millman, I. Genome of hepatitis B virus: Restriction enzyme cleavage and structure of DNA extracted from Dane particles. Proc. Nat. Acad. Sci. *72:* 4597—4601, 1975.
27. Blumberg, B. S. and Millman, I. Vaccine against viral hepatitis and process. U.S. Patent Office No. 3,636,191, 1972.
28. Lustbader, E. D., London, W. T. and Blumberg, B. S. Study design for a hepatitis B vaccine trial. Proc. Nat. Acad. Sci. *73:* 955—959, 1976.
29. Purcell, R. H. and Gerin, J. L. Hepatitis B subunit vaccine: A preliminary report of safety and efficacy tests in chimpanzees. Am. J. Med. Sci. *270:* 395—399, 1975.
30. Hilleman, M. R., Buynak, E. B., Roehm, R. R., Tytell, A. A., Bertland, A. V., and Lampson, S. P. Purified and inactivated human hepatitis B vaccine: Progress report. Am. J. Med. Sci. *270:* 401—404, 1975.
31. Maupas, P. Coursaget, P. Goudeau, A., Drucker, J., Bagros, P. Immunisation against hepatitis B in man. Lancet *1:* 1367—1370, 1976.
32. Buynak, E. B., Roehm, R. R., Tytell, A. A., Bertland, A. U., Lampson, G. P., and Hilleman, M. R. Vaccine against human hepatitis B. J. Am. Med. Assoc. *235:* 2832—2834, 1976.
33. Blumberg, B. S. Australia antigen: The history of its discovery with comments on genetic and family aspects. *In* Viral Hepatitis and Blood Transfusion, edited by G. N. Vyas, H. A. Perkins and R. Schmid, Grune & Stratton, New York, pp. 63—83, 1972.
34. Patterson, C. P., Boyer, K. M., Maynard, J. E. and Kelly, P. C. Epidemic hepatitis in a clinical laboratory. J. Am. Med. Assoc. *230:* 854—857, 1974.
35. Blumberg, B. S., London, W. T., Lustbader, E. D., Drew, J. S. and Werner, B. G. Protection vis-a-vis de l'hépatite B par l'anti-HBs chez des malades hémodialysés. *In* Hépatite a Virus B et Hémodialyse, Flammarion Médecine-Sciences, Paris, pp. 175—183, 1975.
36. London, W. T., Drew, J. S., Lustbader, E. D., Werner, B. G. and Blumberg, B. S. Host response to hepatitis B infection among patients in a chronic hemodialysis unit. Kidney Int. (in press) 1977.
37. London, W. T., Drew, J. S., Blumberg, B. S., Grossman, R. A., and Lyons, P. S. Association of graft survival with host response to hepatitis B infection in patients with kidney transplants. N. Engl. J. Med. *296:* 241—244, 1977.
38. Hesser, J. E., Economidou, J. and Blumberg, B. S. Hepatitis B surface antigen (Australia antigen) in parents and sex ratio of offspring in a Greek population. Human Biol. *47:* 415—425, 1975.
39. Drew, J. S., London, W. T., Lustbader, E. D. and Blumberg, B. S. Cross reactivity between hepatitis B surface antigen and an antigen on male cells. Proc. of the 1977 March of Dimes Birth Defects Conference (in press) 1977.
40. Payet, M., Camain, R., Pene, P. Le cancer primitif du foie, étude critique a propos de 240 cas. Rev. Intern. Hepatol. *4:* 1—20, 1956.
41. Davies, J. N. P. Hepatic neoplasm. The Liver, edited by E. A. Gall, F. K. Mostofi. Baltimore, Williams and Wilkins, pp. 361—369, 1973.
42. Steiner, P. D. and Davies, J. N. P. Cirrhosis and primary liver carcinoma in Uganda Africans. Br. J. Cancer *11:* 523—534, 1957.
43. Blumberg, B. S., Larouzé, B., London, W. T., Werner, B., Hesser, J. E., Millman, I., Saimot, G. and Payet, M. The relation of infection with the hepatitis B agent to primary hepatic carcinoma. Am. J. Path. *81:* 669—682, 1975.
44. Ohbayashi, A., Okochi, K. and Mayumi, M. Familial clustering of asymptomatic carriers of Australia antigen and patients with chronic liver disease or primary liver cancer. Gastroenterol. *62:* 618—625, 1972.
45. Larouzé. B., Blumberg, B. S., London, W. T., Lustbader, E. D., Sankale, M., and Payet, M. Forecasting the development of primary hepatic carcinoma by the use of risk factors. Studies in West Africa. J. Nat. Canc. Inst. *58:* 1557—1561, 1977.

46. Prince, A. M., Metselaar, D. Kafuko, G. W., Mukwaya, L. G., Ling, C. M. and Overby, L. R. Hepatitis B antigen in wild-caught mosquitoes in Africa. Lancet *2:* 247—250, 1972.
47. Smith, J. A., Ogunba, E. O., and Francis, T. I. Transmission of Australia Au(1) antigen by Culex mosquitoes. Nature *237:* 231—232, 1970.
48. Muniz, F. J. and Micks, D. W. The persistence of hepatitis B antigen in *Aedes aegypti*. Mosquito News *33:* 509—511, 1973.
49. Newkirk, M. M., Downe, A. E. R., and Simon, J. B. Fate of ingested hepatitis B antigen in blood-sucking insects. Gastroenterol. *69:* 982—987, 1975.
50. Millman, I., Hutanen, H., Merino, F., Bayer, M. E. and Blumberg, B. S. Australia antigen: Physical and chemical properties. Res. Commun. Chem. Path. Pharm. *2:* 667—686, 1971.
51. Blumberg, B. S., Millman, I., Stunick, A. I. and London, W. T. The nature of Australia antigen and its relation to antigen-antibody complex formation. J. Exp. Med. *134* : 320—329, 1971.
52. London, W. T., Sutnick, A. I., Millman, I., Coyne, V. Blumberg, B. S. and Vierucci, A. Australia antigen and hepatitis: Recent observations on the serum protein polymorphism, infectious agent hypotheses. C.M.A.J. *106:* 480—485, 1972.
53. Blumberg, B. S. Bioethical questions related to hepatitis B antigen. Am. J. Clin. Path. *65:* 848—853, 1976.
54. Krugman, S., Giles, J. P., and Hammond, J. Viral hepatitis, type B (MS2 strain). Studies on active immunization. J. Am. Med. A. *217:* 41—45, 1971.
55. Maugh, T. H. Hepatitis B: A new vaccine ready for human testing. Science *188:* 137—138, 1975.
56. Larouzé, B., London, W. T., Saimot, G., Werner, B. G., Lustbader, E. D., Payet, M., and Blumberg, B. S. Host responses to hepatitis B infection in patients with primary hepatic carcinoma and their families. A case/control study in Senegal, West Africa. Lancet *2:* 534—538, 1976.
57. Blumberg, B. S., Wills, W., Millman, I., and London, W. T. Australia antigen in mosquitoes. Feeding experiments and field studies. Res. Comm. CP *6:* 719—732, 1973.
58. Wills, W., Saimot, G., Brochard, C., Blumberg, B. S., London, W. T., Dechene, R., and Millman, I. Hepatitis B surface antigen (Australia antigen) in mosquitoes collected in Senegal, West Africa. Am. J. Trop. M. *25:* 186—190, 1976.
59. Raunio, V. K., London, W. T., Sutnick, A. I., Millman, I., and Blumberg, B. S. Specificities of human antibodies to Australia antigen. P. Soc. Exp. M. *134:* 548—557, 1970.
60. Werner, B. and London, W. T. Host responses to hepatitis B infection: Hepatitis B surface antigen and host proteins. Ann Int. Med. *83:* 113—114, 1975.
61. Wills, W., Larouzé, B., London, W. T., Blumberg, B. S., Millman, I., Pourtaghra, M., and Coz, J. Hepatitis B surface antigen in West African mosquitoes and bedbugs, Abstract 25th Annual Joint Meeting of the American Society of Tropical Medicine and Hygiene and the Royal Society of Tropical Medicine and Hygiene, Philadelphia, Pennsylvania, November 3—5, 1976.
62. Magnius, L. O. Characterization of a new antigen-antibody system associated with hepatitis B. Clin. Exp. Im. *20:* 209—216, 1975.

D. Carleton Gajdusek

D. CARLETON GAJDUSEK

My scientific interests started before my school years, when as a boy of five years I wandered through gardens, fields and woods with my mother's entomologist-sister, Tante Irene, as we overturned rocks and sought to find how many different plant and animal species of previously hidden life lay before us. We cut open galls to find the insects responsible for the tumors, and collected strange hardening gummy masses on twigs which hatched indoors to fill the curtains with tiny praying mantises, and discovered wasps with long ovipositors laying their eggs into the larvae of wood-boring beetles. In petri dishes we watched some leaf-eating insects succumb to insecticide poison while others survived, and on exciting excursions visited the laboratories and experimental greenhouses of the Boyce Thompson Institute for Plant Research in my hometown of Yonkers, New York, where my aunt, Irene Dobroscky, worked, studying in the 1920's virus inclusions in the cells of leaf-hoppers.

In my first years at school I had problems with my teachers for carrying to school insect-killing jars, correctly labeled "Poison: potassium cyanide". As a grade schoolboy, I met at the Boyce Thompson Institute laboratories the quiet, amused, watchful and guiding eyes of the mathematician and physical chemist, Dr. William J. Youden, who enjoyed letting me play with his hand cranked desk calculator, with his circular or cylindrical slide rules, and with models of crystal lattice structure, and on his laboratory bench where he taught me to prepare colloidal gold solution time color reactions and to manufacture mercuric thiocyanate snake-generating tablets. Before I was ten years old I knew that I wanted to be a scientist like my aunt and my quiet mathematician tutor. I rejected completely, as did my younger brother, Robert, who is now a poet and critic, the interests of our father and maternal grandfather in business, which had made our life style possible.

My life and outlook were greatly influenced by the polyglot immigrant Eastern European communities, adjacent and unwillingly interlaced, living in the carpet, elevator and copper wire manufacturing and sugar refining city of Yonkers, just upstream on the Hudson River from the New York megalopolis and possessing a schoolbook history of a Seventeenth Century Royal Dutch land grant of Indian land to Johng Heer (hence Yonkers) Adrian van der Donck. The cimbalon in our living room, beside the piano, Romanian and Hungarian gypsies who fiddled the *czardas* and *halgatos* at our family festivities and camped in the empty store adjacent to my father's butcher shop, an uninterrupted flow of loud conversation in many tongues, rarely English, and kitchen odors of many Habsburg cuisines filling our crowded expanded-family-filled home, gave me an orthodox and optimistic view of America as a land of change and possibility which I never lost. Below our almost rural hilltop home—our family had "risen"—clustered the factories, churches, shops and two to four family houses of immigrant factory workers and tradesmen in the valleys of the almost obliterated Nepperhan and Tuckahoe Indian-named creeks. In this hollow stood Hungarian, Slovak and Polish Catholic and Russian Orthodox churches and a Presbyterian mission to the factory workers. (This exciting conglomeration of Eastern Europeans has been later displaced by Mediterranean and Caribbean and, still later, Black Americans, all similarly "melting".)

My father, Karl Gajdusek, was a Slovak farm boy from a small village near Senica, who had left home as an adolescent youth to emigrate to America before World War I, alone and without speaking English, to become a butcher in the immigrant communities of Yonkers, where he met and married my mother, Ottilia Dobroczki. Her parents had also come, each alone, as youthful immigrants from Debrecen, Hungary to America. On my father's side we were a family of farmers and tradesmen, vocations which never interested my brother or myself, but my father's temperament for laughter and ribald fun, lust for life in work and play, music, song, dance and food, and above all, conversation, affected us strongly. On my mother's side were the more somber academic and aesthetic aspirations of four university educated first generation American siblings and a heroic interest in fantasy and inquiry, in the classics and culture, nature, nurture and process. Because of my mother's unquenchable interest in literature and folklore, my brother and I were reared listening to Homer, Hesiod, Sophocles, Plutarch and Virgil long before we learned to read.

I was born on September 9, 1923 in the family home we still own, while my maternal grandparents and my mother's youngest sister shared the home. My brother arrived nineteen months later. He and I grew up closely together; for every move I made further into mathematics and the sciences, he moved further into poetry, music and the other arts. In 1930 we traveled to Europe to visit our relatives, mostly those of my father's large family, which he had abandoned twenty years earlier. My brother and I were left for months in my father's birthplace with his old father and the huge remaining family (the squire had sired some twenty five children), while our parents toured European capitals.

Back in America, my early school years were those of great happiness: I liked school and the enchanting family excursions up the Hudson valley were frequent. My Tante Irene was working on problems of economic entomology in the Philippines and South East Asia, and exotic artifacts and natural history specimens, particularly the beautiful giant leafhoppers clad in batik-like patterns, arrived to fascinate me. On her return from the Orient she took me on ever broader excursions to collect insects, to watch the emergence of the seventeen-year cicadas and to attend scientific meetings in the American Museum of Natural History. I became an early habitué of New York city's museums, attending courses on Egyptology at the Metropolitan Museum of Art on schoolday afternoons after my fifth grade classes and at weekend and evening lectures on entomology, geology and botany at the Museum of Natural History.

Today, I and my large family of adopted sons from New Guinea and Micronesia still occupy, on our frequent visits to New York city, our family home in which I was born fifty-three years ago. Here, the boys recently discovered, while installing new attic insulation, daguerreotypes and tintypes of the family taken in towns east of the Danube and in turn-of-the-century New York city and also school notebooks which once belonged to my mother, her siblings, my brother, and myself. From this home, too, we buried both of my maternal grandparents, and my father and mother. On the occasion of my pagan mother's death, the unavoidably close proximity of Slovak Catholic and Russian Orthodox churches, both named Holy Trinity, led to the confusion which resulted in burying her with ministrations of the wrong denomination, which she would have enjoyed, when I attemped to assuage, by asking the funeral director to call in the priest, the pious Roman Catholic relatives of my irreverent father, at whose earlier funeral the Slovak priest had declined to officiate.

I started to read seriously before puberty. Books by Scandinavian authors, Henrik Ibsen and Sigrid Undset, were among the earlier works I read myself. I devoured enthusiastically three biographical works which must have had a profound effect on me: René Vallery-Radot's biography of his father-in-law, Louis Pasteur; Eve Curie's biography of her mother, Marie Curie; and Paul de Kruif's "Microbe Hunters." I then stenciled the twelve names of microbiologists whom de Kruif had selected on the steps leading to my attic chemistry laboratory, where they remain today. At about this time, when I was about ten years old, I wrote an essay on why I planned to concentrate on chemistry, physics, and mathematics, rather than classical biology, in preparation for a career in medicine. Dr. Youden had succeeded in making it clear to me that education in mathematics, physics and chemistry was the basis for the biology of the future.

During the summers of my thirteenth to sixteenth years, I was often working at the Boyce Thompson Laboratories. Under Dr. John Arthur's tutelage, I synthesized and characterized a large series of halogenated aryloxyacetic acids, many previously unsynthesized. The series of new compounds I derived from these failed to yield the fly-killing potency anticipated, but when they were tested several years later for their phytocidal capacity one of my new compounds, 2,4-dichlorophenoxyacetic acid, became the weed killer of commerce; and the Institute based its patent rights to royalties on my boyhood laboratory notebooks—the only venture I have had which involved commerce.

My experiences at the Boyce Thompson, especially with Youden, directed me towards physics at the University of Rochester, where I hoped to fulfill my plan, formulated in boyhood from my readings and teachings of my aunt and Youden, of studying mathematics, physics, and chemistry in preparation for a career in medical research.

From 1940 to 1943 I studied at the University of Rochester under Victor Weisskopf in physics; Curt Stern, Don Charles, David Goddard, Jim Goodwin, in biology; Vladimir Seidel in mathematics; and Ralph Helmkamp in chemistry. In the summer of 1941 I was inspired by the marine embryology course of

Viktor Hamburger's at Woods Hole Marine Biology Laboratories. In those years of my teens I learned to love mountaineering, hiking, canoeing and camping with a passion as great as that for science.

At nineteen to twenty-two years of age while at Harvard Medical School, I worked with John T. Edsall in the laboratory of protein physical chemistry, and with James L. Gamble in his laboratory of electrolyte balance at Boston Children's Hospital. Thereafter, at ages of twenty-five and twenty-six, I worked at Caltech with Linus Pauling and John Kirkwood, where I was also greatly influenced by Max Delbrück, George Beadle, Walter Zechmeister and James Bonner. It was at Caltech that my peers—fellow postdoctoral students and young investigators (Gunther Stent, Jack Dunitz, Elie Wollman, Benoit Mandelbrot, David Shoemaker, John Cann, Harvey Itano, Aage Bohr, Ole Maaloe, Ted Harold, John Fincham, Reinhart Ruge, Arnold Mazur, Al Rich, and others)—had a profound effect on my intellectual development, goals and appreciation of quality in creative life, and on my career. This was the "Golden Age" at Caltech and the many close friends working in several different disciplines, as well as our mentors, have remained mutually stimulating coworkers in science and, above all, lasting personal friends for the past thirty years. With the group of students about Linus Pauling, John Kirkwood, Max Delbrück and George Beadle, I spent many days and evenings in wide-ranging discussions in the laboratories and at the Atheneum, and in even more protracted exchanges on camping and hiking trips to the deserts and mountains of the West, of Mexico and Canada. Max and Mannie Delbrück were often the hosts for our group at their home, and the prime organizers of many of our expeditions. This period of less than two years at Caltech has given me a group of friends who are interested critics of my work, who together with my major teachers in clinical and laboratory investigation, comprise, perhaps unwittingly, the jury whose judgements I most respect.

I had not counted on my captivation with clinical pediatrics. Children fascinated me, and their medical problems (complicated by the effect of variables of varying immaturity, growth, and maturation upon every clinical entity that beset them) seemed to offer more challenge than adult medicine. I lived and worked within the walls of Boston Children's Hospital through much of medical school. Thereafter, I started my postgraduate specialty training in clinical pediatrics which I carried through to Specialty Board qualification, while also working in the laboratory of Michael Heidelberger at Columbia University College of Physicians and Surgeons, while at Caltech, and while with John Enders on postgraduate work at Harvard. I have never abandoned my clinical interests, particularly in pediatrics and neurology, which were nurtured by a group of inspiring bedside teachers: Mark Altschuler, Louis K. Diamond, William Ladd, Frank Ingraham, Sidney Gellis, and Canon Ely at Harvard; Rustin McIntosh, Hattie Alexander, Dorothy Anderson, and Richard Day at Babies Hospital, Columbia Presbyterian Medical Center in New York; Katie Dodd, Ashley Weech, Joe Warkany, and Sam Rappaport at Cincinnati Children's Hospital, and Ted Woodward of Baltimore.

In 1951 I was drafted to complete my military service from John Enders' laboratory at Harvard to Walter Reed Army Medical Service Graduate School as a young research virologist, to where I was called by Dr. Joseph Smadel. I found that he responded to my over-ambitious projects and outlandish schemes with severity and metered encouragement, teaching me more about the methods of pursuing laboratory and field research, and presenting scientific results, than any further theoretical superstructure, which he assumed I already possessed.

From him and from Marcel Baltazard of the Institut Pasteur of Teheran, where I worked in 1952 and 1953 on rabies, plague, arbovirus infections, scurvy and other epidemic disease in Iran, Afghanistan and Turkey, I learned of the excitement and challenge offered by urgent opportunistic investigations of epidemiological problems in exotic and isolated populations. My quest for medical problems in primitive population isolates took me to valleys of the Hindu Kush, the jungles of South America, the coast and inland ranges of New Britain, and the swamps and high valleys of Papua New Guinea and Malaysia, but always with a base for quiet contemplation and exciting laboratory studies with John Enders in Boston, Joe Smadel in Washington, and Frank Burnet in Melbourne. To these teachers I am indebted for guidance and inspiration and for years of encouragement and friendship.

To Joe Smadel I also owe the debt of further sponsorship and encouragement, and recognition of my scientific potential for productive research which led him to create for me several years later a then-unique position as an American visiting scientist at the National Institutes of Health, in the National Institute of Neurological Diseases and Blindness, under Dr. Richard Masland, wherein I could nurture my diverse interests in a selfstyled Study of Child Growth and Development and Disease Patterns in Primitive Cultures. Our Laboratory of Slow, Latent and Temperate Virus Infections grew out of the elucidation of one of our "disease patterns", kuru, and blossomed into a new field of medicine. For about two decades I have enjoyed at the National Institutes of Health the base and haven for our diverse studies in remote parts of the world together with a small group of students and coworkers and many visiting colleagues who have formed the strong team of our endeavor. Here, Marion Poms, Joe Gibbs, Paul Brown, Vin Zigas, Michael Alpers, David Asher and Nancy Rogers have shared these adventures with me through almost two decades.

My boyhood reading, first in Homer, Virgil, and Plutarch, on which we were nurtured by our Classicist-Romanticist Hungarian mother, led, upon the instigation of my poet brother, to my more thorough return to the classics as a young, too-ardent scientist-cum-physician, and to the modern literature of European authors and philosophers, which I had missed in my university days devoted too exclusively to mathematics and the sciences. This reading changed greatly my way of thinking. Particularly, I would have to credit Dostoevsky, Chekhov and Tolstoy; Montaigne, Baudelaire, Rimbaud, Valery and Gide; Shakespeare, Wordsworth, Yeats and Lawrence; Poe,

Whitman and Melville; Ibsen; Goethe, Schiller, Kant, Nietzsche, Kafka and Mann; Saadi and Hafiz.

In 1954 I took off for Australia to work as a visiting investigator with Frank Burnet at the Walter and Eliza Hall Institute of Medical Research in Melbourne from where, between periods of bench work in immunology and virology, I launched studies on child development and disease patterns with Australian aboriginal and New Guinean populations.

In eighteen volumes of some five thousand pages of published personal journals on my explorations and expeditions to primitive cultures, I have told far more about myself and my work since 1957, when I first saw kuru, under the guidance of Vincent Zigas, than one should in a lifetime . . I do not see how I can précis that here.

UNCONVENTIONAL VIRUSES AND THE ORIGIN AND DISAPPEARANCE OF KURU

Nobel Lecture, December 13, 1976

by D. CARLETON GAJDUSEK
National Institutes of Health, Bethesda, Maryland, U.S.A.

Kuru was the first chronic degenerative disease of man shown to be a slow virus infection, with incubation periods measured in years and with a progressive accumulative pathology always leading to death. This established that virus infections of man could, after long delay, produce chronic degenerative disease and disease with apparent heredofamilial patterns of occurrence, and with none of the inflammatory responses regularly associated with viral infections. Soon thereafter, several other progressive degenerative diseases of the brain were likewise attributed to slow virus infections (see Tables 1 and 2). These include delayed and slow measles encephalitis, now usually called subacute sclerosing panencephalitis (SSPE), progressive multifocal leukoencephalopathy (PML), and transmissible virus dementias usually of the Creutzfeldt-Jakob disease (CJD) type. Thus, slow virus infections, first recognized in animals, became recognized as a real problem in human medicine.

Kuru has led us, however, to a more exciting frontier in microbiology than only the demonstration of a new mechanism of pathogenesis of infectious disease, namely the recognition of a new group of viruses possessing unconventional physical and chemical properties and biological behavior far different from those of any other group of microorganisms. However, these viruses still demonstrate sufficiently classical behavior of other infectious microbial agents for us to retain, perhaps with misgivings, the title of "viruses". It is about these unconventional viruses that I would further elaborate.

The group consists of viruses causing four known natural diseases: two of man, kuru and CJD, and two of animals, scrapie in sheep and goats, and transmissible mink encephalopathy (TME) (Table 1). The remarkable unconventional properties of these viruses are summarized in Tables 3 and 4. Because only primate hosts have been available as indicators for the viruses

Table 1. Naturally-occurring slow virus infections caused by unconventional viruses (subacute spongiform virus encephalopathies)

In man:	In animals:
Kuru	Scrapie
Transmissible virus dementia	In sheep
Creutzfeldt-Jakob disease	In goats
Sporadic	Transmissible mink encephalopathy
Familial	
Familial Alzheimer's disease	

Table 2. Slow infections of man caused by conventional viruses

Disease	*Virus*
Subacute post-measles leukoencephalitis	Paramyxovirus—defective measles
Subacute sclerosing panencephalitis (SSPE)	Paramyxovirus—defective measles
Subacute encephalitis	Herpetovirus – Herpes-simplex Adenovirus—Adeno-types 7 and 32
Progressive congenital rubella	Togavirus—rubella
Progressive panencephalitis as a late sequela following congenital rubella	Togavirus—defective rubella
Progressive multifocal leukoencephalopathy (PML)	Papovavirus—JC; SV-40
Cytomegalovirus brain infection	Herpetovirus—cytomegalovirus
Epilepsia partialis continua (Kozhevnikov's epilepsy) and progressive bulbar palsy in USSR	Togavirus—RSSE and other tick-borne encephalitis viruses
Chronic meningoencephalitis in immunodeficient patients	Picornaviruses—poliomyelitis, ECHO virus
Crohn's disease	Unclassified—RNA virus
Homologous serum jaundice	Unclassified—Hepatitis B, Dane particle
Infectious hepatitis	Parvovirus—Hepatitis A Unclassified—Hepatitis B, Dane particle Unclassified—Hepatitis C

Table 3. Atypical physical and chemical properties of the unconventional viruses

1. Resistant to formaldehyde
2. Resistant to β-propiolactone
3. Resistant to ethylenediamine tetraacetic acid (EDTA)
4. Resistant to proteases (trypsin, pepsin)
5. Resistant to nucleases (RNase A and III, DNase I)
6. Resistant to heat (80° C); incompletely inactivated at 100° C
7. Resistant to ultraviolet radiation: 2540 Å
8. Resistant to ionizing radiation (γ rays): equivalent target 150,000 daltons
9. Resistant to ultrasonic energy
10. Atypical UV action spectrum: 2370 Å inactivation = 6 × 2540 Å inactivation
11. Invisible as recognizable virion by electron microscopy (only plasma membranes, no core and coat)
12. No non-host proteins demonstrated

causing human disease (or, more recently, cats (24) and guinea pigs (48) for CJD and mink for kuru (24), but with long incubation periods), it has been impossible to characterize these agents well; knowledge of the properties of unconventional viruses is based mostly on the study of the scrapie virus adapted to mice (39, 60) and hamsters (42, 49, 57). The unusual resistance of the viruses to various chemical and physical agents (items 1 to 9 in Table 3), separate this group of viruses from all other microorganisms. In fact, their resistance to ultraviolet (UV) and ionizing radiation, the atypical UV action spectrum for inactivation, and the failure to contain any demonstrable non-host protein, make these infectious particles unique in the biology of replicating infectious agents, and it is only to the newly-described viroids causing six natural plant diseases [potato spindle tuber disease (7—10, 34), chrysanthemum stunt disease, citrus exocortis disease (57, 58), Cadang-Cadang disease of coconut palms (55), cherry chloratic mottle, and cucumber pale fruit disease] that we must turn for analogy (see Figs. 1a, 1b).

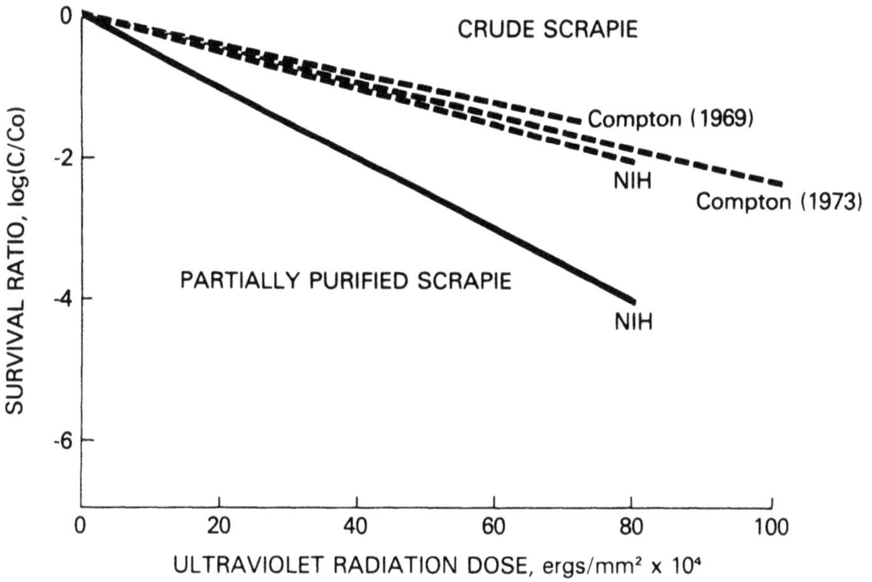

Figure 1. Scrapie virus is unusually resistant to UV inactivation at 2534 to 2540 Å (45,46). This has been interpreted as an indication that it contains no nucleic acid. Recent data from Diener (7,8), however, indicate that the smallest plant virus (potato spindle tuber viroid: PSTV), which is a naked single-stranded RNA of 120,000 daltons, is 90 times more resistant to such UV inactivation than are conventional plant viruses. Since the small infectious nucleic acid of tobacco ring spot virus satellite virus (single-stranded RNA of 75,000 daltons) is 70 times as resistant as are conventional viruses, this high resistance of the two plant viruses is probably because of their small size. The small RNA of PSTV is apparently single-stranded with a circular structure and of such small size that it could code for only about 25 amino acids. Inactivation of scrapie virus by ionizing radiation yields a target size for inactivation equivalent to molecular weight of 150,000 (45). These data, taken with the association of scrapie virus with smooth vesicular membrane during purification and the absence of recognizable virions on electron microscopic study of highly infectious preparations, suggest that the virus is a replicating membrane subunit. It may contain its genetic information in a small nucleic acid moiety incorporated into the plasma membrane. The membrane appears to be the host membrane without altered antigenicity.

1a. Scrapie virus in crude suspensions of mouse brain has been very resistant to UV inactivation at 2540 Å (36,45,46). These three experiments with crude scrapie are in close agreement: NIH (45); Compton A (36); Compton B (45).
Survival ratio is calculated as log C/C_0:

$$\log_{10} \frac{\text{Infectivity titer after irradiation}}{\text{Infectivity titer before irradiation}}$$

Partially purified scrapie (suspension of scrapie mouse brain clarified by two treatments with Genetron in the cold) is somewhat less resistant to UV inactivation, but still much more resistant than other conventional viruses.

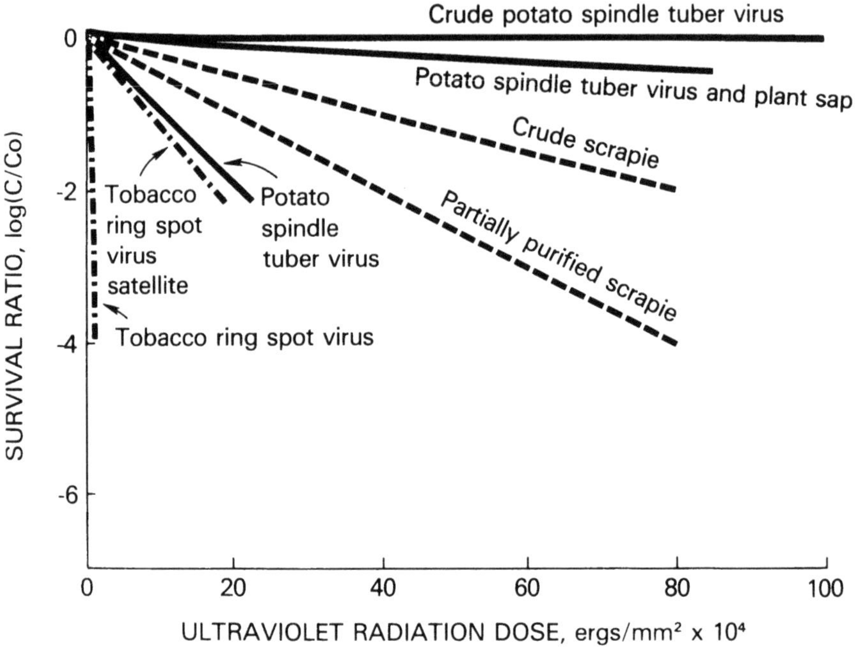

1b. Scrapie inactivation by UV irradiation is compared with that of a conventional plant virus, tobacco ring spot virus, and with the tobacco ring spot virus satellite and potato spindle tuber viroid, both of which contain nucleic acid of molecular weights under 100,000 daltons (7,8). PSTV, as a highly purified nucleic acid, becomes almost totally resistant to UV inactivation (2540 Å) when mixed with clarified normal plant sap, while other viruses placed in this sap are not rendered so resistant. In the crude extract from infected plants the PSTV is almost totally resistant to UV inactivation (7,8). In a second such experiment a different Genetron treated scrapie preparation showed less reduction in UV resistance (45).

UNCONVENTIONAL VIRUSES AS A CAUSE OF THE SUBACUTE SPONGIFORM VIRUS ENCEPHALOPATHIES

Kuru and the transmissible virus dementias have been classified in a group of virus-induced slow infections that we have described as subactute spongiform virus encephalopathies because of the strikingly similar histopathological lesions they induce; and, scrapie and mink encephalopathy appear, both from their histopathology, pathogenesis, and the similarities of their infectious agents, to belong to the same group (Table 1). The basic neurocytological lesion in all of these diseases is a progressive vacuolation in the dendritic and axonal processes and cell bodies of neurons and, to a lesser extent, in astrocytes and oligodendrocytes; an extensive astroglial hypertrophy and proliferation; and, finally, spongiform change or status spongiosus of grey matter (1, 2, 41, 43, 44). These atypical infections differ from other diseases of the human brain which have been subsequently demonstrated to be slow virus infections (Table 2) in that they do not evoke a virus-associated inflammatory response in the brain; they usually show no pleocytosis nor marked rise in protein in the cerebrospinal fluid throughout the course of infection; furthermore, they show no evidence of an immune response to the causative virus and, unlike the situation in the other virus diseases, there are no recognizable virions in electron-microscopic sections of the brain (Table 4).

There are other slow-infections of the central nervous systems which are caused by rumbling nonproductive, even defective, more conventional viruses including measles virus, papovaviruses (JC and SV-40-PML), rubella virus, cytomegalovirus, herpes-simplex virus, adenovirus types 7 and 32, and probably RSSE virus (Table 2). However, unlike these "conventional" viruses the "unconventional" viruses of the spongiform encephalopathies have unusual resistance to ultraviolet radiation and to ionizing radiation (45), to ultrasonication, to heat, proteases and nucleases, and to formaldehyde, β-propiolactone, ethylenediamine tetraacetic acid (EDTA), and sodium desoxycholate (Table 4). They are moderately sensitive to most membrane-disrupting agents such as phenol (90 %), chloroform, ether, urea (6 M), periodate (0.01 M), 2-chloroethanol, alcoholic iodine, acetone, chloroform-butanol, and hypochlorite (0.5—5.0 %) (Table 5). Virions are not recognized on electron microscopic study of infected cells *in vivo* or *in vitro*, nor in highly infectious preparations of virus concentrated by density-gradient banding in the zonal rotor (60). This has led to the speculation that the infectious agents lack a nucleic acid, perhaps are even a self-replicating membrane fragment.

Table 4. Atypical biological properties of the unconventional viruses

1. Long incubation period (months to years; decades)
2. No inflammatory response
3. Chronic progressive pathology (slow infection)
4. No remissions or recoveries: always fatal
5. "Degenerative" histopathology: amyloid plagues, gliosis
6. No visible virion-like structures by electron microscopy
7. No inclusion bodies
8. No interferon production or interference with interferon production by other viruses
9. No interferon sensitivity
10. No virus interference (with over 30 different viruses)
11. No infectious nucleic acid demonstrable
12. No antigenicity
13. No alteration in pathogenesis (incubation period, duration, course) by immuno-suppression or immunopotentiation:
    - (a) ACTH, cortisone
    - (b) cyclophosphamide
    - (c) X-ray
    - (d) antilymphocytic serum
    - (e) thymectomy/splenectomy
    - (f) "nude" athymic mice
    - (g) adjuvants
14. Immune "B" cell and "T" cell function intact *in vivo* and *in vitro*
15. No cytopathic effect in infected cells *in vitro*
16. Varying individual susceptibility to high infecting dose in some host species (as with scrapie in sheep)

Table. 5. Methods of inactivating unconventional viruses

1.	Autoclaving (121° C at 20 p.s.i.; 30 min.)
2.	Hypochlorite ("Clorox"); 0.5—5.0 %
3.	Phenol (90 %)
4.	Alcoholic iodine solution and organic iodine disinfectants
5.	Ether
6.	Acetone
7.	Chloroform or chloroform-butanol
8.	Strong detergents
9.	Periodate (0.01 M)
10.	2-chloroethanol
11.	Urea (6 M)

A major effort in my laboratory has been and is now being directed toward the molecular biological elucidation of the nature and structure of this group of atypical viruses.

The scrapie virus has been partially purified by fluorocarbon precipitation of proteins and density-gradient banding by zonal rotor ultracentrifugation (60). Other semipurified preparations have been made using ultrafiltration and repeated complete sedimentation and washing of the scrapie virus by means of ultrasonication for resuspension of the virus-containing pellets; such resuspended and washed virus has been banded into peaks of high infectivity using cesium chloride, sucrose, and metrizamide density gradients in the ultracentrifuge by Dr. Paul Brown in my laboratory. Sucrose-saline density-gradient banding of scrapie virus in mouse brains produced wide peaks of scrapie infectivity at densities of 1.14 to 1.23. A second smaller peak of high infectivity at density of 1.26 to 1.28 disappeared on filtration of the crude suspension through 200 nm Nucleopore membranes. On electron microscopic examination, fractions of high infectivity (10^7 to 10^8 LD_{50}/ml) revealed only smooth vesicular membranes with mitochondiral and ribosomal debris and no structures resembling recognizable virions. Lysosomal hydrolases (*n*-acetyl-β-D-glucosaminidase; β-galactosidase; acid phosphatase) and mitochondrial marker enzyme (INT-succinate reductase) showed most of their activity in fractions of lower density than in the fractions having high scrapie infectivity (60).

We have confirmed the previously noted resistance of scrapie virus to UV inactivation at 254 nm and UV inactivation action spectrum with a six-fold increased sensitivity at 237 nm over that at 254 or 280 nm (45). This should not be taken as proof that no genetic information exists in the scrapie virus as nucleic acid molecules, since work with the smallest RNA viruses, called viroids, indicates a similar resistance to UV inactivation in crude infected plant-sap preparations. Ultraviolet sensitivity also depends greatly on small RNA size, as has been shown by the high resistance of the purified very small tobacco ring spot satellite virus RNA (about 80,000 daltons) (7, 8). Partial purification of scrapie by fluorocarbon only slightly increases UV sensitivity at 254 nm (Figs. 1a, 1b) (7, 8, 45). Fluorocarbon-purified scrapie was neither inactivated by RNase A nor III nor by DNase I.

On the other hand, the unconventional viruses possess numerous properties in which they resemble classical viruses, and some of these properties suggest far more complex genetic interaction between virus and host than one might expect for genomes with a molecular weight of only 10^5 daltons (Table 6). They are, moreover, not totally resistant to inactivation nor so dangerous that we cannot work safely with them by using appropriate inactivating agents (Table 5). In spite of very unusual resistance to heat, they are rapidly inactivated by temperatures over 85° C. Autoclaving (120° C/20 p.s.i./45 minutes) completely inactivated scrapie virus in suspensions of mouse brain.

Table 6. Classical virus properties of unconventional viruses

1. Filterable to 25 nm average pore diameter (a.p.d.) (Scrapie, TME); 100 nm p.s.i. (kuru, CJD)
2. Titrate "cleanly" (all individuals succumb to high LD_{50} in most species)
3. Replicate to titers of 10^8/g to 10^{12}/g in brain
4. Eclipse phase
5. Pathogenesis: first replicate in spleen and elsewhere in the reticuloendothelial system, later in brain
6. Specificity of host range
7. "Adaptation" to new host (shortened incubation period)
8. Genetic control of susceptibility in some species (sheep and mice for scrapie)
9. Strains of varying virulence and pathogenicity
10. Clonal (limiting dilution) selection of strains from "wild stock"
11. Interference of slow-growing strain of scrapie with replication of fast-growing strain in mice

CONVENTIONAL VIRUSES CAUSING CHRONIC DISEASE BY DEFECTIVE OR NON-DEFECTIVE REPLICATION

The other chronic diseases of man which have been shown to be slow virus infections are all caused by conventional viruses which in no way tax our imagination (Table 2). They comprise a wide spectrum of chronic and so-called degenerative diseases. Within this group of slow virus infections we find diverse mechanisms of viral replication, various modes of pathogenesis, and different kinds of involvement of the immune system.

In SSPE, the offending measles virus is apparently not present as a fully infectious virion, but instead asynchronous synthesis of virus subunits with defective or incomplete virion assembly occurs; only a portion of the virus genome is expressed, and replication is defective (6, 38, 54, 59). In the case of PML, on the other hand, fully assembled and infectious virus particles are produced (52, 64, 66). In fact, electron microscopically monitored suspensions of the virus particles of the JC papovavirus, density banded from human PML brain, shows that fewer defective particles are being produced than in any known *in vitro* system for cultivating papovaviruses, including the SV-40 virus. Thus, these ordinary viruses are causing slow infections by very different mechanisms. In some cases, as with PML, an immune defect is demonstrated in association with the disease: in this case severe immunosuppression, either from natural primary disease (leukemia, lymphoma, sarcoid, etc.), or an iatrogenic immune suppression, as for renal transplantation or cancer chemotherapy.

The Russian Spring-Summer, or tick-borne encephalitis virus in cases of Kozhevnikov's epilepsy (epilepsia partialis continua) in the Soviet Union, Japan and India, and the rubella virus in adolescents with recrudescence of their congenital rubella infection (60a, 63a) appear also to be proceeding with defective virus replication. In chronic recurrent ECHO virus infection of the central nervous system in children with genetic immune defects, and in subactute brain infection with adenovirus types 7 (47) or 32 (55a), wholly infectious virus, as in the case of PML, seems to be produced.

Kuru and CJD, however, belong to a very different category of virus infections in which no involvement of the immune system has been demonstrable, in which there is no inflammatory response (no pleocytosis in the cerebrospinal fluid and no alteration in CSF protein), and in which the causative virus has defied all conventional attempts at virus taxonomy.

In recent years many other slow virus infections causing chronic diseases in animals have been used as models for various human diseases. Some of these are tabulated in Table 7. In these examples, as for the human diseases, many different mechanisms of virus replication or partial replication are involved in the persistent, latent, chronic, recurrent or slow virus infections. In some of these diseases the host genetic composition is crucial to the type of

Table 7. Slow infections of animals caused by conventional viruses

Disease	*Virus*
Visna	Retrovirus—Visna
Maedi (Zoegerziekte)	Retrovirus—Maedi
Progressive pneumonia of sheep (PPS; Montana sheep disease)	Retrovirus—PPS-Visna and Maedi related
Motor neurone disease of mice (mouse ALS)	Retrovirus—type "C"
Lymphocytic choriomeningitis	Arenavirus—LCM
Aleutian mink disease	Parvovirus
Hard-pad disease (old-dog distemper)	Paramyxovirus—distemper
Chronic tick-borne encephalitis (RSSE)	Togavirus—RSSE
Pulmonary adenomatosis of sheep (Jaagsiekte)	Unclassified
Mouse cataract disease	Unclassified—mouse cataract virus
Lactic dehydrogenase elevating virus of mice	?Togavirus—LDV
Equine infectious anemia	Unclassified—EIA virus
Rabies	Rhabdovirus—rabies
NZB mouse hemolytic anemia	Retrovirus
Chronic hydrocephalus in hamsters	Paramyxovirus—mumps Orthomyxovirus—influenza
Spontaneous progressive multifocal leukoencephalopathy in rhesus monkeys	Papovavirus—SV-40

pathogenesis that occurs, as is the age of the host at the time of infection, and the immune system may be involved in different ways; immune complex formation is important in some cases and not in others.

The suspicion has been awakened that many other chronic diseases of man may be slow virus infections (see Table 8). Data have gradually accumulated both from the virus laboratory and from epidemiological studies, which suggest that multiple sclerosis and Parkinson's disease, disseminated lupus erythematosis and juvenile diabetes, polymyositis and some forms of chronic arthritis may be slow infections with a masked and possible defective virus as their causes. The study of kuru was carried on simultaneously with a parallel attack on multiple scleroris, amyotrophic lateral sclerosis, and Parkinson's disease; in addition, other degenerative dementias such as Alzheimer's disease, Pick's disease, Huntington's chorea and parkinsonism-dementia were also studied. Chronic encephalitis, epilepsia partialis continua, progressive supranuclear palsy, and degenerative reactions to schizophrenia are among the other diseases under investigation (16, 22, 25, 62). Our attempts at transmission of these diseases to subhuman primate and non-primate laboratory animals have been unsuccessful; no virus has been unmasked from *in vitro* cultivated tissues from the patients, and no virus etiology has been demonstrated for any of these diseases.

Table 8. Chronic diseases of man of suspected slow virus etiology

Multiple sclerosis	Carcinomatous cerebellar degeneration
Neuromyelitis optica—Devic's syndrome	Tuberous sclerosis
Parkinson's disease	Ataxia telangiectasia
Amyotrophic lateral sclerosis	Progyria
Progressive supranuclear palsy	Schizophrenic dementia
Chronic encephalitis with focal epilepsy	Neurofibromatosis
Alzheimer's disease	Disseminated lupus erythematosis
Pick's disease	Chronic arthritis
Huntington's chorea	Dermatomyositis
Parkinsonism-dementia	Scleroderma
Syringomyelia	Ulcerative colitis
Alper's disease	Juvenile diabetes
Polymyositis	Beget's disease
Papulosis atrophicans maligna (Köhlmeier-Degos)	Sjögren's disease

KURU

Kuru is characterized by cerebellar ataxia and a shivering-like tremor that progresses to complete motor incapacity and death in less than one year from onset. It is confined to a number of adjacent valleys in the mountainous interior of New Guinea and occurs in 160 villages with a total population of just over 35,000 (Figs. 2—4). *Kuru* means shivering or trembling in the Fore language. In the Fore culture and linguistic group, among whom over 80 % of the cases occur, it had a yearly incidence rate and prevalence ratio of

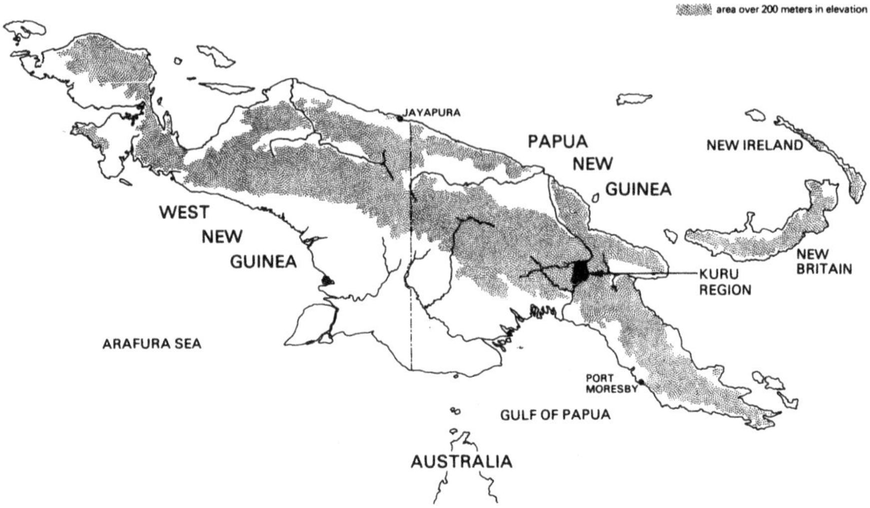

Figure 2. The region in New Guinea from which all kuru patients have come is shown by the irregular black area in the Eastern Highlands Province on the eastern side of the island in Papua New Guinea. It contains more than 35,000 people living in 160 villages (census units) that have experienced kuru. All kuru-affected hamlets lie nestled among rain forest covered mountains from 1,000 to 2,500 m. above sea level.

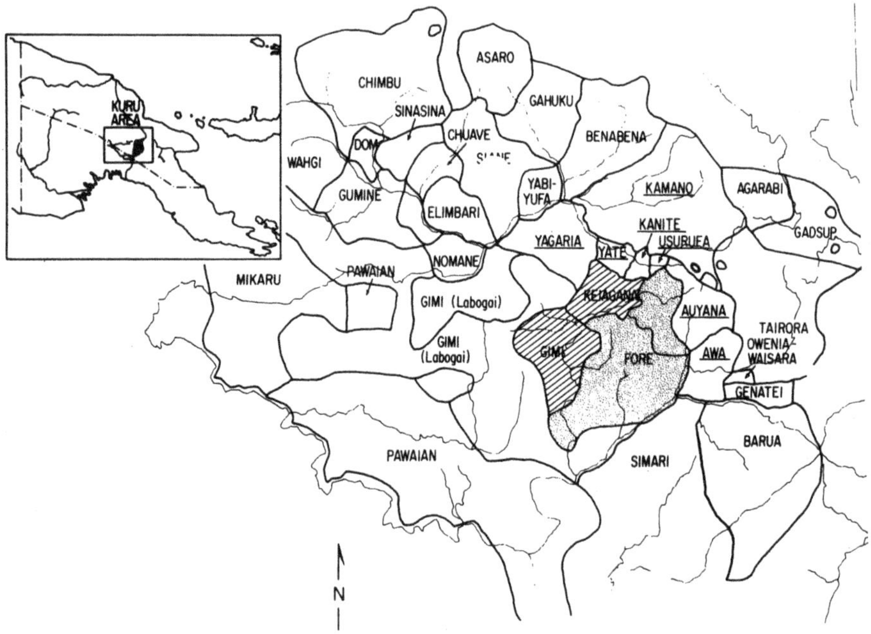

Figure 3. The kuru region in the Eastern Highlands Province of Papua New Guinea showing the cultural and linguistic groups in and surrounding the kuru affected populations.
Inset, upper left: Eastern half of the island of New Guinea showing, in rectangle, area included in the map of larger scale.

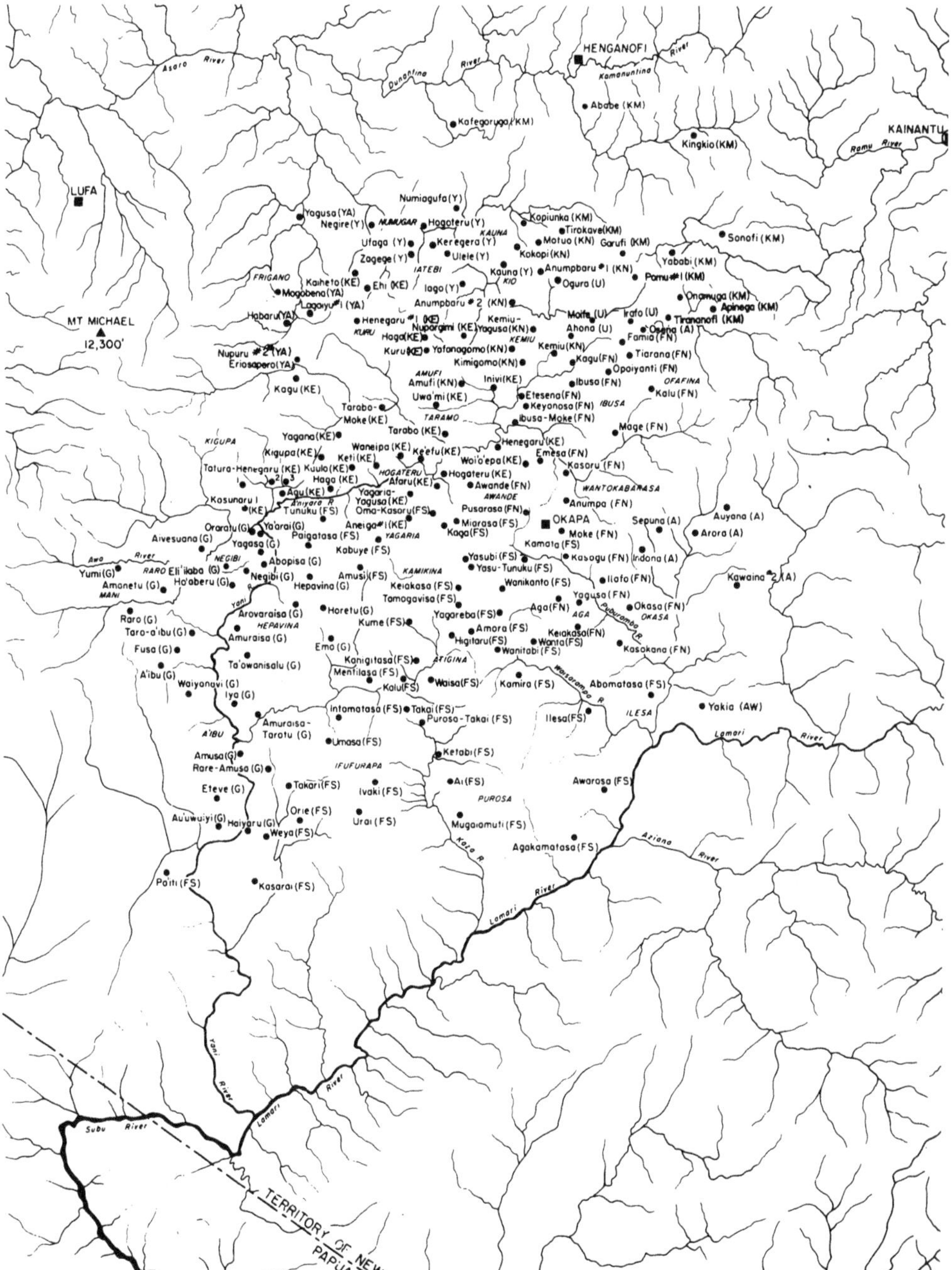

Figure 4. River drainages of the kuru region with superimposed locations of the 160 villages (census units) in which kuru has ever occurred. The cultural and linguistic group of each village is indicated: A Auyana, AW Awa, FN North Fore, FS South Fore, G Gimi, KE Keiagana, KM Kamano, KN Kanite, U Usurufa, Y Yate, YA Yagaria.

about 1 % of the population (Figs. 5a, b). During the early years of investigation, after the first description by Gajdusek and Zigas in 1957 (28), it was found to affect all ages beyond infants and toddlers; it was common in male and female children and in adult females, but rare in adult males (Fig. 6). This marked excess of deaths of adult females over males has led to a male-to-female ratio of over 3:1 in some villages, and of 2:1 for the whole South Fore group (17, 28, 29, 65).

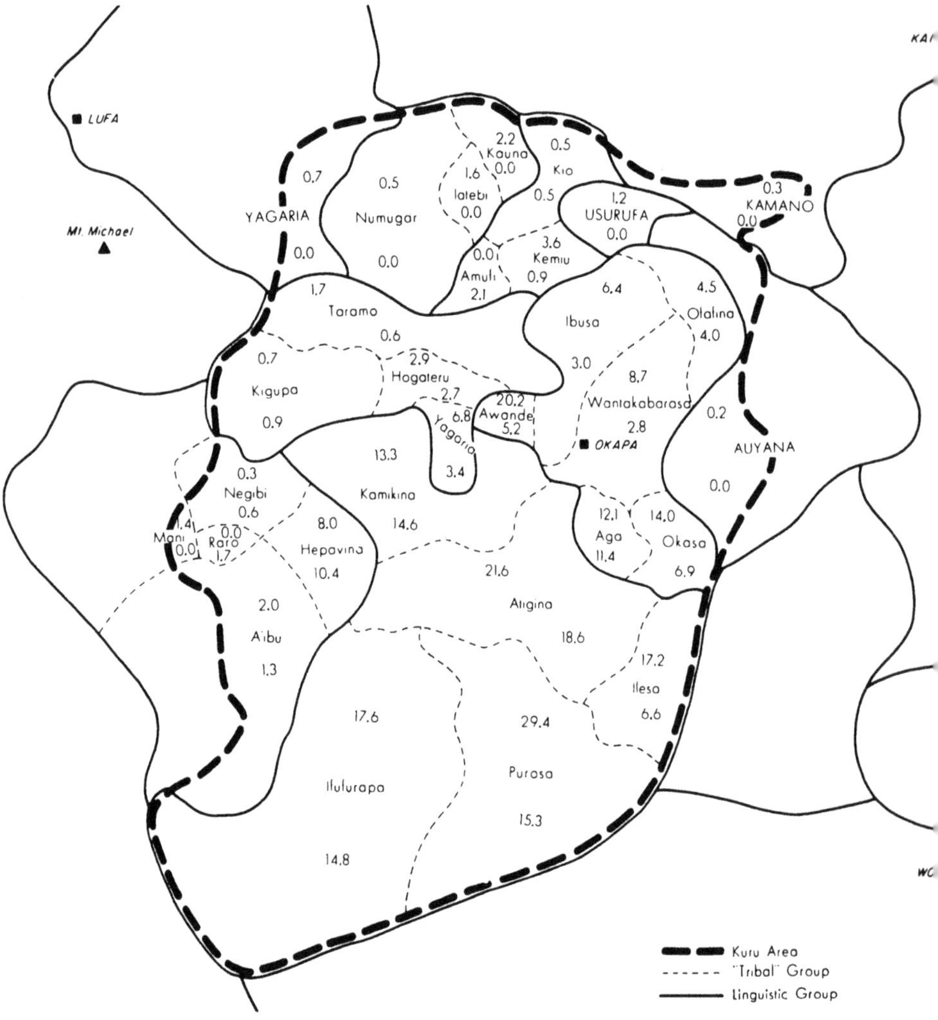

Figure 5. The discovery of kuru coincided with the height of the "epidemic".

5a. Kuru mortality rate in deaths per thousand population per annum in each "tribal" group of the kuru region, 1957—59 and 1961—1963. The numerators of the rates are obtained from the deaths which occurred in the two 3-year periods, the denominators are the populations for 1958 and 1962, respectively. The rates above each name refer to 1957—59, those below to 1961—63.

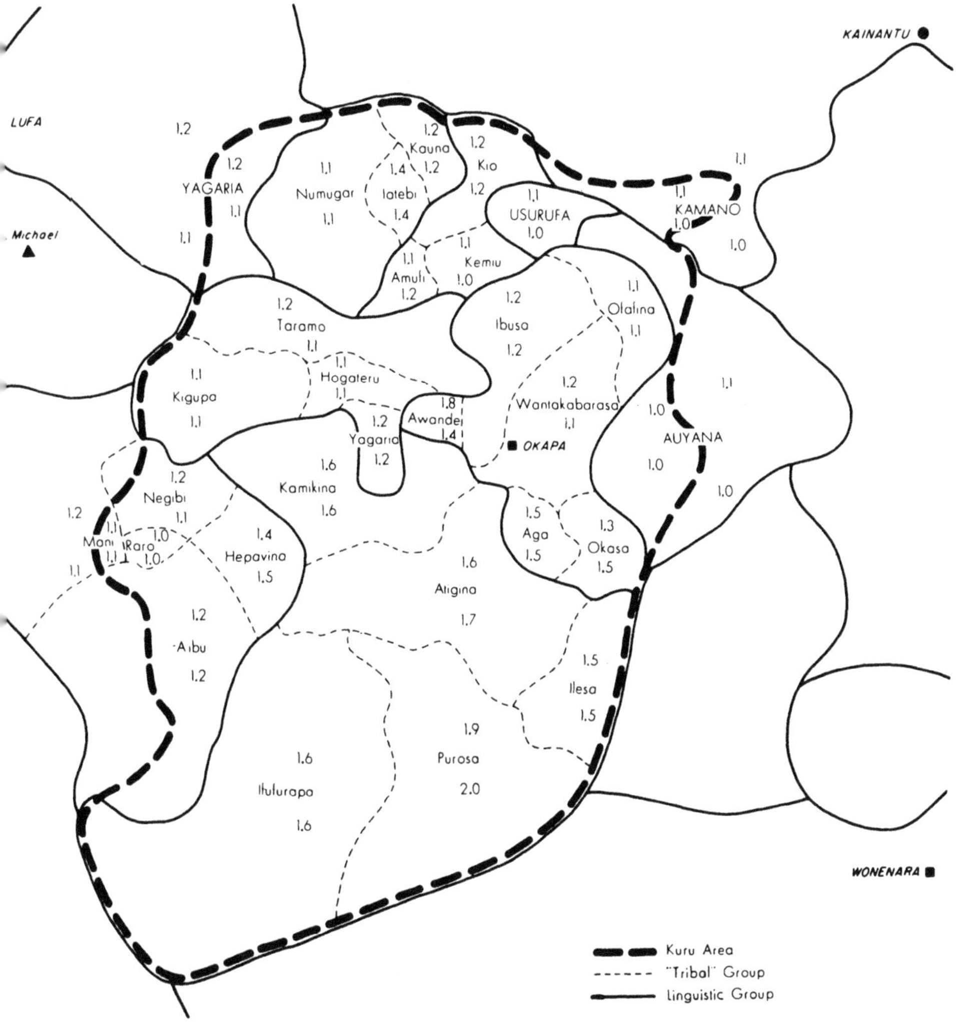

5b. Male: female population ratio in each "tribal" group of the kuru region, 1958 and 1962. The two sets of figures for peripheral groups refer to their portions within and without the kuru region. The ratios above each name refer to 1958, those below to 1962. In these early years of kuru investigation the disease, affecting predominantly females, was causing increasing distortion of the sex ratio.

Kuru has been disappearing gradually during the past 15 years (Fig. 7). The incidence of the disease in children has decreased during the past decade, and the disease is no longer seen in either children or adolescents (Figs. 8 and 9.) This change in occurrence of kuru appears to result from the cessation of the practice of ritual cannibalism as a rite of mourning and respect for dead kinsmen, with its resulting conjunctival, nasal, and skin contamination with highly infectious brain tissue mostly among women and small children (17).

AGE AND SEX DISTRIBUTION OF 1276 STUDIED CASES OF KURU

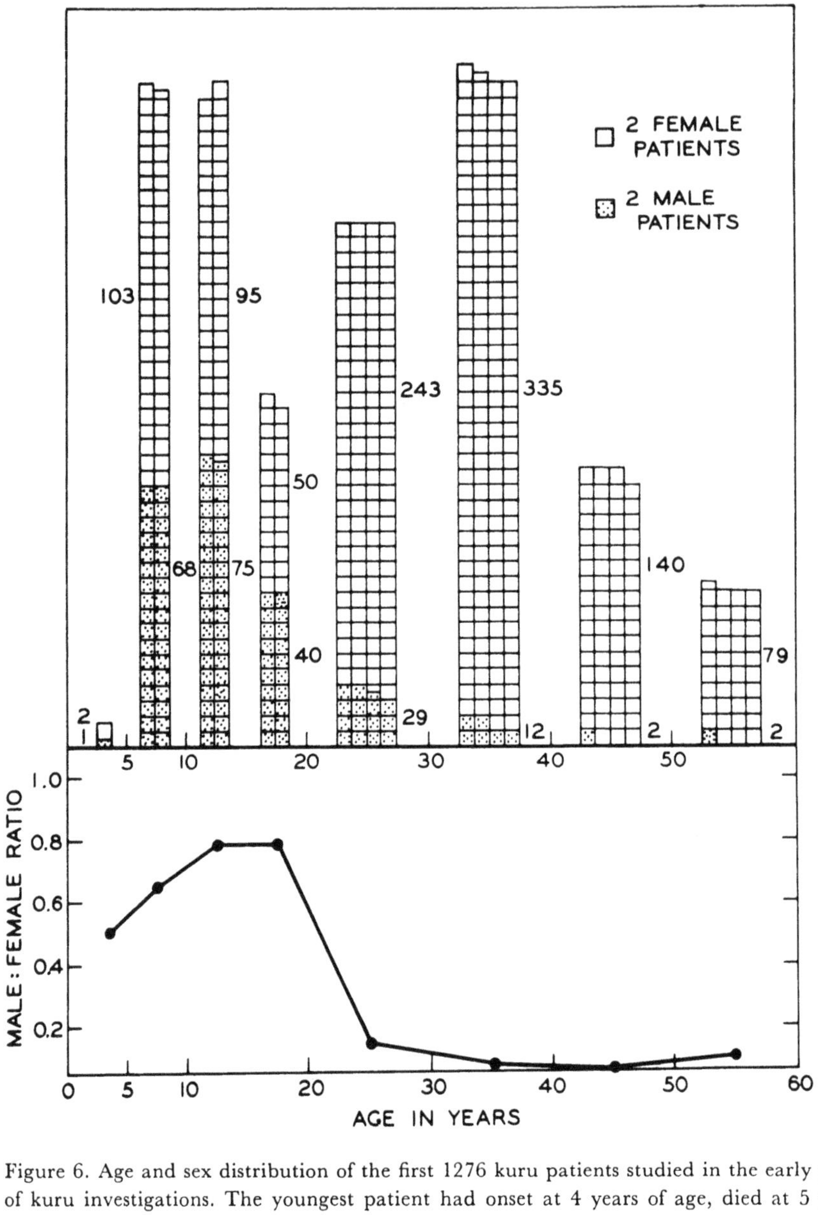

Figure 6. Age and sex distribution of the first 1276 kuru patients studied in the early years of kuru investigations. The youngest patient had onset at 4 years of age, died at 5 years of age.

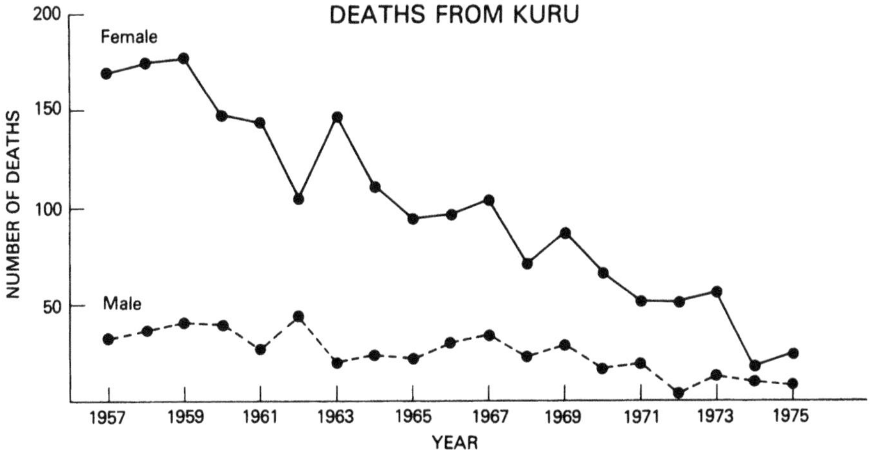

Figure 7. The overall incidence of kuru deaths in male and female patients by year since its discovery in 1957 through 1975. More than 2,500 patients died of kuru in this 17 year period of surveillance, and there has been a slow, irregular decline in the number of patients to one-fifth the number seen in the early years of kuru investigation. The incidence in males has declined significantly only in the last few years, whereas in females it started to decline over a decade earlier. This decline in incidence has occurred during the period of acculturation from a stone age culture in which endocannibalistic consumption of dead kinsmen was practiced as a rite of mourning, to a modern coffee planting society practicing cash economy. Because the brain tissue with which the officiating women contaminated both themselves and all their infants and toddlers contained over 1,000,000 infectious doses per gram, self-inoculation through the eyes, nose, and skin, as well as by mouth, was a certainty whenever a kuru victim was eaten. The decline in incidence of the disease has followed the cessation of cannibalism, which occurred between 1957 and 1962 in various villages.

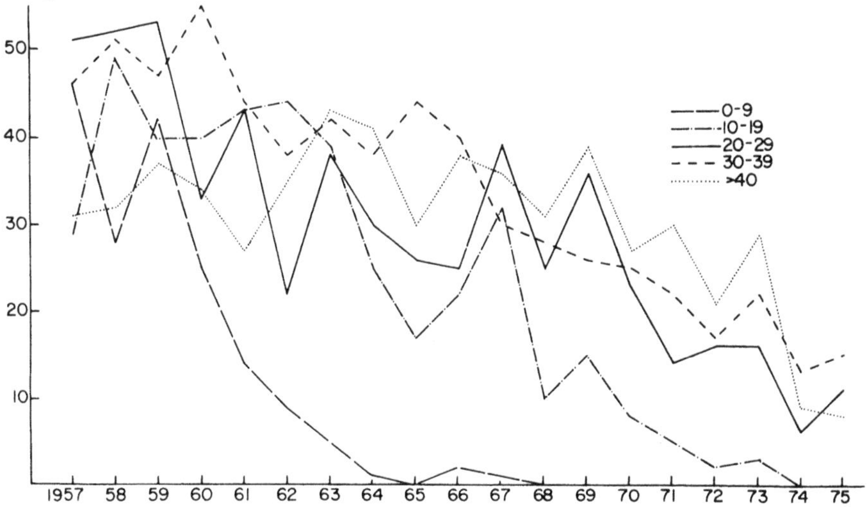

Figure 8. Kuru deaths by age group from 1957 through 1975. The disease has disappeared from the youngest age group (4—9 years) about 5 years before it disappeared in the 10 to 14 year olds, and now it has disappeared in the 15 to 19 year olds. The number of adult patients has declined to less than one-fifth since the early years of investigation. These changes in the pattern of kuru incidence can be explained by the cessation of cannibalism in the late 1950's. No child born since cannibalism ceased in this area has developed the disease.

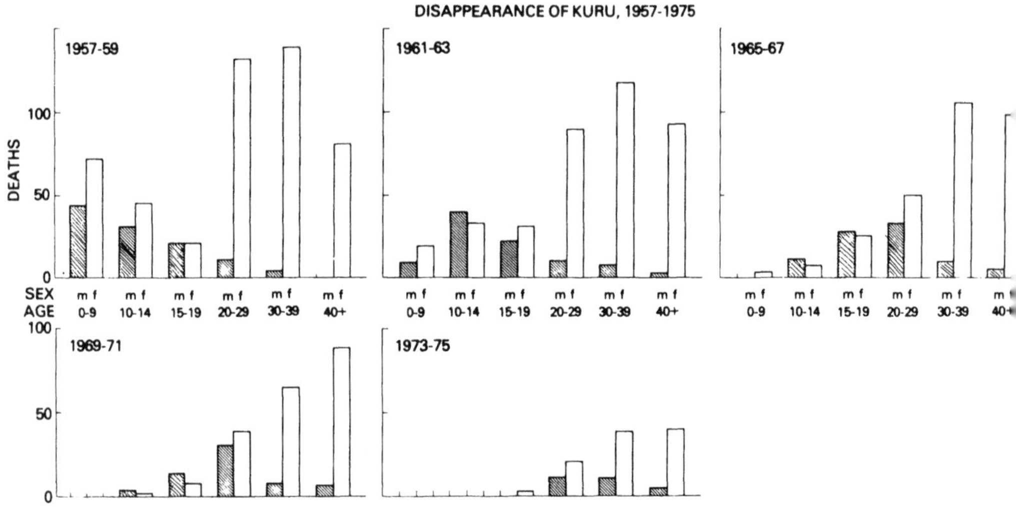

Figure 9. Kuru deaths by age and sex for the years 1957 through 1975 are plotted in 3 year periods, with the exception of those dying in the 1 year intervals between each plot, namely 1960, 1964, 1968, and 1972. These years have been omitted because of irregularities which may have occurred in arbitrarily assigning exact dates of death at the end of the year when dates were not known precisely. The disappearance first in the 4 to 9 year old patients (there were no cases in children under 4 years of age), then in the 10 to 14 year group, and, finally, in the 15 to 19 year group, is clearly shown. No patient under 22 years has died since 1973, and the youngest still-living patient is 24 years old.

The clinical course of kuru is remarkably uniform with cerebellar symptomatology progressing to total incapacitation and death, usually within three to nine months. It starts insidiously without antecedent acute illness and is conveniently divided into three stages: ambulant, sedentary and terminal (Figs. 10—15).

For several years all work on the kuru virus was done using chimpanzees, the first species to which the disease was transmitted (Figs. 16—18) (22, 25). Eventually, other species of nonhuman primates developed the disease: first, several species of New World monkeys with longer incubation periods than in the chimpanzee; and later, several species of Old World monkeys with yet longer incubation periods (Tables 9 and 10) (23, 32). Very recently, we have transmitted kuru to the mink and ferret, the first nonprimate hosts that have proved to be susceptible, although dozens of other species of laboratory, domestic and wild nonprimate and avian hosts have been inoculated without developing disease after many years of observation. We have now extended the nonprimate host range for the subucute spongiform virus encephalopathies, as shown in Table 11.

The virus has been regularly isolated from the brain tissue of kuru patients. It attains high titers of more than 10^8 infectious doses per gram. In peripheral tissue, namely liver and spleen, it has been found only rarely at the time of death, and in much lower titers. Blood, urine, leukocytes, cerebrospinal fluid, and placenta and embryonal membranes of patients with kuru have not yielded the virus.

Figure 10a. Nine victims of kuru who were assembled one afternoon in 1957 from several villages in the Purosa valley (total population about 600) of the South Fore region. The victims included six adult women, one adolescent girl, one adolescent boy, and a prepubertal boy. All died of their disease within 1 year after this photograph was taken.

10b. Five women and one girl, all victims of kuru, who were still ambulatory, assembled in 1957 in the South Fore village of Pa'iti. The girl shows the spastic strabismus, often transitory, which most children with kuru developed early in the course of the disease. Every patient required support from the others in order to stand without the aid of the sticks they had been asked to discard for the photograph.

Figure 11. Six women with kuru so advanced that they require the use of one or two sticks for support, but are still able to go to garden work on their own. In all cases their disease progressed rapidly to death within less than a year from onset.

12a

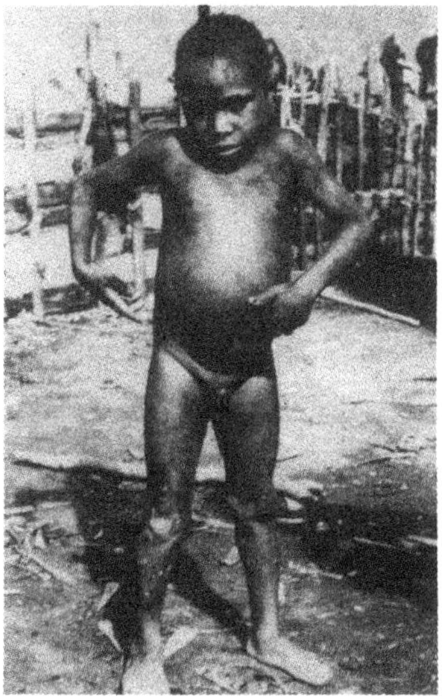

12b

12c

Figure 12. Three Fore boys with kuru in 1957; all three were still ambulatory.

12a. The youngest patient with kuru, from Mage village, North Fore, who self-diagnosed the insidious onset of clumsiness in his gait as kuru at 4 years of age, and died at 5 years of age, several years before his mother developed kuru herself.

12b. A South Fore boy from Agakamatasa village, about 8 years of age, who was caught by the camera in an athetoid movement while trying to stand without support, in the early stage of kuru.

12c. A mid-adolescent youth from Anumpa village, North Fore, who demonstrates the difficulty in standing on one foot associated with the early ambulatory stage of kuru.

13a

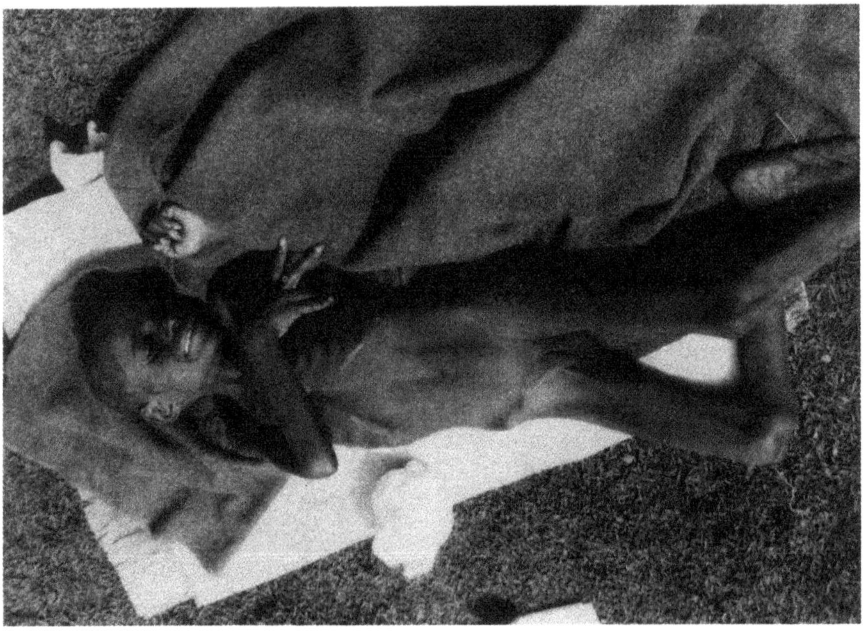

13b

Figure 13. Two Fore children with advanced kuru in 1957. Both had been sedentary for several months and were reaching the terminal stage of the disease.

13a. A girl, about 8 years old, who was no longer able to speak, but who was still alert and intelligent.

13b. A boy, about 8 years of age, who was similary incapacitated after only 3 months of illness.

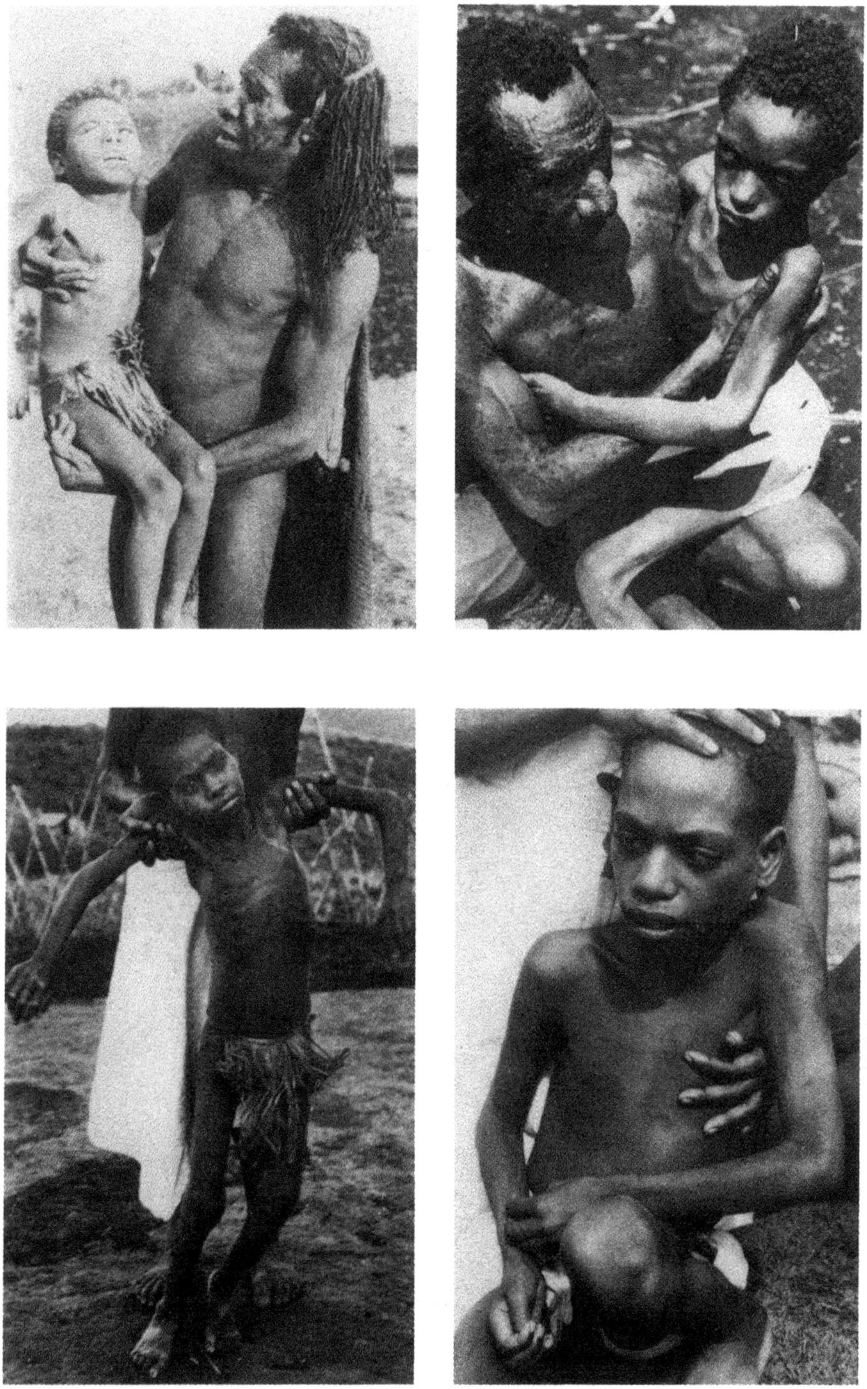

Figure 14. Four preadolescent children, totally incapacitated by kuru in 1957. All had such severe dysarthria that they could no longer communicate by word, but all were still intelligent and alert. All had spastic strabismus. None could stand, sit without support, or even roll over; none had been ill for over six months, and all died within a few months of the time of photography.

15a

15b

15c

Figures 15. Groups of kuru patients in 1957 at the Kuru Research Hospital in Okapa, New Guinea. All died within 1 year of photography. The pictures show many preadolescent child victims of kuru, an age group in which kuru has not occurred in recent years.

15a. Eight kuru patients in the first, or ambulatory, stage of the disease. Five adult women are holding sticks to maintain their balance. Three girls who are still able to walk without the aid of a stick, but with severe ataxia, sit in front of the women.

15b. Eight preadolescent children, four boys and four girls, with kuru. The girl at the far left, in her father's lap, is the same child as that on the left in (a), but is seen 2 months later in the secondary, or sedentary, stage of the disease.

15c. Five children with kuru, two boys in the center, a girl on each side: the adolescent boy supporting the girl on the right is a kuru victim himself, but he is in an earlier stage of the disease. The 4 children requiring support are just passing from the first, or ambulatory, to the second, or sedentary, stage of the disease.

16a

16b

16c

Figure 16a. Chimpanzee with a vacant facial expression and a drooping lip, a very early sign of kuru preceding any "hard" neurological signs. Most animals show this sign for weeks or even months before further symptoms of kuru are detectable other than subtle changes in personality.

16b. Three successive views of the face of a chimpanzee with early kuru drawn from cinema frames. Drooping lower lip is an early sign of kuru.

16c. Face of a normal chimpanzee drawn in three successive views from successive cinema views.

17a

17b

Figure 17a. Chimpanzee with early experimentally-induced kuru eating from floor without use of prehension. This "vacuum cleaner" form of feeding was a frequent sign in early disease in the chimpanzee when tremor and ataxia were already apparent (*From:* Asher, D. M. et al., In: Nonhuman Primates and Human Diseases, W. Montagna and W. P. McNulty, Jr., eds., Vol. 4, 1973, pp. 43—90).

17b. Range of movement in forelimbs in walking: left, normal chimpanzee; right, chimpanzee in stage 2 of experimental kuru. Quantitative assessment was made by studying individual frames of Research Cinema film (24).

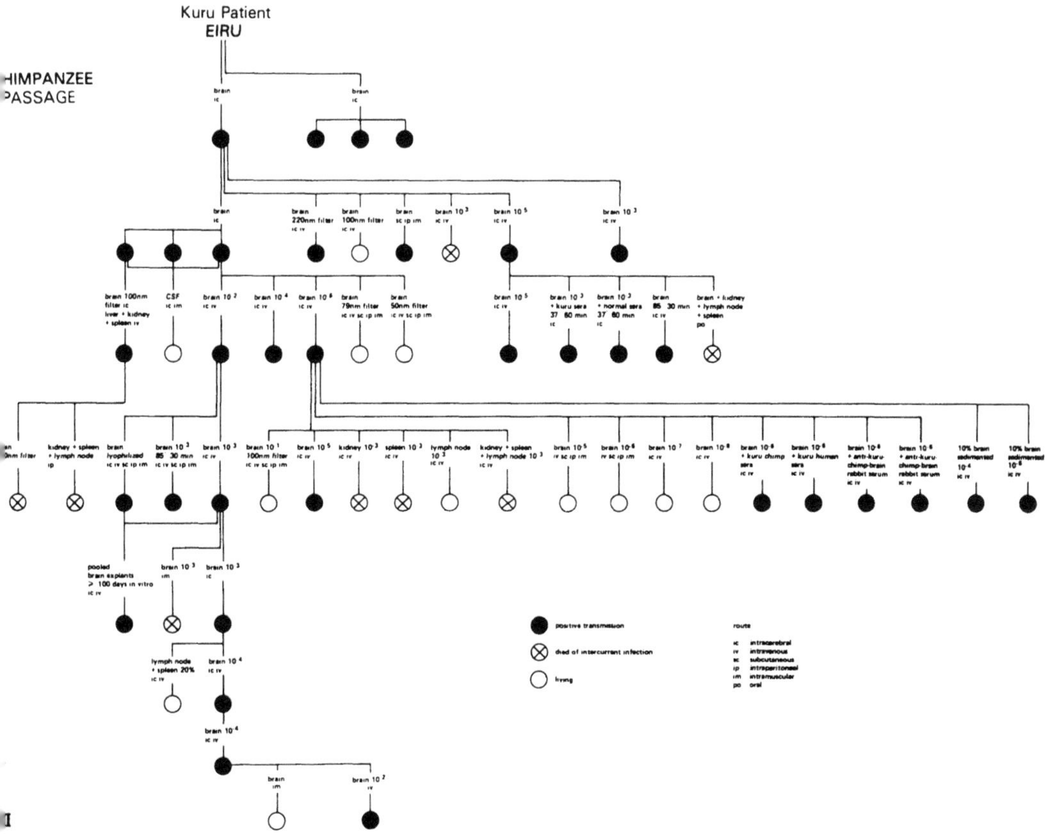

Figure 18. Kuru transmission experiments in chimpanzees, illustrating the early extensive use of this rare and diminishing species and significant curtailment of chimpanzee inoculations after the 4th chimpanzee passage. It was at this time that we discovered that New World monkeys could be used in lieu of the chimpanzee, although they required considerably longer incubation periods. The experiments indicate failure of the agent to pass a 100 nm or smaller filter. They also show the failure of a conventional virus neutralization test, using only 10 infectious doses of kuru virus to neutralize the virus using sera from patients with kuru or from chimpanzees with experimental kuru or antisera made by immunizing rabbits with kuru chimpanzee brain. In these experiments, kidney, spleen and lymph node have not yielded virus, and although chimpanzee brain has had a titer above 10^{-6} by intracerebral inoculation, at 10^{-5} dilutions such brain suspensions inoculated by peripheral routes have not produced disease. In the 3rd passage (on the left), liver, spleen and kidney given intracerebrally, presumably caused disease since 100 nm filtrates of infectious brain have regularly failed to produce the disease; the affected 3rd passage animal had received both inocula.

TRANSMISSIBLE VIRUS DEMENTIAS (CREUTZFELDT-JAKOB DISEASE)

Creutzfeldt-Jacob disease (CJD) is a rare, usually sporadic, presenile dementia found worldwide; it has a familial pattern of inheritance, usually suggestive of autosomal dominant determinations in about 10 % of the cases (Fig. 19). The typical clinical picture includes myoclonus, paroxysmal bursts of high voltage slow waves on EEG, and evidence of widespread cerebral dysfunction. The disease is regularly transmissible to chimpanzees (3, 33), New and Old World monkeys (Tables 9 and 10) and the domestic cat (Tables 11 and 12) (23, 32), with pathology in the animal indistinguishable at the cellular level from that in the natural disease or in experimental kuru (Fig. 20) (3, 43). We have recently confirmed in our laboratory reports of transmission of CJD from human brain to guinea pigs (48, 48a). In spite of a recent convincing report of transmission of CJD from human brain to mice (5, 5a) we have not yet succeeded in transmitting CJD or kuru to mice.

As we have attempted to define the range of illness caused by the CJD virus, a wide range of clinical syndromes involving dementia in middle and late life have been shown to be such slow virus infections associated with neuronal vacuolation or status spongiosus of gray matter and a reactive astrogliosis.

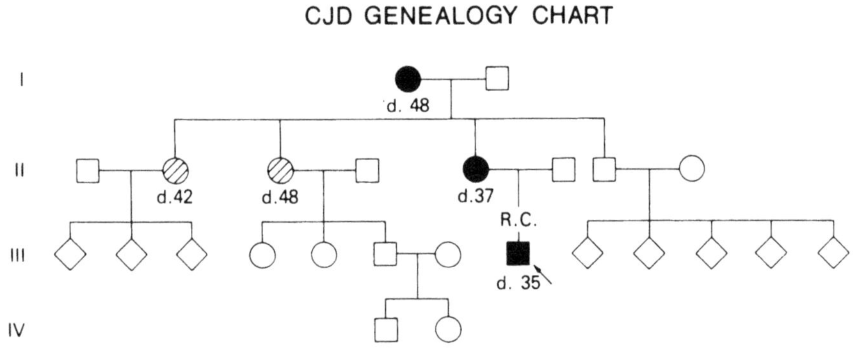

Creutzfeldt-Jakob disease was a rare, almost unknown disease; now cases of transmissible virus dementia are regularly found wherever they are looked for. Better ascertainment of cases, the study of familial aggregations and an unusually high prevalence in Libyan Jews, and the possibility of occupational hazard and transmission through corneal transplant provide promising epidemiological leads to the understanding of the natural history of the disease, and hence to its prevention.

■ Creutzfeldt-Jakob disease confirmed pathologically

▨ Probable Creutzfeldt-Jakob disease

↗ Transmitted to chimpanzee from brain tissue inoculated intracerebrally

Figure 19. Subacute spongiform virus encephalopathy has been transmitted to chimpanzees or New World monkeys from 8 patients with transmissible virus dementias of a familial type. Ten percent of CJD patients have a history of similar disease in kinsmen.

19a. Genealogical chart shows a family with 5 cases of CJD over threee generations, suggesting autosomal dominant inheritance. From patient R. C., the disease has been transmitted to a chimpanzee.

W Family

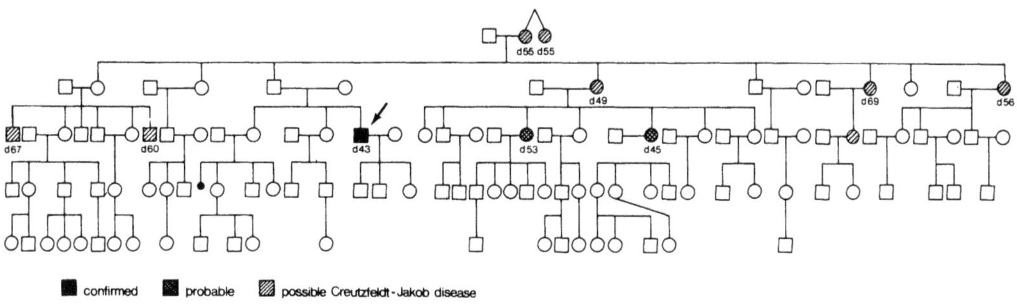

19b. This family has 11 members suffering from CJD-like disease in three generations. From the brain tissue of patient J. W., obtained at autopsy, the disease has been transmitted to a squirrel monkey.

B FAMILY

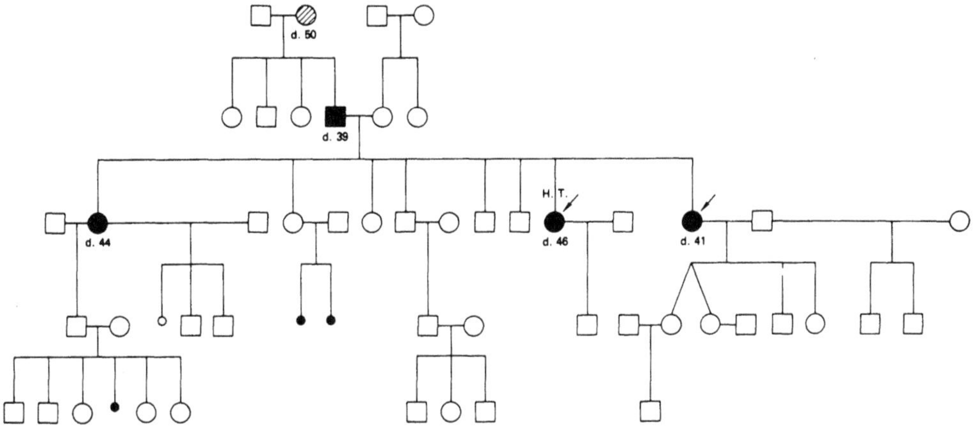

19c. This family has 5 cases of CJD over three generations, again suggesting autosomal dominant inheritance. From the brain tissue of patient H. T., obtained at autopsy, the disease has been transmitted to two squirrel monkeys.

These even include cases that have been correctly diagnosed as brain tumors (glioblastoma, meningioma), brain abscess, Alzheimer's disease, progressive supranuclear palsy, senile dementia, or stroke, or Köhlmeier-Degos disease (27), at some time in their clinical course (51, 62). Hence, the urgent practical problem is to delineate the whole spectrum of subacute and chronic neurological illnesses that are caused by or associated with this established slow virus infection. Because some 14 % of the cases show amyloid plaques akin to those found in kuru, and many show changes similar to those af Alzheimer's disease, in addition to the status spongiosus and astrogliosis of CJD, and because other cases also involve another neurological disease as well as CJD (50, 51, 62), we have started to refer to the transmissible disorder as transmissible virus dementia (TVD).

Since our first transmission of Creutzfeldt-Jakob disease, we have obtained brain biopsy or early postmortem brain tissue on over 200 cases of pathologically confirmed CJD. The clinical, laboratory, and virus investigations of these cases have been summarized in a recent report (62) that extends and updates our earlier report of 35 cases (56). We have been aware of occasional clustering of cases in small population centers, admittedly lacking in natural boundaries, and the unexplained absence of any cases over periods of many years in some large population centers where, at an earlier date, cases were more frequent.

Table 9. Species of laboratory primate susceptible to the subacute spongiform virus encephalopathies

In man	
Kuru	*Apes:* chimpanzee, gibbon *New World monkeys:* capuchin, marmoset, spider, squirrel, woolly *Old World monkeys:* African green, bonnet, cynomolgus macaque, mangabey, rhesus, pig-tailed macaque
Creutzfeldt-Jakob disease	*Apes:* chimpanzee *New World monkeys:* capuchin, marmoset, spider, squirrel, woolly *Old World monkeys:* African green, bushbaby, cynomolgus macaque, mangabey, patas, pig-tailed macaque, rhesus, stump-tailed macaque
In animals	
Scrapie	*New World monkeys:* capuchin, spider, squirrel *Old World monkeys:* cynomolgus macaque, rhesus
Transmissible mink encephalopathy	*New World monkeys:* squirrel *Old World monkeys:* rhesus, stump-tailed macaque

Table 10. Species of laboratory primates susceptible to subacute spongiform encephalopathies

	Incubation periods (in months)			
	Kuru	CJD	Scrapie	TME
Apes				
Chimpanzee (*Pan troglodytes*)	10—82	11—71	(111)	(72+)
Gibbon (*Hylobates lar*)	+(10)	NT	NT	NT
New World monkeys				
Capuchin (*Cebus albifrons*)	10—15	29—34	NT	NT
Capuchin (*Cebus apella*)	11—61	11—47.5	32—35.5	NT
Spider (*Ateles geoffroyi*)	10—85.5	4—50	38	NT
Squirrel (*Saimiri sciureus*)	8—50	5—41	8—63	8—13
Marmoset (*Saguinus sp.*)	1.5—36	18—54	NT	NT
Woolly (*Lagothrix lagothricha*)	33	21	NT	NT
Old World monkeys				
African green (*Cercopithecus aethiops*)	18	33—49.5	(109)	NT
Baboon (*Papia anubis*)	(114)	47.5	NT	NT
Bonnet (*Macaca radiata*)	19—27	(43)	NT	NT
Bushbaby (*Galago senegalensis*)	(104)	16	NT	NT
Cynomolgus macaque (*Macaca fascicularis*)	16	52.5—60	27—72	NT
Patas (*Erythrocebes patas patas*)	(120)	47—60.5	NT	NT
Pig-tailed macaque (*Macaca nemestrina*)	70	+(2)	NT	NT
Rhesus (*Macaca mulatta*)	15—103	43—73	30—37	17—33
Sooty mangabey (*Cercocebus atys*)	+(2)	+(2)—43	NT	NT
Stump-tailed macaque (*Macaca arctoides*)	(120)	60	NT	13
Talapoin (*Cercopithecus talapoin*)	(1+)	64.5	NT	NT

Numbers in parentheses are the number of months elapsed since inoculation, during which the animal remained asymptomatic.

Table 11. Nonprimate hosts for experimental subacute spongiform encephalopathies

In man:	
Kuru	Ferret, mink
Creutzfeldt-Jakob disease	Cat, ferret, guinea pig, ?mouse (5, 5a), hamster (48b)
In animals:	
Scrapie	Gerbil, goat, hamster, mink, mouse, rat, sheep, vole
Transmissible mink encephalopathy	Ferret, goat, hamster, mink, opossum, raccoon, sheep, skunk

Table 12. Creutzfeldt-Jakob disease in cats

Inoculum	Incubation period (months)	Duration (months)
Primary passage		
Human brain	30	2
Serial passage		
Cat brain (passage 1)	19—24	4—5 1/2
Cat brain (passage 2)	18—24	

This geographic and temporal clustering does not apply, however, to a majority of cases and is unexplained by the 10 % of the cases that are familial. Matthews has recently made a similar observation in two clusters in England (50). There are two reports of conjugal disease in which husband and wife died of CJD within a few years of each other (30, 50).

The prevalence of CJD has varied markedly in time and place throughout the United States and Europe, but we have noted a trend toward making the diagnosis more frequently in many neurological clinics in recent years, since attention has been drawn to the syndrome by its transmission to primates (3, 33). For many large population centers of the United States, Europe, Australia, and Asia, we have found a prevalence approaching one per million with an annual incidence and a mortality of about the same magnitude, as the average duration of the disease is 8 to 12 months. Matthews (50) found an annual incidence of 1.3 per million in one of his clusters, which was over 10 times the overall annual incidence for the past decade for England and Wales (0.09 per million). Kahana *et al.* (40) reported the annual incidence of CJD ranging from 0.4 to 1.9 per million in various ethnic groups in Israel. They noted, however, a 30-fold higher incidence of CJD in Jews of Libyan origin above the incidence in Jews of European origin. From recent discussions with our Scandinavian colleagues it is apparent that an annual incidence of at least one per million applies to Sweden and Finland in recent years.

Probable man-to-man transmission of CJD has been reported in a recipient of a corneal graft, which was taken from a donor who was diagnosed retrospectively to have had pathologically confirmed CJD (12). The disease occurred 18 months after the transplant, an incubation period just the average for chimpanzees inoculated with human CJD brain tissue (32, 62). From suspension of brain of the corneal graft recipient we succeeded in transmitting CJD to a chimpanzee although the brain had been at room temperature in 10 % formol-saline for seven months (26a). More recently we learned that two of our confirmed cases of TVD were professional blood donors until shortly before the onset of their symptoms. To date, there have been no transmissions of CJD from blood of either human patients or animals affected with the experimentally transmitted disease. However, we have only transfused two chimpanzees each with more than 300 ml of human whole blood from a different CJD patient

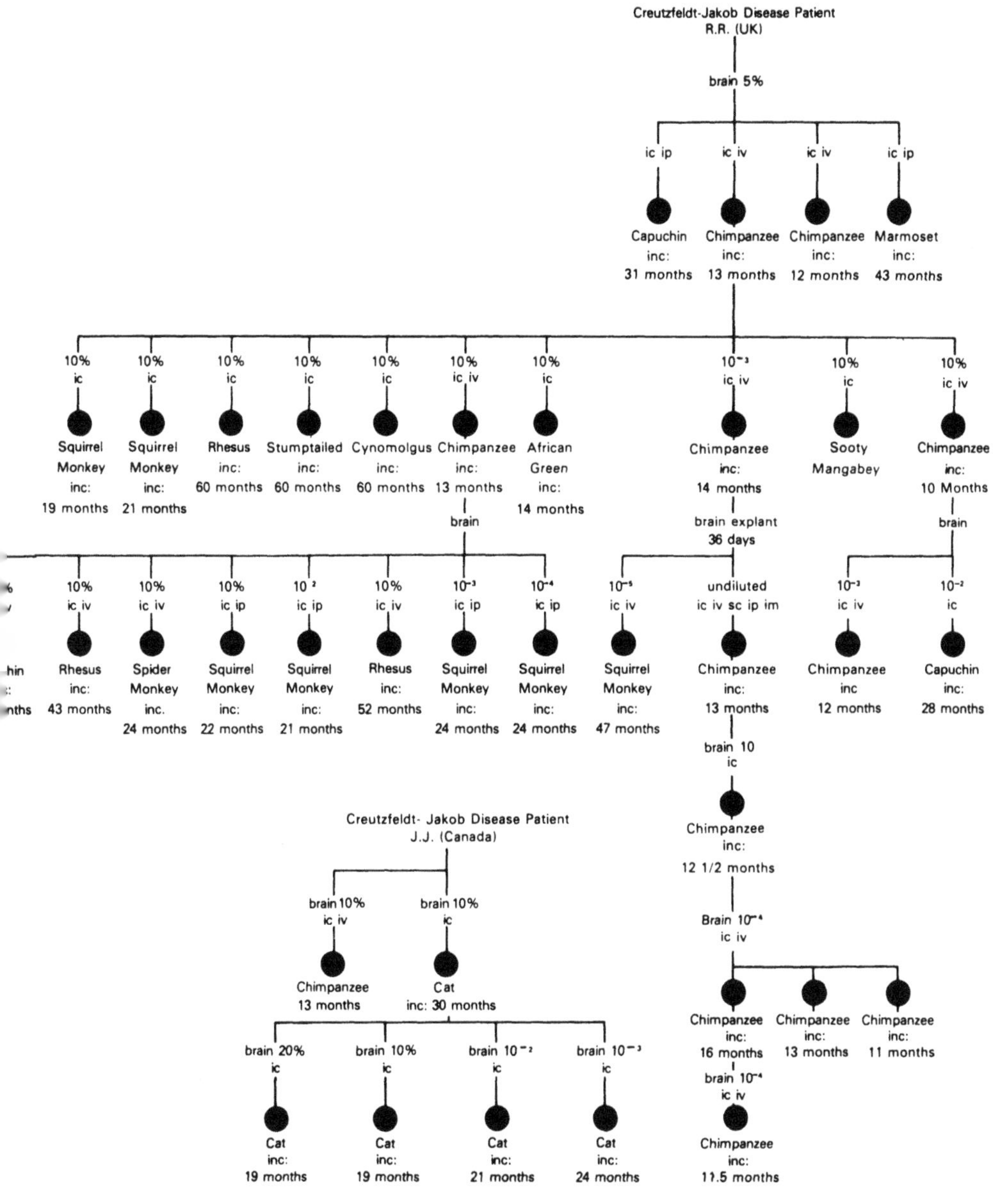

Figure 20. Six serial passages of CJD in chimpanzees, starting with brain tissue from a biopsy of a patient (R. R.) with CJD in the United Kingdom (U. K.). Also shown is transmission of the disease directly from man to the capuchin monkey and marmoset, and from chimpanzee brain to three species of New World monkeys (squirrel, capuchin, spider monkeys), and to six Old World species (rhesus, stumptailed, cynomolgus, African green, pigtailed, and sooty mangabey). Incubation periods in the New World monkeys ranged from 19 to 47 months, and in the Old World monkeys from 43 to 60 months. The pigtailed macaque and the sooty mangabey showed positive CJD pathology when sacrificed without

within the past several months. Finally, the recognition of TVD in a neurosurgeon (27), and more recently in two physicians, has raised the question of possible occupational infection, particularly in those exposed to infected human brain tissue during surgery, or at postmortem examination (61, 63).

The unexpectedly high incidence of previous craniotomy in CJD patients noted first by Nevin *et al.* (51) and more recently by Matthews (50) and by ourselves (62), raises the possibility of brain surgery either affording a mode of entry for the agent or of precipitating the disease in patients already carrying a latent infection. The former unwelcome possibility now seems to be a reality with the probable transmission of CJD to two young patients with epilepsy from the use of implanted silver electrodes sterilized with 70 % ethanol and formaldehyde vapor after contamination from their use on a patient who had CJD. The patients had undergone such electrode implantation for stereotactic electroencephalographic localization of the epileptic focus at the time of correctional neurosurgery (3a).

Two patients with transmissible virus dementias were not diagnosed clinically or neuropathologically as having CJD, but rather as having Alzheimer's disease (62). In both cases the disease was familial: in one (Fig. 21) there were six close family members with the disease in two generations; in the other both the patient's father and sister had died of presenile dementia. The diseases as transmitted to primates were clinically and pathologically typical subacute spongiform virus encephalopathies, and did not have pathological features of Alzheimer's disease in man. More than 30 additional specimens of brain tissue from non-familial Alzheimer's disease have been inoculated into TVD-susceptible primates without producing disease. Therefore, although we

clinical disease. A third passage to the chimpanzee was accomplished using frozen and thawed explanted tissue culture of brain cells that had been growing *in vitro* for 36 days. Using 10^{-3}, 10^{-4}, and 10^{-4} dilutions of brain, respectively, the 4th, 5th, and 6th chimpanzee passages were accomplished. This indicates that the chimpanzee brain contains $\geqslant$50,000 infectious doses per gram, and that such infectivity is maintained in brain cells cultivated *in vitro* at 37° C for at least one month. The lower left shows transmission of CJD from a second human patient (J. T.) to a cat with a 30 month incubation and serial passage in the cat with 19 to 24 month incubation.

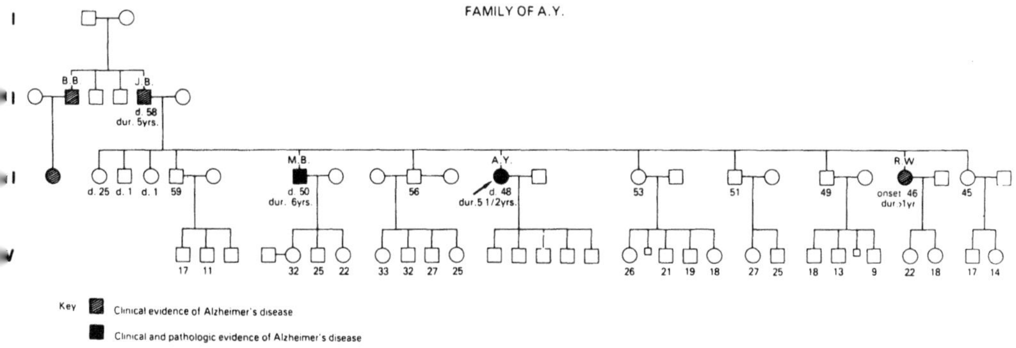

Figure 21a. Y family. Brain tissue obtained from patient A. Y. at biopsy induced subacute spongiform encephalopathy in a squirrel monkey 24 months after intracerebral inoculation. The patient, a 48-year old woman who died after a 68 month course of progressive dementia, quite similar in clinical aspects to the progressive dementia from which her father and brother had died at 54 and 56 years of age, respectively, was diagnosed clinically and neuropathologically as suffering from Alzheimer's disease. Her sister is at present incapacitated by a similar progressive dementia of 4 years' duration. Although the transmitted disease in the squirrel monkey was characterized by severe status spongiosis, none was seen in the patient, although amyloid plaques and neurofibrillary tangles were frequent.

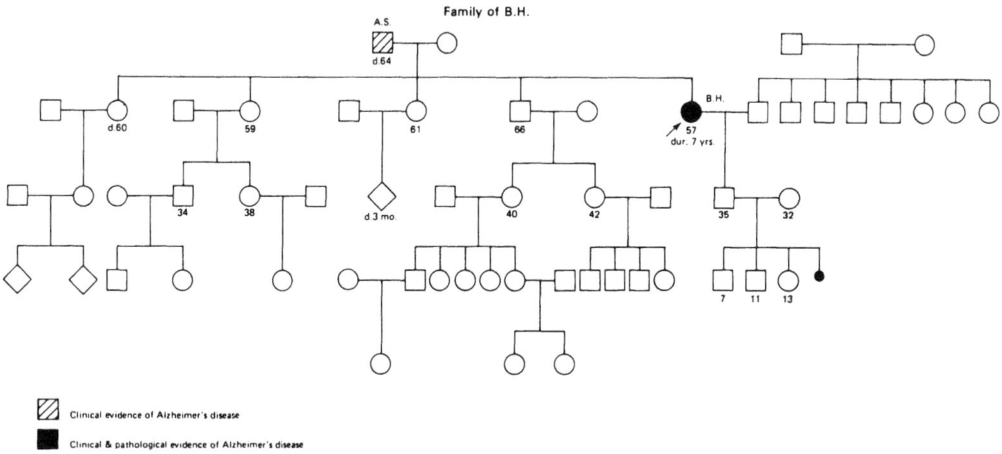

21b. H family. Brain tissue obtained from patient B. H. at surgical biopsy induced subacute spongiform encephalopathy in a squirrel monkey and a capuchin monkey 29 1/2 months and 43 months, respectively, after intracerebral inoculations. The patient, a 57 year old woman, has had slowly progressive dementia and deterioration for the past 7 years. Neuropathological findings revealed abundant neurofibrillary tangles and senile plaques and no evidence of status spongiosis. The patient's father, A. S., had died at age 64 following several years of progressive dementia, behavioral change and memory loss. B. H. is presently alive and institutionalized.

cannot claim to have transmitted the classical sporadic Alzheimer's disease to primates, we are confronted with the anomaly that the familial form of Alzheimer's disease has, in these two instances, transmitted as though it were CJD.

The above findings have added impetus to our already extensive studies of Huntington's chorea, Alzheimer's and Pick's diseases, parkinsonism-dementia, senile dementia, and even "dementia praecox", the organic brain disease associated with late uncontrolled schizophrenia.

SCRAPIE

Scrapie is a natural disease of sheep, and occasionally of goats, that has widespread distribution in Europe, America, and Asia. Affected animals show progressive ataxia, wasting, and frequently severe pruritis. The clinical picture and histopathological findings of scrapie closely resemble those of kuru; this permitted Hadlow (35) to suggest that both diseases might have similar etiologies. As early as 1936, Cuillé and Chelle (5b) had transmitted scrapie to the sheep, and its filterable nature and other virus-like properties had been demonstrated two to three decades ago (26). Because scrapie is the only one of the subacute spongiform virus encephalopathies that has been serially transmitted in mice, much more virological information is available about this agent than about the viruses that cause the human diseases.

Although scrapie has been studied longer and more intensely than the other diseases, the mechanism of its spread in nature remains uncertain. It may spread from naturally infected sheep to uninfected sheep and goats, although such lateral transmission has not been observed from experimentally infected sheep or goats. Both sheep and goats, as well as mice, have been experimentally infected by the oral route. It appears to pass from ewes to lambs, even without suckling; the contact of the lamb with the infected ewe at birth appears to be sufficient, because the placenta itself is infectious (39). Transplacental versus oral, nasal, optic, or cutaneous infection in the perinatal period, are unresolved possibilities. Older sheep are infected only after long contact with diseased animals; however, susceptible sheep have developed the disease in pastures previously occupied by scrapied sheep.

Both field studies and experimental work have suggested genetic control of disease occurrence in sheep. In mice, there is evidence of genetic control of length of incubation period and of the anatomic distribution of lesions, which is also dependent on the strain of scrapie agent used. Scrapie has been transmitted in our laboratory to five species of monkeys (Tables 9 and 10) (23, 31, 32), and such transmission has occurred using infected brain from naturally infected sheep and from experimentally infected goats and mice (Figures 22a, b, c). The disease produced is clinically and pathologically indistinguishable from experimental CJD in these species.

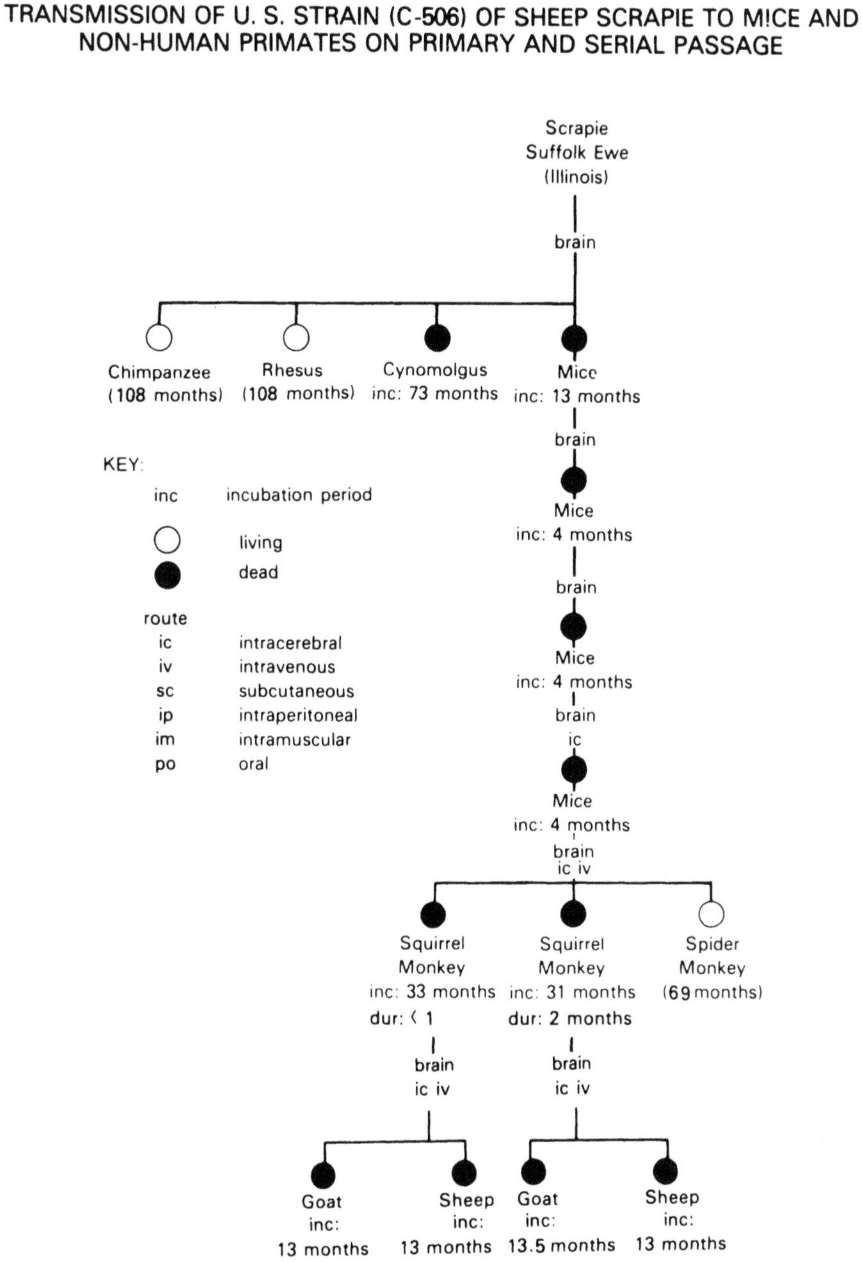

Figure 22. Scrapie has been transmitted to three species of New World monkeys and two species of Old World monkeys (Tables 9, 10).

22a. Transmission of scrapie from the brain of a scrapie-infected Suffolk ewe (C506) in Illinois to a cynomolgus monkey, and from the 4th mouse passage of this strain of scrapie virus to two squirrel monkeys. Incubation period in the cynomolgus was 73 months and in the squirrel monkeys 31 and 33 months. A chimpanzee and a rhesus monkey inoculated 109 months ago with this sheep brain remain well, as does a spider monkey inoculated 70 months ago with brain from the 4th passage of the C506 strain of scrapie in mice.

TRANSMISSION OF GOAT-ADAPTED AND MOUSE-ADAPTED SCRAPIE (COMPTON STRAIN) TO NON-HUMAN PRIMATES

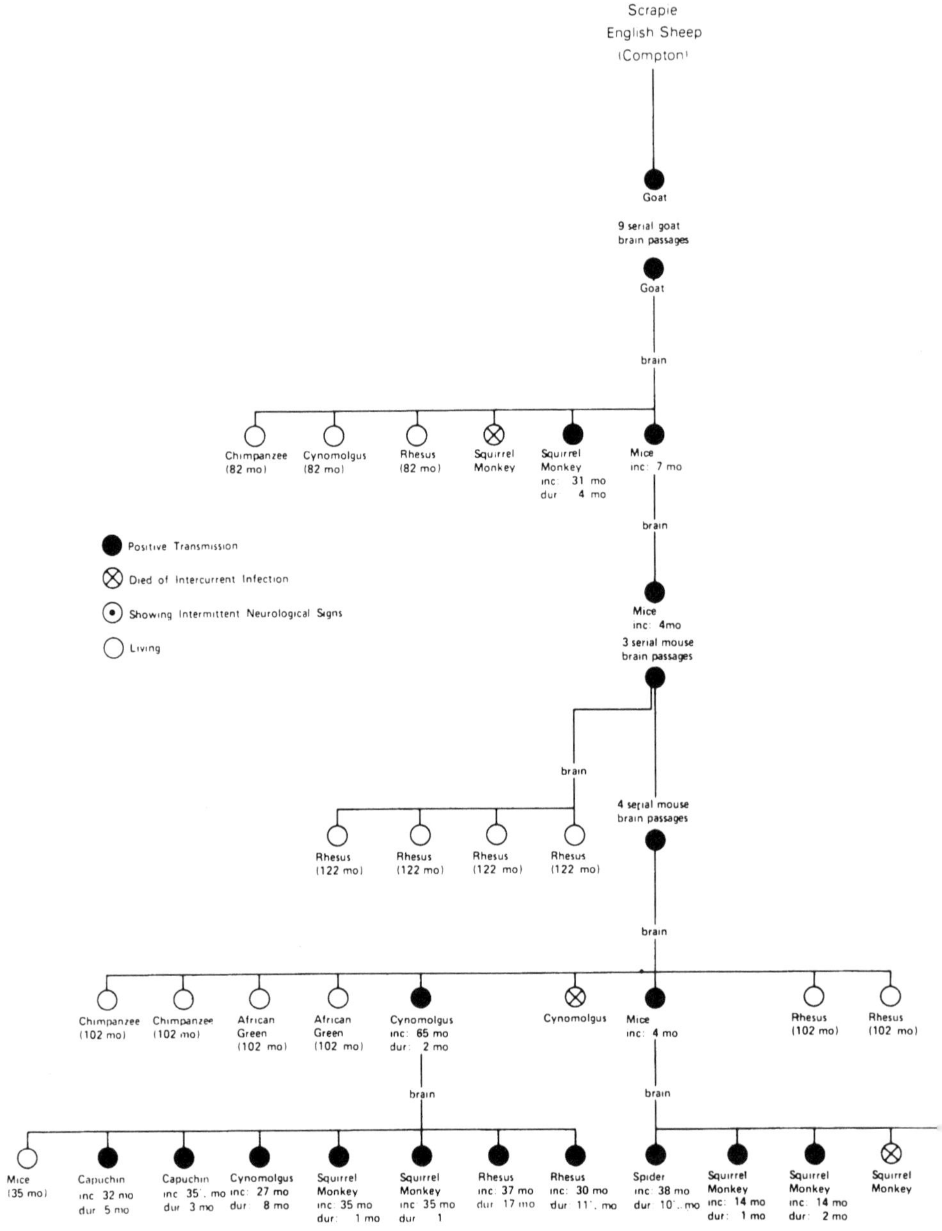

22b. Primary transmission of goat-adapted scrapie (Compton, England strain) to the squirrel monkey and to mice and the transmission of mouse-adapted scrapie to two species of Old World and three species of New World monkeys. Numbers in parentheses are the number of months elapsed since inoculation, during which the animal remained asymptomatic.

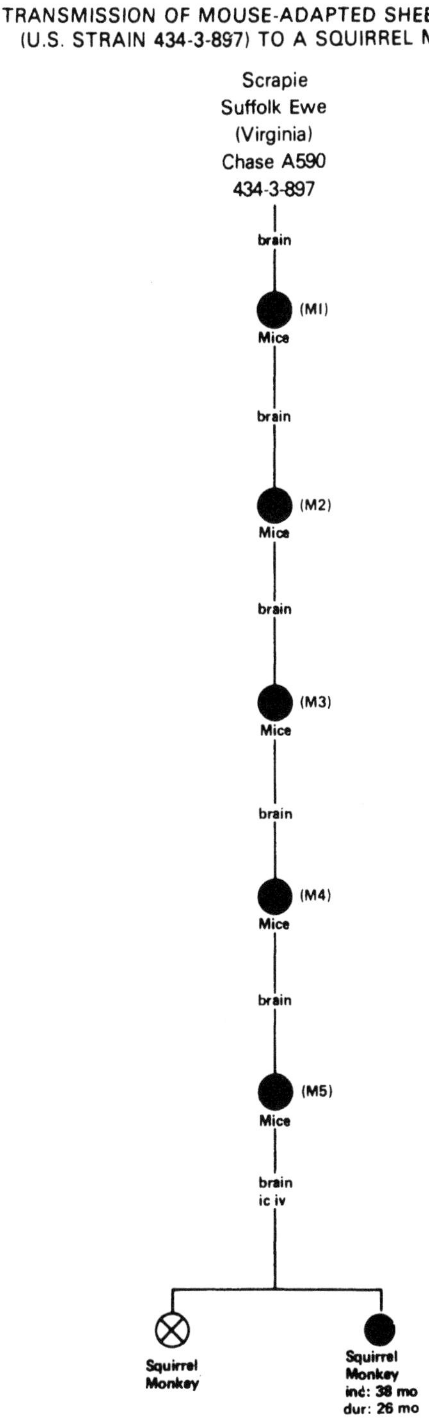

22c. Transmission of mouse-adapted sheep scrapie (U. S. strain 434-3-897) to a squirrel monkey 38 months following intracerebral inoculation with a suspension of scrapie-infected mouse brain containing $10^{7.3}$ infectious units of virus per ml. This animal showed signs of ataxia, tremors and incoordination, and the disease was confirmed histologically. See (b) for an explanation of symbols.

TRANSMISSIBLE MINK ENCEPHALOPATHY

Transmissible mink encephalopathy (TME) is very similar to scrapie both in clinical picture and in pathological lesions. On the ranches on which it developed, the carcasses of scrapie-infected sheep had been fed to the mink; presumably the disease is scrapie. The disease is indistinguishable from that induced in mink by inoculation of sheep or mouse scrapie. Like scrapie, TME has been transmitted by the oral route, but transplacental or perinatal transmission from the mother has not been demonstrated. Physicochemical study of the virus has thus far revealed no differences between TME and the scrapie virus (42, 49).

The disease has been transmitted to the squirrel, rhesus, and stump-tailed monkey (Tables 9 and 10; Fig. 13), and to many nonprimate hosts, including the sheep, goat, and ferret, but has not been shown to transmit to mice (Table 11). In monkeys the illness is indistinguishable from experimental CJD in these species.

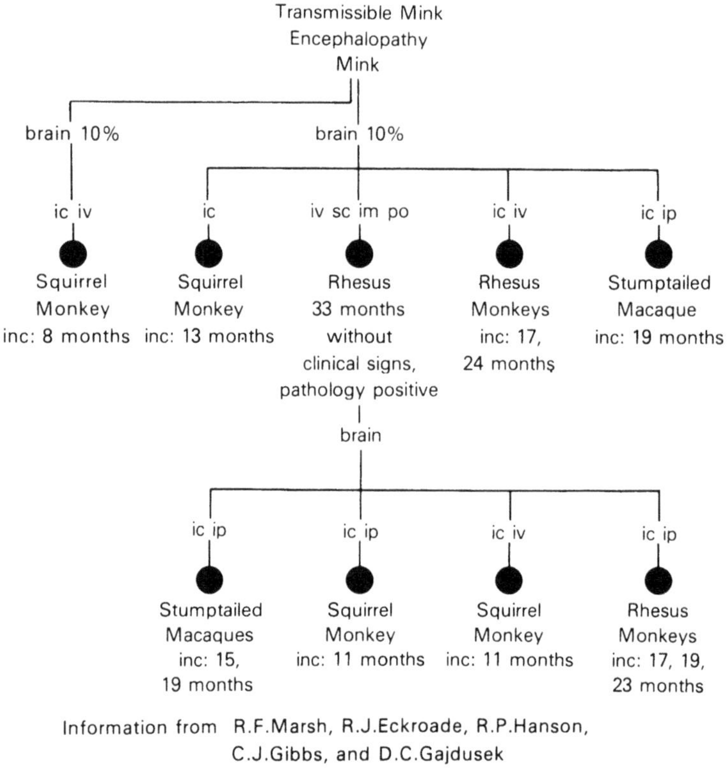

Figure 23. Transmissible mink encephalopathy (TME), a rare disease of American ranch mink, is possibly a form of scrapie. The clinical picture and histopathological lesions attendant in the brain, resemble that of scrapie, and scrapie sheep carcasses were fed to mink on ranches on which TME appeared. The disease is transmissible to sheep, goats, certain rodents and New and Old World monkeys. Illustrative data on the primary transmissions of transmissible mink encephalopathy to one species of New World monkey and two species of Old World monkeys, and serial passage of the virus in squirrel, rhesus and stumptailed monkeys are presented in this Figure. Incubation periods are shown in months that elapsed between inoculation and onset of clinical disease. (Figure includes information from our laboratory and from R. F. Marsh, R. J. Eckroade, and R. P. Hanson.)

ORIGIN AND SPREAD OF KURU

Unanswered crucial questions posed by all of these agents are related to their biological origin and mode of survival in nature. The diseases they evoke are not artificial diseases, produced by researchers tampering with cellular macromolecular structures, as some would have it. They are naturally occurring diseases, for none of which do we know the mode of dissemination or maintenance which is adequate to explain their long-term persistence. For kuru we have a full explanation of the unique epidemiological findings and their change over the past two decades: the contamination of close kinsmen within a mourning family group by the opening of the skull of dead victims in a rite of cannibalism, during which all girls, women, babes-in-arms, and toddlers of the kuru victim's family were thoroughly contaminated with the virus (15, 17, 21). The disease is gradually disappearing with the cessation of cannibalism and has already disappeared in children, with progressively increasing age of the youngest victims (Figs. 7—9, 24, 26). However, this does not provide us with a satisfactory explanation for the origin of kuru. Was it the unlikely event of a sporadic case of worldwide CJD, which in the unusual cultural setting of New Guinea produced a unique epidemic? We now have the report of a spontaneous case of CJD in a 26 year old native Chimbu New Guinean from the Central Highlands, whose clinical diagnosis was proved by

Figure 24. A Fore mother mourning over the body of her dead daughter, who has just died of kuru. The deep decubitus ulcer below her right hip indicates her chronic debility, which is in contrast to her good nutritional state. Men, and already initiated boys, rarely participated in the mourning rite around the corpse, and even more rarely in the dissection and preparation of the kuru victim's flesh for its ritual endocannibalistic consumption.

Figure 25. All cooking, including that of human flesh from diseased kinsmen, was done in pits with steam made by pouring water over the hot stones, or cooked in bamboo cylinders in the hot ashes. Children participated in both the butchery and the handling of cooked meat, rubbing their soiled hands in their armpits or hair, and elsewhere on their bodies. They rarely or never washed. Infection with the kuru virus was most probably through the cuts and abrasions of the skin, or from nose-picking, eye rubbing, or mucosal injury.

light- and electronmicroscopic examination of a brain biopsy specimen (24, 37a). Serial passage of brain in main in successive cannibalistic rituals might have resulted in a change in the clinical picture of the disease, with modification of the virulence of the original agent.

If such spontaneous CJD is not related to the origin of kuru, another possibility might be that the serial brain passage that occurred in this ritual inoculation of brain from successive victims in multiple sequential passages into their kinsmen yielded a new neurotropic strain of virus from some well-known virus. Finally, in view of what occurs in the defective replication of measles virus in patients with SSPE, we must wonder if a ubiquitous or, at least, a well-known virus may not be modified into a defective, incomplete, or highly integrated or repressed agent *in vivo* in the course of its long masked state in the individual host. Such a new breed of virus may no longer be easily recognizable either antigenically or structurally, because of failure of full synthesis of viral subunits or of their assembly into a known virion. Therefore, we may ask if kuru does not contain some of the subunits of a known agent, modified by its unusual passage history (15, 16, 22).

a

b

26c

Figure 26a. An Awa boy just before first stage initiation, while still living in the women's house with his sisters and small pigs. At this age, boys were already well trained in the use of bows and arrows in hunting.

26b. Youthful Awa toxophilite, already a warrior.

26c. Young Awa warriors in their boy's house.

27a

27b

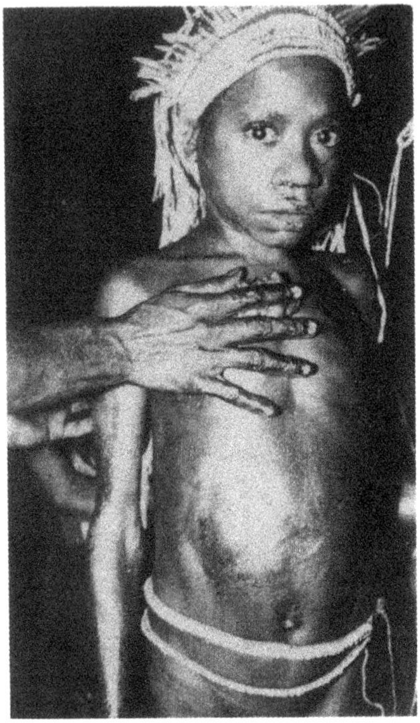

27c

27d

Figure 27. Boys of prepubertal age were removed from the women's houses to enter the *wa'e*, men's house, after elaborate first-stage initiation ceremonies. Thereafter, and for the rest of their lives, they would live, eat, and sleep separately from the women. Married men did not share the houses of their wives, and sexual activity was restricted to daylight in the secluded privacy of the gardens. Three Fore boys are shown in the first stage of initiation in 4 sequences (a—d) during their ceremonial adornment, after having been held in seclusion for several days and having their nasal septa pierced.

27a. Bark strips have been braided into their hair.

27b. Bands of shells of high value to the Fore are fastened to their foreheads.

27c. Their bodies are rubbed with pig grease.

27d. They are given new bark sporans and bows and arrows.

CONJECTURAL NATURAL HISTORY OF THE SUBACUTE SPONGIFORM VIRUS ENCEPHALOPATHIES: HYPOTHETICAL ORIGIN OF CREUTZFELDT-JAKOB DISEASE, KURU, AND TRANSMISSIBLE MINK ENCEPHALOPATHY FROM NATURAL SHEEP SCRAPIE

Scrapie has now been found to cause a disease clinically and neuropathologically indistinguishable from experimental Creutzfeldt-Jakob disease in three species of New World and two species of Old World monkeys (Tables 9 and 10). This disease occurs after either intracerebral or peripheral routes of inoculation. Natural sheep scrapie, as well as experimental goat and mouse scrapie strains of virus have caused disease in the monkeys. The Compton strain of scrapie virus, as a result of such passage through primates, develops an altered host range, for it no longer produces disease in inoculated mice, sheep and goats. A similar situation has been noted to prevail when scrapie is produced in ferrets or mink; the mink or ferret brain virus is no longer pathogenic for mice. This is also true for the virus of natural mink encephalopathy, which, presumably, had its origin in the feeding of scrapie sheep carcasses to mink on commercial mink farms.

Creutzfeldt-Jakob disease or kuru viruses may produce, after over two years of asymptomatic incubation, an acute central nervous disease with death in a few days in the squirrel monkey; even sudden death without previously noted clinical disease has been seen. The same strains of kuru or CJD viruses produce chronic clinical disease in the spider monkey, closely mimicking the human disease, after incubation periods of two years or more. The time sequence of disease progression also mimics that in man, ranging from several months to over a year until death. A single strain of kuru or CJD virus may cause severe status spongiosus lesions in many brain areas, particularly the cerebral cortex in chimpanzees and spider monkeys with minimal or no involvement of the brainstem or spinal cord, whereas in the squirrel monkey this same virus strain may cause extensive brainstem and cord lesions.

From the above findings, it is clear that neither incubation periods nor host range, nor the distribution or severity of neuropathological lesions, can be interpreted as having any significance toward unraveling the possible relationships of the four viruses causing the subacute spongiform virus encephalopathies.

As mentioned earlier, we have found that the prevalence of CJD in the United States and abroad appears to be about one per million whenever extensive neurological survey for cases is instituted. In a study in Israel, an overall prevalence in Jews of Libyan origin is 30 times as high as in Jews of European origin (40). The custom of eating the eyeballs and brains of sheep in the Jewish households of North African and Middle Eastern origin, as opposed to Jewish households of European origin, has understandably given rise to the conjecture that scrapie-infected sheep tissue might be the source of such CJD infection (37).

Figure 28 presents a conjectural schematic natural history of the subacute spongiform virus encephalopathies in which the hypothetical origin of CJD,

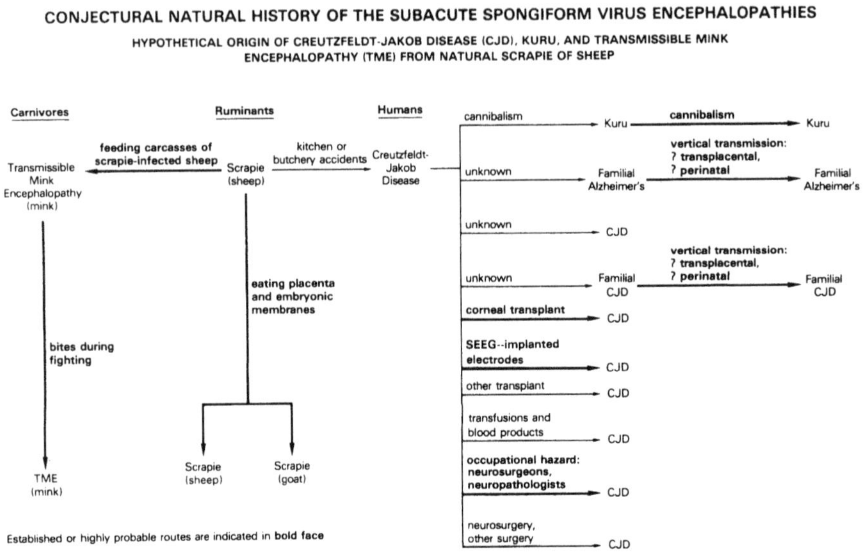

Figure 28.

kuru, and TME from natural scrapie in sheep is proposed with possible routes of transmission indicated. However, such games of armchair speculation provide schemata that cannot yet be tested. They may, nevertheless, have heuristic value. In the absence as yet of proven antigenicity or identified infectious nucleic acid in the agents, neither serological specificity nor nucleic acid homology can be used to answer the compelling question of the relationship between the viruses of kuru, transmissible virus dementia, scrapie, and transmissible mink encephalopathy.

The possibility that the viruses of all four of the subacute spongiform virus encephalopathies are not just closely related agents, but different strains of a single virus which have been modified in different hosts, is easily entertained. The passage of sheep scrapie into other sheep and into goats, at least by the route of feeding of material contaminated with placenta and embryonic membrane (53), and into mink from feeding carcasses of scrapied sheep, are established paths of scrapie transmission. In view of the experimental transmission of scrapie to monkeys, there is serious cause for wonder whether kitchen and butchery accidents involving the contamination of skin and eyes may not be a possible source of CJD in man (36a, 37). We believe that contamination during the cannibalistic ritual was the sole source of transmission of kuru from man to man, and have conjectured above that a spontaneous case of CJD may have given rise to the chain of kuru transmissions (17). The documented case of CJD from corneal transplant (12) suggests that other tissue transplantation may also be a source of infection. It is known that the virus is present in peripheral tissue, as well as in the brain. The case of CJD in a neurosurgeon who had frequently performed autopsies (27), poses a possibility of occupational hazard to the neurosurgeon and neuropathologist (61—63). Finally, the rather frequent report of neurosurgery or other surgery preceding the appearance of CJD, as noted by us (62) and by other workers (50, 51), may indicate that such surgery has been a source of infection, rather than a virus activating incident. This seems to be a real hazard in view of the recent episode of transmission of CJD to two patients from the use of CJD-contaminated electrodes in stereotactic EEG during surgery for epilepsy (3a). The use of formaldehyde for their sterilization was, in view of the resistance of the unconventional viruses to it (26), a very unfortunate choice. The mode of transmission, which at first sight would appear to be vertical in the cases of familial CJD or familial Alzheimer's disease, remains unknown (4, 13, 50, 62). Whether infection is transovarian or occurs in utero or during parturition, or from a milk factor or some other neonatal infection, also remains unknown, although from kuru epidemiological study (i.e., failure to see kuru in children born to kuru-affected mothers since the cessation of cannibalism), we have no evidence for such transmission (17).

PROSPECT

The elucidation of the etiology and epidemiology of a rare, exotic disease restricted to a small population isolate—kuru in New Guinea—has brought us to worldwide considerations that have importance for all of medicine and microbiology. For neurology, specifically, we have considerable new insights into the whole range of presenile dementias and, in particular, to the large problems of Alzheimer's disease and the senile dementias. The implications of vertical transmission of slow virus infections, and of host genetic control of disease expression for all genetic diseases, and the relationship of these slow virus infectious processes to those which may lead to neoplastic transformation, are obvious.

However, the major problems among the degenerative diseases: multiple sclerosis, amyotrophic lateral sclerosis, and parkinsonism remain unsolved, although there are tantalizing laboratory and epidemiological data pointing to the possible role of virus-like agents in these diseases. Perhaps the masked and defective slow infections with conventional viruses such as are seen in PML and SSPE, may provide the best leads for studying these diseases.

The foci of high incidence of amyotrophic lateral sclerosis with associated high incidence of parkinsonism-dementia complex among the Chamorro people on Guam and the Japanese of the Kii Peninsula remain continuing challenges. Our discovery (14) and reevaluation (20) of the very small but very intense focus of such motor neuron disease with associated high incidence of parkinsonism, parkinsonism-dementia, and other peculiar bradykinetic and myoclonic dementia syndromes among the Auyu and Jaqai people in a remote population of West New Guinea, suggests strongly that some common etiological factor may underly the occurrence of all these very different syndromes, as they occur strangely in this one small population and are not found in the much larger surrounding populations.

The models of lysogenicity and of subviral genetically active macromolecular structures from the study of bacterial viruses and bacterial genetics supply ample imaginative framework for an expression of our ideas of possible mechanisms of infectious pathogenesis in man. The unconventional viruses tax even our imagination in relation to molecular biology gained from these studies in bacteria.

For a now-disappearing disease in a small primitive population to have brought us this far is ample reason for pursuing intensively the challenges offered by the still inexplicable high incidence and peculiar profusion of different neurological syndromes, pathologically distinct yet apparently somehow related to each other, which have been discovered in the several small population enclaves (14, 20, 21).

REFERENCES

1. Beck, E., Daniel, P. M., Alpers, M., Gajdusek, D. C., Gibbs, C. J., Jr. and Hassler, R. (1975): Experimental kuru in the spider monkey. Histopathological and ultrastructural studies of the brain during early stages of incubation. Brain 98: 592—620.
2. Beck, E., Daniel, P. M., and Gajdusek, D. C. (1966): A comparison between the neuropathological changes in kuru and scrapie, a system degeneration. In: Proceedings of the Fifth International Congress of Neuropathology, F. Luthy and A. Bischoff, eds., pp. 213—218. Excepta Medica International Congress Series No. 100, Amsterdam.
3. Beck, E., Daniel, P. M., Matthews, W. B., Stevens, D. L., Alpers. M. P., Asher, D. M., Gajdusek, D. C. and Gibbs, C. J., Jr. (1969): Creutzfeldt-Jakob disease: the neuropathology of a transmission experiment. Brain 92: 699—716.

3a. Bernoulli, C., Siegfried, J. Baumgartner, G., Regli, F., Rabinowicz, T., Gajdusek, D. C., and Gibbs, C. J., Jr. (1977): Danger of accidental person-to-person transmission of Creutzfeldt-Jakob disease by surgery. Lancet 1 (8009): 478—479.

4. Bobowick, A., Brody, J. A., Matthews, M. R., Roos, R. and Gajdusek, D. C. (1973): Creutzfeldt-Jakob disease: a case control study. Am. J. Epidemiol 98: 381—394.
5. Brownell, B., Campbell, M. J. and Greenham, L. W. (1975): The experimental transmission of Creutzfeldt-Jakob disease. 51st Annual Meeting, American Association of Neuropathologists, May 30—June 1, New York. Program and Abstracts, #32, p. 46.

5a. Brownell, B., Campbell, M. J., Greenham, L. W. and, Peacock, D. B. (1975): Experimental transmission of Creutzfeldt-Jakob disease. Lancet 2 (7926): 186—187.

5b. Cuillé, J. and Chelle, P.-L. (1936): Pathologie animale la maladie dite tremblante du mouton est-elle inoculable? C. R. Acad. Sci. (D). (Paris) 203: 1552—1554.

6. Dawson, J. R. Jr. (1933): Cellular inclusions in cerebral lesions of lethargic encephalitis. Am. J. Pathol. 9: 7—16.
7. Diener, T. O. (1973): Similarities between the scrapie agent and the agent of potato spindle tuber disease. Ann. Clin. Res. 5: 268—278.
8. Diener, T. O. (1974): Viroids: the smallest known agents of infectious disease. Ann. Rev. Microbiol. 28: 23—29.
9. Diener, T. O. (1976): Towards an understanding of viroid nature and replication. Ann. Microbiol. (Inst. Pasteur) 127A, pp. 7—17.
10. Diener, T. and Hadidi, A. (1977): Viroids. In: Comprehensive Virology, H. Fraenkel-Conrat and R. R. Wagner, eds. Plenum Press, New York. In press.
11. Dubois-Dalcq, M., Rodriguez, M., Reese, T. S., Gibbs, C. J., Jr. and Gajdusek, D. C. (1977): Search for a specific marker in the neural membranes of scrapie mice (a freeze-fracture study). Lab. Invest. 36: 547—553.
12. Duffy, P., Wolf, J., Collins, G., DeVoe, A. G., Steeten, B. and Cowen, D. (1974): Possible person-to-person transmission of Creutzfeldt-Jakob disease. New Engl. J. Med. 299: 692—693.
13. Ferber, R. A., Wiesenfeld, S. L., Roos, R. P., Bobowick, A. R., Gibbs, C. J., Jr. and Gajdusek, D. C. (1974): Familial Creutzfeldt-Jakob disease: transmission of the familial disease to primates. In: Proceedings of the X International Congress of Neurology, A. Subirana, J. M. Espadaler and E. H. Burrows, eds., September 8—15, 1973, Barcelona, pp. 358—380. Excerpta Medica International Congress Series No. 296, Amsterdam.
14. Gajdusek, D. C. (1963): Motor-neuron disease in natives of New Guinea. New Engl. J. Med. 268: 474—476.
15. Gajdusek, D. C. (1972): Spongiform virus encephalopathies. In: Host Virus Reactions with Special Reference to Persistent Agents, G. Dick, ed. J. Clin. Pathol. (Suppl.) 25: 78—83.

16. Gajdusek, D. C. (1973): Kuru and Creutzfeldt-Jakob disease. Experimental models of noninflammatory degenerative slow virus diseases of the central nervous system. Ann. Clin. Res. 5: 254—261.
17. Gajdusek, D. C. (1973): Kuru in the New Guinea Highlands. In: Tropical Neurology, J. D. Spillane, ed., pp. 376—383. Oxford Press, New York.
18. Gajdusek, D. C., ed. (1976): Correspondence on the Discovery and Original Investigations of Kuru. Smadel-Gajdusek Correspondence 1956—1959. National Institutes of Health, Bethesda, Maryland.
19. Gajdusek, D. C. (1977): Urgent opportunistic observations: the study of changing, transient, and disappearing phenomena of medical interest in disrupted primitive human communities. In: Health and Disease in Isolated and Tribal Societies, Julie Whelan, ed. Ciba Foundation Monograph 49 pp. 69—102.
20. Gajdusek, D. C. (1977): Focus of high incidence of motor neuron disease associated with high incidence of Parkinsonism and dementia syndromes in a small population of Awyu New Guineans. New Engl. J. Med. In preparation.
21. Gajdusek, D. C. (1959—1977): Journals 1956—1976, 21 volumes, published in limited edition. National Institutes of Health, Bethesda, Maryland.
22. Gajdusek, D. C. and Gibbs, C. J., Jr. (1973): Subacute and chronic diseases caused by atypical infections with unconventional viruses in aberrant hosts. In: Perspectives in Virology, 8, M. Pollard, ed., pp. 279—311. Academic Press, New York.
23. Gajdusek, D. C. and Gibbs, C. J., Jr. (1975): Familial and sporadic chronic neurological degenerative disorders transmitted from man to primates. In: Primate Models of Neurological Disorders, B. S. Meldrum and C. D. Marsden. eds. Adv. Neurol. 10: 291—317. Raven Press, New York.
24. Gajdusek, D. C. and Gibbs, C. J., Jr. (1975): Slow virus infections of the nervous system and the Laboratories of Slow, Latent and Temperate Virus Infections. In: The Nervous System, D. B. Tower, ed., Vol. 2, The Clinical Neurosciences, T. N. Chase, ed., pp. 113—135.
25. Gajdusek, D. C., Gibbs, C. J., Jr. and Alpers, M. (1966): Experimental transmission of a kuru-like syndrome in chimpanzees. Nature 209: 794—796.
26. Gajdusek, D. C., Gibbs, C. J., Jr. and Alpers, M., eds. (1965): Slow, Latent and Temperate Virus Infections. NINDB Monograph No. 2, National Institutes of Health. PHS Publication No. 1378, U. S. Govt. Printing Office, Washington, D. C., 489 pp.
26a. Gajdusek, D. C., Gibbs, C. J., Jr., Collins, G. and Traub, R. (1976): Survival of Creutzfeldt-Jakob disease virus in formol-fixed brain tissue. New Engl. J. Med. 294: 553.
27. Gajdusek, D. C., Gibbs, C. J., Jr., Earle, K., Dammin, C. J., Schoene, W. and Tyler, H. R. (1974): Transmission of subacute spongiform encephalopathy to the chimpanzee and squirrel monkey from a patient with papulosis atrophicans maligna of Köhlmeier-Degos. In: Proceedings of the X International Congress of Neurology, A. Subirana, J. M. Espadaler and E. H. Burrows, eds., September 8—15, 1973, Barcelona, pp. 390—392. Excerpta Medica International Congress Series No. 319, Amsterdam.
28. Gajdusek, D. C. and Zigas, V. (1957): Degenerative disease of the central nervous system in New Guinea. The endemic occurrence of "kuru" in the native population. New Engl. J. Med. 257: 974—978.
29. Gajdusek, D. C. and Zigas, V. (1959): Kuru: clinical, pathological and epidemiological study of an acute progressive degenerative disease of the central nervous system among natives of the Eastern Highlands of New Guinea. Am. J. Med. 26: 442—469.
30. Garzuly, F., Jellinger, K. and Pilz, P. (1971): Subakute spongiose encephalopathie (Jakob-Creutzfeldt-Syndrom). Klinische-morphologische Analyse von 9 fällen. Arch. Psychiatr. Nervenkr. 214: 207—227.
31. Gibbs, C. J., Jr. and Gajdusek, D. C. (1972): Transmission of scrapie to the cynomolgus monkey (Macaca fascicularis). Nature 236: 73—74.
32. Gibbs, C. J., Jr. and Gajdusek, D. C. (1976): Studies on the viruses of subacute spongiform encephalopathies using primates, their only available indicator. First Inter-American Conference on Conservation and Utilization of American Nonhuman Pri-

mates in Biomedical Research, Lima, Peru, June 2—4. PAHO Scientific Publication No. 317, pp. 83—109. Washington, D. C.

33. Gibbs, C. J., Jr., Gajdusek, D. C., Asher, D. M., Alpers, M. P., Beck, E., Daniel, P. M. and Matthews, W. B. (1968): Creutzfeldt-Jakob disease (subacute spongiform encephalopathy): transmission to the chimpanzee. Science 161: 388—389.
34. Hadidi, A., Jones, D. M., Gillespie, D. H., Wong-Staal, S. and Diener, T. O. (1976): Hybridization of potato spindle tuber viroid to cellular DNA of normal plants. Proc. Nat. Acad. Sci. (USA) 73: 2453—2457.
35. Hadlow, W. J. (1959): Scrapie and kuru. Lancet 2: 289—290.
36. Haig, D. C., Clarke, M. C., Blum, E. and Alper, T. (1969): Further studies on the inactivation of the scrapie agent by ultraviolet light. J. Gen. Virol. 5: 455—457.
36a. Harris, R. A. (1977): A reporter at large: a nice place to live. New Yorker (April 25): 48—91 (citation from page 53).
37. Herzberg, L., Herzberg, B. N., Gibbs, C. J., Jr., Sullivan, W., Amyx, H. and Gajdusek, D. C. (1974): Creutzfeldt-Jakob disease: hypothesis for high incidence in Libyan Jews in Israel. Science 186: 848.
37a. Hornabrook, R. W., and Wagner, F. (1975): Creutzfeldt-Jakob disease. Papua New Guinea Medical Journal 18:226—228.
38. Horta-Barbosa, L., Fuccillo, D. A., London, W. T., Jabbour, J. T., Zeman, W. and Sever, J. L. (1969): Isolation of measles virus from brain cell cultures of two patients with subacute sclerosing panencephalitis. Proc. Soc. Exp. Biol. Med. 132: 272—277.
39. Hunter, G. D., Collis, S. C., Millson, G. C. and Kimberlin, R. H. (1976): Search for scrapie-specific RNA and attempts to detect an infectious DNA or RNA. J. Gen. Virol. 32: 157—162.
40. Kahana, E., Alter, M., Braham, J. and Sofer, D. (1974): Creutzfeldt-Jakob disease: focus among Libyan Jews in Israel. Science 183: 90—91.
41. Klatzo, I., Gajdusek, D. C. and Zigas, V. (1959): Pathology of kuru. Lab. invest. 8: 799—847.
42. Kimberlin, R. H. and Marsh, R. F. (1975): Comparison of scrapie and transmissible mink encephalopathy in hamsters. I. Biochemical studies of brain during development of disease. J. Infect. Dis. 131: 97—103.
43. Lampert, P. W., Gajdusek, D. C. and Gibbs, C. J., Jr. (1972): Subacute spongiform virus encephalopathies. Scrapie, kuru and Creutzfeldt-Jakob disease. Am. J. Pathol. 68: 626—646.
44. Lampert, P., Hooks, J., Gibbs, C. J., Jr. and Gajdusek, D. C. (1971): Altered plasma membranes in experimental scrapie. Acta Neuropathol. (Berlin) 19: 80—93.
45. Latarjet, R., Gajdusek, D. C. and Gibbs, C. J., Jr. (1977): Unusual resistance to UV and ionizing radiation of kuru and scrapie by ionizing radiation. In preparation.
46. Latarjet, R., Muel, B., Haig, D. A., Clarke, M. C. and Alper, T. (1970): Inactivation of the scrapie agent by near-monochromatic ultraviolet light. Nature 227: 1341—1343.
47. Lord, Ann. Sutton, R. N. P. and Corsellis. J. A. N. (1975): Recovery of adenovirus type 7 from human brain cell cultures. J. Neurol. Neurosurg. Psychiat. 38: 710—712.
48. Manuelidis, E. E. (1975): Transmission of Creutzfeldt-Jakob disease from man to the guinea pig. Science 190: 571—572.
48a. Manuelidis, E. E., Kim, J., Angelo, J. N., and Manuelidis, L. (1976): Serial propagation of Creutzfeldt-Jakob disease in guinea pigs. Proc. Nat. Acad. Sci. (USA) 73:223—227.
48b. Manuelidis, E. E., Angelo, J. N., Gorgacz, E. J., and Manuelidis, L. (1977): Transmission of Creutzfeldt-Jakob disease to Syrian hamster. Lancet 1 (8009): 479.
49. Marsh, R. F. and Kimberlin, R. H. (1975): Comparison of scrapie and transmissible mink encephalopathy in hamsters. II. Clinical signs, pathology, and pathogenesis. J. Infect. Dis. 131: 104—110.
50. Matthews, W. B. (1975): Epidemiology of Creutzfeldt-Jakob disease in England and Wales. J. Neurol. Neurosurg. Psychiat. 38: 210—213.
51. Nevin, S., McMenemy, W. H., Behrman, D. and Jones, D. P. (1960): Subacute

spongiform encephalopathy. A subacute form of encephalopathy attributable to vascular dysfunction (spongiform cerebral atrophy). Brain 83: 519—564.

52. Padgett, B. L., ZuRhein, G. M., Walker, D. L., Eckroade, R. J. and Dessel, B. H. (1971): Cultivation of papova-like virus from human brain with progressive multifocal leucoencephalopathy. Lancet 1(7712): 1257—1260.
53. Pattison, I. H., Hoare, M. N., Jebbett, J. N. and Watson, W. A. (1972): Spread of scrapie to sheep and goats by oral dosing with fetal membranes from scrapie affected sheep. Vet. Rec. 99: 465—467.
54. Payne, F. E., Baublis, J. V. and Itabashi, H. H. (1969): Isolation of measles virus from a patient with subacute sclerosing panencephalitis. New Eng. J. Med. 281: 585—589.
55. Randles, J. W., Rillo, E. P. and Diener, T. O. (1976): The viroidlike structure and cellular location of anomalous RNA associated with the Cadang-Cadang disease. Virology 74: 128—129.
55a. Roos, R., Chou, S. M., Rogers, N. G., Basnight, M. and Gajdusek, D. C. (1972): Isolation of an adenovirus 32 strain from human brain in a case of subacute encephalitis. Proc. Soc. Exper. Biol. Med. 139: 73—74.
56. Roos, R., Gajdusek, D. C. and Gibbs, C. J., Jr. (1973): The clinical characteristics of transmissible Creutzfeldt-Jakob disease. Brain 96: 441—462.
57. Semancik, J. S., Marsh, R. F., Geelen, J. L. M. C. and Hanson, R. P. (1977): Properties of the scrapie agent-endomembrane complex from hamster brain. J. Virol. In press.
58. Semancik, J. S. and Vanderwonde, W. J. (1976): Exocortis disease: cytopathic effect on the plasma membrane in association with the pathogenic RNA. Virology 69(2): 719—726.
59. Sever, J. L. and Zeman, W., eds. (1968): Conference on Measles Virus and Subacute Sclerosing Panencephalitis. Neurology 18: 1(Pt. 2), 192 pp.
60. Siakotos, A. N., Bucana, C., Gajdusek, D. C., Gibbs, C. J., Jr. and Traub, R. D. (1976): Partial purification of the scrapie agent from mouse brain by pressure disruption and zonal centrifugation in a sucrose-sodium chloride gradient. Virology 70: 230—237.
60a. Townsend, J. J., Baringer, J. R., Wolinsky, J. S., Malamud, N., Mednick, J. P., Panitch, H. S., Scott, R. A. T., Oshiro, L. S. and Cremer, N. E. (1975): Progressive rubella panencephalitis. Late onset after congenital rubella. New Eng. J. Med. 292: 990—993.
61. Traub, R. D., Gajdusek, D. C. and Gibbs, C. J., Jr. (1974): Precautions in conducting biopsies and autopsies on patients with presenile dementia. J. Neurosurg. 41: 394—395.
62. Traub, R., Gajdusek, D. C. and Gibbs, C. J., Jr. (1977): Transmissible virus dementias. The relation of transmissible spongiform encephalopathy to Creutzfeldt-Jakob disease. In: Aging and Dementia, M. Kinsbourne and L. Smith, eds., pp. 91—146. Spectrum Publishing Inc., Flushing, New York.
63. Traub, R. D., Gajdusek, D. C. and Gibbs, C. J., Jr. (1975): Precautions in autopsies on Creutzfeldt-Jakob disease. Am. J. Clin. Pathol. 64: 417.
63a. Weil, M. L., Itabashi, H. H., Cremer, N. E., Oshiro, L. S., Lennette, E. H. and Carnay, L. (1975): Chronic progressive panencephalitis due to rubella virus simulating subacute sclerosing panencephalitis. New Eng. J. Med. 292: 994—998.
64. Weiner, L. P., Herndon, R. M., Narayan, O., Johnson, R. T., Shah, K., Rubinstein, L. J., Preziosi, T. J. and Conley, F. K. (1972): Isolation of virus related to SV40 from patients with progressive multifocal leucoencephalopathy. New Eng. J. Med. 286: 385—390.
65. Zigas, V. and Gajdusek, D. C. (1957); Kuru: clinical study of a new syndrome resembling paralysis agitans in natives of the Eastern Highlands of Australian New Guinea. Med. J. Australia 2: 745—754.
66. ZuRhein, G. M. and Chou, S. (1968): Papovavirus in progressive multifocal leukoencephalopathy. In: Infections of the Nervous System, H. M. Zimmerman, editor. Research Publication of the Association for Nervous and Mental Diseases, 44: 254—280. Williams and Wilkins, Baltimore, Maryland.

1977
Physiology or Medicine

ROGER GUILLEMIN,
ANDREW V. SCHALLY

"for their discoveries concerning the peptide hormone production of the brain"

and ROSALYN S. YALOW

"for the development of radioimmunoassays of peptide hormones"

THE NOBEL PRIZE FOR PHYSIOLOGY OR MEDICINE

Speech by Professor ROLF LUFT of the Karolinska Medico-Chirurgical Institute
Translation from the Swedish text

Your Majesties, Your Royal Highnesses, Ladies and Gentlemen,
The word "hormones" and associated terms have always stimulated our fantasy. The mystery in connection with hormones has been, from the beginning, equally overwhelming to the researcher and the layman. It is easy to understand why. These were chemical substances with often very powerful actions at concentrations which for a long time seemed so low that they were impossible to measure. However, mystery and belief lead nowhere, at least not in scientific research and medicine. Once one learned to identify the active chemical substances—in this case hormones—and to measure their rate of synthesis, only then did one establish a firm basis for turning fantasy and mystery into reality.

This year's three Nobel laureates in medicine have all made contributions which are outstanding examples of this kind of activity. Rosalyn Yalow's name is for ever associated with her methodology of measuring the presence of hormones in the blood at concentrations as low as one thousand billionths of a gram per milliliter of blood. This was a necessity, since a great many hormones, primarily the so-called protein hormones, are present in the blood in such small quantities. Before Yalow, these hormones could not be determined quantitatively in the blood, and therefore, active research in this field had stagnated.

Rosalyn Yalow and Solomon Berson, her late coworker, discovered by chance that *one* small protein hormone, insulin, following injection into man resulted in a production of antibodies against insulin. All diabetics who receive insulin develop similar antibodies against the administered insulin. The discovery by Yalow and Berson was unacceptable at first—their first scientific paper concerning this observation was even refused publication—since it was commonly believed that proteins as small as these protein hormones were unable to stimulate antibody formation. However, Yalow and Berson did not give up, and furthermore, after a couple of years of intensive work, they presented in 1960 a methodology for the determination of protein hormones in the blood, the fundamental principle of which utilized the ability of these hormones to stimulate antibody formation in man. This methodology, known as the Yalow-Berson method, is genial in all its simplicity, and can even be described in simple terms.

As a result of mixing in a test tube a known quantity of radioactive insulin with a known quantity of antibodies against insulin, a specific amount of the insulin becomes attached to these antibodies. Subsequently, if one adds to this mixture a small amount of blood which contains insulin, the insulin of the blood becomes similarly attached to the antibodies and a certain portion of the radioactive insulin is detached from the antibodies. The higher the concentration of insulin is in the blood sample, the larger is the amount of radioactive insulin that will be detached from the antibodies. The amount of radioactive insulin thus removed can easily be determined, providing an exact measure of the amount of insulin present in the blood sample.

The Yalow-Berson method which makes it possible to determine the exact amounts of all hormones present, represented a real revolution in the field of hormone research. A field where one refers to the time period before Yalow, and the new epoch which began with her achievement. Her methodology and the modifications thereof, subsequently made their triumphant journey far beyond her own field of research, reaching into vast territories of biology and medicine. It has been said that Yalow changed the life of a multitude of researchers within these fields. *Rarely have so many had so few to thank for so much.*

Roger Guillemin and Andrew Schally have also contributed greatly to this field of research, exploring protein hormones. It is justifiable to say that they have uncovered a substantial part of the link between body and soul.

For decades, one has talked about the indivisible homo sapiens, maintaining that our body and soul can not be separated since they form an entity. Emotional and psychic phenomena do influence our bodily functions. Let me give you an example. When American soldiers were sent to the European war scene, thousands of female companions who were left behind, stopped menstruation. They were completely healthy, but the emotional stress had an influence on certain body functions, causing these functions to cease. Through which mechanisms did the psyche thus influence the body?

Psychic phenomena as well as input from the entire body bring about electrical impulses in the brain. This is the language of the nervous system, the brain speaks "electrically". The brain informs some of its centers of what is going on, and these centers relay the message further. Those centers which pass on the information to the hormone producing organs of the body are situated in the midbrain, an area on the base of the brain. Delicate blood vessels in turn connect the midbrain with the pituitary, an important hormone producing gland, often referred to as the hypophysis. This sequence provides the pathway for transmission of information from the surroundings to the brain, to the midbrain, to the pituitary, and thus to all those bodily functions which are influenced and controlled by hormones.

By the mid 1950's it was evident—also here through the contributions of Guillemin and Schally—that the midbrain produces chemical substances which are transported to the pituitary via the delicate blood vessels just mentioned. Once in the pituitary, they determine the exact quantities of the various hypophyseal hormones which must be produced at a given point in time. But which were these substances in the midbrain, evidently passing the information from soul to body?

Guillemin and Schally worked independently in different parts of the U.S.A. together with their large staff of coworkers, trying to isolate one of these chemical substances, and both researchers concentrated on the same substance. Each started with five million pieces taken from the midbrain of sheeps and pigs—half a ton—and in 1969, after years of arduous labor, they each came up with 1 milligram of the purified hormonal substance. *Rarely have so many gained so little from so much.*

Guillemin and Schally were the first to isolate several of the communicating chemical links between the brain and the pituitary, and they also determined their structure and succeeded in synthesizing them.

The discoveries by Guillemin and Schally brought on a revolution in their own field of research. Still other protein hormones have subsequently been isolated from the midbrain, this wondrous organ of control and guidance which today—more than ever—emerges as part of the link between the body (soma) and the soul (brain).

Rosalyn Yalow, Roger Guillemin, Andrew Schally: the road of every scientist is paved by frustration. But some reach the goal they have set up and enjoy the pleasure and excitement of having learned something that no one knew before, and for that enjoy imperishable honor in the learned world. Few ever reach the point at which you have arrived: to undertake a formidable task and to come to a solution, which not only attracts the admiration of your scientific colleagues, but which—in the best spirit of Alfred Nobel—also contains a possibility to understand the structure of human life and human behaviour.

The Karolinska Institute is happy to be able to award you this year's Nobel Prize in Physiology or Medicine for your contributions and congratulates you. May I now ask you to receive the insignia of the Nobel Prize from His Majesty, the King.

ROGER GUILLEMIN

I was born in France on January 11, 1924 in the small town of Dijon, the capital of Burgundy. I was educated there in the public schools and the lycée. I entered medical school in Dijon in 1943 and received the M.D. degree from the Faculté de Médecine of Lyon in 1949,—the two schools were then administratively connected, with the larger school of Lyon granting the degrees. All my medical studies and training were totally clinically oriented, with three years of what we could call rotating internship. There was no laboratory facility of any sort in Dijon, except for gross anatomy. Dark years of no fun youth these were; France had fallen to the Germans in 1940; Dijon was from then on occupied by the German army until liberation days in 1944.

During these five years of medical studies, I had always been interested in endocrinology, probably because two of my best teachers of clinical medicine, P. Etienne-Martin and J. Charpy were themselves interested in what were in those days the early concepts of endocrinology and the beginning logical therapy it appeared to offer. I always hoped that somehow I could one day work in a laboratory. In France you had terminated your medical studies after 5 years of curriculum; you could then practice medicine—which I did for some time. To obtain the degree of Doctor in Medicine you had to write and defend a dissertation, a thesis; that was usually *pro forma.* I decided, however, to write a dissertation for the M.D. degree that I would enjoy, hopefully on some work I could perform in a laboratory.

One day, I learned that Hans Selye would lecture in Paris on his alarm reaction and the endocrinology of the general adaptation syndrome. I went to hear him. The magnetism of the man was extraordinary. I went to talk to him after one of his lectures. A few months later I was in Selye's newly created Institute of Experimental Medicine and Surgery at the University of Montreal, with a modest fellowship from Selye's funds. In one year I completed some experimental work on desoxycorticosterone-induced hypertension in bilaterally nephrectomized rats kept alive for several weeks by peritoneal dialysis; that constituted the material for the thesis necessary in the French system to obtain the M.D. degree, which I obtained in Lyon upon the defense of that dissertation, in 1949. Not much interested in the academism and formalism of a research career within the French system that was then open to me, I returned to Selye's Institute, and three years later eventually obtained a Ph.D. degree in physiology in 1953. In these four years I had learned experimental endocrinology in a remarkable program jointly conducted between McGill University and the University of Montreal. In 1953, I joined the staff of the Department of

Physiology at the Baylor University College of Medicine in Houston, Texas, as a young assistant professor. I taught physiology at Baylor College of Medicine for 18 years, until 1970. While in Selye's department, I had become interested in the problem of the physiological control of the secretion of the pituitary gland as it was involved in the acute response to stress. This was due particularly to friendly contacts with Claude Fortier and to a long visit by Geoffrey W. Harris from London.

I have recounted in some details in a chapter of Volume 2, of *Pioneers in Neuroendocrinology*, J. Meites (ed.), Plenum Press Publ., 1978, how I became more and more involved in the search for the chemical mediators of hypothalamic origin, suspected to control the functions of the pituitary gland; how Schally came to me at Baylor from the laboratory of Murray Saffran at McGill immediately after he had obtained there his Ph.D. degree in biochemistry—as we both thought that we would solve in no time the problem of the nature of CRF (the corticotropin releasing factor); how I went back to France in 1960,—on academic promises that did not materialize, thus returned to Houston in 1963, and later, in 1970, went to the Salk Institute to establish our present Laboratories for Neuroendocrinology. That chapter written in a light anecdotal manner, along with two reviews, one concerning the isolation and characterization of the first of the hypothalamic releasing factor, TRF (in *Vitamins and Hormones, 29*, 1—39, 1971), the other concerning the isolation of the luteinizing hormone releasing factor (in *Am. J. Obs. and Gynecol., 129*, 214—218, 1977) will give the interested reader a good historical description of these early years of, indeed, true pioneering in neuroendocrinology.

I served for 11 years on several advisory groups (Study Sections) of the National Institutes of Health (NIH)—an experience that was as rewarding as it was exhausting—as a member of the Council of the American Endocrine Society from 1969—1973.

I was elected a member of the National Academy of Sciences of the USA, in 1974, a member of the American Academy of Arts and Sciences in 1976. I have been honored by several national and international scientific recognitions: among which The Gairdner International Award, Toronto, Canada, 1974; The Dickson Prize in Medicine, The University of Pittsburgh, Pennsylvania, 1976; the Passano Award in the Medical Sciences, Baltimore, Maryland, 1976; the Lasker Award in Basic Sciences, New York, 1975; and recently the National Medal of Science presented by the President of the USA.

I have received honorary degrees, from the University of Rochester (D.Sc.), 1976; the University of Chicago (D.Sc.), 1977; and the Légion d'Honneur from the French government in 1973.

I consider as major honors to have been asked to deliver numerous memorial lectures, in particular the Harvey Lecture, The Rockefeller University, New York, 1974; the Jane Russell Wilhelmi Memorial Lecture, Emory University, Atlanta, Georgia, 1976; the Geoffrey W. Harris Memorial Lecture, International Congress of Endocrinology, Hamburg, Germany, 1976; The Gregory Pincus Memorial Lecture, The Laurentian Hormone Conference, 1976;

The Herbert M. Evans, Memorial Lecture, University of California in San Francisco, 1977.

In Houston, in Paris, in La Jolla, where I set up shop—sometime simultaneously as in the days of commuting between Paris and Houston—I have had the extraordinary privilege to work with wonderful collaborators some so much more knowledgeable in their own field than I was (or still am), all full of enthusiasm and sharing common ethics of science. The work recognized in this Nobel Prize was a group effort and achievement. I started writing the list of these colleagues, collaborators, students who worked with me, starting in 1953; I stopped when I realized more than one hundred names were involved. Of unique roles and significance in the saga of the hypothalamic hormones in which I was involved, I must call to the lime light Edvart Sakiz, now in Paris, Roger Burgus, now in La Jolla, Wylie Vale who came to me as a graduate student, now in La Jolla, Nicholas Ling and Jean Rivier, both now in La Jolla. They, and their own students, are and will be the future of this expanding field or research.

My wife is a musician of talent and, so far, five of our six children are already in artistic careers or show a definite preference for artistic endeavors; one only, may be a biologist some day. And all that is fine with me. Since 1970 when we came to the Salk Institute, we have lived in La Jolla, a suburb of San Diego, in a Mediterranean house which we have filled, if not overfilled, with many contemporary paintings French and American, sculptures and potteries mostly from pre-Columbian Mexico and also from New Guinea. Several keyboards and string instruments are also part of the enjoyable living environment of that happy house.

PEPTIDES IN THE BRAIN. THE NEW ENDOCRINOLOGY OF THE NEURON

Nobel Lecture, 8 December, 1977

by
ROGER GUILLEMIN
Laboratories for Neuroendocrinology
The Salk Institute, San Diego, California, U.S.A.

PART I

A. *The Existence of Brain Peptides Controlling Adenohypophysial Functions. Isolation and Characterization of Their Primary Molecular Structures*

In the early 1950s, based on the anatomical observations and physiological experimentation from several groups in the USA and Europe, it became abundantly clear that the endocrine secretions of the anterior lobe of the hypophysis—well known by then to control all the functions of all the target endocrine glands, (thyroid, gonads, adrenal cortex) plus the overall somatic growth of the individual—were somehow entirely regulated by some integrative mechanism located in neuronal elements of the ventral hypothalamus (review Harris, 1955). Because of the peculiar anatomy of the junctional region between ventral hypothalamus (floor of the 3rd ventricle) and the parenchymal tissue of the anterior lobe of the pituitary (Fig. 1), the mechanisms involved in this hypothalamic control of adenohypophysial functions were best explained by proposing the existence of some secretory product(s) by some (uncharacterized) neuronal elements of the ventral hypothalamus, the products of which would somehow reach the adenohypophysis by the peculiar capillary vessels observed as if to join the floor of the hypothalamus to the pituitary gland.

That concept was definitely ascertained in simple experiments using combined tissue cultures of fragments of the pituitary gland and of the ventral hypothalamus (Guillemin and Rosenberg, 1955). The search for characterizing the hypothetical hypothalamic hypophysiotropic factors started then. Simple reasoning and early chemical confirmation led to the hypothesis that these unknown substances would be small peptides. After several years of pilot studies involving both biology and relatively simple chemistry in several laboratories in the USA, Europe and Japan, it became clear that characterizing these hypothalamic hypophysiotropic substances would be a challenge of (originally) unsuspected proportions. Entirely novel bioassays would have to be devised for routine testing of a large number of fractions generated by the chemical purification schemes; more sobering still was the realization in the early 1960s that enormous amounts of hypothalamic fragments (from slaughter house animals) would have to be obtained to have available a sufficient quantity of starting material to attempt a meaningful program of chemical isolation. The early pilot studies had indeed shown the hypothalamic sub-

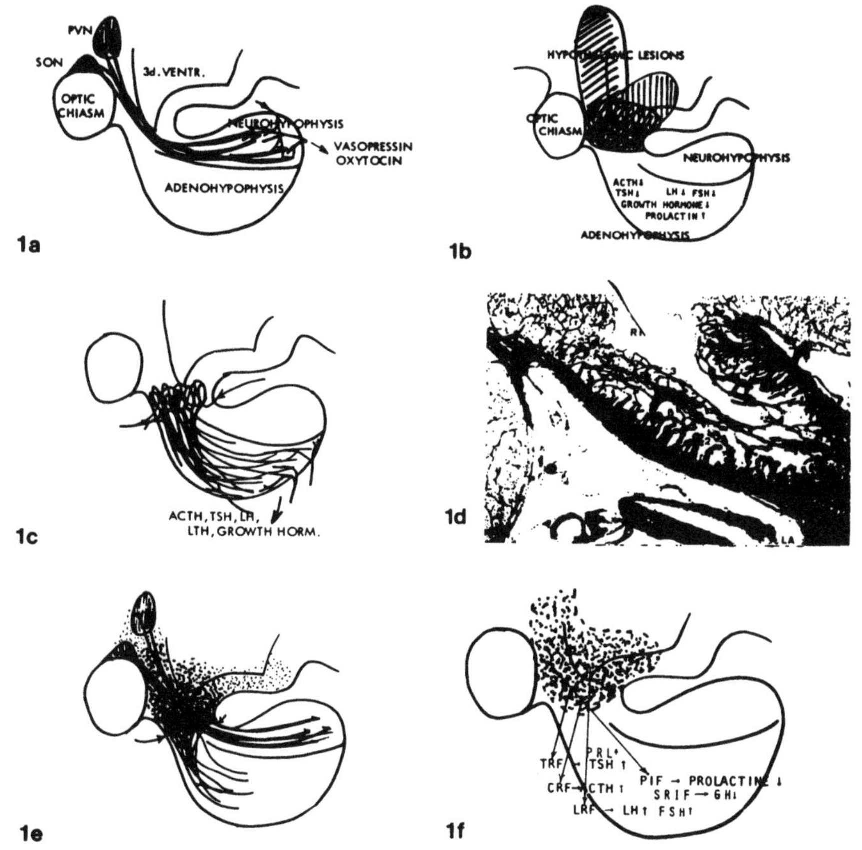

Fig. 1: a) Diagrammatic representation of the pituitary gland and the innervation of the neurohypophysis by nerve fibers from the n.paraventricularis (PVN) and supra-opticus (SON). b) Localized lesions in the hypothalamus produce changes in the pituitary secretion of the various adenohypophysial hormones (increase ↑, or decrease ↓). c) Diagrammatic representation of the hypothalamo-hypophysial portal system. d) Photomicrograph of the hypothalamo-hypophysial portal system after injection with an opaque dye. e) Diagrammatic representation of the hypophysiotropic area. f) Changes in pituitary secretion of various adenohypophysial hormones (increase ↑, or decrease ↓).

stances to be extremely potent and, on the basis of simple assumptions, to be present in each hypothalamic fragment only in a few nanogram quantities. Essentially one, then two groups of investigators approached the problem with enough constancy and resolution to stay with it for the ten years that it took to provide the first of its definitive solution, i.e. the primary structure of one of the hypothalamic hypophysiotropic factors: My own group, then at Baylor College of Medicine in Houston, Texas (with an episode at the Collège de France in Paris), organized the collection over several years of more than 5 million sheep brains, handling in the laboratory more than 50 tons of hypothalamic fragments. Schally and his collaborators, now in the Tulane

University School of Medicine in New Orleans, after he had left my laboratory at Baylor, collected also very large numbers of porcine hypothalamic fragments. Late in 1968, from 300,000 sheep hypothalami, Burgus and I isolated 1.0 mg of the first of these hypothalamic hypophysiotropic peptides, the thyrotropin releasing factor (TRF), the molecule by which the hypothalamus regulates through the pituitary the functions of the thyroid gland (Guillemin et al., 1962).

The following year, after more technical difficulties were overcome, we established the primary structure of ovine TRF by mass spectrometry as that of the deceivingly simple tripeptide pGlu-His-Pro-NH_2. The material of porcine origin was shown by Schally and his collaborators to be identical. The synthetic replicate, rapidly available in unlimited quantities, was shown to be highly potent in all vertebrate species and particularly in man; it is now widely used throughout the world in a highly sensitive test of pituitary function and an early means of detection of pituitary tumors in man.

The isolation and characterization of TRF was the result of an enormous effort. It was also the turning point which separated doubt — and often confusion, from unquestionable knowledge. It was of such heuristic significance, that I can say that neuroendocrinology became an established science on that event.

The characterization of the molecular structure of TRF was achieved in an unconventional manner, which will be briefly recounted here. I have with Burgus and Vale given an extensive technical review of the whole series of events that led to the characterization of ovine TRF (Guillemin et al., 1971); readers interested in the technical as well as historical aspects of these developments will find them in that review.

Purification, Isolation and Characterization of TRF

In January 1969, with the latest supply of highly purified ovine TRF available, —1.0 mg obtained from 300,000 sheep hypothalamus fragments, amino acid analysis of 6 *N* HCl hydrolyzates of this preparation revealed only the amino acids, Glu, His, and Pro, in equimolar ratios and accounting in weight for 81 % of the preparation (theoretical ponderal contribution of His, Pro, and Glu for a tripeptide monoacetate calculates to 86 %) (Burgus & Guillemin, 1970a). Furthermore the ultraviolet (UV), infrared (IR), and nuclear magnetic resonance (NMR) spectra obtained with that preparation of TRF were consistent with those of a polypeptide and upon close examination most of the characteristics of those spectra could be accounted for by the structural features of the amino acids found in the hydrolyzates of TRF. Moreover, the solubility properties and the lack of volatility observed in early attempts to obtain mass spectra or to perform gas chromatography, as well as other analytical data, were consistent with those of a polypeptide; also, the lack of effect of classical proteolytic enzymes, observed earlier, could be related to the particular amino acids observed. With the analyses of the more highly purified material unmistakably showing the amino acids to account for the total weight of the preparation, an earlier hypothesis that TRF could be a heteromeric polypeptide was therefore abandoned in favor of the possibility that it might be a cyclic or a protected peptide, a view compatible with failure to detect an N-terminus (Burgus et al. 1966b; Burgus and Guillemin, 1967, 1970a; Schally et al., 1966c, 1968, 1969) or a C-terminus (Schally et al., 1969; Burgus and Guillemin, 1970a) as well as the resistance of the biological activity to proteases.

The knowledge that the amino acids His, Pro, and Glu not only occurred in equimolar ratio in porcine and ovine TRF but indeed accounted for almost the theoretical total weight of the molecule in the case of ovine TRF, along with the previous knowledge of a lack of an N-terminal amino acid, led us to re-examine (this had been done by Schally in 1968) derivatives of synthetic polypeptides containing equimolar ratios of these amino acids to serve at least as possible models for the methodology to be used in the characterization of ovine TRF. We tested for TRF activity 6 tripeptide isomers containing L-His, L-Pro, and L-Glu synthesized upon our request by Gillessen *et al.* (1970) (containing only the peptides involving the α-carboxyl group of glutamic acid). The tripeptides proved to be devoid of TRF activity, confirming the earlier results of Schally *et al.* (1968 , 1969). Our response to these negative results was, however, different from what had been that of Schally et al (see Schally et al., 1969).

I proposed treating each of the six tripeptides by acetic anhydride in an effort to protect the N-terminus as in natural TRF. The acetylation mixture from one, and only one of the peptides, namely H-Glu-His-Pro-OH, yielded biological activity qualitatively indistinguishable from that of natural TRF. It was active in *in vivo* and *in vitro* assays specific for TRF and its action *in vivo* was blocked by prior injection of the animals with thyroxine (Burgus et al., 1969a). The specific activity of the material obtained was lower (*ca.* 1×10^{-3}) than that of purified natural TRF. The nature of several possible reaction products was considered: mono- or diacetyl-derivatives, polymers of Glu-His-Pro, and cyclic peptide derivatives or derivatives containing pyroglutamic acid (pGlu) as the N-terminus. Subsequently we reported (Burgus et al., 1969a, b) that the major product by weight of this procedure was indeed pGlu-His-Pro-OH. The material was isolated from the reaction mixture and its structure was confirmed by mass spectrometry of the methyl ester and by its identity to authentic pGlu-His-Pro-OH (Gillessen et al., 1970) on TLC, the IR spectrum, as well as similarity of intrinsic biological activity *in vivo*.

This represented the first demonstration of a fully characterized synthetic molecule, based on the known composition of natural TRF, to reproduce the biological activity of a hypothalamic releasing factor.

Several other products present in the acetylation mixture, some possibly having higher specific activity than pGlu-His-Pro-OH, were not characterized. It is of interest that acetyl-Glu-His-Pro-OH obtained by total synthesis (Gillessen et al., 1970) was devoid of TRF activity in the *in vivo* assay at doses up to 250 μg (Burgus et al., 1969b).

Because of the differences between the specific biological activities of pGlu-His-Pro-OH and natural ovine TRF and the different behavior of these two compounds in various chromatographic systems, it was evident that TRF was not pGlu-His-Pro-OH as such. It was the proposal of Burgus based on knowledge of the primary structures of other biologically active polypeptides (vasopressins, oxytocin, gastrins, etc.) that a likely candidate for the structure of the natural material would be pGlu-His-Pro-NH_2 and its synthesis was approached through the simple procedure of methanolysis of the methyl ester, pGlu-His-Pro-OMe (Burgus et al., 1969b, 1970 b, c). The ester, prepared by treatment of the pure synthetic pGlu-His-Pro-OH with methanolic HCl, was purified by partition chromatography and identified as pGlu-His-Pro-OMe, on the basis of its behavior on TLC, its IR spectrum, and by mass spectrometry (Burgus et al., 1970 b, c). It had biological activity *in vitro* and *in vivo* now approaching half of the specific activity of isolated ovine TRF. Ammonolysis of the methyl ester in methanol produced a material which upon partition chromatography gave a small yield of a substance presumably pGlu-His-Pro-NH_2, occurring in a Pauly positive zone separated from the starting material, which had a specific activity *in vivo* or *in vitro* statistically identical to that of ovine TRF. Among the derivatives tested, the properties of native ovine TRF most closely matched that of the amide, failing to separate from the synthetic compound in four different systems of TLC when run in mixtures. The IR spectra of several of the more highly purified preparations of the amide, including pGlu-His-Pro-NH_2 now prepared by total synthesis (Gillessen et al., 1970), were almost identical to that of ovine TRF, showing only minor differences in two regions of the spectra. These new observations, together with the demonstration that the specific activity of the pGlu-His-Pro-NH_2 was not statistically different from that of natural ovine TRF, led us to reconsider (Burgus et al., 1969c; 1970b) an earlier hypothesis (Burgus et al., 1969b) that ovine TRF may have a secondary or tertiary amide on the C-terminal proline, rather than correspond to the primary amide of the tripeptide pGlu-His-Pro.

Availability of large amounts of the synthetic tripeptides made possible a series of experiments with Desiderio, of Horning's group at Baylor, to modify the design of the direct probe of the then available low resolution mass spectrometer and simultaneously to obtain volatile derivatives of the peptides that would give clear mass spectra on only a few micrograms of the peptides. Once this was achieved, evidence was obtained (Burgus et al., 1969c) based on low and high resolution mass spectrometry, that the native ovine TRF preparation originally obtained in late 1968 had been all along essentially homogeneous and had unquestionably the structure pGlu-His-Pro-NH_2. Both synthetic pGlu-His-Pro-NH_2 (Burgus et al., 1969c; 1970b, c) and the highly purified ovine TRF (Burgus and Guillemin, 1970a) were introduced by direct probe into a low resolution mass spectrometer as the methyl or trifluoroacetyl (TFA) derivatives (Fig. 2); all preparations gave volatile materials in the temperature range of 150—200°C ($\leqslant 10^{-6}$ torr). Several mass spectra taken throughout the range of the thermal gradient (7 in the case of the isolated ovine TRF) showed fragmentation patterns corresponding to a single component. Although none of the spectra revealed a molecular ion, fragments

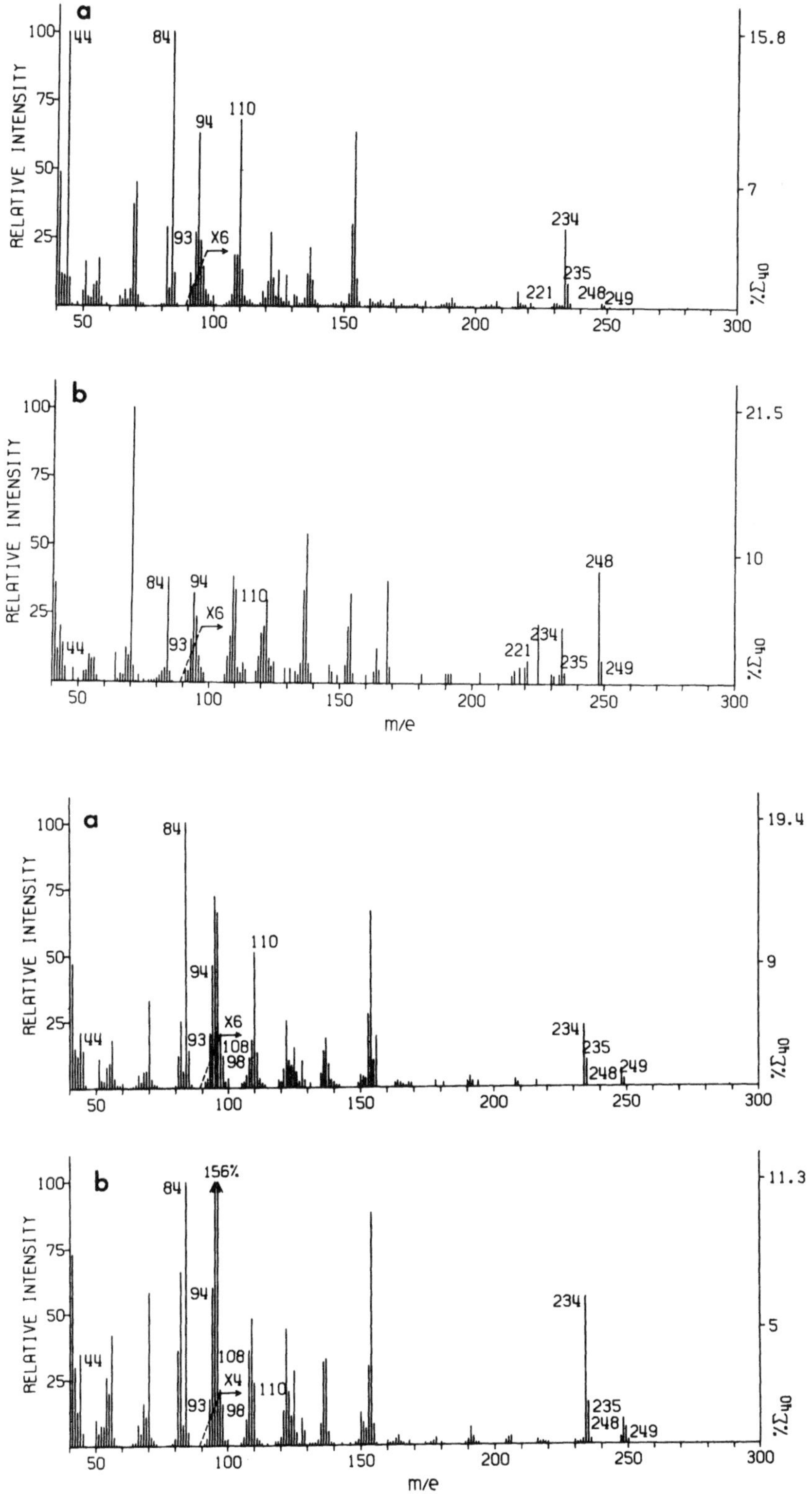

Fig. 2: Upper a, b: Low resolution mass spectra of trifluoroacetylated ovine TRF (a) and synthetic PCA-His-Pro-NH_2 (b).

Lower a, b: Low resolution mass spectra of methylated ovine TRF (a) and synthetic arising from the structures pGlu, methyl-pGlu, His, methyl-His, Pro, Pro-NH_2, $CONH_2$, pGlu-His, and His-Pro-NH_2 were observed. The low resolution mass spectra of the corresponding derivatives of synthetic pGlu-His-Pro-NH_2 and TRF were essentially identical. Fragments arising from unsubstituted pGlu or His were observed in the spectra of both types of derivatives.

The elemental composition of all the fragments, except *m/e* 221, the intensity of which was too weak for it to be observed on the photoplate used, were confirmed by high resolution mass spectroscopy of the methyl derivatives (Burgus et al., 1969c; 1970b).

Thus, the structure of ovine TRF as isolated from the hypothalamus was established as pGlu-His-Pro-NH_2 (Fig. 3). However, we did point out (Burgus et al., 1969c) that the possibility was not excluded that, as opposed to the *isolated* material, the *native* molecule of TRF may occur as Gln-His-Pro-NH_2 either free or conjugated to another structure such as a protein, which would not be necessary for biological activity *in vivo* or *in vitro*. At the time of this lecture we and others are still looking for a hypothetical prohormone of TRF.

The structure of *porcine* TRF was shown in a series of reports by Schally and his collaborators to be compatible with, and finally to be identical with, that of pGlu-His-Pro-NH_2; mass spectrometry was also the method of ultimate proof used by Nair et al. (1970).

STRUCTURE OF HYPOTHALAMIC TRF (OVINE)
WITH FRAGMENTATION POINTS IN MASS SPECTROMETRY

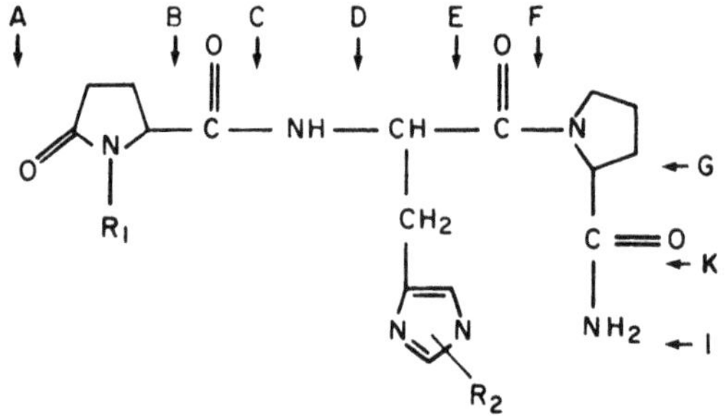

Fig. 3: The primary structure of TRF with indication of the fragmentation points in mass spectrometry. R_1, R_2 represent the methyl derivative prepared for mass spectrometry; in the native molecule, $R_1 = R_2 = H$.

It is most interesting that TRF from two widely different species of mammals should have the same structure and apparently the same specific (biological) activity in similar assays. It was rapidly shown that TRF shows no evidence of species specificity for its biological actions, pGlu-His-Pro-NH_2 being readily active in humans (see Fleischer et al., 1970; Fleischer and Guillemin, 1976).

Purification, Isolation and Characterization of LRF

In the early 1960s, several investigators reported experimental results that were best explained by proposing the existence in crude aqueous extracts of hypothalamic tissues of substances that specifically stimulated the secretion of luteinizing hormone, and that were probably polypeptides (McCann et al., 1960; Campbell et al., 1961; Courrier et al., 1961). The active substance was named LH-releasing factor or LRF. Rapidly following these early observations, preparations of LRF, active at 1 μg per dose in animal bioassays, were obtained by gel filtration and ion-exchange chromatography on carboxymethylcellulose (Gullemin et al., 1963) an observation that was confirmed with similar methods by several investigators (Schally et al., 1968). In spite of the vagaries of the various bioassay methods available, several laboratories reported preparations of LRF of increased potency. Several of these early publications led, however, to contradictory statements regarding purification and separation of LH-releasing factor (LRF), from a follicle-stimulating hormone releasing factor (Schally et al., 1968; Dhariwal et al., 1967; Guillemin 1963).

Two laboratories independently reported in 1971 the isolation of porcine LRF (Schally et al., 1971a) and ovine LRF (Amoss et al., 1971), *both groups* concluding that LRF from either species was a *nonapeptide* containing, on the basis of acid hydrolysis, 1 His, 1 Arg, 1 Ser, 1 Glu, 1 Pro, 2 Gly, 1 Leu, 1 Tyr. Earlier results with the pyrrolidonecarboxylylpeptidase prepared by Fellows and Mudge (1970) had led us to conclude (Amoss et al., 1970) that the N-terminal residue of LRF was Glu in its cyclized pyroglutamic (pGlu) form, as in the case of hypothalamic TRF, (pGlu-His-Pro-NH_2). The total amount of the highly purified ovine LRF that we had isolated from side fractions of the TRF program and that was available for amino acid sequencing was ca. 80 nmol (as measured by quantitative dansylation).

It is to the credit of Schally's group to have first recognized and reported (Matsuo et al., 1971a) that porcine LRF contained one residue of tryptophan (Trp), in addition to the other amino acids earlier observed by acid hydrolysis. On the basis of a series of experiments including enzymatic hydrolysis with chymotrypsin and thermolysin and analysis of the partial sequences of their decapeptide by Edman degradation-dansylation and selective tritiation of C-termini, Matsuo *et al.* (1971b) proposed the sequence pGlu-His-Trp-Ser-Tyr-Gly-Leu-Arg-Pro-Gly-NH_2 for porcine LRF as that best compatible with the partial sequence data. Their studies were carried out with *ca.* 200 nmol of peptide. They also stated that synthesis of that particular sequence had given a material with biological activity. A few weeks later, we reported the synthesis by solid-phase methods of the decapeptide pGlu-His-Trp-Ser-Tyr-Gly-Leu-Arg-Pro-Gly-NH_2; after isolation from the reaction mixture it had quantitatively the full biological activity *in vivo* and *in vitro* of ovine LRF (Monahan et al., 1971).

Shortly thereafter, we reported (Burgus et al., 1971) the amino-acid sequence of ovine LRF obtained on 40 nmoles of peptide by analysis of hydrolysis products after digestion with chymotrypsin or pyrrolidonecarboxylylpeptidase, using Edman-degradation followed by determination of N- and C-termini by a quantitative [^{14}C]-dansylation technique. Confirmation of the positions of some of the amino-acid residues obtained by combined gas chromatographic-mass spectrometric analysis of phenylthiohydantoin (PTH) derivatives (Fales et al., 1971; Hagenmaier et al., 1970) resulting from Edman degradations was described; we also reported results obtained by degradation of the synthetic decapeptide, since they confirmed and clarified some peculiarities observed upon enzymatic cleavage of the native peptide (Burgus et al., 1972).

The amino acid sequence of ovine LRF was established to be pGlu-His-Trp-Ser-Tyr-Gly-Leu-Arg-Pro-Gly-NH_2. It is identical to that of the material of porcine origin.

Of considerable interest was the observation that the synthetic replicate of LRF, now available in large quantities, was shown to stimulate concomitant secretion of the two gonadotropins LH (luteinizing hormone) and FSH (follicle stimulating hormone) in all assay systems *in vivo* and *in vitro* in which it was tested. This confirmed the earlier results obtained with the minute quantities of the isolated ovine or porcine LRF (Amoss et al., 1971; Schally et al., 1971a). In other words, the stimulation of the release of FSH appeared to be inherent to the molecule of LRF — thus throwing considerable doubt on earlier reports (Schally et al., 1966; Schally et al., 1968; Igarashi et al., 1964) claiming to have obtained preparations of LRF free of FSH-releasing activity.

To the day of this lecture, no solid evidence has been produced which could be interpreted to indicate the existence of an FSH-releasing factor as a specific entity, discrete from the decapeptide LRF. Reports by Folkers *et al.* (1969) claiming purification of an FSH-releasing factor are difficult to appreciate in view of the paucity of data offered. Moreover, other evidence is against the existence of an FSH-releasing factor, separate from LRF: all synthetic analogs of LRF made so far, with no exception, can be shown to release LH and FSH with the same ratio of specific activity when related to the activity of LRF in the particular assay involved. Thus, none of these analogs has shown any evidence of dissociated activity for releasing FSH vs. LH. Also, there is increasing evidence that the two gonadotropins (LH and FSH) can be demonstrated (immunocytochemistry) mostly in the same pituitary cell (Moriarty, 1973). It is thus unlikely that one could be released without the other as they appear to be present in the same secretory granules.

Later on, both Schally's group (1971b) and our group (Ling et al., 1973) confirmed the primary structure of porcine and ovine LRF, respectively, using larger quantities of native material.

Purification, Isolation and Characterization of Somatostatin

It has been generally accepted that the control of the pituitary secretion of growth hormone would be exerted by a hypothalamic hypophysiotropic releasing factor, as is now proven to be the case for the secretion of thyrotropin and the gonadotropins. The nature of the postulated hypothalamic releasing factor for growth hormone, however, remains elusive to this day, mostly due to the difficulties and ambiguities of the various assay systems used so far in attempts at its characterization. For instance there is now agreement that the "growth hormone releasing hormone" (GH-RH), isolated on the basis of a bioassay and characterized by Schally *et al.* (1971c) as H-Val-His-Leu-Ser-Ala-Glu-Glu-Lys-Glu-Ala-OH, was actually a decapeptide fragment of the N-terminal of the β-chain of porcine hemoglobin (Veber et al., 1971). The material has never been shown to be active in stimulating secretion of immunoreactive growth hormone. Similarly, biological activity of a tetrapeptide recently reported as a growth hormone releasing factor of porcine origin has not been confirmed by others, including our own laboratory (see Guillemin, 1973).

Searching to demonstrate the presence of this still hypothetical somatotropin releasing factor in the crude hypothalamic extracts used in the isolation of TRF (thyrotropin releasing factor) and LRF (luteinizing hormone releasing factor), we regularly observed that their addition in minute doses ($\leqslant$.001 of a hypothalamic fragment equivalent) to the incubation fluid of dispersed rat pituitary cells in monolayer cultures (Vale et al., 1972a) significantly decreased the resting secretion of immunoreactive growth hormone by the pituitary cells. This inhibition was related to the dose of hypothalamic extract added and appeared to be specific. It was not produced by similar extracts of cerebellum, and the crude hypothalamic extracts that inhibit secretion of growth hormone simultaneously stimulated secretion of LH and TSH. The inhibition of growth hormone secretion could not be duplicated by addition to the assay system of [Arg^8]-vasopressin, oxytocin, histamine, various polyamines, serotonin, catecholamines, LRF, or TRF. We decided to attribute this inhibitory effect on the secretion of growth hormone to a "somatotropin-release inhibiting factor" which we later named *somatostatin*.

Inhibition of secretion of growth hormone by crude hypothalamic preparations had been reported by others (Krulich and McCann, 1969). The active factor possibly involved in these early studies based on various types of assays for growth hormone activity had not been characterized. The results on the inhibition by the hypothalamic extracts of the secretion of immunoreactive growth hormone by the monolayer pituitary cultures were so consistent and easily quantitated that we decided to attempt the isolation and characterization of the hypothalamic factor involved. We realized the possible interest of such a substance in inhibiting abnormally elevated secretion of growth hormone in juvenile diabetes; also, we considered that knowledge of the primary structure of a native inhibitor of the secretion of a pituitary hormone could be of significance in our efforts at designing synthetic inhibitors of the gonadotropin releasing factor LRF.

The starting material was the chloroform-methanol-glacial acetic acid extract of about 500,000 sheep hypothalamic fragments (Burgus et al., 1971; Burgus et al., 1972) used in the program of characterization of the releasing factors for the gonadotropins. The extract (2 kg) had been partitioned in two systems; the LRF concentrate was subjected to ion-exchange chromatography on carboxymethyl cellulose. At that stage, a fraction with growth hormone-release inhibiting-activity was observed well separated from the LRF zone; it was further purified by gel filtration (Sephadex G-25) and liquid partition chromatography (*n*-butanol, acetic acid, water, 4:1:5). Thin-layer chromatography and electrophoresis of the final product showed only traces of peptide impurities. The yield was 8.5 mg of a product containing 75 percent of amino acids by weight; we will refer to this material by the name *somatostatin* which was actually given to it only after it had been fully characterized.

Analysis of amino acids obtained from somatostatin after acid hydrolysis in 6 N HCl-0.5 % thioglycollic acid gave the molar ratios Ala (0.9), Gly (1.1), Cys (0.2), Cys-SS-Cys (1.0), Lys (2.0), Asp (1.0), Phe (3.3), Trp (0.5), Thr (2.0), Ser (0.8), and NH_3 (1.1). Enzymic hydrolysis gave the ratios Ala (0.9), Gly (0.9), Lys (2.0), Phe (3.4) and Trp (0.9); Asn, Thr, and Ser were not well resolved, giving a total of about 3.6 mol/mol of peptide. Edman degradation of the carboxymethylated trypsin digests of somatostatin and mass spectrometry led finally to the final demonstration of the following primary structure for somatostatin: H-Ala-Gly-Cys-Lys-Asn-Phe-Phe-Trp-Lys-Thr-Phe-Thr-Ser-Cys-OH, in the oxidized form (Burgus et al., 1973).

This peptide was reproduced by total synthesis using Merrifield method (see Rivier, 1974). It had the full biological activity of native somatostatin *in vivo* and *in vitro* (Brazeau et al., 1973, Vale et al., 1972b, Brazeau et al., 1974). Of interest was the unexpected observation that the peptide has the full biological activity either in the oxidized form (native) or reduced form.

B. *Purification, Isolation and Characterization of the Endorphins, Opiate-Like Peptides of Brain or Pituitary Origin*

The concept and the demonstration, some years ago, of the existence in the brain of mammalians of (synaptosomal) opiate-receptors (Pert and Snyder, 1973) led to the search of what has been termed the endogenous-ligand(s) of these opiate receptors. The generic name *endorphins* (from endogenous and morphine) was proposed for these (then hypothetical substances) by Eric Simon and will be used here. Sometime in the summer of 1975 I became interested in these early observations. Besides the challenge of characterizing an endogenous substance as the ligand of the brain opiate receptors, we could not ignore that, like morphine, the (then hypothetical) endorphins might stimulate the secretion of growth hormone; the nature of the growth hormone releasing factor was and still remains unknown. I thus decided to engage in the isolation and characterization of the endogenous ligand(s) for the opiate receptors. The isolation of these endogenous ligands of the opiate receptors turned out to be a relatively simple problem to which a solution was provided in less than a couple of months of effort.

Dilute acetic acid-methanol extracts of whole brain (ox, pig, rat) were confirmed to contain substances presumably peptidic in nature, with naloxone-reversible, morphine-like activity in the bioassay using the myenteric-plexus longitudinal muscle of the guinea pig ileum. Evidence of such biological activity in our laboratory was in agreement with earlier results of Hughes (1975), Terenius & Wahlstrom (1975), Teschemacher et al. (1975), and Pasternak et al. (1975). Searching for an enriched source of endorphins in available concentrates from our earlier efforts towards the isolation of CRF, TRF, LRF, somatostatin, I recognized that acetic acid-methanol extracts of porcine hypothalamus-neurohypophysis contained much greater concentrations of the morphine-like activity than extracts of whole brain. From such an extract of approximately 250,000 fragments of pig hypothalamus-neurohypophysis we isolated several oligopeptides (*endorphins*) with opioid activity (Guillemin et al., 1976a; Lazarus et al., 1976; Ling et al., 1976). The isolation procedure involved successively gel filtration, ion exchange chromatography, liquid partition chromatography and high pressure liquid chromatography (Guillemin et al., 1976a; Lazarus et al., 1976; Ling et al., 1976). By that time, had appeared the evidence for the isolation and primary structure of Met^5-enkephalin and Leu^5-enkephalin (Hughes, et al., 1975). Hughes et al. (1975) had also made the remarkable observation of the identity of the amino acid sequence of Met^5-enkephalin with that of the sequence Tyr^{61}-Met^{65} of β-lipotropin, a polypeptide of ill defined biological activity, isolated and characterized in 1964 by C. H. Li et al. (see Li & Chung, 1976). The primary structure of *α-endorphin* was established (Ling et al., 1976; Guillemin et al., 1976a) by mass spectrometry and classical Edman degradation of the enzymatically cleaved peptide and is H-Tyr-Gly-Gly-Phe-Met-Thr-Ser-Glu-Lys-Ser-Gln-Thr-Pro-Leu-Val-Thr-OH (Fig. 4 a, b). The primary structure of *γ-endorphin*

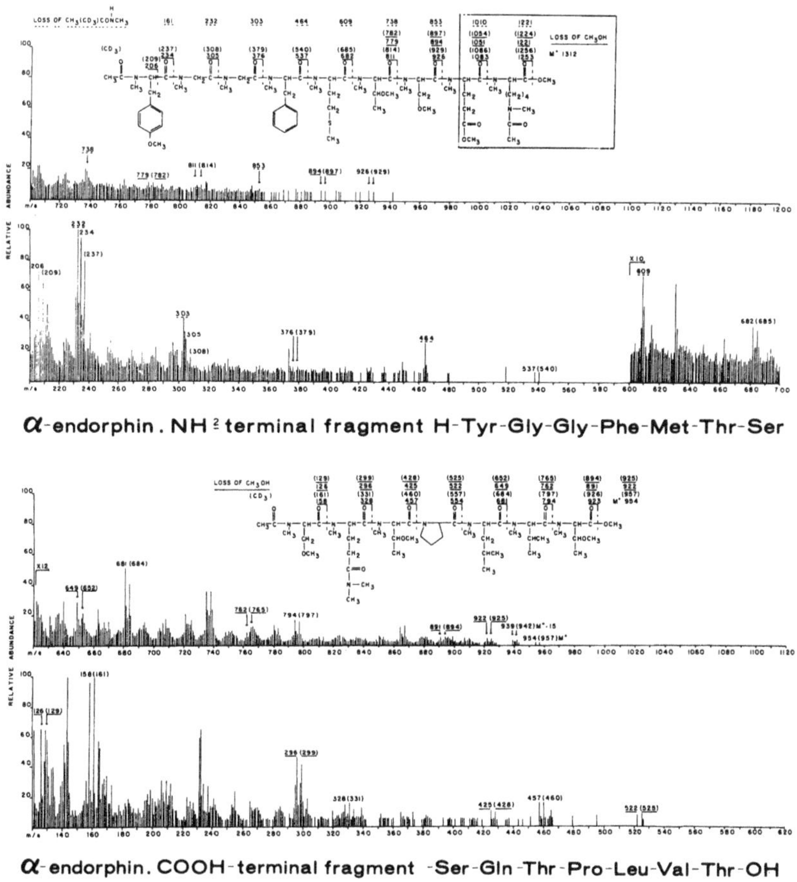

Fig. 4: Mass spectra of α-endorphin after trypsin digestion, acetic and deuterioacetic anhydride acetylation and permethylation. The sequences are: (a) H-Tyr-Gly-Gly-Phe-Met-Thr-Ser-; (b): H-Ser-Gln-Thr-Pro-Leu-Val-Thr-OH.

was similarly established by mass spectrometry and by Edman degradation: γ-endorphin has the same primary structure as α-endorphin with one additional Leu as the COOH-terminal residue in position 17.

Thus, it was obvious that Met-enkephalin is the N-terminal pentapeptide of α- and γ-endorphin, which have respectively the same amino acid sequence as β-lipotropin [61—76] and [61—77]. β-LPH-[61—91], a fragment of β-LPH isolated earlier on the basis of its chemical characteristics (Bradbury et al., 1975; Li and Chung, 1976) was shown also to have opiate-like activity (Bradbury et al., 1976; Lazarus et al., 1976; Cox et al. 1976) and has been named *β-endorphin* (Li and Chung, 1976). Recently we have isolated, from the same starting material of hypothalamus-neurohypophysis origin from which we originally isolated α- and γ-endorphin, two peptides characterized by amino acid composition as β-endorphin (β-LPH[61—91]) and *δ-endorphin* (β-LPH [61—87]). No effort was made to obtain the amino acid sequences of these two samples. The synthetic replicates of these two polypeptides have identical chromatographic behavior in several systems as the native materials.

PART II

Biological Activities of the Hypothalamic Peptides and Synthetic Analogs. Experimental and Clinical Studies

As soon as they were obtained in large quantity from total synthesis, TRF and LRF were extensively studied for their biological activities, both in the laboratory and in clinical medicine. Indeed, the observation was rapidly made that TRF, and later LRF, both characterized only from tissues of ovine and porcine origin, were biologically fully active in all species of vertebrates studied, including man (Fig. 5), the same was to apply for the synthetic replicate of somatostatin, as characterized from ovine brains (Fig. 6).

For early clinical studies with synthetic TRF see Fleischer et al., 1970 and more recently Fleischer and Guillemin, 1976; for early clinical studies with synthetic LRF see Yen et al., 1973; Rebar et al., 1973. Schally's group also published extensively on clinical investigations with either purified native or synthetic TRF and LRF (review in Schally et al., 1973).

Both in the case of TRF and of LRF chemists have prepared large numbers of synthetic analogs of the primary (native) structure for studies of correlation between molecular structure and biological activity. Also, biologists carefully screened these analogs, in the hope that some of them would prove to be antagonists of the native (agonist) releasing factor. This was of particular interest in the case of the gonadotropin releasing factor; a powerful antagonist of LRF would be of considerable interest as a chemical means of controlling or regulating fertility, thus introducing a totally new type of substances for contraception. In 1972, our laboratory reported the first partial agonist/antagonist analogs of LRF, (Vale et al., 1972c). They all had a deletion or a substitution of His^2 or Trp^3 in the (otherwise identical) amino acid sequence of LRF. These were antagonists of low activity and of no possible practical value as clinically significant inhibitors of LRF. They showed, however, that analogs as competitive antagonists of the decapeptide LRF could be prepared.

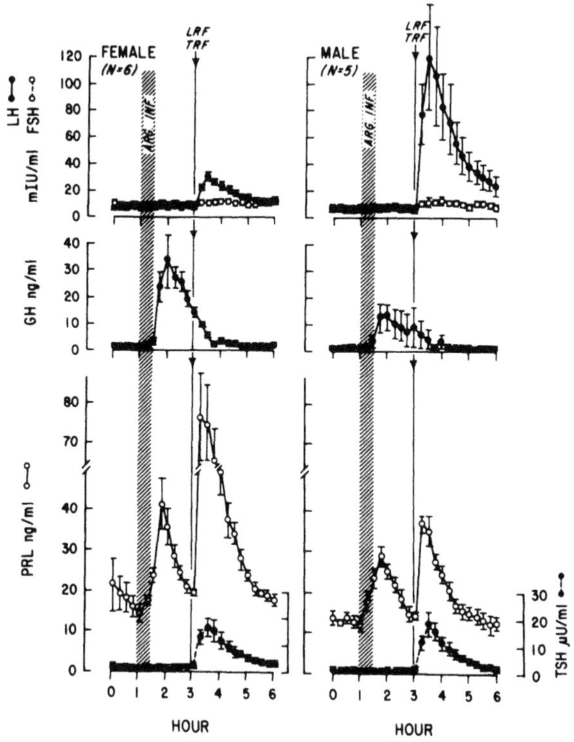

Fig. 5: Testing of the ability of the anterior pituitary to secrete GH, TSH, prolactin (PRL), LH and FSH in normal human subjects. Stimulation of the secretion of GH is achieved by i.v. administration of arginine; stimulation of the secretion of TSH and PRL, LH and FSH is produced by i.v. injection of a solution in saline of synthetic TRF (250 μg) and synthetic LRF (150 μg)—note that arginine infusion stimulates secretion of GH and PRL. All pituitary hormone plasma concentrations measured by radioimmunoassays (from Yen et al.).

To this day, the most potent antagonist-analogs of LRF still have the early deletion or substitution of His^2 or Trp^3 of the amino acid sequence of LRF.

Also, analogs of TRF and LRF with increased potency (over that of the native compound) were expected, searched for, and obtained (see below).

a. *Biological Activity of Thyrotropin Releasing Factor* (*TRF*) *and Luteinizing Hormone Releasing Factor* (*LRF*)

The remarkable observation was originally made by Tashjian et al. (1971) that TRF stimulates the secretion of prolactin by the cloned line GH_3 of pituitary cells. This was confirmed by others and extended to show the observation to be valid also with normal pituitary tissues *in vitro* and *in vivo*, including the human pituitary. TRF can thus be considered as involved in the control of the secretion of thyrotropin and of prolactin. Of the many analogs of TRF which have been synthesized and studied biologically, only one has a significantly increased specific activity over that of the native compound. Described by our group and synthesized by Rivier (1971) a few years ago, it is the analog [3N-Methyl-His]-TRF. Its specific activity is approximately

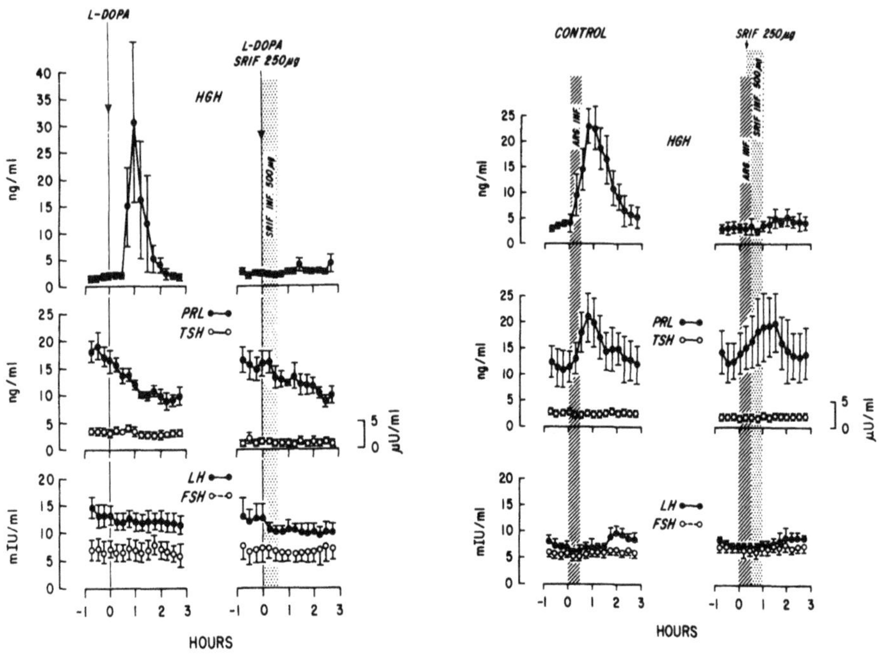

Fig. 6: Effects of the administration of synthetic somatostatin in normal human subjects. There is complete inhibition of the increase in GH secretion normally produced by infusion of arginine or oral administration of L-DOPA, where somatostatin is administered prior to or concurrently with the stimulating agent. Plasma concentrations of pituitary hormones were measured by radioimmunoassays (from Yen et al.).

10 times that of the native molecule, on the secretion of TSH as well as of prolactin. Of the several hundreds of TRF analogs synthesized, none has been found so far to be even a partial antagonist. They are all agonists with full intrinsic activity but variable specific activity; no true antagonist of TRF has been reported.

In contradistinction to the statement above regarding analogs of TRF, antagonist as well as extremely potent agonist analogs of LRF have been prepared by a number of laboratories. There are now available preparations of a series of what we may accurately call "super-LRFs", analogs which have as much as 150 times the specific activity of the native compound. In fact, in certain assays such as ovulation, they may have 1,000 times the specific activity of the native peptide. All the agonist-analogs or super-LRFs possess structural variations around two major modifications of the amino acid sequence of native LRF: They all have a modification of the C-terminal glycine, as originally reported by Fujino *et al.* (1974). The Fujino modification consists of deletion of Gly^{10}-NH_2 and replacement by primary or secondary amide on the (now C-terminal) Pro^9. In addition to the Fujino modification, they have an additional modification at the Gly^6 position by substitution of one of several D-amino acids as originally discovered in our laboratories (Monahan et al., 1973). The most potent of the LRF-analog agonists prepared are [D-Trp^6]-LRF; des-Gly^{10}-[D-Trp^6-Pro^9-N-Et]-LRF, [D-Leu^6, Pro^9-N-Et]-LRF.

In an *in vitro* assay in which the peptides stimulate release of LH and FSH by surviving adenohypophysial cells in monolayer cultures, these analogs of LRF have a specific activity 50 to 100 times greater than that of the synthetic replicate of native LRF. There is no evidence of dissociation of the specific activity for the release of LH from that of FSH. All agonist analogs release LH and FSH in the same ratio (in that particular assay system) as does native LRF. Probably because of their much greater specific activity, when given in doses identical in weight to the reference doses of LRF, the super-LRFs are remarkably long acting. While the elevated secretion of LH (or FSH) induced by LRF is returned to normal in 60 minutes, identical amounts in weight, of [D-Trp^6-des-Gly^{10}]-N-Et-LRF lead to statistically elevated levels of LH up to 24 hours in several in vivo preparations, including man. These analogs are ideal agents to stimulate ovulation (Vilchez-Martinez et al., 1975). Marks and Stern (1975) have reported that these analogs are considerably more resistant than the native structures to degradation by tissue enzymes.

Injection into laboratory animals of large doses of the super-LRFs (i.e. doses of several micrograms/animal while minimal active doses or physiological range are in nanograms/animal) has been recently shown by several groups to have profound anti-gonadotropic effects, both in males and females; moreover, when such large doses of the super-LRFs are injected in the early days of pregnancy in rats, they consistently lead to resorption of the conceptus (see Rivier et al., 1977a); mechanisms involved in these observatons have not been fully clarified as yet but are best explained by the current concepts of negative cooperativity between the peptidic ligands involved and their receptors at the several target-organ sites.

All of the antagonist LRF-analogs as originally found by our group (Vale et al. 1972c) or as later reported by others have deletion or a D-amino acid substitution of His^2. For reasons not clearly understood, addition of the Fujino modification on the C-terminal (Fujino et al., 1974) does not increase the specific activity (as antagonists) of the antagonist analogs. Administered simultaneously with LRF the antagonist analogs inhibit LRF in weight ratios ranging from 3:1 to 15:1. The most potent of these antagonists inhibit activity of LRF not only *in vitro*, but also in various tests *in vivo*. They inhibit the release of LH and FSH induced by an acute dose of LRF; they also inhibit endogenous release of LH-FSH and thus prevent ovulation in laboratory animals. The clinical testing of some of these LRF-antagonists prepared in our laboratory has recently started in collaboration with Yen at the University of California in San Diego.

b. *Biological Activity of Somatostatin*

It is now recognized that somatostatin has many biological effects other than the one on the basis of which we isolated it in extracts of the hypothalamus, *i.e.* as an inhibitor of the secretion of growth hormone (Brazeau et al., 1973). Somatostatin inhibits the secretion of thyrotropin, but not prolactin, normally stimulated by TRF (Vale et al., 1974); it also inhibits the secretion of glucagon, insulin (Koerker et al., 1974), gastrin, secretin, by acting directly on the secretory elements of these peptides. I have recently shown (Guillemin, 1976b) that somatostatin also inhibits the secretion of acetylcholine from the (electrically stimulated) myenteric plexus of the guinea pig ileum probably at a presynaptic locus — thus explaining at least in part the reportedly inhibitory effects of somatostatin on gut contraction, *in vivo* and *in vitro* (Fig. 7).

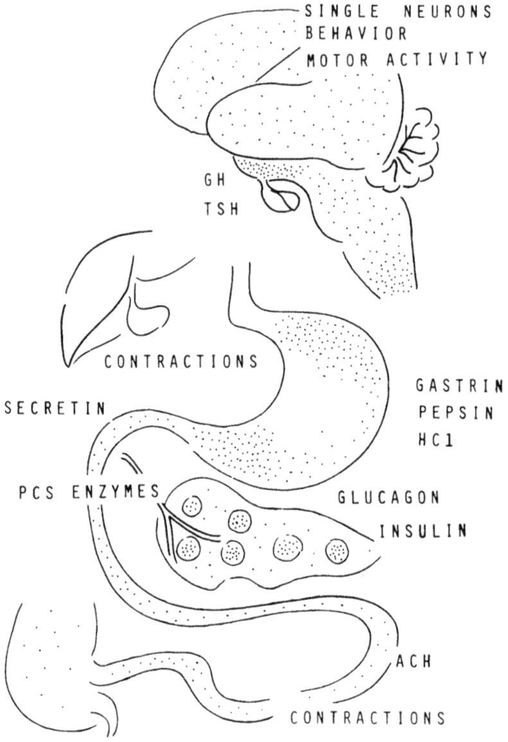

Fig. 7: Multiple locations of somatostatin and multiple effects of somatostatin.

It is also now well recognized that somatostatin is to be found in many locations other than the hypothalamus (Fig. 7), from which we originally isolated it. Somatostatin has been found in neuronal elements and axonal fibers in multiple locations in the central nervous system, including the spinal cord (see Hökfelt et al., 1976). It has been found also in discrete secretory cells of classical epithelial appearance in all the parts of the stomach, gut and pancreas (see Luft et al., 1974; Dubois, 1975) in which it had been first recognized to have an inhibitory effect.

Somatostatin does not inhibit indiscriminately the secretion of all polypeptides or proteins. For instance, as already stated, somatostatin does not inhibit the secretion of prolactin concomitant with that of thyrotropin when stimulated by a dose of TRF; this is true *in vivo* with normal animals or *in vitro* with normal pituitary tissue (see Vale et al., 1974). Somatostatin does not inhibit the secretion of either gonadotropin LH or FSH, the secretion of calcitonin, the secretion of ACTH in normal animals or from normal pituitary tissues *in vitro*; it does not inhibit the secretion of steroids from adrenal cortex or gonads under any known circumstances (see Brazeau et al., 1973). Regarding the secretion of polypeptides or proteins from abnormal tissues of experimental or clinical sources, such as pituitary adenomas, gastrinomas, insulinomas, etc. somatostatin has been shown to be inhibitory according to its normal pattern of activity or being nondiscriminative. The latter must reflect one of the differences between normal and neoplastic tissue. This is in keeping with observation that TRF or LRF can stimulate release of growth hormone from the pituitaries of acromegalic patients though that does not happen with normal tissues.

Clinical studies have confirmed in man all observations obtained in the laboratory. The powerful inhibitory effects of somatostatin on the secretion not only of growth hormone but also of insulin and glucagon have led to extensive studies over the last three years of a possible role of somatostatin in the management or treatment of juvenile diabetes (Figs. 8—9). First of all, the ability of somatostatin to inhibit insulin and glucagon secretion has provided a useful tool for studying the physiological and pathological effects

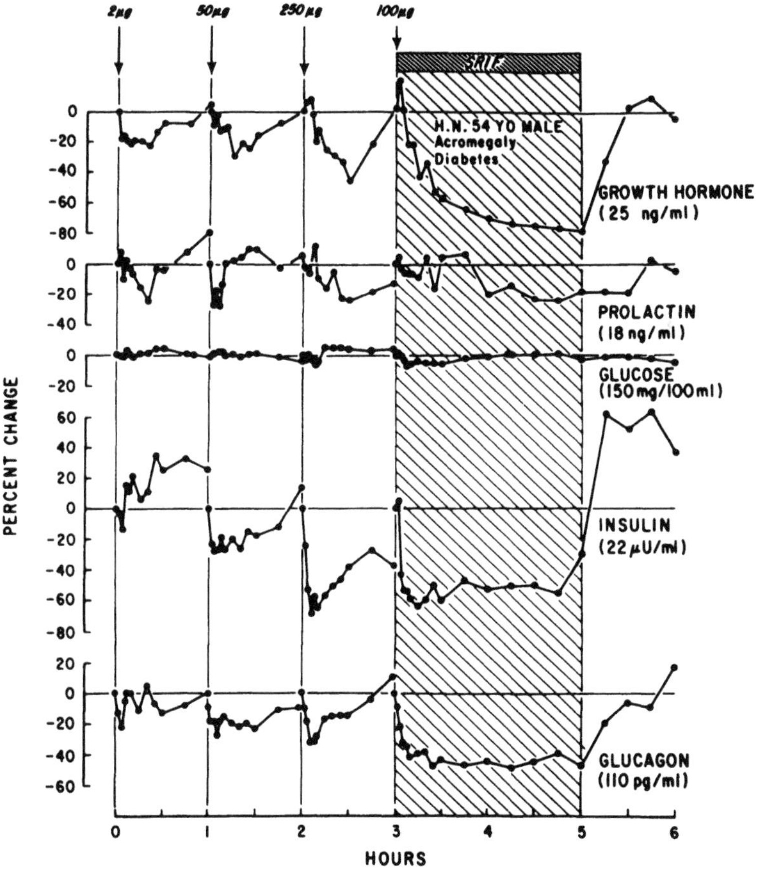

Fig. 8: Effect of multiple doses of somatostatin decreasing the plasma levels of growth hormone, insulin and glucagon in a patient with acromegaly and diabetes. (from Yen et al.).

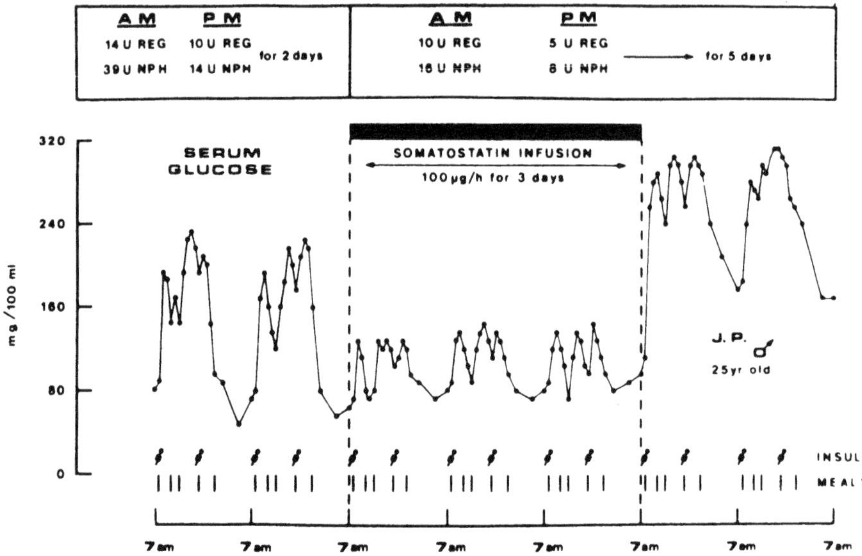

Fig. 9: Juvenile diabetic; improved control of glycemia during infusion of somatostatin and reduced amounts of insulin (from Gerich et al.).

of these hormones on human metabolism. Infusion of somatostatin lowers plasma glucose levels in normal man despite lowering of plasma insulin levels (Alford et al., 1974; Gerich et al., 1974; Mortimer et al., 1974). These observations provided the first clear-cut evidence that glucagon has an important physiological role in human carbohydrate homeostasis. Somatostatin itself has no direct effect on either hepatic glucose production or peripheral glucose utilization, since the fall in plasma glucose levels could be prevented by exogenous glucagon (Gerich et al., 1974).

In juvenile-type diabetics, somatostatin diminishes fasting hyperglycemia by as much as 50% in the complete absence of circulating insulin (Gerich et al., 1974). Although somatostatin impairs carbohydrate tolerance after oral or intravenous glucose challenges in normal man by inhibiting insulin secretion, carbohydrate tolerance after ingestion of balanced meals is improved in patients with insulin-dependent diabetes mellitus through the suppression of excessive glucagon responses (Gerich et al., 1974). The combination of somatostatin and a suboptimal amount of exogenous insulin (which by itself had prevented neither excessive hyperglycemia nor hyperglucagonemia in response to meals) completely prevents plasma glucose levels from rising after meal ingestion in insulin-dependent diabetics (Gerich et al., 1974). Through its suppression of glucagon and growth hormone secretion, somatostatin has also been shown to moderate or prevent completely the development of diabetic ketoacidosis after the acute withdrawal of insulin from patients with insulin-dependent diabetes mellitus (Gerich et al., 1975).

At the moment, clinical studies with somatostatin are proceeding in several clinical centers throughout the world.

From the foregoing description of the ability of somatostatin to inhibit the secretion of various hormones, it would appear that it may be of therapeutic use in certain clinical conditions such as acromegaly, pancreatic islet cell tumors, and diabetes mellitus. With regard to endocrine tumors, it must be emphasized that while somatostatin will inhibit hormone secretion by these tissues, it would not be expected to diminish tumor growth (in view of its locus of action relating to that of c-AMP — see Vale et al., 1972b). Thus, in these conditions it is unlikely that somatostatin will find use other than as a symptomatic or temporizing measure.

In diabetes mellitus, however, somatostatin might be of considerable clinical value. First, it has already been demonstrated that it can acutely improve fasting as well as postprandial hyperglycemia in insulin-requiring diabetics, by inhibiting glucagon secretion. Second, since growth hormone has been implicated in the development of diabetic retinopathy, the inhibition of growth hormone secretion by somatostatin may lessen this complication of diabetes. Finally, through suppression of both growth hormone and glucagon secretion, somatostatin may prevent or diminish the severity of diabetic ketoacidosis and find application in "brittle diabetes." These optimistic expectations must be considered in light of the facts that the multiple effects of somatostatin on hormone secretions and its short duration of action make its clinical use impractical at the present time; moreover, its long-term effectiveness and safety have not been established as yet. Regarding the clinical use of somatostatin, see the recent review by Guillemin and Gerich (1976).

With the considerable interest in somatostatin as a part of the treatment of diabetics, "improved" analogs of somatostatin have been in the mind of clinicians and investigators. Analogs of somatostatin have been prepared in attempts to obtain substances of longer duration of activity than the native form of somatostatin; this has not been very successful so far. Other analogs have been sought that would have dissociated biological activity on one or more of the multiple recognized targets of somatostatin. Remarkable results have recently been obtained. The first such analog so recognized by the group of the Wyeth Research Laboratories was [des-Asn^5]-somatostatin, an analog with approximately 4%, 10% and 1% the activity of somatostatin to inhibit respectively secretion of growth hormone, insulin and glucagon (Sarantakis et al., 1976). While such an analog is not of clinical interest, it showed that dissociation of the biological activities of the native somatostatin on three of its receptors could be achieved. Some of the most interesting analogs with dissociated activities reported so far were prepared and studied by J. Rivier, M. Brown and W. Vale in our laboratories; they are [D-Ser^{13}]-somatostatin, [D-Cys^{14}]-somatostatin and [D-Trp^8, D-Cys^{14}]-somatostatin. When compared to somatostatin, this latest compound has ratios of activity such as 300%, 10%, 100% to inhibit the secretions respectively of growth hormone, insulin and glucagon (Brown et al., 1976). These and other analogs are obviously of much clinical interest and are being so investigated at the moment in several laboratories and clinical centers. An international symposium was recently devoted entirely to the biology and chemistry of somatostatin and analogs (see Raptis, 1978).

c. *Biological Studies with the Endorphins*

1. *Relation of Endorphins to β-lipotropin*

So far, all morphinomimetic peptides isolated from natural sources on the basis of a bioassay or displacement assay for ^{3}H-opiates on synaptosomal preparations, and chemically characterized, have been related to a fragment of the C-terminus of the molecule of β-lipotropin, starting at Tyr^{61}. In the case of Leu^5-enkephalin, the relationship still holds for the sequence Tyr-Gly-Gly-Phe; no β-lipotropin with a Leu residue in position 65 has been observed.

β-LPH has no opioid activity in any of the tests above. Incubation of β-LPH at 37°C with the 10^5 *g* supernatant of a neutral sucrose extract of rat-brain generates opioid activity suggesting the presence of peptidases in the rat brain that could cleave β-LPH to fragments with opioid activity. Thus, β-LPH may be a prohormone for the opiate-like peptides (Lazarus et al. 1976). This would imply that the biogenesis of endorphins may be similar to that of angiotensin with cleaving enzymes available in the central nervous system or in peripheral blood. There is also good evidence by immunocytochemistry (Bloom et al., 1977) and biosynthesis studies (Mains, Eipper and Ling, 1977;

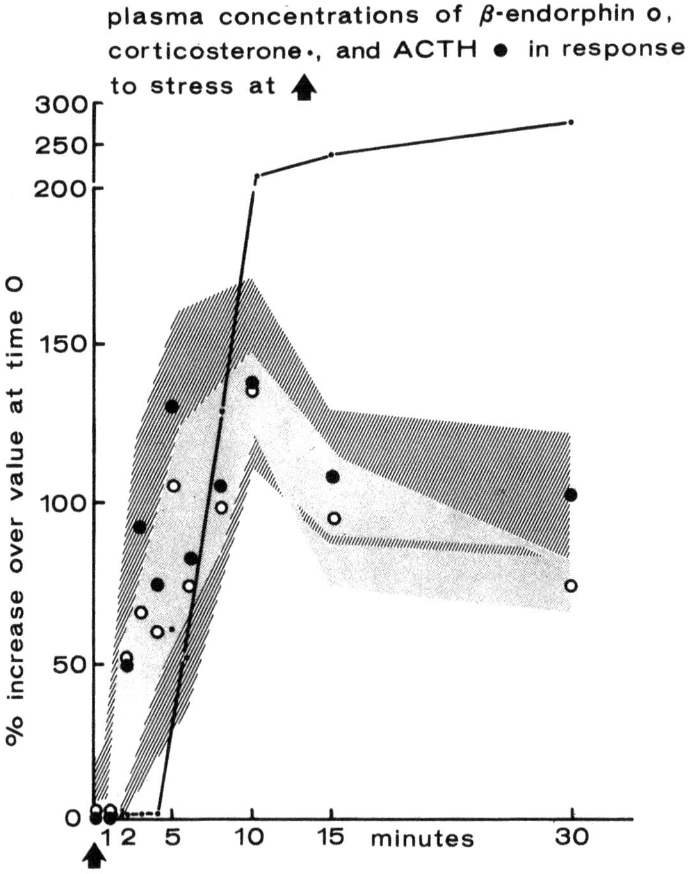

Fig. 10: Concomitant elevation of plasma levels of ACTH and β-endorphin upon acute exposure to stress applied at time 0. (Peptide measured by radioimmunoassays). Solid line shows plasma corticosterone levels measured by fluorometry.

Chrétien et al., 1977) that β-endorphin exists as such and as part of a larger precursor in discrete pituitary cells. We have indeed recently shown (Guillemin et al., 1977b) that β-endorphin and ACTH are secreted simultaneously *in vivo* or *in vitro* in all circumstances tested so far (Fig. 10).

β-LPH [61—63] has no opioid activity at 10^{-4} M; β-LPH[61—64], β-LPH[61—65]-NH_2, (Met $(O)^{65}$)-β-LPH[61—65], β-LPH[61—69], β-LPH-[61—76], β-LPH[61—91] all have opioid activity. β-LPH[61—65]-NH_2, β-LPH[61—65]-NEt and all peptides larger than β-LPH[61—65] have longer duration of biological activity than met-enkephalin in the myenteric plexus bioassay. All these peptides were prepared by solid phase synthesis (see Ling and Guillemin, 1976). β-Endorphin is by far the longest-acting peptide when compared at equimolar ratios with all other fragments of the 61—91 COOH-fragment of β-LPH. In quantitative assays using the myenteric plexus, β-endorphin is approximately 5 times more potent than Met^5-enkephalin; the two analogs of Met^5-enkephalin amidated on the C-terminal residue have also 2—3 times greater specific activity than the free acid form of the peptide, with 95% fiducial limits of the assays overlapping those of β-endorphin. Several analogs of the enkephalins have recently been reported with greater specific activity than the native molecule; all involve substitution with D-amino acids (Pert et al., 1976; Ling et al., 1978).

A series of analogs of the endorphins was synthesized by N. Ling and further purified to high purity. All these peptides have parallel competition curves when studied at 5—6 dose levels in an opiate-displacement assay from rat brain synaptosomes (Lazarus et al., 1976) with the exception of β-LPH-[62—91] which is definitely divergent from the other curves. Comparing the values obtained in the bioassay (myenteric plexus-longitudinal muscle) and the synaptosomal displacement assay, it is obvious that the two assay systems do not give necesssarily identical values.

Of considerable interest are some results observed with the analogs of α, β, γ, δ-endorphins in which a residue of leucine has been substituted for methionine in position 5 from the NH^2-terminus (Ling et al., 1978; Guillemin et al., 1977). [Leu^5]-β-endorphin and [Leu^5]-γ-endorphin are considerably more potent than their native congeners in the brain synaptosome assays, though not in the guinea pig ileum assay. It is tempting to speculate that the brain variety of endorphins might contain a residue of leucine in position 5. Proof of such a hypothesis would require isolation and characterization of such molecules. To this date, no [Leu^{65}]-β-lipotropin has been recognized and characterized. On the other hand, Hughes and collaborators (Hughes et al., 1975) and later Simantov and Snyder (1976) have isolated from brain extracts not only Met^5-enkephalin but also Leu^5-enkephalin. Leu^5-enkephalin might come from an allele of β-lipotropin of brain origin. It is also possible that Leu^5-enkephalin of brain origin is a sub-unit of a larger molecule with no relation to β-lipotropin (other than the common tetrapeptide Tyr-Gly-Gly-Phe). Recent studies in collaboration with Bloom (to be published) and Rossier (1977) have indeed shown remarkable dissociation in the distribution of neurons containing either β-endorphin or enkephalins.

2. *Release of Pituitary Hormones by Endorphins*

One of our original interests in engaging in the isolation and characterization of the endorphins was that the opiate-like peptides might be involved in the secretion of pituitary hormones, particularly growth hormone and prolactin, long known to be acutely released following injection of morphine.

We have shown (Rivier et al., 1977b) that β-endorphin is a potent releaser of immunoreactive growth hormone and prolactin when administered to rats by intracisternal injection. These effects were prevented by prior administration of naloxone. The endorphins are not active directly at the level of the pituitary cells; they show no effect, even in large doses, when added directly to monolayer cultures of (rat) pituitary cells. Thus, the hypophysiotropic effects of the endorphins, like those of the opiate alkaloids, are mediated by some structure in the central nervous system and are not directly at the level of the adenohypophysis. Similar results have been observed by several groups of investigators. We have recently shown that β-endorphin is a potent stimulator of the secretion of vasopressin, possibly acting at a hypothalamic level, since it is not active on the *in vitro* isolated neurohypophysis (Weitzman et al., 1977).

c. *Neuronal Actions of Endorphins and Enkephalins Among Brain Regions*

The existence of endogenous peptides with opiate-like actions suggests that these substances may function as neuromodulators or neurotransmitters in the CNS. Indeed, recent iontophoretic studies have shown that the enkephalins can modify the excitability of a variety of neurons in the CNS. Most neurons tested were inhibited by these peptides (Frederickson and Norris, 1976; Hill et al., 1976; Zieglgansberger et al., 1976), although Renshaw cells responded with an excitation (Davies and Dray, 1976). Studies have recently appeared exploring systematically the sensitivity of neurons to the endorphins or reporting a systematic regional survey of neurons responsive to the peptides (Nicoll et al., 1977).

A surprising finding in that study was the potent excitatory effects of the peptides and normorphine on hippocampal pyramidal cells (Fig. 11). The regional specificity of this excitatory action could be clearly demonstrated with the same electrode by recording from cells in the overlying cerebral cortex and the underlying thalamus during a single penetration. No tachyphylaxis was observed either to the excitatory or inhibitory action of the peptides in any of the regions examined, even though the peptides were often applied repeatedly to the same cell for periods in excess of one hour.

To determine whether the responses observed with the peptides were related to the activation of opiate receptors the specific opiate antagonist, *naloxone*, was administered both by iontophoresis from an adjacent barrel of the microelectrode and by subcutaneous injections. Administered by either route, naloxone antagonized both the excitations and the inhibitions.

All these effects of opiate-like peptides on neuronal activity, taken with biochemical and histochemical evidence for their existence in brain, are consistent with the hypothesis that these peptides are neurotransmitters in

NEURONAL EFFECTS OF OPIOID PEPTIDES

Region (Cell Type)	MET-Enkephalin % Exc.	% Inh.	Beta-Endorphin % Exc.	% Inh.	Normorphine % Exc.	% Inh.
Cerebellum (Purkinje)	18	21	23	23	20	60
	N= 34		N=13		N= 5	
Cerebral Cortex (unidentified)	1	79	25	49	26	52
	N= 58		N=44		N=27	
Brain Stem (Lat. Ret. Nuc. +)	3	47	23	45	10	75
	N=113		N=35		N=20	
Caudate Nuc. (unidentified)	0	83	10	86	9	73
	N= 18		N=35		N=20	
Thalamus (unidentified)	0	100	0	100	0	100
	N= 15		N= 5		N= 4	
Hippocampus (Pyramidal)	90	5	86	7	92	0
	N= 19		N=14		N=12	

Fig. 11: Summary of neuronal effects of opioid peptides and morphine. In each category the total number of cells tested and the percentage of this total that were inhibited or excited is shown (from Nicol et al.).

the CNS. Moreover, I have recently observed in collaboration with Grumbach, Peternelli and Davis that purified synaptosomes of ventral hypothalamus origin (rat brain) release large amounts of immunoreactive β-endorphin when exposed to elevated [KCl] (to be published). When the cells of origin of these peptide-containing fibers have been determined, it may then be possible to proceed with studies into the effects on cellular activity and the secretion of the peptides in order to satisfy more completely the criteria for a neurotransmitter. Crucial points in such future analyses will be the questions of whether the endorphin and enkephalin containing fibers are mutually inclusive systems (recent studies indicate that they are not — Rossier et al., 1977), whether the length of the peptide released by neuronal activity is subject to modulation, and whether intermediate length peptides (such as the α-, γ-, and δ-endorphins) may participate in such modulatory changes. Although the endorphins and β-LPH may be prohormones for Met^5-enkephalin, there are at present no such candidates for Leu^5-enkephalin. The results presented here indicate to us that the cellular roles of endorphin and enkephalin peptides cannot now be generalized across all brain regions where they are found, and that no simple cellular action of any peptide will yield an integrative picture of the way in which opiate alkaloids produce complex analgesic, euphoric, and addictive responses. Involvement of the endorphins in the control of adenohypophysial functions is still a subject for further study at the time of writing this review. So far, no direct hypophysiotropic activities of the endorphin peptides have been clearly demonstrated.

d. *Behavioral effects of endorphins*

The pharmacological properties of endorphins have so far been screened through application of tests *in vitro* or *in vivo* previously used to characterize opiate agonists and antagonists.

When injected into the cerebrospinal fluid, endorphins affect several behavioral and physiological measures, in addition to responses to noxious agents, and each of the peptides exhibits different dose-effect profiles on these measures: β-endorphin induces a marked catatonic state (Fig. 12) lasting for hours (Bloom et al., 1976) at molar doses 1/100 those at which Met^5-enkephalin transiently inhibits responses to noxious agents (Belluzzi et al., 1976; Buscher et al., 1976; Loh et al., 1976). This potent behavioral effect of a naturally occurring substance suggests its regulation could have etiological significance in mental illness.

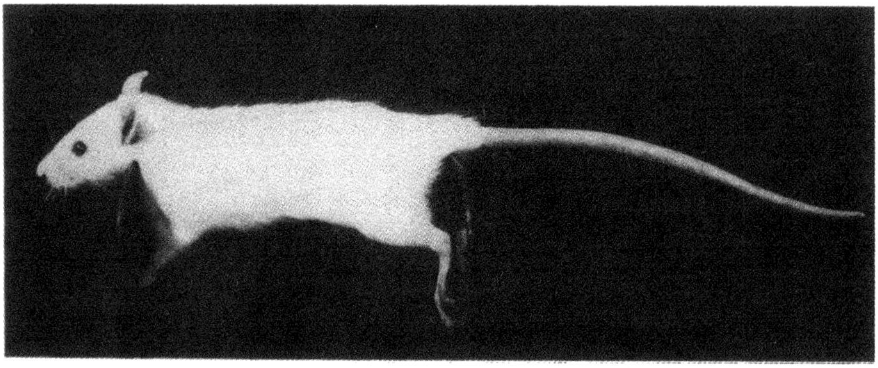

Fig. 12: Thirty minutes after the intracisternal injection of β-endorphin (15×10^{-9} mole) this rat exhibited sufficient rigid immobility to remain totally self-supporting when placed across metal bookends which are in contact only at the upper neck and base of the tail. Such postures were maintained for prolonged periods. Note the erect ears and tail, widely opened eyelids and extended lower limbs.

In terms of molar dose-effectiveness on the various parameters examined, β-endorphin is clearly the most potent substance tested.

Rats given seven daily intracisternal injections of 15×10^{-9} mole of β-endorphin continued to show the full set of responses and duration of action. The catatonic state induced by β-endorphin was not observed with the other endorphin peptides, even at considerably higher doses. At very high doses of α-endorphin, γ-endorphin or Met^5-enkephalin, transient losses of corneal reflexes were observed, and α-endorphin seemed more potent in this regard than either γ-endorphin or Met^5-enkephalin. No significant depressions of responsiveness to tail-pinch or pin-prick stimuli were observed with Met^5-enkephalin, α-endorphin, or γ-endorphin, but such effects (Bradbury et al., 1976; Buscher et al., 1976; Ross et al., 1976) could have been missed by the 5-minute interval after injection and before testing began. In contrast to the syndrome induced by β-endorphin, rats given γ-endorphin showed consistent elevations in rectal temperature (about 2.0°C$\pm$.2°C at 30 minutes after 281×10^{-9} mole), and sometimes exhibitied some degree of hyperresponsivity to sensory testing and handling, although there were individual variations in this response.

All of our observations suggest that normal variations—either qualitative or quantitative—in the homeostatic mechanisms regulating the postulated (Lazarus et al., 1976) conversion of β-LPH as a prohormone to its several endorphin cleavage products could constitute a system fundamentally involved in maintaining "normal" behavior; alterations of the mechanisms normally regulating β-lipotropin-endorphins homeostasis could lead to signs and symptoms of mental illness. Such a potential psychophysiological role of endorphins could logically be testable through the therapeutic administration of available opiate antagonists. This has already been attempted in several clinical centers throughout the world; results obtained have been interpreted differently and, in my own mind, are much too preliminary as yet to warrant any conclusion, positive or negative. The ultimate identification of endorphin-sensitive behavioral events and specific treatment of their dysfunctional states may require the development of more specific "anti-endorphins" than those now available; other naturally occurring brain peptides such as Substance P, have already been reported to be endorphin antagonists in some assay systems (Guillemin et al., 1976c). There is little doubt that the potential significance in any such studies is so great that major efforts in this area both in the laboratory and in clinical studies should be sponsored and pursued without fail.

PART III

Endocrine and Paracrine Secretions of the Brain. Hormones and Cybernins

It has been known for some time that TRF, LRF, and somatostatin originally isolated from extracts of the hypothalamus are actually to be found in non-negligible amounts throughout the central nervous system, including the spinal cord. This was demonstrated first by bioassays and has been amply confirmed by radioimmunoassays. Extrahypothalamic TRF, LRF, or somatostatin have not been isolated and chemically characterized as yet. There is thus no uncontrovertible evidence that the pertinent substances in extra-hypothalamic brain-extracts are identical to those characterized in the hypothalamus. All the current circumstantial evidence is in favor, however, of the identity of the materials in their multiple locations (parallelism of dose-response curves in several bioassays, parallelism of displacement curves in several types of radioimmunoassays, etc. using purified tissue extracts and pure synthetic peptides as reference standards). Ubiquity of these peptides throughout the central nervous system does not imply that they are randomly distributed. Several groups have shown that each of these peptides has a unique distribution pattern as they have been identified by immunocytochemistry in axonal tracts and neuronal bodies in well characterized anatomical formations of the central nervous system (Hökfelt et al., 1977). It was this knowledge of the multiplicity of location of the releasing factors and particularly of somatostatin in the central nervous system that lead a few years ago to the solution of one of the most puzzling dilemmas in this field, and that solution had far-reaching consequences: It was difficult to reconcile the short biological half-life of somatostatin (less than four minutes), when injected in the blood stream with its well established effects on the secretion of glucagon and insulin, with the hypothesis that *hypothalamic* somatostatin could be involved in the physiological control of the secretion of *pancreatic* glucagon and insulin. Luft in Stockholm and I independently wondered whether somatostatin could be delivered to the endocrine pancreas by means other than the general circulation, possibly by nerve fibers, known to innervate the islets of Langerhans. The remarkable observation was then made that in fact the endocrine pancreas of all vertebrates studied so far contain a discrete population of cellular elements containing somatostatin as shown by immunocytochemistry (Luft et al., 1974; Dubois, 1975). The somatostatin-containing cells belong to the D-cells of the endocrine pancreas, long known to the morphologists to be different from the α-cells containing glucagon and the β-cells containing insulin, but for which no specific secretory products had been recognized so far. Moreover, in these early reports a large number of secretory cells containing immunoreactive somatostatin were found throughout the gastrointestinal tract and it has now been shown that somatostatin can inhibit the secretion of gastrin, of secretin, of cholecystokinin, also the secretion of pepsin and HCl by acting directly at the level of the gastric mucosa (Nakaji, et al., 1975). TRF has recently been reported in extracts of the stomach and of the duodenum (Leppaluoto et al. in press). Neurotensin and substance P have also been located in the hypothalamus and throughout the gastrointestinal tract in specific cells and in crude extracts — as has been known in the case of substance P, since 1931 from the work of Gaddum and von Euler. There is now evidence that other peptides originally characterized from extracts of tissues of the gastrointestinal tract can be found and located in the brain; this is the case for gastrin/cholecystokinin, vaso-intestinal peptide (VIP), the gastric inhibitory peptide (GIP) (see review by Pearse and Takor, 1976); this is also true for the endorphins and enkephalins and for several of the small peptides such as bombesin, caerulein, physalamine, isolated years ago from extracts of the skin of several species of frogs. There are remarkable analogies and homologies between the amino acid sequences of several of these peptides of central nervous system origin and gastrointestinal origin, as well as those isolated from frog skin.

These peptides have been found by immunocytochemistry essentially in two types of cells: 1) They are seen in cell bodies and nerve fibers i.e. axonal and dendritic processes of *neurons* in brain, spinal cord, in spinal ganglia and in the myenteric plexus; 2) They are seen also in typical *endocrine cells*, for instance in the pancreatic islets of Langerhans, in the enterochromaffin cells of the gut and the adrenal medulla. Neuroblastomas have been reported to contain high levels of the vaso-intestinal peptide (VIP) (Said and Rosenberg, 1976).

There is already an interesting unifying concept to bring together these rather startling observations. Much credit must go to A.G.E. Pearse for his visionary concept, formulated some ten years ago, of the APUD cells: Pearse observed that neurons and some endocrine cells producing polypeptide hormones shared a set of common cytochemical features and ultrastructural characteristics. APUD is an acronym referring to Amine content and/or Amine Precursor Uptake and Decarboxylation, as common qualities of these cells (Pearse, 1968). The APUD concept postulated that these endocrine cells were derived from a common neuroectodermal ancestor, the transient neural crest. Pearse postulated further that a still larger number of endocrine cells would be eventually found sharing these common properties if one were to explore further in the adult, endocrine tissues derived from the neural crest. Recent observations with refined techniques, particularly the work of Le Douarin on topical chimeras with chromosomal markers, have led Pearse to modify the original APUD concept, and, as we will see, in a remarkable manner. The new evidence regarding the multiple sources of the several peptides mentioned above showed that tissues were involved that were not of neural crest origin; this is particularly true for the peptide-secreting cells of the gut. All these cells have been shown to arise from specialized neuroectoderm (Pearse and Takor, 1976); that is, not only the neural crest but also the neural tube, the neural ridges and the placodes.

The expanded concept now postulates that all peptide hormone-producing cells are derived from the neural ectoderm, as are all neurons. With such ontogenic commonality, it is thus less surprising to recognize the presence of "gastro-intestinal peptides" in the brain, and of "brain peptides" in the gastro-intestinal tract.

ACTH, β-endorphin and growth hormone of pituitary origin, cholecystokinin, secretin, or gastrin of gastrointestinal origin are well recognized hormones which satisfy all the definitions of the word, particularly as it implies their distant action on target cells or organs far removed from the source of these peptides. In the case of the release of TRF, LRF, or somatostatin by hypothalamic neurons at the level of the hypothalamo-hypophysial portal vessels, these hypophysiotropic activities can also be considered as hormonal in nature. Let us note immediately that, while the means of conveyance are indeed blood vessels, the distance travelled by the hypophysiotropic peptides in these vessels is measured by a few mm. until they reach their target pituitary cells. This is very different from the long distance travelled by the classical hormones mentioned above. Moreover, even in the case of the hypothalamic peptides as involved in their hypophysiotropic functions, there is no generally accepted evidence that they enter the general circulation for any length of time and in a physiologically meaningful concentration.

When we consider these same peptides in parts of the brain other than the hypophysiotropic hypothalamus, the situation is altogether different: Both the optic and the electron microscope, combined with immunocytochemistry begin to show evidence of very punctual localizations which imply similarly pùnctual roles; i.e. to be played over distances measured in angstroms. Berta Scharrer has for some time described what she calls peptidergic synapses

(Scharrer, 1975). Moreover, recent studies with antibodies to somatostatin have yielded pictures which have been interpreted by Petrusz et al. (1977) as showing the localization of immunoreactive somatostatin in multiple dendritic endings. Some of these pictures are spectacular (Fig. 13). Their most heuristic interpretation is that each dark point is that of a dendritic contact either with another dendrite or abutting on a specialized locus of the axon or of the soma of recipient neurons. Clearly these recipient neurons do not seem to contain immunoreactive somatostatin. The cells of origin containing and sending the presumptive somatostatinergic terminals have not been characterized as yet. Similar pictures have already been observed in multiple locations in the brain and for several immunoreactive peptides. In this context these peptides which we called hormones earlier do not fit the definition of a hormone any more; they seem to be candidates for the definition of neurotransmitters. Recently Hökfelt (1977) has concluded that some neurons may contain both peptides and one of the catecholamines, a classical neurotransmitter (1977).

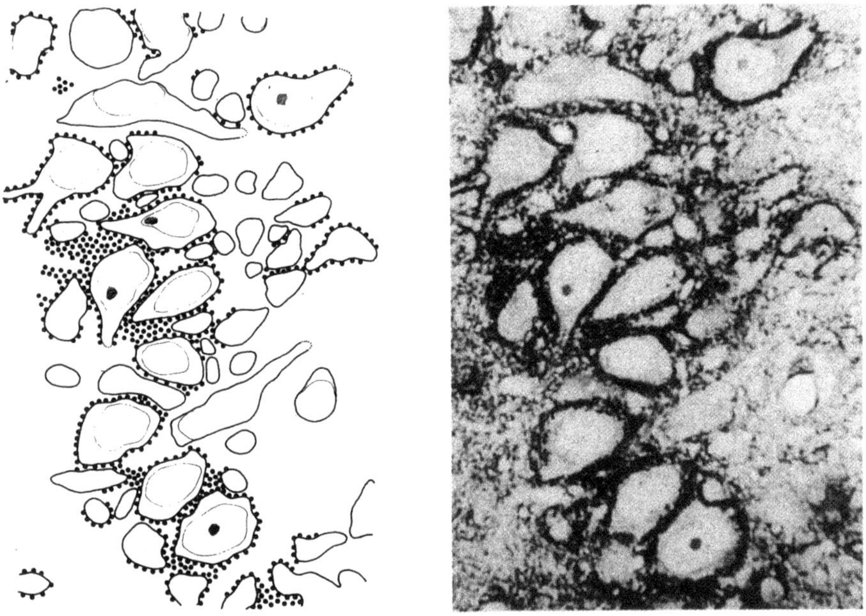

Fig. 13: Presumptive nerve terminals as somato-dendritic terminals around pyramidal cells of the hippocampus rat brain). Immunocytochemistry with antisera to somatostatin (from Petrusz et al.).

It will be obvious by now that we are only at the very beginning of the physiological significance of these peptides in the brain. The local punctual release that we have seen here with its multiple locations and short range of traffic would make them fit what Feyrter had called earlier the paracrine secretory system. Interestingly enough, Feyrter had evolved his concept of paracrine secretion while studying with very simple morphological tools and a great deal of observational acumen the very cells of the gastrointestinal tract and of the pancreas that we know now to secrete the very peptides discussed in this lecture. I have proposed earlier the generic name *cybernin* for these substances; the etymology of the word implying local information. Obviously again, years of investigation are ahead to understand the mechanisms involved in the cell biology and the cell physiology of the neurons as they synthesize, release, respond to and metabolize the many peptides discussed here.

Of the many discoveries ahead of us will be those that will lead us to understand the role of each of these peptides in the brain, not only in their cellular physiology and biochemistry, but also in their significance in the higher functions of the central nervous systems. Though admittedly based on simple and enthusiastic teleology, it is difficult not to hypothesize that these peptides must indeed play some role in the functions of the brain. Once this simple proposal is made, if we recognize that not one word about the existence of these substances in the central nervous system is to be found in any of the classical texts of neuro-psychiatry, one can not but be optimistic that the early observations summarized in this lecture will lead to profound reappraisals of the mechanisms involved in the functions of the normal brain, but also of mental illness.

REFERENCES

Alford, F. P., Bloom, S. R., Nabarro, J. D. N., Hall, R., Besser, G. M., Coy, D. H., Kastin, A. J., and Schally, A. V., Lancet *2*, 974—976, 1974.

Amoss, M., Burgus, R., Ward, D. N., Fellows, R., and Guillemin, R., Endocrinology *86*, A61, 1970.

Amoss, M., Burgus, R., Blackwell, R., Vale, W., Fellows, R., and Guillemin, R., Biochem. Biophys. Res. Commun. *44*, 205—210, 1971.

Belluzzi, J. D., Grant, N., Garsky, V., Sarantakis, D., Wise, C. C., and Stein, L., Nature *260*, 625, 1976.

Bloom, F., Segal, D., Ling, N., and Guillemin, R., Science *194*, 630—632, 1976.

Bloom, F., Battenberg, E., Rossier, J., Ling, N., Leppaluoto, J., Vargo, T., Guillemin, R., Life Science *20*, 43—48, 1977.

Bradbury, A. F., Smyth, D. G., and Snell, C. R., in *Proc. 4th Am. Peptide Symposium*, R. Walter and G. Meienhofer (eds.), pp. 609—615, Ann Arbor Science Publ., 1975.

Bradbury, A. F., Smyth, D. G., Snell, C. R., Birdsall, N. J. M., and Hulme, E. C., Nature (London) *260*, 793—795, 1976.

Brazeau, P., Vale, W., Burgus, R., Ling, N., Butcher, M., Rivier, J., and Guillemin, R., Science *179*, 77—79, 1973.

Brazeau, P., Rivier, J., Vale, W., and Guillemin, R., Endocrinology *94*, 184—186, 1974.

Brown, M., Rivier, J., and Vale, W., Metabolism *25*, 1501—1503, 1976.

Burgus, R., Stillwell, R. N., McCloskey, J. A., Ward, D. N., Sakiz, E., and Guillemin, R., Physiologist *9*, 149, 1966b.

Burgus, R., and Guillemin, R., Fed. Proc., *26*, 255, 1967.

Burgus, R., Dunn, T. F., Desiderio, D., Vale, W., and Guillemin, R., C. R. Acad. Sci. (Paris) *269*, 226, 1969b.

Burgus, R., Dunn, T. F., Ward, D. N., Vale, W., Amoss, M., and Guillemin, R., C. R. Acad. (Paris) Sci. *268*, 2116, 1969a.

Burgus, R., Dunn, T. F., Desiderio, D., and Guillemin, R. C. R., Acad. Sci. (Paris) *269*, 1870, 1969c.

Burgus, R., and Guillemin, R. In *Hypophysiotropic Hormones of the Hypothalamus*, J. Meites (ed.), p. 227. Williams and Wilkins, Baltimore Md. (publ), 1970a.

Burgus, R., Dunn, T. F., Desiderio, D., Ward, D. N., Vale, W., and Guillemin, R., Nature (London) *226*, 321, 1970b.

Burgus, R., Dunn, T. F., Desiderio, D., Ward, D. N., Vale, W., Guillemin, R., Felix, A. M., Gillessen, D., and Studer, R. O., Endocrinology *86*, 573, 1970c.

Burgus, R., Butcher, M., Ling, N., Monahan, M., Rivier, J., Fellows, R., Amoss, M., Blackwell, R., Vale, W., and Guillemin, R., C. R. Acad. Sci. (Paris) *273*, 1611—1613, 1971.

Burgus, R., Butcher, M., Amoss, M., Ling, N., Monahan, M., Rivier, J., Fellows, R., Blackwell, R., Vale, W., and Guillemin, R., Proc. Nat. Acad. Sci. (USA) *69*, 278—282, 1972.

Burgus, R., Ling, N., Butcher, M., and Guillemin, R., Proc. Nat. Acad. Sci. (USA) *70*, 684—688, 1973.

Buscher, H. H., Hill, R. C., Romer, D., Cardinaux, F., Closse, A., Hauser, D., Pless, J., Nature *261*, 423—425, 1976.

Campbell, H. T., Feuer, G., Garcia, J., and Harris, G. W., J. Physiology *157*, 30, 1961.

Chrétien, M., Seidah, N. G., Benjannet, S., Dragon, N., Routhier, R., Motomatsu, T., Crine, P., and Lis, M., Ann. N.Y. Acad. Sci. *297*, 84—105, 1977.

Courrier, R., Guillemin, R., Jutisz, M., Sakiz, E., and Aschheim, P., C. R. Acad. Sci. (Paris) *253*, 922, 1961.

Cox, B. M., Goldstein, A., and Li, C. H., Proc. Nat. Acad. Sci. (USA) *73*, 1821—1823, 1976.

Davies, J., and Dray, A., Nature *262*, 603—604, 1976.

Dhariwal, A. P. S., Watanabe, S., Antunes-Rodrigues, J., and McCann, S., Neuroendocrinology *2*, 294—303, 1967.
Dubois, M., Proc. Nat. Acad. Sci. (USA) *72*, 1340—1343, 1975.
Fales, H. M., Nagai, Y., Milne, G. W. A., Brewer, H. B., Bronzert, R. J., and Pisano, J. J., Anal. Biochem. *43*, 288—299, 1971.
Fellows, R. E. and Mudge, A., Fed. Proc. *30*, 1078, 1970.
Fleischer, N., Burgus, R., Vale, W., Dunn, T., and Guillemin, R., J. Clin. Endocrinol. *31*, 109—112, 1970.
Fleischer, N., and Guillemin, R., in *Peptide Hormones*, J. A. Parsons (ed.), p. 317—335, MacMillan Press, Ltd., N.Y., 1976.
Folkers, K., Enzmann, F., Boler, J., Bowers, C. Y., and Schally, A. V., Biochem. Biophys. Res. Commun. *37*, 123, 1969.
Frederickson, R. C. A., and Norris, F. H., Science *194*, 440—442, 1976.
Fujino, M., Yamazaki, I., Kobayashi, S., Fukuda, T., Shinagawa, S., Nakayama, R., Biochem. Biophys. Res. Commun. *57*, 1248—1256, 1974.
Gerich, J. E., Lorenzi, M., Schneider, V., Karam, J., Rivier, J., and Guillemin, R., New Engl. J. Med. *291*, 544—547, 1974.
Gillessen, D., Felix, A. M., Lergier, W., and Studer, R. O., Helv. Chim. Acta *53*, 63, 1970.
Guillemin, R. and Rosenberg, B., Endocrinology *57*, 599—697, 1955.
Guillemin, R., Yamazaki, E., Gard, D. A., Jutisz, M., and Sakiz, E., Endocrinology *73*, 564, 1963.
Guillemin, R., Yamazaki, E., Jutisz, M., and Sakiz, E., C. R. Acad. Sci. (Paris) *255*, 1018—1020, 1962.
Guillemin, R., Burgus, R., and Vale, W., Vitamins and Hormones *29*, 1—39, 1971.
Guillemin, R., in *Advances in Human Growth Hormone Research*, Vol. 1, S. Raiti (ed.), U.S Gov't Print. Office, DHEW, Publ. No. (NIH) 74—612, p. 139—143, 1973.
Guillemin, R., Ling, N., and Burgus R., C. R. Acad. Sci. (Paris), *282*, 783—785, 1976a.
Guillemin, R., Endocrinology *99*, 1653, 1976b.
Guillemin, R. and Gerich, J., Ann. Rev. of Med. *27*, 379—388, 1976.
Guillemin, R., Ling, N., Burgus, R., Lazarus, L., Psychoneuroendocrinology *2*, 59—62, 1976c.
Guillemin, R., Ling, N., Lazarus, L., Burgus, R., Minick, S., Bloom, F., Nicoll, R., Siggins, G., and Segal, D., Ann. N.Y. Acad. Sci. *297*, 131—156, 1977a.
Guillemin, R., Vargo, T., Rossier, J., Minick, S., Ling, N., Rivier, C., Bloom, F., and Vale, W., Science *197*, 1367—1369, 1977b.
Hagenmaier, H., Ebbighausen, W., Nicholson, G. and Votsch, W., Z. Naturforsch. *25b*, 681—689, 1970.
Hill, R. G., Pepper, C. M., and Mitchell, F. J., Nature *262*, 604—606, 1976.
Hökfelt, T., Elfvin, L. G., Elde, R., Schultzberg, M., Goldstein, M., and Luft, R., Proc. Nat. Acad. Sci. *74*, 3587—3591, 1977.
Hughes, J., Brain Res. *88*, 295—308, 1975.
Hughes, J., Smith, T. W., Kosterlitz, H. W., Fothergil, L. A., Morgan, B. A., and Morris, H. R., Nature (London) *258*, 577—579, 1975.
Igarashi, M., Nallar, R., and McCann, S. M., Endocrinology *75*, 901—907, 1964.
Koerker, D. J., Ruch, W., Chideckel, E., Palmer, J., Goodner, C., Ensinck, J., and Gale, C., Science *184*, 482—484, 1974.
Krulich, L., Dhariwal, A. P. S., and McCann, S. M., Endocrinology *85*, 319, 1969.
Lazarus, L. H., Ling, N., and Guillemin, R., Proc. Nat. Acad. Sci. (USA) *73*, 2156—2159, 1976.
Li, C. H., and Chung, D., Proc. Nat. Acad. Sci. (USA) *33*, 1145—1148, 1976.
Ling, N., Rivier, J., Burgus, R., and Guillemin, R., Biochemistry *12*, 5305—5310, 1973.
Ling, N., and Guillemin, R., Proc. Nat. Acad. Sci. (USA), *73*, 3308—3310, 1976.
Ling, N., Burgus, R., and Guillemin, R., Proc. Nat. Acad. Sci. (USA), *73*, 3942—3946, 1976.
Ling, N., Minick, S., and Guillemin, R., to be published, 1978.

Loh, H. H., Tseng, L. F., Wei, E., Li, C. H., Proc. Nat. Acad. Sci. (USA) *73*, 2895—2896, 1976.
Luft, R., Efendic, S., Hokfelt, T., Johansson, O., and Arimura, A., Med, Biol. *52*, 428—430, 1974.
Mains, R., Eipper, B., and Ling, N., Proc. Nat. Acad. Sci. (USA) *74*, 3014—3018, 1977.
Marks, N., and Stern, F., FEBS Letters *55*, 220—224, 1975.
Matsuo, H., Arimura, A., Nair, R. M. G., and Schally, A. V., Biochem. Biophys. Res. Commun. *45*, 822—827, 1971a.
Matsuo, H., Nair, R. M., Arimura, A., and Schally, A. V., Biochem. Biophys. Res. Commun. *43*, 1334, 1971b.
Monahan, M., Rivier, J., Burgus, R., Amoss, M., Blackwell, R., Vale, W., and Guillemin, R., C. R. Acad. Sci. (Paris) *273*, 205, 1971.
Monahan, M., Amoss, M., Anderson, H., and Vale, W., Biochemistry *12*, 4616—4620, 1973.
Moriarty, G., J. Histochem and Cytochem. *21*, 855—894, 1973.
Mortimer, C. H., et al., Lancet *1*, 697—701, 1974.
McCann, S. M., Taleisnick, S., and Friedman, H. M., Proc. Soc. Exp. Biol. Med. *104*, 432, 1960.
Nair, R. M. G., Barrett, F. J., Bowers, C. Y., and Schally, A. V., Biochemistry *9*, 1103, 1970.
Nicoll, R., Siggins, G., Ling, N., Bloom, F., and Guillemin, R. Proc. Nat. Acad. Sci. (USA) *74*, 2584—2588, 1977.
Nakaji, N. T., Charters, A. C. III, Guillemin, R. C. L., and Orloff, M. J., Gastroenterology *70*: 989, 1976.
Pasternak, G., Goodman, R., and Snyder, S. H., Life Sci. *16*, 1765—1770, 1975.
Pearse, A. G. E., Proc. Roy. Soc. *B170*, 71, 1968.
Pearse, A. G. E., and Takor, T., Clin. Endocrinol. 229s—244s (suppl), 1976.
Petrusz, P., Sar, M., Grossman, G. H., and Kizer, J. S., Brain Research *137*, 181—187, 1977.
Pert, C. B., and Snyder, S. H., Science *179*, 1011—1014, 1973.
Pert, C. B., Pert, A., Chang, J. K., and Fong, B. T., Science *194*, 330—332, 1976.
Raptis, K. (ed.), in *Somatostatin*, Serono Foundation Symposium, Plenum Press Publ., 1978.
Rebar, R., et al., J. Clin. Endocrinol. and Metab. *36*, 10—16, 1973.
Rivier, J., Burgus, R., and Vale, W., Endocrinology *88*: A86, 1971.
Rivier, J., J. Amer. Chem. Soc. *96*, 2986—2992, 1974.
Rivier, C., Rivier, J., and Vale, W., Endocrinology (in press), 1977a.
Rivier, C., Vale, W., Ling, N., Brown, M., and Guillemin, R., Endocrinology *100*, 238—241, 1977b.
Ross, M., Su, T. P., Cox, B. M., and Goldstein, A., in *Opiates and Endogenous Opioid Peptides*, Elsevier, Publ., Amsterdam, 1976.
Rossier, J., Vargo, T., Minick, S., Ling, N., Bloom, F., and Guillemin, R., Proc. Nat. Acad. Sci. (USA), *74*, 5162—5165, 1977.
Said, S. I., Rosenberg, R. N., Science *192*, 907, 1976.
Sarantakis, D., McKinley, W. A., Jaunakais, I., Clark, D. and Grant, N. Clin. Endocrinol *5*, 275s—278s, 1976.
Schally, A. V., Redding, T. W., Barrett, F. J., and Bowers, C. Y., Fed. Proc. *25*, 348, 1966b.
Schally, A. V., Bowers, C. Y., and Redding, T. W., Endocrinology *78*, 726, 1966a.
Schally, A. V., Bowers, C. Y., Redding, T. W., and Barrett, J. F., Biochem. Biophys. Res. Commun. *25*, 165, 1966c.
Schally, A. V., Saito, T., Arimura, A., Muller, E. E., Bowers, C. Y., and White, W. F., Endocrinology *79*, 1087—1094, 1966.
Schally, A. V., Arimura, A., Bowers, C. Y., Kastin, A. J., Sawano, S., and Redding, T. W. Rec. Progr. Horm. Res. *24*, 497, 1968.
Schally, A. V., Redding, T. W., Bowers, C. Y., and Barrett, F., J. Biol. Chem. *244*, 4077, 1969.
Schally, A. V., Arimura, A., Baba, Y., Nair, R. M. G., Matsuo, H., Redding, T. W., Debeljuk, L., and White, W. F., Biochem. Biophys. Res. Commun. *43*, 393—399, 1971a.
Schally, A. V., Nair, R. M. G., Redding, T. W., and Arimura, A., J. Biol. Chem. *246*, 7230—7236, 1971b.

Schally, A. V., Baba, Y., Nair, R. M. G., Nair, E., Bennett, J., J. Biol. Chem. *246*, 6647, 1971c.

Scharrer, B., Am. J. Anat. *143*, 451—456, 1975.

Simantov, R., and Snyder, S. H., Life Sci. *18*, 781—788, 1976.

Tashjian, A. H., Barowsky, N. J., and Jensen, D. K., Biochem. Biophys. Res. Commun. *43*, 516—523, 1971.

Terenius, L., and Wahlstrom, A., Life Sci. *16*, 1759—1764, 1975.

Teschemacher, H., Opheim, K. E., Cox, B. M., and Goldstein, A., Life Sci. 1771—1776, 1975.

Vale, W., Grant, G., Amoss, M., Blackwell, R., and Guillemin, R., Endocrinology *91*, 562—572, 1972a.

Vale, W., Brazeau, P., Grant, G., Nussey, A., Burgus, R., Rivier, J., Ling, N., and Guillemin, R., C. R. Acad. Sci. (Paris) *275*, 2913—2915, 1972b.

Vale, W., Grant, G., Rivier, J., Monahan, M., Amoss, M., Blackwell, R., Burgus, R., and Guillemin, R., Science *176*, 933—934, 1972c.

Vale, W., Rivier, C., Brazeau, P., and Guillemin, R., Endocrinology *95*, 968—977, 1974.

Veber, D. F., Bennett, C. D., Milkowski, J., Gal, G., Denkewalter, R. G., and Hirschmann, R., Biochem. Biophys. Res. Commun. *45*, 235, 1971.

Vilchez-Martinez, J., Coy, D., Coy, E., De la Cruz, A., Niyhi, N., and Schally, A. V., Endocrinology *96*, 354A, 1975.

Weitzman, R., Fisher, D., Minick, S., Ling, R., and Guillemin, R., Endocrinology, *101*, 1643—1646, 1977.

Yen, S. S. C., Rebar, R., Vandenberg, G., Naftolin, F., Judd, H., Ehara, Y., Ryan, K., Rivier, J., Amoss, M., and Guillemin, R., in *Hypothalamic Hypophysiotropic Hormones*, C. Gual and E. Rosenberg (eds.), Excerpta Medica, Amsterdam, pp. 217—229, 1973.

Zieglgansberger, W., Fry, J. P., Herz, A., Moroder, L., and Wunsch, E., Brain Research *115*, 160—164, 1976.

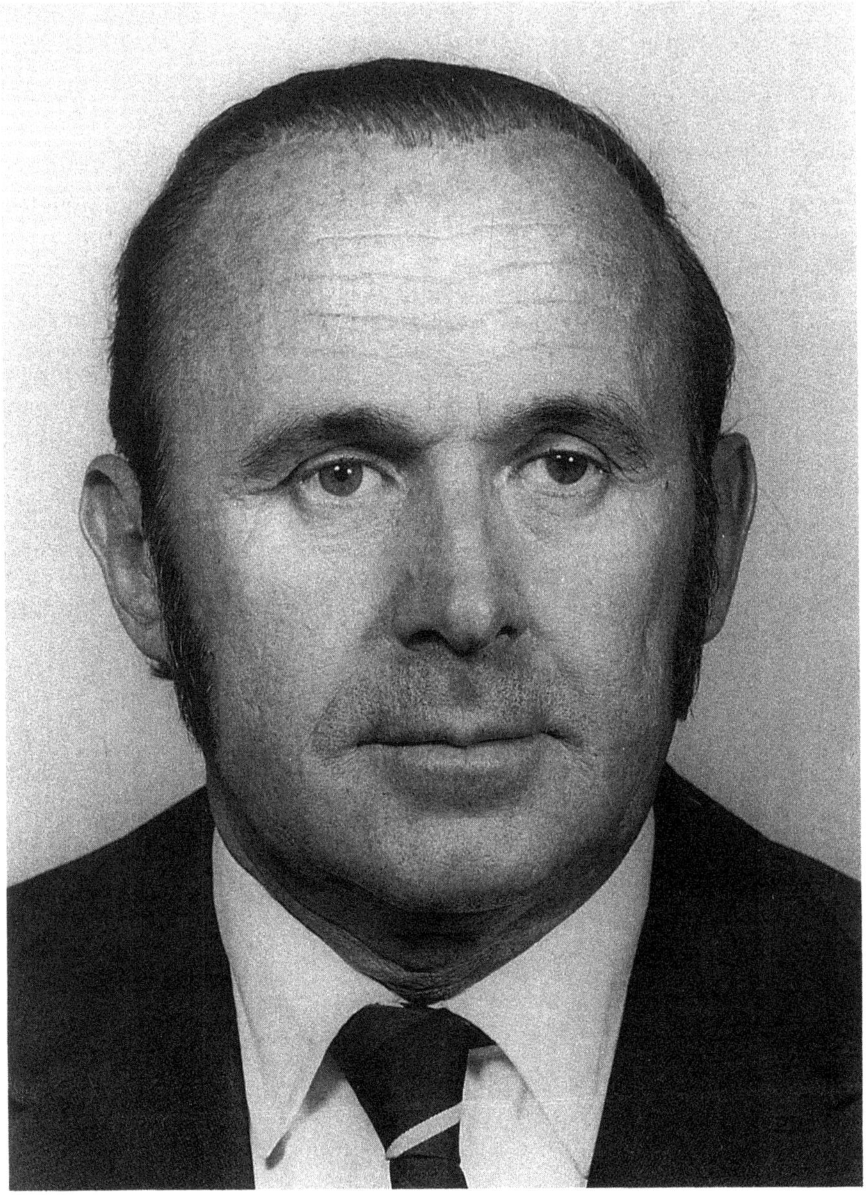

Andrew V. Schally

ANDREW VICTOR SCHALLY

I was born in Wilno, Poland on November 30, 1926, being of Polish, Austro-Hungarian, French and Swedish ancestry. My father, a professional soldier trained in the military academies of Vienna, Austria and St. Cyr, France, had to leave his family when the Second World War broke out to fight with the Allied Forces. My life and outlook were influenced by the harsh childhood which I spent in the Nazi-occupied Eastern Europe, but I was fortunate to survive the holocaust while living among the Jewish-Polish Community in Roumania. I used to speak Polish, Roumanian, Yiddish, Italian and some German and Russian, but I have almost completely forgotten them, and my French in which I used to excell is also now far from fluent. In 1945, I moved via Italy and France to England and Scotland. In spite of post-war economic and nutritional austerity, the United Kingdom seemed like a paradise to me because of the respect for human rights. Since that time, I have always had a profound friendship for the British. I received my high school diploma in Scotland in 1946 and afterwards studied chemistry in London. I adored English and Scottish association football and I even tried out as an inside forward for some English and Scottish football clubs, but since I could not devote enough time to training I never made regular First Division teams. However, since 1946 I have always stayed in excellent physical shape by swimming daily and practicing other sports. "Mens sana in corpore sano" has always truly been my motto. In England I also developed a great liking for classical music, especially Beethoven, Brahms and Liszt.

My interest in medical research started at the age of 23, when I joined the National Institute of Medical Research (NIMR, MRC) Mill Hill, London, England. I was fortunate to work with and be exposed to the stimulating influences of such scientists as Dr. D. F. Elliott, Sir Charles Harington, Dr. R. R. Porter, Dr. A. J. P. Martin, Dr. R. Pitt-Rivers, Dr. J. Gross, Dr. T. S. Work, Dr. H. Fraenkel-Conrat, and Dr. W. Cornforth, several of whom later won Nobel prizes for chemistry or physiology and medicine. Although my position was very junior at Mill Hill, my work was appreciated and this was a source of tremendous satisfaction for me, inasmuch as this recognition came from scientists of such caliber. I learned much in those 2 1/2 years, not only technical expertise but also the philosophy of research and a systematic approach to scientific investigations. These years of instruction (1950—1952) were decisive in providing inspiration, training, and laboratory discipline and profoundly influenced the course of my career. In fact, it was at NIMR, Mill Hill where I endured my "baptism of fire" in medical research and became addicted to it. In May, 1952, I moved to Montreal, Canada where

I was given the opportunity to work and study at McGill University. There I learned endocrinology from the brilliant lectures by Professor D. L. Thomson and from my work with Dr. M. Saffran in the laboratory of experimental therapeutics of the Allan Memorial Institute of Psychiatry headed by Dr. R. A. Cleghorn. The work at this laboratory was devoted to ACTH and adrenal cortical steroids. That period marked the beginning of my interest in the relationship between brain function and endocrine activity, and it was there in 1954 that my involvement in the hypothalamic field began.

In 1955, using *in vitro* systems, Dr. M. Saffran and I demonstrated the presence of corticotropin releasing factor (CRF) in hypothalamic and neurohypophysial tissue. This was the first experimental proof of the existence of hypothalamic hormones regulating pituitary function postulated with prophetic insight by Dr. G. W. Harris. I obtained my doctorate at McGill University in May, 1957, and in September of the same year I was able to secure a position which enabled me to continue my work on CRF at Baylor University College of Medicine in Houston, Texas, where I was associated with Dr. R. Guillemin. My years in Houston (1957—1962) where I was Assistant Professor of Physiology and a Senior Research Fellow of the U.S. Public Health Service were discouraging and frustrating because of problems with the isolation of CRF. Our failure to obtain enough CRF to determine its structure tended to cast doubt on the initial findings. We encountered much skepticism, but I remained unshaken in my confidence in the correctness of the observations on CRF and in the postulation of other hypothalamic hormones regulating anterior piuitary function.

In 1961 I spent about one month at the Institute of Biochemistry in Uppsala with Dr. J. Porath where I gained useful experience in the use of Sephadex and column electrophoresis. I also visited Dr. V. Mutt and the late Professor E. Jorpes in Stockholm, in connection with our collaboration on gastrointestinal hormones, and I was encouraged that they and other astute scientists had confidence in our work and the foresight to appreciate the possible scientific and medical importance of hypothalamic hormones.

I was grateful for the opportunities I was given in the United States, for which I felt a complete allegiance, and in 1962 became a naturalized citizen. When Dr. Joe Meyer, then head of the Veterans Administration (VA) basic research, offered in June, 1962, to set up a VA laboratory devoted to research on the hypothalamus and make me its chief, I accepted since this gave me a clear opportunity to be in complete command of such an effort. The support of a number of individuals, including Dr. E. H. Bresler, then Associate Chief of Staff for Research of the New Orleans VA Hospital, Dr. C. Y. Bowers and Dr. G. Burch of the Department of Medicine of Tulane University School of Medicine, and Dr. W. Locke of the Ochsner Foundation Hospital, was instrumental in helping me establish the laboratory in New Orleans. In December of 1962, I was appointed Chief of the Endocrine and Polypeptide Laboratories at the VA Hospital in New Orleans and Associate Professor of Medicine at Tulane University, and, in 1966, Professor. The earliest members of our 1962 VA-Tulane team were T. W. Redding, W. H. Carter, and M.

Tanaka. They have stayed with me all these years, and without their devoted help we could not have resolved the many problems associated with our work on TRH in 1969, LH-RH in 1971, and porcine somatostatin in 1975. Working in a clinical environment, I became more aware of the need for better diagnosis and treatment of patients than I had been before. It occurred to me early that problems with infertility on the one hand and the necessity for population control on the other would make a breakthrough in the control of reproduction particularly desirable from the standpoint of society, and therefore I became especially interested in reproductive endocrinology. To broaden our knowledge of reproductive processes at the brain level, we studied the central effects of contraceptive steroids and clomiphene. In some of the early studies on LH-RH, before its isolation, we collaborated with one of the pioneers of the hypothalamus and the man I always admired deeply, Professor C. H. (Tom) Sawyer and also with Drs. J. Hilliard, D. Holtkamp, A. Parlow and W. F. White.

It was my good fortune that in 1964 Dr. A. J. Kastin and in 1965 Dr. A. Arimura came to join our laboratory. Dr. Abba Kastin was mainly interested in continuing his work on control of release of MSH and in helping us in clinical work on hypothalamic hormones. He quickly became my best friend and a most efficient collaborator. Dr. Akira Arimura was an experienced physiologist and endocrinologist. Because of his great knowledge, enthusiasm and very hard work, he made great contributions in all phases of our program, and also broadened it with many independent ideas, especially in immunology. Other excellent collaborators at that time included Drs. I. Ishida, A. Kuroshima, T. Saito, and S. Sawano from Japan, and Dr. E. E. Müller from Italy.

All during the period since 1962, I had been hard at work on TRH with Cy Bowers and Tom Redding. In 1966, we reported for the first time the isolation of porcine TRH and determined that it contained three amino acids (glutamic acid, histidine, and proline) in equimolar ratio, but did not take full advantage of this original early finding, as we were preoccupied with parallel studies on reproduction and growth hormone-releasing hormone (GH-RH). However, when R. Burgus and R. Guillemin announced at the 1969 Tucson, Arizona, conference that they also found the same three amino acids in ovine TRH, I realized that we had the right substance. The same year I established the correct amino acid sequence of porcine TRH with Dr. R. M. G. Nair in New Orleans. Subsequently, with help from Drs. F. Enzmann and J. Bøler working in K. Folkers laboratory in Austin, Texas, we were able to determine the structure of porcine TRH and synthesize it. We have shared the credit with R. Burgus, W. Vale and R. Guillemin, who elucidated the structure of ovine TRH at about the same time.

The identification of TRH removed the skepticism surrounding the work on the hypothalamus and I realized that many workers would now be attracted to the field. We therefore redoubled our efforts on LH-RH.

In 1965, in Mexico City, I met Dr. C. Gual of the National Institute of Nutrition who invited me to collaborate with him in the clinical testing of hypothalamic hormones in Mexico. We took advantage of this invitation and in 1968 demonstrated, with Cy Bowers, that preparations of natural TRH are active in humans. Subsequently, again in collaboration with Carlos Gual, Abba Kastin and I established that highly purified porcine LH-RH unequivocally released LH and FSH in men and women under a variety of conditions. It was clear that LH-RH might be useful clinically and this encouraged us to continue the agonizing effort involved in the isolation of this hormone. Although I consider myself an endocrinologist or neuroendocrinologist, with considerable interest in clinical endocrine research and not a biochemist, I personally carried out the isolation work on TRH, LH-RH, somatostatin, and other hormones. Only a person such as myself with strong faith in the presence of these materials would have the patience to go through the many fastidious steps of the isolation procedure, since the effort required in isolating exceedlingly small quantities of gradually purer and purer materials from a crude hypothalamic exctract is so enormous. I was able to isolate a small amount (800 μg) of LH-RH from 160,000 hypothalami and proved it to be a polypeptide. This tiny amount of material was passed to our chemists, Dr. H. Matsuo and Dr. Y. Baba, with suggestions for a structural approach. Since I did not think that amounts of LH-RH on hand would be enough to complete our structural work, I decided to isolate additional amounts of LH-RH. Drs. Matsuo and Baba worked hard and efficiently, and we were able to determine the complete structure of LH-RH with the 800 μg material. After confirming the structure by synthesis, we were in a position to present our findings at the Endocrine Society meeting in San Francisco, California, in June 1971. It was one of the high points in my life to be able to report for the first time the solution to the problem which had preoccupied me and others for so long.

Physiological and subsequently immunological studies with natural and synthetic LH-RH in our laboratory by Drs. A. Arimura, L. Debeljuk, J. Reeves and M. Saito, and with others demonstrated that LH-RH was indeed the physiological hormone. With the synthetic LH-RH readily available, Dr. Kastin and I continued to carry out a variety of clinical studies in Mexico in association with Dr. Gual and later with Drs. A. Zarate and D. Gonzalez-Barcena. I also did parallel clinical tests with Dr. J. Zanartu in Chile and in Argentina with Drs. L. Schwarzstein, N. Aparicio, and the late R. Mancini.

The importance of analogs, particularly with respect to the possibility of developing a new birth control method was uppermost in my mind. I was very fortunate in being able to induce Dr. D. H. Coy, a superb peptide chemist and his wife Esther, also a researcher, to join our laboratory in 1972. More than 300 analogs of LH-RH were synthesized by the Coys with the help of Drs. Y. Hirotsu, K. Nikolics and J. Seprödi in our laboratory between 1972 and 1977. We were particularly interested in stimulatory long-acting superactive analogs for clinical use and in inhibitory analogs which would block LH and FSH release. We were joined in this important work by researchers from many countries. The work of Drs. J. Vilchez from Venezuela, A. de la Cruz from Peru, E. Pedroza from Colombia, and N. Nishi from Japan established in

1976 that the antagonists of LH-RH can indeed completely block ovulation in animals. Very recently with Dr. D. Gonzalez-Barcena in Mexico we showed that these analogs are also active in humans. This of course raises the possibility that such analogs could eventually form the basis of a new birth control method. However, much work is still needed to make my dream of being able to control reproduction at the central level come true.

In 1971, immediately after solving the LH-RH problem, I decided to reinforce our attacks on PIF and GH-RF next, but six years of hard work with Dr. A. Arimura and Drs. J. Sandow from Germany, A. Dupont from Canada and J. Takahara from Japan resulted only in a demonstration that hypothalamic catecholamines and gamma-amino butyric acid (GABA) may be involved in the control of release of prolactin, but did not yet lead to the development of any clinical agents. In our preoccupation with PIF and GH-RH, we did not work on factors inhibiting growth hormone release but after P. Brazeau and collaborators in 1973 announced the isolation and structure of ovine somatostatin, we purified this hormone from porcine hypothalami, determined its structure and synthesized it. We also carried out much physiological and immunological work (some in collaboration with Dr. F. Labrie in Quebec, Canada), as well as clinical work which convinced us of its importance. Particularly important was the establishment of a radioimmunoassay for somatostatin by Dr. Arimura. The clinical work on somatostatin was carried out mainly in England. Brilliant clinicians Professor R. Hall from the Royal Victoria Infirmary in Newcastle-upon-Tyne and Professor G. M. Besser of St. Bartholomew's Hospital in London were our leaders of two clinical teams which also included excellent collaborators such as Drs. A. Gomez-Pan, D. Evered, C. Mortimer, S. R. Bloom, and others. These clinical studies in England (based in part on some of our suggestions) showed that somatostatin inhibits the release of GH, TSH, glucagon, insulin, and gastrin. Basic studies carried out in England in collaboration Dr. A. Gomez-Pan and in Poland with Professor S. Konturek showed that somatostatin also inhibits gastric acid and pepsin secretion, and the release of duodenal hormones, secretin and cholecystokinin. Since the immunological work of Dr. Arimura showed the presence of somatostatin in the pancreas, stomach and intestine, we then suggested that this substance may be involved in the control of secretion not only of the pituitary, but also of the pancreas, stomach and duodenum. Since somatostatin has multiple short-lived effects, Drs. D. H. Coy and C. Meyers are achieving considerable success in the synthesis of long-acting and selective analogs of somatostatin, some of which could be more practical clinical agents.

Also among our present projects is the isolation of all the compounds with PIF activity, of PRH, GH-RH, CRF, and other hypothalamic substances. In addition to authoring or co-authoring many publications, I take satisfaction from the fact that I helped Dr. W. Locke write a book for clinical endocrinologists.

Since much work with hypothalamic hormones and their analogs is being carried out in Latin America and Spain, my ability to communicate in Spanish and Portuguese has aided me greatly, and resulted in the formation of many beautiful friendships. However, the greatest reward for learning Spanish and Portuguese came when, in 1974, in the course of my work in Brazil I met a very charming endocrinologist, Ana Maria de Medeiros-Comaru (M.D.). Our friendship soon deepened into love and led to our marriage.

I have had the satisfaction that my work in the hypothalamus was honored by top U.S., Canadian and Spanish awards: Van Meter Prize of the American Thyroid Association; Ayerst-Squibb Award of the U.S. Endocrine Society; William S. Middleton Award, the highest award of the VA; Charles Mickle Award of the University of Toronto; Gairdner Foundation International Award, Canada; Edward T. Tyler Award; Borden Award of the Association of American Medical Colleges; Albert Lasker Basic Medical Research Award, and the Laude Award, Spain. In 1973 I was made a Senior Medical Investigator by the Veterans Administration, an honor reserved for only a few. When I learned about my Nobel Prize, I was too grateful and too moved to be overcome with joy, but that came a few hours later when my friends from all over the world began to phone or wire. However, I do not feel that these prizes will have an adverse effect on my future productivity. I am still as keen as ever to make new discoveries and useful contributions to endocrinology.

ASPECTS OF HYPOTHALAMIC REGULATION OF THE PITUITARY GLAND WITH MAJOR EMPHASIS ON ITS IMPLICATIONS FOR THE CONTROL OF REPRODUCTIVE PROCESSES

Nobel Lecture, 8 December, 1977

by

ANDREW V. SCHALLY
Veterans Administration Hospital and Tulane University School of Medicine, New Orleans, Lousiana, U.S.A.

I am profoundly grateful for the great honor which has been bestowed upon me in recognition of my research efforts. It is a privilege for me to give an account of my search for the hypothalamic regulatory hormones. Since my work on the hypothalamus has extended over 23 years, it will be necessary to give a somewhat simplified version of it and omit reference to studies which did not contribute directly to my main objective, that is, demonstration of hormonal activity in hypothalamic extracts and the purification, isolation, determination of the structures of hypothalamic hormones and their testing in biological and clinical settings. Also, in order to avoid significant overlapping with Dr. Guillemin's lecture, I will concentrate primarily on the LH-releasing hormone (LH-RH).

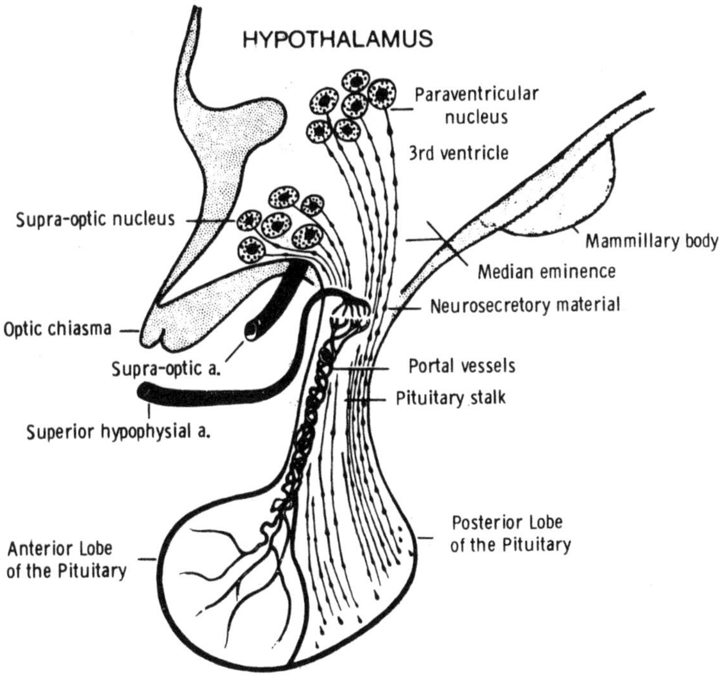

Figure 1. Simplified schematic reconstruction of the hypothalamic-hypophysial nerve tracts and blood supply to illustrate the principles of neurohumoral control of the anterior pituitary gland.

I was attracted to the hypothalamic endocrine field in 1954 while still an undergraduate student at McGill University in Montreal. A decisive stimulus was provided by the formulation by G. W. Harris and others (1) of hypotheses relating to the hypothalamic control of secretion of the anterior pituitary gland (Fig. 1).

Harris and others postulated that neurohumoral substances might originate in the median eminence of the tuber cinereum, reach the anterior lobe by way of the hypophysial portal system, and thus regulate pituitary secretion (1). About the same time Sawyer *et al* (2) demonstrated involvement of the central nervous system in the control of gonadotropin secretion. Without the brilliant work of these men my contributions would not have been forthcoming. It was clear that despite a strong circumstantial case favoring hypothalamic control of the pituitary, the proposition would remain speculative until direct evidence for the existence of specific hypothalamic chemotransmitters controlling release of pituitary hormones could be demonstrated.

In the beginning it was not possible to prove the existence of and isolate hypothalamic hormones because of a lack of specific methods for the detection of their activity. Working on the problem of control of ACTH secretion, M. Saffran and I reached the conclusion that the hypothalamic theory best explained most of the then existent experimental facts. We devised a test system for measuring the release of ACTH using isolated rat anterior pituitary fragments (Fig. 2) (3). This *in vitro* pituitary system was delightfully simple and consisted of exposure of symmetrical portions of the gland to test sub-

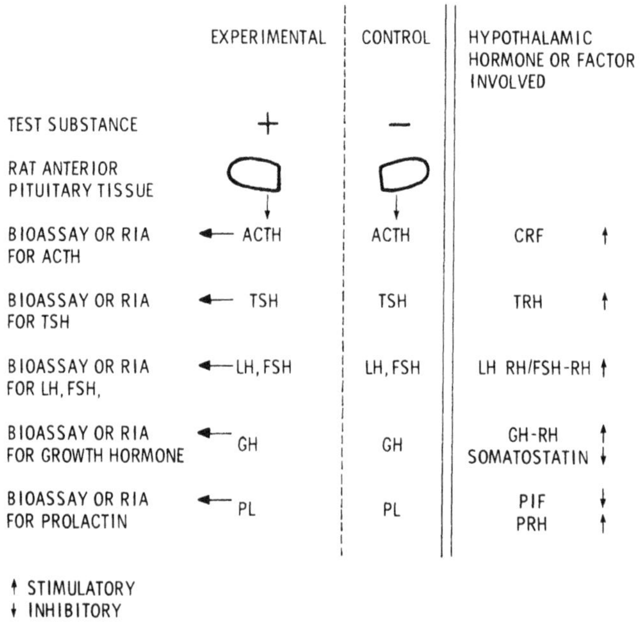

Figure 2. A diagrammatic representation of the *in vitro* test system for the detection of hypothalamic hormones and factors controlling the release of anterior pituitary hormones.

Table 1. CRF Activity of Pig Hypothalamic Preparations Measured by Stimulation of ACTH Release *in vitro* from Isolated Rat Pituitary Fragments.

PREPARATION	% ACTH RELEASED DOSE (μg)	RATIO EXPERIMENTAL TO CONTROL	95% LIMITS
Oxycel non-adsorbed	10	340	220—520
Oxycel non-adsorbed	10	240	130—470
Oxycel adsorbed	10	240	130—430
HOAc insoluble	10	240	90—660

A. V. Schally Ph. D. Thesis, McGill University, April, 1957.
Table 29, p. 91; Also, Schally et al, *Biochem. J.*, Vol. *70*, No. 1, p. 97—103, 1958.

stances. This permitted compensation for any possible indirect effect or contamination with trophic hormones and proved to be of decisive importance not only for demonstrating the existence of the corticotropin releasing factor but also hypothalamic hormones regulating the secretion of TSH, GH, LH, FSH, and prolactin. I still vividly recall the great sense of exaltation when we found that hypothalamic or neurohypophysial extracts added to the anterior pituitary tissue caused an unequivocal increase in the release of ACTH (Table 1). We "knew" then that we had done it, that the existence of a substance which stimulated the release of ACTH had been demonstrated experimentally for the first time (3—5). We named this substance corticotropin releasing factor (CRF). We still apply the term "factor" to those hypothalamic substances whose activity cannot be ascribed to a specific chemical structure. However, for those substances which have had their structures determined and which have been shown to be likely physiological regulators of secretion of respective anterior pituitary hormones, we employ the name "hormone" (Table 2).

In our early attempts to purify CRF we used mainly posterior pituitary powders, since large quantities of hypothalami were not readily available. We obtained evidence that CRF was a polypeptide (5), but despite seven years

Table 2. Hypothalamic hormones or factors controlling the release of pituitary hormones.

	Abbreviation
Corticotropin (ACTH)-releasing factor	CRF
Thyrotropin (TSH)-releasing hormone	TRH
Luteinizing hormone (LH)-releasing hormone/ Follicle-stimulating hormone (FSH)-releasing hormone	LH-RH/FSH-RH
Growth hormone (GH)-release inhibiting hormone	GH-RIH; somatostatin
Growth hormone (GH)-releasing factor	GH-RF
Prolactin release-inhibiting factor	PIF
Prolactin releasing factor	PRF
Melanocyte stimulating hormone (MSH)-release-inhibiting factor	MIF
Melanocyte-stimulating hormone (MSH)-releasing factor	MRF

of intensive effort, two with M. Saffran in Montreal and five with R. Guillemin in Houston, we were unable to isolate enough material for the determination of its structure. However, during that period (1955—1962) new *in vivo* assays for hypothalamic hormones and improved purification methods were developed. Techniques of gel filtration on Sephadex, which I learned at the Institute of Biochemistry in Uppsala with Dr. J. Porath, proved to be of particular value.

Arrangements were also made in 1962, after I moved to New Orleans, for the procurement of hundreds of thousands of hypothalami. Oscar Mayer & Co. generously donated about a million pig hypothalami. This enabled us to undertake a large-scale effort aimed at the purification of adequate amounts of material to permit chemical characterization. In addition to CRF, we systematically investigated purified fractions for the presence of TRH, LH-RH, FSH-RH, GH-RF, PIF and MIF (6—15), since the discovery of CRF opened the way to the demonstration of these other releasing factors.

THYROTROPIN-RELEASING HORMONE (TRH)

Our next great effort was devoted to the isolation and identification of TRH. We first demonstrated the presence of TRH in pig, beef, and human hypothalami using *in vitro* assays based on the release of TSH from rat pituitary glands and *in vivo* assays based on I^{131} release from thyroid glands of mice (6, 12, 13, 15). Then with the help of C. Y. Bowers and T. W. Redding I undertook the purification of bovine and porcine TRH. In 1966 we isolated 2.8 mg of TRH from 100,000 pig hypothalami (16) by Sephadex gel filtration, phenol extraction, CMC chromatography, CCD (Fig. 3), free flow electrophoresis (Fig. 4), and partition chromatography (Fig. 5). One ng of this homogeneous porcine TRH was active in our *in vivo* assay, and *in vitro* 0.01 ng stimulated TSH release (16). We also correctly reported that it had

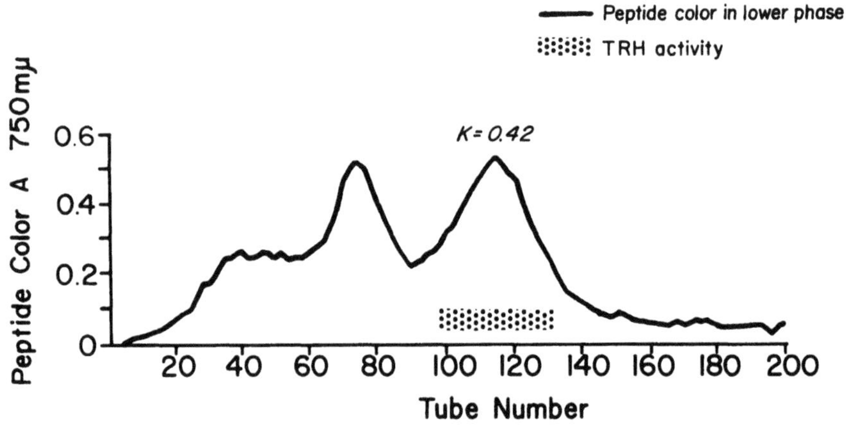

Figure 3. Countercurrent distribution of 1.0 g of porcine TRH from Cm-cellulose in a system of 0.1% acetic acid-1-butanol-pyridine, 11: 5: 3 (v/v). The number of transfers was 400. Peptide analyses were carried out on 50 μl aliquots lower phase. Based on Schally *et al*, *Biochem. Biophys. Res. Commun. 25*: 165, 1966; Schally *et al*, *J. Biol. Chem. 244*: 4077, 1969.

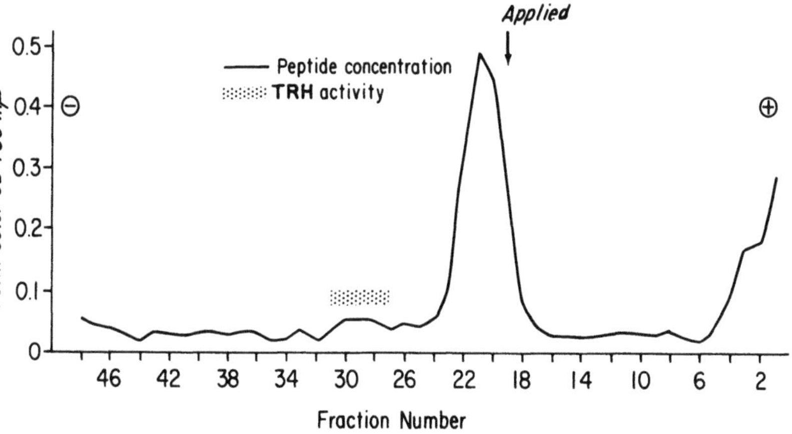

Figure 4. Free-flow electrophoresis of porcine TRH from CCD in pyridine acetate buffer pH 6.1. Conditions: 1800 V, 160 mA, 5°C, 7 hours. Peptide analyses carried out on 50 μl aliquots. Courtesy of Schally *et al*, *Rec. Prog. Hormone Res.*, *Vol. 24*, p. 514, 1968.

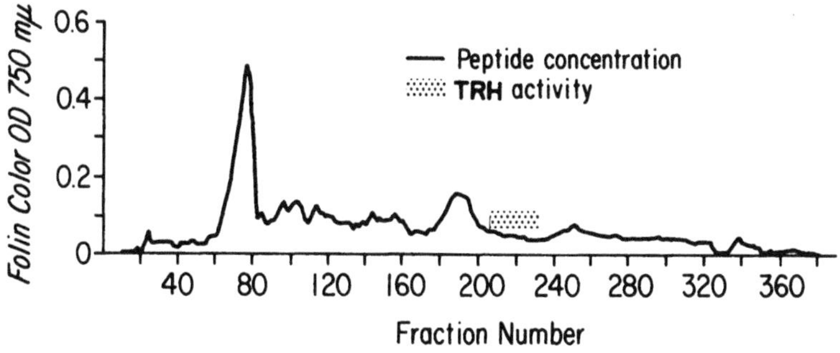

Figure 5. Partition chromatography of porcine TRH from FFE on a column of Sephadex G-25 0.9 × 76 cm. The solvent system consisted of upper phase of n-butanol: acetic acid: water = 4: 1: 5. Fraction size, 1 ml. H.U. volume = 20 ml. Courtesy of Schally *et al*, *Rec. Prog. Hormone Res. Vol. 24*, p. 514, 1968.

three amino acids, glutamic acid, histidine and proline in equimolar ratios (16), which established for the first time that TRH was a peptide. By mass spectra we detected a band due to the diketopiperazine of His-Pro and we also determined that an intact histidine was necessary for full biological activity of TRH, but unfortunately we did not take full advantage of these original early findings (13, 16).

Although the TRH problem could have been solved in 1966, three more years had to elapse for additional technological breakthroughs necessary to determine its precise structure. Since we lacked synthetic capabilities at that time, Merck, Sharp and Dohme Laboratories synthesized for us eight tripeptides containing histidine, proline and glutamic acid or glutamine, one of which was in the correct sequence, Glu-His-Pro. None of these, however, proved to have biological activity (13) and a complete series of possible analogs was not made. Somewhat discouraged by these negative results, I turned my attention to LH-RH, leaving the problem of the structure of TRH to my chemists with whom I was working. However when Burgus and Guillemin announced in

1969 that they found the same three amino acids in ovine TRH as I had three years earlier for porcine TRH (16), my enthusiasm for the program was rekindled and we intensified our efforts.

Fortunately, since I thought that the amount of TRH originally isolated would be insufficient to allow complete determination of structure, I took the precaution of obtaining about five additional milligrams of TRH from 250,000 pig hypothalami (17). Its structure then was systematically investigated by a series of degradation reactions (17, 18). First with R. M. G. Nair we established the correct amino acid sequence in New Orleans (17) and then in a parallel effort between my group and F. Enzmann and J. Bøler working in K. Folkers laboratory in Austin, Texas, we were able to assign the correct structure to porcine TRH and synthesize it (18—20). The structure was based on: 1. the amino acid sequence of TRH established in my laboratory (17); 2. comparison of activity of synthetic analogs of Glu-His-Pro in assays carried out by C. Y. Bowers and, independently by T. W. Redding (19, 20); 3. mass spectra of natural and synthetic preparations (18), and 4. synthetic modification and physico-chemical comparisions of these synthetic analogs and natural TRH (19, 20). Thus, synthetic experiments were carried out on Glu-His-Pro to modify both the amino and the carboxyl ends in order to generate TRH activity. Treatment of the methyl ester of Glu-His-Pro with anhydrous ammonia led predominantly to formation of (pyro)Glu-His-Pro-amide, and

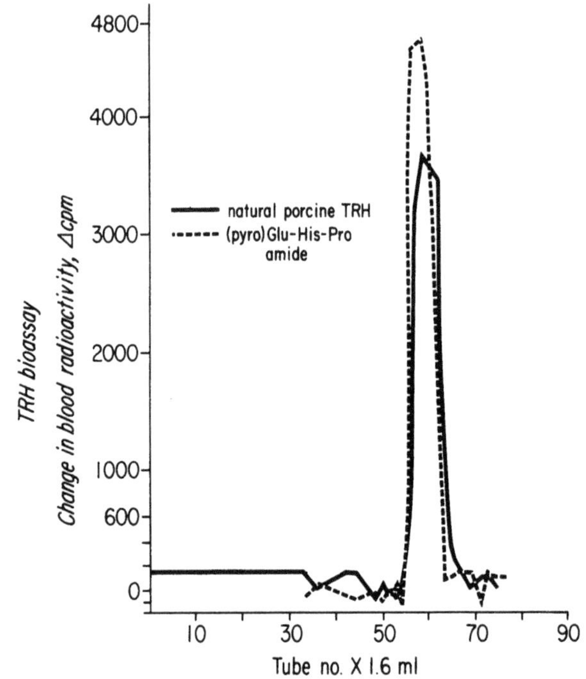

Figure 6. Gel filtration of natural porcine TRH (160 μg) and synthetic (pyro)Glu-His-Pro amide (200 μg) on Sephadex G-25. Column 1.1 × 123 cm. Solvent 0.2 M acetic acid. Fraction size 1.6 ml. The biological activity of effluents was followed by bioassay for TRH. From Schally *et al*, in *Proc. 6th Midwest Conf. on Thyroid and Endocrinology, 1970*, Univ. of Missouri-Columbia Press, p. 42.

to generation of TRH activity (19). Synthetic L-(pyro)Glu-L-His-L-Pro-amide gave R_f values identical to those of natural TRH in 17 chromatographic systems (20). Upon gel filtration on Sephadex G-25 columns in 0.2 M acetic acid, natural porcine TRH and synthetic (pyro)Glu-His-Pro-amide displayed identical migration rates (Fig. 6). The structure of TRH was thus (pyro)Glu-His-Pro-amide, or 2-pyrrolidone-5-caboxylyl-His-Pro-amide (Fig. 7). The biological activity of synthetic TRH was the same as that of natural porcine TRH (21). It is somewhat ironic to realize that had Merck, Sharp and Dohme Laboratories furnished us with Glu(NH_2)-His-Pro-NH_2, it would have partially cyclized to the active (pyro)Glu-His-Pro-NH_2 form after the synthesis, and we would in all probability have solved the problem of TRH structure three years earlier.

The structural work of Burgus and Guillemin (22) on ovine TRH paralleled that of our group and they elucidated the structure of ovine TRH about the same time. Subsequent studies disclosed that bovine and human TRH probably have the same structure as the porcine and ovine hormone.

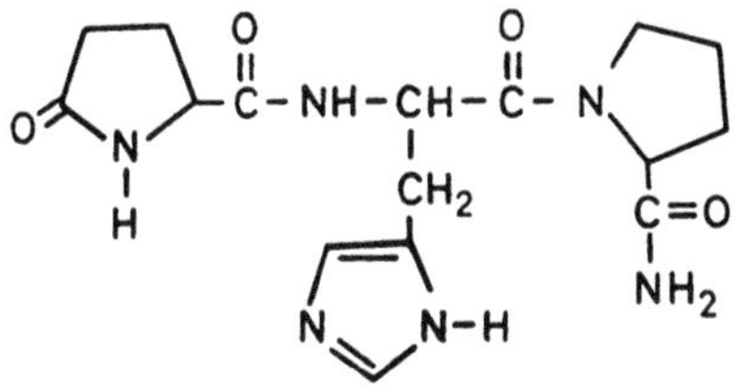

(Pyro) Glu – His —— Pro-NH_2

Figure 7. Molecular structure of thyrotropin-releasing hormone (TRH).

We have conducted various physiological studies since 1962 with natural preparations of TRH (12, 13, 16) and confirmed and extended them by using synthetic TRH, demonstrating that the concentration of TSH in the plasma .increased when TRH was administered intravenously, subcutaneously, intraperitoneally, or orally (21). Later we found that TRH also stimulates prolactin release in sheep (23). A direct action on the pituitary tissue *in vitro* in picogram doses was demonstrated in the pituitaries of rats, sheep, and goats (24). In pituitary tissue cultures, TRH was shown to stimulate the synthesis as well as the release of TSH. A dose-response relationship, both *in vivo* and *in vitro*, was also demonstrated; i.e., increasing doses of TRH caused a progressively greater release of TSH (21, 24). Thyroxine (T_4) and triiodothyronine (T_3) blocked the stimulatory effect of TRH on TSH release (13, 24). This occurred not only *in vivo* but also with pituitary fragments *in vitro* (Table 3), thus confirming that thyroid hormones must exert an action directly on the pituitary gland. We also suggested that among the physiological stimuli that may release TRH is exposure to mild cold (25). In pursuing the characterization of this hormone, we showed that it is rapidly inactivated in the blood stream and studied its excretion and metabolism

Table 3. Inhibition by Triiodothyroxine (T_3) of the *in vitro* Stimulation of TSH Release Induced by 0.5 Nanograms of TRH.

ADDITIONS	TSH ASSAY Change in blood ^{131}I (cpm) at 2 hours ± SE		MEAN Δ cpm	P
	Control, no TRH	Experimental, with TRH		
None	100 ± 55	818 ± 92	718	.01
None	60 ± 25	668 ± 117	608	.01
1 μg T_3	144 ± 27	184 ± 54	41	NS
1 μg T_3	162 ± 23	155 ± 38	−7	NS
1 μg T_3	106 ± 51	166 ± 36	60	NS
1 μg T_3	92 ± 17	118 ± 24	26	NS

NS: not significant ($p > 0.05$).
From: Schally and Redding (1967) (Ref. 25). Courtesy of the Proceedings of the Society for Experimental Biology and Medicine, and Academic Press.

(13, 25). Recent results suggest that TRH in addition to its effect on the pituitary might have central nervous system (CNS) effects, possibly as a neurotransmitter or modulator (26). These studies and subsequent ones by others helped establish the physiological importance of TRH.

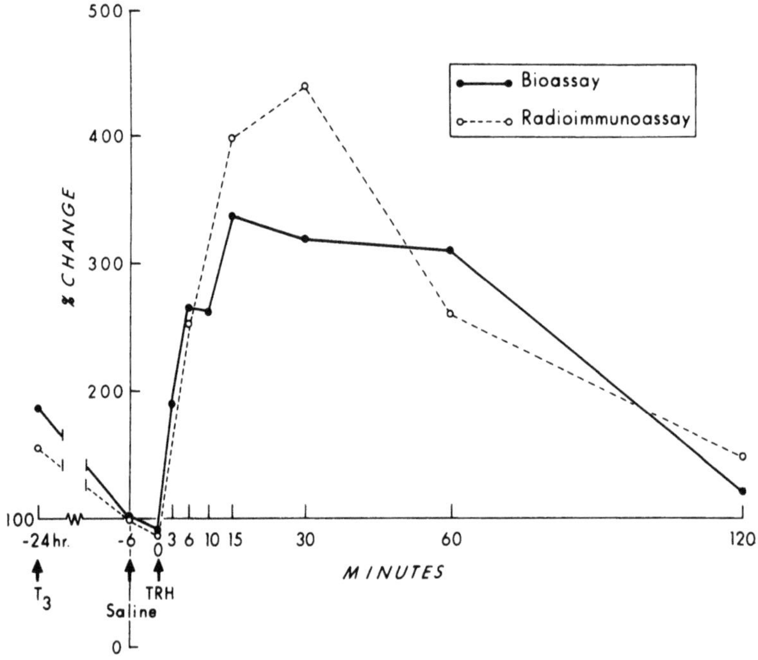

Figure 8. Changes in plasma TSH levels after the injection of TRH to cretins: at −24 h 25 μg T_3 was given orally and at 0 time 300 μg porcine TRH was given iv. Plasma samples were measured by both bioassay and radioimmunoassay. From Bowers, Schally, *et al*, *Endocrinology 86*: 1143, 1970.

The first clinical studies with TRH, carried out in 1967 with Dr. C. Bowers and Dr. C. Gual, showed that natural porcine TRH stimulated TSH release in humans (27) as measured by both bioassay and radioimmunoassay (Fig. 8). After synthetic TRH became available, these findings were confirmed and extended by us and others (28—31). It has been particularly gratifying to me that TRH is useful clinically for the differentiation between hypothalamic and pituitary hypothyroidism and for the diagnosis of mild hyper- and hypothyroidism, since one always enjoys seeing one's work bear fruit clinically.

After identification of TRH, we redoubled our efforts on the LH and FSH releasing hormone.

LUTEINIZING HORMONE- AND FOLLICLE STIMULATING HORMONE-RELEASING HORMONE (LH-RH/FSH-RH)

It has long been known that the reproductive activity of animals is influenced by seasonal and external environmental factors such as nutrition, light and temperature and that aberrations in the menstrual cycles of women can occur as a result of adverse environmental and psychological stimuli and emotional disturbances (31). In the late 1920's, after the involvement of the pituitary in the processes of reproduction was established, systematic investigations were initiated on the link between the hypothalamic region of the central nervous system (CNS) and the secretion of pituitary gonadotropins.

Based on experiments involving electrical stimulation of the hypothalamus, interruption of the blood vessels between the hypothalamus and the anterior pituitary by sectioning the hypophysial stalk, and the transplantation of the pituitary to various sites, Harris proposed the hypothesis of neurohumoral regulation of gonadotropin secretion (1). Sawyer *et al* (2) also adduced evidence for involvement of the CNS in control of the secretion of gonadotropins by demonstrating neuropharmacological stimulation and inhibition of ovulation with centrally acting stimulants and blocking agents.

The question that remained to be resolved was how information perceived in the CNS would be communicated to the pituitary. It was our aim to find that link between the hypothalamus and the pituitary insofar as the control of reproductive functions was concerned.

Although strong evidence for the existence of LH-RH and FSH-RH in hypothalamic extracts of rats and domestic animals was provided in the early 1960's (7, 8, 11, 32—40), it was thought that these activities were due to two different substances. We were able to demonstrate that materials with the properties of peptides purified from beef and pig hypothalami stimulated LH release not only *in vivo* but also *in vitro* (7, 8). The latter was the first demonstration that hypothalamic materials release LH by a direct action on the pituitary. With purified LH-RH at our disposal, we initiated work on how the interaction between LH-RH and sex steroids regulates gonadotropin secretion. At first we postulated that the inhibitory effect of contraceptive steroids on gonadotropin release was exerted mainly on the hypothalamus (41, 42), but subsequently with Drs. A. Arimura, C. H. Sawyer, and J. Hilliard (31, 43, 44) we were able to prove that estrogens, progestins and androgens also suppressed in part the response to LH-RH at the pituitary level. Later, we confirmed by *in vitro* studies this inhibitory effect (negative feedback) of steroids on the pituitary (45). Several laboratories, including ours (31, 44, 46) obtained evidence that estrogens and progesterone can also exert a positive feedback at the pituitary and the hypothalamus, and augment the pituitary responsiveness to LH-RH. These results may be correlated with events in the human menstrual cycle and the estrous cycle of animals. Thus, an increase in estrogen concentration in plasma which precedes the ovulatory surge of LH in animals and women appears to augment the pituitary responsiveness to LH-RH. Conversely, the large amounts of estrogen and progesterone which are secreted

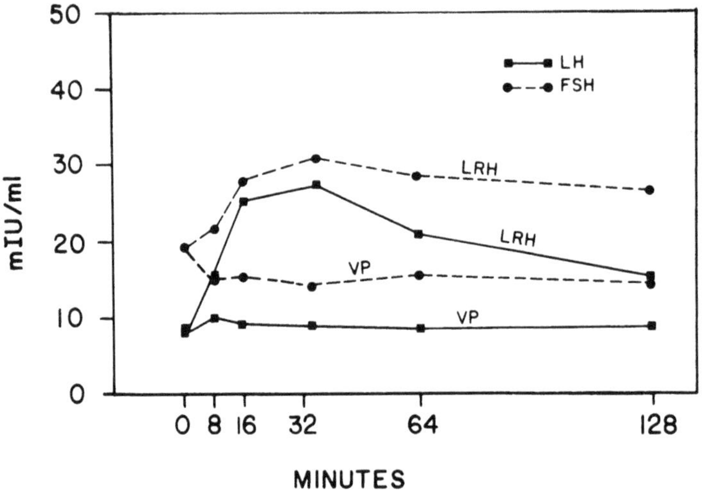

Figure 9. Effects of administration of 0.7 mg iv of porcine LH-RH on the levels of LH and FSH in a normal woman, on day 9 of the menstrual cycle. Based on Kastin, Schally *et al.*, *Amer. J. Obstet. Gynecol. 108*: 177, 1970.

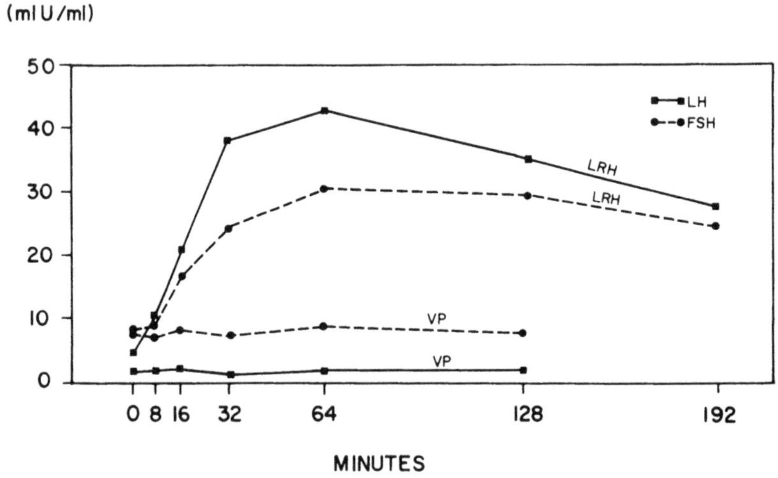

Figure 10. Effects of administration of .7 mg sc of porcine LH-RH on the levels of LH and FSH in a man pretreated with 1.5 mg ethinyl estradiol for 3 days. From Kastin, Schally, *et al*, *Amer. J. Obstet. Gynecol. 108*: 177, 1970.

after ovulation may lower pituitary responsiveness to LH-RH. With the aid of LH-RH, we also found that clomiphene exerts a central effect on the hypothalamus (47).

In view of the relative purity and apparent absence of visible toxicity of porcine LH-RH, I decided to test it in humans. These studies carried out in 1968 and 1969 in collaboration with Dr. A. J. Kastin and Dr. C. Gual in Mexico (48, 49) unequivocally established that highly purified LH-RH released LH and FSH in men and women under a variety of conditions (Figs. 9—10). Realizing that LH-RH might be useful clinically, we intensified our efforts to establish the structure of LH-RH. As in the case of TRH, tens of thousands of hypothalami had to be laboriously extracted, concentrated and purified to obtain enough material for a chemical characterization. The first isolation of 800 μg LH-RH/FSH-RH from ventral hypothalami of 165,000 pigs was achieved by twelve successive purification steps which included extraction with 2 N acetic acid, gel filtration on Sephadex, phenol extraction, chromatography and rechromatography on CM-cellulose, free-flow electrophoresis, countercurrent distribution (CCD), partition chromatography in two different solvent systems, and high voltage zone electrophoresis (Figure 11) (50). During the purification, the LH-RH

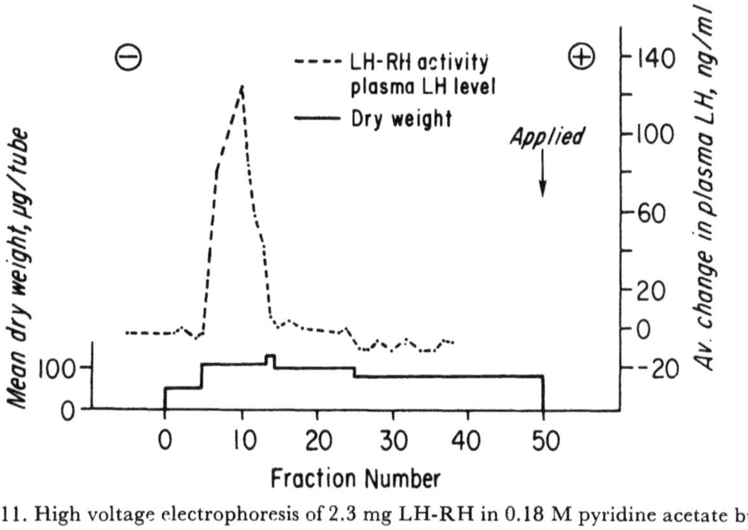

Figure 11. High voltage electrophoresis of 2.3 mg LH-RH in 0.18 M pyridine acetate buffer, pH 6.3. Vertical column 0.9 × 97.6 cm, with external cooling at 5°C, packed with cellulose powder. After the electrophoretic separation at 2570 V; 20 mA for 18 hrs, the column was eluted with buffer and 1.3 ml fractions were collected. From Schallv *et al*, *Piochem. Biophys. Res. Commun. 43*: 393, 1971.

and FSH-RH activities were followed by bioassay *in vitro* and *in vivo*. The LH and FSH released were determined by bioassays and later by radioimmunoassays. Subsequent isolation of 11 mg amounts of LH-RH/FSH-RH from 250,000 pig hypothalami was carried out mainly by the countercurrent distribution technique (Figs. 12—13) (51). In all the isolation steps, the LH-RH activity and FSH-RH activity were located in identical fractions.

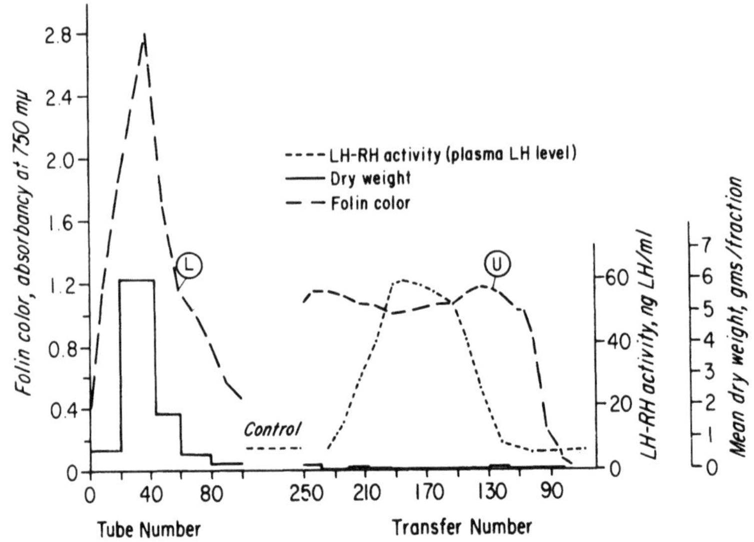

Figure 12. Preparative countercurrent distribution of LH-RH concentrate in a system of 0.1% acetic acid-1-butanol-pyridine, 11: 5: 3 by the single withdrawal method. Phenol extract (179.9 g) was extracted with the distribution solvent (1 liter of lower phase and 500 ml of upper phase) and 171.3 g of material which dissolved was loaded in tubes 0 to 19. One hundred cell train was filled with 50 ml of lower phase and 25 ml of upper phase. Two hundred fifty transfers were performed. Folin-Lowry analyses were carried out on 10 μl of lower phase (L) and 25 μl of upper phase (U). LH-RH activity was determined on 1-μl aliquots of upper phase, equivalent to approximately 0.8 μg dry weight. From Schally *et al*, *Anal. Chem. 43*: 1527, 1961.

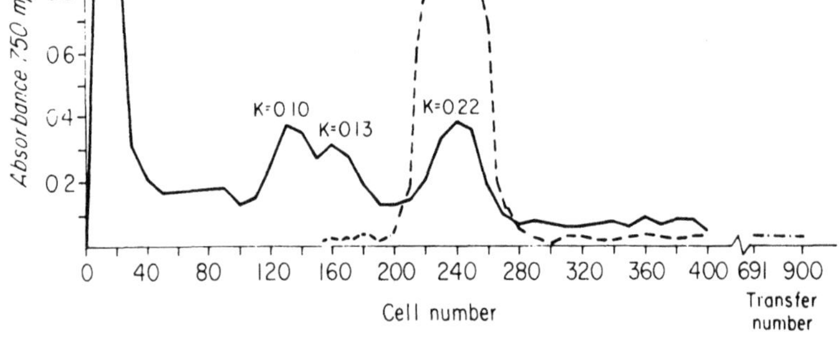

Figure 13. Countercurrent distribution (III) of 55.4 mg of LH-RH from countercurrent distribution (II) in a system of 1-butanol-acetic acid-water, 4: 1: 5 (v/v). Lower phase was 3 ml and upper phase 5 ml. The number of transfers was 900. Folin-Lowry analyses were done on 100-μl aliquots of lower phase. Assays for location of LH-RH activity were carried out on 2 μl of lower phase per rat. From Schally *et al*, *J. Biol. Chem. 246*: 7230, 1971.

They could not be separated by additional partition chromatography in 10 different solvent systems (50, 51). The amino acid composition determined after hydrolysis with 6 N HCl at 110° showed the presence of the following nine amino acids: His 1, Arg 1, Ser 1, Glu 1, Pro 1, Gly 2, Leu 1, and Tyr 1 (50). Since hydrolysis in 6 N HCl leads to destruction of Trp, the analysis for this amino acid was then carried out after acid hydrolysis in the presence of thioglycollic acid or by alkaline hydrolysis and showed the presence of one residue of Trp. Thus, the molecule of LH-RH/FSH-RH consisted of ten amino acids (52). Experiments with proteolytic enzymes showed that LH-RH/FSH-RH was a polypeptide (52). Both LH-RH and FSH-RH activities were simultaneously abolished by incubation with some endopeptidases (chymotropsin, papain, subtilisin, and thermolysin) but not by exopeptidases (leucine aminopeptidase, aminopeptidase M, and carboxypeptidase A and B) (52). Lack of inactivation by the Edman procedure and failure to detect any amino acid by the dansyl method indicated a blocked N-terminus. Inactivation by pyrrolidone carboxylyl peptidase suggested that the N-terminus was occupied by pyroglutamic acid. In the initial structural attack on LH-RH/FSH-RH with Dr. Matsuo and Dr. Baba we utilized the combined Edman-dansyl procedure coupled with the selective tritiation method for C-terminal analyses (53). These procedures were used directly on the digestion products of LH-RH with chymotrypsin and thermolysin without prior separation of fragments. Additional data were provided by high resolution mass spectral fragmentation of LH-RH/FSH-RH. On the basis of these results, we proposed the decapeptide sequence for LH-RH/FSH-RH seen in Fig. 14. The correctness of this structure was confirmed by additional conventional structural analyses involving the separation of chymotryptic fragments (54) after the cleavage of N-terminal pyroglutamyl residue in pyrrolidone carbox-

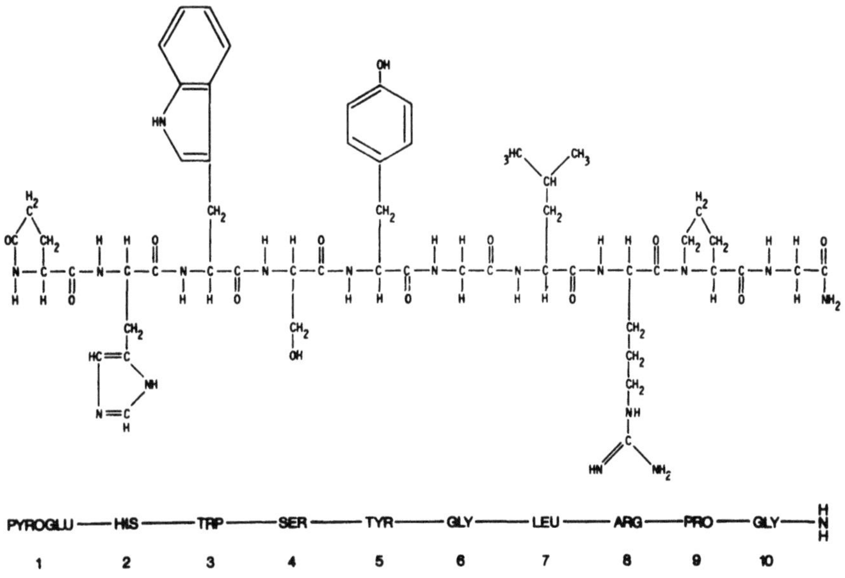

Figure 14. Molecular structure of LH- and FSH-releasing hormone (LH-RH/FSH-RH).

ylyl (PCA) peptidase as well as by synthesis of this material using the solid phase methods (55—57). Synthetic LH-RH/FSH-RH possessed the same properties as the natural material (56, 57). Thus, in rats it stimulated the release of LH and FSH *in vitro* and *in vivo* (58, 59) (Fig. 15). The time courses of LH and FSH release *in vitro* induced by natural or synthetic LH-RH

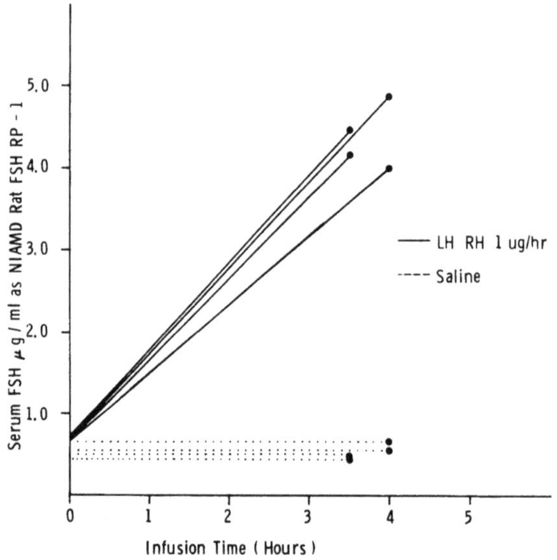

Figure 15. Serum FSH levels in immature male rats after prolonged iv infusion of synthetic LH-RH. Based on Arimura, Debeljuk, and Schally, *Endocrinology 91*: 529, 1972.

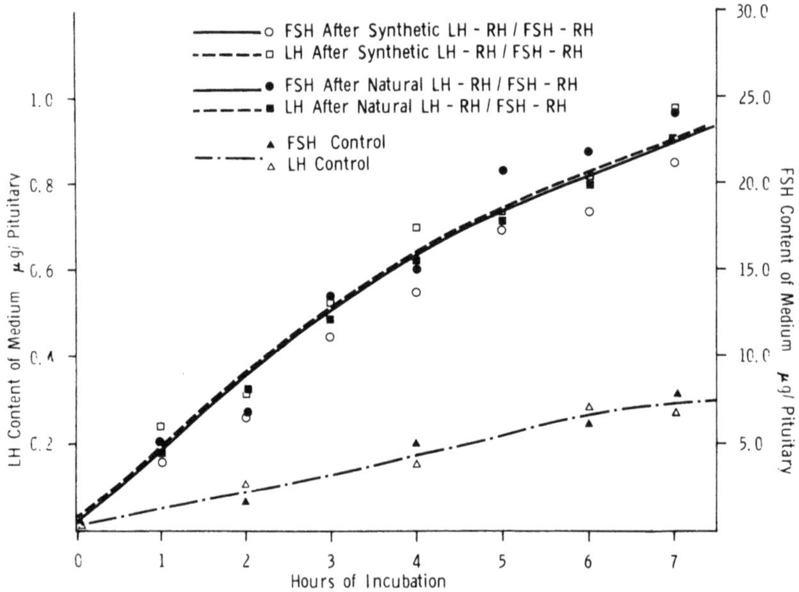

Figure 16. Release of LH and FSH from pituitaries of male rats (10 pituitary halves/beaker in 10 ml KRBG), containing 4 ng of natural or synthetic LH-RH/ml. The ordinates were adjusted to compensate for the content of LH and FSH. LH expressed as NIH-LH-S-17. FSH expressed as NIAMD-RAT-RP-1. From Schally *et al*, *Endocrinology 90*: 1561, 1972.

were identical (Fig. 16) (58). Simultaneous studies demonstrated that in human beings synthetic LH-RH also raised plasma LH and FSH levels (56, 57, 60) (Figs. 17—18).

Because both natural LH-RH and the synthetic decapeptide corresponding to its structure possessed major FSH-RH as well as LH-RH activity, we took the bold step of proposing that one hypothalamic hormone, designated LH-RH/FSH-RH, could be responsible for this dual effect (56, 57). This

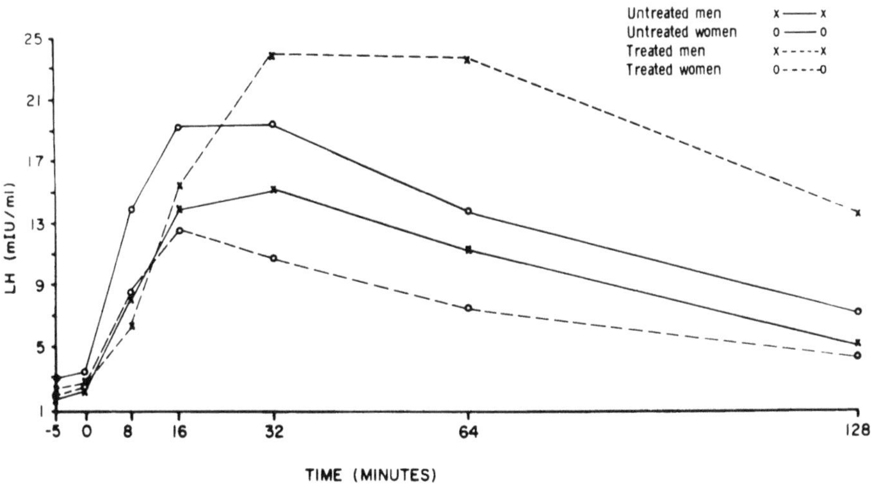

Figure 17. Mean plasma LH levels after iv administration of synthetic LH-RH equivalent to 38 μg to four groups of subjects: untreated men, untreated women, men pretreated with 1.0 mg ethinyl estradiol for 3 days; women pretreated with oral contraceptive Lyndiol for 1 week. From Kastin, Schally, *et al*, *J. Clin. Endocrinol. Metab. 34*: 753, 1972.

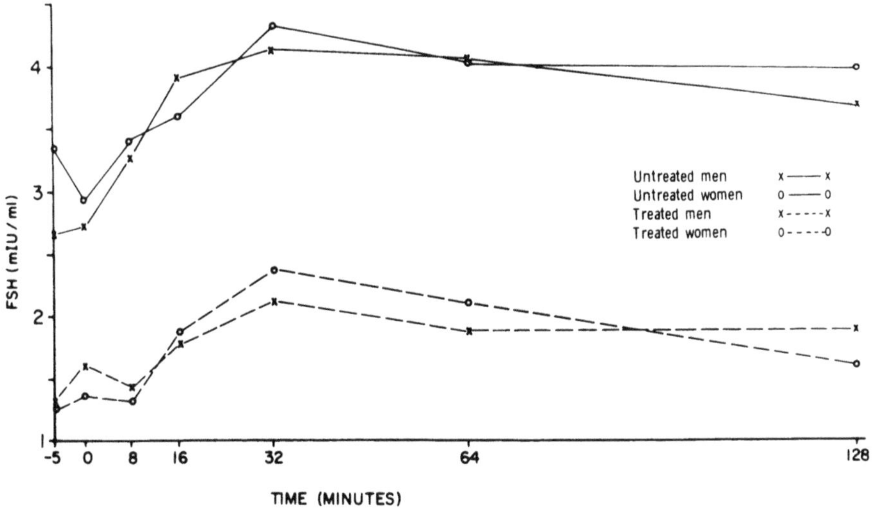

Figure 18. Mean plasma FSH levels after iv administration of synthetic LH-RH equivalent to 38 μg to four groups of subjects: untreated men, untreated women, men pretreated with 1.0 mg ethinyl estradiol for 3 days; women pretreated with oral contraceptive Lyndiol for 1 week. From Kastin, Schally, *et al*, *J. Clin. Endocrinol. Metab. 34*: 753, 1972.

concept is now supported by many physiological as well as immunological data. The LH-RH decapeptide represents the bulk of FSH-RH activity in the hypothalamus and it appears to be the principal FSH releasing hormone. Our subsequent studies in collaboration with Dr. J. Reeves and those of others established that in addition to rats and humans, LH-RH greatly enhances the release of LH and FSH in other mammals, including mice, nutria, rabbits, golden hamsters, mink, spotted skunk, impala, rock hyrax, sheep, cattle, pigs, horses and monkeys (Table 4) (31, 57, 61). In most of these species LH-RH can also induce ovulation. LH-RH was also found to be active in non-mammalian species such as chickens and pigeons, and even in some species of fishes such as brown trout and carp and in amphibia like newts and frogs (61). These studies in mammals, birds, fish, and amphibia indicate that species-specificity does not occur with LH-RH. We also obtained evidence that LH-RH can increase the synthesis of LH and FSH in addition to their release (62) and that prolonged treatment with LH-RH after hypophysectomy and transplantation of the pituitary stimulates spermatogenesis in male rats and follicular development in female rats (63—64). In another study with E. Rennels, we demonstrated that LH-RH increases the extrusion of secretory granules from LH gonadotrophs in rats with persistent estrus (Fig. 19) (65). Studies using synthetic ^{3}H-labelled LH-RH proved that LH-RH is rapidly degraded in blood by enzymatic cleavage of (pyro)Glu-His group and is excreted in the kidneys (66). Its half-life is about four minutes in man.

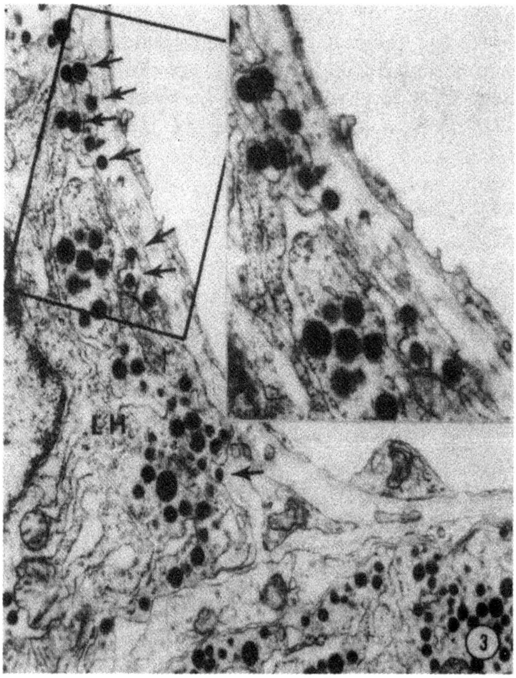

Figure 19. LH-gonadotroph of a persistent estrous rat sacrificed 15 minutes after injection of 200 ng synthetic LH-RH. Arrows indicate massive extrusion of secretory granules into pericapillary space (X 26300; insert X38000). From Shiino, Arimura, Schally, and Rennels, *Zeit. Zellforsch. Mikrosk. Anat. 128*: 152, 1972.

Table 4. Effect of Natural and Synthetic LH-RH/FSH-RH. In Animals and Humans*

Species	Effects
Rat	Release of LH & FSH *in vivo* and *in vitro* (N.S.)
	Stimulation of synthesis of LH & FSH *in vitro* (N.S.)
	Stimulation of spermatogenesis (S)
	Stimulation of follicular maturation (S)
	Ovulation (N.S.)
	Can be given iv, ic, sc, orally, intravaginally, cutaneously (DMSO) and intraventricularly (3rd ventricle)
Mice	Release of LH, ovulation (S) (S.C.)
Golden Hamster	Release of LH *in vivo* (S)
	Ovulation (S)
Nutria	Release of LH *in vivo* (N) (I.V.)
Rock Hyrax	Release of LH *in vivo* (S)
Rabbit	Release of LH *in vivo* (N.S.)
	Ovulation (N.S.)
Mink	Ovulation
Spotted Skunk	Ovulation
Sheep	Release of LH & FSH *in vivo* (N.S.) (I.V., S.C., I.M.)
	Ovulation (S)
Pigs	Release of LH *in vivo* (S) (I.V., I.M.)
	Ovulation (S)
Cattle	Release of LH and FSH *in vivo* (S) (I.M., I.C., S.C.)
	Ovulation (S)
Impala	Release of LH *in vivo* (S)
Horses	FSH release > LH release, ovulation (S) (S.C.)
Monkeys	Release of LH (N.S.)
Humans	Release of LH and FSH (N.S.)
	Stimulation of spermatogenesis (S)
	Ovulation (N.S.)
Pigeons	
Chickens	Premature ovulation (S) and release of LH (I.C., I.V., I.M.)
Fish**	Release of Gonadotropins *in vivo* (N.S.) (I.V.)
Newts	Release of LH & FSH *in vivo* (S)
Frogs	Spermiation (S) (S.C.)

* N. Natural LH RH
S: Synthetic LH RH
** Brown Trout, Carp
Modified from: Schally and Arimura, in: *Clinical Neuroendocrinology* (L. Martini & G. M. Besser, eds) Academic Press

Immunological and Immunohistochemical Studies

Production of antisera to LH-RH by Arimura *et al* (67) and by others permitted the establishment of radioimmunoassays (RIA) and the performance of a variety of immunological studies (68—73). Male rabbits that were actively immunized with LH-RH and had generated its antibodies developed testicular atrophy associated with aspermatogenesis (67). Castrated rats actively immunized with LH-RH showed parallel decreases in serum LH and FSH levels associated with a rise in serum antibody titer to LH-RH. Administration of anti-LH-RH gamma-globulin to castrated rats prevented the rise in serum

LH and FSH levels normally seen after such operation, and the development of castration cells in the pituitary (68). Passive immunization of normal cycling rats or hamsters with LH-RH arrested follicular maturation, prevented the preovulatory surge of LH and FSH, blocked ovulation (Table 5) and reduced serum estradiol levels (69, 70). We also showed that hypothalamic LH-RH is necessary for normal implantation and maintenance of pregnancy since passive immunization with LH-RH in early pregnancy causes a delay in implantation of fertilized ova or termination of pregnancy in rats, depending on the time the antibody to LH-RH is injected (71, 72). Before and during the preovulatory surge of LH-RH release, Arimura *et al* (73) detected by RIA a peak of LH-RH levels in the peripheral plasma in women, and others found it in the blood of rats, sheep, rabbits and monkeys. This indicates that this decapeptide is the mediator responsible for the release of the ovulatory quota of LH. These studies and others clearly established that LH-RH is the main link between the brain and the pituitary gland insofar as reproductive function is concerned.

Table 5. Effect of iv Injection of 1 ml of Anti-LH-RH Serum (No. 742) on Serum LH and FSH Levels and Ovulation in Cycling Rats

TREATMENT	No. of Rats INJECTED	No. of Rats OVULAT-ED	MEAN no. of ova, ± S.E.	MEAN serum LH level ± S.E. (ng/ml)*	MEAN serum FSH level ± S.E. (ng/ml)**
Normal rabbit serum	4	4	12 ± 1.3	58. ±13.2	720 ± 108
Anti-LH-RH	4	0	0	0.8 ± 0.13+	145 ± 21+

One ml of normal rabbit serum or rabbit anti-LH-RH serum was injected iv into cycling rats at 9 a.m. on the day of proestrus and blood was collected at 4:30 p.m. for assays of serum LH and FSH level.

* Expressed in terms of NIH-LH-S-17.

** Expressed in terms of NIAMD-Rat-FSH-RP-1.

\+ $P < 0.01$ by Student's *t* test as compared with the corresponding LH and FSH levels of the sera from the rats which were injected with normal rabbit serum.

From: Arimura, Debeljuk, and Schally, *Endocrinology 95*: 323, 1974.

The availability of antisera to LH-RH made possible various studies on localization of LH-RH by RIA or immunohistochemical methods. The bulk of LH-RH was localized in the median eminence and in the arcuate nucleus, and small but significant amounts were found in the preoptic and suprachiasmatic areas. In studies with Drs. B. Flerkó and G. Sétáló, we found that the pathway of LH-RH-containing nerve fibers in the median eminence of rats coincides with the course of the nerve fibers of the tubero-infundibular tract (74) and that LH-RH is produced in neuronal cell bodies, especially in the medial preoptic and the suprachiasmatic area. However, other studies showed that extrahypothalamic brain areas also contain LH-RH and may be involved in its synthesis. This could suggest that, in addition to being the regulator of the release of LH and FSH, LH-RH might act as a central neuromodulator. LH-RH has indeed been shown to excite sexual behavior in rats (75). This is in agreement with CNS effect of hypothalamic peptides, which I helped A. J. Kastin to demonstrate in 1971 (76, 77).

Analogs of LH-RH

The interest in possible veterinary and medical applications of LH-RH stimulated us and others to synthesize many hundreds of LH-RH analogs. Between 1972 and 1977 our laboratory synthesized more than 300 analogs by the use of rapid solid-phase techniques (61, 78, 79). Our aims were: 1. to develop analogs with prolonged biological activity, so that they would be more useful therapeutically than LH-RH itself; 2. to obtain inhibitory (antagonistic) analogs which could form the basis of new birth control methods. The studies on these peptides have shed much light on the relationship between biological activity and structure. Early results showed that the amino-terminal tripeptide and tetrapeptide fragments of LH-RH as well as the carboxyl-terminal nonapeptide and octapeptide of LH-RH have very little or no LH-RH activity (61, 80). Thus, very active small fragments cannot be obtained from LH-RH. In general, amino acids in position 1 and from 4 to 10 appeared to be involved only in binding to the receptors and in exerting conformational effects. However, histidine and tryptophan probably exert a functional effect in addition to providing receptor-binding capacity, since simple substitutions or deletions in positions 2 or 3 greatly decrease or abolish LH-RH activity. Dr. D. Coy in our laboratory (81—83) and others (84—87) showed that some analogs substituted in position 6, 10 or both are 10—60 times more potent than LH-RH and also possess prolonged activity; of these, the most interesting were [D-Phe^{6}]-LH-RH, [D-trp^{6}]-LH-RH, [D-Ala^{6}, desGly^{10}]-LH-RH ethylamide (EA), [D-Leu^{6}, desGly^{10}]-LH-RH EA, and [D-Ser(Bu^{t})^{6}, desGly^{10}]-LH-RH EA. These superactive LH-RH analogs cause a prolonged release of LH and FSH (78—79, 81—86).

However, it has also been recently demonstrated that chronic treatment with pharmacologic doses of these analogs or with large amounts of LH-RH can cause temporary and reversible impairment of reproductive functions. Thus, chronic administration of 1—10 μg of [D-Leu^{6}, desGly^{10}]-LH-RH EA to mature female rats caused cessation of cycling and atrophy of the ovaries and uterus (88). Prolonged administration of the same analog to male rats resulted in a reduction of testicular LH/HCG receptors and of testosterone levels (89). Pharmacologic doses of LH-RH or superactive analogs block implantation and terminate gestation when given daily postcoitally to rats (90, 91). These paradoxical antifertility effects of LH-RH and its analogs appear to be directly related to hypersecretion of LH and after nidation to functional luteolysis and/or inhibition of progesterone secretion and have caused us and others to initiate investigations on their possible application as pre-coital (male and female) and postcoital contraceptives.

Clinical Uses of LH-RH and Its Superactive Agonistic Analogs

LH-RH has been used diagnostically to determine pituitary LH and FSH reserve. It is not a complete diagnostic tool, but used alone, especially repeatedly, or in combination with the clomiphene test, it may be helpful in differentiating pituitary and hypothalamic causes of hypogonadism (61, 92). LH-RH alone or in combination with HMG or HCG has also been used therapeutically in Mexico, Chile, USA, Israel, Sweden, Japan and other countries (93—96) to induce ovulation in amenorrheic women. The use of LH-RH and its analogs can prevent superovulation and the resultant multiple births which are not uncommon after administration of HMG and/or HCG. LH-RH has also been used in Argentina and England to treat oligospermia and hypogonadotropic hypogonadism in men (97, 98). We participated in many of these studies. Recently, LH-RH given intranasally was successfully used for treatment of cryptorchidism (99). It was determined in collaborative studies carried out in Mexico, Brasil, Japan, England, Spain and Germany (100—105) that single administration of the superactive analogs [D-Ala^6, desGly-NH_2^{10}]—LH-RH EA, [D-Leu^6, desGly-NH_2^{10}]-LH-RH EA, [D-Ser-$(Bu^t)^6$, desGly-NH_2^{10}]-LH-RH EA, or [D-Trp^6]-LH-RH can induce protracted stimulation of the release of LH and FSH lasting as long as 24 hours (Fig. 20). Consequently these analogs should be more convenient and practical to use than LH-RH, which has to be given repeatedly each day (96). Moreover, these analogs are active not only after parenteral but also intranasal (Fig. 21), intravaginal, intrarectal and oral administration if suitable doses are given (61, 103—105). No significant untoward side effects of LH-RH and analogs have been observed. However, in spite of some positive

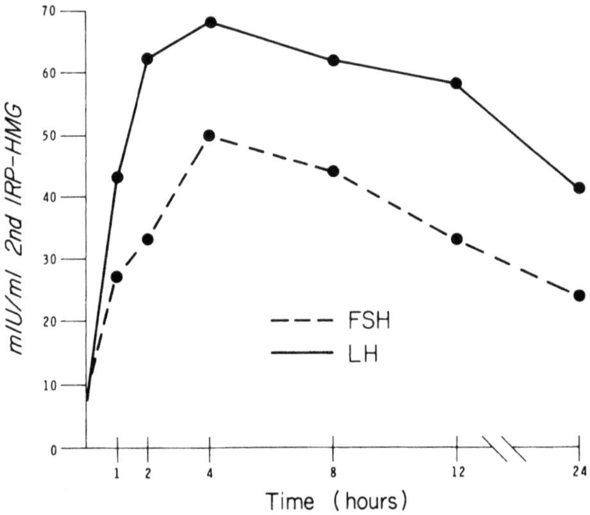

Figure 20. Plasma LH and FSH levels in 34-year-old woman with amenorrhea and galactorrhea after intramuscular administration of 250 μg of D-Leu-6-LH-RH ethylamide. From deMedeiros-Comaru, Rodrigues, Povoa, Franco, Dimetz, Coy, Kastin and Schally, *Internat. J. Fertil. 21*: 239, 1976.

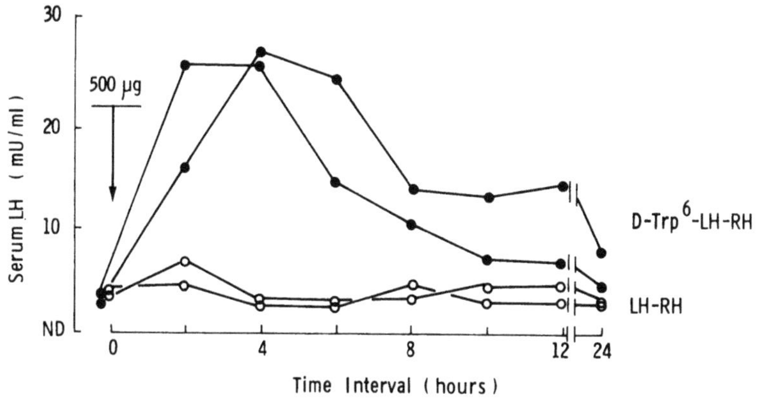

Figure 21. Effect of intranasal administration of 500 μg amounts of LH-RH or D-Trp^6-LH-RH on serum LH levels in two men. Administration carried out in 1 ml saline using pasteur pipette. From collaborative study with Prof. G. M. Besser and Prof. R. Hall.

results, our current knowledge about the use of these analogs for treatment of female and male infertility is inadequate and the therapeutic regimens are largely empirical. Moreover, in view of the paradoxical antifertility effects of large doses of LH-RH and longacting superactive analogs, caution must be exercised in devising clinical protocols. In order to fully exploit the potential of analogs of LH-RH for control of fertility at the level of the brain, we will need further work.

Inhibitory Analogs of LH-RH

The concept of antagonists of LH-RH proposed by us in 1971 (106) was based on the assumption that replacement or deletion of some amino acids in LH-RH might result in analogs possessing features requisite for binding, but lacking those which are necessary for a functional effect. Such analogs would be competitive inhibitors of LH-RH; that is, they would be devoid of LH-RH activity, but by competing for attachment to the receptor site with endogenous LH-RH would lead to a decrease of LH and FSH secretion. From the early inactivation studies on LH-RH (50—52, 57), we surmised the importance of His and Trp for the biological activity of LH-RH. However, the analogs based only on deletion of His or Trp were not very effective antagonists. [DesHis^2, desGly^{10}]-LH-RH EA, made by Dr. D. Coy in our laboratory, was the first LH-RH inhibitor found to be active *in vivo* (107). Incorporation of a D-amino acid in the 6 position, in agreement with original report of Monahan *et al* (85), also improved the inhibitory activity (78, 79). It was then determined by Rees *et al* (108) that replacement of His in position 2 by D-Phe created more effective inhibitors than its deletion. Analogs such as [D-Phe^2, D-Phe^6]-LH-RH and [D-Phe^2, D-Leu^6]-LH-RH were synthesized (78, 79). The former was found to inhibit LH and FSH release for 6—8 hours after injection, and the latter to partially block ovulation in rats in doses of about 6 mg/kg (78, 79). [D-Phe^2, Phe^3, D-Phe^6]-LH-RH was a still more potent inhibitor, since given at noon on the proestrous day it suppressed the preovulatory LH

(Fig. 22) and FSH surge, and completely blocked ovulation (Table 6) (109). The replacement of Trp by D-Trp in position 3 appeared to further increase the potency of inhibitory peptides (110). [D-Phe^2, D-Trp^3, D-Phe^6]-LH-RH (Fig. 23) is both longer-acting (nearly 10 hours in the rat) and more potent than [D-Phe^2, Phe^3, D-Phe^6]-LH-RH. Both these inhibitors also inhibit ovulation in hamsters and rabbits, and suppress LH release in monkeys (79, 111). We have observed that [D-Phe^2, D-Trp^3, D-Phe^6]-LH-RH and the superactive agonist [D-Trp^6]-LH-RH compete with LH-RH for its pituitary plasma membrane receptors, displacing the [^{125}I]-LH-RH more strongly than its parent hormone (Fig. 24) (112). Therefore, both stimulatory and inhibitory analogs of LH-RH may exert their action on the same pituitary plasma membrane receptors as those for LH-RH. Recently, in collaboration with Dr.

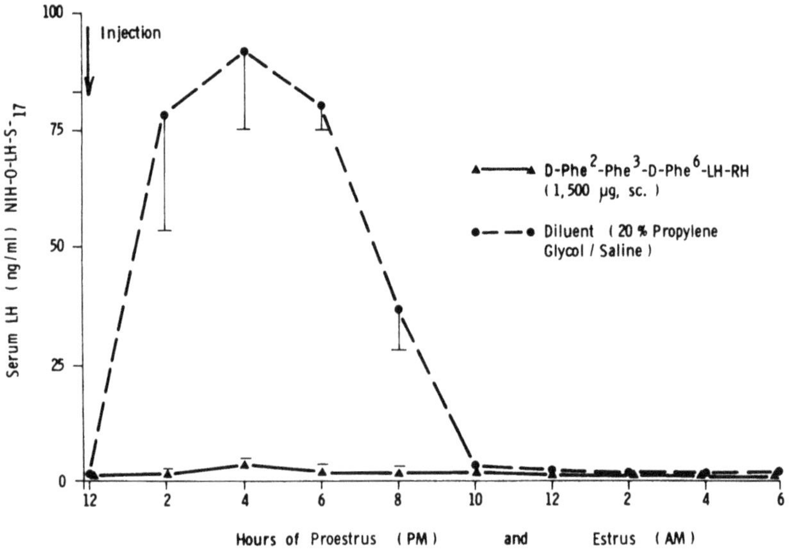

Figure 22. Effect of a single subcutaneous administration of [D-Phe^2, Phe^3, D-Phe^6]-LH-RH (1.5 mg) on the preovulatory surge of LH in proestrous rats. The differences in LH levels between animals treated with diluent (20 percent propylene glycol in saline) and analog were significant at 1400, 1600, 1800 and 2000 hours (P<.01). From de la Cruz, Coy, Vilchez-Martinez, Arimura, and Schally, *Science 191*: 195, 1976.

Table 6. Suppression of Ovulation in Rats by [D-Phe^2, Phe^3, D-Phe^6]-LH-RH

TREATMENT	DOSE (mg)	No. of Animals	No. of Animals Ovulating	No. of Ova (mean ± S.E.)	Suppression (%)	P
Diluent	(x3)	6	6	13.3 ± 0.8	———	———
[D-Phe^2, Phe^3, D-Phe^6]-LH-RH	1 (x3)	5	0	0.0 ± 0.0	100.0	.001

The rats had a 4-day estrus cycle and weighed 202.6 ± 1.6 g. Three subcutaneous injections were administered at 12:00, 14:30, and 17:00 hours (C.S.T.). The diluent was 20 percent propylene glycol in saline.
From: de la Cruz, Coy, Vilchez-Martinez, Arimura, and Schally, *Science 191:* 195, 1976.

p-GLU-D-PHE-D-TRP-SER-TYR-D-PHE-LEU-ARG-PRO-GLY-NH_2

1 2 3 4 5 6 7 8 9 10

D-PHE^2,D-TRP^3,D-PHE^6-LH-RH

Figure 23. Structure of the inhibitory analog [D-Phe^2, D-Trp^3, D-Phe^6]-LH-RH.

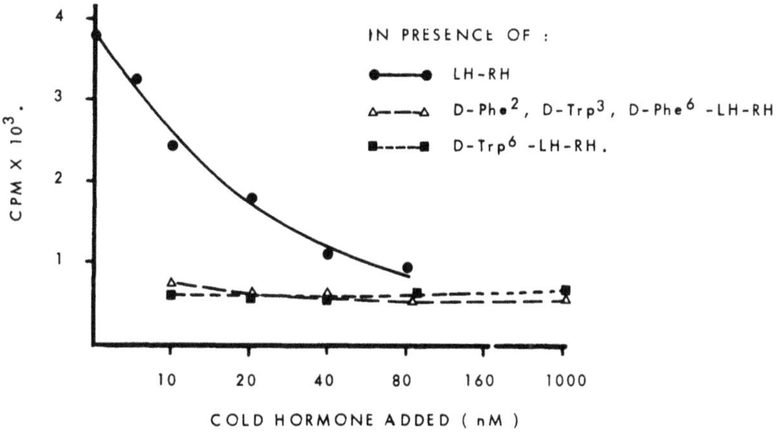

Figure 24. Effects of LH-RH and its analogs on the binding of ^{125}I-LH-RH by pituitary homogenates. The anterior pituitary was homogenized in Hepes buffer pH 7.2. The homogenate equivalent to 1 pituitary in 0.5 ml was incubated with 25 μl of a solution 4.5 nM of ^{125}I-LH-RH (300 μCi/μg) for 30 min and at 4°C in presence of LH-RH, [D-Phe^2, D-Trp^3, D-Phe^6]-LH-RH, or [D-Trp^6]-LH-RH. Each point represents the average of triplicate experiments. From Pedroza, Vilchez-Martinez, Fishback, Arimura, and Schally, *Biochem. Biophys. Res. Commun.*, in press.

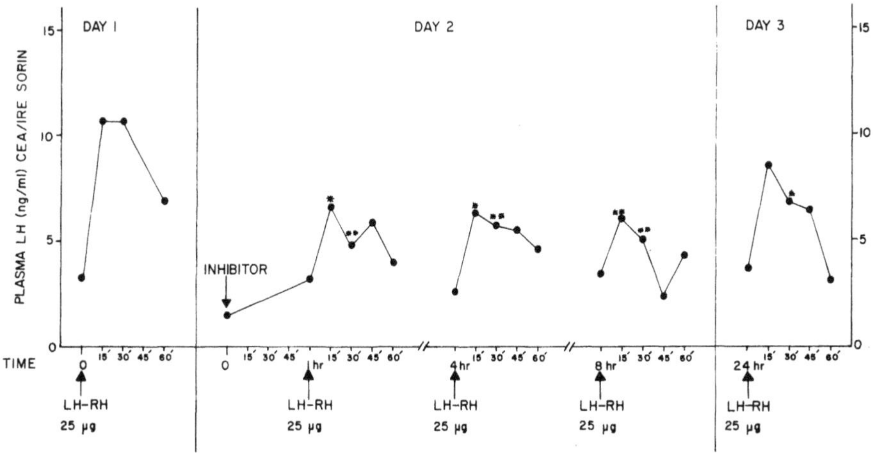

Figure 25. Mean response in serum LH levels of 4 men to 25 μg LH-RH before (control) and after im administration of 90 mg [D-Phe^2, D-Trp^3, D-Phe^6]-LH-RH. Asterisks indicate values significantly ($P<0.01$) different from the value at that time during the control period. From Gonzalez-Barcena, Kastin, Coy, Nikolics, and Schally, *Lancet ii*: 997, 1977.

D. Gonzalez-Barcena, we have determined that [D-Phe^2, D-Trp^3, D-Phe^6]-LH-RH significantly suppressed the release of LH and FSH in response to LH-RH in normal men (Fig. 25) (113). We believe that the progress being made in this area may eventually lead to development of new, safer birth control methods.

GROWTH HORMONE-RELEASE INHIBITING HORMONE (GH-RIH, SOMATOSTATIN)

In 1973, Brazeau *et al* (114) isolated from sheep hypothalami and established the structure of a tetradecapeptide which they named somatostatin, or GH-RIH, which inhibited the release of GH *in vitro* and *in vivo* in rats. The presence of somatostatin in the hypothalamus was first observed by Krulich *et al* (115). Somatostatin was synthesized by several groups, including ours (116, 117). Subsequently, we isolated and determined the structure of porcine somatostatin, showed the primary structures of native porcine and ovine somatostatin to be identical (118), and thus confirmed the existence of this peptide (114) in another species. We also found larger and more basic forms of somatostatin in pig hypothalami (118). These materials are biologically and immunologically active, possess different physico-chemical properties from somatostatin, appear to have several amino acids including arginine attached to the N-terminus, and may represent precursors of somatostatin. We have also found high concentrations of somatostatin in extracts of pancreas, stomach and duodenum of the rat, as well as two types of immunoreactive somatostatin (119). In agreement with parallel studies by others, we found with Hall *et al* (92, 120) and Besser *et al* (121) that somatostatin inhibits the secretion of pituitary GH and TSH in human beings. A physiological role for somatostatin in the regulation of GH and TSH secretion is supported by our observations with Dr. A. Arimura that passive immunization with anti-somatostatin elevates basal GH levels, prevents the stress-induced decrease of GH in rats (122) and increases the TSH response to TRH (123). In collaborative clinical studies in England, also with Profs. Hall and Besser, parallel to those of others, we then determined that somatostatin suppresses the secretion of glucagon and insulin in humans (124). In joint investigations with Dr. S. Konturek we later established that somatostatin affects the exocrine pancreas as well, since it reduced the secretin-induced secretion of pancreatic fluid and bicarbonate (125). With Bloom *et al* (126), Gomez-Pan *et al* (127) and Konturek *et al* (128) we made the original observations that somatostatin decreases the circulating levels of gastrin in men and dogs, and that it also exerts a direct antisecretory effect on gastric parietal and peptic cells, since it inhibits pentagastrin-induced gastric acid and pepsin secretion in cats (Fig. 26) and dogs. These studies established for the first time that this hormone can exert exocrine, as well as endocrine, effects. In our work with Konturek *et al* (125), it was also determined that somatostatin inhibited the release of secretin and cholecystokinin/pancreozymin from the duodenal mucosa.

Dr. Arimura in our laboratory was the first to generate antisera to somatostatin and to establish a RIA for this hormone (129). These antisera were used by us and Hökfelt *et al* (130) for the localization of immunoreactive somatostatin in the brain, including hypothalamus, D-cells of pancreas, the gastrointestinal mucosa and other tissues by immunocytochemical methods.

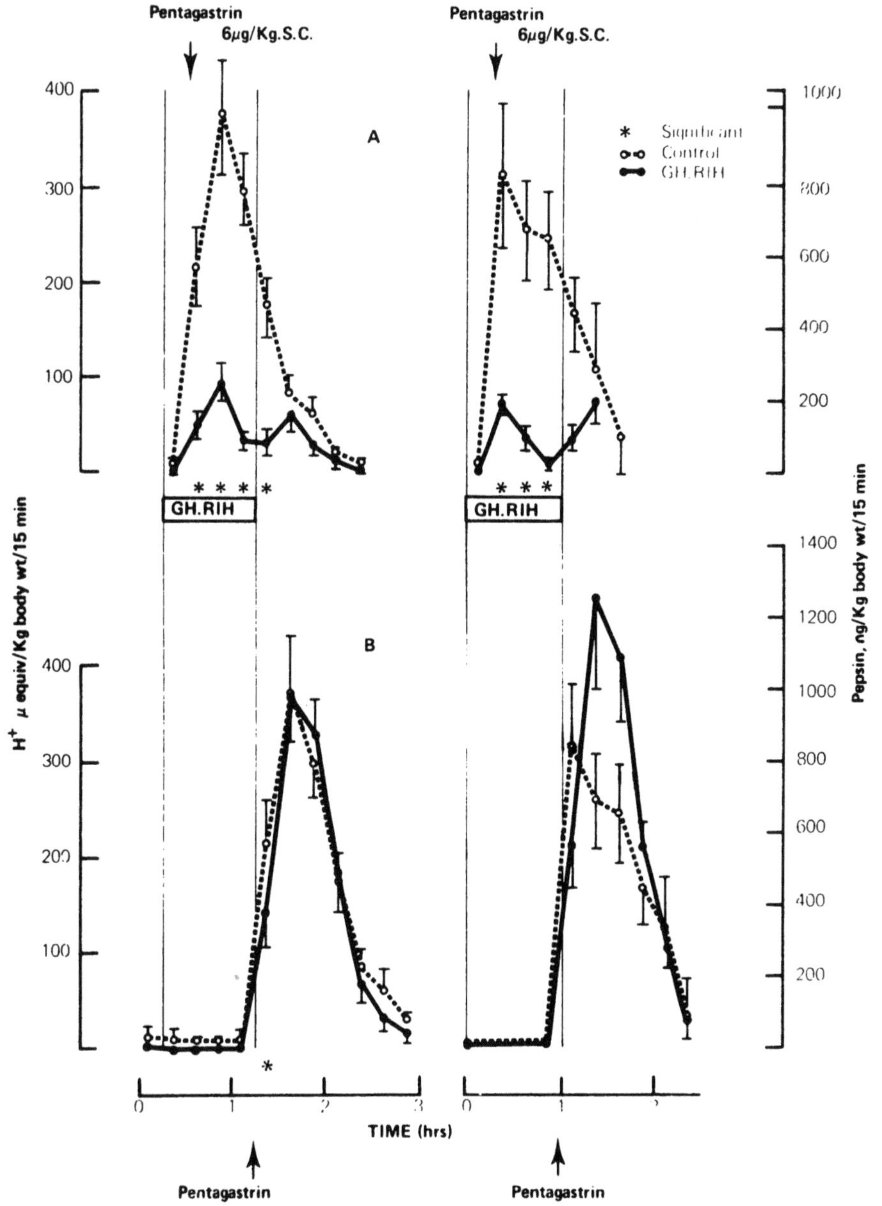

Figure 26. The mean acid and pepsin responses to pentagastrin of cats with gastric fistulae in control experiments (○— ——○) compared in (A) with those when pentagastrin was injected during the infusion of GH-RIH (10 µg/kg/hr) (●———●) and in (B) with those when the injection of pentagastrin was preceded by the infusion of GH-RIH (●———●). Responses are expressed as the mean ± S.E.M.. Asterisk indicates significant difference between the means at least at the 5% level. From Gomez-Pan, Reed, Albinus, Shaw, Hall, Besser, Coy, Kastin, and Schally, *Lancet i*: 888, 1975.

These studies support the view that somatostatin plays a role in the regulation not only of the pituitary but also of the pancreas, duodenum and stomach.

Somatostatin itself is of little therapeutic value because it has multiple actions and a short biological half-life. Attempts are therefore continuing by us and others to produce analogs of somatostatin with prolonged activity and the ability to inhibit the release of some or only one hormone. We showed that [D-Ala^2, D-Trp^8]-somatostatin (131) has a potency 20 times greater than somatostatin on inhibition of GH release, but is only three times as potent in inhibiting pentagastrin-induced gastric acid secretion (132). Meyers *et al* (133) in our laboratory, and others (134), synthesized [D-Cys^{14}]-somatostatin and [D-Trp^8, D-Cys^{14}]-somatostatin which selectively inhibited GH and glucagon release more than insulin secretion. [D-Trp^8, D-Cys^{14}]-somatostatin has a ratio of 22:1 for the selective inhibition of glucagon over insulin, 100:1 for that of GH over insulin (133) and 3:1 for that of GH over gastric acid. Since potent analogs of somatostatin with selective activities can be prepared and promising results with long-acting analogs have already been realized in our laboratory, it is possible that future analogs may be useful in the treatment of such disorders as acromegaly, diabetic retinopathy, juvenile diabetes, peptic ulcers, and other diseases.

CONCLUSIONS AND PERSPECTIVES

At the inception of my scientific career, the concept of hypothalamic control of anterior pituitary function was in its formative stage. It was my good fortune to have arrived on the scene at such a crucial time and to have helped place it on the solid foundation on which it now rests. At present, the validity of this concept stands proven by the isolation, structural identification, and synthesis of three hypothalamic regulatory hormones. The presence of at least six other hypothalamic hormones which stimulate or inhibit the release of pituitary hormones from the pituitary gland is now reasonably well-established. It is likely that additional hypothalamic hormones will be found. Many clinical applications are now well-established and more will come. The information gathered from both animal and human studies with natural and synthetic TRH, LH-RH and somatostatin, has provided us with new understanding and even more importantly, I believe, opened vast new vistas for probing ever more deeply into these marvelously integrated systems of which the animate world is composed.

In any case, I hope that my work will be of practical use to humanity, and that I will be able to make new contributions in this field in the years to come.

ACKNOWLEDGMENTS

Basic studies done by us since 1962 and quoted here would not have been possible without the generous support of the Veterans Administration and the National Institutes of Health.

REFERENCES

1. Green, J. D., and Harris, G. W.: The neurovascular link between the neurohypophysis and adenohypophysis. *J. Endocrinol. 5*: 136—146, 1947.
2. Sawyer, C. H., Everett, J. W., and Markee, J. E.: A neural factor in the mechanism by which estrogen induces the release of luteinizing hormone in the rat. *Endocrinology 41*: 218—233, 1949.
3. Saffran, M., and Schally, A. V.: The release of corticotrophin by anterior pituitary tissue *in vitro*. *Canad. J. Biochem. Physiol 33*: 408—415, 1955.
4. Schally, A. V.: *In vitro* studies on the control of the release of ACTH. Ph. D. Thesis, McGill University, April, 1957.
5. Schally, A. V., Saffran, M., and Zimmerman, B.: A corticotrophin releasing factor: partial purification and amino acid composition. *Biochem. J. 70*: 97—103, 1958.
6. Schally, A. V., Bowers, C. Y., and Locke, W.: Neurohumoral functions of the hypothalamus. *Am. J. Med. Sci. 248*: 79—101, 1964.
7. Schally, A. V., and Bowers, C. Y.: *In vitro* and *in vivo* stimulation of the release of luteinizing hormone. *Endocrinology 75*: 312—320, 1964.
8. Schally, A. V., and Bowers, C. Y.: Purification of luteinizing hormone releasing factor from bovine hypothalamus. *Endocrinology 75*: 608—614, 1964.
9. Schally, A. V., Kuroshima, A., Ishida, Y., Redding, T. W., and Bowers, C. Y.: The presence of prolactin inhibiting factor (PIF) in extracts of beef, sheep and pig hypothalami. *Proc. Soc. Exp. Biol. Med. 118*: 350—352, 1965.
10. Schally, A. V., Steelman, S., and Bowers, C. Y.: Effect of hypothalamic extracts on the release of growth hormone *in vitro*. *Proc. Soc. Exp. Biol. Med. 119*: 208—212, 1965.
11. Kuroshima, A., Ishida, Y., Bowers, C. Y., and Schally, A. V.: Stimulation of release of follicle stimulating hormone by hypothalamic extract *in vitro* and *in vivo*. *Endocrinology 76*: 614—619, 1965.
12. Bowers, C. Y., Redding, T. W., and Schally, A. V.: Effect of thyrotropinreleasing factor (TRF) of ovine, bovine, porcine and human origin on thyrotropin release *in vitro* and *in vivo*. *Endocrinology 77*: 609—616, 1965.
13. Schally, A. V., Arimura, A., Bowers, C. Y., Kastin, A. J., Sawano, S., and Redding, T. W.: Hypothalamic neurohormones regulating anterior pituitary function. In: *Recent Progress in Hormone Research* (E. B. Astwood, ed.), *24*: 497—590, Academic Press, New York, 1968.
14. Kastin, A. J., and Schally, A. V.: MSH activity in pituitaries of rats treated with hypothalamic extracts. *Gen. Comp. Endocrinol. 7*: 452—456, 1966.
15. Schally, A. V., Müller, E. E., Arimura, A., Bowers, C. Y., Saito, T., Redding, T. W., Sawano, S., and Pizzolato, P.: Releasing factors in human hypothalamic and nuerohypophysial extracts. *J. Clin. Endocrinol. Metab. 27*: 755—762, 1967.
16. Schally, A. V., Bowers, C. Y., Redding, T. W., and Barrett, J. F.: Isolation of thyrotropin releasing factor (TRF) from porcine hypothalamus. *Biochem. Biophys. Res. Commun. 25*: 165—169, 1966.
17. Schally, A. V., Redding, T. W., Bowers, C. Y., and Barrett, J. F.: Isolation and properties of porcine thyrotropin releasing hormone. *J. Biol. Chem. 244*: 4077—4088, 1969.
18. Nair, R. M. G., Barrett, J. F., Bowers, C. Y., and Schally, A. V.: Structure of porcine thyrotropin-releasing hormone. *Biochemistry 9*: 1103—1106, 1970.
19. Folkers, K., Enzmann, F., Bøler, J., Bowers, C. Y., and Schally, A. V.: Discovery of modification of the synthetic tripeptide sequence of the thyrotropin releasing hormone having activity. *Biochem. Biophys. Res. Commun. 37*: 123—126, 1969.
20. Bøler, J., Enzmann, F., Folkers, K., Bowers, C. Y., and Schally, A. V.: The identity of chemical and hormonal properties of the thyrotropin releasing hormone and pyroglutamyl-histidine-proline amide. *Biochem. Biophys. Res. Commun. 37*: 705—710, 1969.
21. Bowers, C. Y., Schally, A. V., Enzmann, F., Bøler, J., and Folkers, K.: Porcine thyrotropin releasing hormone is (Pyro)Glu-His-Pro(NH_2). *Endocrinology 86*: 1143—1153, 1970.

22. Burgus, R., Dunn, T. F., Desiderio, D., Ward, D. N., Vale, W., and Guillemin, R.: Characterisation of ovine hypothalamic hypophysiotropic TSH-releasing factor. *Nature 226*: 321—325, 1970.
23. Debeljuk, L., Arimura, A., Redding, T., and Schally, A. V.: Effect of TRH and tri-iodothyronine on prolactin release in sheep. *Proc. Soc. Exp. Biol. Med. 148*: 421—423, 1973.
24. Schally, A. V. and Redding, T. W.: *In vitro* studies with thyrotropin releasing factor. *Proc. Soc. Exp. Biol. Med. 126*: 320—325, 1967.
25. Redding, T. W., and Schally, A. V.: Studies on thyrotropin-releasing hormone (TRH) activity in peripheral blood. *Proc. Soc. Exp. Biol. Med. 131*: 420—425, 1969.
26. Jackson, I. M. D., and Reichlin, S.: Thyrotropin-releasing hormone (TRH): distribution in hypothalamic and extrahypothalamic brain tissues of mammalian and submammalian chordates. *Endocrinology 95*: 854—862, 1974.
27. Bowers, C. Y., Schally, A. V., Hawley, D. W., Gual, C., and Parlow, A. F.: Effect of thyrotropin releasing factor in man. *J. Clin. Endocrinol. Metab. 28*: 978—982, 1968.
28. Anderson, M. S., Bowers, C. Y., Kastin, A., Schalch, D. S., Schally, A. V., Snyder, P. J., Utiger, R. D., Wilber, J. F., and Wise, A. J.: Synthetic thyrotropin releasing hormone (TRH): a potent stimulator of thyrotropin secretion in man. *New Engl. J. Med. 285*: 1279—1283, 1971.
29. Hershman, J. M., and Pittman, J. A., Jr.: Response to synthetic thyrotropin releasing hormone in man. *J. Clin. Endocrinol. Metab. 31*: 457—460, 1970.
30. Hall, R., Amos., J., Garry, R., and Buxton, R. L.: Thyroid stimulating hormone response to synthetic thyrotrophin-releasing hormone in man. *Brit. Med. J. 2*: 274—277, 1970.
31. Schally, A. V., Arimura, A., and Kastin, A. J.: Hypothalamic regulatory hormones. *Science 179*: 341—350, 1973.
32. McCann, S. M., Taleisnik, S., and Friedman, H. M.: LH-releasing activity in hypothalamic extracts. *Proc. Soc. Exp. Biol. Med. 104*: 432—434, 1960.
33. Harris, G. W.: In *Control of Ovulation* (C. A. Villee, ed.), p. 56 Pergamon Press, Inc., London, 1961.
34. Campbell, H. J., Feuer, G., Garcia, J., and Harris, G. W.: The infusion of brain extracts into the anterior pituitary gland and the secretion of gonadotrophic hormone. *J. Physiol. (London) 157*: 30P—31P, 1961.
35. Courrier, R., Guillemin, R., Jutisz, M., Sakiz, E. , and Aschheim, P.: Presence dans un extrait d-hypothalamus d'une substance qui stimule la secretion de l'hormone antehypophysaire de luteinisation (LH). *C. R. Acad. Sci. 253*: 922—927, 1961.
36. Nikitovitch-Winer, M. B.: Induction of ovulation in rats by direct intrapituitary infusion of median eminence extracts. *Endocrinology 70*: 350—358, 1962.
37. Endröczi, E., and Hilliard, J.: Luteinizing hormone releasing activity in different parts of rabbit and dog brain. *Endocrinology 77*: 667—673, 1965.
38. Igarashi, M., and McCann, S. M.: A hypothalamic FSH-releasing factor. *Endocrinology 74*: 446—452, 1964.
39. Campbell, H. J., Feuer, G., and Harris, G. W.: The effect of intrapituitary infusion of median eminence and other brain extracts on anterior pituitary gonadotrophic secretion. *J. Physiol. (London) 170*: 474—486, 1964.
40. Mittler, J. C. and Meites, J.: *In vitro* stimulation of pituitary FSH release by hypothalamic extract. *Proc. Soc. Exp. Biol. Med. 117*: 309—313, 1964.
41. Schally, A. V., Carter, W. H., Saito, M., Arimura, A., and Bowers, C. Y.: Studies on the site of action of oral contraceptive steroids. I: the effect of anti-fertility steroids on plasma LH levels and on the response to luteinizing hormone-releasing factor in rats. *J. Clin. Endocrinol. Metab. 28*: 1747—1755, 1968.
42. Schally, A. V., Parlow, A. F., Carter, W. H., Saito, M., Bowers, C. Y., and Arimura, A.: Studies on the site of action of oral contraceptive steroids. II: plasma LH and FSH levels after administration of antifertility steroids and luteinizing hormone-releasing hormone (LRH) as merasured by bioassay and radioimmunoassays. *Endocrinology 86*: 530—541, 1970.

43. Arimura, A., and Schally, A. V.: Progesterone suppression of LH-releasing hormone-induced stimulation of LH-release in rats. *Endocrinology 87*: 653—657, 1970.
44. Hilliard, J., Schally, A. V. and Sawyer, C. H.: Progesterone blockade of the ovulatory response to intrapituitary infusion of LH-RH in rabbits. *Endocrinology 88*: 730—736, 1971.
45. Schally, A. V., Redding, T. W., and Arimura, A.: Effect of sex steroids on pituitary responses to LH and FSH-releasing hormone *in vitro*. *Endocrinology 93*: 893—902, 1973.
46. Arimura, A., and Schally, A. V.: Augmentation of pituitary responsiveness to LH-releasing hormone (LH-RH) by estrogen. *Proc. Soc. Exp. Biol. Med. 136*: 290—293, 1971.
47. Schally, A. V., Carter, W. H., Parlow, A. F., Saito, M., Arimura, A., Bowers, C. Y., and Holtkamp, D. E.: Alteration of LH and FSH release in rats treated with clomiphene or its isomers. *Am. J. Obstet. Gynecol. 107*: 1156—1167, 1970.
48. Kastin, A. J., Schally, A. V., Gual, C., Midgley, A. R., Jr., Bowers, C. Y., and Diaz-Infante, A. Jr.: Stimulation of LH release in men and women by hypothalamic LH-releasing hormone purified from porcine hypothalami. *J. Clin. Endocrinol. Metab. 29*: 1046—1050, 1969.
49. Kastin, A. J., Schally, A. V., Gual, C., Midgley, A. R., Jr., Bowers, C. Y., and Gomez-Perez, E.: Administration of LH-releasing hormone to selected human subjects. *Am. J. Obstet. Gynecol. 108*: 177—182, 1970.
50. Schally, A. V., Arimura, A., Baba, Y., Nair, R. M. G., Matsuo, H., Redding, T. W., Debeljuk, L , and White, W. F.: Isolation and properties of the FSH and LH-releasing hormone. *Biochem. Biophys. Res. Commun. 43*: 393—399, 1971.
51. Schally, A. V., Nair, R. M. G., Redding, T. W., and Arimura, A.: Isolation of the LH and FSH-releasing hormone from porcine hypothalami. *J. Biol. Chem. 246*: 7230—7236, 1971.
52. Schally, A. V., Baba, Y., Matsuo, H., Arimura, A., and Redding, T. W.: Further studies on the enzymatic and chemical inactivation of hypothalamic follicle-stimulating hormone-releasing hormone. *Neuroendocrinology 8*: 347—358, 1971.
53. Matsuo, H., Baba, Y., Nair, R. M. G., Arimura, A., and Schally, A. V.: Structure of the porcine LH- and FSH-releasing hormone. I. The proposed amino acid sequence. *Biochem. Biophys. Res. Commun. 43*: 1334—1339, 1971.
54. Baba, Y., Matsuo, H., and Schally, A. V.: Structure of porcine LH- and FSH-releasing hormone. II. Confirmation of the proposed structure by conventional sequential analysis. *Biochem. Biophys. Res. Commun. 44*: 459—463, 1971.
55. Matsuo, H., Arimura, A., Nair, R. M. G., and Schally, A. V.: Synthesis of the porcine LH- and FSH-releasing hormone by the solid-phase method. *Biochem. Biophys. Res. Commun. 45*: 822—827, 1971.
56. Schally, A. V., Arimura, A., Kastin, A. J., Matsuo, H., Baba, Y., Redding, T. W., Nair, R. M. G., Debeljuk, L., and White, W. F.: The gonadotropin-releasing hormone: one polypeptide regulates the secretion of luteinizing and follicle stimulating hormones. *Science 173:* 1036—1038, 1971.
57. Schally, A. V., Kastin, A. J., and Arimura, A.: Hypothalamic FSH and LH-regulating hormone: structure, physiology and clinical studies. *Fertility and Sterilitiy 22*: 703—721, 1971.
58. Schally, A. V., Redding, T. W., Matsuo, H., and Arimura, A.: Stimulation of the FSH and LH release *in vitro* by natural and synthetic LH and FSH releasing hormone. *Endocrinology 90*: 1561—1568, 1972.
59. Arimura, A., Debeljuk, L., and Schally, A. V.: Stimulation of FSH release *in vivo* by prolonged infusion of synthetic LH-RH. *Endocrinology 91*: 529—532, 1972.
60. Kastin, A. J., Schally, A. V., Gual, C., and Arimura, A.: Release of LH and FSH after administration of synthetic LH-releasing hormone. *J. Clin. Endocrinol. Metab. 34*: 753—756, 1972.
61. Schally, A. V., Kastin, A. J., and Coy, D. H.: Edward T. Tyler Prize Oration-LH-releasing hormone and its analogues: recent basic and clinical investigations. *Internat. J. Fertil. 21*: 1—30, 1976.

62. Redding, T. W., Schally, A. V., Arimura, A., and Matsuo, H.: Stimulation of release and synthesis of luteinizing hormone (LH) and follicle stimulating hormone (FSH) in tissue cultures of rat pituitaries in response to natural and synthetic LH and FSH releasing hormone. *Endocrinology 90*: 764—770, 1972.
63. Debeljuk, L., Arimura, A., Shiino, M., Rennels, E. G., and Schally, A. V.: Effect of chronic treatment with LH-RH in hypophysectomized pituitary grafted male rats. *Endocrinology 92*: 921—930, 1973.
64. Arimura, A., Debeljuk, L., Shiino, M., Rennels, E. G., and Schally, A. V.: Follicular stimulation by chronic treatment with synthetic LH-releasing hormone in hypophysectomized female rats bearing pituitary grafts. *Endocrinology 92*: 1507—1514, 1973.
65. Shiino, M., Arimura, A., Schally, A. V., and Rennels, E. G.: Ultrastructural observations of granule extrusion from the rat anterior pituitary cells after the injection of LH releasing hormone. *Zeit. Zellforsch. Mikrosk. Anat. 128*: 152—161, 1972.
66. Redding, T. W., Kastin, A. J., Gonzalez-Barcena, D., Coy, D. H., Coy, E. J., Schalch, D. S., and Schally, A. V.: The half-life, metabolism and excretion of tritiated luteinizing hormone-releasing hormone (LH-RH) in man. *J. Clin. Endocrinol. Metab. 37*: 626—631, 1973.
67. Arimura, A., Sato, H., Kumasaka, T., Worobec, R. B., Debeljuk, L., Dunn, J. D., and Schally, A. V.: Production of antiserum to LH-RH associated with marked atrophy of the gonads in rabbits: characterization of the antibody and development of a radioimmunoassay for LH-RH. *Endocrinology 93*: 1092—1103, 1973.
68. Arimura, A., Shiino, M., de la Cruz, K. G., Rennels, E. G., and Schally, A. V.: Effect of active and passive immunization with LH-RH on serum LH, FSH levels, and the ultrastructure of the pituitary gonadotrophs in castrated male rats. *Endocrinology 99*: 291—303, 1976.
69. Arimura, A., Debeljuk, L., and Schally, A. V.: Blockade of preovulatory surge of gonadotropins LH and FSH and of ovulation by anti-LH-RH serum in rats. *Endocrinology 95*: 323—325, 1974.
70. de la Cruz, A., Arimura, A., de la Cruz, K. G., and Schally, A. V.: Effect of administration of anti-LH-RH serum on gonadal function during the estrus cycle in the hamster. *Endocrinology 98*: 490—497, 1976.
71. Arimura, A., Nishi, N., and Schally, A. V.: Delayed implantation caused by administration of sheep immunogamma globulin against LH-RH in the rat. *Proc. Soc. Exp. Biol. Med. 152*: 71—75, 1976.
72. Nishi, N., Arimura, A., de la Cruz, K. G., and Schally, A. V.: Termination of pregnancy by sheep anti-LH-RH gamma globulin in rats. *Endocrinology 98*: 1024—1030, 1976.
73. Arimura, A., Kastin, A. J., Schally, A. V., Saito, M., Kumasaka, T., Yaoi, Y., Nishi, N., and Ohkura, K.: Immunoreactive LH-releasing hormone in plasma: midcycle elevation in women. *J. Clin. Endocrinol. Metab. 38*: 510—513, 1974.
74. Sétáló, G., Vigh, S., Schally, A. V., Arimura, A., and Flerkó, B.: LH-RH-containing neural elements in the rat hypothalamus. *Endocrinology 96*: 135—142, 1975.
75. Moss, R., and McCann, S. M.: Induction of mating behavior in rats by luteinizing hormone-releasing factor. *Science 181*: 177—179, 1973.
76. Plotnikoff, N. P., Kastin, A. J., Anderson, M. S., and Schally, A. V.: DOPA potentiation by a hypothalamic factor, MSH releasing inhibiting hormone. *Life Sci. 10*: 1279—1283, 1971.
77. Kastin, A. J., Plotnikoff, N. P., Schally, A. V., and Sandman, C. A.: Endocrine and CNS effects of hypothalamic peptides and MSH. In: *Reviews of Neuroscience, Vol. 2* (S. Ehrenpreis and I. J. Kopin, eds). pp. 111—148, Raven Press, New York, 1976.
78. Coy, D. H., Coy, E. J., and Schally, A. V.: Structure activity relationship of the LH and the FSH releasing hormone. In: *Research Methods in Neurochemistry, Vol. 3* (N. Marks and R. Rodnight, Eds.), pp. 393—406, Plenum Press, New York, 1975.
79. Schally, A. V., and Coy, D. H.: Stimulatory and inhibitory analogs of luteinizing hormone-releasing hormone (LH-RH). In: *Hypothalamic Peptide Hormones and Pituitary Regulation* (J. C. Porter, ed.), pp. 99—121, Plenum Press, New York, 1977.

80. Schally, A. V., Arimura, A., Carter, W. H., Redding, T. W., Geiger, R., König, W., Wissman, H., Jaeger, G., Sandow, J., Yanaihara, N., Yanaihara, C., Hashimoto, T.,. and Sakagami, M.: Luteinizing hormone-releasing hormone (LH-RH) activity of some synthetic polypeptides. I. fragments shorter than decapeptide. *Biochem. Biophys. Res. Commun.* *48*: 366—375, 1972.
81. Coy, D. H., Coy, E. J., Schally, A. V., Vilchez-Martinez, J. A., Hirotsu, Y., and Arimura, A.: Synthesis and biological properties of [D-Ala-6, des-Gly-NH_2^{10}]-LH-RH-ethylamide, a peptide with greatly enhanced LH- and FSH-releasing activity. *Biochem. Biophys. Res. Commun.* *57*: 335—340, 1974.
82. Vilchez-Martinez, J. A., Coy, D. H., Arimura, A., Coy, E. J., Hirotsu, Y., and Schally, A. V.: Synthesis and biological properties of [D-Leu-6]-LH-RH and [D-Leu-6, desGly-NH_2^{10}]-LH-RH ethylamide. *Biochem. Biophys. Res. Commun.* *59*: 1226—1232, 1974.
83. Coy, D. H., Vilchez-Martinez, J. A., Coy, E. J., and Schally, A. V.: Analogs of luteinizing hormone-releasing hormone with increased biological activity produced by D-amino acid substitutions in position 6. *J. Med. Chem.* *19*: 423—425, 1976.
84. Fujino, M., Kobayashi, S., Obayashi, M., Shinagawa, S., Fukuda, T., Kitada, C., Nakayama, R., Yamazaki, I., White, W. F., and Rippel, R. H.: Structure-activity relationships in the C-terminal part of luteinizing hormone releasing hormone (LH-RH). *Biochem. Biophys. Res. Commun.* *49*: 863—869, 1972.
85. Monahan, M. W., Amoss, M. S., Anderson, H. A., and Vale, W.: Synthetic analogs of the hypothalamic luteinizing hormone releasing factor with increased agonist or antagonist properties. *Biochemistry* *12*: 4616—4620, 1973.
86. Fujino, M., Fukuda, T., Shinagawa, S., Kobayashi, S., Yamazaki, I., Nakayama, R., Seely, J. H., White, W. F., and Rippel, R. H.: Synthetic analogs of luteinizing hormone releasing hormone (LH-RH) substituted in position 6 and 10. *Biochem. Biophys. Res. Commun.* *60*: 406—413, 1974.
87. König, W., Sandow, J., and Geiger, R.: In: *Peptides: Chemistry, Structure and Biology* (R. Walter and J. Meienhofer, eds). Ann Arbor Science Publishers, Ann Arbor, *4*: 357—366, 1977.
88. Johnson, E. S., Gendrich, R. L., and White, W. F.: Delay of puberty and inhibition of reproductive processes in the rat by a gonadotropin-releasing hormone agonist analog. *Fertil. Steril.* *27*: 853—860, 1976.
89. Auclair, C., Kelly, P. A., Labrie, F., Coy, D. H., and Schally, A. V.: Inhibition of testicular luteinizing hormone receptor level by treatment with a potent luteinizing hormone-releasing hormone agonist or human chorionic gonadotropin. *Biochem. Biophys. Res. Commun.* *76*: 855—862, 1977.
90. Corbin, A., Beattie, C. W., Yardley, J., and Foell, T. J.: Post-coital contraceptive effects of an agonistic analogue of luteinizing hormone releasing hormone. *Endocrine Res. Commun.* *3*: 359—376, 1976.
91. Arimura, A., Pedroza, E., Vilchez-Martinez, J., and Schally, A. V.: Prevention of implantation by [D-Trp^6]-LH-RH in the rat: comparative study with the effects of large doses of HCG on pregnancy. *Endocrine Res. Commun.*, *4*: 357—366, 1977.
92. Hall, R., and Gomez-Pan, A.: The hypothalamic regulatory hormones and their clinical applications. In: *Advances in Clinical Chemistry, Vol. 18*, pp. 173—212, Academic Press, New York, 1976.
93. Zarate, A., Canales, E. S., Schally, A. V., Ayala-Valdes, L., and Kastin, A. J.: Successful induction of ovulation with synthetic luteinizing hormone-releasing hormone in anovulatory infertility. *Fertil. Steril.* *23*: 672—674, 1972.
94. Zañartu, J., Dabacens, A., Kastin, A. J., and Schally, A. V.: Effect of synthetic hypothalamic hormone (FSH/LH-RH) releasing gonadotropic hormones in anovulatory sterility. *Fertil. Steril.* *25*: 160—169, 1974.
95. Figueroa Casas, P. R., Badano, A. R., Aparicio, N., Lencioni, L. J., Berli, R. R., Badano, H., Biccoca, C., and Schally, A. V.: Hypothalamic gonadotropin-releasing hormone (LH-RH) in the treatment of anovulatory sterility. *Fertil. Steril.* *26*: 549—553, 1975.

96. Nillius, S. J., and Wide, L.: Gonadotrophin-releasing hormone treatment for induction of follicular maturation and ovulation in amenorrheic women with anorexia nervosa. *Brit. Med. J. 3*: 405—408, 1975.
97. Mortimer, C. H., McNeilly, A. S., Fisher, R. A., Murray, M. A. F., and Besser, G. M.: Gonadotrophin-releasing hormone therapy in hypogonadal males with hypothalamic or pituitary dysfunction. *Brit. Med. J. 4*: 617—621, 1974.
98. Aparicio, N. J., Schwarzstein, L., Turner, E. A. de, Turner, D., Mancini, R., and Schally, A. V.: Treatment of idiopathic normogonadotropic oligoasthenospermia with synthetic LH-RH. *Fertil. Steril. 27*: 549—555, 1976.
99. Illig, R., Kollmann, F., Borkenstein, M., Kuber, W., Exner, G. U., Kellerer, K., Lunglmayr, L., and Prader, A.: Treatment of cryptorchidism by intranasal synthetic luteinizing-hormone releasing hormone. *Lancet ii*: 518—520, 1977.
100. Soria, J., Zarate, A., Canales, E. S., Ayala, A., Schally, A. V., Coy, D. H., Coy, E. J., and Kastin, A. J.: Increased and prolonged LH-RH/FSH-RH activity of synthetic [D-Ala^{6},desGly-NH_2^{10}]-LH-RH ethylamide in normal women. *Am. J. Ob. Gyn. 123*: 145—146, 1975.
101. deMedeiros-Comaru, A. M., Rodrigues, J., Povoa, L. C., Franco, S., Dimetz, T., Coy, D. H., Kastin, A. J., and Schally, A. V.: Clinical studies with long-acting superactive analogs of LH-RH in women with secondary amenorrhea. *Internat. J. Fertil. 21*: 239—245, 1976.
102. Jaramillo Jaramillo, C., Perez Infante, V., Lopez Macia, A., Charro Salgado, A., Coy, D. H., and Schally, A. V.: Serum LH, FSH, and testosterone response to the administration of a new LH-RH analog, D-Trp^{6}-LH-RH in normal men. *Internat. J. Fertil. 22*: 77—84, 1977.
103. Gonzalez-Barcena, D., Kastin, A. J., Schalch, D. S., Coy, D. H., and Schally, A. V.: Prolonged elevation of luteinizing hormone (LH) after intranasal administration of an analogue of LH releasing hormone. *Fertil. Steril. 27*: 1246—1249, 1976.
104. Gonzalez-Barcena, D., Kastin, A. J., Coy, D. H., Schalch, D. S., Miller, M. C. III, Escalante-Herrera, A., and Schally, A. V.: Stimulation of luteinizing hormone release after oral administration of an analogue of LH-releasing hormone. *Lancet ii*: 1126—1128, 1975.
105. Saito, M., Kumasaka, T., Yaoi, Y., Nishi, N., Arimura, A., Coy, D. H., and Schally, A. V.: Stimulation of LH and FSH secretion by [D-Leu^{6}, Des-Gly^{10}-NH_2]-LH-RH ethylamide after subcutaneous, intravaginal and intrarectal administration to women. *Fertil. Steril. 28*: 240—245, 1977.
106. Schally, A. V., and Kastin, A. J.: Stimulation and inhibition of fertility through hypothalamic agents. *Drug Therapy 1* (11): 29—32, 1971.
107. Coy, D. H., Vilchez-Martinez, J. A., Coy, E. J., Arimura, A., and Schally, A. V. A peptide inhibitor of luteinizing hormone-releasing hormone (LH-RH). *J. Clin. Endocrinol. Metab. 37*: 331—333, 1973.
108. Rees, R. W. A., Foell, T. J., Chai, S., and Grant, N.: Synthesis and biological activities of analogs of the luteinizing hormone-releasing hormone (LH-RH) modified in position 2. *J. Med. Chem. 17*: 1015—1019, 1974.
109. de la Cruz, A., Coy, D. H., Vilchez-Martinez, J. A., Arimura, A., and Schally, A. V.: Blockade of ovulation in rats by inhibitory analogs of luteinizing hormone-releasing hormone. *Science 191*: 195—197, 1976.
110. Coy, D. H., Vilchez-Martinez, J. A., and Schally, A. V.: Structure-function studies on LH-RH. In: *Peptides, 1976* (A. Loffet, ed.), pp. 463—469, Editions de l'Universite de Bruxelles, Bruxelles, 1977.
111. Phelps, C. P., Coy, D. H., Schally, A. V., and Sawyer, C. H.: Blockade of LH release and ovulation in the rabbit with inhibitory analogues of luteinizing hormone-releasing hormone. *Endocrinology 100*: 1526—1532, 1977.
112. Pedroza, E., Vilchez-Martinez, J. A., Fishback, J., Arimura, A., and Schally, A. V.: Binding capacity of LH-RH and its analogue for pituitary receptor sites. *Biochem. Biophys. Res. Commun., 79*: 234—238, 1977.

113. Gonzalez-Barcena, D., Kastin, A. J., Coy, D. H., Nikolics, K., and Schally, A. V.: Suppression of gonadotropin release in man by an inhibitory analogue of LH-releasing hormone. *Lancet ii*: 997—998, 1977.
114. Brazeau, P., Vale, W., Burgus, R., Ling, N., Butcher, M., Rivier, J., and Guillemin, R.: A hypothalamic polypeptide that inhibits the secretion of pituitary growth hormone. *Science 179*: 77—79, 1973.
115. Krulich, L., Dhariwal, A. P. S., and McCann, S. M.: Stimulatory and inhibitory effects of purified hypothalamic extracts on growth hormone release from rat pituitary *in vitro*. *Endocrinology 83*: 783—790, 1968.
116. Rivier, J., Brazeau, P., Vale, W., Ling, N., Burgus, R., Gilon, C., Yardley, J., and Guillemin, R.: Synthèse totale par phase solide d'un tetradecapeptide ayant les propriétés chimiques et biologiques de la somatostatine. *C. R. Acad. Sci. Paris 276*: 2737—2740, 1973.
117. Coy, D. H., Coy, E. J., Arimura, A., and Schally, A. V.: Solid phase synthesis of growth hormone-release inhibiting factor. *Biochem. Biophys. Res. Commun. 54*: 1267—1273, 1973.
118. Schally, A. V., Dupont, A., Arimura, A., Redding, T. W., Nishi, N., Linthicum, G. L., and Schlesinger, D. H.: Isolation and Structure of growth hormone-release inhibiting hormone (somatostatin) from porcine hypothalami. *Biochemistry 15*: 509—514, 1976.
119. Arimura, A., Sato, H., Dupont, A., Nishi, N., and Schally, A. V.: Abundance of immunoreactive somatostatin in the stomach and the pancreas. *Science 189*: 1007—1009, 1975.
120. Hall, R., Besser, G. M., Schally, A. V., Coy, D. H., Evered, D., Goldie, D. J., Kastin, A. J., McNeilly, A. S., Mortimer, C. H., Phenekos, C., Tunbridge, W. M. G., and Weightman, D.: Action of growth hormone-release inhibitory hormone in healthy men and in acromegaly. *Lancet ii*: 581—584, 1973.
121. Besser, G. M., Mortimer, C. H., Carr, D., Schally, A. V., Coy, D. H., Evered, D., Kastin, A. J., Tunbridge, W. M. G., Thorner, M. O., and Hall, R.: Growth hormone release inhibiting hormone in acromegaly. *Brit. Med. J. 1*: 352—355, 1974.
122. Arimura, A., Smith, W. D., and Schally, A. V.: Blockade of the stress-induced decrease in blood GH by anti-somatostatin serum in rats. *Endocrinology 98*: 540—543, 1976.
123. Arimura, A. and Schally, A. V. Increase in Basal and TRH-Stimulated Secretion of TSH by Passive Immunization with Antiserum to Somatostatin in Rats. *Endocrinology, 98*: 1069—1072, 1976.
124. Mortimer, C. H., Tunbridge, W. M. G., Carr, D, Yeomans, L, Lind, T., Bloom, S. R., Kastin, A., Mallinson, C. N., Besser, G. M., Schally, A. V., and Hall, R.: Effects of growth-hormone release-inhibiting hormone on circulating glucagon, insulin, and growth hormone in normal, diabetic, acromegalic and hypopituitary patients. *Lancet i*: 697—701, 1974.
125. Konturek, S. J., Tasler, J., Obtulowicz, W., Coy, D. H., and Schally, A. V.: Effect of growth hormone-release inhibiting hormone on hormones stimulating exocrine pancreatic secretion. *J. Clin. Invest. 58*: 1—6, 1976.
126. Bloom, S. R., Mortimer, C. H., Thorner, M. O., Besser, G. M., Hall, R., Gomez-Pan, A., Roy, V. M., Russell, R. C. G., Coy, D. H., Kastin, A. J., and Schally, A. V.: Inhibition of gastrin and gastric-acid secretion by growth-hormone release-inhibiting hormone. *Lancet ii*: 1106—1109, 1974.
127. Gomez-Pan, A., Reed, J. D., Albinus, M., Shaw, B., Hall, R., Besser, G. M., Coy, D. H., Kastin, A. J., and Schally, A. V.: Direct inhibition of gastric acid and pepsin secretion by growth hormone-release inhibiting hormone in cats. *Lancet i*: 888—890, 1975.
128. Konturek, S. J., Tasler, J., Cieszkowski, M., Coy, D. H., and Schally, A. V.: Effect of growth hormone release-inhibiting hormone on gastric secretion, mucosal blood flow and serum gastrin. *Gastroenterology 70*: 737—741, 1976.
129. Arimura, A., Sato, H., Coy, D. H., and Schally, A. V.: Radioimmunoassay for GH-release inhibiting hormones. *Proc. Soc. Exp. Biol. Med. 148*: 784—789, 1975.

130. Hökfelt, T., Efendic, S., Hellerström, C., Johansson, O., Luft, R., and Arimura, A.: Cellular localization of somatostatin in endocrine-like cells and neurons of the rat with special references to the A_1-cells of the pancreatic islets and to the hypothalamus. *Acta Endocrinologica 80(Suppl. 200)*: 5—41, 1975.
131. Coy, D. H., Coy, E. J., Meyers, C., Drouin, J., Ferland, L., Gomez-Pan, A., and Schally, A. V.: Structure-function studies on somatostatin (abstract). In: *Program 58th Meeting of the Endocrine Society, San Francisco, June 23—25, 1976, p. 209.*
132. Brown, M. P., Coy, D. H., Gomez-Pan, A., Hirst B. H., Hunter, M., Meyers, C. A., Reed, J. D., Scally A. V. and Shaw, B. Structure-activity relationships of eighteen somatostatin analogues on gastric secretion. *J. Physiol. 277*: 1—14, 1978.
133. Meyers, C., Arimura, A., Gordin, A., Fernandez-Durango, R., Coy, D. H., Schally, A. V., Drouin, J., Ferland, L., Beaulieu, M., and Labrie, F.: Somatostatin analogs which inhibit glucagon and growth hormone more than insulin release. *Biochem. Biophys. Res. Commun. 74*: 630—636, 1977.
134. Brown, M., Rivier, J., and Vale, W.: Somatostatin: Analogs with selected biological activities. *Science 196*: 1467—1469, 1977.

Rosalyn S. Yalow

ROSALYN S. YALOW

I was born on July 19, 1921 in New York City and have always resided and worked there except for 3 1/2 years when I was a graduate student at the University of Illinois.

Perhaps the earliest memories I have are of being a stubborn, determined child. Through the years my mother has told me that it was fortunate that I chose to do acceptable things, for if I had chosen otherwise no one could have deflected me from my path.

My mother, née Clara Zipper, came to America from Germany at the age of four. My father, Simon Sussman, was born on the Lower East Side of New York, the Melting Pot for Eastern European immigrants. Neither had the advantage of a high school education but there was never a doubt that their two children would make it through college. I was an early reader, reading even before kindergarten, and since we did not have books in my home, my older brother, Alexander, was responsible for our trip every week to the Public Library to exchange books already read for new ones to be read.

By seventh grade I was committed to mathematics. A great chemistry teacher at Walton High School, Mr. Mondzak, excited my interest in chemistry, but when I went to Hunter, the college for women in New York City's college system (now the City University of New York), my interest was diverted to physics especially by Professors Herbert N. Otis and Duane Roller. In the late '30's when I was in college, physics, and in particular nuclear physics, was the most exciting field in the world. It seemed as if every major experiment brought a Nobel Prize. Eve Curie had just published the biography of her mother, Madame Marie Curie, which should be a must on the reading list of every young aspiring female scientist. As a Junior at college, I was hanging from the rafters in Room 301 of Pupin Laboratories (a physics lecture room at Columbia University) when Enrico Fermi gave a colloquium in January 1939 on the newly discovered nuclear fission—which has resulted not only in the terror and threat of nuclear warfare but also in the ready availability of radioisotopes for medical investigation and in hosts of other peaceful applications.

I was excited about achieving a career in physics. My family, being more practical, thought the most desirable position for me would be as an elementary school teacher. Furthermore, it seemed most unlikely that good graduate schools would accept and offer financial support for a woman in physics. However my physics professors encouraged me and I persisted. As I entered the last half of my senior year at Hunter in September 1940 I was offered what seemed like a good opportunity. Since I could type, another of my physics professors, Dr. Jerrold Zacharias, now at Massachusetts Institute of Technology, obtained a part time position for me as a secretary to Dr. Rudolf Schoenheimer, a leading biochemist at Columbia University's College of Physicians and Surgeons (P&S). This position was supposed to provide an entrée for me into graduate courses, via the backdoor, but I had to agree to take stenography. On my graduation from Hunter in January 1941, I went to business school. Fortunately I did not stay there too long. In mid-February I received an offer of a teaching assistantship in physics at the University of Illinois, the most prestigious of the schools to which I had applied. It was an achievement beyond belief. I tore up my stenography books, stayed on as secretary until June and during the summer took two tuition-free physics courses under government auspices at New York University.

In September I went to Champaign-Urbana, the home of the University of Illinois. At the first meeting of the Faculty of the College of Engineering I discovered I was the only woman among its 400 members. The Dean of the Faculty congratulated me on my achievement and told me I was the first woman there since 1917. It is evident that the draft of young men into the armed forces, even prior to American entry into the World War, had made possible my entrance into graduate school.

On the first day of graduate school I met Aaron Yalow, who was also beginning graduate study in physics at Illinois and who in 1943 was to become my husband. The first year was not easy. From junior high school through Hunter College, I had never had boys in my classes, except for a thermodynamics course which I took at City College at night and the two summer courses at NYU. Hunter had offered a physics major for the first time in September 1940 when I was an upper senior. As a result my course work in physics had been minimal for a major—less than that of the other first year graduate students. Therefore at Illinois I sat in on two undergraduate courses without credit, took three graduate courses and was a half-time assistant teaching the freshman course in physics. Like nearly all first-year teaching assistants, I had never taught before—but unlike the others I also undertook to observe in the classroom of a young instructor with an excellent reputation so that I could learn how it should be done.

It was a busy time. I was delighted to receive a straight A in two of the courses, an A in the lecture half of the course in Optics and an A- in its laboratory. The Chairman of the Physics Department, looking at this record, could only say "That A- confirms that women do not do well at laboratory work". But I was no longer a stubborn, determined child, but rather a stubborn, determined graduate student. The hard work and subtle discrimination were of no moment.

Pearl Harbor on December 7, 1941 brought our country into the war. The Physics Department was becoming decimated by loss of junior and senior faculty to secret scientific work elsewhere. The campus was filled with young Army and Navy students sent to the campus by their respective Services for training. There was a heavy teaching load, graduate courses, an experimental thesis requiring long hours in the laboratory, marriage in 1943, war-time housekeeping with its shortages and rationing, and in January 1945 a Ph.D. in Nuclear Physics. My thesis director was Dr. Maurice Goldhaber, later to become Director of Brookhaven National Laboratories. Support and encouragement came from the Goldhabers. Dr. Gertrude Goldhaber, his wife, was a distinguished physicist in her own right, but with no University position because of nepotism rules. Since my research was in nuclear physics I became skilled in making and using apparatus for the measurement of radioactive substances. The war was continuing. I returned to New York without my husband in January 1945 since completion of his thesis was delayed and I accepted a position as assistant engineer at Federal Telecommunications Laboratory, a research laboratory for ITT—the only woman engineer. When the research group in which I was working left New York in 1946, I returned to Hunter College to teach physics, not to women but to returning veterans in a pre-engineering program.

My husband had come to New York in September 1945. We established our home in an apartment in Manhattan, then in a small house in the Bronx. It and a full-time teaching position at Hunter were hardly enough to occupy my time fully. By this time my husband was in Medical Physics at Montefiore Hospital in the Bronx. Through him I met Dr. Edith Quimby, a leading medical physicist at P&S. I volunteered to work in her laboratory to gain research experience in the medical applications of radioisotopes. She took me to see "The Chief", Dr. G. Failla, Dean of American medical physicists. After talking to me for a while, he picked up the phone, dialed, and I heard him say "Bernie, if you want to set up a radioisotope service, I have someone here you must hire." Dr. Bernard Roswit, Chief of the Radiotherapy Service at the Bronx Veterans Administration Hospital and I appeared to have no choice; Dr. Failla had spoken.

I joined the Bronx VA as a part time consultant in December 1947, keeping my position at Hunter until the Spring Semester of 1950. During those years while I was teaching full-time, I equipped and developed the Radioisotope Service and started research projects together with Dr. Roswit and other physicians in the hospital in a number of clinical fields. Though we started with nothing more than a janitor's closet and a small grant to Dr. Roswit from a veterans' group, eight publications in different areas of clinical investigation resulted from this early work. The VA wisely made a commitment to set up Radioisotope Services in several of its hospitals around the country because of its appreciation that this was a new field in which research had to proceed pari passu with clinical application. Our hospital Radioisotope Service was one of the first supported under this plan.

In January 1950 I chose to leave teaching and join the VA full time. That Spring when he was completing his residency in internal medicine at the Bronx VA, Dr. Solomon A. Berson and I met and in July he joined our Service. Thus was to begin a 22 year partnership that lasted until the day of his death, April 11, 1972. Unfortunately, he did not survive to share the Nobel Prize with me as he would have had he lived.

During that period Aaron and I had two children, Benjamin and Elanna.

We bought a house in Riverdale, less than a mile from the VA. With sleep-in help until our son was 9, and part-time help of decreasing time thereafter, we managed to keep the house going and took pride in our growing children: Benjamin, now 25, is a systems programmer at the CUNY Computer Center; Elanna, now 23, is a third year doctoral candidate in Educational Psychology at Stanford University. She has just married Daniel Webb and is with us on part of her honeymoon.

But to return to the scientific aspects of my life, after Sol joined our Service, I soon gave up collaborative work with others and concentrated on our joint researches. Our first investigations together were in the application of radioisotopes in blood volume determination, clinical diagnosis of thyroid diseases and the kinetics of iodine metabolism. We extended these techniques to studies of the distribution of globin, which had been suggested for use as a plasma expander, and of serum proteins. It seemed obvious to apply these methods to smaller peptides, i.e., the hormones. Insulin was the hormone most readily available in a highly purified form. We soon deduced from the retarded rate of disappearance of insulin from the circulation of insulin-treated subjects that all these patients develop antibodies to the animal insulins. In studying the reaction of insulin with antibodies, we appreciated that we had developed a tool with the potential for measuring circulating insulin. It took several more years of work to transform the concept into the reality of its practical application to the measurement of plasma insulin in man. Thus the era of radioimmunoassay (RIA) can be said to have begun in 1959. RIA is now used to measure hundreds of substances of biologic interest in thousands of laboratories in our country and abroad, even in scientifically less advanced lands.

It is of interest from this brief history that neither Sol nor I had the advantage of specialized post-doctoral training in investigation. We learned from and disciplined each other and were probably each other's severest critic. I had the good fortune to learn medicine not in a formal medical school but directly from a master of physiology, anatomy and clinical medicine. This training was essential if I were to use my scientific background in areas in which I had no formal education.

Sol's leaving the laboratory in 1968 to assume the Chairmanship of the Department of Medicine at the Mount Sinai School of Medicine and his premature death 4 years later were a great loss to investigative medicine. At my request the laboratory which we shared has been designated the Solomon A. Berson Research Laboratory so that his name will continue to be on my papers as long as I publish and so that his contributions to our Service will be memoralized. At present my major collaborator is a young, talented physician, Dr. Eugene Straus, who joined me in 1972, first as a Fellow, then as Research Associate and now as Clinical Investigator.

Through the years Sol and I together, and now I alone, have enjoyed the time spent with the "professional children", the young investigators who trained in our laboratory and who are now scattered throughout the world, many of whom are now leaders in clinical and investigative medicine. In the training in my laboratory the emphasis has been not only in learning our research techniques but also our philosophy. I have never aspired to have, nor do I now want, a laboratory or a cadre of investigators-in-training which is more extensive than I can personally interact with and supervise.

The laboratory since its inception has been supported solely by the Veterans Administration Medical Research Program and I acknowledge with gratitude its confidence in me and its encouragement through the years. My hospital is now affiliated with The Mount Sinai School of Medicine where I hold the title of Distinguished Service Professor. I am a member of the National Academy of Sciences. Honors which I have received include, among others: Albert Lasker Basic Medical Research Award; A. Cressy Morrison Award in Natural Sciences of the N.Y. Academy of Sciences; Scientific Achievement Award of the American Medical Association; Koch Award of the Endocrine Society; Gairdner Foundation International Award; American College of Physicians Award for distinguished contributions in science as related to medicine; Eli Lilly Award of the American Diabetes Association; First William S. Middleton Medical Research Award of the VA and five honorary doctorates.

RELEVANT SELECTED BIBLIOGRAPHY

ROSALYN S. YALOW

1. Berson, S. A., Yalow, R. S., Bauman, A., Rothschild, M. A. and Newerly, K.: Insulin-I^{131} metabolism in human subjects: Demonstration of insulin binding globulin in the circulation of insulin-treated subjects. J. Clin. Invest. 35: 170, 1956.
2. Berson, S. A. and Yalow, R. S.: Quantitative aspects of reaction between insulin and insulin-binding antibody. J. Clin. Invest. 38: 1996, 1959.
3. Yalow, R. S. and Berson, S. A.: Assay of plasma insulin in human subjects by immunological methods. Nature 184: 1648, 1959.
4. Yalow, R. S. and Berson, S. A.: Immunoassay of endogenous plasma insulin in man. J. Clin. Invest. 39: 1157, 1960.
5. Yalow, R. S. and Berson, S. A.: Plasma insulin concentration in non-diabetic and early diabetic subjects. Diabetes 9: 254, 1960.
6. Roth, J., Glick, S. M., Yalow, R. S. and Berson, S. A. Hypoglycemia: A potent stimulus to secretion of growth hormone. Science 140: 987, 1963.
7. Yalow, R. S., Glick, S. M., Roth, J. and Berson, S. A.: Radioimmunoassay of human plasma ACTH. J. Clin. Endocrinol. Metab. 24: 1219, 1964.
8. Berson, S. A. and Yalow, R. S.: Parathyroid hormone in plasma in adenomatous hyperparathyroid, uremia, and bronchogenic carcinoma. Science 154: 907, 1966.
9. Walsh, J. H., Yalow, R. S. and Berson, S. A.: Detection of Australia antigen and antibody by means of radioimmunoassay techniques. J. Inf. Dis. 121: 550, 1970.
10. Yalow, R. S. and Berson, S. A.: Size and charge distinctions between endogenous human plasma gastrin in peripheral blood and heptadecapeptide gastrins. Gastroenterology 58: 609, 1970.
11. Yalow, R. S. and Berson, S. A.: Characteristics of "Big ACTH" in human plasma and pituitary extracts. J. Clin. Endocrinol. Metab. 36: 415, 1973.
12. Silverman, R. and Yalow, R. S.: Heterogeneity of parathyroid hormone: Clinical and physiologic implications. J. Clin. Invest. 52: 1958, 1973.
13. Yalow, R. S. and Wu, N.: Additional studies on the nature of big big gastrin. Gastroenterology 65: 19, 1973.
14. Gewirtz, G. and Yalow, R. S.: Ectopic ACTH production in carcinoma of the lung. J. Clin. Invest. 53: 1022, 1974.
15. Coslovsky, R. and Yalow, R. S.: Influence of the hormonal forms of ACTH on the pattern of corticosteroid secretion. Biochem. Biophys. Res. Commun. 60: 1351, 1974.
16. Yalow, R. S., Hall, K. and Luft, R.: Radioimmunoassay of Somatomedin B: Application to clinical and physiologic studies. J. Clin. Invest. 55: 127, 1975.
17. Yalow, R. S. and Gross, L.: Radioimmunoassay for intact Gross mouse leukemia virus. Proc. Natl. Acad. Sci. USA 73: 2847, 1976.

RADIOIMMUNOASSAY:

A Probe for Fine Structure of Biologic Systems

Nobel Lecture, 8 December, 1977

by

ROSALYN S. YALOW

Veterans Administration Hospital, Bronx, New York, U.S.A. and

The Mount Sinai School of Medicine, City University of New York, New York, U.S.A.

To primitive man the sky was wonderful, mysterious and awesome but he could not even dream of what was within the golden disk or silver points of light so far beyond his reach. The telescope, the spectroscope, the radio-telescope — all the tools and paraphernalia of modern science have acted as detailed probes to enable man to discover, to analyze and hence better to understand the inner contents and fine structure of these celestial objects.

Man himself is a mysterious object and the tools to probe his physiologic nature and function have developed only slowly through the millenia. Becquerel, the Curies and the Joliot-Curies with their discovery of natural and artificial radioactivity and Hevesy, who pioneered in the applicaton of radioisotopes to the study of chemical processes, were the scientific progenitors of my career. For the past 30 years I have been committed to the development and application of radioisotopic methodology to analyze the fine structure of biologic systems.

From 1950 until his untimely death in 1972, Dr. Solomon Berson was joined with me in this scientific adventure and together we gave birth to and nurtured through its infancy radioimmunoassay, a powerful tool for determination of virtually any substance of biologic interest. Would that he were here to share this moment.

Radioimmunoassay came into being not by directed design but more as a fall-out from our investigations into what might be considered an unrelated study. Dr. I. Arthur Mirsky had hypothesized that maturity-onset diabetes might not be due to a deficiency of insulin secretion but rather to abnormally rapid degradation of insulin by hepatic insulinase (1). To test this hypothesis we studied the metabolism of ^{131}I-labeled insulin following intravenous administration to non-diabetic and diabetic subjects (2). We observed that radioactive insulin disappeared more slowly from the plasma of patients who had received insulin, either for the treatment of diabetes or as shock therapy for schizophrenia, than from the plasma of subjects never treated with insulin (Fig. 1). We suspected that the retarded rate of insulin disappearance was due to binding of labeled insulin to antibodies which had developed in response to administration of exogenous insulin. However classic immunologic techniques were not adequate for the detection of antibodies which we presumed were likely to be of such low concentration as to be non-precipitating. We therefore introduced radioisotopic methods of high sensitivity

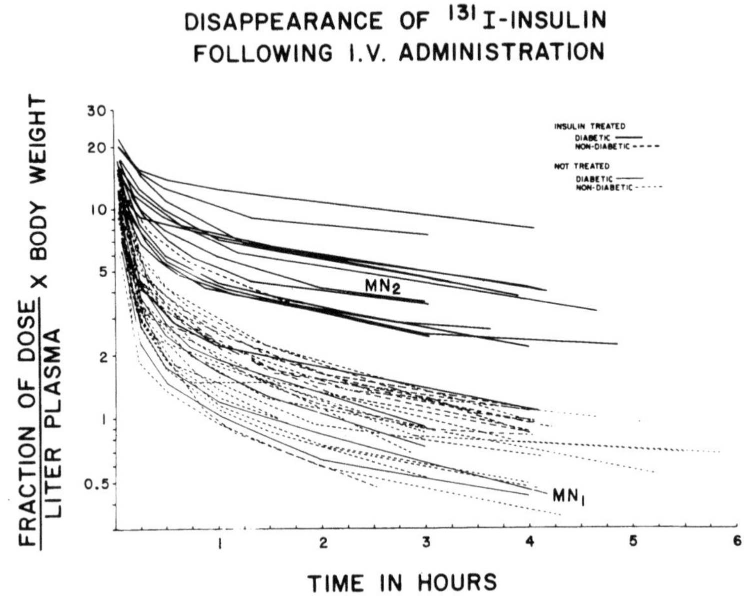

Fig. 1. Trichloractic acid precipitable radioactivity in plasma as a function of time following intravenous administration of ^{131}I-insulin to insulin-treated and non-insulin-treated subjects. The disappearance was retarded in the insulin-treated subjects irrespective of whether they had received the hormone for treatment of diabetes or for shock therapy for schizophrenia. The retarded rate is a consequence of binding to insulin antibodies generated in response to administration of animal insulins. Note the slower disappearance from the plasma of MN after 4 months of insulin therapy (curve MN_2) than prior to such therapy (curve MN_1). (Data reproduced from Ref. 2.)

for detection of soluble antigen-antibody complexes. Shown in Fig. 2 are the electrophoresis patterns of labeled insulin in the plasma of controls and insulin-treated subjects. In the insulin-treated patients the labeled insulin is bound to and migrates with an inter beta-gamma globulin. Using a variety of such systems we were able to demonstrate the ubiquitious presence of insulin-binding antibodies in insulin-treated subjects (2). This concept was not acceptable to the immunologists of the mid 1950's. The original paper describing these findings was rejected by Science and initially rejected by the Journal of Clinical Investigation (Fig. 3). A compromise with the editors eventually resulted in acceptance of the paper, but only after we omitted "insulin antibody" from the title and documented our conclusion that the binding globulin was indeed an antibody by showing how it met the definition of antibody given in a standard textbook of bacteriology and immunity (3). Our use of radioisotopic techniques for studying the primary reaction of antigen with antibody and analyzing soluble complexes initiated a revolution in theoretical immunology in that it is now generally appreciated that peptides as small as vasopressin and oxytocin are antigenic in some species and that the equilibrium constants for the antigen-antibody reaction can be as great as 10^{14} liters per mole, a value up to 10^{8} greater than the highest value predicted by Pauling's theory of 1940 (quoted in 4).

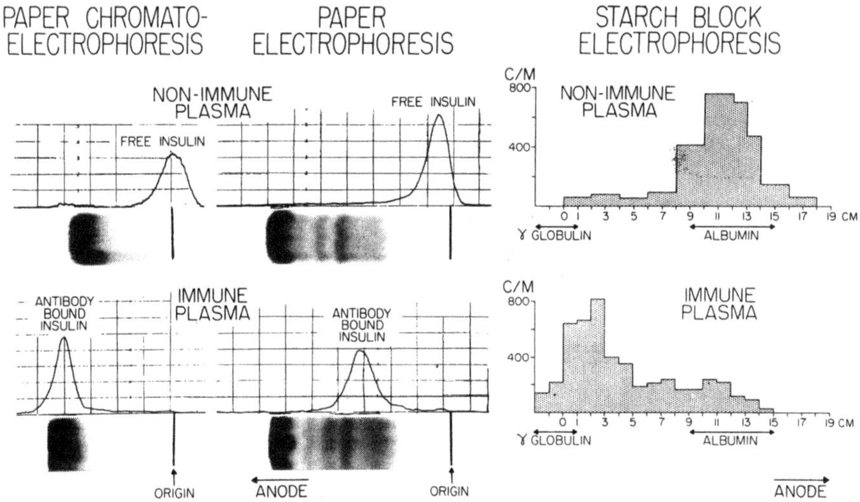

Fig. 2. ^{131}I-insulin was added to the plasmas of insulin-treated (bottom) and untreated (top) human subjects and the mixtures were applied to a starch block (right) or to paper strips (middle) for electrophoresis or to paper strips for hydrodynamic flow chromatography combined with electrophoresis (left). After completion of electrophoresis, segments were cut out of the starch block for assay of radioactivity and the paper strips were assayed in an automatic strip counter. The zones of migration of albumin and γ-globulin were identified on the starch block by running samples containing ^{131}I-albumin and ^{131}I-γ-globulin on the same block. (Starch block reproduced from ref. 2; paper strips reproduced from Berson and Yalow, 1962, Ciba Found. Colloq. Endocrinol. 14, 182—201.)

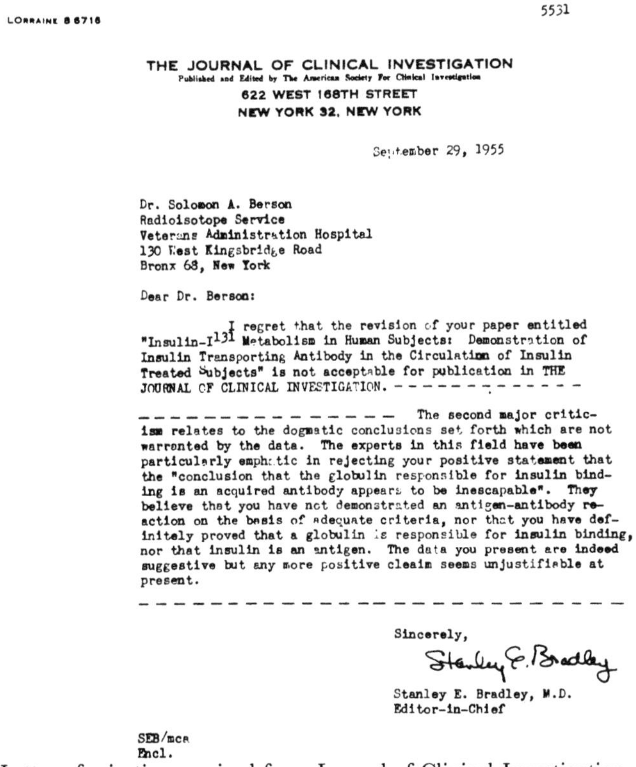
LORRAINE 8 6716

5531

THE JOURNAL OF CLINICAL INVESTIGATION
Published and Edited by The American Society For Clinical Investigation
622 WEST 168TH STREET
NEW YORK 32, NEW YORK

September 29, 1955

Dr. Solomon A. Berson
Radioisotope Service
Veterans Administration Hospital
130 West Kingsbridge Road
Bronx 68, New York

Dear Dr. Berson:

I regret that the revision of your paper entitled "Insulin-I^{131} Metabolism in Human Subjects: Demonstration of Insulin Transporting Antibody in the Circulation of Insulin Treated Subjects" is not acceptable for publication in THE JOURNAL OF CLINICAL INVESTIGATION. - - - - - - - - - - - - - -

- - - - - - - - - - - - - - - The second major criticism relates to the dogmatic conclusions set forth which are not warranted by the data. The experts in this field have been particularly emphatic in rejecting your positive statement that the "conclusion that the globulin responsible for insulin binding is an acquired antibody appears to be inescapable". They believe that you have not demonstrated an antigen-antibody reaction on the basis of adequate criteria, nor that you have definitely proved that a globulin is responsible for insulin binding, nor that insulin is an antigen. The data you present are indeed suggestive but any more positive claim seems unjustifiable at present.

- -

Sincerely,

Stanley E. Bradley

Stanley E. Bradley, M.D.
Editor-in-Chief

SEB/mca
Encl.

Fig. 3. Letter of rejection received from Journal of Clinical Investigation.

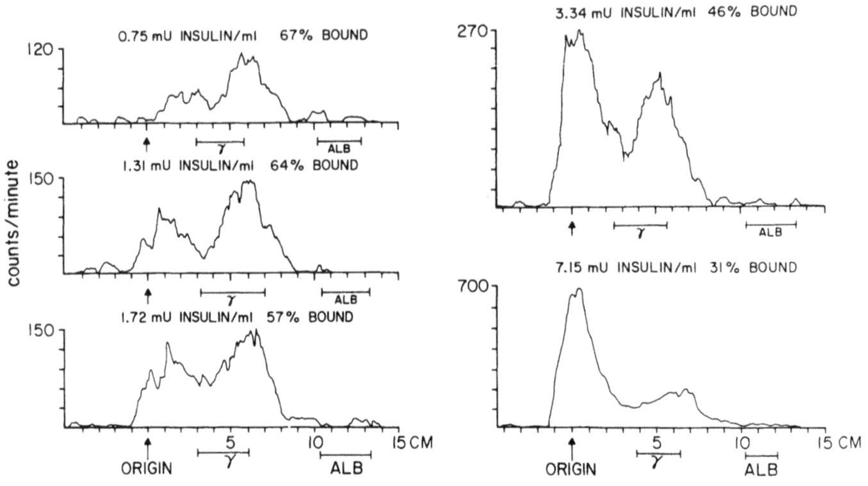

Fig. 4. Paper electrophoretograms showing the distribution of ^{131}I-insulin between that bound to antibody (migrating with serum protein) and that free (remaining at site of application) in the presence of increasing concentrations of labeled insulin. The antibodies were from an insulin-treated human subject. (Data reproduced from Ref. 2.)

In this paper we also reported that the binding of labeled insulin to a fixed concentration of antibody is a quantitative function of the amount of insulin present (Fig. 4). This observation provided the basis (5) for the radioimmunoassay of plasma insulin. However investigations and analysis which lasted for several years and which included studies on the quantitative aspects of the reaction between insulin and antibody (6) and the species specificity of the available antisera (7) were required to translate the theoretical concepts of radioimmunoassay into the experiments which led first to the measurement of plasma insulin in rabbits following exogenous insulin administration (8) and finally in 1959 to the measurement of insulin in unextracted human plasma (9).

Radioimmunoassay (RIA) is simple in principle. It is summarized in the competing reactions shown in Fig. 5. The concentration of the unknown unlabeled antigen is obtained by comparing its inhibitory effect on the binding of radioactively labeled antigen to specific antibody with the inhibitory effect

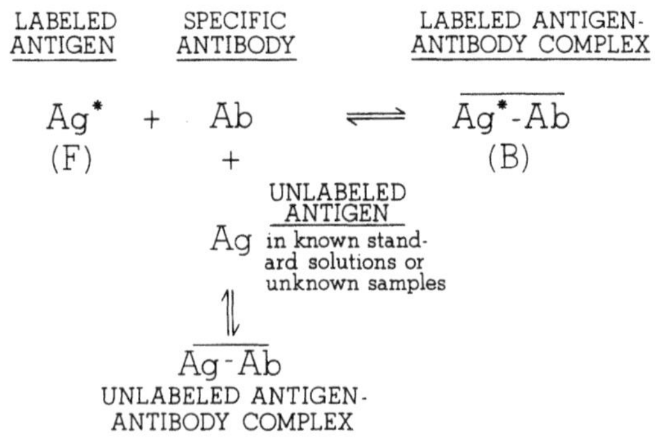

Fig. 5. Competing reactions that form the basis of radioimmunoassay (RIA).

of known standards (Fig. 6). The sensitivity of RIA is remarkable. As little as 0.1 pg gastrin/ml of incubation mixture, i.e., 0.05 picomolar gastrin, is readily measurable. RIA is not an isotope dilution technique, with which it has been confused, since there is no requirement for identical immunologic or biologic behavior of labeled and unlabeled antigen. The validity of RIA is dependent on identical immunologic behavior of antigen in unknown samples with the antigen in known standards. The specificity of immunologic reactions can permit ready distinction, for instance, between corticosterone and cortisol, steroids which differ only in the absence of or presence of respectively a single hydroxyl residue. There is no requirement for standards and unknowns to be identical chemically or to have identical biologic behavior. Furthermore it has been demonstrated that at least some assays can be clinically useful, even though they cannot be properly validated due to lack of immunologic identity between standards and the sample whose concentration is to be determined.

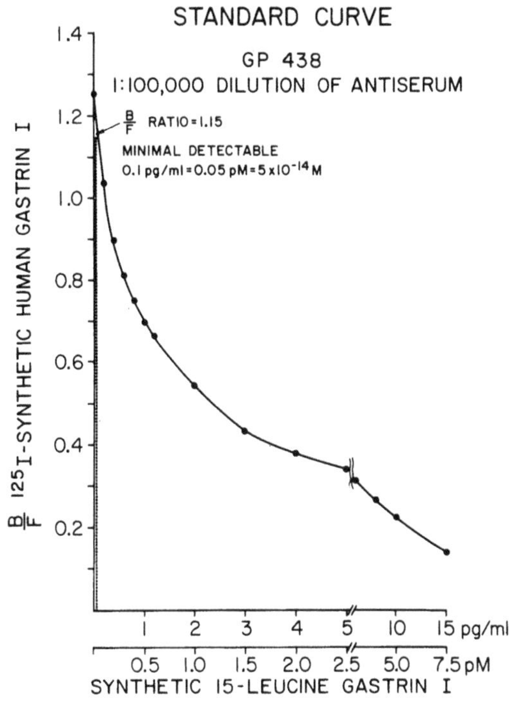

Fig. 6. Standard curve for the detection of gastrin by RIA. Note that as little as 0.2 pg gastrin/ml incubation mixture (0.1 picomolar) is readily detectable.

The RIA principle is not limited to immune systems but can be extended to other systems in which in place of the specific antibody there is a specific reactor or binding substance. This might be a specific binding protein in plasma, a specific enzyme or a tissue receptor site. Herbert and associates (10, 11) first demonstrated the applicability of competitive radioassay to the measurement of vitamin B_{12} in a liver receptor assay using ^{60}Co-vitamin B_{12} and intrinsic factor as the binding substance. However it remained for Rothenberg in our laboratory (12) and Ekins (13) to develop assays for serum vitamin B_{12} using this principle. Ekins (14) and later Murphy (15) employed thyroxine binding globulin as the specific reactor for the measurement of serum thyroxine.

It is not necessary that a radioactive atom be the "marker" used to label the antigen or other substance which binds to the specific reactor. Recently there has been considerable interest in employing as "markers" enzymes which are covalently bound to the antigen. Although many variations of competitive assay have been described, RIA has remained the method of choice and is likely to remain so at least in those assays which require high sensitivity. The receptor site assays for the peptide hormones have the presumed advantage of measuring biologic activity but are generally at least 10-to 100-fold less sensitive than RIA. Enzyme marker assays have several disadvantages; the most important is that the steric hindrance introduced into the antigen-antibody reaction because of the presence of the enzyme molecule almost inevitably decreases the sensitivity of the assay.

Two decades ago, when bioassay procedures were in the forefront, the first presentation on the potential of hormonal measurements by radioimmunoassay (16) went virtually unnoticed. Somewhat more interest was generated by the demonstration in 1959 of the practical application of radioimmunoassay to the measurement of plasma insulin in man (9). It became evident that the sensitivity and simplicity of radioimmunoassay permitted ready assay of hundreds of plasma samples, each as small as a fraction of a milliliter, and made possible measurement not only of single blood samples, as had been performed on occasion with in vivo bioassay, but also of multiple samples, thus permitting study of dynamic alterations in circulating insulin levels in response to physiologic stimuli (9, 17). Nonetheless in the early 60's the rate of growth of radioimmunoassay was quite slow. Only an occasional paper other than those from our laboratory appeared in prominent American journals of endocrinology and diabetes before 1965 (Fig. 7). But by the late 60's RIA had become a major tool in endocrine laboratories and more recently it has expanded beyond the research laboratory into the nuclear medicine and clinical

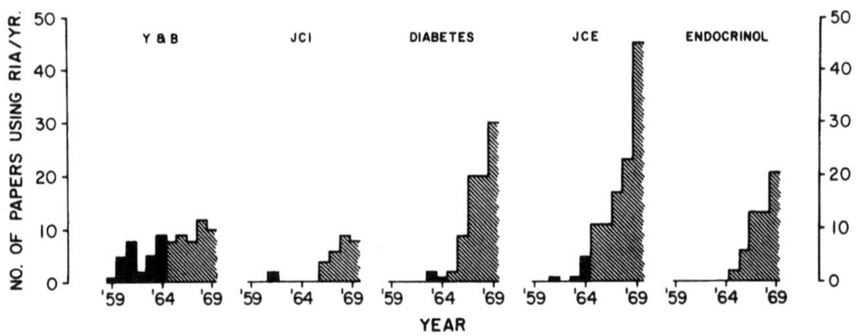

Fig. 7. Number of papers using radioimmunoassay published by Yalow and Berson (Y and B, left) and by all others in American journals of endocrinology and diabetes through 1969. Papers before 1965 are shown in black; 1965 and later are cross-hatched. (JCI, Journal of Clinical Investigation; JCE, Journal of Clinical Endocrinology; Endocrinol, Endocrinology.)

laboratories. It has been estimated (18) that in 1975, in the United States alone, over 4000 hospital and non-hospital clinical laboratories performed radioimmunoassays of all kinds, almost double the number of a year or two earlier and the rate of increase appears not to have diminished in the past two years. The technical simplicity of RIA and the ease with which the reagents may be obtained have enabled its extensive use even in scientifically under-developed nations.

The explosive growth of RIA has derived from its general applicability to many diverse areas in biomedical investigation and clinical diagnosis. A representative but incomplete listing of substances measured by RIA is given in Figure 8.

SUBSTANCES MEASURED BY RADIOIMMUNOASSAY

PEPTIDAL HORMONES

PITUITARY HORMONES
- Growth hormone
- Adrenocorticotropic hormone (ACTH)
- Melanocyte stimulating hormone (MSH)
  - α-MSH
  - β-MSH
- Glycoproteins
  - Thyroid stimulating hormone (TSH)
  - Follicle stimulating hormone (FSH)
  - Luteinizing hormone (LH)
- Prolactin
- Lipotropin
- Vasopressin
- Oxytocin

CHORIONIC HORMONES
- Human chorionic gonadotropin (HCG)
- Human chorionic somatomammotropin (HCS)

PANCREATIC HORMONES
- Insulin
- Glucagon
- Pancreatic Polypeptide

CALCITROPIC HORMONES
- Parathyroid hormone (PTH)
- Calcitonin (CT)

GASTROINTESTINAL HORMONES
- Gastrin
- Secretin
- Cholecystokinin (CCK)
- Vasoactive intestinal polypeptide (VIP)
- Gastric inhibitory polypeptide (GIP)

VASOACTIVE TISSUE HORMONES
- Angiotensins
- Bradykinins

RELEASING AND RELEASE INHIBITING FACTORS
- Thyrotropin releasing factor (TRF)
- LHRF
- Somatostatin

OTHER PEPTIDES
- Substance P
- Endorphins
- Enkephalins

NON-PEPTIDAL HORMONES

THYROIDAL HORMONES
- Thyroxine (T_4)
- Triiodothyronine (T_3)
- Reverse T_3

STEROIDS
- Aldosterone
- Corticosteroids
- Estrogens
- Androgens
- Progesterones

PROSTAGLANDINS

BIOLOGIC AMINES
- Serotonin
- Melatonin

NON-HORMONAL SUBSTANCES

DRUGS & VITAMINS
- Cardiac glycosides
- Drugs of Abuse
- Psychoactive Drugs
- Antibiotics
- CNS Depressants
- Vitamin A, Folic acid

CYCLIC NUCLEOTIDES

ENZYMES
- C_1 esterase
- Fructose 1, 6 diphosphatase
- Plasminogen, Plasmin
- Chymotrypsin, Trypsin
- Carbonic anhydrase isoenzymes
- Aldose reductase
- Carboxypeptidase B
- Pancreatic elastase

VIRUSES
- Hepatitis associated antigen
- Murine Leukemia viruses (Gross, Rauscher, Moloney)
- Mason-Pfizer monkey virus

TUMOR ANTIGENS
- Carcinoembryonic antigen
- α-Fetoprotein

SERUM PROTEINS
- Thyroxine binding globulin
- IgG, IgE, IgA, IgM
- Properdin
- Fibrinogen
- Apolipoprotein B
- Myoglobin
- Myelin Basic Protein

OTHER
- Intrinsic factor
- Rheumatoid factor
- Hageman factor
- Neurophysins
- Staphylococcal β-Enterotoxin

Fig. 8. Partial listing of peptidal and non-peptidal hormones and other substances measured by radioimmunoassay.

The exquisite sensitivity, specificity and comparative ease of RIA especially now that instrumentation and reagents are so readily and universally available, have permitted assay of biologically significant materials where measurements were otherwise difficult or impossible. Only if we can detect and measure can we begin really to understand, and herein lies the major contribution of RIA as a probe for insight into the function and perturbations of the fine structure of biologic systems.

For the first decade following the development of RIA and its first application to the measurement of plasma insulin in man, primary emphasis was given to its importance in endocrinology. The ability to measure in the presence of billion-fold higher concentrations of plasma proteins the minute concentrations (10^{-10} to 10^{-12} M) of peptide hormones in plasma with the high specificity characteristic of immunologic reactions has provided greatly increased accuracy of diagnosis of pathologic states which are characterized by hormonal excess or deficiency. It has provided virtually all the information now known about the regulation of hormonal secretion and the interrelationships among hormones and has contributed greatly to our understanding of the mechanisms of hormonal release and of hormonal physiology in general. More recently, as perhaps will be discussed by Drs. Guillemin and Schally, it has been applied to investigations of the potential role of the hypothalamic releasing and release inhibiting factors; studies which have been made easier by RIA of the hormones they control as well as of the factors themselves. Over the past few years, RIA has had an important role in the discovery of new forms of hormones in blood and in tissue. These include the larger hormonal forms — proinsulin (19), big gastrin (20—22), proparathyroid hormone (23, 24), big ACTH (25, 26), etc., and the hormonal fragments — the biologically inactive COOH-terminal parathyroid hormone fragment (27, 28) among others. These studies have generated new insights concerning the biosynthesis of the peptide hormones.

Let us now consider some examples from our laboratory of older and of newer diverse applications of RIA. Proper interpretation of plasma hormone levels in clinical diagnosis requires a clear understanding of the factors involved in the regulation of hormonal secretion. Generally, such secretion is stimulated by some departure from the state of biologic "homeostasis" that the hormone is designed to modulate. A representative model for one such system is shown in Fig. 9. Regulation is effected through the operation of a feed-back control loop which contains the hormone at one terminus and, at the other, the substance which it regulates and by which it is in turn regulated. Gastrin secretion increases gastric acidity, which then suppresses secretion of antral gastrin. Modulation of this system can be effected by a number of

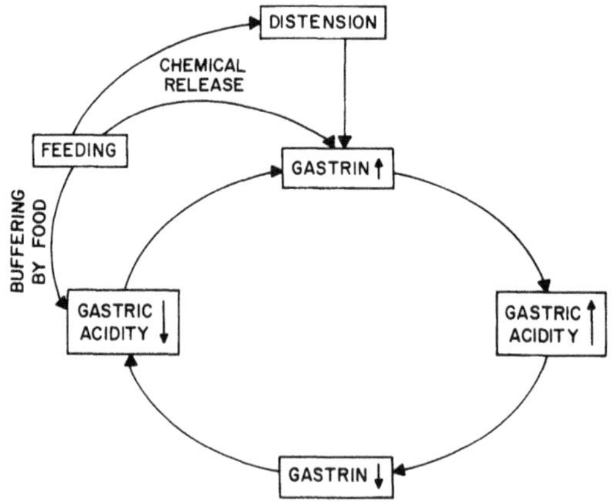

Fig. 9. Feed-back control loop for gastrin regulation of gastric acidity: effect of feeding.

factors, perhaps the most important of which is feeding. Feeding promotes gastrin release directly through a chemical effect on antral cells and indirectly through gastric distension and through the buffering action of food (Fig. 9).

In Figure 10 are compared basal gastrin concentrations in patients with pernicious anemia (PA), in patients with Zollinger-Ellison syndrome (ZE) and in a group of patients we have diagnosed as having non-tumorous hypergastrinemic hyperchlorhydria (NT-HH) (29—31). Gastrin levels are generally considerably higher in each of the three groups than the 0.1 ng/ml considered to be the upper limit for normal subjects. However the reasons are different. Patients with PA have gastric hypoacidity. Since gastric hydrochloric acid normally suppresses gastrin secretion, the continued absence of acid and the repeated stimulation by feeding eventually produces secondary hyperplasia of gastrin-producing cells. The high level of plasma gastrin in PA is quite appropriate in view of the absence of the inhibitory effect of HCl on the secretion of antral gastrin.

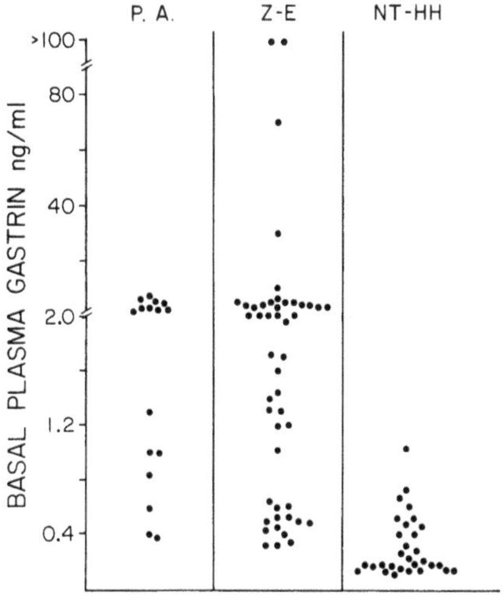

Fig. 10. Basal plasma gastrin concentrations in gastrin hypersecretors, i.e., patients with pernicious anemia (PA), Zollinger-Ellison syndrome (ZE) and non-tumorous hypergastrinemic hyperchlorhydria (NT-HH). Most control subjects without known gastrointestinal disease have basal levels less than 0.1 ng/ml. (Data reproduced in part from Ref. 29—31.)

The elevated values in ZE and NT-HH are inappropriate since these patients have marked hyperacidity and the feed-back mechanisms which should suppress gastrin secretion have failed. How does one distinguish between patients with a gastrin-secreting tumor (ZE) and those whose inappropriate gastrin secretion appears to be due to overactivity of the gastrin-secreting cells of the gastrointestinal tract (NT-HH)? Accurate diagnostic differentiation between these diseases is essential because procedures appropriate for their treatment are so markedly different that diagnostic error might be fatal. Some ZE patients have levels higher than those ever achieved by the non-tumorous group. However in the region of overlap the distinction between them is readily made on the basis of responsiveness to various provocative agents. Patients with ZE respond to a calcium challenge (2 mg Ca^{++}/kg intravenously) or to a secretin challenge (4 U/kg intravenously) with a dramatic increase in plasma gastrin but fail to respond to a test meal; for patients with NT-HH the reverse is true (Fig. 11).

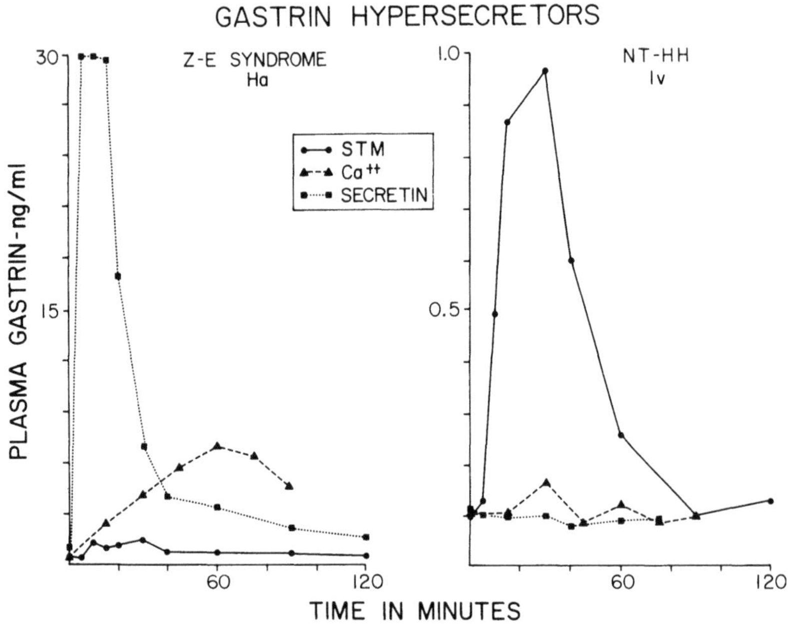

Fig. 11. Plasma gastrin concentrations in the fasting state and in response to three provocative stimuli in gastrin hypersecretors; patient Ha (left) has ZE; subject Iv (right) is in the non-tumorous (NT-HH) group. (Reproduced from Ref. 31.)

Thus, in the application of radioimmunoassay to problems of hypo- or hypersecretion we should seldom rely on a single determination of plasma hormone. Generally, to test for deficiency states, plasma hormonal concentrations should be measured not only in the basal state but also following administration of appropriate physiologic or pharmacologic stimuli. When hypersecretion is suspected and high hormonal concentrations are observed, one must determine whether the level is appropriate or inappropriate and whether the hormonal secretion is autonomous or can be modulated by appropriate physiologic or pharmacologic agents.

Studies such as these are now common in endocrinology and would not have been possible without RIA.

During the past decade our concepts of the chemical nature of peptide hormones and their modes of synthesis have changed dramatically. This change is due in large part to observations based on RIA which have demonstrated that many, if not all, peptide hormones are found in more than one form in plasma and in glandular and other tissue extracts. These forms may or may not have biologic activity and may represent either precursor(s) or metabolic products(s) of the well-known, well-characterized, biologically active hormone. Their existence has certainly introduced complications into the interpretation of hormonal concentration as measured by RIA, and as measured by bioassay as well. A typical example of the work in this area is the current interest in the heterogeneity of gastrin.

Investigations concerning the possible heterogeneity of gastrin were stimulated by considerations in comparative endocrinology, in that the immunochemical heterogeneity of parathyroid hormone (27) and the demonstration of a precursor form for insulin, proinsulin, (19) spurred the search for heterogeneous forms of gastrin as soon as a radioimmunoassay for gastrin (29) had been developed with sufficient sensitivity to permit fractionation of plasma in a variety of physicochemical systems and assay of the immunoreactivity in the various fractions.

Several analytical methods were used to elucidate the nature of plasma gastrin. Quite unexpectedly it appeared that the major component of immunoreactive gastrin in the fasting state of patients with hypergastrinemia was a peptide clearly different from heptadecapeptide gastrin (HG), a 17 amino acid peptide that had earlier been purified and sequenced by Gregory and Tracy (32, 33). The newly discovered peptide eluted between insulin and proinsulin on Sephadex G50 gel filtration, in contrast with HG which eluted after insulin (Fig. 12). This peptide had an electrophoretic mobility on starch gel just greater than serum albumin, which is about half that of HG (20, 21). Characterization in other physical chemical systems helped verify that this

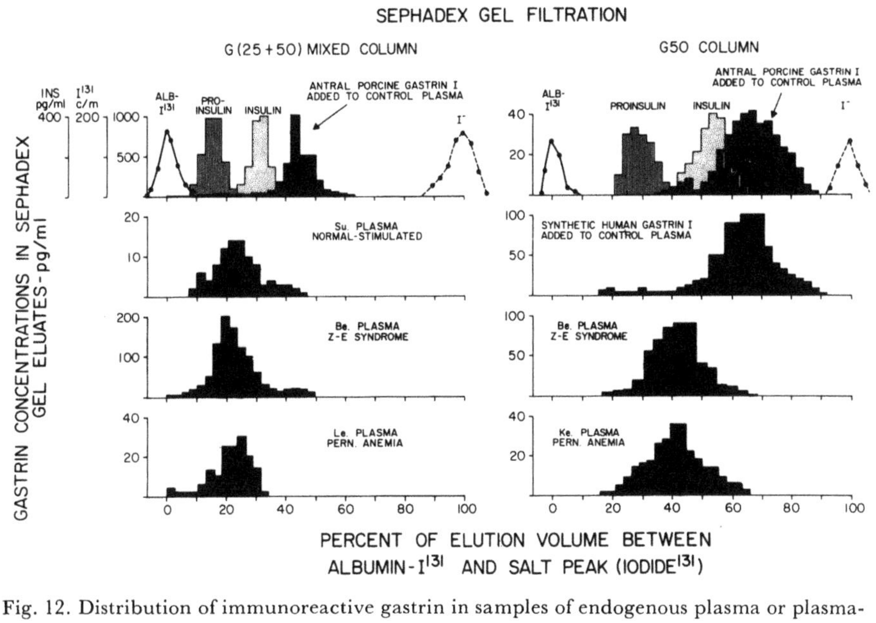

Fig. 12. Distribution of immunoreactive gastrin in samples of endogenous plasma or plasma-gastrin mixtures added to columns of Sephadex G-50 (right), or mixtures of G-50 and G-25 (left) for gel filtration. The zones of elution of the marker molecules are shown in the top frames. (Reproduced from Ref. 20.)

peptide was indeed larger and more basic than HG. In advance of its further characterization we called this new form big gastrin (BG). Both gastrins were found in extracts of a ZE tumor as well as in extracts of the antrum and proximal small bowel (22). We further demonstrated that HG could be generated by tryptic digestion of BG, with no significant change of total immunoreactivity. We predicted that the larger form was composed of the smaller form linked at its amino terminal end to a lysine or arginine residue of another peptide (21).

Our predictions based on measurement of picogram to nanogram amounts of immunoreactive gastrin in the presence of billion-fold higher concentrations of other proteins stimulated Gregory and Tracy to purify and chemically characterize this material. Soon thereafter they succeeded in isolating, both from a ZE tumor and from hog antral extracts, pairs of gastrin peptides with physico-chemical behavior similar to that we had described for BG (34, 35). They then demonstrated that BG is a 34 amino acid peptide with two lysine residues adjacent to the N-terminal residue of HG (35).

Unlike proinsulin which is virtually devoid of biologic activity (36), the in vivo administration of immunochemically identical amounts of BG and HG resulted in the same integrated acid output in a dog (reported in 21). However, the turnover time for BG is prolonged 3 to 5 fold longer than for HG (37, 38). Therefore, following administration of equivalent doses the plasma levels of BG are approximately 3 to 5 fold greater than that of HG. It is evident that the observed heterogeneity introduces complications into bioassay, as well as into immunoassay, since biologic activity as defined by the traditional dose-response method is certainly different from that defined by plasma concentration-response data in the case of the gastrins or any other groups of biologically active related peptides with different turnover times.

Our discovery of the immunochemical heterogeneity of parathyroid hormone (27) and Steiner's discovery of proinsulin (19) just over a decade ago initiated a revolution in concepts of biosynthesis of the peptide hormones. The original suggestion that a major function of proinsulin in biosynthesis was to facilitate disulfide bond formation (39) could not prognosticate that virtually all peptidal hormones, including those which consist simply of a linear peptide chain, appear also to have larger precursor forms. In many, but not all, peptide hormones the smaller peptide is joined into the larger form by two basic residues (gastrin, insulin, etc.). A notable exception to this rule is cholecystokinin (CCK). In the case of CCK and its COOH-terminal octapeptide (CCK-8), both of which are biologically active, cleavage to the smaller form occurs at the COOH-terminal side of a single arginine residue (40). As is discussed below, both forms are found in the tissues of origin.

At present, a decade after the concept of heterogeneity was developed and in spite of an enormous body of descriptive data in this field, we still do not know very much about the rules or reasons for this precursor-product synthetic scheme. Is the synthesis of the peptide hormones in a form in which they are linked to another peptide essential only for their proper storage or release or is some other mechanism involved? What are the enzymes involved in the conversion process? Are the converting enzymes hormone specific or species specific? Is conversion effected only in the secreting tissue, or is there peripheral conversion from inactive to active form? What is the role of the part of the precursor molecule which is discarded after biosynthesis? Finding the answers to these and related questions will keep many of us busy for quite a while.

Since investigations concerned with the brain peptides as well as RIA have enjoyed prominence this year it is relevant to combine the two and discuss some applications of RIA to the understanding of peptides in the brain. Much interest has been generated recently in the finding that several peptides are common to the brain and the gastrointestinal tract. A determination of the location and concentration of these peptides has usually depended on immunologic techniques. The finding by Vanderhaeghen et al (41) of a new peptide in the vertebrate central nervous system that reacts with antibodies against gastrin has been confirmed by Dockray (42), who suggested that the brain peptide resembled cholecystokinin (CCK)-like peptides more closely than it did gastrin-like peptides. We extended these studies and demonstrated that the peptides in the brain are not from the gastrin family or simply CCK-like, but are in fact intact cholecystokinin (CCK) and its COOH-terminal octapeptide (CCK-8) (43—45). These observations depended on the use of two antisera with different immunochemical specificities. One was prepared in a goat by immunization with porcine CCK (pCCK). For all practical purposes this antiserum does not crossreact with CCK-8 or the gastrins, big or little, in spite of their sharing a common COOH-terminal pentapeptide. The other antiserum was prepared by immunization of a rabbit (Rabbit B) with the COOH-terminal gastrin tetrapeptide amide. With this antiserum the crossreactivities of pCCK and of CCK-8 are virtually identical on a molar basis. Using the Rabbit B antiserum, we have observed that in all animal species studied the immunoreactive content as measured in the CCK-8 assay was about five-fold greater in gut extracts than in brain extracts (Table 1). However, the concentrations in the gut and brain extracts were comparable among the different species and did not change significantly on tryptic digestion (Table 1).

Sephadex gel filtration and assay in the CCK-8 system of the brain and gut extracts of the pig, dog and monkey generally revealed two peaks of comparable size, one with an elution volume resembling that of CCK and the other with an elution volume like CCK-8 (Fig. 13). A minor void volume peak whose significance has not yet been determined was also generally observed. Although there was no change in immunoreactivity following prolonged tryptic digestion (Table 1) there was complete conversion of all immunoreactivity to a peptide resembling CCK-8 (Fig. 13).

In the same monkey and dog extracts in which CCK-like material was present in about the same concentration as in the pig extracts we failed to detect immunoreactivity with the anti-pCCK serum (Table 1). The extracts of the gut and brain of the pig contained comparable molar amounts of CCK when measured with either antiserum (Table 1). The failure to detect intact CCK in dog and monkey brain and gut extracts, which were proven to have

Table 1. Immunoreactive content of brain and gut extracts

SPECIES	ORGAN	Goat 1 ASSAY μg pCCK equivalent/ml		Rabbit B ASSAY μg CCK-8 equivalent/ml	
		BEFORE TRYPSIN	AFTER TRYPSIN	BEFORE TRYPSIN	AFTER TRYPSIN
Pig (2)*	Brain	$0.80 \pm 0.05^{+}$	ND	0.20 ± 0.01	0.15 ± 0.01
	Gut	1.80 ± 0.1	ND	0.60 ± 0.05	0.50 ± 0.03
Monkey (1)	Brain	ND	ND	0.05 ± 0.01	0.05 ± 0.02
	Gut	ND	ND	0.40 ± 0.05	0.35 ± 0.05
Dog (2)	Brain	ND	ND	0.10 ± 0.01	0.10 ± 0.01
	Gut	ND	ND	0.70 ± 0.02	0.70 ± 0.05

( )* no of animals
ND = not detected
$^{+}$ mean $\pm$ standard error of the mean of multiple assays

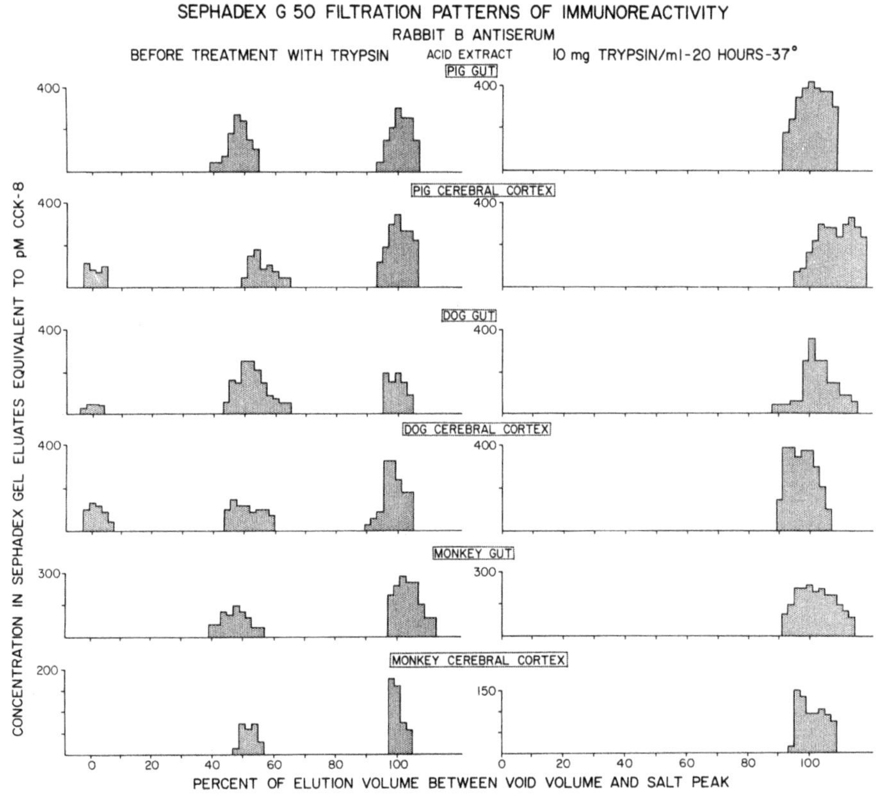

Fig. 13. Immunoreactivity in eluates following Sephadex G50 gel filtration was determined using an antiserum which reacts identically on a molar basis with intact porcine cholecystokinin (pCCK) and its COOH-terminal octapeptide (CCK-8). Purified pCCK has an elution volume midway between the void volume and the salt peak and CCK-8 coelutes with the salt peak in this system. Shown are the patterns for pig, dog and monkey brain and gut extracts before (left) and after (right) prolonged tryptic digestion. Note the complete conversion to a CCK-8-like peptide with no loss in immunoreactivity (Data reproduced from Ref. 45.)

this hormone when measured in the CCK-8 assay, forms the basis for our prediction based on RIA that there are major differences between pig and the other animal cholecystokinins in the amino terminal portion of the molecule. Since this portion of the molecule is not directly involved in its biologic action, it is not surprising that the amino acid sequences in this region of the molecule have diverged during the course of evolution. As yet the amino acid sequences of CCK from animals other than a pig have not been reported. Just as our predictions based on RIA stimulated Gregory and Tracy to purify and chemically characterize big gastrin, we hope our predictions of the nature of the amino terminal portion of CCK will encourage chemical verification by others.

Where in the brain is CCK found? Its concentration is highest in the cerebral cortex (43). Our immunohistochemical studies (Fig. 14) suggest that CCK-8, at least, appears to be concentrated in the cortical neurons (44).

The finding of peptides resembling CCK and CCK-8 in the central nervous system raises intriguing questions about their physiologic function particularly with respect to their potential roles as satiety factors. The observation of Gibbs et al (46, 47) that injection of purified CCK or CCK-8 evoked satiety, although pentagastrin and secretin did not, has suggested negative feedback from the gastrointestinal tract as the causative mechanism. These studies of

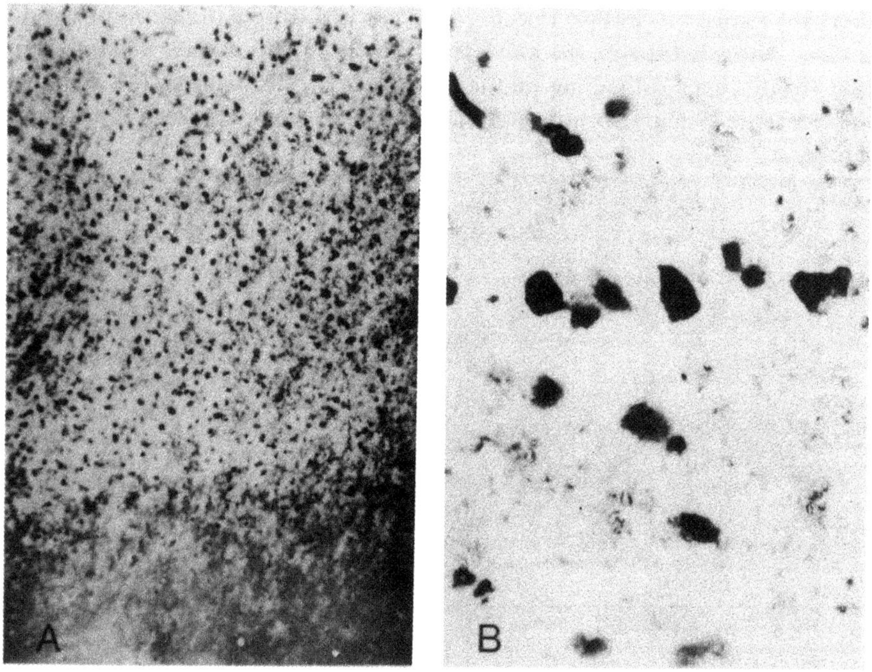

Fig. 14. Left: Low-power photomicrograph of rabbit cerebral cortex (frontal lobe). The tissue was stained by the immunoperoxidase technique using rabbit B antiserum in a 1 : 10 dilution. Staining of individual cell bodies can be seen in all layers of cortical grey matter and diffuse staining can be seen at the bottom in subcortical white matter (X33). Right: Higher-power photomicrograph showing staining of cell bodies in cortical grey matter. (X208) (Data reproduced from Ref. 44.)

Gibbs et al (46, 47) confirm the earlier work of Schally et al (48) who had shown that enterogastrone, a gut extract undoubtedly rich in CCK, inhibited eating by fasted mice. The finding that CCK peptides appear to be endogenous in the brain suggests a more direct role for them as neuroregulators.

It is now commonly accepted that there are a group of peptides such as somatostatin (49, 50), substance P (51), vasoactive intestinal peptide (52) and cholecystokinin or its C-terminal octapeptide (43—45) which are found both in the gastrointestinal tract and in the central nervous system. Some evidence has also been presented that peptide hormones such as β lipotropin, ACTH and peptides structurally related to them, initially thought to be of pituitary origin, are found widely distributed in the brain in extrahypothalamic regions (53—56). We had considered the possibility that the finding of a pituitary hormone, such as ACTH in the brain of the rat might be a consequence of the small dimensions of its brain. Therefore, we recently undertook to study the distribution of ACTH in the brains not only of rodents such as the rat and rabbit but also of animals with large brains such as the dog, monkey and man (57). We observed that the dimensions within which ACTH is found is about the same for all five of these species but that the particular anatomical regions which contain ACTH depend on the brain size (57). Thus, ACTH is widely distributed in the brain of the rat, but is found in the brain only in the hypothalamic regions of primates (Table 2) (Fig. 15). Since there is no reason to assume that the synthetic mechanism is different in small brained animals than in the primates, we believe that these studies suggest that the pituitary is likely to be the sole site of synthesis of ACTH and that the hormone is found in other cranial sites due to mechanisms other than synthesis.

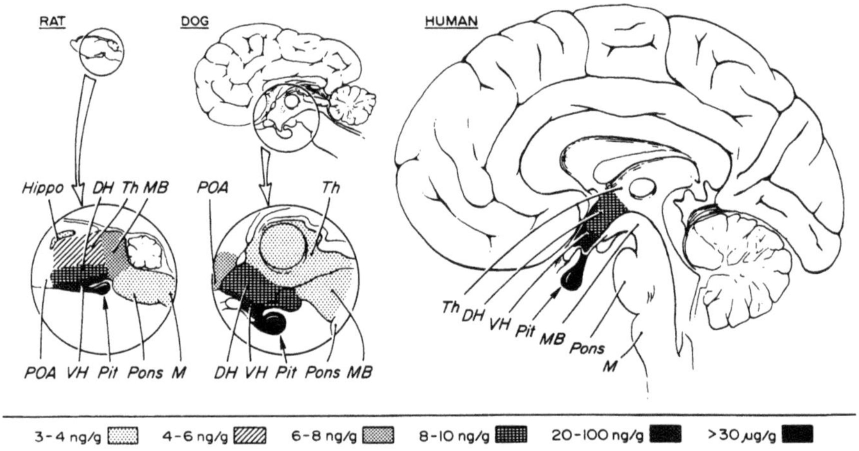

Fig. 15. Distribution of immunoreactive ACTH in the pituitary and brain of several animal species. The rat, dog and human brains are drawn to scale. The regions shown in circles have been enlarged to show in better detail the concentrations of ACTH in the brain of the rat and dog. ACTH was not detectable in regions shown in white. Abbreviations: Pit, pituitary gland; VH, ventral hypothalamus; DH, dorsal hypothalamus; POA, preoptic area; MB, midbrain; M, medulla; TH, thalamus; Hippo, hippocampus. (Reproduced from Ref. 57.)

Table 2. Regional distribution of ACTH in brains of several mammalian species ACTH Concentration (ng/g wet weight)

Brain Region	Human	Monkey			Dog			Rabbit			Rat Pool
		Craniot.	Sacrifice	Autopsy	1	2	3*	P	G	J*	n = 6
Hypothalamus								100	100	100	14
Ventral	33	—	76	10	28	26	17	—	—	—	40
Dorsal	10	4	4	3	—	5	11	—	—	—	10
Thalamus	ND†	ND	ND	ND	2	4	3	20	20	12	4
Preoptic Area	ND	ND	ND	ND	—	8	7	—	—	7	5
Amygdala	ND	ND	ND	ND	3	4	4	6	5	6	3
Hippocampus	ND	ND	ND	ND	ND	ND	ND	ND	ND	ND	3
Striatum	ND	ND	ND	ND	ND	ND	ND	ND	ND	ND	ND
Midbrain	ND	ND	ND	ND	4	3	2	20	21	15	7
Pons	ND	ND	ND	ND	ND	ND	ND	10	8	10	4
Medulla	ND	ND	ND	ND	ND	ND	ND	3	6	10	3
Cerebral Cortex	ND	ND	ND	ND	ND	ND	ND	ND	ND	ND	ND
Cerebellum	ND	ND	ND	ND	ND	ND	ND	ND	ND	ND	ND

* Dissected frozen
† ND — not detectable, <1 ng/g

The presence of pituitary hormones in the brains of commercially prepared hypophysectomized rats has been taken as evidence for de novo synthesis of pituitary hormones in the brain (54—56, 58). We also have observed that in these animals residual pituitary tissue is rarely detected upon visual inspection of the sella (57). Nonetheless although there is an immediate decrease in stress-stimulated ACTH release in hypophysectomized rats, after two months the plasma ACTH concentrations can be stress-stimulated to about 80% of the level found in intact rats (Fig. 16). It would appear therefore that visual

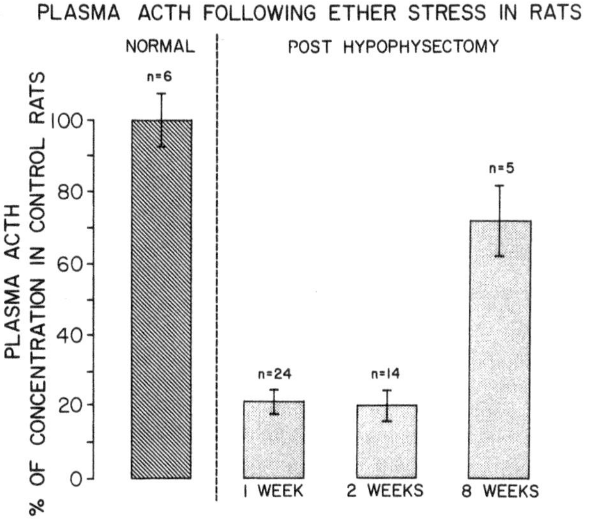

Fig. 16. Plasma ACTH following ether stress in control and hypophysectomized rats. (Reproduced from Ref. 57.)

inspection of the sella is not sufficient to insure that the hypophysectomy has been total. Scrapings from the sella have been shown to contain ACTH amounting to almost 5% that of the normal pituitary (57). This represents a considerable residual source of ACTH since the hypothalamic ACTH content is only about a fraction of a percent of that of the pituitary. Thus, even in these "hypophysectomized" rats we believe that residual pituitary fragments are the source of the brain ACTH (57).

At the present state of our knowledge, we consider it most likely that hormones known to be synthesized in the pituitary are synthesized only there and are transported to the brain by one or more mechanisms; perhaps by retrograde flow along the portal vessels or by leakage into the basal cistern. In addition, there is another group of peptides common to, and likely to have been synthesized in, the gastrointestinal tract and the central nervous system. We leave to others to determine where in this schema is the source of the enkephalins.

The examples chosen come from a sampling of studies in endocrinology since my Nobel citation specifically deals with the application of RIA in this subspecialty. Nonetheless RIA is rapidly growing beyond the borders of endocrinology, its first home.

RIA has already added a completely new dimension to the identification and measurement of pharmacologically active substances in plasma and tissue— and the list of compounds for which such assays are available is growing rapidly (Fig. 8). In general since the molar concentrations of drugs at pharmacologic levels are high compared, for instance, to the concentration of the peptide hormones in body fluids, achieving adequate sensitivity is not likely to be difficult. However, the requirements for the specificity of RIA of drugs merit some consideration. Structurally related compounds or metabolites may have significant immunoreactivity with some antisera but not with others and may or may not constitute a problem, depending on the purpose of the assay. For instance, if the clinical problem relates to the toxicity of a particular drug, then the question as to whether or not the assay measures only the biologically active form is relevant. If the question relates simply to whether or not a drug had been taken surreptitiously, then the reactivity of metabolites or variation of the immunoreactivity with the exact form of the drug may be irrelevant.

The application of RIA to the measurement of enzymes is a field of increasing interest. The very great sensitivity of RIA permits measurement of enzyme levels much lower than that possible by the usual catalytic methods. It permits direct assay of the enzyme rather than only its effects and is not influenced by inhibitors or activators of enzyme systems or variations in substrates. That in the same system one can with RIA measure both an enzyme and its proenzyme and other inactive forms has both advantages and disadvantages, depending on the problem under investigation. It must be appreciated that many enzymes may be species specific and biologic activity need not parallel immunologic activity. At present, RIA seems likely to complement rather than to replace catalytic methods for enzymatic analysis.

There is another field in which the potential application of RIA is in its infancy. My crystal ball—or intuition—tells me that in the '80's the impact of RIA on the study of infectious diseases may prove as revolutionary as its impact on endocrinology in the '60's. A start has already been made in virology. RIA of hepatitis-associated-antigen (59, 60) has become the method of choice for testing for infected blood in Red Cross and other blood banks in the United States where transfusion-transmitted hepatitis has been a significant public health problem. The recent description of a RIA for intact murine leukemia virus (61) with sufficient sensitivity to detect virus in 0.5 μl of blood or of tissue extracts from animals with viral induced or spontaneous leukemia gives us a tool with which we may be able to determine where and in what concentration a virus resides during the period from infection to the time when the fully developed pathologic manifestations of the disease are present. Recently we have developed a sensitive and specific RIA for some constituent of tuberculin purified protein derivative (PPD) (62) which is shed into culture medium in vitro or in vivo by growing Mycobacterium tuberculosis. We have already reported (62) earlier detection of growth of tubercle bacilli in culture medium than is possible by other means and we envision its applicability to rapid and early detection of bacterial growth in biologic fluids. We anticipate that this preliminary work will lead the way to the more extensive use of RIA in bacteriology.

Infectious diseases have become less prominent as causes of death and disability in regions of improved sanitation and adequate supplies of antibiotics. Nonetheless they remain a major public health problem throughout the world and simple inexpensive methods of identifying carriers of disease would facilitate eradication of these diseases. RIA is likely to provide those methods and one can anticipate its fuller exploitation in this virtually untapped field.

The first telescope opened the heavens; the first microscope opened the world of the microbes; radioisotopic methodology, as exemplified by RIA, has shown the potential for opening new vistas in science and medicine.

REFERENCES

1. Mirsky, I. A. 1952. "The Etiology of Diabetes Mellitus in Man." Recent Progr. Horm. Res. 7, 437.
2. Berson, S. A., R. S. Yalow, A. Bauman, M. A. Rothschild and K. Newerly. 1956. "Insulin-I^{131} Metabolism in Human Subjects: Demonstration of Insulin Binding Globulin in the Circulation of Insulin-Treated Subjects." J. Clin. Invest. 35, 170—190.
3. Topley, W. W. C. and G. S. Wilson. 1941. The Principles of Bacteriology and Immunity. Williams and Wilkins Co.
4. Day, E. D. 1966. Foundations of Immunochemistry. Williams and Wilkins Co.
5. Berson, S. A. and R. S. Yalow. 1957. "Kinetics of Reaction Between Insulin and Insulin-Binding Antibody." J. Clin. Invest. 36, 873.
6. Berson, S. A. and R. S. Yalow. 1959. "Quantitative Aspects of Reaction Between Insulin and Insulin-Binding Antibody." J. Clin. Invest. 38, 1996—2016.
7. Berson, S. A. and R. S. Yalow. 1959. "Species-Specificity of Human Anti-Beef Pork Insulin Serum." J. Clin. Invest. 38, 2017—2025.
8. Berson, S. A. and R. S. Yalow. 1958. "Isotopic tracers in the study of diabetes". Advances in Biological and Medical Physics. Academic Press. pp. 349—430.
9. Yalow, R. S. and S. A. Berson. 1959. "Assay of Plasma Insulin in Human Subjects by Immunological Methods." Nature 184, 1648—1649.
10. Herbert, V. 1959. "Studies on the Role of Intrinsic Factor in Vitamin B_{12} Absorption, Transport, and Storage." Am. J. Clin. Nutr. 7, 433—443.
11. Herbert, V., Z. Castro and L. R. Wasserman. 1960. "Stoichiometric Relation Between Liver-Receptor, Intrinsic Factor and Vitamin B_{12}." Proc. Soc. Exp. Biol. Med. 104, 160—164.
12. Rothenberg, S. P. 1961. "Assay of Serum Vitamin B_{12} Concentration Using Co^{57}-B_{12} and Intrinsic Factor." Proc. Soc. Exp. Biol. Med. 108, 45—48.
13. Barakat, R. S. and R. P. Ekins. 1961. "Assay of Vitamin B_{12} in Blood." Lancet 2, 25—26.
14. Ekins, R. P. 1960. "The Estimation of Thyroxine in Human Plasma by an Electrophoretic Technique." Clin. Chim. Acta 5, 453—459.
15. Murphy, B. E. P. 1964. "Application of the Property of Protein-Binding to the Assay of Minute Quantities of Hormones and Other Substances." Nature (Lond.) 201, 679—682.
16. Berson, S. A. 1957. Resume of Conference on Insulin Activity in Blood and Tissue Fluids. Editors: R. Levine and E. Anderson. National Institutes of Health, Bethesda, Maryland. p. 7.
17. Yalow, R. S. and S. A. Berson. 1960. "Immunoassay of Endogenous Plasma Insulin in Man." J. Clin. Invest. 39, 1157—1175.
18. Zucker, B. 1976. Laboratory Management. The Medical Div. of the United Business Publications, Inc. pp. 35—38.
19. Steiner, D. F., D. Cunningham, L. Spigelman and B. Aten. 1967. "Insulin Biosynthesis: Evidence for a Precursor." Science 157, 697.
20. Yalow, R. S. and S. A. Berson. 1970. "Size and Charge Distinctions Between Endogenous Human Plasma Gastrin in Peripheral Blood and Heptadecapeptide Gastrins." Gastroenterology 58, 609—615.
21. Yalow, R. S. and S. A. Berson. 1971. "Further Studies on the Nature of Immunoreactive Gastrin in Human Plasma." Gastroenterology 60, 203—214.
22. Berson, S. A. and R. S. Yalow. 1971. "Nature of Immunoreactive Gastrin Extracted from Tissues of Gastrointestinal Tract." Gastroenterology 60, 215—222.
23. Kemper, B., J. F. Habener, J. T. Potts, Jr. and A. Rich. 1972. "Proparathyroid Hormone: Identification of a Biosynthetic Precursor to Parathyroid Hormone." Proc. Nat. Acad. Sci. 69, 643—647.
24. Cohn, D. V., R. R. MacGregor, L. L. H. Chu, J. R. Kimmel and J. W. Hamilton. 1972. "Calcemic Fraction-A: Biosynthetic Peptide Precursor of Parathyroid Hormone." Proc. Nat. Acad. Sci. 69, 1521—1525.

25. Yalow, R. S. and S. A. Berson. 1971. "Size Heterogeneity of Immunoreactive Human ACTH in Plasma and in Extracts of Pituitary Glands and ACTH-Producing Thymoma." Biochem. Biophys. Res. Commun. 44, 439—445.
26. Yalow, R. S. and S. A. Berson. 1973. "Characteristics of 'big ACTH' in Human Plasma and Pituitary Extracts." J. Clin. Endocrinol. Metab. 36, 415—423.
27. Berson, S. A. and R. S. Yalow. 1968. "Immunochemical Heterogeneity of Parathyroid Hormone in Plasma." J. Clin. Endocrinol. Metab. 28, 1037—1047.
28. Silverman, R. and R. S. Yalow. 1973. "Heterogeneity of Parathyroid Hormone: Clinical and Physiologic Implications." J. Clin. Invest. 52, 1958—1971.
29. Yalow, R. S. and S. A. Berson. 1970. "Radioimmunoassay of Gastrin." Gastroenterology 58, 1—14.
30. Berson, S. A., J. H. Walsh and R. S. Yalow. 1973. Frontiers in Gastrointestinal Hormone Research. Almqvist & Wiksell, Stockholm. pp. 57—66.
31. Straus, E. and R. S. Yalow. 1975. Gastrointestinal Hormones. Editor: J. C. Thompson. Univ. Texas Press, Austin. pp. 99—113.
32. Gregory, R. A. and H. J. Tracy. 1964. "The Constitution and Properties of Two Gastrins Extracted from Hog Antral Mucosa: I. The Isolation of Two Gastrins from Hog Antral Mucosa." Gut 5, 103—114.
33. Gregory, R. A., H. J. Tracy and M. I. Grossman. 1966. "Isolation of Two Gastrins from Human Antral Mucosa." Nature 209, 583.
34. Gregory, R. A. and H. J. Tracy. 1972. "Isolation of Two 'Big Gastrins' from Zollinger-Ellison Tumour Tissue." Lancet 2, 797—799.
35. Gregory, R. A. and H. J. Tracy. 1973. "Big Gastrin." Mt. Sinai J. Med. 40, 359—364.
36. Lazarus, N. R., J. E. Panhos, T. Tanese, L. Michaels, R. Gutman and L. Recant. 1970. "Studies on the Biological Activity of Porcine Proinsulin." J. Clin. Invest. 49, 487.
37. Straus, E. and R. S. Yalow. 1974. "Studies on the Distribution and Degradation of Heptadecapeptide, Big, and Big Big Gastrin." Gastroenterology 66, 936—943.
38. Walsh, J. H., H. T. Debas and M. I. Grossman. 1974. "Pure Human Big Gastrin: Immunochemical Properties, Disappearance Half-Time, and Acid-Stimulating Action in Dogs." J. Clin. Invest. 54, 477—485.
39. Steiner, D. F., J. L. Clark, C. Nolan, A. H. Rubenstein, E. Margoliash, F. Melani and P. E. Oyer. 1970. Pathogenesis of Diabetes Mellitus. Nobel Symposium 13. Editors: E. Cerasi and R. Luft. Almqvist & Wiksell, Stockholm, Sweden. pp. 57—78.
40. Jorpes, J. E. and V. Mutt. 1973. Methods in Investigative and Diagnostic Endocrinology, Part III — Non-Pituitary Hormones. Editors: S. A. Berson and R. S. Yalow. North-Holland Publishing Co., Amsterdam. pp. 1075—1080.
41. Vanderhaeghen, J. J., J. C. Signeau and W. Gepts. 1975. "New Peptide in the Vertebrate CNS Reacting with Antigastrin Antibodies." Nature 257, 604—605.
42. Dockray, G. J. 1976. "Immunochemical Evidence of Cholecystokinin—like Peptides in Brain." Nature 264, 568—570.
43. Müller, J. E., E. Straus and R. S. Yalow. 1977. "Cholecystokinin and Its C-terminal Octapeptide in the Pig Brain." Proc. Nat. Acad. Sci. 74, 3035—3037.
44. Straus, E., J. E. Muller, H-S. Choi, F. Paronetto and R. S. Yalow. 1977. "Immunohistochemical Localization in Rabbit Brain of a Peptide Resembling the C-terminal Cholecystokinin Octapeptide." Proc. Nat. Acad. Sci. 74, 3033—3034.
45. Straus, E. and R. S. Yalow. 1978. "Species Specificity of Cholecystokinin in Gut and Brain of Several Mammalian Species." Proc. Nat. Acad. Sci. 75, 486—489.
46. Gibbs, J., R. C. Young and G. P. Smith. 1973. "Cholecystokinin Decreases Food Intake in Rats." J. Comp. Physiol. Psychol. 84, 488—495.
47. Gibbs, J., R. C. Young and G. P. Smith. 1973. "Cholecystokinin Elicits Satiety in Rats with Open Gastrics Fistula." Natue 245, 323—325.
48. Schally, A. V., T. W. Redding, H. W. Lucien and J. Meyer. 1967. "Enterogastrone Inhibits Eating by Fasted Mice." Science 157, 210—211.
49. Brownstein, M., A. Arimura, H. Sato, A. V. Schally and J. S. Kizer. 1975. "The Regional Distribution of Somatostatin in the Rat Brain." Endocrinology 96, 1456—1461.

50. Hokfelt, T., S. Efendic, C. Hellerstrom, O. Johansson, R. Luft and A. Arimura. 1975. "Cellular Localization of Somatostatin in Endocrine-like Cells and Neurons of the Rat with Special References to the A_1-Cells of the Pancreatic Islets and to the Hypothalamus." Acta Endocrinol. 80 (Suppl. 200), 1—41.
51. Leeman, S. E., E. A. Mroz and R. E. Carraway. 1977. Peptides in Neurobiology. Editor: H. Gainer. Plenum Press, New York. pp. 99—144.
52. Bryant, M. G., J. M. Polak, I. Modlin, S. R. Bloom, R. H. Alburquerque and A. G. E. Pearse. 1976. "Possible Dual Role for Vasoactive Intestinal Peptide as Gastrointestinal Hormone and Neurotransmittal Substance." Lancet 1, 991—993.
53. Simantov, R., M. J. Kuhar, G. R. Uhl and S. H. Snyder. 1977. "Opioid Peptide Enkephalin: Immunohistochemical Mapping in Rat Central Nervous System." Proc. Nat. Acad. Sci. 74, 2167—2171.
54. Cheung, A. L. and A. Goldstein. 1976. "Failure of Hypophysectomy to Alter Brain Content of Opioid Peptides (Endorphins)." Life Sci. 19, 1005—1008.
55. Krieger, D. T., A. Liotta and M. J. Brownstein. 1977. "Presence of Corticotropin in Brain of Normal and Hypophysectomized Rats." Proc. Nat. Acad. Sci. 74, 648—652.
56. Krieger, D. T., A. Liotta and M. J. Brownstein. 1977. "Presence of Corticotropin in Limbic System of Normal and Hypophysectomized Rats." Brain Res. 128, 575—579.
57. Moldow, R. and R. S. Yalow. 1978. "Extrahypophysial Distribution of Corticotropin as a Function of Brain Size." Proc. Nat. Acad. Sci. 75, 994—998.
58. Hong, J. S., T. Yang, W. Fratta and E. Costa. 1977. "Determination of Methionine Enkephalin in Discrete Regions of Rat Brain." Brain Res. 134, 383—386.
59. Walsh, J. H., R. S. Yalow and S. A. Berson. 1970. "Radioimmunoassay of Australia Antigen." Vox Sanguinis 19, 217—224.
60. Walsh, J. H., R. S. Yalow and S. A. Berson. 1970. "Detection of Australia Antigen and Antibody by Means of Radioimmunoassay Techniques." J. Inf. Dis. 121, 550—554.
61. Yalow, R. S. and L. Gross. 1976. "Radioimmunoassay for Intact Gross Mouse Leukemia Virus." Proc. Nat. Acad. Sci. 73, 2847—2851.
62. Straus, E. and R. S. Yalow. 1977. "Radioimmunoassay for Tuberculin Purified Protein Derivative." Clin. Res. 25, A384.

1978
Physiology or Medicine

WERNER ARBER, DANIEL NATHANS and HAMILTON O. SMITH

"for the discovery of restriction enzymes and their application to problems of molecular genetics"

THE NOBEL PRIZE FOR PHYSIOLOGY OR MEDICINE

Speech by Professor PETER REICHARD of the Karolinska Medico-Chirurgical Institute
Translation from the Swedish text

Your Majesties, Your Royal Highnesses, Ladies and Gentlemen,

"Their research opens up the possibility to copy human beings in the laboratory, to construct geniuses, to massproduce workers, or to create criminals." This is a quotation from the presentation on Swedish television of this year's laureates in medicine. The presentation was not made as a joke. Let me for now, however, leave this Frankenstein-fixation of the news-media. Reality is remarkable enough, without such excursions into science fiction.

The discoveries of this year's laureates mark the beginning of a new era of genetics. Genetics started as a science more than 100 years ago with the experiments of Gregor Mendel who showed that our heritage is packaged into genes. Each gene directs a particular function and is faithfully propagated from generation to generation. The second era of genetics started about 30 years ago when Avery succeeded in transferring with DNA a hereditary property from one bacterium to another. Thus, genetics became molecular, and our concept of both genes and their functions acquired a chemical basis. We realized that the gene is a piece of DNA, and that DNA contains the genetic code for the synthesis of specific proteins. During this period many fundamental discoveries were made in molecular genetics, as witnessed by the fact that six Nobel prizes in medicine were awarded to scientists working in this field in the last 20 years.

These discoveries originated mostly in experiments with bacteria and viruses, but usually the results could be directly extrapolated to man. However, man depends on many biological processes directed by genes, which do not take place in microorganisms. So we must ask how do our genes direct the development of a single fertilized egg cell into a complete individual with many different organs? What mechanism forces the cells in one organ to retain their specialized functions? We know that disturbances in normal development give rise to diseases and malformations. During the 1950s and 60s scientists worked very hard to answer these questions but apparently knocked on a closed door. This door has now been opened wide by our laureates, whose discoveries started the third era of genetics.

The difficulties in this field of research were mainly due to the large amount of information contained in our genes and to the enormous length of the DNA molecule. We can compare the DNA of a single human cell with a book containing all the information for the development and function of the cell. The text written on one page of this book might then correspond to one gene containing all the information necessary for the synthesis of one protein. The whole book consists of 1 million pages and would occupy about 100 meters in a book shelf. The whole book is faithfully copied at each cell division. One mistake in one letter on a single page may result in disease or death. Changes in the text can be caused by the action of chemicals or viruses and this may result in cancer, malformations or hereditary diseases. The scientist wants to be able to read the book and to localize and identify any misprints. He first tries to find the correct page with the interesting text, but in doing so he realizes that the pages of the book are glued together. How can he separate the pages without destroying the text?

Restriction enzymes are the tools which make it possible to open the sealed book. *Werner Arber* discovered these enzymes in the early 1960s when he analyzed an apparently obscure phenomenon in bacteria, discovered 10 years earlier by Bertani and Weigle, called host-controlled modification. In a series of simple but elegant experiments Arber showed that this phenomenon was caused by a change in DNA and apparently served to protect the host from foreign genes. Foreign DNA is degraded, and Arber postulated that bacteria contain restriction enzymes with the capacity to recognize and bind to recurring structural elements of DNA. At these locations the DNA-helix is severed: the pages of the book are separated.

Hamilton Smith verified Arber's hypothesis. He purified one restriction enzyme and showed that it could cleave foreign DNA. He determined the chemical structure of the regions of DNA which were severed by the enzyme and discovered certain rules which later could be applied to other restriction enzymes. Today maybe 100 such enzymes are known. They all cleave DNA, each at different, defined regions. With their aid, these giant molecules can be dissected into well-defined segments which subsequently can be used for structural investigations or in genetic experiments.

The last step in this development was taken by *Dan Nathans.* He pioneered the application of restriction enzymes in genetics and his work has been a source of inspiration for scientists all over the world. He constructed the first genetic map using restriction enzymes by cleaving the DNA from a monkey virus. The methodology devised by him for this purpose was later used by others to construct increasingly more complicated maps. Today we can write the complete chemical formula for the genes of the monkey virus that Nathans started to investigate.

The application of restriction enzymes has revolutionized the genetics of higher organisms and completely changed our ideas of the organisation of their genes. In contrast to the DNA of bacteria, the DNA of higher organisms is not a contiguous structure coding for one protein. Instead, genes contain "quiet" regions alternating with regions containing the genetic code. Restriction enzymes have also been used for genetic engineering. With their aid we can selectively remove parts of the genetic material and transplant genes into a foreign background. In this way genes from higher organisms have been transferred to bacteria, and in certain cases such bacteria can be used to produce human hormones. In the near future we can expect many products of medical importance to be synthesized.

These experiments gave rise to the earlier mentioned fears of copying human beings in the laboratory. Such fears are due to a complete ignorance of the content of genetics and the nature of man. A similar misunderstanding once caused the distortion of Darwinian evolution theory to social Darwinism. I would like to illustrate this with the following example which I take from the geneticist Dobzhansky: "Whereas birds, bats and insects became fliers by evolving genetically for millions of years, man has become the most powerful flier of all, by constructing flying machines, not by reconstructing his genotype".

Dr. Arber, Dr. Nathans, Dr. Smith: The discovery of restriction enzymes started off an avalanche in molecular genetics. Their application made possible the detailed chemical analysis of the organisation of the genetic material, and this has in particular in higher organisms given unexpected but far reaching results. At long last we are in a position to tackle successfully the basic problem of cell differentiation. Your work has pioneered this development. On behalf of the Karolinska Institute I wish to convey to you our warmest congratulation and I now ask you to receive the prize from the hands of his Majesty the King.

WERNER ARBER

I was born on June 3rd, 1929 in Gränichen in the Canton of Aargau, Switzerland, where I went to the public schools until the age of 16. I then entered the gymnasium at the Kantonsschule Aarau where I got a B-type maturity in 1949. From 1949 to 1953 I studied towards the diploma in Natural Sciences at the Swiss Polytechnical School in Zurich. It is in the last year of this study that I made my first contacts with fundamental research, when working on the isolation and characterisation of a new isomer of Cl^{34}, with a halflife of 1.5 seconds.

On the recommendation of my professor in experimental physics, Paul Scherrer, I took an assistantship for electron microscopy at the Biophysics Laboratory at the University of Geneva in November 1953. This laboratory was animated by Eduard Kellenberger and it had two prototype electron microscopes requiring much attention. In spite of spending many hours to keep the microscope "Arthur" in reasonable working condition, I had enough time not only to help developing preparation techniques for biological specimens in view of their observation in the electron microscope, but also to become familiar with fundamental questions of bacteriophage physiology and genetics, which at that time was still a relatively new and unknown field. My first contribution to our journal club concerned Watson and Crick's papers on the structure of DNA.

In the 1950's the Biophysics Laboratory at the University of Geneva was lucky enough to receive each summer for several months the visit of Jean Weigle. He was the former professor of experimental physics at the University of Geneva. After having suffered a heart attack, he had left Geneva to become a researcher at the Department of Biology of the California Institute of Technology in Pasadena. There, he had been converted to a biologist under the influence of Max Delbrück and had chosen to study bacteriophage λ. This is why the first electron micrographs of phage λ were made in Geneva. Stimulated by Jean Weigle we soon turned our interests also to other properties of λ, and the study of defective λ prophage mutants became the topic of my doctoral thesis.

In the summer of 1956, we learned about experiments made by Larry Morse and Esther and Joshua Lederberg on the λ-mediated transduction (gene transfer from one bacterial strain to another by a bacteriophage serving as vector) of bacterial determinants for galactose fermentation. Since these investigators had encountered defective lysogenic strains among their transductants, we felt that such strains should be included in the collection of λ prophage mutants under study in our laboratory. Very rapidly, thanks to the stimulating help by Jean Weigle and Grete Kellenberger, this turned out to be extremely fruitful. We could indeed show that λ-mediated transduction is based on the formation of substitution mutants, which had replaced a part of the phage genes by genes from the bacterial chromosome. This made the so-called λgal phage derivatives so defective that they were not able any longer to propagate as a virus. In fact, one of the at first sight rather frustrating observation was that lysates of λgal, which indeed could still cause the infected host cell to lyse as does wild type phage λ, did not contain any structural components of λ (phage particles, heads or tails) discernible in the electron microscope. This was the end of my career as an electron microscopist and in chosing genetic and physiological approaches I became a molecular geneticist.

After my Ph. D. exam in the summer of 1958 I had the chance to receive an offer to work at the University of Southern California in Los Angeles with Joe Bertani, a former collaborator of Jean Weigle. Several years before, Bertani had isolated and characterised another bacteriophage of *E. coli*, P1. Phage P1 rapidly had become a very welcome tool of bacterial geneticists, since it gives general transduction, i.e. any particular region of the host chromosome gets at some low frequency wrapped into P1 phage particles if P1 multiplies in a cell, and this enables the geneticists to carry out linkage studies of bacterial genes. While working as a research associate with Bertani, I received P1 at first hand which enabled me to study phage P1-mediated transduction of monomeric and dimeric λ prophage genomes as well as of the fertility plasmid F.

In the meantime, my Ph. D. thesis on λgal, although written in French, had been read, or, what is perhaps more essential, understood in its conclusions by many leading microbial geneticists.

This may be the reason why I received offers to spend additional postdoctoral time in several excellent laboratories. On the other hand, I had remained in close contact with Eduard Kellenberger, and he urged me to come back to Geneva in order to lead an investigation on radiation effects on microorganisms. As a compromise, I decided to return to Geneva at the beginning of 1960, but only after having spent several very fruitful weeks at each of the laboratories of Gunther Stent in Berkeley, Joshua Lederberg in Stanford and Salvador Luria at the Massachusetts Institute of Technology, Cambridge.

At the end of the 1950's, a special credit had been voted for by the Swiss Parliament for research in atomic energy, including radiation effects on living organisms. Eduard Kellenberger felt that important contributions to the latter questions could be expected from studies with microorganisms, and he had therefore submitted a research proposal which found approval by the granting agency, the Swiss National Science Foundation. The project could bring insight into the nature of radiation damage to genetic material and its repair mechanisms, as well as of the stimulation of genetic recombination by radiation. These topics had already engaged the attention of Jean Weigle and Grete Kellenberger for a number of years.

One of the first experiments after my return to Geneva was to render *E. coli* B and its radiation resistant strain B/r sensitive to phage λ. The first step to accomplish this was easy thanks to a hint received from Esther Lederberg to look for cotransduction of the Mal^+ and λ^S characters. However, the strains thus obtained still did not allow an efficient propagation of λ. Very rapidly I realized that this was due to host-controlled modification, a phenomenon described for λ and *E. coli* strains seven years earlier by Joe Bertani and Jean Weigle. However, I was not satisfied to know how to overcome this barrier. I was also anxious to know how the restriction of phage growth and the adaptation of λ to the new host strain worked. When I started investigations on the mechanisms of host-controlled modification, I did not of course imagine that this sidetrack would keep my interest for many years. Otherwise I might not have felt justified to engage in this work because of its lack of direct relevance to radiation research. However, a lucky coincidence rapidly dissipated these concerns. At the same time, Grete Kellenberger had looked at the fate of DNA from irradiated phage λ upon infection of host bacteria: part of it was rapidly degraded after injection into the host. And so was the DNA from unirradiated phage λ used to measure adsorption and DNA injection into restrictive bacterial strains! This phenomenon became the topic of Daisy Dussoix's doctoral thesis, who very carefully not only studied the DNA degradation of phage that was not properly modified, but who also tried to detect parallels between the fate of unmodified DNA in restrictive conditions and of irradiated DNA in normal host cells.

Within about one year of study, it had become clear that strain-specific restriction and modification directly affected the DNA, without however causing mutations. It soon also became obvious that restriction and modification were properties of the bacterial strains and acted not only on infecting bacteriophage DNA, but also on cellular DNA as manifested in conjugation experiments. These findings were reported by myself and Daisy Dussoix for the first time to the scientific community during the First International Biophysics Congress held in Stockholm in the summer of 1961. In a more extended version I presented them in 1962 to the Science Faculty of the University of Geneva as my work of habilitation as privatdocent. This work earned me in the same year the Plantamour-Prévost prize of the University of Geneva.

At a time before the Swiss Universities received direct financial help from the federal government, the Swiss National Science Foundation awarded "personal grants" to qualified researchers to allow them to guide projects of fundamental research at a Swiss University. I was lucky to benefit from such a support form 1965 to 1970. These years were devoted to hard work to consolidate the preliminary data and the concepts resulting from them, and to extend the acquired notions, in particular with regard to the mechanisms of modification by nucleotide methylation, with regard to the genetic control of restriction and modification and with regard to the enzymology and molecular mechanisms of these reactions.

This work would not have been possible without a very fruitful help by a large number of collaborators in my own laboratory and of colleagues working on related topics in their own laboratories. I was extremely lucky to receive in my laboratory in the basement of the Physics Institute of the University of Geneva a number of first class graduate students, postdoctoral fellows and senior scientists. It is virtually impossible to list them all in this context, but my warmest collective thanks go to all of them. In 1964 Bill Wood laid out a solid basis for the genetics of the restiction and modification systems *Eco*K and *Eco*B. Later, Stuart Linn, profiting from his fruitful contacts with Bob Yuan and Matt Meselson, who worked in the USA on the enzymology of *Eco*K restriction, set the basis for in vitro studies with *Eco*B restriction and modification activities. These studies culminated in the final proof that modification in *E. coli* B and K is brought about by nucleotide methylation. This concept had found its first experimental evidence during my two months' visit in 1963 with Gunther Stent at the University of California in Berkeley. Several years later Urs Kühnlein, a Ph. D student, and John Smith, working for various lengths of time with us, succeeded in careful in vivo and in vitro measurements on methylation to validate and extend the earlier conclusions. Their experiments also brought important conclusions with regard to the concept of the sites of recognition on the DNA for the restriction and modification enzymes.

As an illustration that my work has not always been easy and accompanied by success, I would like to refer to my long, fruitless and thus largely unpublished attempts to find experimental evidence for the diversification of restriction and modification systems in the course of evolution. Systems *Eco*K and *Eco*B form a closely related family as judged from genetic and functional studies. Another family is formed by restriction and modification systems *Eco*P1 and *Eco*P15. One could expect that mutations affecting the part of the enzymes responsible for recognition of the specificity site on the DNA might result in new members of the family, recognizing new specificity sites on DNA. We have in vain spent much time in search for such evolutionary changes both after mutagenization and after recombination between two members of the same family of the above mentioned systems. That the basic idea for this search was good was recently shown by Len Bullas, Charles Colson and Aline van Pel (J. Gen. Microbiol. *95,* 166–172, 1976) who encountered such a new system in their work with *Salmonella* recombinants.

In 1965 I was promoted extraordinary professor for molecular genetics at the University of Geneva. Not only did I always enjoy a continued contact with the students, but I also considered teaching as a welcome obligation to keep my scientific interests wide. Although we had a few excellent students in our laboratories, the teaching of molecular genetics at the University of Geneva in the 1960's suffered a bit from a lack of interest by the young generation. This might have been related to a more general lack of public interest for this field, which was perhaps due to the economic structure of the city of Geneva and its environments. These, at that time perhaps more subconscious concerns, might have helped me to accept in 1968 an offer for a professorship at the University of Basel, since I felt that more general interest would be given to molecular genetics in this city with a long tradition of biomedical research at its industries.

I started my new appointment at the University of Basel in October 1971 after having spent one year as a visiting Miller Research Professor at the Department of Molecular Biology of the University of California in Berkeley. In Basel, I was one of the first persons to work in the newly constructed Biozentrum, which houses several University Departments, in particular those of Biophysics, Biochemistry, Microbiology, Structural Biology, Cell Biology and Pharmacology. This diversity within the same house largely contributes to fruitful collaborative projects and it helps to keep horizons broad both in research and teaching. Additional contributions to this goal come from contacts with other nearby University Institutes as well as with the private research Institutions in the city.

Since my coming to Basel, I devoted relatively little of my time to further studies on restriction and modification mechanisms. Not that I have lost my interest in them. On the contrary, I was fortunate to be able to set up a junior group which under the leadership of Bob Yuan and more recently of Tom Bickle, became rapidly quite independent, and it continues to be very successful in its investigations on the more detailed aspects of the molecular mechanisms of restriction and modification. This allowed me to turn my main interests back to other mechanisms affecting either positively or negatively the exchange of genetic material. For a number of years Nick Gschwind, a Ph. D. student, and Dorothea Scandella, a postdoctoral fellow, explored two other mechanisms found in some *E. coli* strains or mutants and affecting more specifically than restriction and modification systems particular steps in the propagation of bacteriophage λ.

For the last several years I have turned my principal interests to the intriguing activities of insertion elements and transposons, which by their actions on genetic rearrangements, seem to be the main driving forces of evolution in microorganisms. Because of their independence on extended nucleotide homologies these forces bring about exchange of largely unrelated genetic materials. Our postdoctoral workers Katsutoshi Mise, Shigeru Iida and Jürg Meyer brought important contributions to the understanding of these phenomena, mainly by the use of the bacteriophage P1 genome as a natural vector of transposable elements. But general knowledge on this to my mind extremely important field is still very scarce and deserves continued attention.

Solid notions on naturally occurring genetic exchange between organisms that are not directly related will also form a good basis for a scientific evaluation of conjectural risks of in vitro recombinant DNA research. Since this research largely makes use of restriction enzymes, although it in no way fully depends on them, I consider it a personal obligation to contribute to the best of my abilities to the solution of questions which arose in the scientific and public debate on this research in the last few years. I see two ways to reach this goal. The first is scientific and tends as just stated to better understand what nature does in its nonhomologous genetic exchange. The second is rather political and it consists in actions to stimulate continued awareness of responsibility to work with a maximum of care in all scientific investigations, which should, however, be allowed to be done under optimal academic freedom.

A curriculum vitae would be incomplete without reference to my private life. I am fortunate to have found a continued support and steady encouragement by my family, in particular by my parents, and, since we became married in 1966, by my wife Antonia. In response to their interest and understanding for my scientific activities, I have tried to give them my personal affection needed for a harmonious life. Our two daughters Silvia and Caroline were born in 1968 and in 1974, respectively. When Silvia learned that I had been honored by the Nobelprize she not only wanted to know what this is, but also why I was chosen as a Laureate. After explaining her in simple terms the basic concepts of the mechanisms of restriction enzymes, she, after some reflection, reexpressed this message in her own terms by a tale, which in the meantime has found wide diffusion around the world. It might thus be justified to finish this curriculum vitae by its reproduction:

"The tale of the king and his servants

When I come to the laboratory of my father, I usually see some plates lying on the tables. These plates contain colonies of bacteria. These colonies remind me of a city with many inhabitants. In each bacterium there is a king. He is very long, but skinny. The king has many servants. These are thick and short, almost like balls. My father calls the king DNA, and the servants enzymes. The king is like a book, in which everything is noted on the work to be done by the servants. For us human beings these instructions of the king are a mystery.

My father has discovered a servant who serves as a pair of scissors. If a foreign king invades a bacterium, this servant can cut him in small fragments, but he does not do any harm to his own king.

Clever people use the servant with the scissors to find out the secrets of the kings. To do so, they collect many servants with scissors and put them onto a king, so that the king is cut into pieces. With the resulting little pieces it is much easier to investigate the secrets. For this reason my father received the Nobel Prize for the discovery of the servant with the scissors".

PROMOTION AND LIMITATION OF GENETIC EXCHANGE

Nobel Lecture, 8 December, 1978

by

WERNER ARBER

Department of Microbiology, Biozentrum, University of Basel, Basel, Switzerland

Exchange of genetic material has widely been observed in practically all living organisms. This suggests that genetic exchange must have been practised since a long time ago, perhaps ever since life has existed. The rules followed by nature in the exchange of genetic information are studied by geneticists. However, as long as the chemical nature of the genetic material remained unknown, genetics remained a rather abstract branch of the biological sciences. This gradually began to change after Avery et al. (1944) had identified DNA as the carrier of genetic information. Their evidence found an independent support by Hershey and Chase (1952), and it was accepted by a majority of biologists by 1953 when Watson and Crick (1953) presented their structural model of DNA. Hence it was clear 25 years ago that very long, filamentous macromolecules of DNA contained the genes. As is usual in fundamental research, the knowledge acquired pointed to a number of new important questions. Among them were those on the structure and function of genes, but also those on the molecular mechanisms of exchange of genetic material.

It is at that time, in the fall of 1953, that I joined more or less by chance a small group of investigators animated by Jean Weigle and Eduard Kellenberg. One of their main interests concerned the mechanisms of genetic recombination. Feeling that the time was not ripe to carry out such studies on higher organisms, they had chosen to work with a bacterial virus, the nowadays famous bacteriophage lambda (λ). It is interesting to see today how knowledge acquired in work with phage λ should later strongly influence other research in molecular genetics. In this lecture I would like to trace back to the origin of some discoveries made in the work with λ and point to their importance for subsequent investigations. But let met first define in more general terms what I mean by genetic exchange.

Escherichia coli and other bacteria carry all their genes on a single, very long DNA molecule, except for occasional cases when bacteria have one or several additional, much shorter DNA molecules, called plasmids, endowed with the ability of autonomous replication. A bacterial strain harbouring besides its chromosome a fertility plasmid F can at times donate by conjugation a copy of its F plasmid to a recipient strain. The F plasmid then establishes itself in the recipient cell as an autonomous plasmid, and it will be propagated in its new environment. The donor strain has thus exchanged genetic information with the recipient strain, and in this case the genetic material transferred had not existed in the recipient strain before the conjugation. We also note that the exchanged material replicates autonomously, and it does not need to be integrated into the bacterial chromosome. Therefore, this is an example of reassortment of DNA molecules.

This situation contrasts with the one encountered in the so-called general recombination. Here two individuals exchange homologous genetic information. An exemple for this is also seen in bacterial conjugation. With a low probability, the fertility plasmid F can integrate into the host chromosome. In conjugation, the resulting strain (=Hfr strain, for high frequency of recombination) transfers a relatively long segment of a copy of its own chromosome to the F^- recipient cell. Maintenance of the information acquired by the recipient depends on its integration into the recipient chromosome, and this integration usually follows the rules of general recombination. This means that recombination depends on the finding of homologous sections on the two interacting genomes; and homology means identity in the nucleotide sequences, with an allowance made for rare exceptions to this rule at sites of mutations, which in fact allow the geneticist to explore these phenomena. Therefore, an exconjugant recombinant genome is a hybrid having received part of its information from one, and part from the other parent. The total information content of the hybrid is the same as that of each parent. The same rule holds true in general recombination between two bacteriophages of the same strain, and it was in work with bacteriophage λ that physical exchange between the two parental genomes was experimentally demonstrated to occur in general recombination (Kellenberger et al., 1961; Meselson and Weigle, 1961).

We know that the molecular mechanism of general recombination is quite complex and depends on a number of specific gene products. Some of these proteins also carry out key functions in DNA replication as well as in DNA repair.

The studies of bacteriophage lysogeny in the 1950's ripened the concept that other mechanisms of molecular exchange between DNA molecules must exists. When bacteriophage λ infects a bacterial host cell it can either reproduce vegetatively to yield a progeny of phage particles or it can lysogenize the host cell. In the latter situation the infected most survives, and it will accept the λ genome as a part of its own chromosome. This is similar to what we have discussed as reassortment of DNA molecules. However, the λ prophage, as is called the λ genome carried in a lysogenic cell, does not replicate autonomously and its maintenance depends on its integration into the host chromosome, which usually occurs at a site close to the genes determining galactose (Gal) fermentation (Wollman, 1953; Lederberg and Lederberg, 1953; Jacob, 1955). Lysogenic bacteria can be induced to phage production, and in this process the prophage gets excised again from the host chromosome. Morse et al. (1956 a,b) observed that phage lysates obtained by such induction of λ-lysogenic Gal^+ bacteria were able to render Gal^- bacteria Gal^+. This phenomenon is called specialized phage-mediated transduction. The authors mentioned that some of the Gal^+ tranductants obtained did not produce plaque forming phage upon induction, although these bacteria were immune to superinfection with λ, a property usually displayed by λ-lysogenic bacteria. At that time I studied λ prophage mutants with defects in genes expressed in the cycle of vegetative phage reproduction. Therefore, Gal^+ transductants as just described were a welcome enrichment of my materials to be studied.

Let me now show what I still consider a simple, straightforward experiment published in my Ph. D. thesis exactly 20 years ago (Arber, 1958). A phage λ lysate transducing the Gal^+ characters at high frequency (HFT lysate) was used to infect a Gal^- bacterial strain at various multiplicities of infection of phage particles per cell, and the surviving bacteria were tested for their Gal and lysogenicity character (Fig. 1).

As expected, the overall probability of an infected cell to become lysogenic remained constant in the range of multiplicities of infection below 1, i.e. the number of normal-lysogenic bacteria linearly dropped with decreasing amounts of phage added (curve 3 of Fig. 1). This curve is exactly paralleled by the one (curve 1) representing Gal^+ transductants found to be immune to superinfection with λ, but which produced no plaque forming phage upon induction. In contrast, the number of Gal^+ transductants which both were λ-immune and produced plaque forming phage upon induction (curve 2) is proportional to the square of the multiplicity of infection (Arber, 1958; Arber et al., 1957).

The interpretation given to these observations was that the HFT lysate used was a mixed population of λ phage particles: (a) normal λ phages and (b) λgal transducing phages which were defective in their capacity to reproduce serially and thus to form plaques, but which were still able to lysogenize even after single infection, although they did so with reduced probability. This interpretation found support in a number of additional experiments, which I would not like to rediscuss now. In summary, by 1958 it was shown that in the excision of λ prophage from the bacterial chromosome errors could sometimes produce aberrant phage genomes having acquired a segment from the host genome and having deleted from the λ genome a segment carrying essential genes for phage reproduction. A molecular model to explain both precise λ excision and the illegitimate formation of λgal was drawn by Campbell (1962), who had also brought very important experimental contributions to this field. The analysis of a large number of independently produced λgal genomes made it clear that recombination within DNA molecules, and by extrapolation also between DNA molecules, occurs sometimes at more or less randomly chosen sites, and hence not selected on the basis of extended regions of homology. Obviously, the likelihood for such recombinants to be viable is relatively small, and nature seems to limit their production to a level several orders of magnitude below the level of general recombination. In evolution, however, this kind of illegitimate recombination may be of great importance.

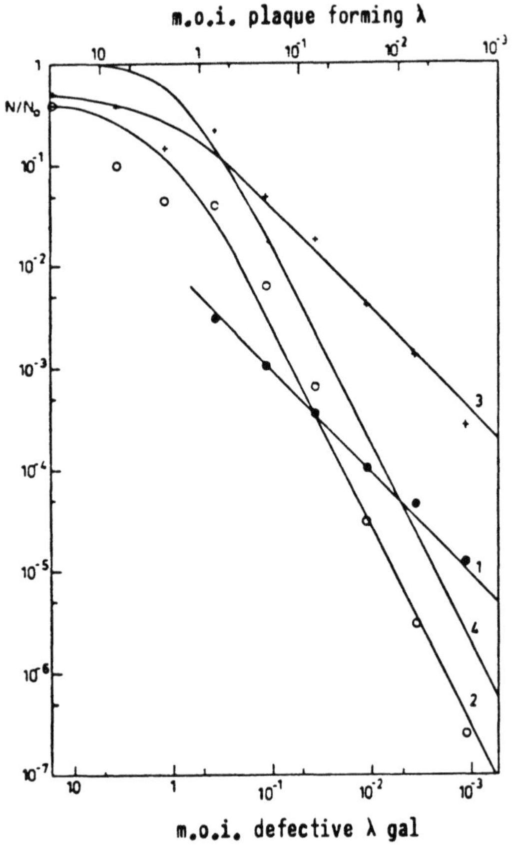

Fig. 1: Transduction and lysogenization of the *galT galK* strain W3350 of *E. coli* K12 by an HFT lysate.

The HFT (high frequency of transduction) lysate used was a phage stock composed of 4.4×10^{10} plaque forming λ phage particles and an estimated (from electron microscopical counts and physiological experiments) 2.3×10^{10} λgal phage particles, which were concluded from this and additional experiments to be defective in vegetative growth and partially affected in lysogenization. Aliquots of the host bacteria were infected at various multiplicities (m.o.i.) with the HFT lysate and then spread on EMB galactose indicator plates containing anti-λ serum. Colonies grown after incubation were tested by replication for λ-immunity and for the ability to produce plaque forming λ upon induction.

N_0 = number of recipient bacteria in infection mixture
N = number of recipient bacteria from the infection mixture found to be (1) Gal^+, λ-immune, not producing plaque forming λ, hence carrying a λgal prophage; (2) Gal^+, λ-immune, producing plaque forming λ and λ gal (as verified with selected subclones), hence being doubly lysogenic for λ and λgal; (3) Gal^-, λ-immune, producing plaque forming λ, hence carrying a λ prophage. (4) represents the calculated fraction of bacteria simultaneously infected with at least one λ and one λgal (From Arber, 1958).

In the meantime, molecular geneticists have learned to isolate *in vivo* derivatives of λ able to transduce practically any desired segment of the host chromosome, and this work has greatly facilitated detailed structural and functional studies of several selected *E. coli* genes. In addition, these studies pointed the way to more recently undertaken approaches to produce *in vitro* recombinational hybrids between λ (or other vector DNA molecules) and DNA fragments from any chosen origin.

The studies of λ lysogeny and of the defective nature of λgal, which could be complemented by helper phage infection, also influenced work on animal and plant viruses. The knowledge acquired with λ was taken as a model and this turned out to be extremely fruitful. We indeed know today that many situations similar to that of λ exist, where viruses are found integrated into the host chromosome and where viruses or fragments thereof can be shown to be defective but activatable by superinfection with exogenous helper virsus.

Let me now return to the process of integration and precise excision of the λ prophage. As already stated, this site specific recombination was explained in a model devised by Campbell (1962). A long and careful study on this system has culminated a few years ago by its demonstration *in vitro*. Some of the few enzymes needed for the process are contributed by the λ phage itself and others by the host bacteria (Nash, 1977). We also know the nucleotide sequences at which the interaction between the λ genome and the bacterial chromosome occurs (Landy and Ross, 1977). These show homology over a stretch of 15 nucleotide pairs, but it has been shown that this length is not sufficient for efficient integration. Rather, the considerably longer, nonhomologous flanking segments play additional key roles in the interaction. This is the system of site specific recombination on which our knowledge is the most advanced.

The demonstration of recombinational events occurring independently of extended nucleotide homology, be they site specific or at random, brought up the question on possible limitations set by nature to such exchange which might be considered rather undesirable for the life of a cell.

Before two DNA molecules of different origin can interact with each other directly they must be brought into proximity, into the same compartment. Nature has certainly set up a number of mechanistic barriers such as membranes to limit free diffusion of genetic materials. On the other hand, we also know that a number of mechanisms exist wich precisely allow the transfer of DNA from one cell to another, and that sometimes this exchange occurs between cells that are not directly related. May I recall that some bacterial conjugation plasmids have a relatively wide host range, and so do some bacteriophage strains able to transduce segments of the host chromosome, either by the already described specialized transduction, or also by the mechanism of general transduction, in which upon maturation the phage wrongly packages a segment of the host chromosome instead of its own phage genome. However, it is also clear that the host range is always limited by the need for specific cell surface interactions, and this seems to hold also for the penetration of free DNA into bacterial cells in the process known as transformation. On the other hand, bypass mechanisms have been demonstrated, e.g. that a phage genome is transferable by bacterial conjugation (Jacob and Wollman, 1956) or that a conjugation plasmid is transduced by bacteriophage (Arber, 1960).

Sometimes the host range of a bacterial virus can become extended due to a mutation in one of the phage genes. In contrast to this situation, host-controlled variation (or modification, as it is now generally called) of bacteriophage, first described in the early 1950's (Luria and Human, 1952; Bertani and Weigle, 1953; Anderson and Felix, 1952; Ralston and Krueger, 1952) presented the puzzling situation that upon growth on different host strains, a virus could adapt to propagate on a new host without this ability being maintained upon backgrowth on the old host. Hence the adaption could not find its explanation by a mutation in the phage genome.

I became interested in these phenomena in 1960 and decided to look at the mechanisms of host-controlled modification of bacteriophage λ. The two host strains of my choice were a pair of *E. coli* strains, K12 (shortly called K), and its Pl-lysogenic derivative K(P1). A few years before, Lederberg (1957) had shown that the P1 prophage determines a system of host-controlled modification. Restriction of λ·K (phage grown on K) by K(P1) bacteria is quite strong: λ·K forms plaques on K(P1) host bacteria with an efficiency of 2×10^{-5} only (Fig. 2). In contrast, phage adapted to K(P1) grows with full efficiency on both K and K(P1). However, as is characteristic of host-controlled modification, when λ·K(P1) serves as inoculum for the growth of a multi-cycle stock of λ on strain K, the resulting phage behaves exactly as the original λ·K. I wanted to know how fast this re-adaptation occurs. Therefore, I grew λ·K(P1) on strain K for just one lytic cycle, taking care to inactivate all non-adsorbed λ with anti-λ serum, which was then removed by washing. The result was striking. The one cycle progeny grew on the restrictive host K(P1) with an efficiency of between 3×10^{-3} and 10^{-2} instead of 2×10^{-5}. Since the burst size per singly infected

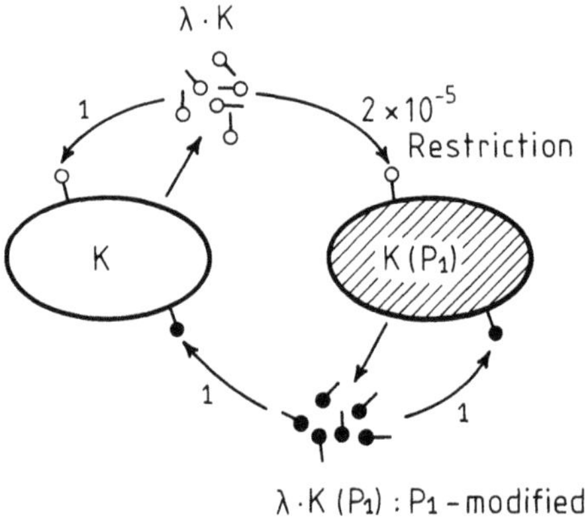

Fig 2: Pl-specific restriction and modification as detected by growth of bacteriophage λ.

Numbers give the efficiency of plating of phage variants λ·K and λ·K(Pl) on the hosts K and K(Pl) as indicated by the arrows.

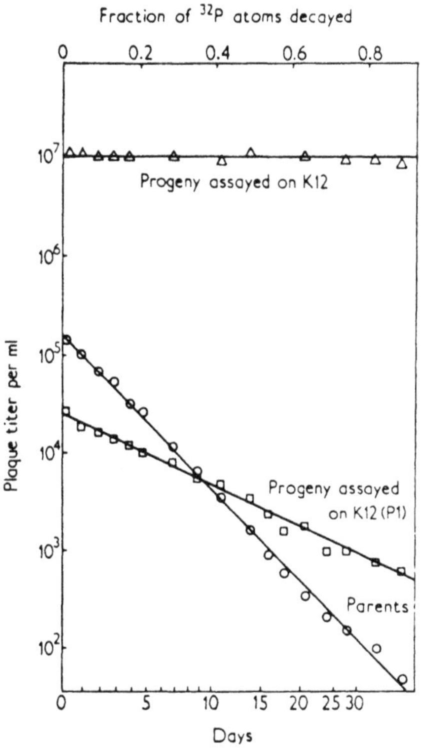

Fig. 3:

Joint transfer of parental DNA and parental Pl-specific modification of heavily ^{32}P labeled λ·K(Pl) into the phage progeny produced after infection at a multiplicity of 0.006 phages per cell of the non-radioactive host strain K. See text for further explanations (From Arber and Dussoix, 1962).

host cell ranged between 100 and 200 λ particles, this result suggested that about one progeny phage per cell had "inherited" the parental P1-specific modification (Arber and Dussoix, 1962; Arber, 1962). We were convinced that this was transferred from the infecting parental phage particle. But was it a diffusible internal phage protein or was it perhaps carried on the parental DNA molecule?

That the second possibility is the correct explanation became clear in the following experiment, which I would like to show, also for historical reasons. It had been demonstrated in the 1950's that DNA carrying radio-isotopes loses its viability as a function of the radioactive decay (Hershey et al., 1951; Stent and Fuerst, 1955). Therefore, phage heavily loaded in its DNA with ^{32}P undergoes suicide upon storage. To my knowledge, the experiment shown in Figure 3 is the only important application of this rather special technique, which did allow us to trace parental DNA molecules in the course of replication at a time before density labeling methods had made their proof. A stock of heavily ^{32}P-labeled λ·K(P1) was prepared and carefully purified. One aliquot was immediately used for one cycle of growth on non-radioactive K bacteria and the phage progeny was stored and assayed from day to day. Another aliquot of the parental phage was directly stored and assayed from day to day. It is seen in Figure 3 that the viability of this parental phage disappeared exponentially as a function of the ^{32}P decay. In contrast, the bulk of the progeny phage grown on non-radioactive cells for one cycle was perfectly stable as revealed upon assay on K. However, those phages in the one cycle progeny able to grow on K(P1) were inactivated, and their inactivation was about half as rapid as that of the parental phage (Arber and Dussoix, 1962). These results indicated that P1-specific modification carried by the parental λ·K(P1) phage is transferred together with a parental DNA strand into the population of progeny phage particles. It must be noted that fragmentation of parental λ DNA molecules can occur by recombination with progeny λ DNA molecules in the course of the intracellular phage growth, but it affects at most half of the parental input. Such recombinants with less than semiconserved parental DNA would not grow on K(P1), and their slow inactivation due to ^{32}P decay would not be detectable upon assay on K, since such phage particles represent a small minority of the progeny population.

About the same time Grete Kellenberger, who worked in the same laboratory, and Daisy Dussoix, a Ph. D. student, studied the breakdown of DNA from irradiated phage λ upon infection of normal host bacteria. We wondered if the mechanisms of P1-specific restriction and of inactivation of phage caused by irradiation had anything in common. For this reason the fate of λ DNA is restrictive host bacteria was investigated, and we could demonstrate that in the infection of K(P1) bacteria with λ·K phage an important fraction of the phage DNA was rapidly degraded (Dussoix and Arber, 1962). No DNA breakdown was seen in the λ·K-infected K bacteria.

The implication of these early findings, that host-controlled modification affected DNA, although the phenomenon could not be explained as a mutation, found rapidly additional support. We also realised that the phenomenon does not directly depend on phage λ used in the studies and that it affects any other DNA in the same way as λ DNA, e.g. bacterial DNA in conjugation (Arber, 1962). Hence restriction and modification (R-M) systems can be looked at as serving as defence mechanisms against the uptake of foreign DNA and restriction to be brought by nucleolytic activity.

It took us a while to find out how bacteria can protect their own DNA against their restriction nucleases. They do so by postreplicative nucleotide methylation at the sites serving the R–M systems for specificity recognition (Arber, 1965; Smith et al., 1972; Kühnlein and Arber, 1972).

Interestingly, the R–M systems *Eco*P1, *Eco*K and *Eco*B with which the fundamental genetic and physiological experiments were carried out do not cleave the DNA precisely at the sites used for recognition (Horiuchi and Zinder, 1972). This points to rather complex molecular mechanisms by which these enzymes act. Careful investigations have already revealed important aspects of them (Arber et al., 1975; Yuan et al., 1975; Bickle et al., 1978), so that these systems can serve as models in investigations of other nucleic acid protein interactions, particularly those showing regional rather than site specificity. I think in particular at some as yet poorly understood recombination phenomena pointing to regionally increased probability of interchange.

Other restriction enzymes, as is now well known, do cleave unmodified DNA at the recognition site, which is specific for each particular R–M system. I think that my colleagues Hamilton Smith and Daniel Nathans will discuss aspects of the mechanisms of these enzymes and of their application to studies of structure and function of DNA. Let me therefore just mention what is relevant with regard to *in vivo* genetic exchange. Since restriction enzymes have been widely used in *in vitro* recombination of DNA molecules it is of interest to see that these enzymes can also trigger recombination *in vivo* (Chang and Cohen, 1977). Hence, as in other biological activities with ability to catalyze antagonistic reactions, the restriction enzymes in question can both inhibit genetic exchange as well as promote it to some degree. This recombination does of course not depend on major nucleotide homologies on the interacting DNA molecules, but only on the existence of recognition sites for the enzymes determined by 4 to 6 nucleotide pairs in general.

The discovery of still another type of genetic exchange not based on nucleotide homology has also its roots in work with phage λ. Peter Starlinger was among those fascinated by the explanation of how λgal phages were formed. In the early 1960's he and his collaborators had fruitfully extended the knowledge on the randomness of the illegitimate recombination by genetically determining the endpoints of the bacterial material picked up by λgal. This formed a part of their studies on the structure and function of the galactose operon of *E. coli*. In the course of their work they encountered unorthodox mutations with strongly polar effects and several other unexpected properties. Further investigations of the nature of these mutations, as well as of similar mutations isolated in other laboratories, finally revealed the existence of what is now known as Insertion Sequences or IS elements (Starlinger and Saedler, 1976; Bukhari et al., 1977).

It would be premature to list general properties of IS elements and related structural entities. It is clear, however, that these elements of the size of about 600 to 2000, or more, nucleotide pairs can be found in *E. coli* and in other bacteria in one or more copies carried at a number of different chromosomal sites. Spontaneously, such an IS element can show up at a site not previously occupied by it. This event is called transposition, although it remains unclear if the element really jumps from one location on the chromosome to another, or if a resident IS element prepares a new copy in view of its integration at a new site. On the other hand, it is clear that IS elements generally can indeed excise, either precisely or imprecisely, from a given site. Another often encountered property of IS elements is their ability to form deletions starting at one of the ends of the IS element and extending to a perhaps randomly chosen site at a distance of sometimes several genes. All these events are rare indeed, but occur at measurable frequencies of perhaps 10^{-4} to 10^{-8} per cell division, depending on the system studied. They must be enzymatically determined, and it is likely that the rate limiting factors are usually repressed. Presumably for this reason, no enzymological *in vitro* studies of these mechanisms have yet been successful. It is still a guess that one or several of the genes determining the activities of IS elements are located on the IS element itself, which is also supposed to have specific sites involved in the events of transposition, excision and deletion formation. Some IS elements have been shown to contain regulatory signals for gene expression.

Most of the IS elements described so far in the literature were chance isolates, and it remains largely unknown how many different IS elements are carried e.g. in *E. coli* K12, nor does one yet have good ideas on the host range of particular IS elements. We have started to look for answers to these questions by the use of a large plasmid, the bacteriophage P1 prophage, to trap transposing IS elements inside the *E. coli* cell. Interestingly, an important proportion of spontaneous P1 prophage mutations affecting the functions of vegetative phage growth is explained by the incorporation of IS elements, which must originate from the host chromosome. On the other hand, several IS elements were also found by chance carried in genomes of P1 derivatives not affected in their functions of vegetative growth. Preliminary studies indicate that many of the IS elements isolated are independent of each other. This should allow us to establish a library of transposable IS elements isolated from host strains of bacteriophage P1. Hybridisation studies with these IS elements and DNA from various origin is then expected to shed light on the question of the host range of particular IS elements (Arber et al., 1979).

IS elements have also been shown to mediate the exchange of more extended DNA segments. Transposons are DNA segments flanked by identical IS elements or at least repeated sequences. One of the important features of a transposon is that it can insert as a unit into another chromosome. For example, an r-determinant element of 23 kb length, carrying the genes for resistance to several antibiotics and originally identified as a constituant of R plasmids, has been shown to transpose into phage P1, from this into the *E. coli* chromosome and from the chromosome into another bacteriophage genome (Arber et al., 1979). This clearly shows that under natural conditions relatively long DNA segments can translocate onto a transferable vector DNA molecule, such as a viral genome or a conjugative plasmid. And the same element at some later time in another host cell can transpose into a cellular chromosome. In principle, there is no limit set for genes to be picked up at one time or another on a transposable element, since the elements flanking a transposon can also transpose independently and thus by chance give rise to the formation of new transposons.

IS elements and related repeated sequences also give rise to cointegration of two DNA molecules, as well as to the dissociation of a single DNA molecule into two. Chromosomal integration and excision of F and R plasmids is just one example. Finally a few additional mechanisms contributing to genetic rearrangement and diversity should also be mentioned: gene inversion, gene amplification and the formation of short partial duplications. These mechanisms seem also to be driven by IS elements flanking the genes involved.

In this lecture I have tried to show that the deeper we penetrate in the studies of genetic exchange the more we discover a multitude of mechanisms either acting as promotors of exchange or acting to set limits to it, and some do both. On purpose I discussed only procaryotes and did so largely by taking examples from *E. coli* and its phages and plasmids.

I am aware and puzzled by the roles that site specific exchanges may play in the ontogeny of higher organisms and at the level of the RNA in gene expression.

I gave some thought on the possible reasons why *E. coli* bacteria might have set up such a multitude of systems involved in the genetic exchange which for some reasons must be vital for them. I must confess that I did not find out why, besides trivial answers such as "serving in repair processes" or "evolutionary driving forces" for the promoting activities, and "species isolation" or "genetic stability" for the activities keeping genetic exchange within limits. More intensive research is needed to understand the apparent complexity of nature. But one important notion already obtained might be good to keep in mind: in spite of possessing a multitude of natural mechanisms to promote exchange between genetic materials of unrelated origin, *E. coli* and other living organisms have succeeded to accomplish a relatively high overall stability in their genetic make-up.

Acknowledgements
In the course of the past 25 years I was very fortunate to benefit from a large number of highly qualified collaborators and I received additional stimulation from even more colleagues. An important part of my research found continued support by the Swiss National Science Foundation.

REFERENCES

Anderson, E.S. and Felix, A. Nature *170,* 492–494 (1952).

Arber, W. Arch. Sci. (Genève) *11,* 259–338 (1958).

Arber, W. Virology *11,* 273–288 (1960).

Arber, W. Path. Microbiol. *25,* 668–681 (1962).

Arber, W. J. Mol. Biol. *11,* 247–256 (1965).

Arber, W. and Dussoix, D. J. Mol. Biol. *5,* 18–36 (1962).

Arber, W., Iida, S., Jütte, H., Caspers, P., Meyer, J. and Hänni, Ch. Cold Spring Harbor Symp. Quant. Biol. *43,* (in press, 1979).

Arber, W., Kellenberger, G. and Weigle, J.J. Schweiz. Z. allg. Pathol. Bakteriol. *20,* 659–665 (1957).

Arber, W., Yuan, R. and Bickle, T.A. Proc. of the 9th FEBS Meeting Budapest 1974. Post-Synthetic Modification of Macromolecules Vol. 34, pp. 3–22 (1975). Eds. F. Antoni & A. Farago.

Avery, O.T., MacLeod, C.M. and McCarthy, M. J. Exp. Med. *79,* 137–158 (1944).

Bertani, G. and Weigle, J.J. J. of Bact. *65,* 113–121 (1953).

Bickle, T.A., Brack, C. and Yuan, R. Proc. Natl. Acad. Sci. (USA) *75,* 3099–3103 (1978).

Bukhari, A.I. and Shapiro, J.A. and Adhya, S.L. DNA Insertion Elements, Plasmids, and Episomes. 1977 by Cold Spring Harbor Laboratory, USA.

Campbell, A.M. Adv. Genet. *11,* 101–145 (1962).

Chang, S. and Cohen, S.N. Proc. Natl. Acad. Sci. (USA) *74,* 4811–4815 (1977).

Dussoix, D. and Arber, W. J. Mol. Biol. *5,* 37–49 (1962).

Hershey, A.D. and Chase, M. J. Gen. Physiol. *36,* 39–56 (1952).

Hershey, A.D., Kamen, M.D., Kennedy, J.W. and Gest, H. J. Gen. Physiol. *34,* 305–319 (1951).

Horiuchi, K. and Zinder, N.D. Proc. Natl. Acad. Sci. (USA) *69,* 3220–3224 (1972).

Jacob, F. Virology, *1,* 207–220 (1955).

Jacob, F. and Wollman, E.L. Ann. Inst. Pasteur *91,* 486–510 (1956).

Kellenberger, G., Zichichi, M.L. and Weigle, J.J. Proc. Natl. Acad. Sci. (USA) *47,* 869–878 (1961).

Kühnlein, U. and Arber, W. J. Mol. Biol. *63,* 9–19 (1972).

Landy, A. and Ross, W. Science, *197,* 1147–1160 (1977).

Lederberg, S. Virology *3,* 496–513 (1957).

Lederberg, E.M. and Lederberg, J. Genetics *38,* 51–64 (1953).

Luria, S.E. and Human, M.L. J. of Bact. *64,* 557–569 (1952).

Meselson, M. and Weigle, J.J. Proc. Natl. Acad. Sci. (USA) *47,* 857–868 (1961).

Morse, M.L., Lederberg, E.M. and Lederberg, J. Genetics *41,* 142–156 (1956a).

Morse, M.L., Lederberg, E.M. and Lederberg, J. Genetics *41,* 758–779 (1956b).

Nash, H.A. C.T. Microbiol. Immunol. *78,* 171–199 (1977).

Ralston, D.J. and Krueger, A.P., Proc. Soc. Exptl. Biol. Med. *80,* 217–220 (1952).

Smith, J.D., Arber, W. and Kühnlein, U. J. Mol. Biol. *63,* 1–8 (1972).

Starlinger, P. and Saedler, H. Current Topics in Microbiology and Immunology, *75,* 111–152 (1976).

Stent, G.S. and Fuerst, C.R. J. Gen. Physiol. *38,* 441–458 (1955).

Watson, J.D. and Crick, F.H.C. Nature *171,* 737–738 (1953).

Wollman, E.L. Ann. Inst. Pasteur *84,* 281–293 (1953).

Yuan, R., Bickle, T.A., Ebbers, W. and Brack, C. Nature *256,* 556–560 (1975).

Daniel Nathans

DANIEL NATHANS

My parents came to the United States in the early years of this century as part of a wave of Russian Jewish immigrants seeking freedom and opportunity in the New World. My mother, Sarah Levitan, came to America when she was 18. My father, Samuel, rebelling against an orthodox family, left home in his midteens and made his way to the United States a few years later. They were married in Philadelphia in 1910. As the last of their nine children I was born in 1928 in Wilmington, Delaware, on the eve of the great depression. Soon after, my father lost his small business and was for some time unemployed. Our house was cold and leaky, and (I learned later) my parents sometimes went hungry. Yet they generally managed to retain their good humor and certainly their hopes for their children. I have only fond memories of this period, no doubt due to the special attentions of an affectionate family.

My education began in the public schools of Wilmington. During most of these years, from about age 10, I also worked at some job or other after school, on weekends, and in the summer months. Following in the footsteps of my brothers and sisters, I went on to the University of Delaware, where I studied chemistry, philosphy, and literature. Although I enjoyed science and mathematics, what I remember most vividly is a small, stimulating circle of professors and students (including a number of veterans just back from the war), interested in philosophy and politics. To my father my interest in natural science meant "medicine", and becoming a physician also seemed more attractive to me than any other alternative I knew about. So I applied to medical school and received a scholarship at Washington University in St. Louis. Washington University turned out to be a lucky choice. The faculty was scholarly and dedicated–and accessible to students. A wonderful summer of research with Oliver Lowry, Professor of Pharmacology, convinced me that a career in medical research and teaching suited me better than medical practice. After getting an M. D. degree in 1954, I went to the Columbia–Presbyterian Medical Center in New York for an internship in medicine with Robert Loeb, a masterful clinician and medical scientist. That was one of the most valuable years of my life. The glimpses of human strength and frailty that a physician sees are with me still. I spent two more years at Columbia as a medical resident, interrupted by service as a Clinical Associate at the National Institutes of Health in Bethesda. During my years in Bethesda, I married Joanne Gomberg, and our son Eli was born.

While at the National Institutes of Health I developed an interest in the biosynthesis of proteins as a result of a study with Michael Potter and John Fahey of myeloma protein formation in plasmacytoma cells. This led me to Fritz Lipmann's laboratory at the Rockefeller Institute in 1959. Here I identified the bacterial "elongation factors" involved in the addition of amino acids to growing peptide chains, worked on the mechanism of action of puromycin as an inhibitor of this step (with Amos Neidle), and in a collaborative study with Norton Zinder, demonstrated that RNA from a bacterial virus directed the synthesis by cell extracts of viral coat protein. During those years in the invigorating atmosphere of Lipmann's laboratory and the Rockefeller Institute, I learned a geat deal, and Lipmann's artistry made a lasting impression on me. I also found out that I liked biochemical research and that I could do it. The intention of returning to a department of medicine was abandoned, and I accepted a position at Johns Hopkins University School of Medicine in the Department of Microbiology, headed by Barry Wood, an inspiring former teacher at Washington University. During the years in New York our son Jeremy was born, and soon after our move to Baltimore in 1962, my wife gave birth to our youngest son, Ben.

In Baltimore I became head of a one-man "Division of Genetics" which gradually developed substance with the recruitment of Hamilton Smith, Bernard Weiss, Kenneth Berns, Thomas Kelly, and recently, John Morrow. My initial research at Hopkins was a continuation of studies on the *in vitro* translation of bacteriophage RNA, particularly its regulation by phage coat protein and the location of genes by translation of fragments of the RNA. My co-workers during these years were Yoshiro Shimura, Max Oeschger, Gerardo Suarez, Robb Moses, Kathleen Eggen, Roy Schmickel, Herbert Kaizer, Marilyn Kozak, and Susan Polmar.

In the mid 60's I became interested in viral tumorigenesis and spent the first half of 1969 learning about animal cells and viruses at the Weizmann Institute of Science in Rehovot, Israel, with Leo Sachs and Ernest Winocour. That spring a letter from Hamilton Smith telling me about the restriction endonuclease he had discovered in *Hemophilus influenzae* aroused my interest in the possibility of using restriction enzymes to dissect the genomes of DNA tumor viruses. Back in Baltimore in the summer and fall of 1969, Stuart Adler and I surveyed the known restriction enzymes for their ability to cleave the DNA of Simian Virus 40, one of the simplest animal viruses that transform cultured cells to tumorigenicity. Using fragments of Simian Virus 40 DNA produced by Smith's enzyme and by similar enzymes discovered subsequently, Kathleen Danna and George Sack constructed a cleavage map of the viral DNA. With this map in hand, other co-workers proceeded to localize viral genes and template functions along the molecule, to construct deletion mutants and later point mutants at pre-selected restriction sites, and to analyse the genomes of naturally arising variants of the virus. Associates in these later studies were Elena Nightingale, Ching-Juh Lai, Theresa Lee, William Brockman, Mary Gutai, Walter Scott, Nicholas Muzyczka, and David Shortle; and collaborators from other laboratories were George Khoury, Malcolm Martin, Kathleen Rundell, and Peter Tegtmeyer.

As I look back on the last few decades of my life, I am struck by the good fortune that came my way. Throughout my schooling there was an abundance of opportunity and encouragement. Several of my teachers were remarkable individuals who had a lasting influence on me. At every stage of my career I have had interesting and cordial colleagues, some of whom are close friends. My field of research is as exciting to me as ever, and it remains essentially a "cottage industry" effort. I have had talented students who are a source of much enjoyment, and I anticipate more to come as their careers develop. And most important, my wife and sons have created in our home an atmosphere of joy and harmony, so essential to everything else.

VITA

Born 30 October, 1928, to Samuel and Sarah (Levitan) Nathans in Wilmington, Delaware, U.S.A.

Married 4 March, 1956, to Joanne Gomberg. Three children: Eli, Jeremy, Ben.

B.S. in Chemistry (1950), University of Delaware, Newark, Delaware.

M.D. (1954), Washington University, St. Louis, Missouri.

Intern (1954–1955) and Resident (1957–1959) in Medicine, Columbia–Presbyterian Medical Center, New York.

Clinical Associate (1955–1957), National Institutes of Health, Bethesda, Maryland.

Guest Investigator (1959–1962), Rockefeller Institute for Medical Research, New York.

Faculty member (1962– present), The Johns Hopkins University School of Medicine, Baltimore, Maryland. Since 1972, Boury Professor and Director of the Department of Microbiology.

American Cancer Society Scholar (1969), Weizmann Institute of Science, Rehovot, Israel.

National Academy of Sciences' U. S. Steel Foundation Award in Molecular Biology (1976).

Fellow, American Academy of Arts and Sciences (1977).

RESTRICTION ENDONUCLEASES, SIMIAN VIRUS 40, AND THE NEW GENETICS

Nobel Lecture, 8 December, 1978

by

DANIEL NATHANS

Department of Microbiology, Johns Hopkins University School of Medicine, Baltimore, Maryland, U.S.A.

Introduction

Some 35 years ago the study of heredity took a chemical turn when Avery and his colleagues discovered that deoxyribonucleic acid–DNA–is the "transforming principle" that converts bacteria from one genotype to another. Watson and Crick's structural model of DNA then provided the basis for investigating its role as the hereditary material. There followed rapid advances in the biochemistry of DNA replication, gene expression, and regulation of genes. Recently, fresh impetus has been given to the study of genetic mechanisms, particularly in higher organisms, by several methodological developments that have opened a new approach to the analysis of chromosomes: site-specific cleavage of DNA by restriction endonucleases and electrophoretic fractionation of the resulting fragments; recombination, cloning, and amplification of DNA segments from any source; rapid methods for determining the nucleotide sequence of DNA; site-directed *in vitro* mutagenesis; synthesis of polydeoxynucleotides of pre-determined sequence; and the ability to introduce cloned, functioning genes into prokaryotic or eukaryotic cells. As a result of these developments even chromosomes which are largely inaccessible to classical genetic methods can now be analysed piece by piece in chemical detail. Genes and signals can be altered at pre-selected sites, and the functional effect of such alterations determined. And active, synthetic genes can be constructed *in vitro* by recombination or by chemical synthesis.

Many investigators have contributed to the "new genetics". Contributions from my own laboratory resulted from our studies of a model eukaryotic chromosome, that of a small mammalian tumor virus. I became interested in tumor viruses in the mid 1960's when I was asked to give a lecture on this subject to Johns Hopkins medical students. Although I had been working with an RNA coliphage (a bacterial virus) for some years, I knew very little about animal viruses. As I reviewed the tumor virus literature, I was impressed by the fact that simple viruses had a profound and permanent effect on the growth of cells in culture or in a living animal. Here was a microcosm of regulatory mechanisms related to the development of the virus itself and to the growth of animal cells, including neoplastic cells. At least some of these mechanisms appeared approachable with the tools of molecular genetics that had been so successfully used with bacterial viruses. Of course all of this was appreciated by a number of people in the tumor virus field, but to me it was an exhilarating revelation. I decided to take a leave of absence in order to explore experimental approaches to understanding viral tumorigenesis, and gradually to wind up my work on the RNA bacteriophage.

At the beginning of 1969 I went to the Weizmann Institute of Science in Israel, where I worked with Ernest Winocour and Leo Sachs and had a chance to read and think without interruption. During that spring I received a letter from a colleague in Baltimore, Hamilton Smith, telling me about the enzyme he had discovered in the bacterium *Hemophilus influenzae* that had the biochemical properties of a restriction endonuclease. This aroused my interest immediately in the possibility that restriction endonucleases were "trypsins and chymotrypsins for DNA" and prompted me to review the literature on bacterial restriction and modification, beginning with the initial observations of Luria & Human (1) and Bertani & Weigle (2). From the incisive work of Arber and his colleagues on the molecular genetics of DNA restriction and modification (3–5), and the biochemical characterization of purified restriction enzymes by Meselson & Yuan (6) and Smith & Wilcox (7), it seemed likely (as first suggested by Arber) that restriction enzymes could be used to digest DNA molecules into specific fragments, just as specific proteolytic enzymes are used to fragment proteins. If the genomes of DNA tumor viruses could be dissected in this way, and if individual fragments of viral DNA could be isolated, one might be able to determine by chemical mapping which segments of the genome were responsible for the various biological activities of the virus, an approach analogous to that Shimura and I had taken earlier to determine the location of genes along the RNA of a bacterial virus (8).

I had already decided that the small papovavirus, Simian Virus 40 (SV40) (9), was the most tractable tumor virus to work with. This virus is a non-enveloped, icosahedral particle with a diameter of about 40 nm (Fig. 1). Its genome is a ring of duplex DNA with only about 5000 nucleotide pairs (10)—equivalent to a few genes—present as a typical eukaryotic minichromosome (11, 12). Despite its paucity of genetic information, SV40 seemed to have all the biological properties of immediate interest: it grew in the nucleus of monkey cells in culture (13), and it caused heritable changes in the growth of rodent cells, i.e., it "transformed" them to tumorigenicity (14). As an initial experiment, I planned to survey the known restriction endonucleases for their ability to cleave SV40 DNA. On my return to Baltimore in the summer of 1969, DNA in hand, our dissection of the SV40 chromosome began.

Cleavage of SV40 DNA by Restriction Endonucleases

For our initial survey of restriction endonucleases Stuart Adler, working with me in the summer and fall of 1969, prepared restriction enzymes from *Escherichia coli* strains B (17), K (6), and K(P1) (18), and he obtained *Hemophilus* enzyme from Smith. To our delight, the *E. coli* B enzyme and the P1 enzyme each cleaved the SV40 DNA circle once, yielding full length

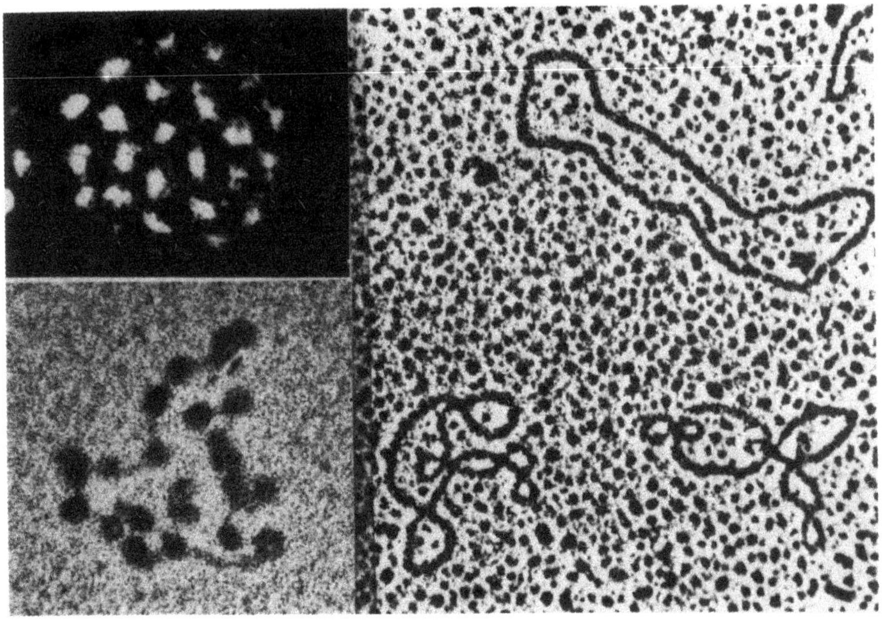

Fig. 1: Electron micrographs of an SV40 virus particle, upper left (15); an SV40 chromosome with typical nucleosomes, lower left (12); and free SV40 DNA, right. The DNA micrograph shows two molecules of form I DNA (covalently closed, circular duplex DNA) and one molecule of form II DNA (nicked, relaxed circular duplex DNA) (16).

linear molecules, and the *Hemophilus* enzyme cleaved SV40 DNA several times (19). However, the *E. coli* K enzyme did not attack SV40 DNA at all. We tentatively concluded that SV40 DNA has no sequences recognized by the K enzyme, that the B and P1 enzymes opened the SV40 circle at a unique site specific for each enzyme, and the *Hemophilus* endonuclease cut the viral DNA at several specific sites. Later we were surprised to find that the *Eco* B* restriction endonuclease, a complex ATP and S-adenosylmethionine-dependent enzyme (17) ("Class I"enzyme), does not break SV40 DNA at a specific site, even though the enzyme cuts each molecule once (20–22). Therefore, *Eco* B and similar enzymes would not be useful for our purpose. However, Smith's *Hin* d endonuclease, a structurally simpler enzyme, not dependent on ATP or S-adenosylmethionine (23), was shown by Kelly and Smith to break DNA at a specific nucleotide sequence (24), and by Kathleen Danna and me to generate specific, electrophoretically separable fragments from SV40 DNA (25) (see Fig. 2). Smith's enzyme turned out to be a mixture of two different restriction endonucleases (*Hin*

Fotnote

* The restriction enzyme nomenclature (100) is based on a three letter abbreviation of the name of the host organism followed by a strain designation and enzyme number where required (e.g., *Eco* B for *E. coli* strain B; *Hin* dIII for *Hemophilus influenzae* strain d, enzyme III).

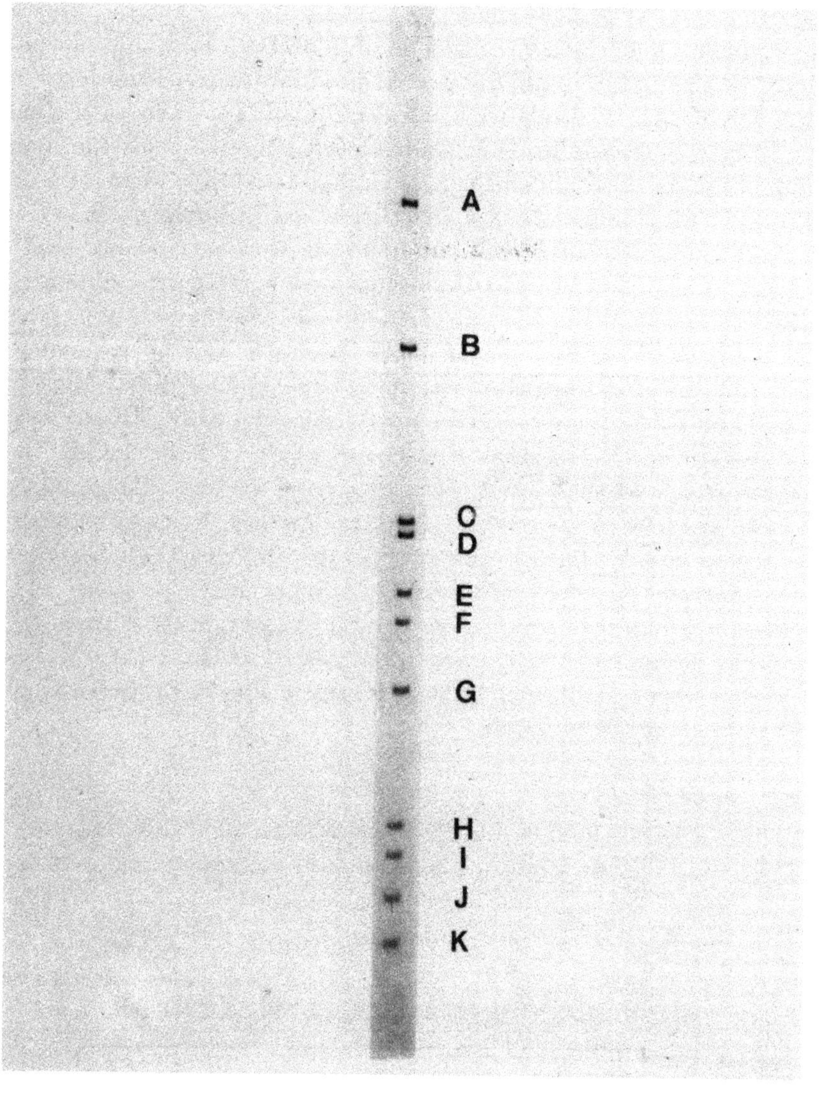

Fig. 2: Autoradiogram of ^{32P}-SV40 DNA after digestion with the *Hin* d enzyme of Smith and electrophoresis from top to bottom in 4 % polyacrylamide gel (25). The largest fragment is near the top (A), and the smallest is near the bottom of the gel (K).

dII and *Hin* dIII) (26, 27) each of which gave a characteristic electrophoretic pattern of fragments from SV40 DNA. Subsequently, newly discovered cleavage site-specific restriction enzymes of the *Hemophilus* type ("Class II" enzymes), over one hundred of which are now known (28), were used to cut SV40 DNA, each yielding its own distinctive digest pattern when the fragments were visualized by electrophoresis in acrylamide or agarose gels. Thus, digestion of DNA by Class II restriction enzymes followed by gel electrophoresis appeared to yield homogeneous fragments derived from specific regions of the genome.

Cleavage Map of the SV40 Chromosome

To use fragments generated by restriction of SV40 DNA for mapping viral functions, we needed to locate the precise positions of restriction sites in the viral DNA, i.e., to construct a "cleavage map" for each restriction enzyme. This was accomplished by Danna and George H. Sack, who first determined the size of fragments in a given digest and then their order in the circular SV40 genome (25, 27). The size of each fragment was determined initially by its relative yield and/or by electron microscopic length measurements, and later, by electrophoretic mobility relative to standards. The order of fragments in the viral genome was determined by electrophoretic analysis of isolated partial digest products and by sequential digestion with different restriction enzymes (Table 1). Our initial cleavage map was based on sites of cleavage by *Eco* RI (29), *Hin* dII + III, and *Hpa* I + II (30) (Figure 3a). The single *Eco* RI site (22, 31) was designated the zero coordinate, and map units were expressed as fractional genome length from that site in an arbitrary direction around the cleavage map. With this map as a reference, sites of cleavage of SV40 DNA by other restriction enzymes have been localized in a number of laboratories, yielding the detailed map shown in Figure 3b (32). As seen in the Figure, the circular SV40 genome can be opened at any one of several different sites by single-cut enzymes, and small or large fragments can be prepared from virtually any part of the molecule.

Nucleotide sequence Map

The ultimate chemical map of a DNA molecule is its nucleotide sequence. The availability of small specific fragments and corresponding cleavage

TABLE 1

ORDER OF *Hin* FRAGMENTS:
ANALYSIS OF PARTIAL DIGESTION PRODUCTS AND *Hpa* FRAGMENTS

Initial fragment (% of SV40 DNA)	*Hin* digest products	Overlapping fragment order										
12	G,J		J	G								
12	F,K	F										K
13	E,K										E	K
22	B,G			G	B							
40	B,F,G,J,K	F	J	G	B							K
43	B,F,G,H,I,J	F	J	G	B	I	H					
51	A,C,D,E							A	C	D	E	
20 (*Hpa*-C)	B,I				B	I						
37 (*Hpa*-B)	A,H,C						H	A	C			
40 (*Hpa*-A)	D,E,F,G,J,K	F	J	G								
	Order:	F	J	G	B	I	H	A	C	D	E	K

Note: Partial *Hin* digest fragments or *Hpa* fragments were recovered from electrophoresis gels and were redigested to completion with *Hin* dII + III. The redigestion products were identified by electrophoretic mobility (27).

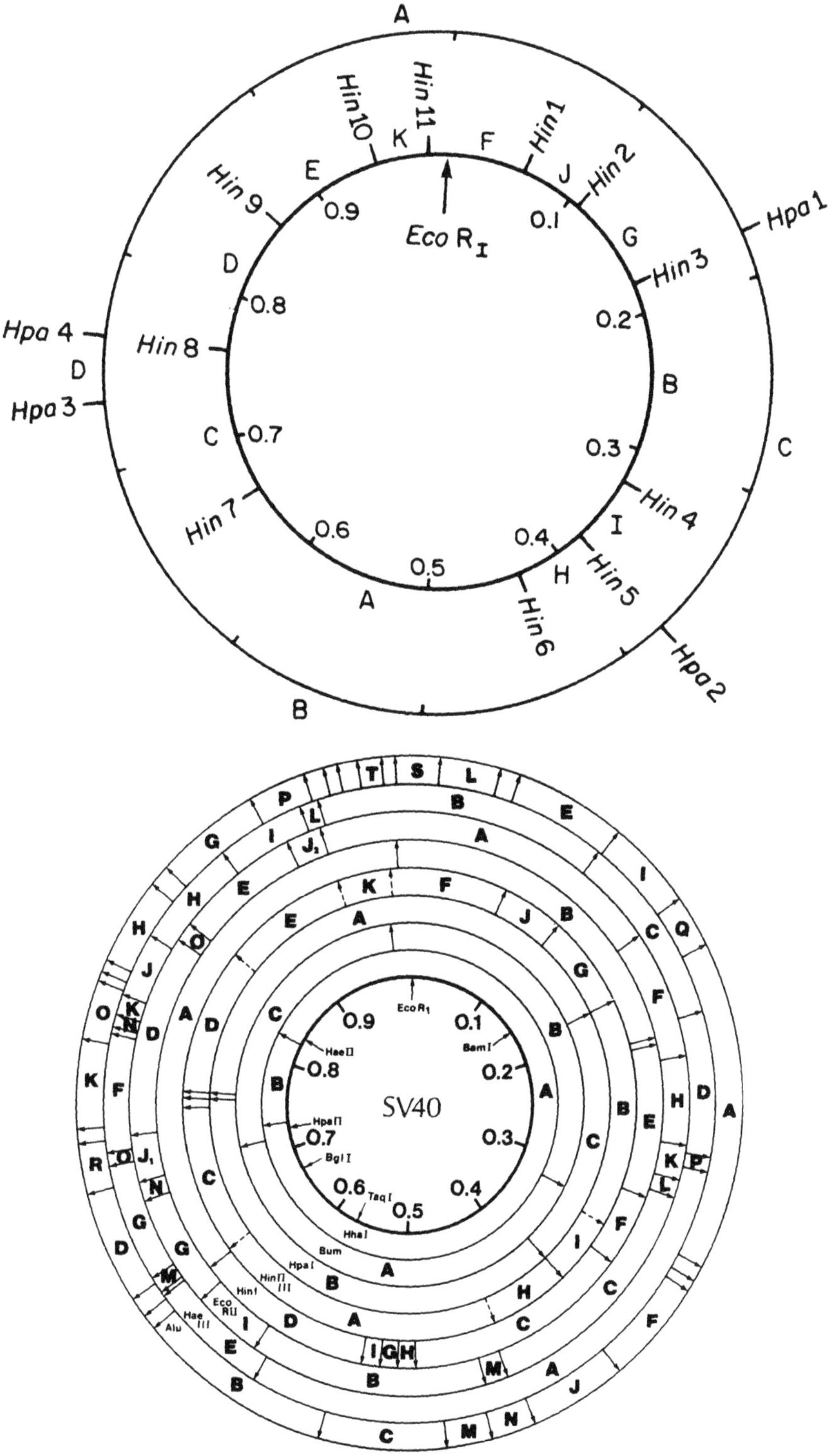

Fig. 3: Cleavage maps of SV40 DNA: a) the initial cleavage map (27), and b) a more recent map. (See (32) and (78) for references to positioning of cleavage sites.)

maps made such an analysis feasible. Soon after our isolation of restriction fragments of SV40 DNA, two groups became interested in carrying out nucleotide sequence analysis of the fragments—S. M. Weissman's laboratory in New Haven and W. Fiers' in Ghent. We were pleased to cooperate in the initial phase of their important work. At the outset, RNA transcripts of isolated fragments were used for sequencing. Later, with the development of rapid DNA sequencing methods by Sanger and Coulson (33) and by Maxam and Gilbert (34), direct sequencing of DNA fragments completed the analysis much sooner than originally expected, to the benefit of all investigators in this field. I am grateful to Sherman Weissman and Walter Fiers for making their sequence data freely available as the work progressed. Their nucleotide sequence map (35, 36), consisting of 5226 nucleotides, provides exact positions for each of the restriction sites in the cleavage map, and allows precise localization of genes and signals in the SV40 genome, as illustrated below.

Functional Map of SV40

The cleavage map and later the sequence map of the SV40 genome served as a framework for identifying functional elements of the viral DNA, for example, the origin and terminus of DNA replication, templates for viral messenger RNA's, and the positions of structural genes.

The origin and terminus of SV40 DNA replication were localized with respect to restriction sites in the DNA by pulse-labelling experiments (37) analogous to those of Dintzis on the rate and direction of globin biosynthesis (38), and by electron microscopic analysis of replicating SV40 DNA (39). The pulse-labelling experiments were carried out by exposing SV40-infected cells to ^{3}H-thymidine for a time period approximating that required for one round of viral DNA replication. DNA molecules whose replication was completed during this time interval were isolated and digested with a restriction endonuclease, and the amount of radioactivity in each restriction fragment determined. If there is a unique replication origin and terminus, fragments derived from the segment of the molecule synthesized last will be most highly labeled, and fragments derived from that segment synthesized first will have the least radioactivity. From the results (illustrated in Fig. 4) we could infer that SV40 DNA replication does begin at a unique site, approximately at map coordinate 0.67, proceeds bidirectionally around the circular genome, and terminates about 180° from the origin at about map coordinate 0.17. Similar experiments carried out with SV40 deletion mutants indicate that whereas the origin is at a fixed position and therefore must be determined by a structural feature of the DNA, the termination point is not fixed, but appears to represent the junction of the two growing forks opposite the origin (40).

Viral messenger RNA's were mapped in collaboration with George Khoury and Malcolm Martin (41, 42) and by Sambrook et al (43) by hybridization to restriction fragments of SV40 DNA. In summary, viral mRNA present in infected cells prior to the onset of viral DNA replication ("early" RNA)

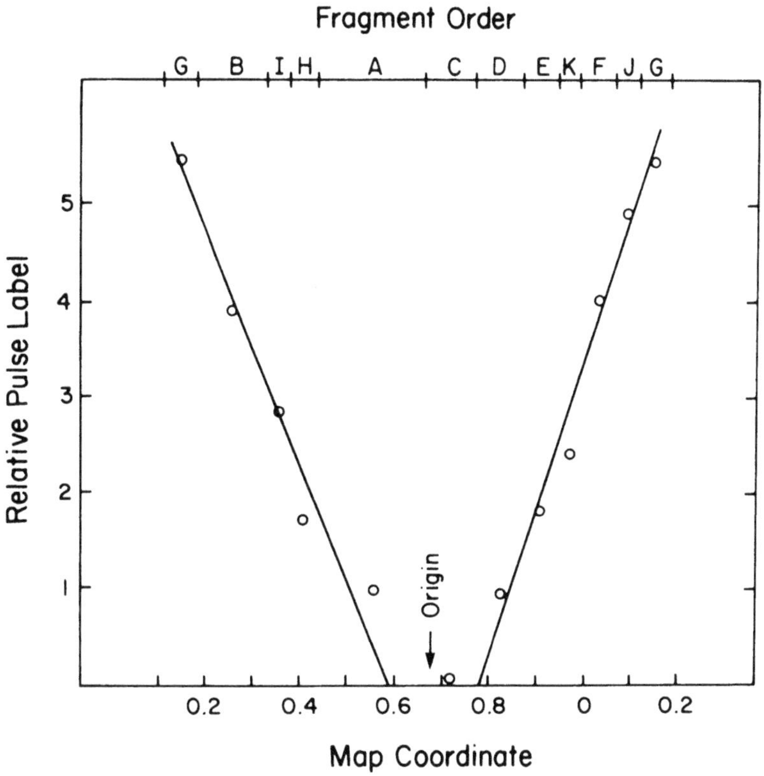

Fig. 4: Distribution of radioactivity in *Hin* fragments of pulse-labeled, newly completed molecules of SV40 DNA extracted from infected cells (37). The circular genome is shown in a linear form, with fragment G duplicated at the ends.

was derived from about half the genome (between map coordinates 0.17 and 0.67) by counterclockwise transcription. "Late" mRNA, i.e., the RNA that appears after the onset of viral DNA replication, was derived from the other half of the genome by clockwise transcription (Fig. 5a). Viral mRNA's from transformed cells were derived from the early genomic segment plus adjacent late regions (44). A particularly intriguing finding was the position of the replication origin between the start of the early and late genome regions, suggesting a regulatory coupling between replication and transcription (Fig. 5a).

Recent major refinements in the analysis of mRNA (45, 46) have allowed much more precise mapping of individual SV40 messengers, down to the nucleotide sequence level in some cases, including the nucleotide positions of segments spliced out of initial transcripts (46). Some of these more recent findings are summarized in the map shown in Fig. 5b, taken from a paper of Weissman and his co-workers (46).

To locate *structural genes* of SV40 on the cleavage map, Ching-Juh Lai and I (48) and Mathei et al (49) determined the mutational sites of temperature-sensitive (ts) mutants of SV40, isolated and characterized by

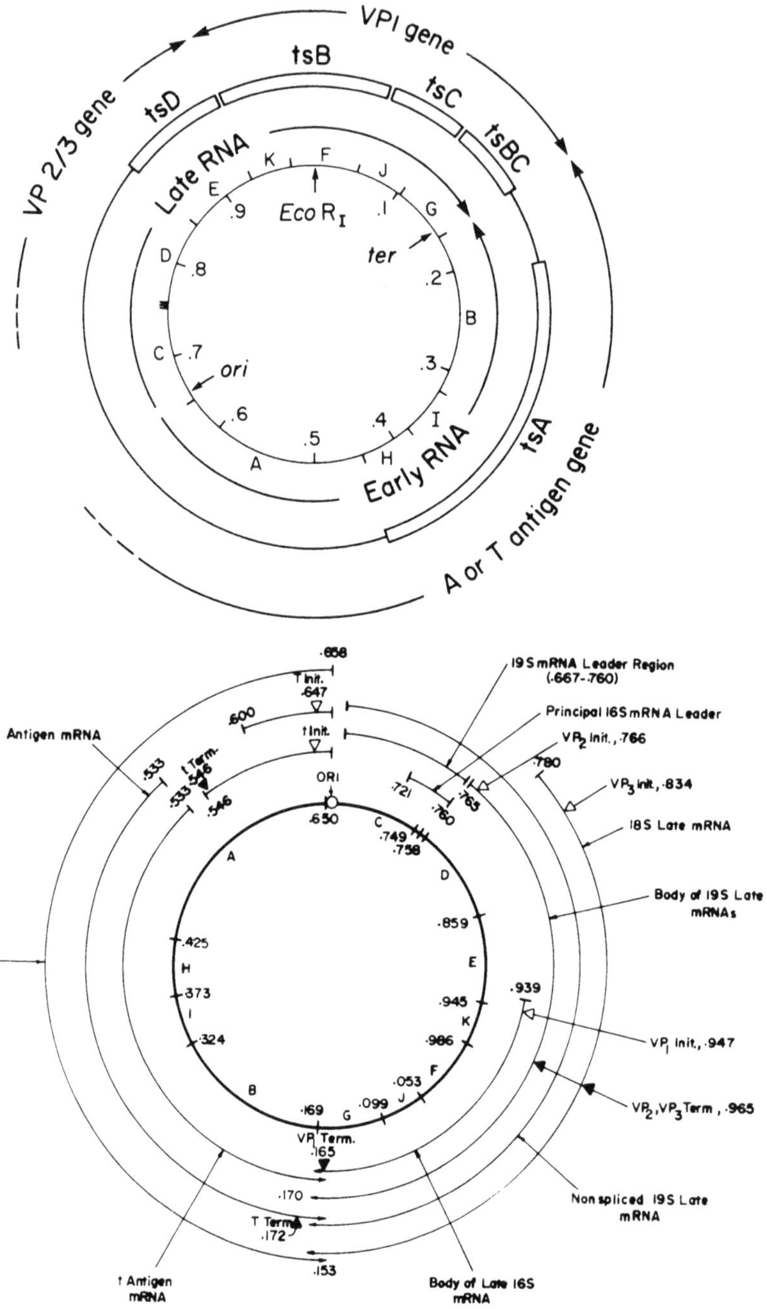

Fig. 5: Maps of viral functions relative to restriction sites: a) an initial map localizing the origin *(ori)* and terminus *(ter)* of DNA replication, early and late mRNA, the direction of transcription, and mutational sites of ts mutants (47). VP1, major viral capsid protein; VP2 and 3, minor viral capsid proteins. b) a recent map based on nucleotide sequence analysis of DNA and mRNA's (46). (Nucleotide position 1 of Reddy et al (35) is near the origin of replication, (here shown at the top of the circle). The nucleotide positions of segments coding for proteins are as fo⁰ (35, 36): t antigen, residues 5081 to 4559; T antigen, residues 5081 to 4837 and 4490 to 2612; VP1, residues 1423 to 2508; VP2, residues 480 to 1535; VP3, residues 834 to 1535.)

Tegtmeyer (50, 51) and Chou & Martin (52), by an adaptation of the "marker rescue" procedure devised for coliphage ΦX 174 (53, 54) (Fig. 6). In this method a single strand circle of mutant DNA is annealed with a single strand restriction fragment derived from wild type SV40 DNA to form a partial heteroduplex. Inside infected monkey cells the partial heteroduplex is repaired to form a duplex circle that has a mismatched base pair if the fragment overlaps the mutational site. By mismatch correction or replication a wild type genome is generated, and is scored by its ability to grow into a plaque under conditions where the mutant virus does not. By using a series of restriction fragments with each mutant to be mapped, we could determine which fragment overlapped a given mutational site. Since the position of each fragment in the cleavage map was known, we could localize any given ts mutation, and hence the genes in which the mutation resides, in the viral chromosome.

Lai's results are summarized in Fig. 5a. All of the tsA mutants (which are defective in initiation of viral DNA replication and in transformation (51, 52, 55, 56)) mapped in the early region of the genome between coordinate 0.20 and 0.43; tsB, C, and BC mutants (which are defective in a viral structural protein (50, 52, 57)) mapped between coordinates 0.94 and 0.17; and tsD mutants, (which are defective in a second viral structural protein (58)), mapped between 0.86 and 0.94 map units. Extensive segments of the genome were mutationally "silent". From the mapping of viral mRNA's and identification of their *in vitro* translation products (59–62), and from an analysis of deletion mutants of SV40 (see below), it is now

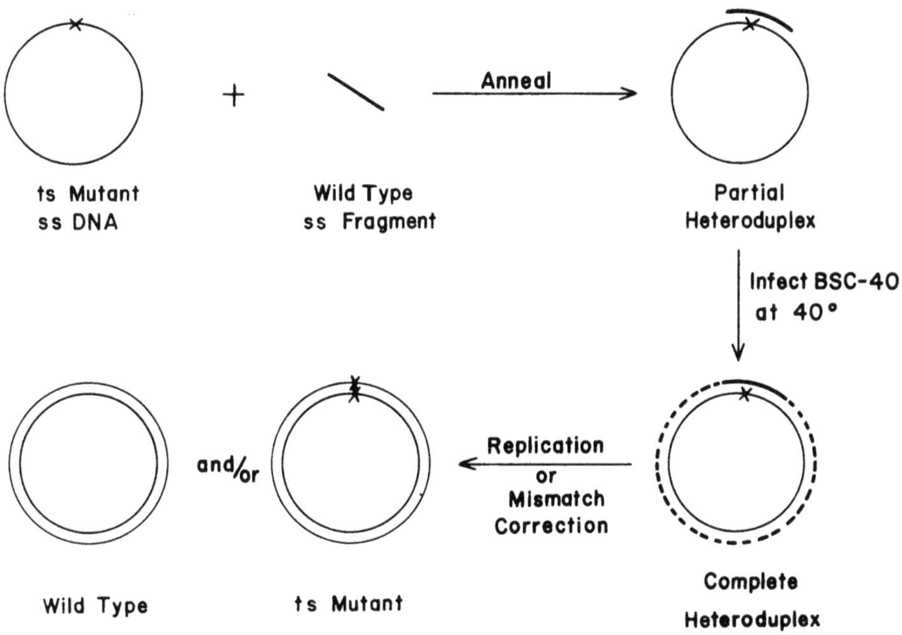

Fig. 6: Mapping of SV40 mutants by marker rescue (48). See text for a description of each step. x, mutational site in the DNA.

known that the A gene codes for SV40 tumor or T antigen (also known as the "A protein"), the B/C gene codes for the major virus structural protein (VP1), and the D gene codes for the overlapping minor virion proteins (VP 2 and 3). As shown in Fig. 5b, nuclotide sequence data has subsequently allowed precise localization of each of these genes, including overlapping in phase "late" sequences coding for the two minor structural proteins and overlapping in-phase "early" sequences coding for a second early protein, the so-called small t antigen (35, 36).

In vitro Construction of SV40 Mutants

The mutants just described were isolated by classical genetic techniques, namely by random mutagenesis and selection of desired phenotypes. As indicated in Fig. 5a, they covered only about half of the SV40 genome. With the advent of site-specific restriction endonucleases, it became possible to take a more active approach to mutational analysis of a DNA genome by creating mutations *in vitro* at preselected sites in the molecule. Site-selection is based on restriction enzyme cleavage of one or both strands of the DNA, and mutations result from enzymatic or chemical modification at or near restriction sites. From DNA thus modified, individual mutants can be isolated without the need for phenotype selection. Given functional and chemical maps of the genome, interesting regions can be selected for perturbation to determine the effect of such changes on the function of genetic elements or gene products. In the case of SV40, to which these methods have been applied most extensively, a series of mutants has been generated with deletions or base substitutions at predetermined sites in the viral genome. These are proving useful in the identification and characterization of gene products and of regulatory signals in the DNA.

Constructed deletion mutants

In general, SV40 deletion mutants are constructed by enzymatic opening of the circular genome to form slightly shortened linear molecules, followed by transfection of cells with the linear DNA (63, 64). Fig. 7 illustrates some of the ways to form linear molecules missing a small segment of the genome. When used to transfect cultured monkey cells, such linear molecules form covalently closed circles within the cell that are missing nucleotide sequences at the joint (Fig. 7). The cyclization process itself (the enzymatic mechanism of which is not understood) leads to variable loss of nucleotides from the ends of the transfecting molecule, thus generating an array of "extended" deletion mutants (63). (To avoid the formation of extended deletions, linear molecules can be cyclized enzymatically *in vivo* prior to infection of cells). If the overall loss of DNA does not remove a sequence essential for virus reproduction, deletion mutants can be isolated simply by selecting individual virus plaques arising in the infected cell monolayer (64). However, if the deletion of DNA leads to unconditional loss of function, the mutant must be isolated and propa-

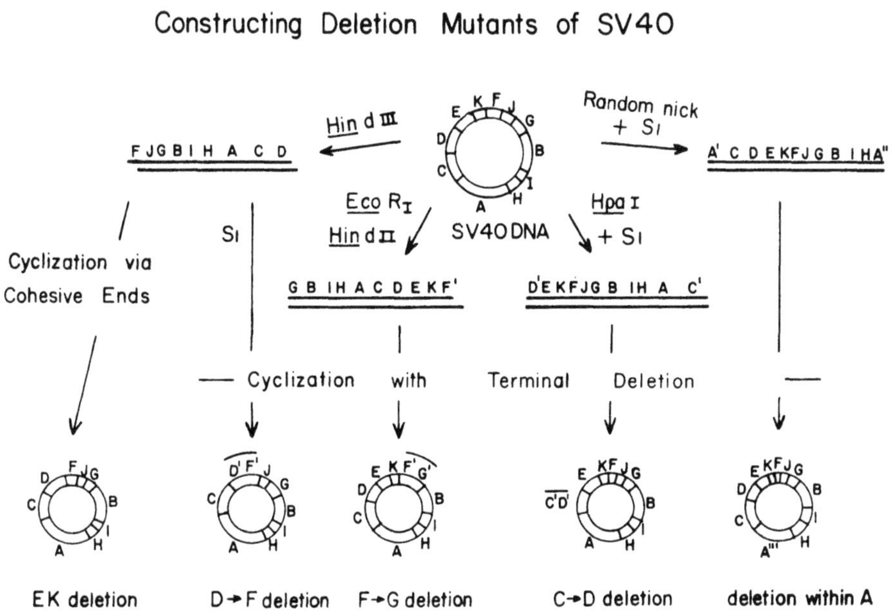

Fig. 7: Generation of deletion mutants of SV40 by enzymatic excision of nucleotides from the DNA followed by cell-mediated cyclization with terminal deletion (63, 64). Restriction enzymes are used to linearize the DNA at specific sites. Non-selective opening of the DNA (far right) can be used to produce a random set of deletions. S1 refers to a single strand specific nuclease (8), which nibbles the ends of linear DNA.

gated in the presence of a complementing "helper virus" (65, 66). Once a mutant is cloned, i.e., isolated in homogeneous form, the postion and extent of the deletion in its DNA can be determined by restriction enzyme analysis and subsequent nucleotide sequence determination.

SV40 deletion mutants have been particularly useful in identifying structural gene products, e.g., the T antigens found in SV40-infected or-transformed cells (67, 68); in locating non-essential parts of the DNA (69); in more precise localization of the origin of replication (70, 71); and in defining those regions required for cell transformation (72, 73, 74). Also important in localizing functions along the SV40 genome are related experiments on transformation by restriction fragments of SV40 DNA (75) and on the activity of microinjected fragments (76), and the studies of adeno-SV40 hybrid viruses containing SV40 DNA segments (77). As a result of these various investigations it became clear that the early region of the SV40 genome codes for the T antigens, as noted earlier, and that this region (plus immediately adjacent sequences) is sufficient for viral DNA replication and for cell transformation.

SV40 mutants with base substitutions at pre-selected sites. Mutants of SV40 with single base pair changes at pre-selected restriction sites have been constructed by David Shortle, using local chemical mutagenesis, as illustrated in Fig. 8 (78). In this procedure, viral DNA is incised in one strand with a restriction enzyme, the "nick" is converted to a small gap with an exonuclease, and bases exposed by the gap are then modified by

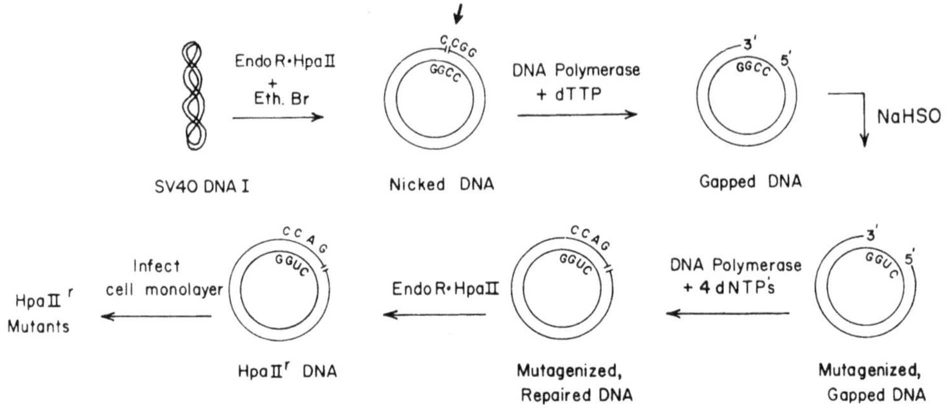

Fig. 8: Outline of the local mutagenesis procedure (78). See text for a description of each step. (Eth Br, ethidium bromide.)

reaction with a single-strand specific mutagen, such as sodium bisulfite, which deaminates cytosine to uracil. When such a bisulfite-treated gap is repaired either *in vitro* or inside a cell, a U-A pair is generated in place of the original C-G pair. Thus, an entire base pair substitution occurs, and the mutation cannot be reversed by cellular enzymes. If (as illustrated in Fig. 8) there is only one site in the genome for the restriction enzyme used to make the initial scission, the mutagenized, repaired DNA can be exposed to the same enzyme to eliminate those molecules that have escaped mutagenesis within the restriction site. When such enzyme-resistant molecules are used to infect cell monolayers, the majority of resulting virus clones contain mutants that have lost the enzyme site (78). Recent extensions of the local mutagenesis procedure have broadened the range of site selection considerably, so that many parts of a DNA molecule can be targeted for mutagenesis (79).

Constructed Regulatory Mutants of SV40

The local mutagenesis method just described has been used by Shortle and by Daniel Di Maio to construct mutants with single base pair substitutions within regulatory sequences of the viral DNA in and around the origin of replication (81). In one set of experiments SV40 DNA was nicked with restriction endonuclease *Bgl I,* which cuts the viral DNA once within a long symmetric sequence or "palindrome" at about map coordinate 0.67 (35), corresponding to the map position of the replication origin (Fig. 9). The *Bgl* I-nicked DNA was then gapped and locally mutagenized to generate *Bgl* I-resistant mutants. Nucleotide sequence analysis of several of these mutants revealed, in each case, a single base pair substitution within the palindrome (82) (Fig. 9). What is most interesting about these mutants is the effect of each base pair change on the rate of viral DNA replication (Table 2). A G/C to A/T change at position 5161, which forms the axis of symmetry of the palindrome, has no effect on the rate of DNA replication;

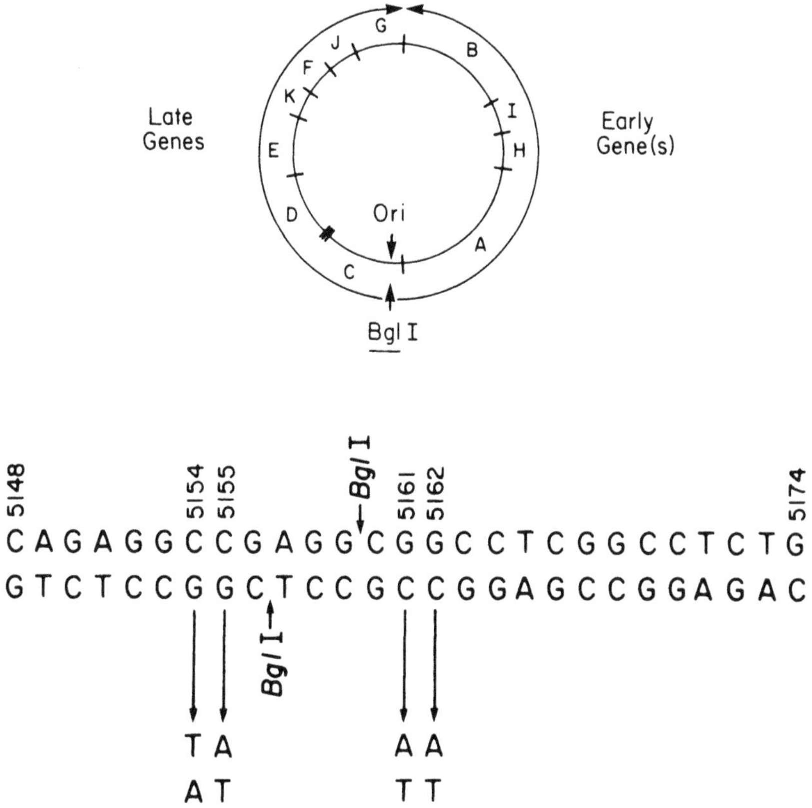

Fig. 9: Positions of base pair substitutions within the palindromic nucleotide sequence in DNA from *Bgl* I-resistant mutants of SV40 constructed by local mutagenesis, as described in the text. *Above* is shown the *Hin* cleavage map of the SV40 genome and the map position of the origin of DNA replication and of the *Bgl* I cleavage site. *Below* is the palindromic sequence at this map position, numbered as in (35). Nucleotide pair 5161 is the axis of symmetry of the palindrome, and the *Bgl* I cleavage sites are as indicated. The single base pair substitution in each of four phenotypically distinct mutants (at nucleotides 5154, 5155, 5161, or 5162) is indicated by an arrow.

TABLE 2

REPLICATION ORIGIN MUTANTS OF SV40

Base pair change	Plaque morphology	DNA replication
5161 G/C → A/T	wild type	normal
5162 G/C → A/T	small	decreased
5154 C/G → T/A	small at 32°	decreased at 32°
	wild type at 40°	normal at 40°
5155 C/G → A/T	small, sharp	increased

Note: Properties of mutants with base substitutions at the origin of viral DNA replication (82). See Figure 9 for nucleotide positions.

a G/C to A/T change at position 5162 causes a marked decrease in the DNA replication rate; a C/G to T/A change at position 5154 leads to a "cold-sensitive" replication phenotype (i.e., reduced at low temperature); and a C/G to A/T change at position 5155 causes an increased rate of DNA replication. Appropriate tests indicate that the replication-defective mutants have abnormlities in a *cis* element controlling the rate of viral DNA replication. The mutational alterations therefore serve operationally to define the origin sequence.

Our interpretation of the altered rates of mutant DNA replication is based on previous evidence of the involvement of the SV40 T antigen in initiation of viral DNA replication (51) and the preferential binding of this protein to a segment of SV40 DNA including that shown in Fig. 9 (83). The postulated first step in the replication of SV40 DNA is the specific binding of T antigen to the origin signal (Fig. 10). From the properties of origin mutants it appears that a single base pair change in the signal alters the binding site, leading to a change in the amount of T antigen bound or in the activity of the complex. In the cold-sensitive mutants, binding may be less efficient at 32° that at higher temperatures either because of a

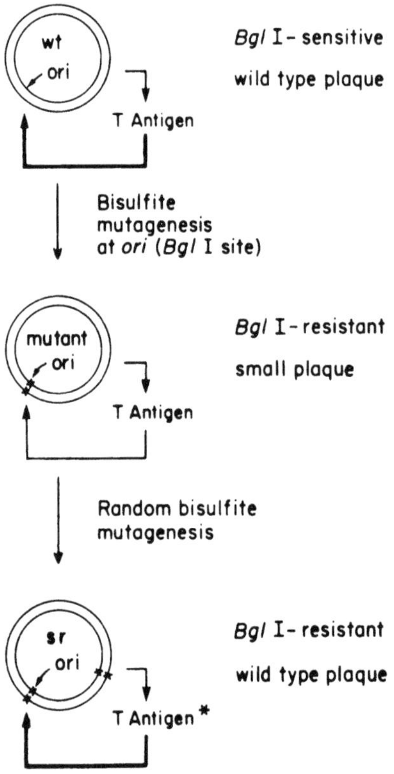

Fig. 10: Model of SV40 T antigen binding to the origin of replication in wild type (wt), *ori* mutant, and second-site revertant (sr) DNA. The thickness of the arrow from T antigen to *ori* reflects the hypothesized extent of binding. x, mutational site; T antigen *, mutant T antigen. On the right the phenotypes of wt, mutant, and second-site revertant viruses are noted.

temperature-dependent change in the binding site or as result of a change in the secondary structure of the T antigen. Recently, second-site revertants of one of the replication-defective origin mutants have been isolated. The mutation responsible for the reversion maps in the gene for T antigen. Therefore, these revertants may be producing T antigens that recognize the mutant origin sequence more efficiently than does the wild type T antigen (Fig. 10). Such double mutants could represent new viral replicons useful for biochemical investigations of T antigen functions.

A striking feature of SV40 origin mutants is that many are conditionally defective. Especially frequent is the cold-sensitive phenotype. This kind of temperature-dependence of specific DNA-protein interactions is well known from *in vitro* studies, e.g., in the case of bacterial RNA polymerase binding to DNA promotor sequences (84), and may be a general property of regulatory protein-nucleic acid interactions that could be exploited to isolate mutants with sequence changes within many different controlling elements in DNA or RNA.

Analysis of More Complex Chromosomes

The methods used to dissect the tiny genome of SV40 are directly applicable to more complex DNA molecules that can be isolated in homogeneous form (85): large viral chromosomes, plasmids, or DNA from cellular organelles. Even certain genes in mammalian DNA, whose complexity is some one million times that of the SV40 genome, have been mapped by restriction enzyme cleavage, using the sensitive detection method devised by Southern (86). However, the completely general application of restriction enzymes to the analysis of cellular chromosomes depends on recombinant techniques for cloning and amplifying individual DNA fragments from complex mixtures (87–91), and on the ability to introduce active genes back into living cells (92–95). These advances have opened the genome of every organism to the type of chemical and functional analysis I have described for SV40. Interesting findings have already emerged, for example the discontinuity of genes in eukaryotes (e.g., 96–98) and the mobility of gene segments during development (99); and experiments are underway to identify regulatory elements in cloned cellular DNA. In time it should be possible to make out the basic regulatory mechanisms used by plant and animal cells, and eventually to understand some of the complex genetic programs that govern the growth, development, and specialized functions of higher organisms, including man.

ACKNOWLEDGMENTS

It is a pleasure to acknowledge the contributions of my associates who carried out the research described in this article and shared its excitement with me: Stuart P. Adler, Kathleen J. Danna, Theresa N. H. Lee, George H. Sack, Jr., William W. Brockman, Elena O. Nightingale, Ching-Juh Lai, Walter A. Scott, Nicholas Muzyczka, Mary W. Gutai, David R. Shortle,

James Pipas, Sondra Lazarowitz, and Daniel DiMaio. I am grateful also to my colleagues, Hamilton O. Smith and Thomas J. Kelly, Jr., for their sound critique and generous help over a period of many years; to George Khoury, Malcolm A. Martin, and Peter Tegtmeyer for enjoyable collaborations; and to the U.S. National Cancer Institute, the American Cancer Society, and the Whitehall Foundation for research support.

REFERENCES

1. Luria, S. E. and Human, S. L. J. Bacteriology *64,* 557, 1952.
2. Bertani, G. and Weigle, J. J. J. Bacteriology *65,* 113, 1953.
3. Arber, W. and Dussoix, D. J. Mol. Biol. *5,* 18, 1962.
4. Dussoix, D. and Arber, W. J. Mol. Biol. *5,* 37, 1962.
5. Arber, W. Ann. Rev. Microbiol. *19,* 365, 1965.
6. Meselson, M. and Yuan, R. Nature, *217,* 1110, 1968.
7. Smith, H. O. and Wilcox, K. Fed. Proc. (Abstr.) Amer. Soc. Exp. Biol. *28,* 465, 1969.
8. Shimura, Y., Kaizer, H. and Nathans, D. J. Mol. Biol. *38,* 453, 1968.
9. Sweet, B. H. and Hilleman, M. R. Proc. Soc. Exp. Biol. Med. *105,* 420, 1960.
10. Crawford, L. V. and Black, P. H. Virology *24,* 388, 1964.
11. Germond, J. E., Hirt, B., Oudet, P., Gross-Bellard, M. and Chambon, P. Proc. Natl. Acad. Sci. (USA) *72,* 1843, 1975.
12. Griffith, J. Science *187,* 1202, 1975.
13. Minowada, J. and Moore, G. E. Exp. Cell Res. *29,* 31, 1963.
14. Shein, H. M. and Enders, J. F. Proc. Natl. Acad. Sci. (USA) *48,* 1164, 1962.
15. Anderer, F. A., Schlumberger, H. D., Koch, M. A., Frank, H. and Eggers, H. J. Virology *32,* 511, 1967.
16. Vinograd, J. and Lebowitz, J. J. Gen. Physiol. *49,* 103, 1966.
17. Linn, S. and Arber, W. Proc. Natl. Acad. Sci. (USA) *59,* 1300, 1968.
18. Haberman, A., Heywood, J., Meselson, M. Proc. Nat. Acad. Sci. (USA) *69,* 3138, 1972.
19. Adler, S. P. and Nathans, D. Fed. Proc. (Abstr.) Amer. Soc. Exp. Biol. *29,* 725, 1970.
20. Adler, S. P. and Nathans, D. Biochim. Biophys. Acta *299,* 177. 1973.
21. Horiuchi, K. and Zinder, N. D. Proc. Natl. Acad. Sci. (USA) *69,* 3220, 1972.
22. Morrow, J. F. and Berg, P. Proc. Natl. Acad. Sci. (USA) *69,* 3365, 1972.
23. Smith, H. O. and Wilcox, K. W. J. Mol. Biol. *51,* 379, 1970.
24. Kelly, T. J., Jr., and Smith, H. O. J. Mol. Biol. *51,* 393, 1970.
25. Danna, K. J. and Nathans, D. Proc. Natl. Acad. Sci. (USA) *68,* 2913, 1971.
26. Roy, P. H. and Smith, H. O. J. Mol. Biol. *81,* 445, 1973.
27. Danna, K. J., Sack, G. H., Jr., and Nathans, D. J. Mol. Biol. *78,* 363, 1973.
28. Roberts, R. J. Gene, *4,* 183, 1978.
29. Hedgpeth, J., Goodman, H. M., and Boyer, H. Proc. Natl. Acad. Sci. (USA) *69,* 3448, 1972.
30. Gromkova, R. and Goodgal, S. H. J. Bacteriol. *109,* 987, 1972.
31. Mulder, C. and Delius, H. Proc. Natl. Acad. Sci. (USA) *69,* 3215, 1972.
32. Kelly, T. J., Jr. and Nathans, D. Adv. Virus Res. *21,* 85, 1977.
33. Sanger, F. and Coulson, A. R. J. Mol. Biol. *94,* 441, 1975; Sanger, F., Nicklen, S., and Coulson, A. R. Proc. Natl. Acad. Sci. (USA) *74,* 5463, 1977.
34. Maxam, A. and Gilbert, W. Proc. Natl. Acad. Sci. (USA) *74,* 560, 1977.
35. Reddy, V. B., Thimappaya, B., Dhar, R., Subramanian, K. N., Zain, B. S., Pan, J., Ghosh, P. K., Celma, M. L. and Weissman, S. M. Science *200,* 494, 1978.
36. Fiers, W., Contreras, R., Haegeman, G., Rogiers, R., Van de Voorde, A., Van Heuverswyn, H., Van Herreweghe, J., Volckgert, G., and Ysebaert, M. Nature *273,* 113, 1978.
37. Danna, K. J. and Nathans, D. Proc. Natl. Acad. Sci. (USA) *69,* 3097, 1972.
38. Dintzis, H. M. Proc. Natl. Acad. Sci. (USA) *47,* 247, 1961.
39. Fareed, G. C., Garon, C. F. and Salzman, N. P. J. Virol. *10,* 484, 1972.
40. Lai, C. J. and Nathans, D. J. Mol. Biol. *97,* 113, 1975.
41. Khoury, G., Martin, M. A., Lee, T. N. H., Danna, K. J. and Nathans, D. J. Mol. Biol. *78,* 377, 1973.
42. Khoury, G., Martin, M. A., Lee, T. N. H. and Nathans, D. Virology *63,* 263, 1975.
43. Sambrook, J., Sugden, B., Keller, W., and Sharp, P. A. Proc. Natl. Acad. Sci. (USA) *70,* 3711, 1973.
44. Khoury, G., Howley, P., Nathans, D. and Martin, M. A. J. Virol. *15,* 433, 1975.

45. Berk, A. J. and Sharp, P. A. Proc. Natl. Acad. Sci. (USA) *75,* 1274, 1978.
46. Ghosh, P. K., Reddy, V. B., Swinscoe, J., Lebowitz, P. and Weissman, S. M. J. Mol. Biol. *126,* 813, 1978.
47. Nathans, D. The Harvey Lectures, Series 70, p. 111, 1976.
48. Lai, C. J. and Nathans, D. Virology *60,* 466, 1974; *66,* 78, 1975.
49. Mantei, N., Boyer, H. W., and Goodman, H. M. J. Virol. *16,* 754, 1975.
50. Tegtmeyer, P. and Ozer, H. J. J. Virol. *8,* 516, 1971.
51. Tegtmeyer, P. J. Virol. *10,* 591, 1972.
52. Chou, J. Y. and Martin, R. G. J. Virol. *13,* 1101, 1974.
53. Hutchison, C. A., III, and Edgell, M. H. J. Virol. *8,* 181, 1971.
54. Weisbeek, P. J. and Van de Pol, J. H. Biochim. Biophys. Acta *224,* 328, 1970.
55. Kimura, G. and Dulbecco, R. Virology *52,* 529, 1973.
56. Kimura, G. and Dulbecco, R. Virology *49,* 394, 1972.
57. Martin, R. M. and Chou, J. J. Virol. *15,* 599, 1975.
58. Robb, J. A. and Martin, R. G. J. Virol. *9,* 956, 1972.
59. Prives, C. L., Aviv, H., Gilboa, E., Revel, M., and Winocour, E. Cold Spring Harbor Symposium *39,* 309, 1974.
60. Prives, C. L., Gilboa, E., Revel, M., and Winocour, E. Proc. Natl. Acad. Sci. (USA) *74,* 457, 1977.
61. Greenblatt, J. F., Allet, B. and Weil, R., and Ahmad-Zadeh, C. J. Mol. Biol. *108,* 361, 1976.
62. Paucha, E., Mellor, A., Harvey, R., Smith, A. E., Hewick, R. W., and Waterfield, M. D. Proc. Natl. Acad. Sci. (USA) *75,* 2165, 1978.
63. Lai, C. J. and Nathans, D. J. Mol. Biol. *89,* 179, 1974.
64. Carbon, J., Shenk, T. and Berg, P. Proc. Natl. Acad. Sci. (USA) *72,* 1392, 1975.
65. Brockman, W. W. and Nathans, D. Proc. Natl. Acad. Sci. (USA) *71,* 942, 1974.
66. Mertz, J. E. and Berg, P. Virology *62,* 112, 1974.
67. Rundell, K., Tegtmeyer, P., Ozer, H. L., Lai, C. J. and Nathans, D. J. Virol. *21,* 636, 1977.
68. Crawford, L. V., Cole, C. N., Smith, A. E., Paucha, E., Tegtmeyer, P., Rundell, K., and Berg, P. Proc. Natl. Acad. Sci. (USA) *75,* 117, 1978.
69. Shenk, T. E. Carbon, J. and Berg, P. J. Virol. *18,* 644, 1976.
70. Subramanian, K. N. and Shenk, T. Nucleic Acids Research *5,* 3635, 1978.
71. DiMaio, D. and Nathans, D. Manuscript in preparation.
72. Scott, W. A., Brockman, W. W. and Nathans, D. Virology *75,* 319, 1976.
73. Sleigh, M. J., Topp, W. C., Hanich, R. and Sambrook, J. F. Cell *14,* 79, 1978.
74. Bouck, N. Beales, N., Shenk, T., Berg, P. and DiMayorca, G. Proc. Natl. Acad. Sci. (USA) *75,* 2473, 1978.
75. Graham, F. L., Abrahams, P. J., Mulder, C. Heijneker, H. L., Warnaar, S. O., de Vries, F. A. J., Fiers, W. and van der Eb, A. J. Cold Spring Harbor Symposium *39,* 637, 1974.
76. Mueller, C., Graessmann, A. and Graessmann, M. Cell *15,* 579, 1978.
77. Lewis, A. M., Jr., Breeden, J. H., Wewerka, Y. L., Schnipper, L. E., and Levine, A. S. Cold Spring Harbor Symposium *39,* 651, 1974.
78. Shortle, D. and Nathans, D. Proc. Natl. Acad. Sci. (USA) *75,* 2170, 1978.
79. Shortle, D. Pipas, J., Lazarowitz, S., DiMaio, D., and Nathans, D. in "Genetic Engineering" Plenum Press, N. Y., 1979, p. 73.
80. Ando, T. Biochim. Biophys. Acta *114,* 158, 1966.
81. Shortle, D. and Nathans, D. Cold Spring Harbor Symposium *43,* 1978, in press.
82. Shortle, D. and Nathans, D. J. Mol. Biol., 1979, in press.
83. Tjian, R. Cell *13,* 165, 1978.
84. Chamberlin, M. J. in *RNA Polymerase* ed. by Losick and Chamberlin, Cold Spring Harbor Laboratory, 1976.
85. Nathans, D. and Smith, H. O. Ann. Rev. Biochem. *44,* 273, 1975.
86. Southern, E. J. Mol. Biol. *98,* 503, 1975.
87. Jackson, D. A., Symon, R. H., and Berg, P. Proc. Natl. Acad. Sci. (USA) *69,* 2904, 1972.

88. Lobban, P. E. and Kaiser, A. K. J. Mol. Biol. *78*, 453, 1973.
89. Cohen, S. N., Chang, A. C. Y., Boyer, H. W., and Helling, R. B. Proc. Natl. Acad. Sci. (USA) *70*, 3240, 1973.
90. Morrow, J. F., Cohen, S. N., Chang, A. C. Y., Boyer, H. W., Goodman, H. M. and Helling, R. B. Proc. Natl. Acad. Sci. (USA) *71*, 1743, 1974.
91. Wensink, P. C., Finnegan, D. J., Donelson, J. E. and Hogness, D. S. Cell *3*, 315, 1974.
92. Brown, D. D. and Gurdon, J. B. Proc. Natl. Acad. Sci. (USA) *74*, 2064, 1977.
93. Struhl, K. and Davis, R. W. Proc. Natl. Acad. Sci. (USA) *74*, 5255, 1977.
94. Hamer, D. H., Davoli, D., Thomas, C. A., Jr., and Fareed, G. C. J. Mol. Biol. *112*, 155, 1977.
95. Wigler, M., Pellicer, A., Silverstein, S., and Axel, R. Cell *14*, 725, 1978.
96. Brack, C. and Tonegawa, S. Proc. Natl. Acad. Sci. (USA) *74*, 5652, 1977.
97. Glover, D. M., and Hogness, D. S. Cell *10*, 167, 1977.
98. Tilghman, S. M., Tiemeier, D. C., Seidman, J. G., Peterlin, B. M., Sullivan, M., Maizel, J. V. and Leder, P. Proc. Natl. Acad. Sci (USA) *75*, 725, 1978.
99. Hozumi, N. and Tonegawa, S. Proc. Natl. Acad. Sci (USA) *73*, 3628, 1976.
100. Smith, H. O. and Nathans, D. J. Mol. Biol. *81*, 419, 1973.

Hamilton O. Smith

HAMILTON O. SMITH

My mother and father each came from simple country backgrounds, but both showed an early inclination for scholarly pursuits. They eventually met as school teachers in a local Panama City, Florida high school and were married in 1929. The following year, my father was appointed Assistant Professor of Education at the University of Florida at Gainesville, and in that year my brother was born. In 1931, my father went on leave to Columbia Universtity in New York City to complete his doctoral work in education. I was born there on August 23, 1931 while he was a graduate student. Though the family commuted annually between New York City and Gainesville over the next five years, I retain the strongest memories of our life in the city. In particular are recollections of life in a small, intimate apartment, walks in the city parks, and quiet evenings spent with my mother and father who entertained us with arithmetic problems and a small Gilbert chemistry set.

In 1937, our family moved to Champaign-Urbana, Illinois. My father had joined the faculty of the Department of Education at the University of Illinois and was to spend the major part of his academic career there until retirement. My entire boyhood was spent in this small midwestern academic community. Despite the fact that our life in Urbana spanned the late depression years and World War II, the community continued to function pretty much as if untouched by world events. At home, an atmosphere of intense intellectualism was maintained. My father was perpetually working and writing. At the same time, my mother struggled to establish herself as a writer, but she was to remain frustrated in her ambitions. However, she, in particular, imbued us with a respect and desire for the creative life.

My brother and I received private French lessons during our pre-teen years. I began piano lessons at age eight and my brother took up violin. We studied with a talented musical family, the Fosters and Sonderskovs. I was in no way gifted and found practice to be a chore until one memorable day when I was about age thirteen. On that day, a friend introduced me to the local music shop and by chance I picked up a recording of Artur Rubinstein playing Beethoven's Pathetique Sonata. I had been struggling with the piece for sometime, but had never appreciated its dramatic beauty. Listening to Rubinstein's magnificent performance for the first time was a truly awakening experience. From then on I became a devoted pupil and music lover.

My boyhood friends were mostly sons of university faculty. Our interests included football, basketball, music, chemistry, electricity, and electronics.

My brother and I spent many hours in our basement laboratory stocked with supplies purchased from our paper route earnings. We attended University High School, a superb small college preparatory school with an array of exceptionally talented students drawn largely from university faculty families. To my knowledge, two Nobel Laureates are counted among "Uni-High's"graduates, as well as numerous successful professionals, and no less than three current professors at Johns Hopkins. I completed high school in three years largely due to a wonderful science teacher, Wilbur E. Harnish, who allowed me to complete chemistry and physics during the two summers preceding ninth grade. Two other teachers at "Uni-High" influenced my development profoundly: Vynce Hines, who taught me the beauties and rigor of plane geometry and Miles C. Hartley, who gave me a sound foundation in algebra.

After completing high school, I matriculated at the University of Illinois, majoring in mathematics for which I had a flair but no deep talent. I had not yet decided on a particular field of science. My brother, who was considerably more gifted in the abstract areas than I, was studying theoretical physics, but this did not appeal strongly to me, nor was I interested in pure chemistry. At that time biology, as taught, was largely descriptive. It was not especially appealing for one brought up on "real" science. However, during my sophomore year, my brother introduced me to a book on mathematical modeling of central nervous system circuits by a biophysicist named Rashevsky. It caught my interest and I began reading about the nervous system. I continued this interest after transferring to the University of California at Berkeley in 1950. There, for the first time, I found courses in cell physiology, biochemistry, and biology that interested me. I recall in particular at that time, a guest lecture by George Wald describing his studies of retinal biochemistry. I was converted overnight into an avid student of visual physiology. It had become clear that mathematics, while providing an excellent basic training, was not my real interest. With a broadening appreciation of biology and a budding interest in human visual- and neurophysiology, I decided to apply to medical school.

In 1952, I began my studies at the Johns Hopkins University Medical School in Baltimore, Maryland. I was immediately caught up in the excitement of a new kind of life, and without firm commitments to any particular area of research, I was to continue a fairly conventional medical career for several years. I received my M. D. degree from Hopkins in 1956 and proceeded to Barnes Hospital in St. Louis for a medical internship. There I experienced for the first time a true feeling of freedom and independence. In my second month of internship I met Elizabeth Anne Bolton, a young nursing student. She was from a family of doctors and engineers, had been born in Spain, reared in Mexico City, and had come to the States for college and nurses' training. We immediately liked each other, and a few months later, were married.

In July, 1957, I was called up in the Doctor's Draft, and rather than seek any of several avenues of deferment, decided it was an opportune time to be done with my service obligation. I chose the Navy and we received a two year assignment in San Diego, California. It was a relaxed and easy time for us after so many years of schooling. For the first time in my life I was faced with greatly reduced demands on my time and the problem of idleness. I began to search for ways to occupy myself. A report of the then new research in human chromosomal aberrations caught my interest. Soon I was reading textbooks on genetics. Because of my mathematical background, I delved deeply into the population genetics of Sewell Wright and Ronald Fisher.

In 1959, with the Navy service completed, my wife and I moved to Detroit, Michigan with our one-year old son to begin my medical residency training at the Henry Ford Hospital. There I found a well-stocked library that included "Bacteriophage" by Mark Adams, the first issues of the Journal of Molecular Biology containing the classical Jacob and Monod paper describing the operon model for gene regulation, and two collections of papers by Adelberg and Stent. I suddenly became aware of the beautiful work of the "phage school" and of Watson and Crick and DNA. After many years of haphazard searching for the "right" area of research, I knew I had found it.

In 1962, armed with a N.I.H. postdoctoral fellowship, I began my research career with Myron Levine in the Department of Human Genetics at the University of Michigan in Ann Arbor. Mike was a geneticist studying *Salmonella* Phage P22 lysogeny. The choice to work with him, while governed more by expediency than by considered planning, turned out to be most fortuitous. Mike was an easy-going young investigator with a solid phage background and well established among the phage crowd. He allowed me just the right blend of independence and encouragement. Together we carried out a series of studies demonstrating the sequential action of the P22 *C*-genes which controlled lysogenization. In 1965, we discovered the gene controlling prophage attachment, now known as the *int* gene. By 1967, I had published this work and had carried out a study of defective transducing particles formed after induction of *int* mutant prophage. During 1966–67, Mike took a sabbatical year with Werner Arber in Geneva and through correspondence, I learned for the first time about Arber's remarkable work on restriction and modification phenomenon in bacteria.

In 1967, I came to Johns Hopkins as an Assistant Professor of Microbiology and have remained there since. My research work includes studies of restriction and modification enzymes, enzymology of genetic recombination, mechanism of bacterial transformation, and genetic regulation in prokaryotes and eukaryotes. In 1975–76, as a Guggenheim Fellow, I collaborated with Max Birnstiel at the Unversity of Zurich in Switzerland on histone gene arrangement and sequence. It was a superbly enriching year for both myself and for my family.

My wife, Elizabeth, is artistically inclined, enjoys a variety of "Handarbeit", sings in a church choir, enjoys classical music, and is a moderately proficient linguist (English, Spanish, German, and French). My major non-scientific diversions are classical music and piano. We have four sons and a daughter, none of whom currently indicate a strong interest in science.

Principal works:

Smith, H. O. and Wilcox, K. W. A restriction enzyme from *Hemophilus influenzae*. I. Purification and general properties. J. Mol. Biol. *51,* 379 (1970).

Kelly, T. J., Jr. and Smith, H. O. A restriction enzyme from *Hemophilus influenzae*. II. Base sequence of the recognition site. J. Mol. Biol. *51,* 393 (1970).

Roy, P. H. and Smith, H. O. The DNA methylases of *Hemophilus influenzae* Rd. I. Purification and properties. J. Mol. Biol. *81,* 427 (1973).

Roy, P. H. and Smith, H. O. The DNA methylases of *Hemophilus influenzae* Rd. II. Partial recognition site base sequences. J. Mol. Biol. *81,* 445 (1973).

NUCLEOTIDE SEQUENCE SPECIFICITY OF RESTRICTION ENDONUCLEASES

Nobel Lecture, 8 December, 1978

by

HAMILTON O. SMITH
Department of Microbiology, The Johns Hopkins University School of Medicine Baltimore, Maryland, U. S. A.

INTRODUCTION

In the past seven to eight years we have witnessed the development of a new DNA technology that has fundamentally altered our approach to modern genetics. The basic ingredients of this new technology are the cleavage-site-specific restriction enzymes: a special class of bacterial endonucleases that can recognize specific nucleotide sequences in duplex DNA and produce double-stranded cleavages. Using a collection of these enzymes, each with its own particular sequence specificity, DNA molecules may be cleaved into unique sets of fragments useful for DNA sequencing, chromosome analysis, gene isolation, and construction of recombinant DNA. The latter, combined with the concept of molecular cloning, has given birth to the new field of genetic engineering, and from this are expected many new and exciting medical and research applications.

My own role in these developments occurred primarily in the period of 1968–1970 when my colleagues and I made the chance discovery of the first of the cleavage-site-specific restriction enzymes. I should like to briefly describe this work in historical context as it leads naturally into the main part of my lecture describing our present knowledge of restriction and modification enzymes. I shall not go into the many applications as these have been reviewed elsewhere (1). However, I should like to describe in some detail the use of these enzymes as model systems for studying sequence-specific protein-DNA interactions since this is one of our major research interests.

Restriction and Modification in Bacteria: the Discovery of Restriction Enzymes

The observations leading to the discovery of restriction enzymes span a period of nearly two decades and constitute a prime example of how basic research on an apparently insignificant bacteriological phenomenon has had unexpectedly far-reaching implications. The story begins in the early 1950s with some observations by Luria and Human (1a) and Bertani and Weigle (2) concerning the curious behavior of phage grown on two different strains of bacteria. Phage propagated on one strain were found to grow poorly (were "restricted") on the second, and vice-versa. However, a few phage always escaped restriction and could then grow well on the new host. They apparently had acquired some type of host-specific modification that then protected them from the restriction effects of the host. The biochemical basis of this phenomenon remained a mystery until the early 1960s when Werner Arber and co-workers were able to show that host-specific modifications was carried on the phage DNA (3), and that restriction was associated with degradation of the phage DNA (4). In a remarkably prophetic review in 1965, Arber postulated the existence of site-specific restriction enzymes and suggested that modification might be produced by hostspecific DNA methylases (5). Thus, the notion became established that each restriction and modification (R–M) system in bacteria was composed of two enzymes with identical specificity: a restriction endonuclease that recognized short nucleotide sequences and cleaved DNA, and a modification enzyme that recognized the same sequence and modified it to protect against cleavage. In this way, the host cell DNA would be protected but foreign DNA entering from outside with improper modification would be cleaved and destroyed.

Although the existence of restriction enzymes was predicted with confidence by 1965, it was not until early 1968 that Linn and Arber (6) actually found an activity in extracts of *E. coli* B with the expected properties of such an enzyme. At the same time, Meselson and Yuan (7) reported more extensive experiments with a highly purified restriction endonuclease from *E. coli* K. Using sucrose gradient centrifugation, the latter demonstrated that their enzyme cleaved unmodified phage λ DNA into large fragments while modified DNA remained undegraded. An unusual feature of the enzyme was its requirement for the cofactors S-adenosylmethionine, ATP, and Mg^{2+}. Meselson and Yuan assumed from Arber's work that their enzyme was attacking the λ DNA at fixed sites, but were unable to confirm this by sucrose gradient analysis of the fragment species.

It is now known that restriction enzymes of *E. coli* B and K are examples of a class of restriction enzymes that do not cleave DNA at specific sites, although this fact was not appreciated for several years. Such Class I enzymes are complex, multimeric proteins that generally require ATP, S-adenosylmethionine and Mg^{2+} as cofactors and function both as restriction endonucleases and as modification methylases (8). Although they recognize specific sites in the DNA, they cleave randomly at a considerable distance from the recognition site (9, 10, 11). Because of this property, they have not proven useful as enzymatic tools for DNA analysis.

Discovery of a Cleavage-Site-Specific Restriction Endonuclease in Haemophilus influenzae Rd

In early Spring of 1968, I read the Meselson and Yuan paper with great interest. Their work imparted, in a very explicit way, a sense of biochemical reality to Arber's observations. I had at that time recently joined the faculty of the Department of Microbiology at Johns Hopkins and, with a young graduate student named Kent Wilcox, was just beginning to explore genetic recombination in *Haemophilus influenzae,* strain Rd, an efficiently transformable bacterium that we had been introduced to by Roger Herriott of the School of Hygiene. Some of our experiments employed a viscometer as a particularly sensitive measure of endonucleolytic cleavage of DNA by cell extracts. Other experiments involved recovery of donor DNA from cells after uptake. In one such experiment we happened to use labeled DNA from phage P22, a bacterial virus I had worked with for several years before coming to Hopkins. To our surprise, we could not recover the foreign DNA from the cells. With Meselson's recent report in our minds, we immediately suspected that it might be undergoing restriction, and our experience with viscometry told us that this would be a good assay for such an activity. The following day, two viscometers were set up, one containing P22 DNA and the other, *Haemophilus* DNA. Cell extract was added to each and we began quickly taking measurements. As the experiment progressed, we became increasingly excited as the viscosity of the *Haemophilus* DNA held steady while the P22 DNA viscosity fell. We were confident that we had discovered a new and highly active restriction enzyme. Furthermore, it appeared to require only Mg^{2+} as a cofactor, suggesting that it would prove to be a simpler enzyme than that from *E. coli* K or B. From that point on, other work in the laboratory was shelved while we turned our full attention to the isolation and study of the new enzyme.

After several false starts and many tedious hours with our laborious, but sensitive viscometer assay, Wilcox and I (12) succeeded in obtaining a purified preparation of the restriction enzyme. We next used sucrose gradient centrifugation to show that the purified enzyme selectively degraded duplex, but not single-stranded, P22 DNA to fragments averaging around 100 base pairs in length, while *Haemophilus* DNA present in the same reaction mixture was untouched. No free nucleotides were released during the reaction, nor could we detect any nicks in the DNA products. Thus, the enzyme was clearly an endonuclease that produced double-strand breaks and was specific for foreign DNA. Since the final (limit) digestion products of foreign DNA remained large, it seemed to us that cleavage must be site-specific. This proved to be case and we were able to demonstrate it directly by sequencing the termini of the cleavage fragments.

Sequencing the Recognition Site

We began our sequencing efforts in late 1968 using a method that had been worked out by Bernard Weiss and Charles C. Richardson at Harvard. The method involved labeling the 5'-termini of DNA with radioactive phosphorus using T4 polynucleotide kinase and ^{32}P gamma-labeled ATP, followed by digestion with pancreatic DNase and either venom phosphodiesterase to yield terminal nucleotides, or exonuclease I to yield terminal dinucleotides (13). These could then be separated by electrophoresis and identified by comparison with known marker nucleotides.

Weiss had come to Hopkins in the Fall of 1967 and occupied a neighboring laboratory in the Microbiology Department. He instructed us in the procedure and supplied us with both the kinase and the ^{32}P-gamma-labeled

ATP. We started our sequencing using restriction enzyme digests of phage T7 DNA, a fortuitous choice as we later learned. In our first experiment, we found it necessary to treat with alkaline phosphatase to remove a terminal 5'-phosphoryl group from the cleavage fragments in order to obtain labeling; thus cleavage of the DNA chain produced 3'-hydroxyl, 5'-phosphoryl termini. We next examined the terminal nucleotides and found terminal ^{32}P label appearing only in dGMP and dAMP. Thus, our enzyme was specific!

We were ready by early 1969 to proceed to the dinucleotide level. Unfortunately, just at this most exciting stage of the work, Wilcox received his draft notice from the Army and was forced to discontinue the work so that he could complete his formal requirements for the Master's degree. Meanwhile, I began occupying myself, for the space of several months, with the laborious preparation of the dinucleotide standards that would be necessary for identification of the terminal dinucleotides. I also prepared a supply of exonuclease I from a side fraction of a large DNA polymerase preparation generously given to me by Paul Englund. Using the exonuclease I and the standards, I proceeded to show by the Weiss and Richardson method that the terminal dinucleotide was either (5')pGpA or (5')pApA, further confirming the remarkable cleavage specificity of our restriction enzyme. I believed that extension of the analysis beyond this point was possible, but standard oligonucleotides with which to identify the longer terminal species were unavailable.

About this time, Thomas J. Kelly, Jr. joined my laboratory, and in a series of discussions we worked out an approach, using the newly available isotope ^{33}P, that was to prove successful. In this approach, T7 DNA was uniformly ^{33}P-labeled, cleaved with the restriction endonuclease, 5'-terminally labeled with the second isotope, ^{32}P, and then digested to oligomers using pancreatic DNase. The products were fractionated according to length by ion-exchange chromatography, and we then analyzed the oligonucleotides of each size class electrophoretically. Two ^{32}P-terminal, ^{33}P-uniformly labeled species were obtained at the dimer and trimer level, but at least six species out of a possible eight were identified at the tetramer level, so that specificity was lost at that point. The dimer and trimer species were eluted from the electrophoretic strip, digested with venom phosphodiesterase and the ^{33}P-labeled nucleotides identified. In this way, the 5'-terminal dinucleotide was again confirmed as pPu-A and the trinucleotide was found to be pPu-A-C.

Three possible sequence arrangements could account for our result depending on whether cleavage was "even" or "staggered." To resolve this, we used micrococcal nuclease to release the 3'-terminal dinucleoside monophosphate and this turned out to be unique and complementary to the 5'-terminal dinucleotide. Thus, our enzyme recognized the 2-fold rotationally symmetrical six nucleotide sequence: ...(5')G–T–Py↓ Pu–A–C(3')... ...(3')C–A–Pu↑Py–T-G(5')... and produced an even duplex cleavage as indicated by the arrows (14). Based on the expected occurrence of this sequence in random DNA, the enzyme would be expected to cut once every 1024 base pairs: a value in good accord with our previous observations. The most interesting feature of the sequence was its symmetry, and we speculated that this might have important implications for the restriction enzyme structure (see later sections).

In retrospect the proof of cleavage specificity was clearly our most important result. It had not been shown for the *E. coli* K and B enzymes, and, in fact, could not have been shown since these were randomly cleaving Class I enzymes. Our enzyme belonged to a different, and as we shall see, a much larger class of restriction enzymes. Such Class II enzymes (8) are cleavage-site-specific and require only Mg^{2+} as cofactor. Later studies have revealed that they are relativly simple proteins, existing typically as dimers or tetramers of a single polypeptide chain (15, 16, 17), and their corresponding modification methylases are separate proteins that exist in some cases as monomers (18). However, in 1970 when we completed the sequence work, only Class I methylases had been studied *in vitro* (8), so we turned next to the isolation of a modification methylase from *H. influenzae* Rd.

Modification Methylases in H. influenzae Rd

While we did not hesitate to call our cleavage-site-specific endonuclease a restriction enzyme—after all it was specific for foreign DNA—formal objections existed to that classification unless a modification enzyme of the same specificity could be found. In the absence of such a modifying enzyme to protect host sequences against cleavage, it would be necessary to postulate total absence of the sequence in the cell chromosome (a rather remote possibility) or, alternatively, a compartmentalization of the activity. To allay these objections, and to satisfy our own curiosity, Paul Roy (a graduate student) and I undertook a survey of the DNA methylases in *H. influenzae* Rd in late 1970 (19, 20).

We first established that *H. influenzae* Rd like many other strains of bacteria (21), contains a small percentage of methylated bases in its DNA: 5-methylcytosine occurs once per about 8 000 bases, and N-6-methyladenine is found once per about 280 bases. We realized that much of this methylation might be unrelated to R–M systems and that several methylases could be present. Arber had shown that in *E. coli* the majority of DNA methylation was not associated with R–M systems; in *E. coli* B, as little as 5 % was so involved (22). With this in mind, we adopted a general approach designed to reveal the total DNA methylases of the cells. Proteins from a crude cell extract were chromatographed on phosphocellulose and assayed for ability to transfer [^{3}H]methyl groups from labeled S-adenosylmethionine onto salmon sperm DNA or T7 DNA. In this way, four DNA adenine methylases were detected.

One of these methylases protected T7 DNA from cleavage by our restriction enzyme; and conversely, the sites for this enzyme in salmon sperm DNA were destroyed by predigestion with our restriction enzyme preparation. These two results together indicated that both the restriction enzyme and the methylase shared common DNA recognition sites. As a further proof that we had isolated the modification methylase, we analyzed 3' and 5' nearest neighbors to the ^{3}H-methylated adenine residues produced by the enzyme. The results gave as the partial sequence for the methylase, the trinucleotide (5')Pu−A−C, in direct agreement with our restriction enzyme sequence.

One additional and unexpected observation came out of our methylase studies. In the salmon sperm DNA experiment in which we predigested with our restriction enzyme preparation, methyl acceptor ability was also lost for one of the other DNA methylases. This particular methylase was not active on T7 DNA unlike our restriction enzyme. It appeared that a second restriction enzyme, corresponding in specificity to this methylase, was contained in our endonuclease preparation. Our interpretation was confirmed by separation of these two restriction activities in Nathans' laboratory, and by a letter from Kenneth Murray of Edinburgh, Scotland reporting that he had purified the new enzyme and determined its recognition site sequence as (5')pA−A−G−C−T−T(23). At this point it is interesting to recall our choice of T7 DNA for the sequencing work. Since, unbeknownst to us, we had worked with a mixture of two enzymes, it was indeed fortuitous that only one was active on T7 DNA.

Search for New Restriction Enzymes

Through the work of Nathans and colleagues (24), beginning in 1969, in which they applied cleavage-site-specific restriction endonucleases to the analysis of the SV40 tumor virus genome, it became clear that these enzymes were valuable tools for DNA analysis. Their work provided an early stimulus to the search for new enzymes of differing specificities. An important consequence of our work with the *Haemophilus* restriction endonuclease was the realization that such enzymes could be readily detected in bacteria by purely biochemical procedures. This was especially true after the introduction of gel electrophoresis by Nathans for analysis of DNA restriction cleavage fragments (24) and the introduction of ethidium bromide as a fluorescent stain for DNA by Sharp *et al.* (25). Armed wih easy and specific assays, other laboratories were soon reporting new restriction endonucleases, first in *E. coli* (26) and other *Haemophilus* species (25, 27, 28), and then in a variety of other bacteria. Richard J. Roberts of the Cold Spring Harbor Laboratories, who had a special interest in their possible use for DNA sequencing, spearheaded the drive to isolate new enzymes. Bacteria available from the American Type Culture Collection were systematically examined for cleavage-site-specific endonucleases that would digest phage lambda and other viral DNAs so as to produce discrete bands on electrophoretic gels. Many other laboratories joined in the effort, and today some 150 enzymes with nearly 50 different cleavage specificities are known (29).

A Catalog of Restriction Enzymes and Their Specificities

In Table I is a current list of known restriction endonuclease cleavage specificities grouped according to type of recognition sequence. The list is taken from Roberts (29) and includes for each sequence only the prototype enzyme name*. In many cases, other enzymes recognizing the same sequence are known. These have been called isoschizomers by Roberts and are given in his complete listing (29); e. g. isoschizomers of *Hind*II are *Chu*II, *Hinc*II, and *Mnn*I. It is important to note that among a group of isoschizomers, cleavage position within the site may vary; e. g., *Sma*I

TABLE 1. CATALOG OF RESTRICTION ENDONUCLEASE SEQUENCE SPECIFICITIES†

Symmetric (N = 6)

AvaIII ATGCAT
BalI TGG|CCA
BamHI G|GATC*C
BclI T|GATCA
BglII A|GATCT
ClaI ATCGAT
EcoRI G|AA*TTC
HindIII A*|AGCTT
HpaI GTT|AAC
KpnI GGTAC|C
MstI TGCGCA
PstI CTGCA|G
PvuI CGATCG
PvuII CAG|CTG
SmaI CCC|GGG
SacI GAGCT|C
SacII CCGC|GG
SalI G|TCGAC
XbaI T|CTAGA
XhoI CTC|GAG

Degenerate symmetric (N = 6)

AccI GT|(A/C)(G/T)AC
AvaI C|PyCGPuG
HaeI (A/T)GG|CC(T/A)
HaeII PuGCGC|Py
HgiAI G(T/A)GC(T/A)|C
HindII GTPy|PuA*C

Symmetric (N = 5)

AsuI G|GNCC
AvaII G|G(A/T)CC
BbvI GC(T/A)GC
EcoRII |CC*(A/T)GG
HinfI G|ANTC

Asymmetric (N = 4,5)

MnlI CCTC cleavage 5 to 10 bases 3′ to site
HgaI GACGCNNNNN| (3′)
 CTGCTNNNNNNNNNN| (5′)
HphI GGTGANNNNNNNN| (3′)
 CCACTNNNNNNN| (5′)
MboII GAAGANNNNNNNN| (3′)
 CTTCTNNNNNNN| (5′)
 GATGC
StaNI AGACC (3′)
†EcoP1 TCTGG (5′) cleavage 24 to 26 bases 3′ to site

Symmetric (N = 4)

AluI AG|CT
FnuDII CG|CG
HaeIII GG|C*C
HhaI GC*G|C
HpaII C|C*GG
MboI |GATC
TaqI T|CGA

Symmetric methylated (N = 4)

DpnI GAmTC

Symmetric (N = 7)

EcaI GGTNACC

† Compiled from data published by R. J. Roberts (29). Names of host organisms and references for enzymes and sequences are listed in his article.
† From Reiser, J., personal communication
Sequences are 5′ ⟶ 3′. They should be visualized as duplexes although the sequence of only one strand is given. Vertical lines represent cleavage positions. Asterisks (*) indicate bases modified by the corresponding modification enzymes. An "m" represents a methyl group. Only the prototype enzyme name is given for each sequence specificity; isoschizomers exist for many sequences and are given in the Roberts reference. Bases in parenthesis indicate that either base may occupy the position; e.g., *Acc*I recognizes GT(A/C)(G/T)AC which signifies the following sequence possibilities: GTAGAC, GTATAC, GTCGAC, GTCTAC. (The first and last sequences are the same in duplex form.) Pu is purine and Py is pyrimidine.

cleaves (5') C–C–C↓G–G–G while *Xma*I cleaves (5') C↑C–C–G–G–G.

Nucleotide sequences of recognition sites have in most cases been determined by analysis of oligonucleotides released from the 3' or 5' labeled termini of cleavage fragments in a manner analogous to that first used for *Hind*II (14). Recently a simple method for palindromic sequences has been devised. It depends on comparing digest fragments of ΦX174 and SV40 DNA produced by a given restriction enzyme with a table of possible fragments predicted by computer analysis of all tetra, penta, and hexanucleotide palindromes in these DNAs (31). Usually a unique sequence assignment is possible. For enzymes that cleave outside of their recognition site, identification can be made by analysis of the location of mutations that remove the site and by determining the position of bases, which when modified by a specific methylase or by certain chemical agents such as dimethylsulfate, inactivate the site (32).

Sites are classified according to whether they show 2-fold rotational symmetry (palindromes) or are asymmetric. Among the symmetric sites are 20 perfect hexanucleotide sites, 6 degenerate hexanucleotide sites, 5 pentanucleotide sites with a central degeneracy, 7 perfect tetranucleotide sites, 1 hexanucleotide site with a central degeneracy, and 1 tetranucleotide site requiring a methylated adenine (see below). Each of the tetranucleotide sites can also be found as a central tetranucleotide in one or more of the hexanucleotide sites. The degenerate hexanucleotide sites, while losing strict structural symmetry at the degenerate position, retain a basic overall symmetry and are probably recognized as symmetrical by the enzymes (see below). When a complete degeneracy exists as in the *Hinf*I sequence, G–A–N–T–C, symmetry is not lost since no discriminating enzyme contacts are made with the degenerate base.

*Dpn*I is unique in that it recognizes the sequence, G-Am-T-C, only when it contains the methylated adenine. It is difficult to rationalize this reversal of the normal role of methylation since other *Diplococcus pneumoniae* strains carry the more conventional restriction enzyme, *Dpn*II, recognizing unmethylated G-A-T-C (33).

Among the 5 restriction enzymes recognizing asymmetric sites, 1 recognizes a tetranucleotide site and 4 recognize pentanucleotides. These enzymes cleave asymmetrically at a distance of 5 to 10 nucleotides 3' to the recognition sequence. *Hga*I deserves special comment since it generates

***Restriction endonucleases derive their names from an R–M system nomenclature (30) that uses an italicized three-letter abbreviation for the host organism followed by a fourth letter for strain where necessary and a roman numeral to indicate each R–M system in the organism. For example, *Hind*II is the name of the R–M system from which our original restriction endonuclease comes. Restriction enzymes are indicated as endonuclease R followed by the system name, and similarly, modification enzymes are designated methylase M followed by the system name; endonuclease R·*Hind*II and methylase M·*Hind*II. Most often a shorter form R·*Hind*II or M·*Hind*II is used, and when only restriction enzymes are being considered, they carry just the system name, i. e., *Hind*II, *Hind*III, *EcoRI*, etc.**

DNA cleavage fragments with cohesive termini that have a high probability of specific reunion with the original complementary partner; thus a small genome could be cut into several fragments that would religate only in original order (34).

Three main cleavage modes are observed by enzymes with symmetric recognition sites: even breaks (e.g. *Hind*II), staggered breaks generating 3'-single-stranded cohesive termini (e.g. *Pst*I), and staggered breaks generating 5'-single-stranded cohesive termini (e.g. *Hind*III). Each of these types of termini has found special uses in recombinant DNA work. So far, all the enzymes examined cleave so as to produce 3'-hydroxyl, 5'-phosphoryl termini.

Mechanism of Nucleotide Sequence Recognition

Restriction and modification enzymes, because of their variety and relative structural simplicity, provide excellent model systems for study of sequence-specific DNA-protein interactions. We have had an interest in this area for some time, and I should like to present, in a general way, our approach to this problem as well as possible directions for future research in the area.

The majority of R-M system recognition sites possess 2-fold rotational symmetry. Two basic recognition mechanisms are possible for these sites: *symmetric* recognition involving bilaterial symmetric contacts in a duplex site and *asymmetric* recognition involving a set of nonsymmetric contacts (Fig. 1). For single-stranded sites, only the asymmetric mechanism can apply. An important consideration then, is whether or not restriction enzymes or their corresponding modification methylases can act on single-stranded sites.

Most restriction endonucleases appear to require duplex sites, as originally demonstrated for *Hind*II (14). A few enzymes, for example, *Hae*III, *Hha*I, *Sfa*I, *Mbo*I, and *Hinf*I act slowly on single-stranded DNAs (35, 36, 37), but this is now thought to be due to formation of transient duplexes (38). It is probable that most of the restriction enzymes employ a symmetric recognition mechanism. This is based on several arguments. First, since hemimethylated sites are generated during replication, the recognition process must be responsive to methylation on either strand. This is most easily achieved by bilaterial, symmetric protein-DNA contacts at the methylation positions within the duplex site. Second, from the standpoint of genetic economy it is less expensive to specify a protein monomer site recognizing n/2 bases than one recognizing n bases (Fig. 1). Finally, the *Eco*RI endonuclease exists as dimers and tetramers of a single 28,500 dalton subunit, and under physiological conditions, cleaves both strands of a duplex site in one binding event (18). It seems likely from symmetry considerations that such a dimeric or tetrameric structure will prove to be the rule for other enzymes.

Modification methylases may recognize sites in a fashion quite different from the restriction enzymes. Some of these enzymes appear capable of

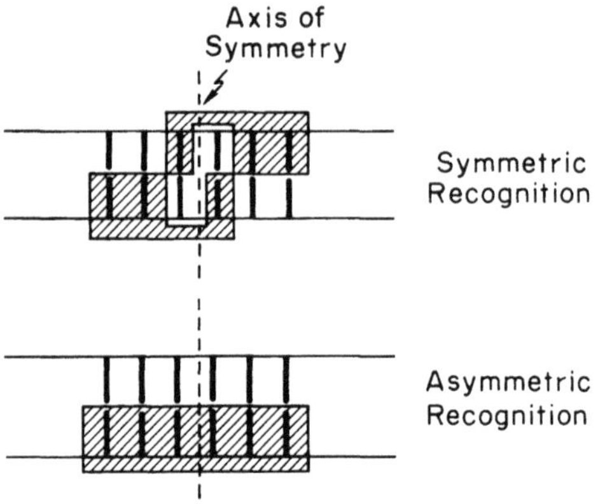

Fig. 1: Two different ways in which restriction and modification enzymes may interact with two-fold rotationally symmetrical nucleotide sequences in order to achieve recognition. In the symmetric recognition model, the protein possesses subunits arranged in two-fold rotational symmetry and each identical half interacts with a minimum of n/2 nucleotides. In the asymmetric model, the protein is assumed to have an asymmetric structure and must interact with at least one nucleotide at each base pair position for a total of n nucleotides. The simplest case, where all the interactions are on one strand, is shown.

acting on purely single-stranded sites, implying an asymmetric recognition process (Fig.1). Michael B. Mann, in my laboratory (39), has shown that M·*Hha*I methylates C residues in the random copolymer, poly(dN-acetyl G, dC) which is unable to form any Watson–Crick base pairing (based on absence of a thermal melting transition), and that M·*Hha*I and M·*Hpa*II methylate poly(dX, dC) which also shows no thermal melting transition. M· *Hae*III and M·*Hpa*II also methylate denatured salmon sperm DNA to the same total extent as native DNA, although at half the rate. A lower rate and extent (30 %) was achieved with M·*Hha*I. These observations support the notion that these methylases can act on single-stranded sites with preservation of specificity. This implies that discriminatory interactions need involve only the bases on one strand. Rubin and Modrich (18) have shown that the *Eco*RI methylase is a functional monomer of molecular weight 39,000 that transfers methyl groups to each strand of the *Eco*RI site in individual catalytic events that are interrupted by dissociation from the site. On theoretical grounds, an asymmetric recognition mechanism is reasonable for the modification methylases because, as pointed out by the above authors, the usual *in vivo* hemimethylated duplex substrate is inherently asymmetric.

Turning again to the restriction endonucleases, it was early proposed that recognition of a symmetric site might depend on some unusual structure of the site. Kelly and I (14) initially suggested the enzymes might interact with open (melted) sites (Fig. 2) because at that time we felt there was insufficient opportunity for base specific interactions in the helical grooves. Meselson *et al.* (41) proposed that symmetric sites might transiently form cruciform structures with special features that would promote enzyme recognition (Fig. 2). Both open and cruciform struc-

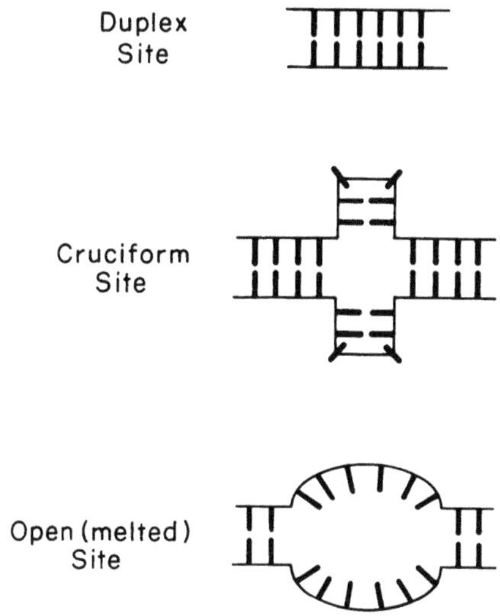

Fig. 2: Representations of three possible structures of a recognition site that might facilitate specific protein-DNA interaction.

tures are energetically unfavorable, and there are no compelling theoretical reasons to favor them. We now accept, as a result of the *lac* repressor-operator studies (42, 43) and recent structural analyses of base pairs (40), that the base groups exposed in the major and minor grooves of helical DNA are sufficient for discriminatory interactions. Therefore, to simplify the discussion, it will be assumed that sites are recognized while in the helical configuration.

The restriction enzyme *Hha*I was chosen by Michael Mann and myself for initial studies because the site, (5')pG-C-G↓C is particularly simple and can easily be synthesized in alternating polymer form. We have chosen chemical modification of the bases as an approach to determining those groups in the major or minor grooves that play a role in recognition. Effects on catalytic activity rather than binding are most easily measured and have been used in our studies, although we acknowledge that each may provide somewhat different information.

We have depended heavily on the analysis of Seeman *et al.* (40) for interpretation of our results. These authors compare the various potential sites for discriminatory protein-DNA contacts in the major and minor grooves of the different base pairs. A G·C base pair, the only kind in the *Hha*I site, is shown in Fig. 3. The major and minor grooves may be visualized as divided into outer and central regions. In the central major groove, the 06 atom of guanine is hydrogen bonded to the amino N4 of cytosine. The outer major groove contains the N7 atom of guanine and C5 hydrogen atom of cytosine. The central minor groove contains the 2-amino (N2) group of guanine. The outer minor groove contains the N3 atom of guanine and the O2 atom of cytosine. (The latter is hydrogen bonded to the 2-amino group of guanine). A top view of the major groove

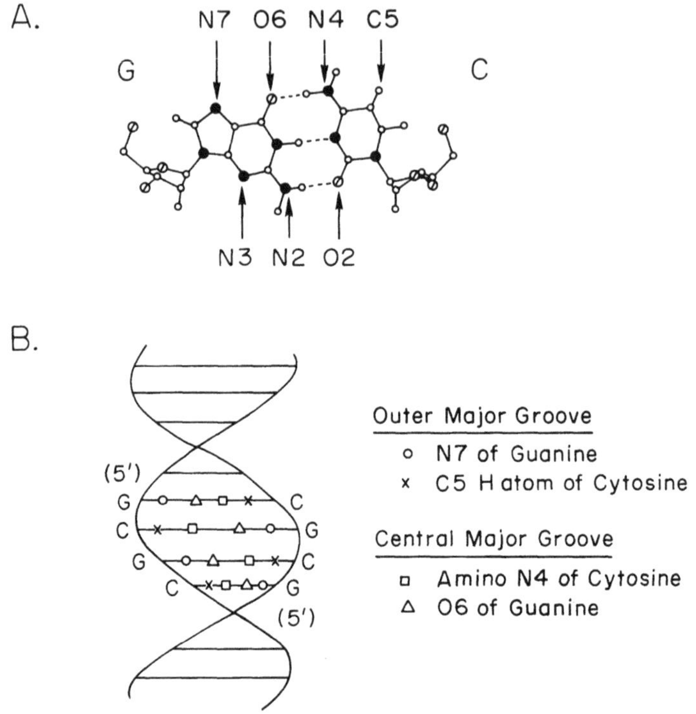

Fig. 3: Potential major and minor groove discriminatory protein-DNA interaction positions in the *Hha*I restriction endonuclease recognition site. A. A stereochemical drawing of a G·C base pair in DNA (adapted from Seeman *et al.* (40). Potential atoms for interaction are indicated. B. A rough sketch of a section of helical DNA containing an *Hha*I site. Base pairs are indicated by horizontal bars. The view is from above the major groove, and approximate positions of interacting atoms are shown.

in the *Hha*I site is shown in Fig. 3 with potential atoms for interaction diagramatically represented.

The *Hha*I modification methylase transfers methyl groups from S-adenosylmethionine onto the 5-position of the internal cytosines (situated between the two guanines) in the *Hha*I site and protects against cleavage by the *Hha*I endonuclease (39). Since the methyl groups probably interfere sterically, we infer that contacts between protein and DNA must take place at these two outer major groove positions of the duplex site. Mann and I have also shown that methylation introduced on the 5-position of the external cytonsines inhibits cleavages by R·*Hha*I. Therefore, it is likely that these are also closely fitted by the enzyme. In another experiment, we used dimethylsulfate to introduce methyl groups on the N7 positions of guanine in an *Hha*I-site positioned 20 bases from the 5' terminus of a ΦX174RF DNA fragment of known sequence. After this treatment, the fragment was digested with *Hha*I endonuclease, and fractionated by gel electrophoresis into cleaved and uncleaved molecules. These were then treated by Maxam and Gilbert (44) sequencing methods so as to cleave at the methylated position and analyzed by electrophoretic gels. Bands representing G's in the site were greatly increased in intensity in the gel tract representing uncleaved molecules and absent from the gel tract representing cleaved molecules. We concluded that methylation of the N7 position of any G residue in the site conferred protection against cleavage. Again, the effect is likely to be steric, and we infer that the enzyme closely fits these positions in the outer major groove of the site (Fig. 2B). In summary, we have been able to demonstrate that methylation at any of eight positions in the outer portions of the major groove inhibits cleavage. Study of the central major groove groups by this approach is more difficult because modifications often destroy helical pairing.

To examine minor groove interactions, we looked at activity on alternating poly(dI-dC). Inosine contains a H-atom in place of the 2-amino group of guanine in the central minor groove. A dI-dC base pair mimics a dA-dT base pair when viewed from the minor groove. R·*Hha*I cleaves alternating poly(dI-dC) efficiently, thus the 2-amino group is not essential for discrimination, and more specifically, plays no role in discrimination of A·T from G·C base pairs. The central minor groove thus seems not to be occupied. Interaction is still possible in the outer positions of the minor groove where the N3 atom of purines and the O2 atom of pyrimidines are exposed. However, since these are both electron-rich hydrogen bond acceptors and occupy strictly similar positions regardless of base pair type, they are not considered likely for discriminatory interactions (40). We have tentatively concluded then that *Hha*I endonuclease occupies the major groove and derives all discriminatory contacts from groups in the central and outer major groove positions.

It is likely that *Hha*I and other restriction endonucleases also interact with the sugar-phosphate backbone within a site. We suggest that stabilizing interactions of this sort, and also non-discriminatory outer minor groove interactions, may extend to adjacent nucleotides to either side of a site. These interactions could explain two observations. First is the size effect. Greene *et al.* (45) found that the affinity of *Eco*RI endonuclease for the symmetric octanucleotide (5') pT-G-A-A-T-T-C-A, containing a central *Eco*RI recognition site is 200 times less than for the *Eco*RI site in SV40 DNA. We found similarly that a symmetric decanucleotide containing terminal *Hpa*II sites is not detectably cleaved by *Hpa*II endonuclease, but addition of nucleotides to the end restores the site (46). Second is the finding that some sites are cut preferentially, depending on external sequence context (47, 48), suggesting that weak contacts are made at neighboring nucleotides outside of the site.

Recognition of Degenerate Sites

Degenerate sites are not strictly symmetrical by structural criteria. *Acc*I endonuclease recognizes the site (5')G-T-(A/C)-(G/T)-A-C which exists as four combinations of the degenerate nucleotides: G-T-A-G-A-C, G-T-A-T-A-C, G-T-C-G-A-C, and G-T-C-T-A-C. The first and last combinations are asymmetric. Yet it is very appealing to think of the enzyme as interacting with each of these sequences in a similar way so as to preserve symmetry. The discrimination rules of Seeman *et al.* (40) allow for this possibility. They describe several potential positions for major and minor groove interaction with each of the Watson–Crick base pairs. A protein-DNA interaction, e.g. a single hydrogen-bond directed to one of the positions, is insufficient to allow discrimination between all the base pairs, although two interactions can be sufficient. Thus, a restriction or modification enzyme making only a single contact at symmetrically placed base pairs could allow a degeneracy, i.e., an ambiguity in recognition.

According to this scheme, four types of degeneracy appear to be possible: (A/G)-(T/C) (Pu-Py type), (A/C)-(G/T), (A/T)-(T/A), and (G/C)-(C/G). The Pu-Py type degeneracy could arise from outer major groove contact directed toward the purine N7 atom. The (A/C)-(G/T) degeneracy could result from a single interaction directed to the central major groove amino N4 of cytosine and amino N6 of adenine position, or to the carbonyl oxygen of thymine or guanine, since these pairs of groups occupy similar positions in the specified degeneracy. The (A/T)-(T/A) degeneracy could result from a central minor groove interaction since the C2 hydrogen atom of adenine occupies a sterically similar position in each A·T orientation. Finally, the (G/C)-(C/G) degeneracy could result from interaction in the central minor groove with the 2-amino group of guanine which is in a sterically similar position for each orientation of the G·C pair. Inspection of the six degenerate hexanucleotide sequences in Table 1 reveals that, of the four predicted degeneracies, only (G/C)-(C/G) has not yet been found.

Among the symmetrical pentanucleotide sites, two are completely degenerate at the middle nucleotide position implying either absence of protein-DNA interaction or possibly non-discriminatory outer minor groove contacts. Three of the pentanucleotide sites contain a middle (A/T) nucleotide degeneracy. This is compatible with a single interaction directed to the C2 hydrogen atom of adenine, which is inherently symmetrical since it falls almost directly on the dyad axis of the site. Other types of degeneracy in the middle nucleotide position of pentanucleotide sites appear less likely.

The general agreement between the above predicted and observed degeneracies further reinforces the notion that restriction enzymes accomplish nucleotide sequence recognition through major and minor groove interactions.

Relaxation of Sequence Specificity

Several DNA enzymes, e.g., terminal deoxynucleotidyl transferase (49) and pancreatic DNase (50), show changes in specificity according to species of divalent cation and ionic conditions in the reaction mixture. Some restriction endonucleases appear to be similarly affected. *Eco*RI endonuclease cleaves the sequence (5')G-A-A-T-T-C in a reaction mixture containing 100m M Tris-Cl (pH 7.3), 50 mM NaCl, and 5 mM $MgCl_2$. When the conditions are changed to 25 mM Tris-Cl (pH 8.5), 2 mM $MgCl_2$, the

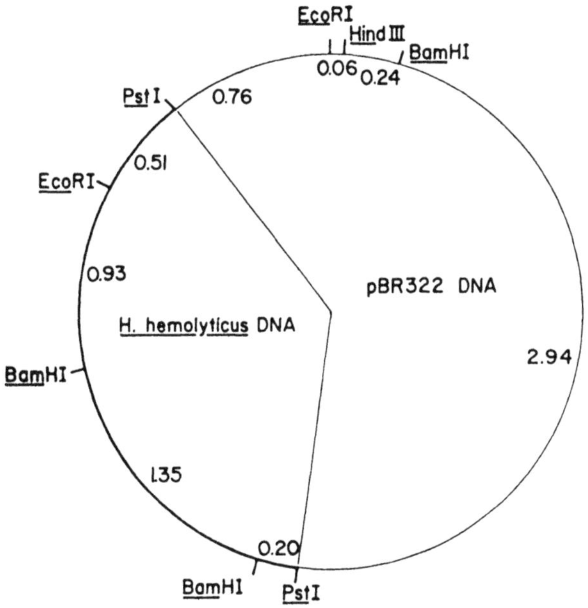

Fig. 4: A restriction enzyme cleavage map of pDI10 DNA. Distances are in kilobases.

specificity is lowered to the central tetranucleotide sequence (5')A-A-T-T (51). However, the enzyme retains a strong preference for the canonical site; extensive digestion is required to achieve cleavage at the new tetranucleotide sites, and there is great variability among them in regard to cleavage rate. The latter presumably reflects the degree of relatedness to the canonical site.

Hsu and Berg (52) obtained decreased specificity with *Eco*RI by substituting Mn^{2+} for Mg^{2+}. They also noted relaxed specificity with *Hin*dIII, but not *Hpa*II, in the presence of Mn^{2+}. This appears to be a promising area for more investigation.

Cloning R-M System Genes

Detailed studies of restriction and modification enzymes require quantities of pure enzyme. However, enzymes are often obtained in poor yield from source bacteria. Because of this we have begun to explore the possibilities of cloning various R-M system genes as a means to achieve enzyme overproduction. Using this approach, Michael Mann and Nagaraja Rao, in my laboratory, have recently cloned the *Hha*II system from *Haemophilus haemolyticus* in the *E. coli*-pBR322 host-vector system using a "shotgun" approach (53). Total chromosomal DNA was cleaved with *Pst*I endonuclease and inserted into the *Pst*I cloning site of the plasmid, located in the ampicillin resistance gene, by means of a GC-extension procedure developed by Rougeon *et al* (54). After transfection into an r^-m^- *E. coli* host (HB 101), tetracycline-resistant recombinant clones were tested for acquisition of a new restriction phenotype using phage λ. A single such clone was found

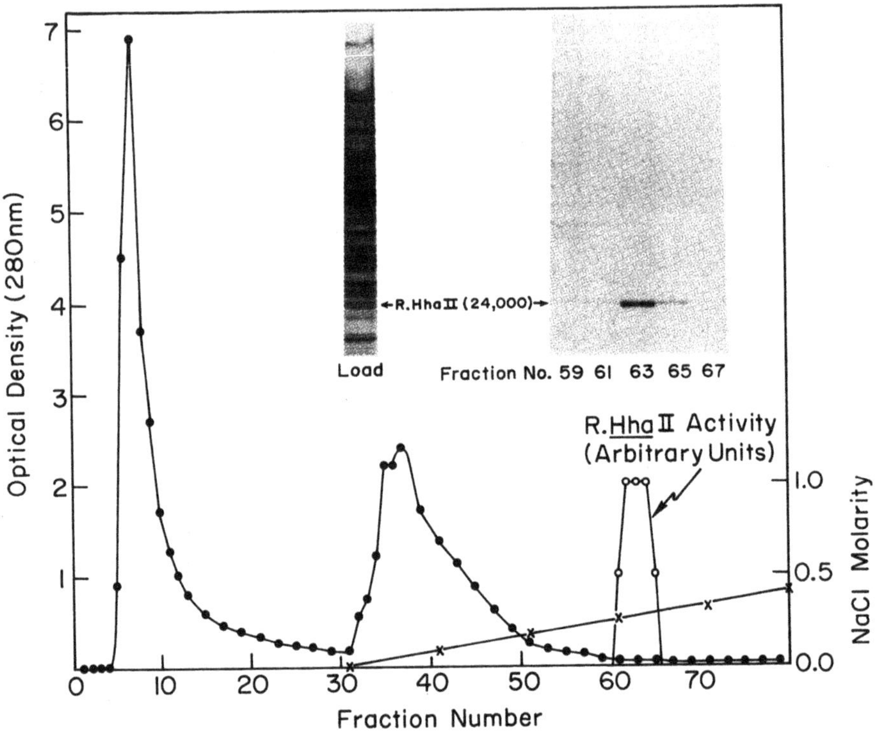

Fig. 5: Single-stranded DNA-agarose affinity chromatogtaphy of *Hha*II endonuclease. A crude extract from about 10 gm of thermally induced KJ34 cells was processed to remove nucleic acids and then chromatographed on a single-stranded DNA agarose column (2.5×20 cm) using a 1,000 ml gradient essentially as described by Mann *et al.* (53). Fractions (approximately 12 ml) were collected and assayed for protein by absorbance at 280 nm and for R·*Hha*II activity by gel electrophoresis of lambda DNA digestion products. The endonuclease units represent estimates of the percent of complete digestion. Protein species in load and in peak endonuclease fractions were examined by SDS-polyacrylamide gel electrophoresis.

among 1400 tested. The recombinant plasmid, pDI10, recovered from this clone contained a 3.0 kb DNA insert flanked by *Pst*I sites. A cleavage map is shown in Fig. 4. The HB101 clone carrying pDI10 exhibits classical restriction and modification behavior with phage λ (e.o.p.$\sim 10^{-7}$) and several other phages.

The pDI10 DNA efficiently re-transfects new HB101 cells suggesting that methylation is expressed well in advance of restriction. To account for this, we have suggested that the methylase might act as a positive regulator for expression of the restriction gene. By this scheme, methylase would initially be occupied with methylation of host chromosomal sites, becoming free to induce restriction enzyme only after its job was complete. This is only one possible scheme to explain the apparent sequential action of these genes. Study of the regulation should prove interesting.

To increase plasmid copy number and consequent enzyme overproduction, the 3.0 kb DNA fragment was excised from pDI10 DNA and transferred into a second plasmid vector, pKC16, a hybrid of pBR322 and phage λ containing a thermally-inducible λ-replication region (55), to yield a new hybrid plasmid pDI21. Using a clone containing pDI21, a 20 minute 42° C treatment raises plasmid copy number and enzyme yield several fold over that obtainable with pDI10. The restriction endonuclease and modification methylase were purified from crude extracts of this clone by single-stranded DNA-agarose affinity chromatography. Typical results for the endonuclease are shown in Fig. 5. This essentially one-step procedure yielded an active fraction showing a single major protein band of about 24,000 daltons by SDS-polyacrylamide gel electrophoresis (Fig. 5). The purified endonuclease gives a DNA cleavage pattern on ΦX174RF DNA identical to *Hin*fI with the sequence specificity (5')pG-A-N-T-C. The DNA methylase from the clone protects against both *Hha*II and *Hin*fI cleavage.

We believe that future developments in the field of restriction and modification enzymes will depend heavily on gene cloning, both for enzyme overproduction and for genetic studies. There is a great advantage to having the *Hha*II genes on a small segment of DNA that can be propagated and expressed in *E. coli.* The genes are easily accessible for genetic studies in the new host, whereas this would be difficult or impossible in the original *Haemophilus* strain. The DNA segment is small enough to be readily sequenced, thus providing direct information on gene arrangement, regulatory sequences, and protein amino acid sequences. The latter will be valuable for future crystallographic studies of enzyme structure, a goal which must be achieved if we are to fully understand the nature of the protein-DNA interactions involved in nucleotide sequence recognition.

REFERENCES

1. Nathans, D. and Smith, H.O. (1975) Ann. Rev. Biochem. *44*, 273–293.
1a. Luria, S.E. and Human, S.L. (1952) J. Bacteriol. *64*, 557–569.
2. Bertani, G. and Weigle, J.J. (1953) J. Bacteriol. *65*, 113–121.
3. Arber, W. and Dussoix, D. (1962) J. Mol. Biol. *5*, 18–36.
4. Dussoix, D. and Arber, W. (1962) J. Mol. Biol. *5*, 37–49.
5. Arber, W. (1965) Ann. Rev. Microbiol. *19*, 365–378.
6. Linn, S. and Arber, W. (1968) Proc. Nat. Acad. Sci. USA *59*, 1300–1306.
7. Meselson, M. and Yuan, R. (1968) Nature *217*, 1110–1114.
8. Boyer, H.W. (1971) Ann. Rev. Microbiol. *25*, 153–176.
9. Horiuchi, K. and Zinder, N.D. (1972) Proc. Nat. Acad. Sci. USA *69*, 3220–3224.
10. Adler, S.P. and Nathans, D. (1973) Biochem. Biophys. Acta *299*, 177–188.
11. Murray, N.E., Batten, P.L., and Murray, K. (1973) J. Mol. Biol. *81*, 395–407.
12. Smith, H.O. and Wilcox, K.W. (1970) J. Mol. Biol. *51*, 379–391.
13. Weiss, B. and Richardson, C.C. (1967) J. Mol. Biol. *23*, 405–417.
14. Kelly, T.J. Jr. and Smith, H.O. (1970) J. Mol. Biol. *51*, 393–409.
15. Modrich, P. and Zabel. D. (1976) J. Biol. Chem. *251*, 5866–5874.
16. Greene, P.J., Betlach, M.C., Goodman, H.M., and Boyer, H.W. (1974) Methods Mol. Biol. 7, 87–111.
17. Smith, L.A. and Chirikjian, J.G., personal communication.
18. Rubin, R.A. and Modrich, P. (1977) J. Biol. Chem. *252*, 7265–7272.
19. Roy, P. H. and Smith, H. O. (1973) J. Mol. Biol. *81*, 427–444.
20. Roy, P.H. and Smith, H.O. (1973) J. Mol. Biol. *81*, 445–459.
21. Vanyushin, B.F., Belozersky, A.N., Kokurina, N.A., and Kadirova, D.X. (1968) Nature *218*, 1066–1067.
22. Arber, W. and Linn, S. (1969) Ann. Rev. Biochem. *38*, 467–500.
23. Old, R., Murray, K., and Roizes, G. (1975) J. Mol. Biol. *92*, 331–339.
24. Danna, K. and Nathans, D. (1971) Proc. Nat. Acad. Sci. USA *68*, 2913–2917.
25. Sharp, P.A., Sugden, B. and Sambrook, J. (1973) Biochemistry *12*, 3055–3063.
26. Yoshimori, R. (1971) Ph.D. thesis, Univ. of Calif., San Francisco.
27. Middleton, J.H., Edgell, M.H., and Hutchison, C.A.III. (1972) J. Virol. *10*, 42–50.
28. Gromkova, R. and Goodgal, S.H. (1972) J. Bacteriol. *109*, 987–992.
29. Roberts, R.J. (1978) Gene, *4*, 183–193.
30. Smith, H.O. and Nathans, D. (1973) J. Mol. Biol. *81*, 419–423.
31. Fuchs, C., Rosenvold, E.C., Honigman, A., and Szybalski, W. (1978) Gene *4*, 1–23.
32. Kleid, D., Humayun, Z., Jeffrey, A., and Ptashne, M. (1976) Proc. Nat. Acad. Sci. USA *73*, 293–297.
33. Lacks, S. and Greenberg, B. (1975) J. Biol. Chem. *250*, 4060–4066.
34. Brown, N.L. and Smith, M. (1977) Proc. Nat. Acad. Sci. USA *74*, 3213–3216.
35. Blakesley, R.W. and Wells, R.D. (1975) Nature *257*, 421–422.
36. Horuichi, K. and Zinder, N.D. (1975) Proc. Nat. Acad. Sci. USA *72*, 2555–2558.
37. Godson, G.N. and Roberts, R.J. (1976) Virology *73*, 561–567.
38. Blakesley, R.W., Dodson, J.B., Nes, I.F., and Wells, R.D. (1977) J. Biol. Chem. *252*, 7300–7306.
39. Mann, M.B. and Smith, H.O. (1978) in Transmethylation, eds. E. Usdin, R. T. Borchardt, C. R. Creveling (Elsevier/North Holland) pp 483–493.
40. Seeman, N.C., Rosenberg, J.M., and Rich, A. (1976) Proc. Nat. Acad. Sci. USA *73*, 804–808.
41. Meselson, M., Yuan, R., and Heywood, J. (1972) Ann. Rev. Biochem. *41*, 447–466.
42. Adler, K., Beyrenther, K., Fanning, E., Geisler, N., Gronenborn, B., Klemm, A., Müller-Hill, B., Pfahl, M., and Schuitz, A. (1972) Nature *237*, 322–327.
43. Gilbert, W., Maxam, A., and Mirzabekov, A. (1976) in Control of Ribosome Synthesis, Alfred Benson Symposium IX, eds. Kjeldgaard, N.O., Maaloe, O. (Munksgaard, Copenhagen) pp. 139–148.

44. Maxam, A. and Gilbert, W. (1977) Proc. Nat. Acad. Sci. USA *74,* 560–564.
45. Greene, P.J., Poonian, M.S., Mussbaum, A.L., Tobias, L., Garfin, D.E., Boyer, H.W., and Goodman, H.M. (1975) J. Mol. Biol. *99,* 237–261.
46. Mann, M.B. and Smith, H.O. (1977) Nucl. Acids Res. *4,* 4211–4221.
47. Thomas, M. and Davis, R.W. (1975) J. Mol. Biol. *91,* 315–328.
48. Smith, H.O. and Birnstiel, M. (1976) Nucl. Acids Res. *3,* 2387–2398.
49. Roychondhury, R., Jay, E., and Wu, R. (1976) Nucl. Acids Res. *3,* 863–877.
50. Melgar, E. and Goldthwait, D.A. (1968) J. Biol. Chem. *243,* 4409–4416.
51. Polisky, B., Greene, P., Garfin, D.E., McCarthy, B.J., Goodman, H.M., and Boyer, H.W. (1975) Proc. Nat. Acad. Sci. USA *72,* 3310–3314.
52. Hsu, M. and Berg, P. (1978) Biochemistry *17,* 131–138.
53. Mann, M.B., Rao, N.R., and Smith, H.O. (1978) Gene *3,* 97–112.
54. Rougeon, F., Kourilsky, P., and Mach, B. (1975) Nucl. Acids Res. *2,* 2365–2378.
55. Rao, N.R. and Rogers, S.G. (1978) Gene *3,* 247–263.

1979
Physiology or Medicine

ALAN M. CORMACK and GODFREY N. HOUNSFIELD

"for the development of computer assisted tomography"

THE NOBEL PRIZE FOR PHYSIOLOGY OR MEDICINE

Speech by Professor TORGNY GREITZ of the Karolinska Medico-Chirurgical Institute.
Translation from the Swedish text.

Your Majesties, Your Royal Highnesses, Ladies and Gentlemen,

Neither of this year's laureates in physiology or medicine is a medical doctor. Nevertheless, they have achieved a revolution in the field of medicine.

It is sometimes said that this new X-ray method that they have developed—computerized tomography—has ushered medicine into the space age. Well, sometimes, art can adumbrate reality. In his epic poem about the space ship "Aniara," Nobel Prizewinner in Literature Harry Martinson tells how, one day, the mimarobe, the computer guardian, "... by means of Mima's formula cycles, phase by phase ... saw into the transtomies ..." and was able to "... see through everything as though it were glass ..."

There, in a single stanza, the poet has captured the essential characteristics and elements of computerized tomography. In addition to X-ray tubes and radiation detectors, the method requires a Mima—that is, a powerful computer; it also calls for a mathematical method, perhaps based on the formula cycles of Fourier transforms, phase by phase; and it produces almost unbelievably clear images of transtomies—that is, cross-sectional views through the human body.

The analogy with the epic poem can be taken even further. The mimarobe remarks: "At this discovery, I went nearly mad ..." Few medical achievements have received such immediate acceptance and met with such unreserved enthusiasm as computerized tomography. It literally swept the world. But the enormous procurement costs involved caused some observers to wonder about the mental health of the health-services sector. Indeed, in the United States, a moratorium on computerized tomography was suggested.

What, then, lay behind this spectacular success? Well, to understand something of the background, let us go back to 1895, the year Röntgen discovered X-rays. The very first X-ray photograph Röntgen ever took—of his wife's hand—indicated, at one and the same time, both the potential and the limitations of conventional X-ray technique. The bones of the hand can be seen, but the complex anatomy of soft tissues—of muscles, tendons, blood vessels, and nerves—does not register.

This inability to distinguish between density differences in the various soft tissues is one of the fundamental limitations of conventional X-ray technique. It means that, in an ordinary X-ray picture, essentially all we can discern are bones and gas-filled spaces. It is the air in the lungs, for example, that enables us to study the lungs' structure and the shape of the heart.

Conventional X-ray technique has two additional shortcomings that are eliminated by computerized tomography.

One such shortcoming is that structures in three-dimensional space overlap in a conventional, two-dimensional, X-ray photograph. What we see is a shadow play—a play, alas, with far too many actors on the stage. It becomes difficult to discern the villain.

The other limitation is that X-ray film cannot indicate any absolute values for variations in tissue density.

Allan Cormack became aware of this latter drawback when, as a young physicist, he was asked to calculate radiation dosages in cancer therapy at Groote Schuur Hospital in Cape Town. He found that the methods then being used—methods based on conventional X-ray examination techniques—were extremely imprecise.

Cormack realized that the problem of obtaining precise values for the tissue-density distribution within the body was a mathematical one. He found a solution and was able, in model experiments, to reconstruct an accurate cross-section of an irregularly shaped object. This was reported in two articles, in 1963 and 1964. Cormack's cross-section reconstructions were the first computerized tomograms ever made—although his "computer" was a simple desktop calculator.

Cormack realized that his method could be used to produce precise X-ray images and so-called positron-camera pictures of cross-sectional "slices" of the body. However, no apparatus for practical diagnostic application of these procedures was constructed.

One probable reason for Cormack's difficulty in arousing interest in his experiments was that the computers of the time were incapable of executing—within a reasonable amount of time—the enormous calculations the procedure required.

It was Godfrey Hounsfield who brought Cormack's predictions to fruition. Hounsfield is indisputably the central figure in computerized tomography. Wholly independently of Cormack, he developed a method of his own for computerized tomography and constructed the first clinically usable computerized tomograph—the EMI scanner, which was intended for examinations of the head.

Publication of the first clinical results in the spring of 1972 flabbergasted the world. Up to that time, ordinary X-ray examinations of the head had shown the skull bones, but the brain had remained a gray, undifferentiated fog. Now, suddenly, the fog had cleared. Now, one could see clear images of cross-sections of the brain, with the brain's gray and white matter and its liquid-filled cavities. Pathological processes that previously could only be indicated by means of unpleasant—indeed, downright painful—and not altogether risk-free examinations could now be rendered visible, simply and painlessly—and as clearly defined as in a section from an anatomical specimen.

Today, computerized tomography is an established method for the examination of all the organ systems of the body. The method's greatest significance, though, is in the diagnosis of neurological disorders. Since nearly one out of three persons suffers from some disease or disorder of the central nervous system—usually of the brain—during his lifetime, computerized tomography means increased certainty in diagnosis and greater precision in treatment for literally millions of patients.

Cormack and Hounsfield have ushered in a new era in diagnostics. Now they, as well as others inspired by their pioneering contributions, are at work developing yet newer methods for the production of images of cross-sections in the body. In those images, we will be able to discern not only structure, but also function; physiology, or biochemistry. In this, new voyages of discovery are being prepared: voyages into man's own interior, into inner space.

Allan Cormack and Godfrey Hounsfield! Few laureates in physiology or medicine have, at the time of receiving their prizes, to the degree that you have, satisfied the provision in Alfred Nobel's will that stipulates that the prizewinner "shall have conferred the greatest benefit on mankind." Your ingenious new thinking has not only had a tremendous impact on everyday medicine; it has also provided entirely new avenues for medical research. It is my task and my pleasure to convey to you the heartiest congratulations of the Karolinska Institute and to ask you now to receive your insignia from His Majesty, the King.

Allan Cormack

ALLAN CORMACK

My parents went from the north of Scotland to South Africa shortly before World War I. My mother had been a teacher and my father was an engineer with the Post Office. I was born in Johannesburg in 1924, the youngest of three children. My family moved around the country quite a lot, as did the families of many civil servants, but after my father's death in 1936 we settled in Cape Town. There I attended the Rondebosch Boys High School and my interests outside my academic work were debating, tennis, and to a lesser extent, acting. I became intensely interested in astronomy and devoured the popular works of astronomers such as Sir Arthur Eddington and Sir James Jeans, from which I learnt that a knowledge of mathematics and physics was essential to the pursuit of astronomy. This increased my fondness for those subjects.

At that time the prospects for making a living as an astronomer were not good, so on going to the University of Cape Town, I followed in the footsteps of my father and brother and started to study electrical engineering. I was fortunate in that a new engineering curriculum had just been introduced by the then Head of the Electrical Engineering Department, Professor B. L. Goodlet. While serving with Mountbatten in the Far East he had seen the value for engineering of a better grounding in physics and mathematics than had previously been the case, and the new curriculum contained a lot of physics and mathematics. After a couple of years I abandoned engineering and turned to physics. At the University of Cape Town I spent most of my spare time mountaineering either on Table Mountain which was almost our back yard, or on the lovely mountain ranges of the Western Cape Province, and what spare time was not spent on climbing was spent listening to music.

After completing my Bachelor and Masters degrees at Cape Town I went to St. John's College, Cambridge, as a Research Student. I worked at the Cavendish Laboratory under Prof. Otto Frisch on problems connected with He^6. While I made some progress on these problems I did not complete them because of the following circumstances. I had met an American girl, Barbara Seavey, in Dirac's lectures on quantum mechanics, and a year and a half later I wanted to marry her, but I was broke. An inquiry at the Physics Department at Cape Town elicited not only the information that there was a vacancy there, but also a telegram offering me a position as Lecturer. So in 1950 I returned to Cape Town with a bride but no cyclotron, and so no further work on He^6.

Working on nuclear physics in Cape Town was lonely because there were very few nuclear physicists in the country, and the nearest one was six hundred miles away. However Professor R. W. James, head of the Physics Department and my mentor as a student, gave me my head and I learnt a lot and published a few papers. In 1956 I by chance became interested in a problem that is now known as CAT-scanning, but that story will be told elsewhere in this volume.

On my first Sabbatical leave it seemed only reasonable that since my wife had willingly come out to the wilds of Africa with me that I should go to the wilds of America with her. In addition the United States was a very good place to do research, and Harvard was a particularly good place to be in, so I spent my Sabbatical at the Harvard cyclotron doing experiments on nucleon-nucleon scattering with Professors Norman Ransey and Richard Wilson and then graduate student Joseph Palmieri. This was the beginning of a long and happy association with the people at the Harvard Cyclotron amongst whom I must mention particularly its present Director, Andreas Koehler.

While on this Sabbatical leave I was offered a position at Tufts University by the then Chairman of the Physics Department, Professor Julian K. Knipp. I accepted the offer and, except for a brief return to South Africa and a couple of Sabbatical leaves, I have been there ever since, progressing up the academic ladder and being Chairman of the Physics Department from 1968 to 1976. My main interest for most of this time was in nuclear and particle physics and I pursued the CT-scanning problem only intermittently, when time permitted. In 1963 and 1964 I published the results of this work, but as there was practically no response I continued my normal course of research and teaching. In the period 1970–72, I became aware of a number of developments in, or related to CT-scanning, and since then I have devoted much of my time to these problems.

Apart from a little swimming and sailing in the summer, I lead a rather sedentary life, spending a lot of time reading. Since my first discussions of ecological problems with Professor John Day around 1950 and since reading Konrad Lorenz's "King Solomon's Ring", I have become increasingly interested in the study of animals for what they might teach us about man, and the study of man as an animal. I have become increasingly disenchanted with what the thinkers of the so-called Age of Enlightenment tell us about the nature of man, and with what the formal religions and doctrinaire political theorists tell us about the same subject. I recently read Edward Wilson's book "On Human Nature", and after this hectic two months of my life culminates in Nobel Week, I look forward to tackling his "Sociobiology".

My wife and I have three children–Margaret, Jean and Robert. Since 1957 we have lived in the town of Winchester, Mass. which I appreciate for still being governed by that unique New England experiment in democracy: a (limited) Town Meeting and and a Board of Selectmen. We enjoy the amenities of New England, particularly summers near, in, and on Lake Winnepesaukee, New Hampshire.

EARLY TWO-DIMENSIONAL RECONSTRUCTION AND RECENT TOPICS STEMMING FROM IT

Nobel Lecture, 8 December, 1979

By
ALLAN M. CORMACK
Physics Department, Tufts University, Medford, Mass., U.S.A.

In 1955 I was a Lecturer in Physics at the University of Cape Town when the Hospital Physicist at the Groote Schuur Hospital resigned. South African law required that a properly qualified physicist supervise the use of any radioactive isotopes, and since I was the only nuclear physicist in Cape Town, I was asked to spend 1 1/2 days a week at the hospital attending to the use of isotopes, and I did so for the first half of 1956. I was placed in the Radiology Department under Dr. J. Muir Grieve, and in the course of my work I observed the planning of radiotherapy treatments. A girl would superpose isodose charts and come up with isodose contours which the physician would then examine and adjust, and the process would be repeated until a satisfactory dose-distribution was found. The isodose charts were for homogeneous materials, and it occurred to me that since the human body is quite inhomogeneous these results would be quite distorted by the inhomogeneities—a fact that physicians were, of course, well aware of. It occurred to me that in order to improve treatment planning one had to know the distribution of the attenuation coefficient of tissues in the body, and that this distribution had to be found by measurements made external to the body. It soon occurred to me that this information would be useful for diagnostic purposes and would constitute a tomogram or series of tomograms, though I did not learn the word "tomogram" for many years.

At that time the exponential attenuation of X- and gamma-rays had been known and used for over sixty years with parallel sided homogeneous slabs of material. I assumed that the generalization to inhomogeneous materials had been made in those sixty years, but a search of the pertinent literature did not reveal that it had been done, so I was forced to look at the problem ab initio. It was immediately evident that the problem was a mathematical one which can be seen from Fig. 1. If a fine beam of gamma-rays of intensity I_0 is incident on the body and the emerging intensity is I, then the measurable quantity $g = \ln(I_0/I) = \int_L f ds$, where f is the variable absorption coefficient along the line L. Hence if f is a function in two dimensions, and g is known for all lines intersecting the body, the question is: "Can f be determined if g is known?". Again this seemed like a problem which would

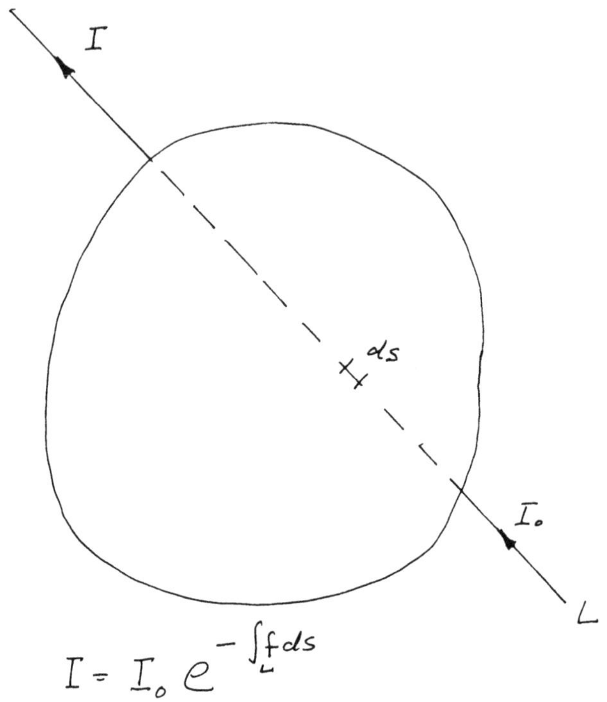

Fig. 1

have been solved before, probably in the 19th Century, but again a literature search and enquiries of mathematicians provided no information about it. Fourteen years would elapse before I learned that Radon had solved this problem in 1917. Again I had to tackle the problem from the beginning. The solution is easy for objects with circular symmetry for which f = f(r), r being the radius. One has Abel's equation to solve, and its solution

$$f(r) = -\frac{d}{dr}\left[\frac{r}{\pi}\int_r^\infty \frac{g(s)ds}{s\sqrt{s^2-r^2}}\right] = -\frac{dI(r)}{dr}, \tag{1}$$

has been known since 1825. In 1957 I did an experiment in Cape Town on a circularly symmetrical sample consisting of a cylinder of aluminum surrounded by an annulus of wood. The results are shown in Fig. 2. Here I(r) is plotted against r and the constant slopes indicate the constant values of the absorption coefficient in wood and aluminum. Even this simple result proved to have some predictive value for it will be seen that the three points nearest the origin lie on a line of a slightly different slope from the other points in the aluminium. Subsequent inquiry in the machine shop revealed that the aluminum cylinder contained an inner peg of slightly lower absorption coefficient than the rest of the cylinder.

r	$I_1(r)$	$I_2(r)$	I(r)
0			2.9938
.5	1.9889	.7915	2.7694
1.0	1.5949	.9640	2.5589
1.5	1.3618	.9614	2.3295
2.0	1.1810	.8886	2.0697
2.5	1.0083	.7900	1.7983
3.0	.8763	.6660	1.5423
3.5	.7480	.5448	1.2928
4.0	.6272	.4148	1.0420
4.5	.4659	.3456	.8115
5.0	.2934	.2862	.5796
5.5	.2353	.2521	.4871
6.0	.2122	.2166	.4288
6.5	.1982	.1828	.3756
7.0	.1757	.1458	.3215
7.5	.1553	.1130	.2683
8.0	.1326	.0791	.2117
8.5	.1141	.0485	.1629
9.0	.0883	.0174	.1057
9.5	.0507	.0028	.0535
10.0	.0074	.0000	.0074
10.5	.0008	.0000	.0008
11.0	.0000	.0000	.0000

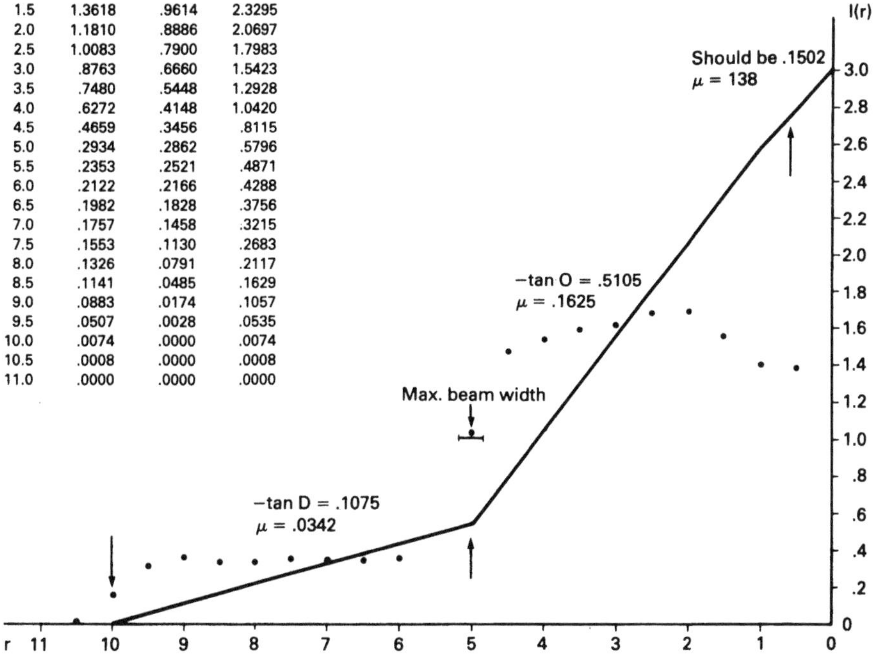

Fig. 2

Further work occurred intermittently over the next six years. Using Fourier expansions of f and g, I obtained equations like Abel's integral equation but with more complicated kernels, and I obtained results like equation (1), but with more complicated integrands. However, they were also integrals from r to ∞, a point which I shall return to. These integrals were not too good for data containing noise, so alternative expansions were developed which dealt with noisy data satisfactorily. By 1963 I was ready to do an experiment on a phantom without circular symmetry with the apparatus shown in Fig. 3. The two cylinders are two collimators which contain the source and the detector, and which permit a 5mm beam of Co^{60} gamma-rays to pass through the phantom which is the disc in between them. At that time I was approached by an undergraduate, David Hennage, who wanted to know whether I had a numerical problem for him to work on so that he could learn FORTRAN and learn how to use a computer. Since he had not had experience of dealing with data in which the principal source of error was statistical, this seemed a good chance for him to learn that too by helping take the data. Final data were taken over two days in the summer of 1963, the calculations were done, and the results are shown in Figs. 4 a and b. Figure 4 a shows the phantom, and

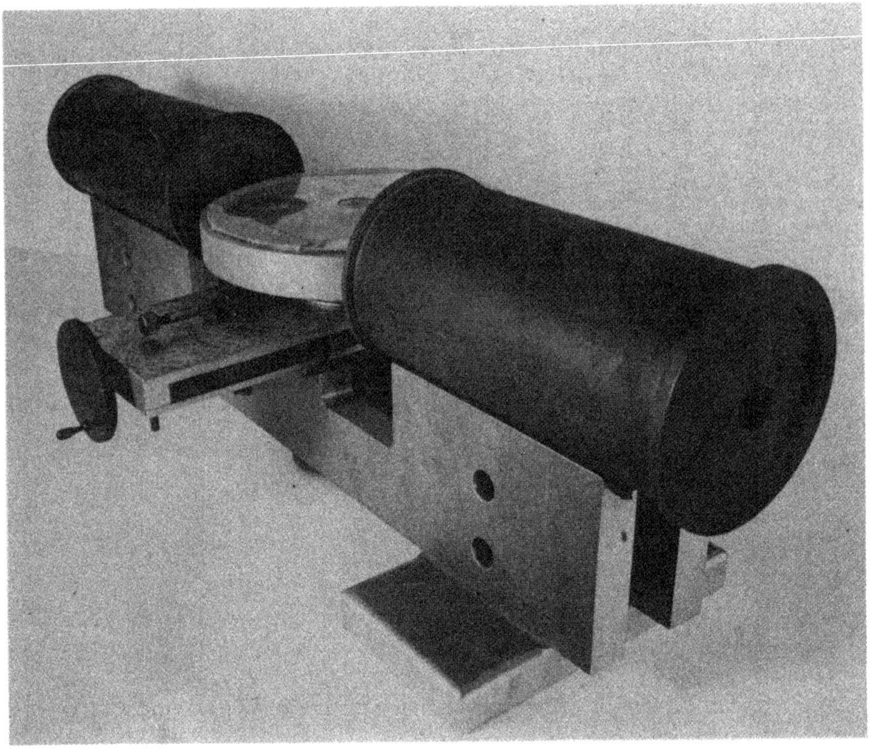

Fig. 3

this was quite a hybrid. The outer ring of aluminum represents the skull, the Lucite inside it represents soft tissue, and the two aluminum discs represent tumors. The ratio of the absorption coefficients of aluminum and lucite is about 3, which is very much greater than the ratio of the absorption coefficients of abnormal and normal tissue. The ratio of 3 was chosen in an attempt to attract the attention of those doing what would now be called emission scanning using positron emitting isotopes, a subject then in its infancy. This was about the ratio between the concentration of radioisotope in abnormal tissue and muscle on the one hand and in normal tissue on the other. The mathematics of emission scanning is of course the same as transmission scanning after an obvious correction for absorption. The right side of the slide shows the results plotted as a graph of attenuation coefficient as a function of distance along the line OA. Similar graphs were made along other lines. The full curves are the true values, the dots are the calculated values, and the agreement is quite good. The results could have been presented on a grey scale on a scope, but the graphs were thought to be better for publication.

Publication took place in 1963 and 1964 (1, 2). There was virtually no response. The most interesting request for a reprint came from a Swiss Centre for Avalanche Research. The method would work for deposits of

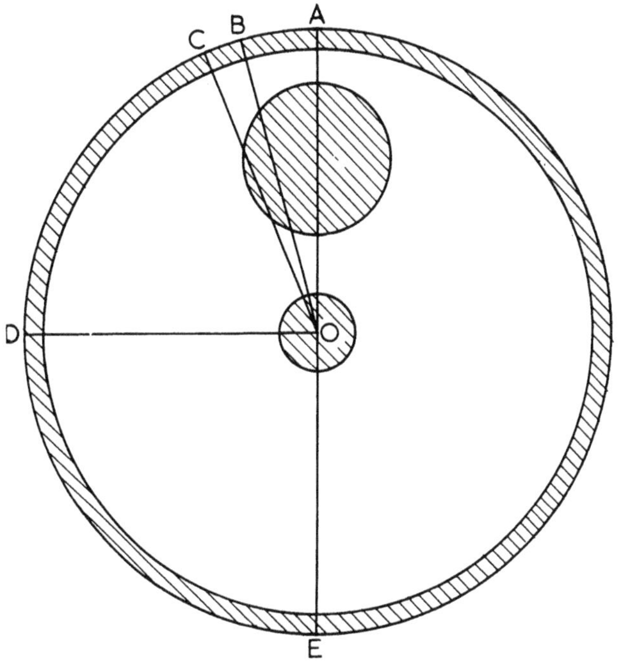

Fig. 4 a

snow on mountains if one could get either the detector or the source into the mountain under the snow!

My normal teaching and research kept me busy enough so I thought very little about the subject until the early seventies. Only then did I learn of Radon's (3) work on the line integral problem. About the same time I

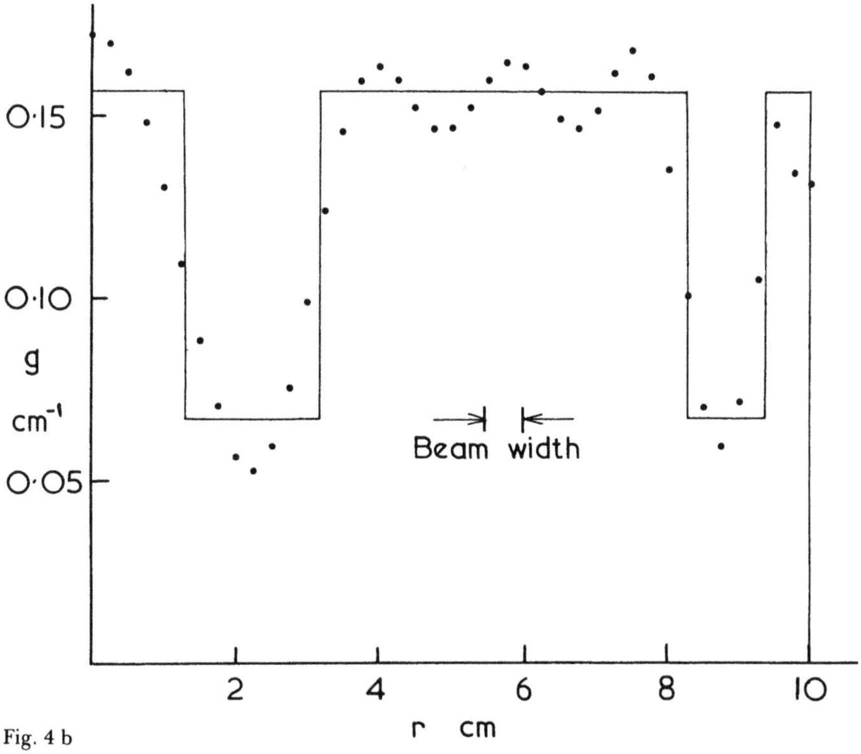

Fig. 4 b

learnt that this problem had come up in statistics in 1936 at the Department of Mathematical Statistics here in Stockholm where it was solved by Cramér and Wold (4). I also learnt of Bracewell's work (5) in radioastronomy which produced the same form of solution as Cramér and Wold had found, of De Rosier and Klug's (6) (1968) work in electron microscopy, and of the work of Rowley (7) (1969) and Berry and Gibbs (1970) in optics. Last, but by no means least, I first heard of Hounsfield and the EMI-scanner, about which you will be hearing shortly. Since 1972 I have been interested in a number or problems related to the CT-scanning problem or stemming from it, and I would like to tell you about some of them.

As I mentioned earlier the solution the Radon's problem which I found was in terms of integrals which extend from the radius at which the absorption coefficient is being found to the outer edge of the sample. What this means is that if one wishes to find the absorption coefficient in an annulus one needs only data from lines which do not intersect the hole in the annulus. This is known as the Hole Theorem and it is an exact mathematical theorem which can be proved very simply. However, as I have mentioned, when one applies my integrals to real data containing noise, the noise from the outer observations is amplified badly as one goes deeper and deeper into the sample, so the integrals as they stand are not useful in practice. This process of working from the outside in is known as "peeling the onion", and several algorithms for peeling the onion have been tried. All show the same feature, namely, that noise from the outer data propagates badly as one goes deeper and deeper into the sample. However, to the best of my knowledge, no one has produced a closed form solution to Radon's problem which includes the Hole Theorem and from which it can be *proved* that noise from the outside *necessarily* propagates badly into the interior. (Doyle and I published a closed form solution to Radon's problem which turned out to be wrong. A retraction will be published soon.) (9)

A solution containing the hole theorem for which noise did *not* propagate into the interior would be extremely valuable for CT-scanning. First it would reduce the dose administered to a patient by not passing X-rays through a region of no interest, i.e. the hole. Second it could be used to avoid regions in which sharp changes in absorption coefficient occur. These sharp changes necessarily produce overshoots which cloud the final image. In an extreme case consider a patient who by accident or by some medical procedure has in his body a piece of metal. This distorts a CT scan of a plane containing the metal beyond recognition. The hole theorem could be used to avoid the piece of metal. Less extremely, an interface between bone and soft tissue introduces overshoots which distort the image of the soft tissue. The hole theorem could be used to avoid the bone and the distortion.

Here then is a question which needs an answer. An unfavorable answer would leave us no worse off than we now are. A favorable answer would result in an improvement of imaging.

In my 1963 paper I had mentioned, but not very favorably, that protons could be used for scanning instead of X-rays. In 1968 my friend and colleague Andreas Koehler started producing some beautiful radiograms (10) using protons, and thinking about his work removed some of the doubts I had previously had about using protons for tomography.

The distinction between protons and X-rays is illustrated in Fig. 5. The number-distance curve for X-rays is the familiar exponential curve. For protons, there is a slow decrease caused by nuclear absorption followed by a very sharp fall-off. Most of the protons penetrate a distance into the material then all that are left stop in a short distance at what is called the range of the protons. It is this sharp fall-off which makes protons very sensitive to small variations in thickness or density along their paths. A comparison of the two curves at the range is a rough measure of this sensitivity to thickness or density changes. Others were thinking about the same thing and Crowe et al (11) at Berkeley produced a tomogram using alpha-particles in 1975, using the not inconsiderable arsenal of equipment at a high-energy physics laboratory. Koehler and I (12) made a tomogram of a simple circularly symmetrical sample to show two things:

1. that very simple equipment could be used, and
2. that we could easily detect 1/2% density differences with that equipment.

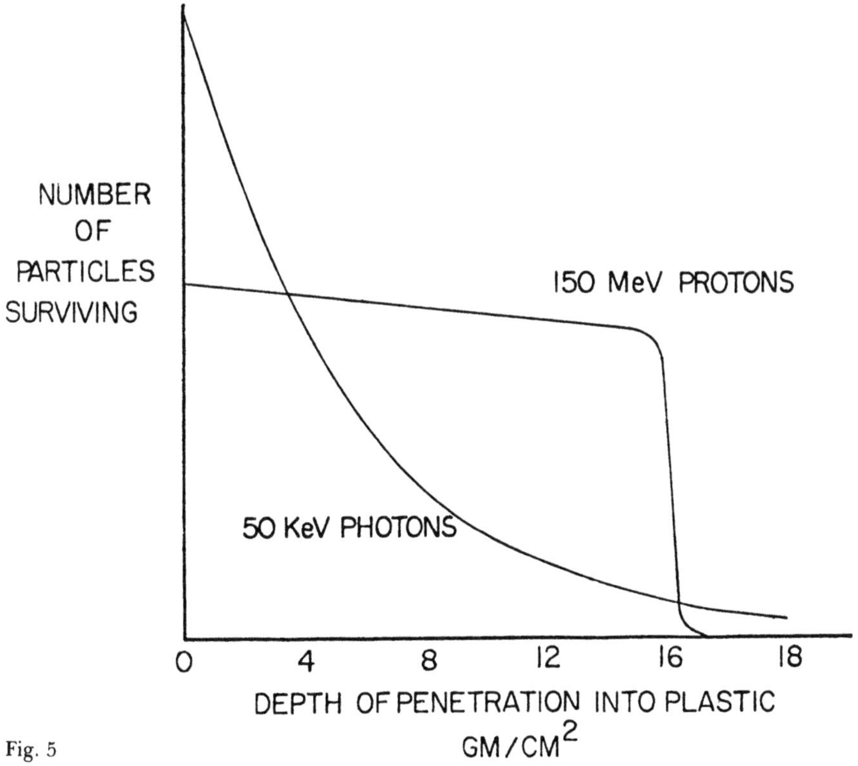

Fig. 5

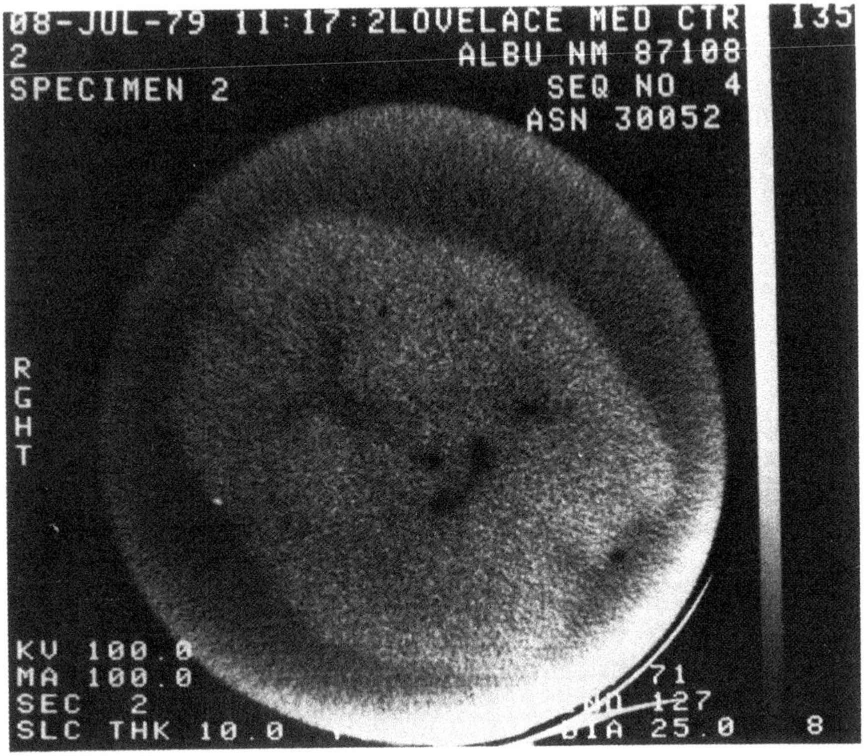
08-JUL-79 11:17:2LOVELACE MED CTR
135
2
ALBU NM 87108
SPECIMEN 2
SEQ NO 4
ASN 30052
RGHT
KV 100.0
MA 100.0
SEC 2
SLC THK 10.0
71
127
25.0
8

Fig. 6 a

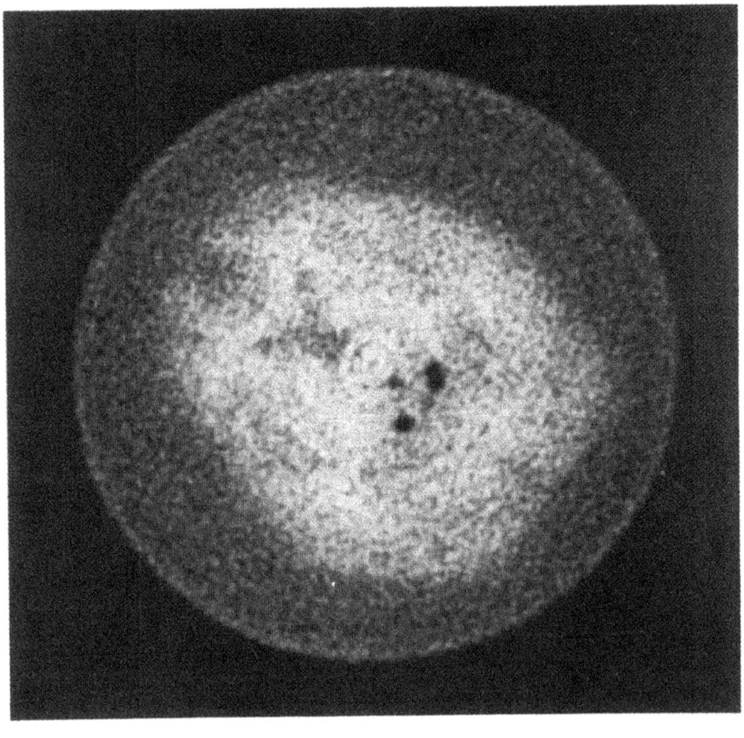

Fig. 6 b

As it happened, we were able to detect density differences of about 0.1% or less caused by machining stresses in our Lucite sample and diffusion of water into and out of the Lucite (13). More recently this work has been continued by Hanson and Steward (14) and their co-workers at Los Alamos, and I am able to show you some of their results. They have recently made tomograms of human organs. Fig. 6a is an X-ray scan of a normal brain made with a Δ2020 scanner at a dose of about 9 rad. Dr. Steward (14) has observed and made density measurements which show that water and electrolyte are lost by white tissue within four hours after death so density differences between grey and white matter disappear. Fig. 6b is a proton scan of the same sample made at a dose of 0.6 rad, and it is roughly as good as the X-ray scan. Figs. 7a and b show, respectively, an X-ray scan at a dose of 9 rad and a proton scan at 0.6 rads of a heart with a myocardial infarction fixed in formalin. According to Dr. Steward, fixing in formalin does not alter the relative stopping powers of the tissues of the heart. Again the pictures are of about the same quality. The time for taking the data was about an hour, but the LAMF accelerator is pulsed and is on for only 6% of the time. A continuous beam would have produced the same data in about four minutes – roughly the same time as the original EMI-scanner. I would like to acknowledge my indebtedness to Hanson and Steward *et al.* for making their results available to me.

Why use protons? First there is the question of dose. The dose in Hanson and Steward's work about 0.6 rad as distinct from to 9 rads required for the X-ray scans. This is in agreement with theoretical calculations (15) that the dose required for proton scans is five to ten times less than for X-ray scans for the same amount of information. Second, the mechanism by which protons are brought to rest is different from the mechanisms by which X-rays are absorbed or scattered, so one ought to see different things by using the different radiations, particularly the distribution of hydrogen. Koehler and I have indeed found a reversal of contrast beween proton and X-ray scans on two occasions, and this is expected from theory.

You know how some people fuss about the high cost of CT-scanners, so you can imagine what they would say if one suggested that the X-ray tube in the scanner should be replaced by a much more expensive cyclotron! Of course they would be right, but only if proton scanning is viewed in the narrow context of diagnostics. Instead of looking at the number-distance curve which I showed when comparing X-rays and protons, let us look at the ionization-distance curve which is the thing that matters for therapy. This is shown in Fig. 8. You will see that the curve is peaked (the Bragg peak) near the end of the range and then falls off very sharply, so that there is virtually no ionization beyond the end of the range. Both of these facts are of importance in therapy as was pointed out by R. R. Wilson (16) in 1946, and as has been shown by treatments of a number of different conditions at Uppsala, Harvard and Berkeley. In fact, protons are far superior to X-rays for the treatment of some conditions. So one can

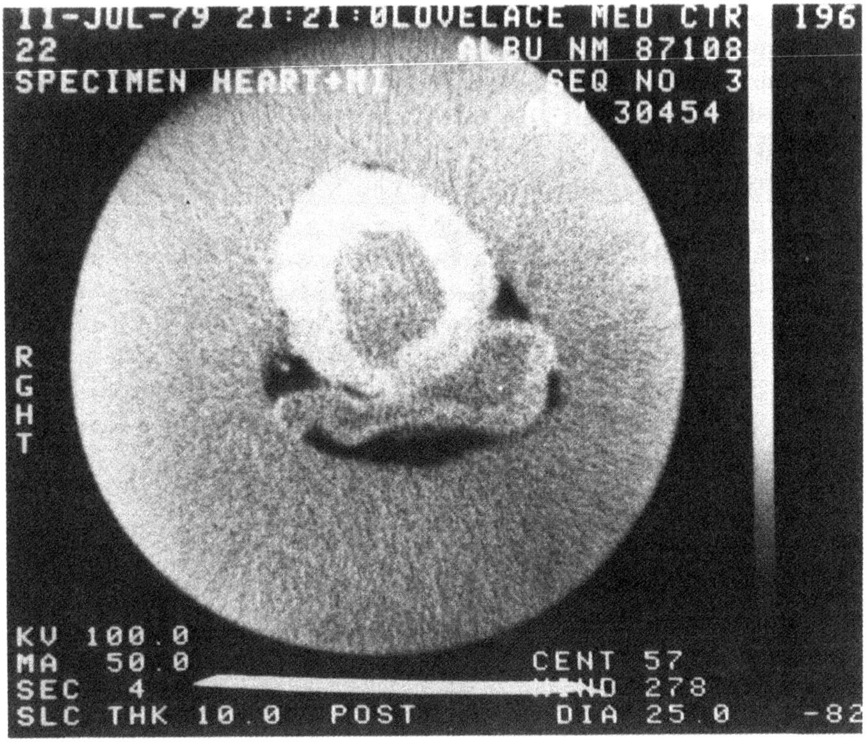
11-JUL-79 21:21:0LOVELACE MED CTR
196
22
NM 87108
SPECIMEN HEART
SEQ NO 3
30454
RGHT
KV 100.0
MA 50.0
SEC 4
SLC THK 10.0 POST
CENT 57
278
DIA 25.0
-82

Fig. 7 a

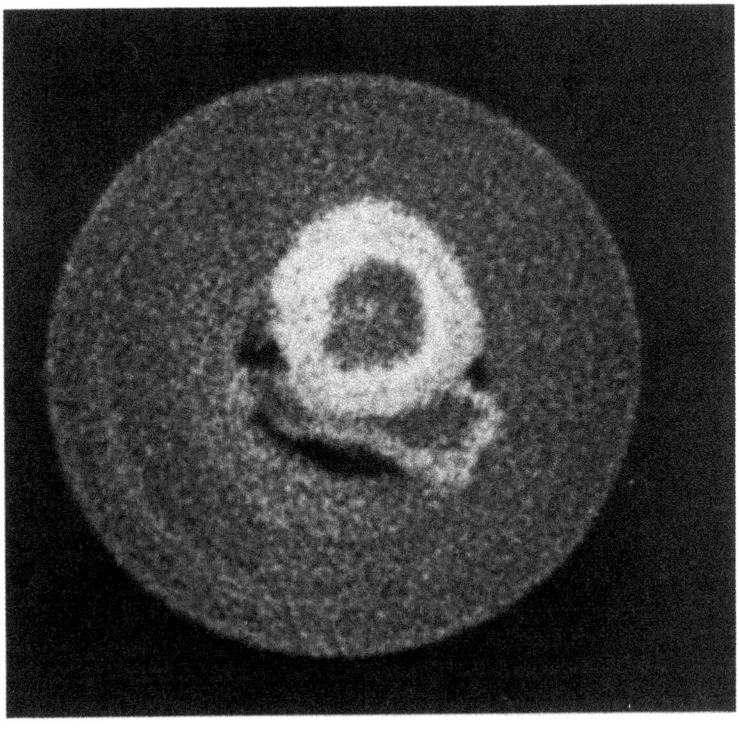

Fig. 7 b

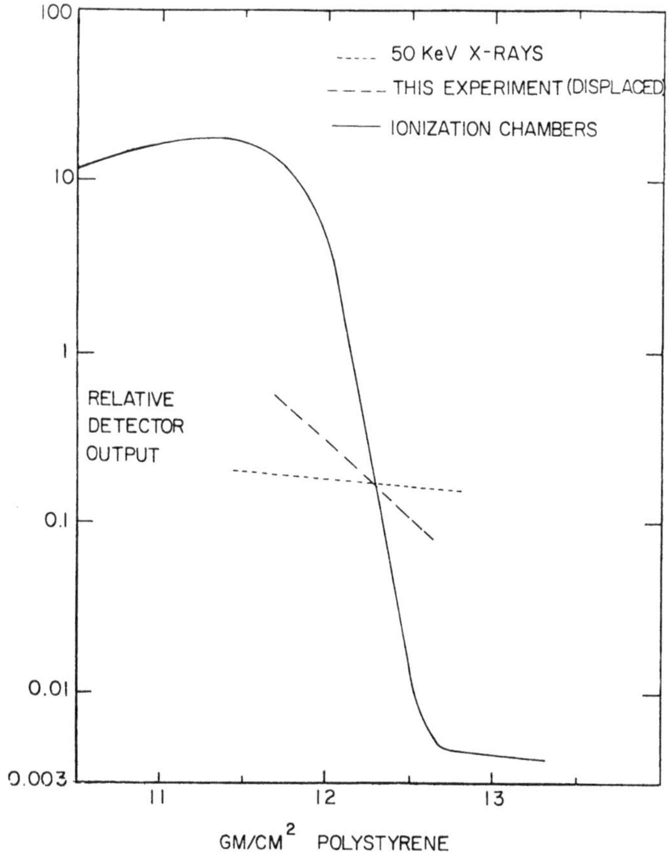

Fig. 8

envisage a large metropolitan area as having a 250 MeV proton accelerator with a number of different ports, say ten, at which patients could be treated simultaneously (17). If one of these ports was devoted to proton tomography, the marginal cost of such tomography would not be great. In exploring these possibilities it is essential that diagnostic radiologists and therapeutic radiologists work together.

Most recently I have been interested in some mathematical topics which I would like to discuss without mathematical details. I have presented Radon's problem in the form in which it is usually given in CT-scanning, namely, how does one recover a function in a plane given its line integrals over all lines in the plane. This can be generalized to three-dimensions as can be seen in Fig. 9. This shows a sphere in which a function is defined, and a plane intersecting that sphere. Suppose that the given information is the integrals of the function over all planes intersecting the sphere, then the questions is "Can the function be recovered from these integrals?" Radon provided an affirmative answer and a formula for finding the function. It turns out that this formula is simpler than in the case of two-dimensions, and it has applications in imaging using Nuclear Magnetic.

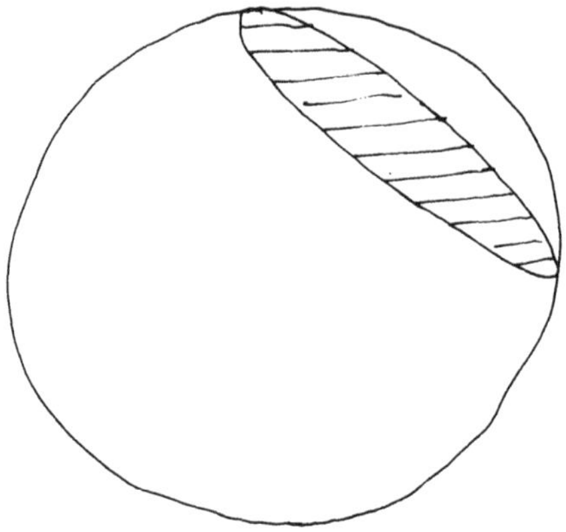

Fig. 9

Resonance (N.M.R.) rather than X-rays. The generalization to n-dimensions follows in a straightforward way for Euclidean space and was given by Radon. The few mathematicians who knew of Radon's work before the nineteen sixties applied his results to the theory of partial differential equations. They also generalized his results to integration over circles or spheres of constant radius and to certain ellipsoids. Generalizations to spaces other than Euclidean were made, for example, by Helgason (18) and Gelfand *et al* (19) in the sixties.

In my 1964 paper I gave a solution to the problem of integrating over circles of variable radius which pass through the origins as shown in Fig. 10. Recently Quinto and I (20) have given much more detailed results on this problem generalized to n-dimensional Euclidean space, and we have applied these results to obtain some theorems about the solutions of Darbox's partial differential equation. There is an intimate connection

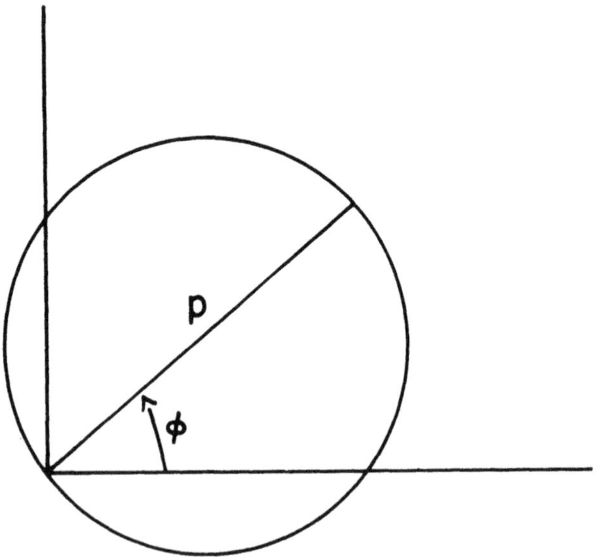

Fig. 10

between our results and Radon's results, and we are presently attempting to find more general results which relate the solutions of Radon's problem for a family of surfaces to the solution of Radon's problem for another family of surfaces related to the first in a particular way.

What is the use of these results? The answer is that I don't know. They will almost certainly produce some theorems in the theory of partial differential equations, and some of them may find application in imaging with N.M.R. or ultrasound, but that is by no means certain. It is also beside the point. Quinto and I are studying these topics because they are interesting in their own right as mathematical problems, and that is what science is all about.

Of the many people who have influenced me beneficially, I shall name only a few. The late Professor R. W. James F.R.S. taught me a great deal, not only about physics, when I was a student and a young Lecturer in his Department in Cape Town. Andreas Koehler, Director of the Harvard Cyclotron Laboratory, has provided me with friendship, moral support, and intellectual stimulation for over twenty years. On the domestic side are my parents, now deceased, and my immediate family. My wife Barbara and our three children have not only put up with me, they have done so in a loving and supportive way for many years. To these people and others unnamed I shall be grateful to the end of my life.

REFERENCES

1. Cormack, A. M., J. Applied Physics *34,* 2722 (1963)
2. Cormack, A. M. J. Applied Physics *35,* 2908 (1964)
3. Radon, J. H., Ber. Sachs. Akad. Wiss. *69,* 262 (1917)
4. Cramér, H., and Wold, H., J. London Math. Soc. *11,* 290 (1936)
5. Bracewell, R. N., Austr. J. Phys., *9,* 198 (1956)
6. De Rosier, D. J., and Klug, A., Nature *217,* 30 (1968)
7. Rowley, P. D., J. Opt. Soc. Amer. *59,* 1496 (1969)
8. Berry, M. V., and Gibbs, D. F., Proc. Roy. Soc. A*314,* 143 (1970)
9. Cormack, A. M., and Doyle, B. J., Phys. Med. Biol. *22,* 994 (1977); Cormack, A. M., Phys. Med. Biol., to be published
10. Koehler, A. M., Science *160,* 303 (1968)
11. Crowe, K. M., Budinger, T. F., Cahoon, J. L., Elischer, V. P., Huesman, R. H., and Kanstein, L. L., Lawrence Berkeley Report LBL-3812 (1975)
12. Cormack, A. M., and Koehler, A. M., Phys. Med. Biol. *21,* 560 (1976)
13. Cormack, A. M., Koehler, A. M., Brooks, R. A., Di Chiro, G., Int. Symp. on C. A. T., National Institutes of Health, (1976)
14. Hanson, K., Steward, R. V., Bradbury, J. N., Koeppe, R. A., Macek, R. J., Machen, D. R., Morgado. R. E., Pacciotti, M. A., Private Communication
15. Koehler, A. M., Goitien, M., and Steward, V. W., Private Communication, and Huesman, R. H., Rosenfeld, A. H., and Solmitz, F. T, Lawrence Berkeley Report LBL-3040 (1976)
16. Wilson, R. R., Radiology *47,* 487 (1946)
17. Cormack, A. M., Physics Bulletin *28,* 543 (1947)
18. Helgason, S., Acta. Math. *113,* 153 (1965)
19. Gelfand, I. M., Graev, M. I., and Shapiro, Z. Y., Functional Anal. Appl. *3.* 24 (1969)
20. Cormack, A. M., Quinto, E. T., to appear in Trans. Amer. Math. Soc.

Godfrey Hounsfield

GODFREY NEWBOLD HOUNSFIELD

I was born and brought up near a village in Nottinghamshire and in my childhood enjoyed the freedom of the rather isolated country life. After the first world war, my father had bought a small farm, which became a marvellous playground for his five children. My two brothers and two sisters were all older than I and, as they naturally pursued their own more adult interests, this gave me the advantage of not being expected to join in, so I could go off and follow my own inclinations.

The farm offered an infinite variety of ways to do this. At a very early age I became intrigued by all the mechanical and electrical gadgets which even then could be found on a farm; the threshing machines, the binders, the generators. But the period between my eleventh and eighteenth years remains the most vivid in my memory because this was the time of my first attempts at experimentation, which might never have been made had I lived in a city. In a village there are few distractions and no pressures to join in at a ball game or go to the cinema, and I was free to follow the trail of any interesting idea that came my way. I constructed electrical recording machines; I made hazardous investigations of the principles of flight, launching myself from the tops of haystacks with a home-made glider; I almost blew myself up during exciting experiments using water-filled tar barrels and acetylene to see how high they could be waterjet propelled. It may now be a trick of the memory but I am sure that on one occasion I managed to get one to an altitude of 1000 feet!

During this time I was learning the hard way many fundamentals in reasoning. This was all at the expense of my schooling at Magnus Grammar School in Newark, where they tried hard to educate me but where I responded only to physics and mathematics with any ease and moderate enthusiasm.

Aeroplanes interested me and at the outbreak of the second world war I joined the RAF as a volunteer reservist. I took the opportunity of studying the books which the RAF made available for Radio Mechanics and looked forward to an interesting course in Radio. After sitting a trade test I was immediately taken on as a Radar Mechanic Instructor and moved to the then RAF-occupied Royal College of Science in South Kensington and later to Cranwell Radar School. At Cranwell, in my spare time, I sat and passed the City and Guilds examination in Radio Communications. While there I also occupied myself in building large-screen oscilloscope and demonstration equipment as aids to instruction, for which I was awarded the Certificate of Merit.

It was very fortunate for me that, during this time, my work was appreciated by Air Vice-Marshal Cassidy. He was responsible for my obtaining a grant after the war which enabled me to attend Faraday House Electrical Engineering College in London, where I received a diploma.

I joined the staff of EMI in Middlesex in 1951, where I worked for a while on radar and guided weapons and later ran a small design laboratory. During this time I became particularly interested in computers, which were then in their infancy. It was interesting, pioneering work at that time: drums and tape decks had to be designed from scratch. The core store was a relatively new idea which was the subject of considerable experiment. The stores had to be designed and then plain-threaded by hand (causing a few frightful tangles on occasions). Starting in about 1958 I led a design team building the first all-transistor computer to be constructed in Britain, the EMIDEC 1100. In those days the transistor, the OC72, was a relatively slow device, much slower than valves which were then used in most computers. However, I was able to overcome this problem by driving the transistor with a magnetic core. This increased the speed of the machine so that it compared with that of valve computers and brought about the use of transistors in computing earlier than had been anticipated. Twenty-four large installations were sold before increases in the speed of transistors rendered this method obsolete.

When this work finished I transferred to EMI Central Research Laboratories, also at Hayes. My first project there was hardly covered in glory: I set out to design a one-million word immediate access thin-film computer store. The problem was that after a time it was evident that this would not be commercially viable. The project was therefore abandoned and, rather than being immediately assigned to another task I was given the opportunity to go away quietly and think of other areas of research which I thought might be fruitful. One of the suggestions I put forward was connected with automatic pattern recognition and it was while exploring various aspects of pattern recognition and their potential, in 1967, that the idea occurred to me which was eventually to become the EMI-Scanner and the technique of computed tomography.

The steps in my work between this initial idea and its realisation in the first clinical brain-scanner have already been well documented. As might be expected, the programme involved many frustrations, occasional awareness of achievement when particular technical hurdles were overcome, and some amusing incidents, not least the experiences of travelling across London by public transport carrying bullock's brains for use in evaluation of an axperimental scanner rig in the Laboratories.

After the initial experimental work, the designing and building of four original clinical prototypes and the development of five progressively more sophisticated prototypes of brain and whole body scanner (three of which went into production) kept me fully occupied until 1976. Since then I have been able to broaden my interest in a number of projects which are currently in hand in the Laboratories, including further possible advances in CT technology and in related fields of diagnostic imaging, such as nuclear magnetic resonance.

As a bachelor, I have been able to devote a great deal of time to my general interest in science which more recently has included physics and biology. A great deal of my adult life has centred on my work, and only recently did I bother to establish a permanent residence. Apart from my work, my greatest pleasures have been mainly out-of-doors, and although I no longer ski I greatly enjoy walking in the mountains and leading country rambles. I am fond of music, whether light or classical, and play the piano in a self-taught way. In company I enjoy lively way-out discussions.

COMPUTED MEDICAL IMAGING

Nobel Lecture, 8 December, 1979
By
GODFREY N. HOUNSFIELD
The Medical Systems Department of Central Research Laboratories EMI, London, England

In preparing this paper I realised that I would be speaking to a general audience and have therefore included a description of computed tomography (CT) and some of my early experiments that led up to the development of the new technique. I have concluded with an overall picture of the CT scene and of projected developments in both CT and other types of systems, such as Nuclear Magnetic Resonance (NMR).

Although it is barely 8 years since the first brain scanner was constructed, computed tomography is now relatively widely used and has been extensively demonstrated. At the present time this new system is operating in some 1000 hospitals throughout the world. The technique has succesfully overcome many of the limitations which are inherent in conventional X-ray technology.

When we consider the capabilities of conventional X-ray methods, three main limitations become obvious. Firstly, it is impossible to display within the framework of a two-dimensional X-ray picture all the information contained in the three-dimensional scene under view. Objects situated in depth, i. e. in the third dimension, superimpose, causing confusion to the viewer.

Secondly, conventional X-rays cannot distinguish between soft tissues. In general, a radiogram differentiates only between bone and air, as in the lungs. Variations in soft tissues such as the liver and pancreas are not discernible at all and certain other organs may be rendered visible only through the use of radio-opaque dyes.

Thirdly, when conventional X-ray methods are used, it is not possible to measure in a quantitative way the separate densities of the individual substances through which the X-ray has passed. The radiogram records the *mean* absorption by all the various tissues which the X-ray has penetrated. This is of little use for quantitative measurement.

Computed tomography, on the other hand, measures the attenuation of X-ray beams passing through sections of the body from hundreds of different angles, and then , from the evidence of these measurements, a computer is able to reconstruct pictures of the body's interior.

Pictures are based on the separate examination of a series of contiguous cross sections, as though we looked at the body separated into a series of thin "slices". By doing so, we virtually obtain total three-dimensional information about the body.

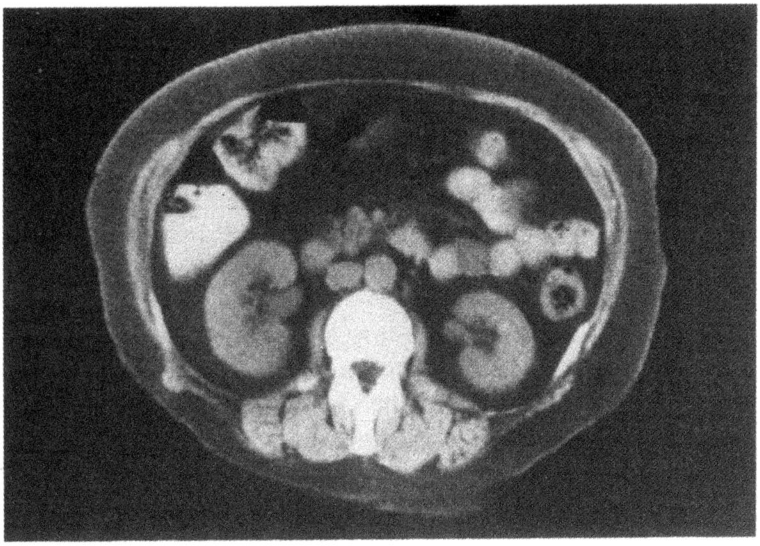

Fig. 1. CT scan taken through the kidneys.

However, the technique's most important feature is its enormously greater sensitivity. It allows soft tissue such as the liver and kidneys to be clearly differentiated, which radiographs cannot do. An example is shown in Fig. 1.

It can also very accurately measure the values of X-ray absorption of tissues, thus enabling the nature of tissue to be studied.

These capabilities are of great benefit in the diagnosis of disease, but CT additionally plays a role in the field of therapy by accurately locating, for example, a tumour so indicating the areas of the body to be irradiated and by monitorig the progress of the treatment afterwards.

It may be of interest if I describe some of the early experiments that led up to the development of CT.

Some time ago I investigated the possibility that a computer might be able to reconstruct a picture from sets of very accurate X-ray measurements taken through the body at a multitude of different angles. Many hundreds of thousands of measurements would have to be taken, and reconstructing a picture from them seemed to be a mammoth task as it appeared at the time that it would require an equal number of many hundreds of thousands of simultaneous equations to be solved.

When I investigated the advantages over conventional X-ray techniques however, it became apparent that the conventional methods were not making full use of all the information the X-rays could give.

On the other hand, calculations showed that the new system used the data very efficiently and would be two orders of magnitude more sensitive than conventional X-rays. For this reason I hoped that it would be possible to distinguish between the various tissues of the body, although I could not find any literature which suggested that such X-ray absorption differences existed.

THE EARLY TESTS

I decided to do some lab experiments with gamma rays to test if the system would work. The equipment was very much improvised. A lathe bed provided the lateral scanning movement of the gamma-ray source, and sensitive detectors were placed on either side of the object to be viewed which was rotated 1° at the end of each sweep. The 28,000 measurements from the detector were digitized and automatically recorded on paper tape. After the scan had been completed this was fed into the computer and processed.

Many tests were made on this machine, and the pictures were encouraging despite the fact that the machine worked extremely slowly, taking 9 days to scan the object because of the low intensity gamma source. The pictures took 2½ hours to be processed on a large computer. The results of the processing were received on paper tape and these were brought to the laboratory and caused to modulate a spot of light on a cathode ray tube point by point, in front of a camera. As paper tape was used this was a slow process and it took at best two hours to produce a photograph. Clearly, nine days for a picture was too time-consuming, and the gamma source was replaced by a more powerful X-ray tube source, which reduced the scanning time to nine hours (Fig. 2). From then on, much better pictures were obtained; these were usually of blocks of perspex. A preserved specimen of a human brain was eventually provided by a local hospital museum and we produced the first picture of a brain to show grey and white matter (Fig. 3).

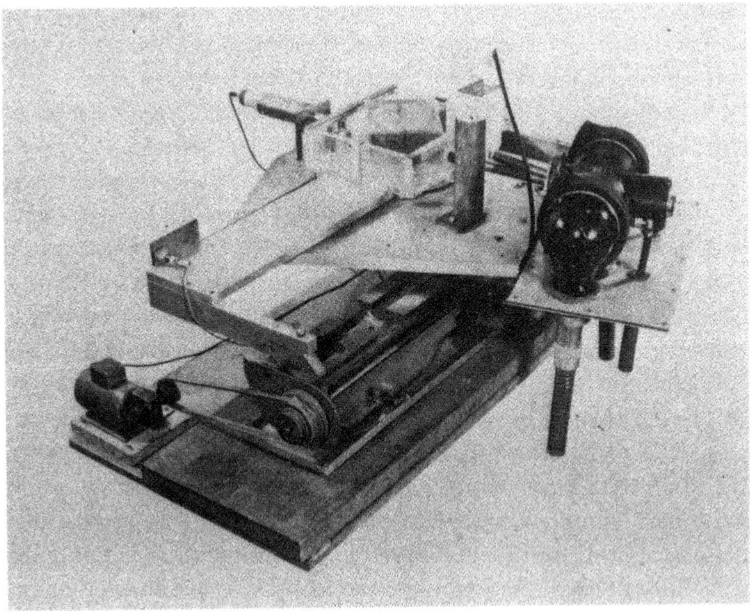

Fig. 2. Laboratory machine, showing X-ray tube and detector traversing along a lathe bed across a human brain. At the end of the stroke the brain would be rotated 1° and the traverse would be repeated.

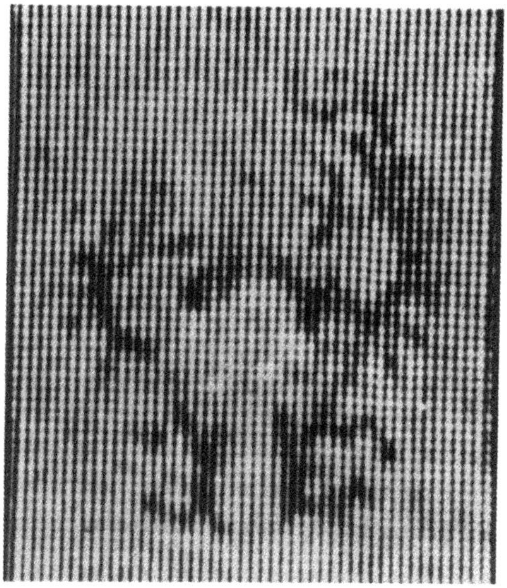

Fig. 3. Picture of the first brain scanned on Laboratory machine Fig. 2.

Disappointingly, further analyses revealed that the formalin used to preserve the specimen had enhanced the readings, and had produced exaggerated results. Fresh bullock's brains were therefore used to cross-check the experiments, and although the variations in tissue density were less pronounced, it was confirmed that a large amount of anatomic detail could be seen. In parallel, tests were carried out on sections through pigs

Fig. 4. First clinical prototype brain scanner installed at Atkinson Morley's Hospital, London.

in the area of the kidneys, and this work also produced most encouraging results. Although the speed had been increased to one picture per day, we had a little trouble with the specimen decaying while the picture was being taken, so producing gas bubbles, which increased in size as the scanning proceeded.

At this point in time, we found that we could see brain and body tissues clearly but we were still very worried as to whether tumours would show up at all. Unless it could do this, the machine would be of very little use. To test this, we had to build a much faster and more sophisticated machine that would scan the brains of living patients in a hospital (Fig. 4).

In 1972 the first patient was scanned by this machine. She was a woman who had a suspected brain lesion, and the picture showed clearly in detail a dark circular cyst in the brain (Fig. 5). From this moment on, as more patients were being scanned, it became evident that the machine was going to be sensitive enough to distinguish the difference between normal and diseased tissue.

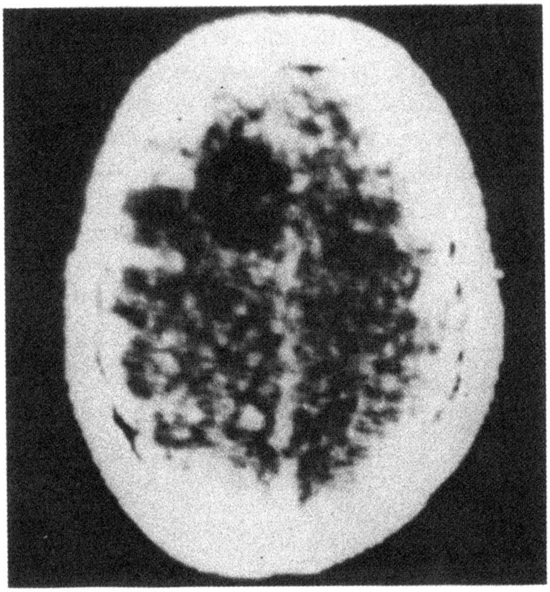

Fig. 5. First clinical picture obtained from prototype machine.

Applying the principles to scanning the body seemed to be the next logical step, a larger and faster scanner was designed capable of taking high resolution pictures of the body in 18 seconds. Fig. 1 is a typical picture taken by this machine.

Since then, there has been a tendency to construct more complicated machined with a scan time of 3 seconds or less (Fig. 6).

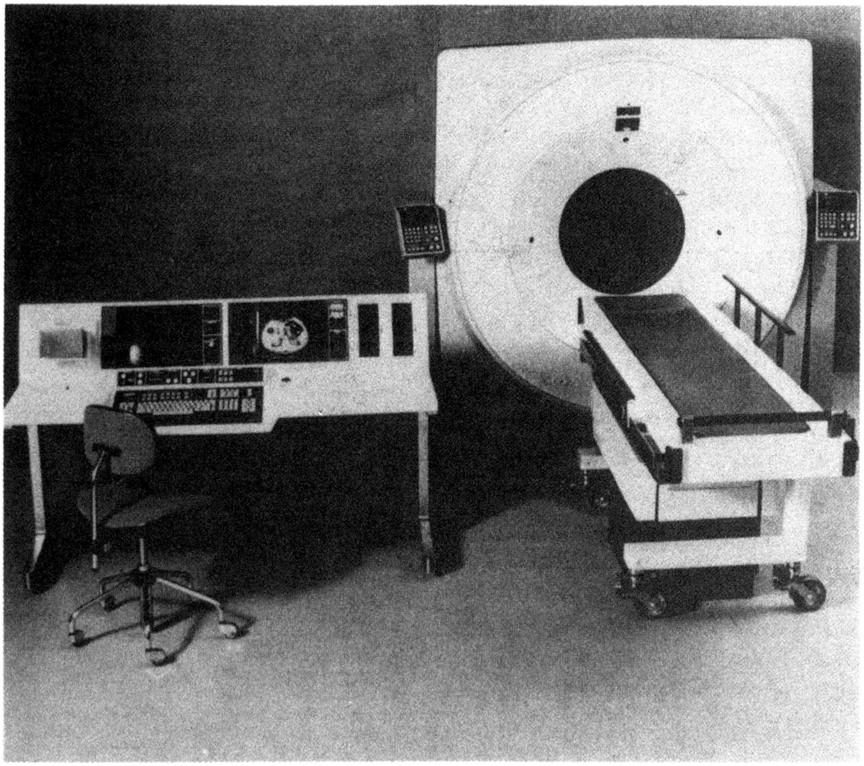

Fig. 6. A machine capable of scanning at a rate of 3 sec/picture.

PRINCIPLES OF THE TECHNIQUE

There are three types of CT machine currently in use (Fig. 7). Basically all three systems use different methods of scanning the patient but end up by taking approximately the same pattern of readings, namely sets of readings across the patient either as parallel sets or into the form of a fan. These are taken at a multitude of different angles (Fig. 8).

The system shown in Fig. 7a, translates across the body, each detector taking parallel sets of readings, at the same time as it rotates around the body. It could have 30 detectors and take 18 seconds to scan a picture.

In the second system (Fig. 7b) the sets of readings taken are in the form of a fan. It does not translate across the body but only rotates around it. This system usually has approximately 300–500 detectors but is faster and can take a picture in three seconds. The detectors in this picture have to be accurately stabilised.

In the third system (Fig. 7c) the detectors are assembled in a fixed circle and only the X-ray tube sweeps around the body, taking a fan of readings as it does so. This system requires 700 – 1 000 detectors and it also

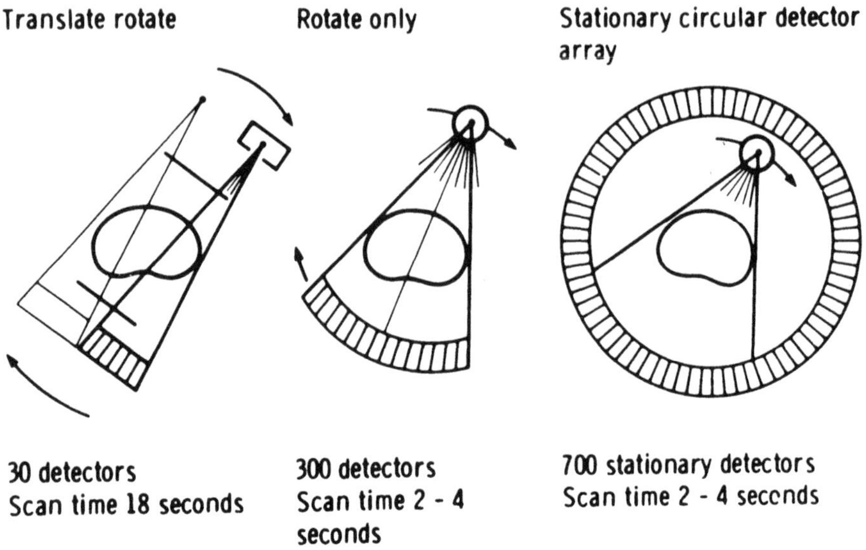

Fig. 7. Three different methods of scanning the patient.

takes a picture in three seconds. It is not necessary to stabilise detectors in this case.

The whole effect of this motion is to take approximately one million accurate absorption measurements through the body in the form of a number of sweeps across (or projections through) the body. These are taken at all angles through the slice, thus providing us with an enormous amount of information about the composition of the slice (Fig. 8). The

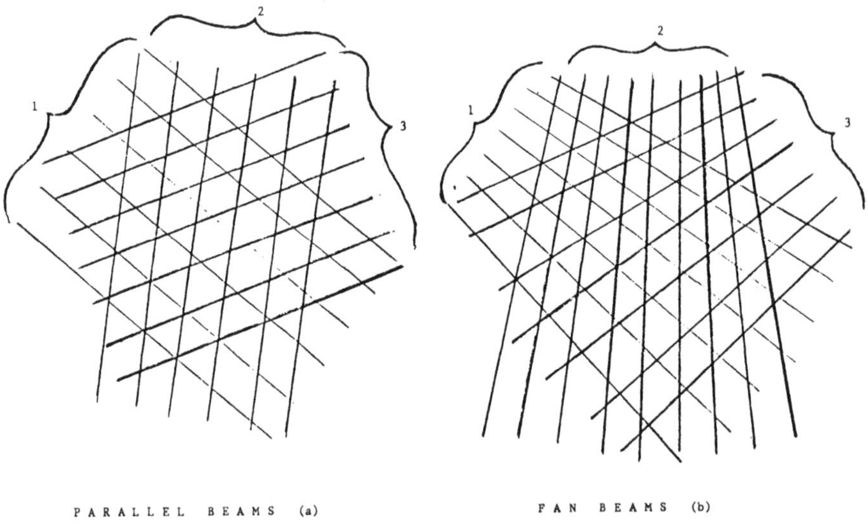

Fig. 8. Illustrating two arrangements of beams for scanning the patient. Usually there are more than 500 readings taken in one scan across the patient and the angles betweens scans advance approx. 1/3°.

readings from the detectors are fed into a computer which derives the absorption of the material in the path of the X-ray beam.

$$\text{Absorption} = \log \frac{\text{Intensity X-rays}}{\text{Detector reading}}$$

The absorption coefficients of the various substances within each square millimetre of the slice can be reconstructed from the readings (i.e. from approximately 1 000 back projections each of some 500 readings). They can be displayed as grey tones on a picture or printed out to an accuracy of approx. 1/4 % with respect to water.

Originally the method of picture reconstruction, I like to think, was attained by common-sense practical steps. Most of the available mathematical methods at the time were of an idealised nature and rather impractical.

FACTORS WHICH GOVERN THE IMPROVED SENSITIVITY OF THE SYSTEM

Many will be aware of the conventional tomogram which also images a slice through the body. This is achieved by blurring the image of the material on the picture either side of the slice, by moving both the X-ray tube and photographic plate in opposite directions while the picture is being taken.

If one makes a comparison of CT with conventional tomography (Fig. 9a) it is clear that in the latter only a short path of the beam (1/10th of its length AB) passes through the slice to be viewed, collecting useful information. The other 9/10ths of the beam pass through material on either side of the slice, collecting unwanted information which will produce artefacts on the picture. Referring to Fig. 9b in computed tomography, the X-ray beam passes along the full length of the plane of the slice via its edges, and thus the measurements taken by it was 100 % relevant to that slice and that slice alone. They are not affected by the materials lying on either side of the section. The material inside the CT slice is seen as a mesh of variables, which is intersected by all the beam paths at a multitude of different angles (Fig. 10). As the absorption measured by each beam path is the sum of all the mesh squares it passes through, the solution of the mesh variables is possible. If the X-ray beam is confined to the slice and there are no external variables, the entire information potential of the X-ray beam is therefore used to the full.

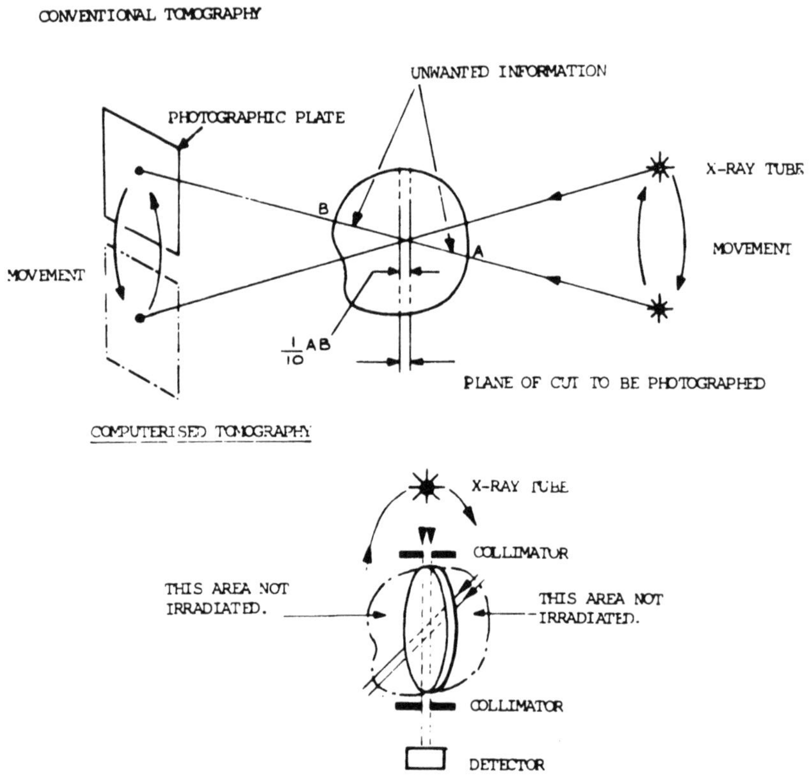

Fig. 9. Illustrating the improved efficiency of information collection when conventional tomography and CT are compared.

Note that in the case of CT the X-ray beam passes through only the material required to be viewed.

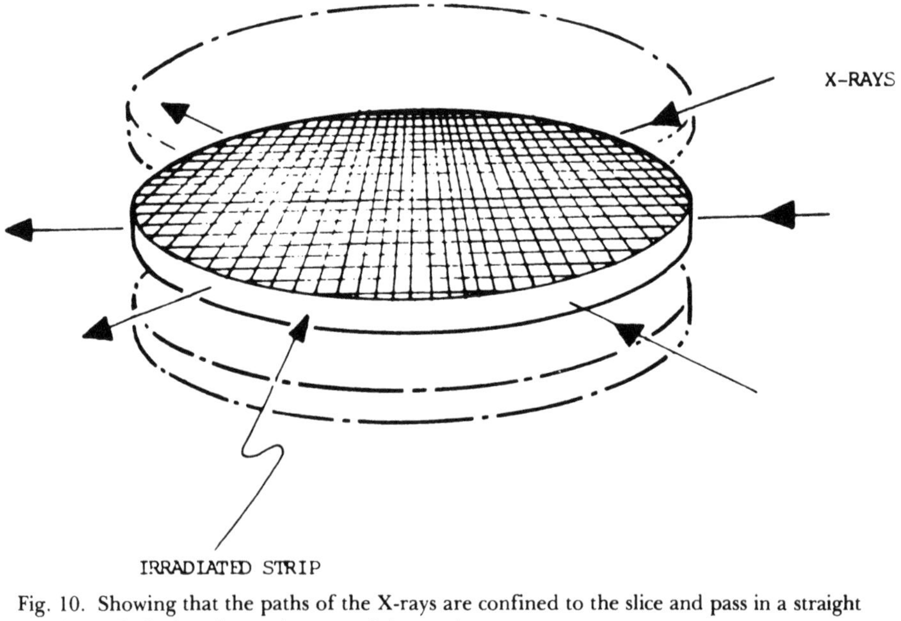

Fig. 10. Showing that the paths of the X-rays are confined to the slice and pass in a straight line through the imaginary elements of the mesh.

ACCURACY

The scale shown (Fig. 11) demonstrates the accuracy to which the absorption values can be ascertained on the picture. It shows the whole range of the machine, from air (–1 000) at the bottom of the scale, to bone at the top of the scale, covering some 1 000 levels of absorption either side of water, which has been chosen to be zero at the centre. (This is done for convenience, as the absorption of water is close to that of tissue). To obtain readings which relate to true absorption, 1 000 must be added to these readings, making air zero, and water would then be +1 000.

The range of tones between black and white seen on the picture can be restricted to a very small part of the scale. This "window" can be raised or lowered according to the absorption value of the material we wish to compare: for example, it must be raised to see the tissue of the heart or lowered to see detail within the air of the lung. The sensitivity can be increased by reducing the "window" width, where the absorption difference between the liver and other organs can be more clearly differentiated.

Let us now consider to what accuracy one can ascertain the absorption values of CT pictures. The clarity of the picture (Fig. 1), and hence the accuracy to which one can measure absorption values, is impaired by a

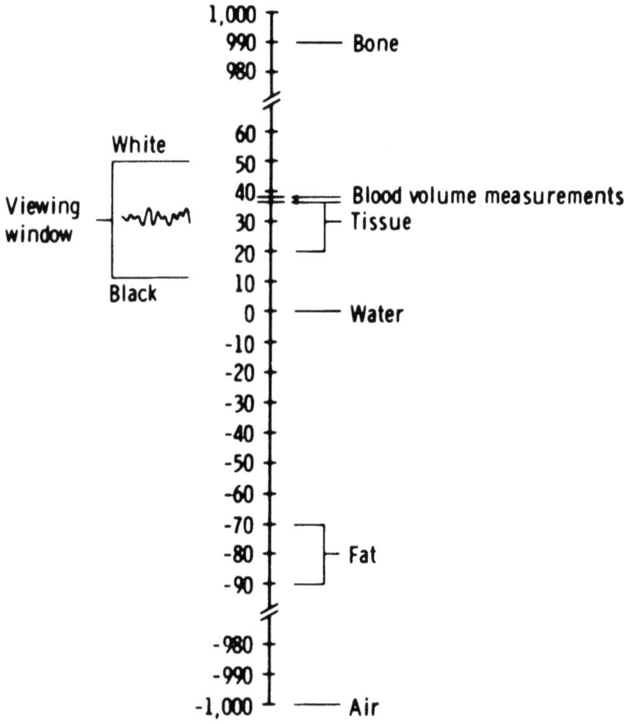

Fig. 11. Demonstrating the accuracy to which absorption values can be ascertained on the CT picture.

mottled appearance (or grain) which unfortunately is fundamental to the system. It is caused by there being a limited number of photons arriving at the detectors after penetrating the body. This results in a statistical spread between readings and is a situation that must be accepted. A typical spread would be a standard deviation of 1/2 % on tissue. (Displayed on a 320x320 matrix).

Present CT methods use very nearly all the available photon information that can be extracted from the X-ray beam, and we must therefore deduce that there is little room for further improvement in grain reduction. However, for industrial uses there are no X-ray dose constraints to be considered. The improvements to picture grain would be proportional to dose for a particular picture resolution.

THE RELATIONSHIP BETWEEN RESOLUTION AND PICTURE NOISE (OR GRAIN)

The study of picture noise reveals a rather important fact. The picture noise concentrates mainly at the high frequencies, there being very little noise at the lower frequencies. In other words, if we reduce the resolution of the picture by filtering out the higher frequency components, the remaining low frequencies will then be very small in amplitude, enabling the sensitivity of the machine to be increased without undue noise appearing on the picture. There is therefore a "trade off" between noise and resolution.

$$\text{Noise} \propto (\text{resolution})\ 3/2$$

During the development of the whole-body CT scanner, it became clear that the availability of an accurate cross-sectional picture of the body, the CT "slice", would have an important effect on the precision and implementation of radiotherapy treatment planning. For many years had existed an imbalance in the degree of precision of various aspects of the chain of events which make up a course of treatment by radiation therapy. For example, the linear accelerator, now regarded as the preferred treatment equipment for radical radiotherapy, can deliver X-ray beams with a precision of approximately ±1.00 mm in terms of spatial "aiming". It can also achieve accuracies of the order of 1 % in most of the other essential parameters of its operation. CT provides us with accurate measurements for aligning these beams.

In the past, radiation treatment planning has been a very lengthy procedure. Now with the aid of CT therapy planning computer programs, we can position the therapy beams automatically with precision in a few minutes. The system is linked to a CT diagnostic display console and a colour display monitor which shows the radiation isodose distributions overlaid on the basic CT scan itself (Fig. 12). The scan is used as the "patient input" to the system and areas of interest such as tumour, bone, lung or sensitive organs are outlined by an interactive light-pen.

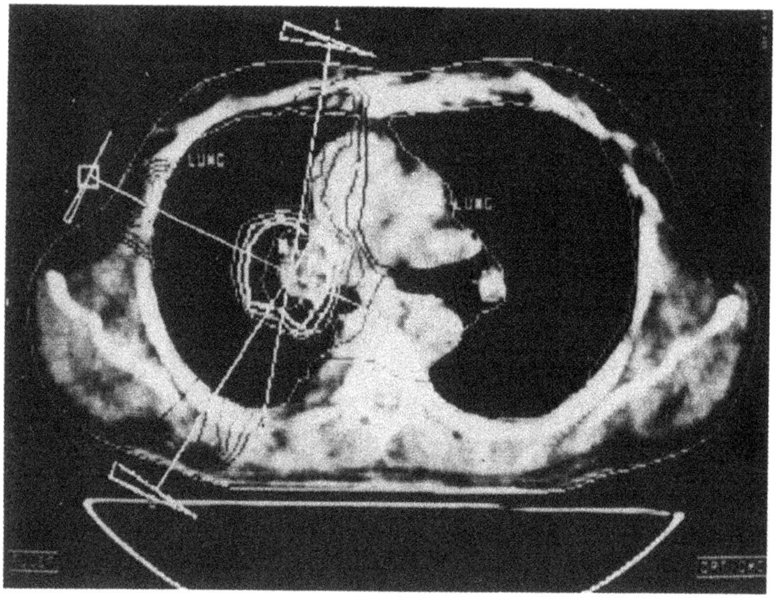

Fig. 12. Computer calculated isodose contours for therapy treatment.

Radiation beams and their computed isodose contours appear as overlays on the colour screen, and the lines of dose level are given different colours to aid the assessment of the plan.

The system uses the CT density numbers for the calculation of the effect of inhomogeneities in the path of radiation beams, although the absorption coefficients have to be corrected for the X-ray energy to be used in the actual treatment.

After areas to be irradiated and those to be avoided (such as the spine) have been outlined by means of a light pen, the position of the beam (shown Fig. 12 as T-shaped lines) can either be rotated around the tumour by means of the same light pen or the optimum position of the beams can be chosen automatically by program. This takes into account the CT numbers on the picture, the size and position of the tumour and the areas outlined by the operator which needs to be avoided. It automatically adjusts position and strength of the beams in order to choose the optimum contour of irradiation across the body.

Fig. 13 shows a series of pictures demonstrating the regression of a seminoma (a very radio-sensitive tumour). The radiation is applied at a low level while the tumour is large, but as the lesion regresses the smaller area is taken advantage of and larger doses are then applied. In this particular case the tumour was completely removed by accurate intense radiation.

These are, however, early days in the applications of CT to radiotherapy planning. It remains to be seen whether the increase in precision really results in a better five-year survival rate. In the opinion of many leading oncologists, evidence already points to such improvements, but it will be the next five years which provide the answers to these questions.

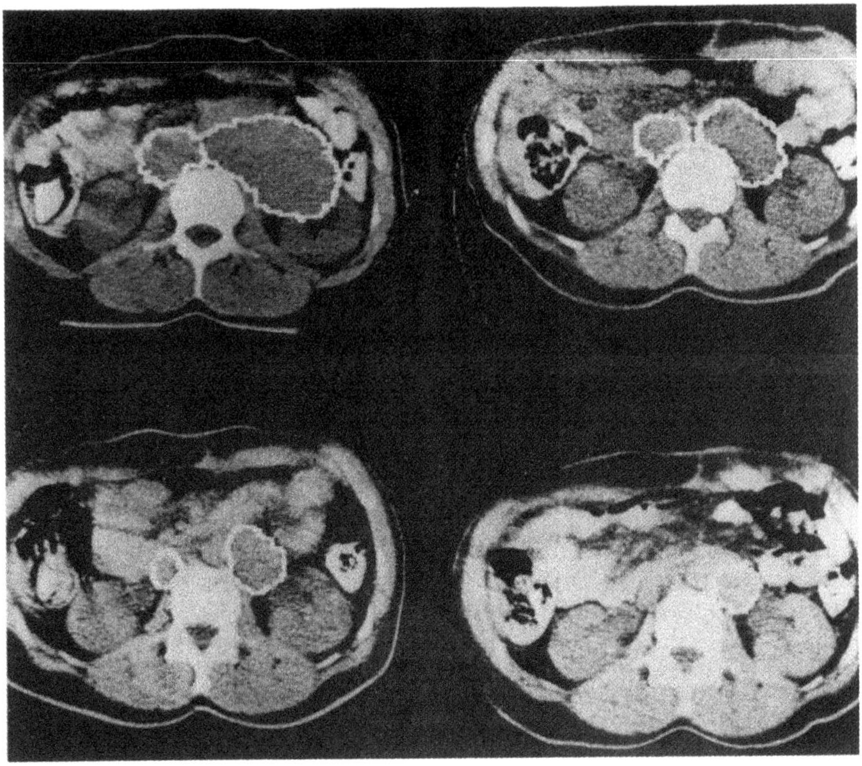

Fig. 13. Demonstrating the regression of a seminoma after four stages of therapy treatment.

WHAT IMPROVEMENTS SHOULD WE EXPECT TO SEE IN THE FUTURE?

Various attempts have been made to achieve useful pictures of the heart.

The time available for taking a picture of the heart is obviously longer than one heart beat. Some experiments were conducted some time ago using conventional CT machines but in which the traverse of the detectors was synchronised to the heart beat via an electro-cargiograph, passing over the heart in diastole (when the heart movement is at a minimum). Fig. 14 shows a picture from the experiment.

The heart chambers can be discerned by a little intravenous injected contrast media.

Another approach is being made at the Mayo clinic, Rochester, America, where a large machine is being constructed with 27 X-ray tubes designed to fire sequentially. It is hoped to take a sequence of pictures in a fraction of a second during one heart beat. However, the complexity and cost may rule out such a machine being used world-wide.

A further promising field may be the detection of the coronary arteries. It may be possible to detect these under special conditions of scanning.

Fig. 15 is an example of present high resolution CT scans of the spine using contrast media. It is more than likely that machines in the future will

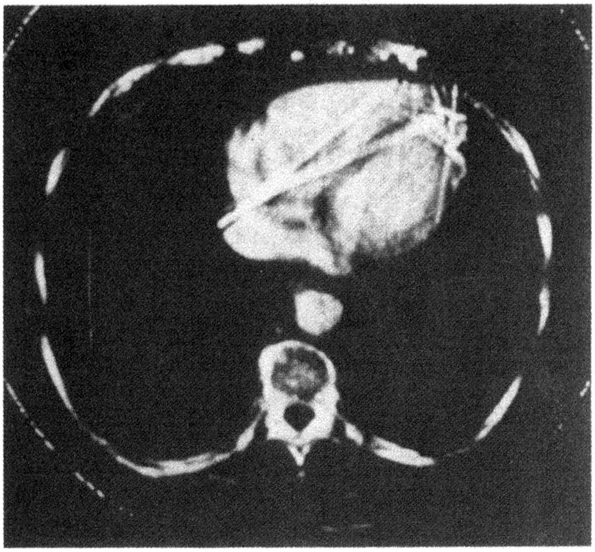

Fig. 14. Scanning of the heart with detectors synchronised to the heart-beat passing over the heart in diastole. (The line artefacts are streaks caused by the wire of a pace-maker).

be designed to provide considerably higher resolution than shown in this picture. Such machines would take up many of the present uses of conventional radiography but would do the job considerably better. They would have the added bonus of having more sensitive detectors than does film.

As all the information on the body is stored in three dimensions, it is possible therefore to display the object at any angle; this allows it to be examined by rotating it around on the screen. The views seen around the organ to be examined may reveal information that hitherto could have been missed, when it was viewed normally in one fixed plane, normal to the axis of the body.

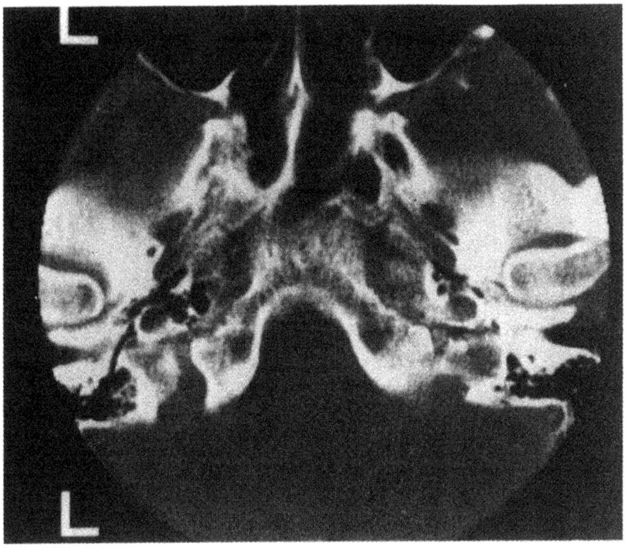

Fig. 15 Higher resolution picture of the base of the skull.

NMR IMAGING

So far, it has been demonstrated that a picture can be reconstructed from X-ray sweeps forming sets of line integrals taken through the body at a multitude of different angles.

However, there are methods other than those using X-rays which are capable of measuring tissue variation, such as the use of protons, neutrons or nuclear magnetic resonance. These can be caused to generate sets of measurements of line integrals at different angles across the body, and in a similar manner a picture can be reconstructed from them.

Fig. 16 shows what is believed to be the first picture taken of a human head using Nuclear Magnetic Resonance. It was taken in 1978 by a team led by Dr. Hugh Clow and Dr. Ian Young of EMI Central Research Laboratories, England. Since then considerable improvements have been made to both head and body pictures and progress is continuing.

The principle of NMR Spectroscopy was well known in the 1960's and was first suggested for imaging by Lauterbur in 1973. It is a new and quite different form of imaging, in which the radio frequencies emitted by the hydrogen nucleus can be measured after they have been excited in a particular way by radio frequency. The nucleus of special interest is the proton of water molecules within the body.

I would now like to compare the merits of CT with those of NMR. Before doing this I must first describe the principles on which NMR operates.

When hydrogen protons are placed in a magnetic field they will precess (or "wobble") around the field direction just as a spinning top precesses around its vertical gravitational field. This precession occurs at a definite frequency, known as the Larmor frequency, and is proportional to the magnetic field intensity.

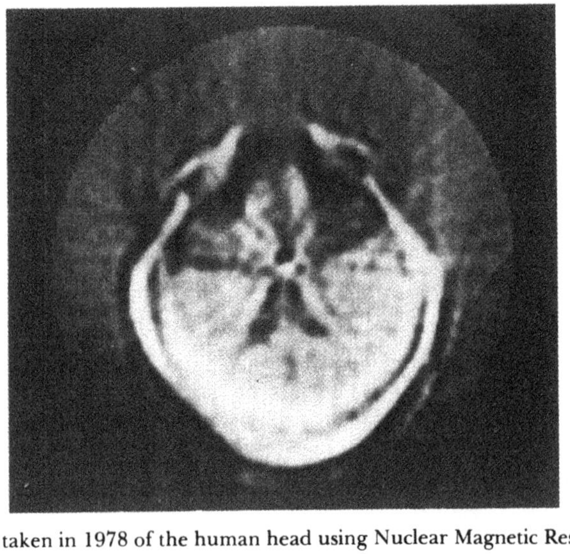

Fig. 16. Picture taken in 1978 of the human head using Nuclear Magnetic Resonance.

The usual NMR procedure for imaging is to apply a strong magnetic field along the body to be studied. (Figs. 17a & 17b). After a short period of time, the nuclei will align with their magnetic movements along the field. A radio frequency tuned to the precession frequency of the hydrogen nucleus is then applied at right angles to the main field by means of a set of coils at the side of the body (Fig. 17c). This causes some of the hydrogen nuclei to precess—all keeping in step. After the radio frequency field has been switched off, the nuclei will continue to precess in phase, generating a similar radio frequency which can be picked up in receiver coils placed at the side of the body (Fig. 17d); these signals detect the water content of the body. It will take some time for the precession to die away, as the nuclei

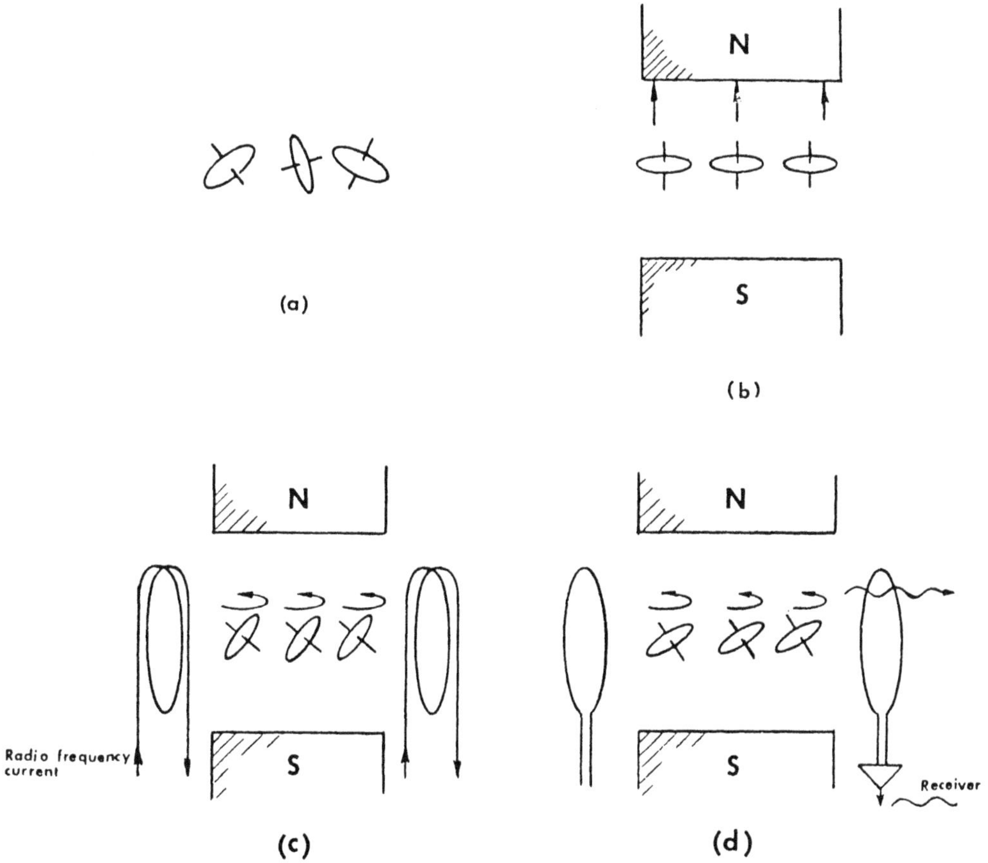

Fig. 17 a) Representation of nucleus without magnetic field applied.
b) Nucleus with field applied.
c) Radio frequency field applied.
d) Radio frequency switched off – receiver on.

again realign themselves with the magnetic field. The measurement of this time is important as it gives some information about the nature of the tissue under investigation. It approximates to an exponential decay of the order of "tenths of a second". (It is known as T_1 or the spin lattice relaxation time).

METHOD OF PRODUCING AN NMR PICTURE

The procedures above refer only to a method of tissue detection. In order to produce a picture which maps the difference of tissue within the body it is necessary to independently measure small volumes of material across it. In NMR imaging this is done by applying a small magnetic field gradient across the body in addition to the main uniform field. The frequency of the nuclei, being dependent upon the magnetic field, will resonate at different frequencies across the body according to the magnitude of the field gradient present (Fig. 18). In one method of NMR imaging, the frequencies received in the coil can be separated (by Fourier analysis) and the whole spectrum of frequencies will represent a series of line integrals across the body, each frequency representing the amount of hydrogen nuclei resonating along that particular line. As a comparison with CT this is equivalent to one X-ray sweep across the body at a particular angle.

A number of "sweeps" can be repeated at different angles by rotating the gradient field, and sufficient data can be built up to reconstruct a picture in similar way to that in which a CT picture is constructed.

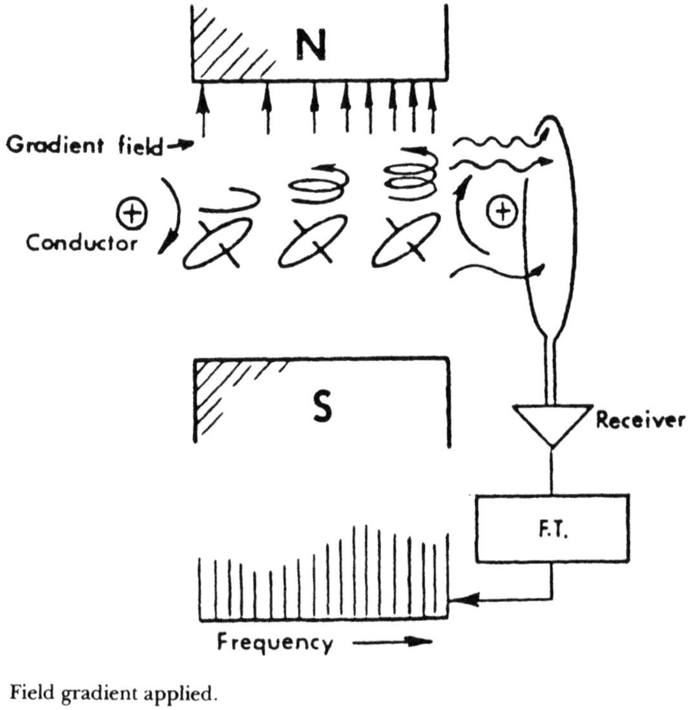

Fig. 18. Field gradient applied.

COMPARISONS OF NMR WITH CT

It is early to predict with certainty the levels of accuracy of a future NMR system, but one could speculate in the following way to illustrate its possible advantages.

Fig. 19a illustrates approximately the number of levels about "noise" one would expect CT to be able to discriminate in tissue when an average scan is taken.

It detects mainly one variable – density. It also detects a minor one – atomic number, which relies upon the photo-electric effect for separation. The discrimination of iodine from tissue by subtraction of two pictures at different X-ray energies is the most used example.

In comparison, Fig. 19b shows the variables one would expect to detect in NMR. It may be possible to separate more than one decay time (the figure indicates that these may be three). If the sensitivity above noise is such that it is possible to select three levels in each of the three decay variables, there should be 27 possible permutations to characterise the tissue to be displayed as against the nine levels of tissue discrimination usually displayed on CT picture (Fig. 19a). This illustrates the major advantage of NMR but is only speculative.

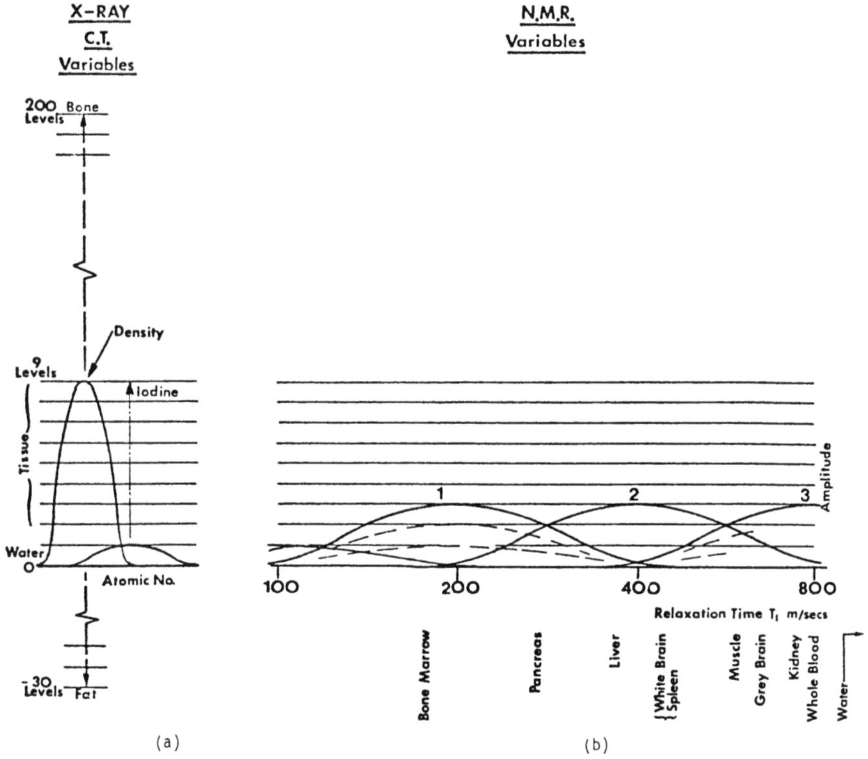

Fig. 19. Comparison between CT and NMR – showing the variables one would expect to find in tissue in both systems.

However, CT has other advantages:

Fat in CT is very easily discriminated with respect to tissue (over 30 levels). For comparison, it is poorly discriminated in NMR.

So far there are indications that the picture resolution on CT is considerably better than in NMR, and a picture can be scanned in a considerably shorter time (2 to 3 secs.). With NMR it is usual to scan the patient over at least a minute or more to collect sufficient data for good tissue discrimination. It must be remembered that each NMR, T_1 measurement is very time consuming, as long relaxation times (tenths of a second) need to be measured, and a few hundred sets of these readings are required to reconstruct one picture. On the other hand, as far as is known, NMR is non-invasive and the long periods of scanning time that the patient may be subjected to should do him no harm.

However, a systems design of NMR has not yet stabilised. There still exist many different avenues which may in the future reveal improvements in both speed and sensitivity.

Despite the fact that it is not yet clearly known what the parameters are that it is measuring, it is felt that NMR has great possibilities for indicating more about the chemistry of the tissue being selected. It is a technique which could image water concentration, its impurities and its binding to macro-molecules in or between the cells of the body. It remains to be seen how this extra information can help diagnoses.

Work on NMR is still in its early stages and has a long way to go. At the present time, the two techniques of CT and NMR should perhaps be seen not as potential competitors but rather as complementary techniques that can exist side by side; NMR providing us whith information on the chemical composition of the tissue, and CT providing us with a means of visualising its position and shape.

May I thank all those at EMI Research Laboratories who have helped me so much in the past in my work – especially Steven Bates who worked on some of the early experiments.

SELECTED REFERENCES

1. Ambrose J: Computerised transverse axial scanning (tomography). II Clinical application. Br J Radiol 46: 1023–1047, 1973.
2. Ambrose J: Computerised X-ray scanning of the brain. J Neurosurg 40: 679–695, 1974.
3. Brooks RA, Di Chiro RA: Theory of Image reconstruction in computed tomography, Radiology 117: 561–572, 1975.
4. Cho ZH, Chan JK, Hall El, Kruger RP, McCaughey DG: A comparative study of 3-D image reconstruction algorithms with reference to number of projections and noise filtering, in Transactions of Nuclear Science, IEEE Catalog no. NS-22 (1), 1975, pp344–358.
5. Hounsfield GN: Computerised transverse axial scanning (tomography) I. Description of system. Br J Radiol 46: 1016–1022, 1973.
6. Pullen BR, Rutherford RA and Isherwood I: 1976 Medical Images: Formation Perception and Measurement. Proceedings of the Seventh L.H. Gray Conference (20–38) University of Leeds 1976. The Institute of Physics, John Wiley and Sons.

1980
Physiology or Medicine

BARUJ BENACERRAF, JEAN DAUSSET and GEORGE D. SNELL

"for their discoveries concerning genetically determined structures on the cell surface that regulate immunological reactions"

THE NOBEL PRIZE FOR PHYSIOLOGY OR MEDICINE

Speech by Professor GEORG KLEIN of the Karolinska Medico-Chirurgical Institute.
Translation from the Swedish text.

Your Majesty, Your Royal Highnesses, Ladies and Gentlemen,

Even the longest journey starts with a single step, the old Chinese have said. The first step of the long journey that has led the three Laureates in Medicine to us tonight was taken in regions that were far from each other. None of the three knew that they were on their way, towards the same chromosome, or, more precisely, the same gene region within one chromosome, now known to influence immune functions in various ways. It is a large region, it can be called a supergene. The system is very ancient: all vertebrates have it: thus, it has been highly conserved during evolution. Its stability across species barriers is in remarkable contrast to its many thousandfold variability within each species, giving rise to a kaleidoscopic pattern that makes all human beings individually distinct, with identical twins as the only exception.

Where did the journey start? George Snell became interested in the genetics of cancer during the 1930s. At that time, the Jackson Laboratory in Bar Harbor, Maine had just succeeded to produce the first highly inbred strains of mice, after more than a decade of brother-sister mating. Within each strain, every mouse had the same genetic constitution, like identical twins. Experiments were performed on the role of genetic factors for the development of cancer. In the same context, tumor cells were also transplanted from cancerous to healthy mice. Snell found that transplanted tumors grew progressively in all mice of the same strain, but were rejected in foreign strains. Crossing experiments showed that transplanted tumor cells could only grow if the donor and the recipient shared certain dominant genes. In the absence of such identity, the tumor cells were killed by a host cell, called the killer lymphocyte. Snell realized at an early stage that the reaction was not limited to cancer cells: the transplantability of *normal* tissues was regulated by the same genes. Snell called them "histocompatibility genes" or *H-genes.* The mouse has at least 80 different H-genes. All of them are not equally important. Some give stronger reactions than others. The *strongest* gene that played the most important role for rejection was called *H-2.* Even the most highly malignant tumor cell cannot escape the rejection reaction induced by a foreign H-2 component, as a rule.

The analysis of the H-2 system by Snell became a monumental masterpiece in mammalian genetics that has laid the foundations of a new science: transplantation immunology and immunogenetics.

At the time when *Dausset* started his activitities in this field, it was already clear that humans reject foreign grafts by the same type of immune mechanism as mice. Since human beings cannot be studied experimentally, nor are they inbred, it was thought that it will take many decades before human H-genes will be identified and mapped. Dausset worked on something quite different. He found that patients who received many blood transfusions produced antibodies that killed white blood cells. At first, he thought that this was an autoimmune reaction, i. e. that the patients reacted against their own white cells. However, this did not fit with the fact that the white cells of the blood donors were killed, but the cells of the recipient remained unharmed. Dausset realized immediately that he had encountered a previously unknown type of genetic variation between people. On the basis of family analyses he could show that the variation was determined by a single genetic system, localized to a single chromosome. It was designated *HLA* and was found to be analogous with H-2 in the mouse. At this point, the paths of Snell and Dausset converged. Research on mice and humans became mutually complementary. One starts to speak about *MHC,* or *major histocompatibility complex,* as a common name for the large gene region and finds that MHC has a closely similar structure in all mammalians. It is also realized that only a fraction of its many components are known: the antibodies of Snell and Dausset have only identified two of its important milestones. It was *Benacerraf's* work that has brought in a third, very important region, located between the two milestones. Like his colleagues, he also started in a seemingly distant area: the antibody response of guinea pigs and the interplay between different cells within the immune system. He found that the immune response against certain substances varied greatly between different guinea pig strains. This was due to a previously unknown group of genes, localized within the MHC complex, designated as *immune responsiveness* or *Ir genes.* They were found to influence the ability of different cell types to cooperate within the "immunological orchestra". Some Ir genes help different cell types to collaborate in order to bring about a certain response wereas others suppress reactions that would otherwise get out of control.

The major histocompatibility complex has acquired *great medical and biological significance.* HLA typing is now indispensable for all forms of tissue and organ *transplantation.* The rapid practical application of the research results is a direct consequence of the exemplary international cooperation organized by Dausset in the form of the "histocompatibility workshops" where research workers from all countries meet in the laboratory to compare their results, exchange reagents and agree on the nomenclature. Now, the most compatible donor-recipient combinations can be readily identified with the help of the data banks that contain the typing information in a language comprehensible for all.

As a more unexpected byproduct of this activity, a strong relationship was found between certain *HLA-types* and some *diseases,* including a rare disease of the spine, juvenile diabetes, multiple sclerosis, a chronic skin disease, etc. The reasons for these associations are not understood, but they further emphasize the great significance of the MHC-region for normal development and function.

Perhaps the most interesting question concerns the role of the MHC system in the *normal* organism. Why does it exist, why has it been retained with such tenacity and in all its complexity during evolution? Protection from foreign tissue grafts, an artefact of our time, is certainly *not* the reason. The answer must be sought in the importance of the MHC system for the cooperation between different cells in the organism and for the ability of the immune system to distinguish between normal cells of the body that should not be exposed to an immune rejection, and changed cells that must be eliminated because they threaten the integrity of the organism. Viral infection, cancerous transformation and perhaps even the normal physiological ageing of cells can be mentioned as examples. The MHC-system provides an extraordinarily sensitive surveillance system to detect cells with changed membranes; it also provides a mechanism to kill cells that are becoming alienated from their community in one way or another. The rejection of foreign grafts is then merely an unavoidable byproduct.

In the early 1950's, I heard George Snell say that he could count the number of research workers in the world who understood the H-2 system without using all his fingers. The development of this field from an outlandish rejection reaction in inbred mice to the mighty supergene system of today that all immunologists, cancer research workers and many virologists and developmental biologists meet in their daily work is one of the most exciting chapters in the enormous building of modern biology.

Drs. Snell, Dausset and Benacerraf! Starting from three different directions, your long journey has led you, after many adventures, to the same supergene area, the major histocompatibility complex, and through it to this happy event tonight. You have been responsible for turning what at first appeared as an esoteric area of basic research on inbred mice into a major biological system of the greatest significance for the understanding of cell recognition, immune responses and graft rejection. We have the rare esthetic pleasure of seeing a series of fundamental discoveries, coupled with immediate applications in clinical medicine. I am very happy to have the privilege of expressing the congratulations of the Nobel Assembly at the Karolinska Institute and to ask you to receive your Nobel Prize from the hands of His Majesty the King.

Barry Benacerraf

BARUJ BENACERRAF

I was born in Caracas, Venezuela, on October 29, 1920 of Spanish-Jewish ancestry. My father, a self-made business man, was a textile merchant and importer. He was born in Spanish Morocco, whereas my mother was born and raised in French Algeria and brought up in the French culture. When I was five years old, my family moved to Paris where we resided until 1939. My primary and secondary education was in French which had a lasting influence on my life. The second World War caused our return to Venezuela, where my father continued to have a thriving business. It was decided that I should pursue my education in the United States, and we moved to New York in 1940. I registered at Columbia University in the School of General Studies, and graduated with a Bachelor of Science Degree in 1942, having also completed the pre-medical requisites for admission to Medical School. By that time, I had elected to study biology and medicine, instead of going into the family business, as my father would have wanted. I did not realize, however, that admission to Medical School was a formidable undertaking for someone with my ethnic and foreign background in the United States of 1942. In spite of an excellent academic record at Columbia, I was refused admission by the numerous medical schools I applied to and would have found it impossible to study medicine except for the kindness and support of George W. Bakeman, father of a close friend, who was then Assistant to the President of the Medical College of Virginia in Richmond. Learning of my difficulties, Mr. Bakeman arranged for me to be interviewed and considered for one of the two remaining places in the Freshman class. I was accepted and began my medical studies in July 1942. While in medical school, I was drafted into the U.S. Army with the other medical students, as part of the wartime training program, and naturalized American citizen in 1943. I greatly enjoyed my medical studies, which at the Medical College of Virginia were very clinically oriented. I received what I considered to be an excellent medical education in the relatively short time of three war years. This busy time was rendered very happy by my marriage in 1943 to Annette Dreyfus, a French student, also a refugee from Paris, whom I had met at Columbia University. I trained as an intern at Queens General Hospital in New York City in 1945 and was commissioned First Lieutenant in the U.S. Army Medical Corps in 1946. After the usual six weeks of basic training at Fort Sam Houston, Texas, I was shipped to Germany with several thousand other physicians. I was happy to be assigned to France, first in Paris, then in Nancy, where my wife had joined me. I stayed there nearly two years, as the head of a medical unit where I enjoyed practising what today would be called community medicine. I was discharged in 1947 and, motivated by intellectual curiosity, decided upon a career in medical research at a time when such a choice was not fashionable. My interest was directed, from my medical student days, to Immunology, and particularly to the mechanism of hypersensitivity. I had suffered from bronchial asthma as a child and had developed a deep curiosity in allergic phenomena. I sought the advice of many scientists, among whom René Dubos at Rockefeller University, John Enders at Harvard Medical School, and Jules Feund at the Public Health Research Institute in New York, to whom I had been recommended by members of the faculty in Richmond. I was strongly urged to work with a dynamic young immunochemist, Elvin Kabat, whose laboratories were at the Neurological Institute, Columbia University School of Physicians and Surgeons. Following an interview with Elvin Kabat, who offered me a Fellowship in his laboratory, I started my research career in February, 1948. Training with Elvin Kabat was one of the significant experiences in my development as a scientist. Elvin Kabat is a hard task-master with rigorous standards and an absolute respect for the quantitative approach to science. He felt that if a phenomenon could not be quantitated, it did not deserve to be studied. He taught me Immunochemistry and basic Immunology, but more importantly, I learned the significance of experimental proof, the need for intellectual honesty and scientific integrity. I was fortunate also that my first two years as a scientist were very productive and my initial goal of understanding experimental hypersensitivity mechanisms was in part fulfilled. My life for the next six years was very much influenced by family considerations. A daughter, Beryl, was born in 1949, and my parents had returned from Venezuela to their home in Paris. My father had suffered a severe stroke and was now a cripple. My wife's family also lived in Paris. The attraction of moving to France and settling close to our respective families was very strong. Accordingly, we moved to Paris in mid−1949 and I accepted a position in Bernard Halpern's laboratory at the Broussais Hospital. This position permitted me also to make frequent trips to Venezuela where my father's business interests now required my personal involvement. During this period I was privileged to form a close relationship with a young Italian scientist who had also joined Halpern's laboratory, Guido Biozzi. For six years we operated as a team and engaged in the study of reticuloendothelial function in relation to immunity. We developed the techniques to study the clearance of particulate matter from the blood by the RES, and formulated the equations that govern this process in mammalian organisms. After six years in Paris, I began to realize that as a foreigner to France, in spite of my French education, I would experience continuous difficulties in pursuing a scientific career and establishing an independent laboratory. This was made painfully clear to me by the chief of the laboratory, Dr. Halpern. The significance of this message was heightened by my unhappy discovery that I could not find another laboratory in Paris in 1956 that would give me a chance to work and establish myself. I decided therefore to return to the United States. I am deeply grateful to Lewis Thomas who offered me an appointment as Assistant Professor of Pathology at New York University School of Medicine and helped me develop my own laboratory and research support. I returned to my earlier studies on hypersensitivity mechanisms, but this time also developed an interest in cellular as well as humoral hypersensitivity. From 1956 to 1961, I worked on cellular hypersensitivity with Philip Gell, immune complex diseases with Robert McCluskey and Pierre Vassalli, anaphylactic hypersensitivity with Zoltan Ovary, tumor specific immunity with Lloyd Old, and the structure of antibodies, in relation with their specificity, with Gerald Edelman. The years at New York University were very happy ones, and it was soon apparent that I had made the correct choice in returning to the United States. The scientific atmosphere at New York University during that period was particularly favorable to the development of Immunology. Numerous immunologists worked enthusiastically and interacted profitably: among these were Jonathan Uhr, Jeanette Thorbecke, Edward Franklin, Victor Nussenzweig, in addition to Robert McCluskey and Zoltan Ovary mentioned earlier. This is the time when I started to teach research fellows and students and realized that the training of young scientists was one of my most valuable and rewarding experiences. Later I chose "The Training of Scientists" as the topic of my presidential address to The American Association of Immunologists. Among the young immunologists with whom I had the pleasure and privilege to work at New York University are: Lloyd Old, William Paul, Ira Green, Victor Nussenzweig, Michael Lamm, Pierre Vassalli, Stanley Cohen, Jeanette Thorbecke, Fred Kantor, Gregory Siskind, Stuart Schlossman, Kurt Bloch, Bernard Levine, Francois Kourilsky, Ted Brunner, and Takeshi Yoshida. During this period also I managed a New York bank, the Colonial Trust Company, which had been bought by my family and associates from Venezuela. However, the success of my laboratory made me realize that I had to choose between a scientific career and my business interests. I made the decision to devote myself solely to my laboratory and my students and to curtail my business career, as I felt the challenges were far greater in my chosen profession. This is precisely the time when I initiated the studies in Immunogenetics that resulted in my being awarded the Nobel Prize in Medicine. I made the observation that random bred animals immunized with antigens with restricted heterogeneity, such as hapten conjugates of poly-L-lysine distribute themselves into two groups, responders and nonresponders. I sensed that this was an important phenomenon. I determined that responsiveness to these or other similar antigens is controlled by dominant autosomal genes termed immune response (Ir) genes. This was the beginning of a long and complex story that led to our understanding of the manner in which these genes, located in the major histocompatibility complex of mammals, exercise their function and determine immune responsiveness. By then I had become Professor of Pathology at New York University. The opportunity, however, arose at the request of John Seal to assume the Directorship of the Laboratory of Immunology of the National Institute of Allergy and Infectious Disease in Bethesda, where I moved in 1968 together with William Paul and Ira Green. Such a laboratory offered very attractive facilities and precious inbred guinea pig strains essential to my work in immunogenetics. Much of the insight on the mechanism of Ir gene function has indeed been obtained in that laboratory, from experiments of William Paul, Ira Green, Alan Rosenthal, Ethan Shevach, and Ronald Schwartz, with the systems I developed.

In 1970, Dean Robert Ebert offered me the Chair of Pathology at Harvard Medical School. I moved to Harvard because I missed the University environment and more particularly the stimulating interaction with the eager, enthusiastic, and unprejudiced young minds of the students and fellows. At Robert Ebert's request, we initiated an interdepartmental immunology graduate program at Harvard Medical School which has developed very successfully under the stewardship of my colleague, Emile Unanue. At Harvard, I have continued my work on immune response genes and their role in the regulation of specific immunity with David Katz, Martin Dorf, Judith Kapp, Carl Pierce, Ronald Germain and Mark Greene. We also determined the role of immune response genes in the control of immune suppression phenomena with the help of Patrice Debré, Judith Kapp, and Carl Waltenbaugh; we analyzed the specificity of cytolytic T lymphocyte in relation to Ir gene function with Steven Burakoff and Robert Finberg and demonstrated how alloreactivity arises as a consequence of the commitment of T lymphocytes to recognize antigen in the context of autologous MHC gene products.

While reaching these scientific goals, I was elected President of the American Association of Immunologists in 1973, President of the American Society for Experimental Biology and medicine in 1974, President of the International Union of Immunological Societies in 1980. I was elected to the American Academy of Arts and Sciences in 1972, the National Academy of Science, U.S.A. in 1973, and I was appointed President of the Sidney Farber Cancer Institute in 1980. I have received the following awards:

R.E. Dyer Lecture of National Institutes of Health 1969
Rabbi Shai Schacknai Lectureship and Prize in Immunology and Cancer Research, Hebrew University of Jerusalem 1974
T. Duckett Jones Memorial Award of The Helen Hay Whitney Foundation 1976
Honorary Degree of Doctor of Medicine, University of Geneva, Switzerland 1980
Waterford Biomedical Science Award 1980

My work has been generously and continuously supported since 1957 by the National Institute of Allergy and Infectious Diseases, and for the last decade also by the National Cancer Institute. I am very grateful for their enlightened support to me and my associates, which made our work possible. I am also particularly indebted to my many students and associates who have contributed so much to our common goal and whom I hold responsible in the largest measure for my achievements.

THE MAJOR HISTOCOMPATIBILITY COMPLEX IN MAN — PAST, PRESENT, AND FUTURE CONCEPTS

Nobel Lecture, 8 December, 1980

by

JEAN DAUSSET

University of Paris VII

Institute of Research in Blood Diseases, Paris, France

As George Snell (*1*) so rightly said, the supergene, the major histocompatibility complex (MHC), is like a page from the nature book read outside of the context.

Today this context is beginning to be better understood. We would here like to recall the evolution of concepts regarding these molecular structures found in the membrane of cells. First, attention was centered on the almost botanical description of their genetic polymorphism. Then the spotlight was turned, for several years, on their importance in transplantation. More recently, their role in the immune response has become more and more apparent. This, however, is probably not the last stage in our search. We, as well as others (*2–5*), have suggested that the essential function of these structures resides in self-recognition. These structures are, in fact, the identity card of the entire organism.

We will discuss these four viewpoints successively. Far from being mutually exclusive, they are landmarks in the stages of our thought process as we have gained deeper knowledge of the subject.

First concept: Polymorphism and Linkage Disequilibrium

Polymorphism: Since Landsteiner's discovery of the first genetic polymorphism in man, knowledge of polymorphic genes has not ceased to increase and will continue to increase with DNA hybridization techniques. Most of these systems, however, are pauci-allelic and more often than not have one very frequent allele, one that is more infrequent, and a few variants. None of these can be compared with the extreme polymorphism of genes in the MHC of vertebrates, and particularly in the human lymphocyte antigen (HLA) complex

The definition of this polymorphism began to emerge in three laboratories: in ours where the first antigen Mac (HLA-A2) was defined (*6*), in Van Rood's laboratory (*7*) with the 4a 4b series (Bw4, Bw6), and Rose Payne's and W. Bodmer's (*8*) with the two alleles HLA-A2 and -A3. Then, thanks to an intense international effort that has spanned more than 15 years and included eight workshops, the web began to be disentangled. The importance of this international effort, launched by Amos in 1964 [see (*9*)], and followed by other workshops perceive antigens on the surface of cells and therefore in the nature of immunogenicity.

I propose to give a historical account of how our present understanding of T cell immunity and of T cell immune regulation has evolved with particular emphasis on the genes of the MHC and the molecules for which they code that regulate essential immune mechanisms.

Carrier Function and the Specificity of T Lymphocytes

The pioneering experiments of Landsteiner (21) established that antibodies can be produced against any type of molecule provided it is presented to the immune system coupled to an immunogenic carrier molecule. The determinants against which antibodies can be made were termed "haptens" and "carriers" the essential immunogenic molecules required to initiate immune responses. Landsteiner's experiments implied the existence of a complex process involving the recognition of a "carrier" function by an entity distinct from antibody to initiate immune responses.

Spurred by Landsteiner's observations of the "carrier" effect, Gell and I (22) investigated the specificity of cellular immune responses to haptenprotein conjugates. We noted a fundamental difference between the specificity of cellular immune reactions and of antibodies. Immune cells displayed classical "carrier" specificity in contrast to antibodies which can be largely hapten specific. This was latter shown to be a general property of T cell mediated immune responses (23). Moreover, we also demonstrated another critical difference between the type of determinants reactive with antibodies and with T cells. Extensive denaturation of protein antigens capable of decreasing drastically reactivity with specific antibody had little effect on the ability of such proteins to initiate or elicit delayed type sensitivity (DTH) to the intact molecules (24). This indicated again that T and B lymphocytes may not be specific for the same determinants, and that T cells react preferentially with sequential determinants on proteins. These observations were confirmed by Schirrmacher and Wigzell (25) and by Ishizaka et al. (26).

The Discovery of Immune Response Genes

The identification of the genes which determine biological phenomena and the study of the control they exert on these phenomena has proven to be the most successful approach to a detailed understanding of the mechanism of biological processes. Some of the most significant advances in molecular biology have relied upon the methodology of genetics. The same statement may be made concerning our understanding of immunological phenomena.

Immunologists had not infrequently observed that certain individuals are weak responders to selected antigens. The complexity of most antigens and the marked heterogeneity of the antibody response did not encourage a genetic analysis of specific immune responsiveness. However, when synthetic polypeptides with relatively restricted structural heterogeneity were synthtesized (27), the appropriate antigens were available to immunologists to study the genetic requirements for immunogenicity. The response of outbred guinea pigs to hapten conjugates of the poly-L-lysine homopolymer (DNP-PLL) was the first specific immune response documented to be under the control of a single dominant autosomal gene (28). We introduced the terms "responders" and "nonresponders" to distinguish animals possessing or not possessing the gene, and the gene responsible was referred to as an immune response or Ir gene. Fortunately, two inbred strains of guinea pigs developed originally by Sewell Wright were available at the National Institute of Allergy and Infectious Diseases, strain 2 and strain 13. Strain 2 animals responded to DNP-PLL and strain 13 guinea pigs did not, whereas $(2 \times 13)F_1$ were responders. The phenomenon was extended to other polypeptide antigens (Table 1), the random copolymers of L-glutamic acid and L-lysine (GL), L-glutamic acid and L-alanine (GA) and L-glutamic acid and L-tyrosine (GT) (29).

The response to conventional antigens, weak isologous antigens (30) or foreign protein antigens, administered at limiting immunizing doses (31, 32) to ensure response to only the most immunogenic determinants, is under similar control of individual Ir genes.

The phenomenon was extended to other experimental species. McDevitt and Sela demonstrated the Ir gene control of the response of inbred mice to a very interesting set of branched copolymers synthesized by Sela, (T,G)-A–L, (H,G)-A–L, and (Phe,G)-A–L which differed only in one of the amino acids on the side chain (33). The responses to these copolymers were under the control of distinct Ir genes. In collaboration with Maurer we also demonstrated Ir gene control of the response of inbred mice to linear random copolymers of L-amino acids (34). Genetic control of immune responsiveness was also reported in rats (35, 36), and rhesus monkeys (37), illustrating the generality of this phenomenon for different antigens and in different species.

Linkage of Ir Genes to the Major Histocompatibility Complex. Mapping of the Genes and Gene Complementation

The availability of inbred strains of mice and guinea pigs permitted the rapid mapping of Ir genes. McDevitt and Chinitz (38) made the exciting finding that responsiveness of inbred mice to (T,G)-A–L, (H,G)-A–L, and (Phe,G)-A–L could be predicted on the basis of their H-2 genotype. The linkage of murine Ir genes with the H-2 complex was confirmed for numerous antigens by many laboratories and is appropriately considered one of the distinctive features of specific Ir genes (39). A summary of the data is shown in Fig. 1 (40). Identical linkage between guinea pig Ir genes and MHC specificities in that species was documented in our laboratory (41). The strategy employed in these experiments is illustrated in Table 1. The genes for the responses to PLL, GA and BSA were observed to be linked to the locus controlling the major histocompatibility complex of strain 2 guinea pigs. Similarly the GT gene and the genes controlling responsiveness to limiting doses of DNP-GPA were found to be linked to the major H locus of strain 13 guinea pigs. Linkage of Ir genes to the MHC of the rat (35, 36) and rhesus monkey (37) was also established, illustrating the general significance of the finding. In contrast, Ir genes were shown not to be linked to the structural genes for the H chain of immunoglobulins (39).

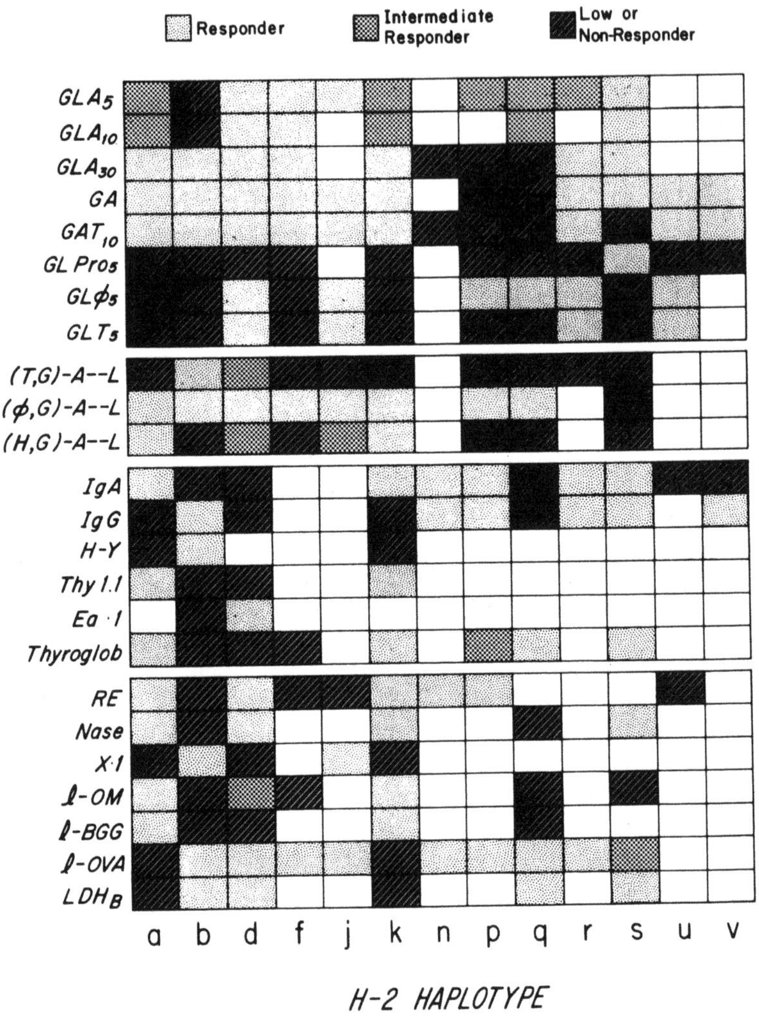

Figure 1. Immune responsiveness to linear random and branched copolymers of L amino acids, to isologous antigens, and to foreign antigen administered at limiting immunizing doses, is determined by the H-2 haplotype.

The availability of congenic resistant mouse strains developed by Snell (42) and of strains with documented recombinant events within the H-2 complex permitted McDevitt, Deak, Shreffler, Klein, Stimpfling and Snell (43) to map the murine Ir-1 locus controlling responsiveness to (T,G)-A−L to a new region of the mouse H-2 complex termed the I region (Fig. 2).

Mapping of individual murine Ir genes by several laboratories (reviewed in 40) revealed that most Ir genes map in I-A, a smaller number map in I-B whereas responsiveness to some antigens map in both I-A and I-E. The latter cases deserve to be discussed in some detail as they provide the genetic basis for the molecular identification of Ir gene products, to be discussed in another section. Whereas most immune responses investigated are under the control of

Table 1. Inheritance of Specific Ir Genes and of the Major Histocompatibility Locus of Strain 2 and Strain 13 Guinea Pigs by (2 x 13)F_1 and Backcross Animals

Antigens	Strain 2	Strain 13	(2×13)F_1	(2×13))F_1×13 50 % *	(2×13))F_1×13 50 %	(2×13)F_1×2 50 %	(2×13)F_1×2 50 %
DNP-PLL GL	+**	−**	+	+	−		
GA	+	−	+	+	−		
GT	−	+	+			+	−
BSA 0.1 μg	+	−	+	+			
HSA 1 μg	+	−	+				
DNP-BSA 1 μg	+	−	+	+	−		
DNP-GPA 1 μg	−	+	+			+	−
Major H locus							
strain 2	+**	−	+	+	−		
strain 13	−	+	+			+	−

* Column identifies the same group of backcross animals.
(**) + indicates responsiveness and presence of major histocompatibility specificities; − indicates nonresponsiveness and absence of major histocompatibility specificities of the inbred strains.
From B. Benacerraf in Ann. Immunol. (Inst. Pasteur) 125c, 143 (1974).

single loci, complementation of Ir genes for the response to certain antigens is observed in rare cases. Thus Dorf and I showed that the response to the terpolymer of L-glutamic acid, L-lysine and L-phenylalanine (GLØ) is determined by two Ir genes which complement in both the cis and trans configura-

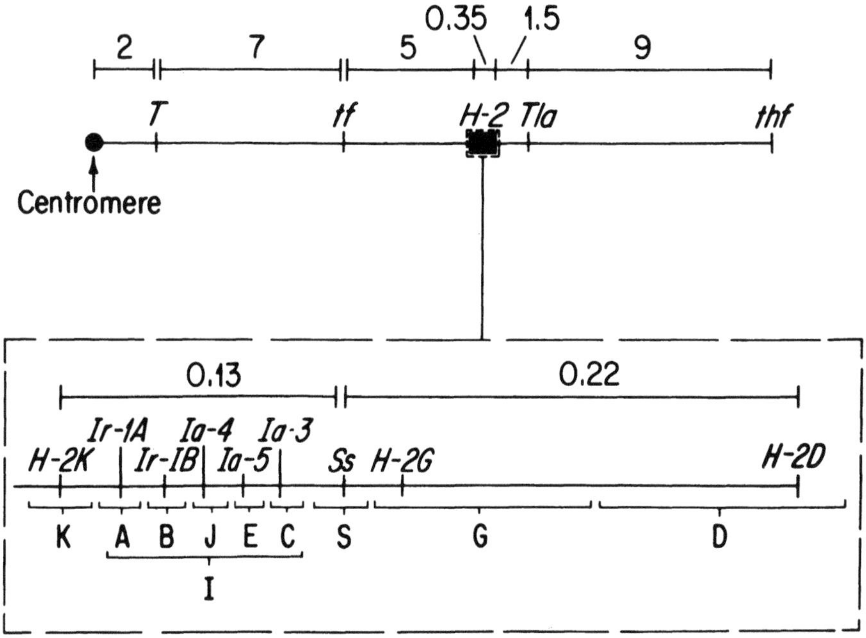

Figure 2. Genetic map of the H-2 complex showing the various loci and the subregions of I. Note that Ir genes have been mapped in I-A, I-B and I-E.

Table 2. Complementation of α and β Ir Genes for Responses to GLØ

Strain	H-2 haplotype	H-2 region formulae								GLØ response
		K	I-A	I-B	I-J	I-E	I-C	S	D	(% binding ± S. E.)
B10	b	b	b	b	b	b	b	b	b	1 ± 3
B10.BR	k	k	k	k	k	k	k	k	k	5 ± 3
(B10×B10.BR)F_1	b/k	b/k	b/k	b/k	b/k	b/k	b/k	b/k	b/k	68 ± 16
B10.S	s	s	s	s	s	s	s	s	s	−1 ± 1
B10.D2	d	d	d	d	d	d	d	d	d	61 ± 5
B10.A	a	k	k	k	k	k	d	d	d	4 ± 2
3R	i3	b	b	b	b	k	d	d	d	59 ± 7
5R	i5	b	b	b	k	k	d	d	d	73 ± 5
18R	i18	b	b	b	b	b	b	b	d	5 ± 2
7R	t2	s	s	s	s	s	s	s	d	4 ± 2
9R	t4	s	s	?	k	k	d	d	d	71 ± 10
A.TL	t1	s	k	k	k	k	k	k	d	4 ± 3
B10.HTT	t3	s	s	s	s	k	k	k	d	77 ± 7
			β			α				

Note:

Vertical bar indicates position of crossing-over.

GLØ genes are indicated by underlines.

Adapted from Dorf, M. E. and Benacerraf, B. in Proc. Natl. Acad. Sci. (USA) 72, 3671 (1975).

tion to permit a response to GLØ to develop (44). These genes which we termed α and β map in the I-E and I-A subregion (Fig. 2), respectively (Table 2). Possession of either α or β genes alone does not confer responsiveness to GLØ which require the presence of both genes. Response to several other antigens follows the pattern of the GLØ response (45).

We shall discuss later the evidence that Ir gene complementation for GLØ responses reflects the molecular complementation of the α and β subunits of the Ia glycoprotein. This molecule must be expressed on the surface of macrophages and B lymphocytes for the response to GLØ. In the case of this Ia molecule, the α and β chains will be shown to be coded respectively in I-E and I-A. When Ir genes map in a single region such as I-A, distinct α and β subunits, the Aα and Aβ chains, are coded in the same A subregion.

Ia Molecules and Histocompatibility Antigens

Taking advantage of the existence of mouse and guinea pig strains which differ solely at the I region of their MHC such as the ATL and ATH strains of mice and the guinea pig strain 2 and strain 13, attempts were made to produce antibodies specific for the Ir gene products by cross-immunization with lymphoid tissue. Alloantisera prepared in this manner by Shreffler and David (46), Klein and Hauptfeld (47) and McDevitt and associates (48) in mice and Schwartz, Paul and Shevach (49) in guinea pigs reacted with alloantigens termed Ia (immune response-associated) antigens expressed on B lymphocytes and a significant fraction of macrophages (50). A detailed study by Shreffler and David (46) of the specificities detected by anti-Ia antisera revealed the considerable polymorphism of these molecules.

Cullen et al (51, 52) studied the structure of murine Ia antigens expressed on B lymphocytes, and analyzed the membrane antigens specifically reactive with anti-Ia antibodies. Such antibodies bound glycoproteins from B cells composed of an α and a β chain with molecular weights of 33,000 and 28,000 daltons respectively.

Similarly, 13 anti-2 and 2 anti-13 reciprocal alloantisera detected homologous Ia molecules with corresponding α and β chain subunits on guinea pig macrophages and B lymphocytes (49,53).

A graphic representation of an Ia molecule is shown in Fig. 3 and compared with a classical transplantation antigen of the MHC expressed on all cells and comprised of a 45,000 dalton polymorphic chain associated with β_2 microglobulin (54).

An analysis of the immunological properties of the highly polymorphic Ia molecules on macrophages and B lymphocytes revealed that these products stimulate the alloreactive proliferation of unprimed clones of T lymphocytes in an *in vitro* test termed the mixed leukocyte reaction (MLR) (55). The ability of Ia bearing cells to stimulate MLR responses is effectively blocked by anti-Ia antibodies (56). I region differences and Ia molecules on cells stimulate strong graft versus host reactions (57) and vigorous homograft rejections (58).

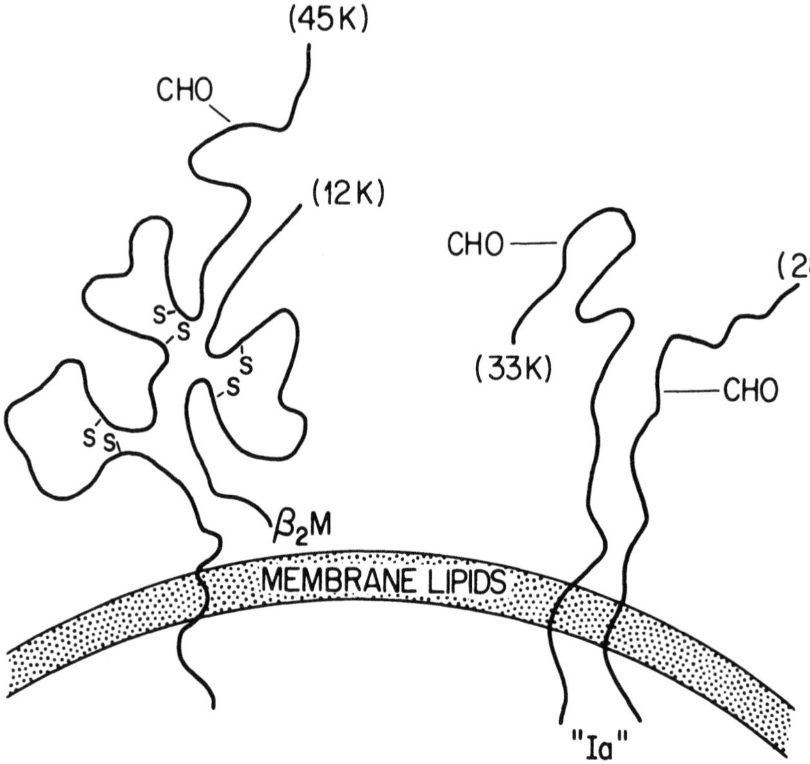

Figure 3. Graphic representation of the chain structure of an Ia molecule, compared with the structure of a histocompatibility antigen. Essentially comparable results were obtained for the mouse H-2 and the human HLA and the guinea pig GPLA complexes.

Function of Ir Genes

The study or Ir gene function contributed to our understanding of the intricate regulatory mechanisms evolved by T cells and macrophages to regulate specific immune responses. Experiments were initially designed to identify the cells of the immune system in which Ir genes are expressed and the nature of the process they control. H-linked Ir genes were shown to determine both humoral and cellular immune responses (28). A further analysis revealed that the genes control the recognition of the "carrier" molecule as an immunogen (59), a property of T lymphocytes. Thus, responder guinea pigs which make anti-DNP antibody upon immunization with DNP-PLL are equally able to make anti-benzylpenicilloyl (BPO) antibody to BPO-PLL whereas nonresponder guinea pigs to DNP-PLL do not (59). Similarly, nonresponder animals who failed to make anti-DNP antibody to DNP-PLL, make anti-DNP antibody when immunized with DNP conjugates of a conventional antigen. Moreover, Ira Green in my laboratory made the significant observation that the DNP-PLL genetic defect could be bypassed and nonresponder animals induced to form anti-DNP-PLL antibodies if DNP-PLL is treated as a macromolecular hapten and administered coupled to an immunogeneic carrier such as ovalbumin (60). Consistent with the critical role of the carrier in cellular immunity, the genetic defect for cellular immunity was not bypassed and the nonresponder guinea pigs immunized with DNP-PLL-ovalbumin did not develop delayed type sensitivity to DNP-PLL in spite of making large amounts of anti-DNP-PLL antibodies. Dunham, Unanue and I (61) then verified the presence of B cells with antibody receptors for nonimmunogeneic polypeptides in the spleens of nonresponder mice. We concluded from these experiments that the process governed by specific H-linked Ir genes controls T cell immune responses and affects antibody production only as a result of the need of helper T cells for B cell responses. In agreement with this conclusion, H-linked Ir genes were shown to control only the response to T dependent antigens (39). T independent responses which result from the direct activation of B lymphocytes by antigen are not under H-linked Ir gene control.

The involvement of Ir genes in T cell responses could result from either: 1) the expression of Ir genes in T cells and their coding for the T cell receptor, or 2) the expression of Ir genes in macrophage and B cells and their role in determinant selection, antigen presentation and T cell-B cell interaction. The latter alternative was shown to be correct in every respect. Shevach and Rosenthal (62), working with the guinea pig systems we developed, made use of the finding that primed T cell clones proliferate *in vitro* and incorporate H^3 thymidine when presented with antigen by antigen pulsed macrophages. $(2 \times 13)F_1$ guinea pigs were immunized with two antigens, DNP-GL, controlled by a strain 2 Ir gene, and GT controlled by a strain 13 Ir gene (Table 1). Their T cells were exposed to DNP-GL or GT on macrophages of 2, 13 or F_1 origin. The results were unequivocal. Primed $(2 \times 13)F_1$ T cells responded to DNP-GL on strain 2 or F_1 macrophages but not on strain 13 macrophages. In contrast, the same primed cell populations responded to GT on strain 13 or F_1 but not strain 2 macrophages (Fig. 4). These experiments were extended in mice by Sredni, Matis, Lerner, Paul and Schwartz (63) using a GLØ specific T cell line cloned from a responder B10.A(5R) mouse. Such GLØ specific clone lines only proliferated when presented GLØ by antigen presenting cells (macrophages) from high responder mice [B10.A(5R) or $(B10.A \times B10)F_1$] expressing both Ir-GLØ α and β genes in the same cell (Table 3). The need for Ia bearing macrophages for T cell stimulation was further documented in our laboratory by Germain and Springer (64). Treatment of antigen presenting cells with monoclonal anti-Ia antibody and complement abolished the ability of the cells to present antigen for proliferative responses to primed T cells (Table 4).

Another approach to the role of Ir genes in the presentation of antigen to T cells by macrophages involved the use of anti-Ia antisera without complement to block antigen presentation. The original experiments were carried out in guinea pigs by Shevach, Paul and Green (65) and later in mice by Schwartz and associates in collaboration with our laboratory (66). T cells from $(2 \times 13)F_1$ guinea pigs primed to DNP-GL and GT were exposed to DNP-GL or GT *in vitro* together with $(2 \times 13)F_1$ macrophages and alloantisera directed to 2 or 13 Ia specificities. Anti-2 antisera blocked only the response to DNP-GL and not to GT whereas anti-13 antisera blocked the response to GT but not to

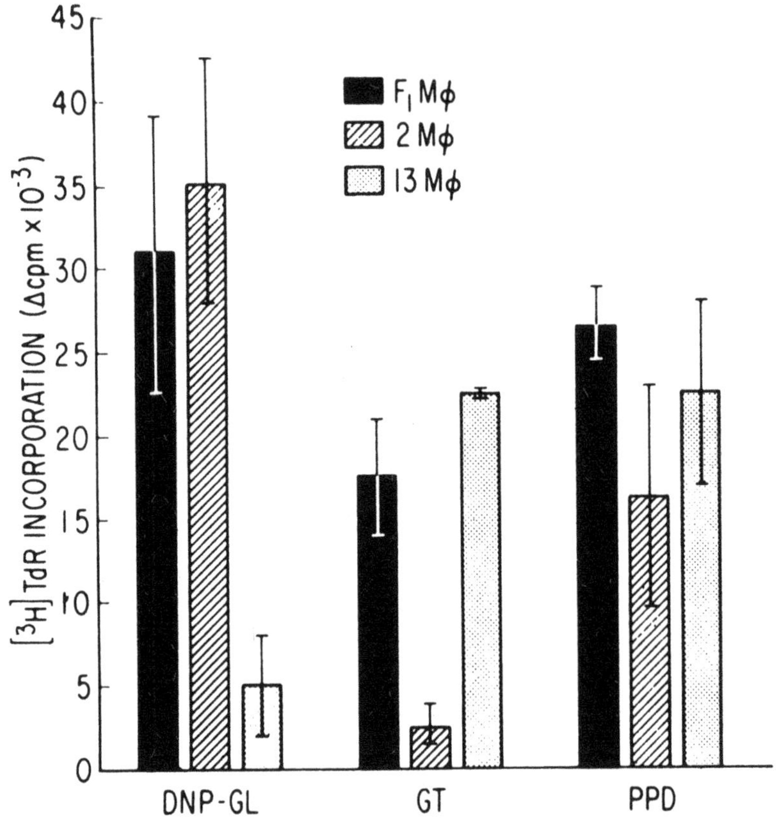

Figure 4. Ir genes are expressed in Ia bearing macrophages. (2 x 13)F_1 T cells from guinea pigs primed to both DNP-GL (to which strain 2 responds) and GT (to which strain 13 responds) were cultured *in vitro* with strain 2, strain 13, or F_1 macrophages pulsed with GT or DNP-GL. The proliferative responses are recorded as the incorporations of ^3H thymidine into DNA. Adapted from Shevach, E. M. and Rosenthal, A. S. in J. Exp. Med. *138*, 1213 (1973).

DNP-GL (Table 5). These experiments led to the important conclusion that: 1) Ir genes are expressed on antigen presenting cells with the morphology of macrophages, and 2) T cells detect antigen on the surface of antigen presenting cells and are specific for foreign antigens perceived in the context of autologous Ia molecules.

Determinant selection for T cell responses clearly results from such a process. Thus, even in the case where two inbred strains are responders to the same T dependent antigen the studies of Barcinski and Rosenthal (67) on the immune response of guinea pig to insulin revealed that strains 2 and 13 respond to distinct determinants; strain 2 T cells respond to a determinant on the A chain of insulin (A_8-A_9-A_{10}), whereas in strain 13 guinea pigs, the response is directed to sequential determinants on the B chain of insulin involving the histidine at position 10 (68). Similar data concerning determinant selection in other antigens was reported by Berzofsky et al (69) and Kipps et al in our laboratory (70).

Table 3. Stimulation of GLØ Specific B10.A(5R) T Cell Clones Requires Antigen Presenting Cells (Macrophages) Expressing Both α and β Ir-GLØ Gene Products in the Same Cell

Antigen (GLØ) presenting cells	Ir-GLØ alleles		Proliferative response by B10.A(5R) (CPM ± SEM)	
	α	β	Clone 6.2	Clone 6.4
none			67 ± 28	173 ± 52
B10.A(5R)	+	+	7,447 ± 61	9,243 ± 1,774
B10.A	+	−	60 ± 15	67 ± 15
B10	−	+	70 ± 10	87 ± 20
B10.A+B10	±	±	73 ± 22	63 ± 23
$(B10.A \times B10)F_1$	+	+	10,177 ± 1,492	9,497 ± 1,514

Note:
Clones 6.2 and 6.4 were selected from GLØ primed T cells of B10.A(5R) responder mice. 5×10^3 cells from those cloned lines were stimulated with 100 μg/ml GLØ in the presence of irradiated antigen presenting spleen cells. Stimulation was assayed by measuring the incorporation of ^{3}H-thymidine.
Adapted from Sredni, B., Matis, L. A., Lerner, E. A., Paul, W. E. and Schwartz, R. H. in J. Exp. Med. *153*, 677, 1981.

Table 4. Treatment with Monoclonal Anti-Ia Antibody and Complement Eliminates Macrophages Required for GAT Induced T Cell Proliferation

Responding cells (a)	Added antigen presenting cells (b)	Response (ΔCPM)
C treated	none	53,656
anti-Ia + C treated	none	722
anti-Ia + C treated	γR spleen	36,146
anti-Ia + C treated	α-Thy 1 + C treated γR spleen	56,505

Note:
a. 4×10^5 nylon passed lymph node T cells from GAT-CFA primed BALB/c mice, treated with C alone, or M5/114 + C, then cultured with or without 100 μg/ml GAT for 3 days, pulsed for 18 hr with ^{3}H-thymidine, harvested, and counted.
b. 3×10^5 1500R γ-irradiated syngeneic spleen cells.

Table 5. Anti-Ia Alloantisera Block Antigen Presentation to Primed T Cells

Responding cells	Antigen	Antisera added	Increased DNA synthesis
(2×13) F_1 T cells	DNP-GL	none	++++
”	DNP-GL	anti-2	+
”	DNP-GL	anti-13	++++
”	GT	none	++++
”	GT	anti-2	++++
”	GT	anti-13	+

Note:
$(2 \times 13)F_1$ guinea pig T cells were primed to DNP–GL and GT. Their response to DNP–GL is blocked by 13-anti-2 antisera and their response to GT is blocked by 2-anti-13 antisera. These alloantisera are specific for Ia antigens on strain 2 and 13 respectively.
Adapted from Shevach, E. M., Paul, W. E., and Green, I., J. Exp. Med. *136*, 1207 (1972).

The genetic restrictions dictated by I region controlled antigen presentation to T cells can also be observed when attempts are made to transfer delayed type sensitivity. I had made the puzzling observation with Paul and Green that delayed type sensitivity to DNP-PLL in random bred guinea pigs could only be adoptively transferred to recipients that were also responders to this antigen (71). Using congenic resistant inbred strains of mice, Miller et al (72) later showed that the successful transfer of delayed type reactivity requires I region identity between the sensitized T cell and the recipient mice which provide the antigen presenting macrophage when the test antigen is injected. Moreover, as expected, sensitized cells from (responder × nonresponder)F_1 mice did not transfer DTH to nonresponder recipients lacking the antigen presenting cells (73).

At the same time as the N.I.H. group documented the importance of Ir genes and Ia molecules in antigen presenting cells and their critical role in the presentation of antigen to T cells, experiments were carried out by Katz, Hamaoka, Dorf and me (74) and by Kindred and Shreffler (75) demonstrating the role of I region genes in the control of T cell-B cell interactions in antibody responses.

We devised a double adoptive transfer protocol whereby hapten specific B cells and carrier specific T cells from either the same parental strain or distinct parental strains were transferred to irradiated F_1 recipient mice prior to secondary challenge (74). The results were unequivocal. Carrier specific helper T cells and hapten primed B cells need to share I region genes for antibody response to develop to the hapten-carrier conjugate. Successful T cell-B cell interactions were observed between F_1 T cells and parental B cells, or parental T cells and F_1 B cells which only need to share one haplotype for successful responses provided both strains are responsers to the carrier antigen used (Table 6).

Table 6. I Region Genes Restrict T-B Cell Cooperative Interactions

H-2 haplotype of hapten-primed B cells	H-2 haplotype of carrier-primed T cells	Secondary responses in F_1 irradiated recipients
a	a	++++
b	b	++++
a	b	–
b	a	–
b	$(a \times b)F_1$	++++
a	$(a \times b)F_1$	++++
$(a \times b)F_1$	a	++++
$(a \times b)F_1$	b	++++

Note:
Carrier primed T cells were adoptively transferred to $(a \times b)F_1$ recipients which were irradiated; then anti-Thy 1 and C treated, hapten primed spleen cells (B cells) were adoptively transferred to the same $(a \times b)F_1$ recipients. The animals were challenged with the hapten-protein conjugates and the secondary anti-hapten response measured as an indication of T-B cell cooperative interactions.
Adapted from Katz, D. H., Hamaoka, T., Dorf, M. E. and Benacerraf, B. in Proc. Natl. Acad. Sci. (USA) *70,* 2624 (1973).

When an antigen under Ir gene control was used such as the copolymer GLT, (responder × nonresponder)F_1 T cells specific for GLT helped the responder but not the nonresponder hapten specific B cells when challenged with DNP-GLT (76), indicating the critical role of Ir gene expression in B cells, in T cell-B cell interactions (Fig. 5).

The need for I region identity for T cell-B cell interactions was confirmed by Sprent (77) and Kappler and Marrack (78) with different systems. More recently, Chiller, working with clonally derived antigen specific helper T cell lines, observed the same I region requirement for successful T-B cell interaction.

The data of Singer and Hodes (79) indicate that in certain experimental conditions where the antibody response involves solely unprimed B cells of the Ly b5 phenotype, antibody responses may be helped by T cells, across I differences. It is clear, however, that the majority of the responses of primed B cells require the type of I region controlled T cell-B cell interaction discussed earlier. The specificity of the interaction for Ia is determined by the specificity

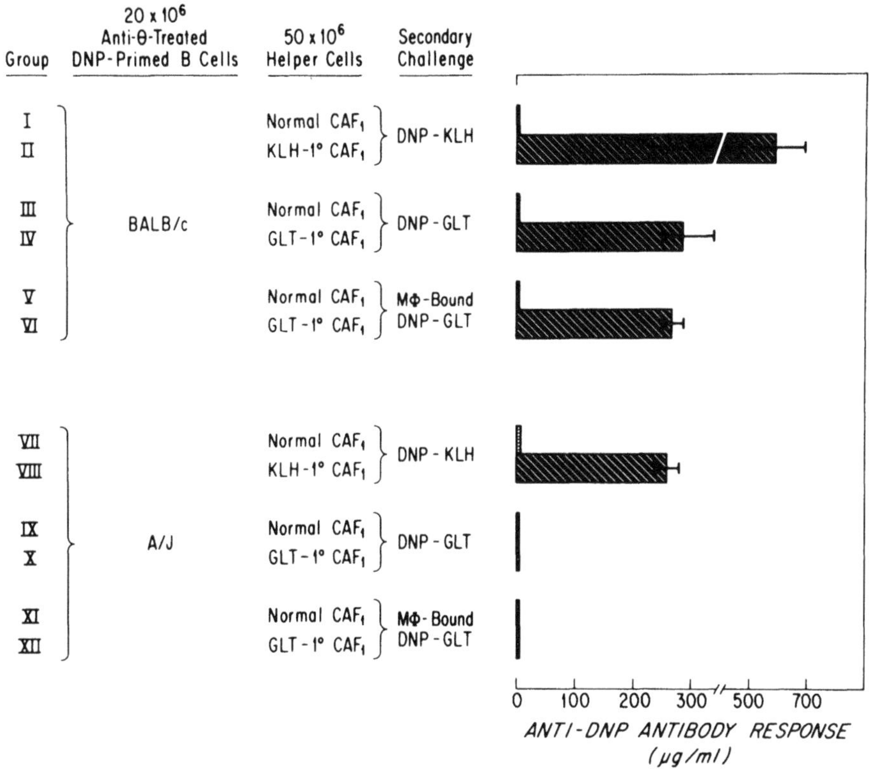

Figure 5. Failure of GLT primed CAF_1 T cells to cooperate with nonresponder hapten primed parental A/J B cells, in contrast with the ability of KLH primed T cells to cooperate equally well with both parental A/J and parental BALB/c hapten primed B cells for anti-DNP secondary responses.

From Katz, D. H., Hamaoka, T., Dorf, M. E., Maurer, P. H., and Benacerraf, B. in J. Exp. Med. *138*, 734 (1973).

of the T cell clones stimulated when antigen is originally presented by the Ia bearing macrophages.

What is the mechanism of I region controlled T cell-B cell interaction and Ir gene function at this level? The data is not as definitive as in the case of macrophage-T cell interaction discussed earlier. I feel, nevertheless, that substantial evidence exists for the view that murine Ly 1^+T cells are specific for antigen perceived in the context of Ia molecules on antigen presenting cells. The cells are stimulated to proliferate and differentiate into DTH or helper T cells. The helper T cells will in turn interact with Ia bearing B cells which bound antigen through their immunoglobuin receptors. The helper T cells deliver their differentiating signal by interacting with antigen and Ia molecules on B cells in a similar manner as on antigen presenting cells.

The Ia Molecules are the Ir Gene Products

There is now substantial and very convincing evidence for the view that Ia molecules are the Ir gene products and determine specific immune responsiveness to thymus dependent antigens.

1) Ia molecules, the surface glycoproteins composed of α and β chains, expressed primarily on a population of macrophages and B lymphocytes, are coded for in precisely the same subregion of I in which Ir genes map: I-A and I-E.

2) Anti-Ia antisera and particularly monoclonal anti-I-A antibodies specifically block *in vitro* responses by interacting with Ia molecules on antigen presenting cells.

3) The mapping of the structural genes coding for the α and β chains of Ia molecules in the I region indicates an intimate correlation between -chain structure and the control of the response to GLØ by complementing α and β genes in I-A and I-E.

The structural analysis of Ia molecules and of their component chains in the mouse was carried out by several laboratories using the techniques of 2 dimensional gel electrophoresis and peptide mapping to analyse the basis of polymorphisms.

Jones et al (80) made the fundamental observation that, in strains bearing the appropriate H-2 haplotype, a gene in the I-E subregion, controls the cell surface expression of an Ia molecule, whose polymorphic determinants are largely controlled by the I-A subregion. Cook, Vitetta, Uhr and Capra (81) and later Silver et al (82) confirmed these findings and demonstrated that the I-E subregion controls the synthesis of the Ia Eα chain which on the cell surface is noncovalently associated with the β chain determined in the I-A subregion (Figs. 2 and 6). Strains with the $H\text{-}2^b$ haplotype in the I-E subregion fail to synthesize this α chain and as a consequence the corresponding Ia molecule is not expressed on the cell membrane, although the β chain coded in I-A is synthesized and found in the cytoplasm.

The genetic control of the polypeptide chains of this class of Ia molecules by I-E and I-A correlates completely with the Ir gene complementation observed in the response to GLØ discussed earlier (44). Moreover those strains that

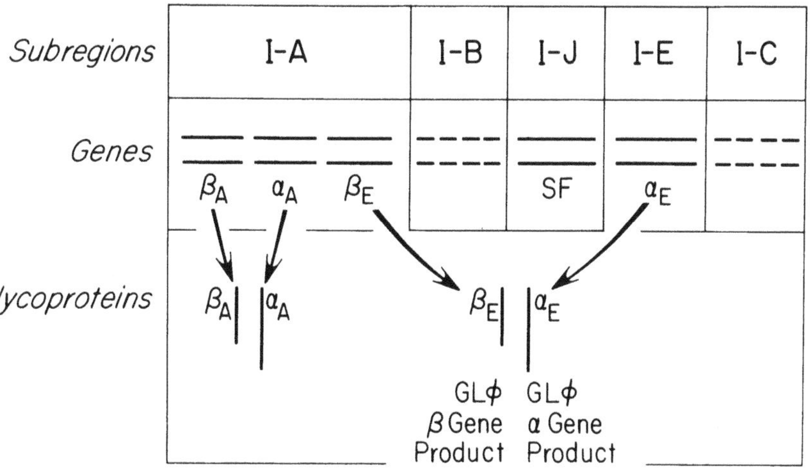

Figure 6. Model for the genetic and structural basis of Ir gene complementation in the response to GLØ. The products of the α and β Ir genes required for GLØ presentation are postulated to consist, respectively, of an Ia.7 bearing α chain (designated αE) encoded in the I-E subregion and of a β chain (designated βE) encoded in the I-A subregion which interact selectively to form a functional Ia molecule on the cell membrane of the antigen presenting cells.

exhibit a responder α gene at I-E synthesize an α chain controlled by this locus. It is indeed fortuitous that we called the I-E gene α and the I-A gene β at a time when we did not know that they determined, respectively, the α and β chains of the corresponding Ia molecule.

The availability of cloned lines of antigen specific T cells and of monoclonal anti-Ia antibody provided still more conclusive evidence for the identity between Ia molecules and Ir gene products. Sredni, Schwartz and associates (63) cloned a GLØ specific T cell line derived from a B10.A(5R) responder mouse which was stimulated only by GLØ presented on B10.A(5R) or (B10 × B10.A)F_1 presenting cells (Table 3). The *in vitro* response of this clone to GLØ was specifically blocked by a monoclonal anti-Ia antibody specific for the conformational determinants (83) on the Ia molecules resulting from the interaction of the I-E coded α chain with the I-A coded β chain.

The other major murine Ia molecule has both α and β chains coded for in I-A. The possibility still exists for genetic complementations at the molecular level corresponding with Ir gene complementation in animals heterozygous at I-A, which is precisely what occurs. Such complementation is more difficult to detect and depends upon clonal analysis of T cells specific for an antigen the response to which is controlled at I-A, such as the terpolymer GAT. Cloned T cell lines specific for GAT were selected by Fathman et al (84) and by Sredni et al (63) from a (B10.A ×B10)F_1 mouse immunized with GAT. Some of these clones responded to GAT when presented by B10.A macrophages, other clones responded to GAT on B10 macrophages and a third type of clone responded to the antigen only when presented on (B10.A × B10)F_1 macrophages (Table 7). We can conclude that the response to GAT in (B 10.A × B10)F_1 mice is determined by three types of genetically distinct I-A coded Ia molecules which,

Table 7. Three Different Types of MHC Restriction of GAT-Specific T Cell Colonies from $(B10.A \times B10)F_1$ Mice Primed to GAT

Colony no.	Proliferative response to GAT on spleen cells from			
	B10.A	B10	$(B10.A \times B10)F_1$	H-2 restriction
1	++++	–	++++	B10.A
4	++++	–	++++	B10.A
8	++++	–	++++	B10.A
12	++++	–	++++	B10.A
2	–	++++	++++	B10
3	–	++++	++++	B10
5	–	++++	++++	B10
9	–	++++	++++	B10
10	–	++++	++++	B10
13	–	++++	++++	B10
6	–	–	++++	$(B10.A \times B10)F_1$
11	–	–	++++	$(B10.A \times B10)F_1$
14	–	–	++++	$(B10.A \times B10)F_1$

Note:
$(B10.A \times B10)F_1$ mice were immunized with GAT; their T lymphocytes stimulated *in vitro* with GAT and cloned in soft agar. 2×10^4 T cells from each colony were stimulated with 100 μg GAT in the presence of antigen presenting cells from (B10, B10.A or $B10.A \times B10)F_1$ mice.
Adapted from Sredni, B., Matis, L. A., Lerner, E. A., Paul, W. E. and Schwartz, R. H. in J. Exp. Med. *153*, 677, 1981 press.

together with antigen, specifically select the three types of clones stimulated. The extent to which these three Ia molecules interact with the same determinant on the GAT antigen or with different ones has not been ascertained.

Significance of Ir Gene Specificity

We have made considerable progress in our understanding: 1) of Ir gene function in antigen presenting cells and in B cells, 2) of the identity of Ia molecules and Ir gene products, and 3) of the commitment of murine T cells with the Ly 1^+ phenotype (85) to react with autologous Ia molecules and antigens. But an important issue remains unresolved which concerns the process by which the specificity of Ir gene function is imparted, i.e., why certain Ia molecules on antigen presenting cells determine T cell response to some antigens and not to others. The issue can also be presented in other terms, i.e. what mechanism determines the development of T cell clones with combined specificity for autologous Ia molecules and selected antigens.

T cells bear receptors coded at least in part by the immunoglobulin H chain linkage group as shown by Binz and Wigzell (86), Eichmann (87) and our laboratory (88–90). Ir genes do not need to be expressed in T cells for responses to occur. T cells become committed to host MHC specificities as they differentiate. Thus, nonresponder parental T cells can be turned into responder T cells by being developed in (responder × nonresponder)F_1 irradiated recipients (73, 91). Such T cells respond to the putative antigen, if it is presented on responder macrophages bearing the appropriate Ia molecules with which the T cells interacted during differentiation.

Two types of hypotheses have been formulated to account for Ir gene controlled restrictions. von Boehmer, Haas and Jerne (92) proposed that T cells generate their repertoire for foreign antigens from their receptors for autologous MHC antigens and that Ir gene defects reflect the absence of clones bearing receptors for certain antigens, based upon the restriction placed on the repertoire by the commitment of T cells to a particular set of autologous MHC antigens. As T cells differentiate and are selected to react with different MHC antigens in different individuals, different H-linked Ir gene defects result.

An alternative hypothesis was proposed independently by Rosenthal (93) and myself (94). It postulates:

1) that Ia molecules are capable of reacting selectively with certain amino acid sequences on protein antigens,

2) that such a selective interaction in antigen presenting cells results in the formation of an Ia molecule-antigen complex reactive with T cell clones differentiated to bear receptors for autologous Ia and antigen (Fig. 7).

A limited number of such binding sites on a relatively small number of Ia molecules can generate from available antigens an almost unlimited number of determinants specifically recognized by T cells. The size of the binding site on the Ia molecule for the antigen or its fragment should encompass a limited number of amino acids in order not to impose undue restrictions on the system and to permit a given site to bind to a great variety of foreign proteins. The likelihood that such a sequence is present in a given protein varies inversely with the size of the sequence and is considerable for a postulated size involving three or at most four amino acids. The location of the binding sites will vary in different proteins. This dictates the antigenic determinants with which T cells react in conjunction with Ia molecules. A given Ia molecule could thus react with a large number of antigens and yet impose unigenic restriction to immune responsiveness.

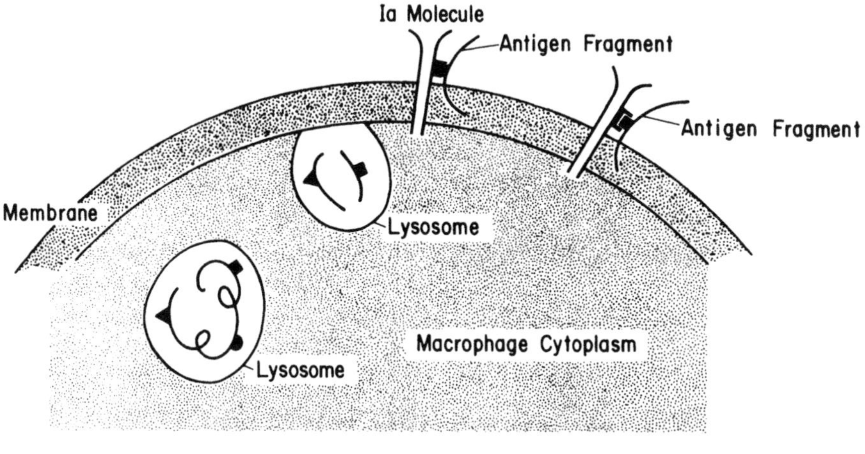

Figure 7. Graphic representation of the specific interaction proposed between Ia molecule and antigen fragment on the surface of antigen presenting cells (macrophages), required for specific interaction with T cells.

The identification of the amino acid sequences critical for immunogenicity may provide some indication of the size of the postulated site of interaction with the Ia molecules. The elegant studies of Schlossman and associates (95, 96) on the immunogenicity of DNP-oligo-L-lysines for strain 2 guinea pigs is very informative. The smallest oligolysine polymer which is immunogenic has seven lysines. However, a peptide containing only four lysines and a sequence of L-alanines, terminated with a DNP-lysine, is equally immunogenic in strain 2 guinea pigs, although the specificity of the response is different. It would appear therefore that the critical interaction site in this antigen for strain 2 guinea pigs may consist of at most 4 lysines. The data on the immunogenicity of insulin A chain for strain 2 guinea pigs and of B chain for strain 13 guinea pigs (68) and on the precise amino acids responsible suggest also an interaction site involving 3 or at most 4 amino acids.

Although the second hypothesis appears more compatible with the available findings, definite evidence of Ia molecule-antigen interaction is lacking. Some recent experiments of Nepom and Germain in our laboratory may also be interpreted to indicate a necessary interaction between Ia molecule and antigen for binding of the complex to T cells. We have indeed observed that when F1 T cells specific for antigens under Ir gene control such as GAT or GLØ are stimulated to proliferate by antigen, they selectively bind autologous Ia molecules of precisely the type which determined responsiveness to the antigen. Considering that some of the T cells in the population studied should have been specific for the allelic Ia molecule, the selective binding observed may indeed imply a requisite interaction between Ia molecules on macrophages and antigen.

I Region Control of T Cell Suppressor Responses

Selected antigens such as the terpolymer GAT induce preferentially suppressor T cells in certain nonresponder mouse strains, which contribute to the unresponsiveness observed (97). These T cells, adoptively transferred, suppress the anti-GAT antibody response to GAT coupled to an immunogenic carrier. Preferential suppressor T cell responses were also observed for other antigens such as the copolymer GT (98) by us and hen egg lysozyme by Sercarz and associates (99).

The ability to develop specific suppressor T cells also proved to be controlled in the I region of the murine H-2 complex (Table 8). The analysis of the genes responsible, of their products, and of the processes involved in the generation of specific suppressor T cells has not yet permitted a definitive understanding of these complex phenomena. As in the case of I region control of T cell responses, immunogenicity of a complex antigen is determined by the determinants it bears. Apparently certain determinants in mice of the appropriate H-2 haplotype induce selectively suppressor T cell responses (99, 100). Moreover, the presence of determinants which stimulate suppressor T cells preferentially may result in suppression of responses to other determinants on the antigen which otherwise would stimulate helper T cells and thereby antibody responses.

In addition to this type of I region control of suppressor T cell responses

Table 8. H-2 (I region) Control of Suppressor T Cell Responses to the Copolymer GT

Strain	H-2	% suppression	p values
A/J	a	0	<0.4
A.By	b	0	<0.3
C57BL/J	b	0	<0.1
129/J	b	0	<0.1
BALB/c	d	80	<0.000001
DBA/2	d	81	<0.001
D1.C	d	76	<0.002
A.CA	f	100	<0.00009
SJL	s	72	<0.000001
A.SW	s	68	<0.001
CAF_1	a/d	74	<0.0001

Note:
100 µg GT was administered intraperitoneally; 3 days later the experimental and matched control groups were immunized with GT coupled to the immunogenic carrier, methylated bovine serum albumin, and the antibody responses compared to determine the suppression elicited by GT preimmunization.
Dominant GT specific suppression can be generated in mice with $H\text{-}2^d$, $H\text{-}2^f$ and $H\text{-}2^s$ but not the $H\text{-}2^a$ or $H\text{-}2^b$ haplotypes. The responsible genes mapped in the I region. In transfer experiments the GT suppression observed was shown to be mediated by suppressor T cells.
Adapted from Debré, F., Kapp. J., Dorf, M., and Benacerraf, B. in J. Exp. Med. *142*, 1447 (1975).

which is specific and determinant oriented, the I region affects suppressor T cell responses by coding for specificities expressed on all suppressor T cells. Murphy et al (101), and Tada et al (102, 103) discovered that a new subregion of I, the I-J, controls alloantigens expressed only on suppressor T cells. All the T cells in the suppressor T cell circuits bear I–J coded specificities. Moreover, antigen specific suppressor factors were extracted from antigen specific suppressor T cells (104, 105). Such factors were shown to stimulate the generation of suppressor T cells (106) and to bear on the same molecule determinants coded by the I–J subregion as well as idiotypic determinants coded for by the immunologlobulin heavy chain linkage group (88, 89). The structure of suppressor factor and the nature of the cellular interactions which result in suppressor T cell responses are currently under investigation in several laboratories.

The Functional Specificity of T Cells for the Antigens of the MHC – The Origin of Alloreactivity

In a preceding section we discussed the commitment of helper and DTH T cells to react with antigen and autologous Ia molecules. An analogous commitment of cytolytic T cells (CTL) to histocompatibility antigens of the MHC has been demonstrated by Zinkernagel and Doherty (107), by Shearer et al (108) and by Bevan (109) illustrating the general nature of the commitment of T lymphocytes to react with antigen only on cell surfaces and in relation with gene products of the MHC.

Zinkernagel and Doherty (107) demonstrated that CTL from mice immune to Lymphocytic choriomeningitis virus (LCM) only lyse LCM infected target cells which share H-2 antigens with the killer cell. The MHC loci concerned map at either K or D of the H-2 complex. Thus, CTL recognize antigen in the context of the K or D histocompatibility antigens, like helper cells react with antigen and Ia molecules.

The evolutionary significance of these restrictions and of the role played by MHC antigens becomes readily apparent when we consider that T cell immune responses are primarily responsible for monitoring self and nonself on cell surfaces. T cells need to determine when an autologous cell becomes malignant or virally infected and must be destroyed. This surveillance function is optimally performed if a large number of T lymphocyte clones are specialized to detect small variants on MHC molecules. Taking advantage of this process, T cells have also evolved the capacity to regulate immune responses as a consequence of their ability to recognize antigen on cell membranes. Because unregulated immune responses can be very harmful, we have developed highly specific T cell mediated mechanisms of immune regulation which require the recognition by T cells of clones of other immune cells bearing antigen.

Because of the two types of MHC specificities exhibited by helper and cytolytic T lymphocytes, two types of Ir gene defects can be observed in CTL responses. A major type of Ir gene defect maps in the I region (92, 110) and concerns the generation of helper cells, as in the case of antibody. The other type maps in K (111) or D and reflects the ability of CTL clones to react with antigens on cell surfaces as they are presented in relation with K or D gene products.

The major topic of this paper has been the specificity of T cells for autologous MHC antigens and the manner in which foreign antigens are perceived by T cells in the context of MHC gene products. We postulate that MHC antigens have evolved for precisely this function. Yet they have been originally discovered by Gorer et al (112) and identified in a different context as the major antigens, within a species, responsible for alloreactivity and the rejection of allografts, a phenomenon which is of limited evolutionary value. The issue of the origin of alloreactivity can now be appropriately addressed as it appears to be closely related to the process whereby T cells become committed as a class to reactivity with autologous MHC antigens during differentiation.

Jerne (113) proposed a theory which was further elaborated by ourselves (114, 115) to explain the generation in the thymus of T cells specific for autologous MHC antigens. According to the theory, in the first stage T cells initially specific for self MHC gene products are selected in the thymus to differentiate and proliferate. Then, in a second stage, only those T cells which bear low affinity receptors for self MHC antigens are allowed to mature and leave the thymus as functional T cells. Such T cells, having low reactivity for self MHC antigens, have concomitantly high affinity for variants of self MHC antigens. These variants appear to be the same or similar to the allogeneic MHC antigens expressed in the same species. Weaker affinity for xenogeneic MHC antigens would thus be expected. Simultaneously and independently these T cells develop recognition for determinants on conventional thymus dependent antigens.

The high degree of reactivity to MHC antigens which constitute the polymorphic population encountered in the same species (i.e. alloantigens) and the lower reactivity to xenogeneic MHC antigens may be attributed to the fact that low affinity receptors for self MHC antigens are expected to react optimally with allogeneic MHC antigens, but much less so with xenogeneic antigens. This would account for the paradox that the strongest T cell responses are not elicited by antigens further removed phylogenetically from the responder. Two predictions from this theory are: (a) that clones of T cells induced by xenogeneic MHC antigens should be highly cross-reactive with allogeneic MHC antigens, even to the extent that they may demonstrate a heteroclitic specificity. This has indeed been demonstrated by Burakoff et al. when mouse anti-rat CTL were shown to be comprised of clones cross-reactive with allogeneic target cells (116) (Table 9). (b) Alloreactive T cells should be expected to be highly cross-reactive with modified syngeneic cells. This was also shown to be the case when we observed considerable cross-reactivity by alloreactive cells for TNP conjugated target cells syngeneic to the responder (114) (Table 10).

Table 9. Mouse Cytolytic T Lymphocytes Elicited by Rat Stimulator Cells Cross-React Extensively with Murine Allogeneic Target Cells

Stimulator	% Specific ^{51}Cr release Targets		
	Lewis	B10.BR	B10.D2
Lewis	78	41	15
	76	56	42
	66	75	30
	43	64	38
	59	64	42

Note:
C57BL/6 (H-2^b) spleen cells were sensitized with rat spleen cells of the Lewis strain. The CTL were assayed for specific ^{51}Cr release on xenogeneic stimulator rat cells or on allogeneic B10.BR (H-2^k) or B10.D2 (H-2^d) mouse cells. Identical results were obtained with ACI and BN stimulators.
Adapted from Burakoff, S. J, Ratnofsky, S. E., and Benacerraf, B. in Proc. Natl. Acad. Sci. (USA) *74*, 4572 (1977).

Table 10. Cytolysis of Syngeneic TNP-Modified Targets by Allogenically Stimulated Cytolytic T Cells

Responder	Stimulator	% Specific ^{51}Cr release targets		
		EL4-TNP (H-2^b)	EL4	P815 (H-2^d)
B6 (H-2^b)	DBA/2 (H-2^d)	32	3	80

Note:
Adapted from Lemonnier, F., Burakoff, S. J., Germain, R. N. and Benacerraf, B. in Proc. Natl. Acad. Sci. (USA) *74*, 1229 (1978).

Since the T cell repertoire for MHC specificities is normally determined by the self MHC antigens of the thymus, we should expect the T cell repertoire to vary according to the MHC of the thymus in which T cells differentiate. Recent experiments utilizing radiation chimeras by Zinkernagel and associates (117) and Bevan (118) have demonstrated this to be the case.

The postulate that alloreactivity results from T cells differentiating in the thymus that are strongly reactive for variants of self MHC antingens leads to the expectation that immunization with virally infected syngeneic cells should result in the stimulation of T cell clones resctive with the virally infected syngeneic cells used to immunize and also reactive with uninfected allogeneic target cells.

Finberg et al in our laboratory have recently shown that immunization of BALB/c (H-2^d) mice with Sendai coated syngeneic cells stimulates CTL which lyse Sendai coated BALB/c target cells but also lyse uncoated H-2^b, H-2^q, H-2^k, H-2^s and H-2^r allogeneic target cells to an appreciable degree (119) (Tabell 11). We further demonstrated by the cold target inhibition technique that the same clones that lysed BALB/c coated Sendai targets also crossreactively lysed the allogeneic targets (Fig. 8). Furthermore, it was observed that separate CTL clones lysed each of the different allogeneic targets. In addition, there was significantly less lysis of target cells bearing the H-2^q haplotype than of target cells bearing the H-2^k or H-2^s haplotypes. This latter finding suggests that the association of Sendai virus antigens with the H-2^d gene products of BALB/c mice creates determinants which are more crossreactive with H-2^k and H-2^r than with H-2^q gene products. Using cloned T cell lines, the alloreactivity of CTL specific for virally infected syngeneic cells was confirmed by von Bochmer et al (120) and extended by Sredni and Schwartz (121) to T cells reactive with autologous I region products and foreign antigens. B10.A T cells specific for DNP-ovalbumin (DNP-ova) were cloned on DNP-ova pulsed macrophages. Such cloned lines proliferated specifically when exposed to DNP-ova on syngeneic macrophages.Some of these clones could also be stimulated to proliferate by H-2^s allogeneic macrophages in the absence of DNP-ova (Table 12). Therefore the same clone selected on DNP-ova pulsed B10.A macrophages reacted identically to DNP-ova pulsed B10.A macrophages or B10.S macro-

Table 11. Mouse Cytolytic T Cells (CTL) Specific for Sendai Infected Syngeneic Target Cells Also Lyse Noninfected Allogeneic Target Cells

Specificity of CTL	Target cells		% Specific ^{51}Cr release
BALB/c anti-BALB/c-Sendai	B10.D2-Sendai	(H-2^d)	78
(H-2^d)	B10	(H-2^b)	37
	B10.G	(H-2^q)	12
	B10.B2	(H-2^k)	38
	B10.RIII	(H-2^r)	46
	B10.S	(H-2^s)	28
	B10.D2	(H-2^d)	1

Note:
Adapted from Finberg, R. S., Burakoff, S. J., Cantor, H. and Benacerraf, B. in Proc. Natl. Acad. Sci. (USA) *75*, 5145 (1978).

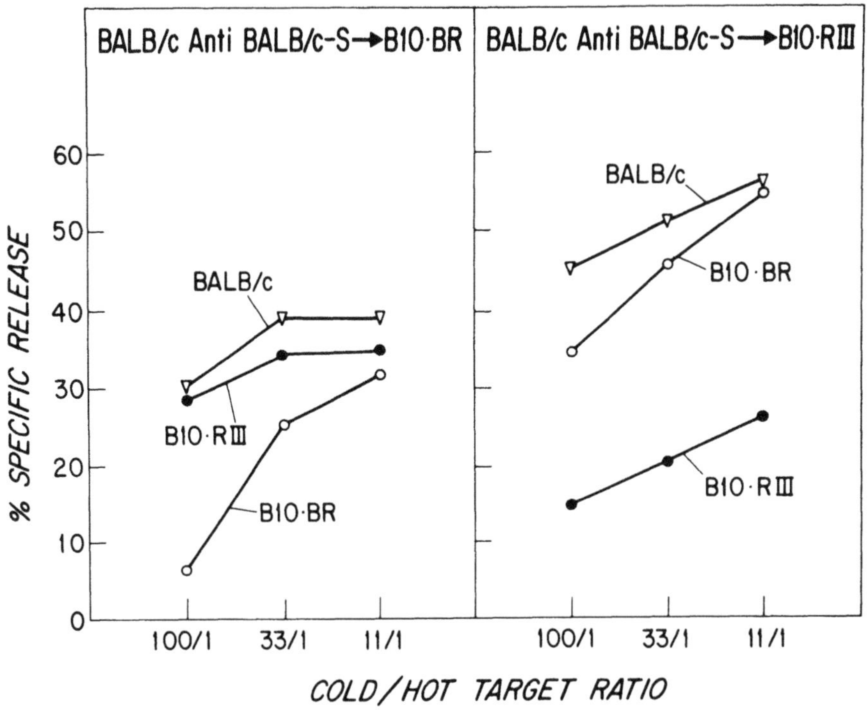

Figure 8. Mature cytolytic T cells specific for Sendai infected syngeneic target cells crossreactively lyse noninfected allogeneic target cells. Using cold target inhibition of lysis by CTL, this experiment illustrates that distinct populations of Sendai specific CTL lyse different allogeneic targets. Adapted from Finberg, R. S., Burakoff, S. J., Cantor, H., and Benacerraf, B., in Proc. Natl. Acad. Sci. (USA) *75*, 5145 (1978).

phages without antigen. In both cases the ability to stimulate mapped in the I-A subregion demonstrating that alloreactivity to I region antigens also arises as a consequence of the commitment of T cells to autologous MHC specificities.

The remaining issue concerns the precise nature of the T cell receptor and how the specificity for foreign antigens and MHC coded molecules is concomi-

Table 12. B10.A Clones Specific for Antigen and Self Ia May Also Be Selectively Alloreactive

Source of spleen cells	H-2	Proliferative response (CPM)	
		Clone 5	Subclone 5.6
B10.A	a	150	180
B10.A + DNP-OVA	a	14,700	39,200
B10	b	130	233
B10.D2	d	236	410
B10.S	s	15,900	36,300

Note:
A colony of DNP-OVA specific proliferating T lymphocytes was derived from lymph node cells of a B10.A mouse immunized with DNP-OVA. A cloned line was derived which was subcloned. Such cells at a concentration of 10^4 or 2×10^4 show reactivity both to DNP-OVA or B10.A cells and to B10.S without antigen.

Adapted from Sredni, B. and Schwartz, R. H., Immunol. Rev., *54*, 187, 1981.

tantly maintained. On the basis of idiotypic and genetic evidence, the variable regions of immunoglobulin heavy chains appears to be responsible for both the specificity directed to MHC and the specificity for foreign determinants, when analyzed independently. The problem still remains whether T cells have one receptor or two coupled receptors and whether one or two $V_H$regions are involved. Moreover, the significance of I region coded determinants on antigen-specific, idiotype bearing regulatory products on T cells must be clarified. I am not tempted to guess at the answer considering the present availability of cloned lines of specific T cells, and of T cell hybrids. A definitive answer should be forthcoming from the laboratory, and the genes coding for the T cell receptors will soon be identified.

Conclusions

The evolutionary significance of the commitment of T cells to MHC antigens should be assessed from several vantage points. From the point of view of the individual concerned, the existence of such a broadly polymorphic system to determine specific responsiveness and suppression will unescapably result in individuals with different immunological potential to a given challenge. Some will clearly be at greater risk, whereas others will be better prepared to resist to certain infectious agents, and it is not surprising that immunological diseases are linked to the MHC. As far as the species is concerned, this polymorphic defense system results in a very significant survival advantage to unforeseen challenges and a better possibility for the immune system to adapt to evolutionary pressures.

As biologists we contemplate with admiration and awe the wondrous array of sophisticated cell interactions and recognitions evolved in the T cell immune system which must be a model for other similarly complex biological systems of highly differentiated organisms.

ACKNOWLEDGEMENTS

I would like to express my deep appreciation and my profound affection for my numerous students and associates who have shared with me in the toil and should now equally share in the honor. I am also deeply grateful to the National Institute of Allergy and Infectious Diseases which supported my work faithfully and generously since 1957, and housed me from 1968 to 1970, and to the National Cancer Institute which supported my program for the last ten years. I owe also a debt of gratitude to New York University Medical School and to Harvard Medical School which provided the stimulating academic environment without which our work could not have progressed.

REFERENCES

1. Ehrlich, P. Proc. Roy. Soc. B. 66, 424, 1900.
2. Gowans, Y. L., McGregor, D. D., and Cowen, D. M. Nature 196, 651, 1962.
3. Burnet, F. M. The clonal selection theory of acquired immunity. Cambridge University Press, 1959.
4. Nossal, G. J. V. Brit. J. Exp. Path. 41, 89, 1960.
5. Green, I., Vassalli, P., Nussenzweig, V., and Benacerraf, B. J. Exp. Med. 125, 511, 1967.
6. Kohler, G., and Milstein, C. Eur. J. Immunol. 6, 511, 1976.
7. Miller, J. F. A. P., and Mitchell, G. F. J. Exp. Med. 131, 675, 1970.
8. Hood, L., Loh, E., Hubert, J., Barstad, P., Eaton, B., Early, P., Fuhrman, J., Johnson, N., Kronenberg, M., and Schilling, J. Cold Spring Harbor Symp. Quant. Biol. 41, 817, 1976.
9. Kabat, E. A., Wu, T. T., Bilowsky, H. U.S. Dept. of H.E.W., Public Health Service, National Institutes of Health Publication #80–2008, 1979.
10. Tonegawa, S., Maxam, A. M., Tizard, R., Bernard, O., and Gilbert, W. Proc. Nat. Acad. Sci. USA 75, 1485, 1978.
11. Seidman, J. G., Leder, A., Edgell, M. H., Polsky, F., Tilghman, S. M., Tiemeier, D. C., and Leder, P. Proc. Natl. Acad. Sci. USA 75, 3881, 1978.
12. Miller, J. F. A. P. Nature (London) 195, 1318, 1962.
13. Good, R. A., Dalmasso, A. P., Martinez, C., Hicher, O. K., Pierce, J. C., and Papermaster, B. W. J. Exp. Med. 116, 773, 1962.
14. Warner, N. L., Szenberg, A., and Burnet, F. M. Austr. J. Exp. Biol. Med. Sci. 40, 373, 1962.
15. Mitchison, N. A. Eur. J. Immunol. 1, 18, 1971.
16. Rajewsky, K., Schirrmacher, B., Nase, S., and Jerne, N. K. J. Exp. Med. 129, 1131, 1969.
17. Katz, D. H., and Benacerraf, B. Adv. Immunol. 15, 1, 1972.
18. Gershon, R. K. Contemp. Top. Immunobiol. 3, 1, 1974.
19. Kapp, J. A., Pierce, C. W., Schlossman, S., and Benacerraf, B. J. Exp. Med. 140, 648, 1974.
20. Unanue, E. R., and Askonas, B. A. J. Exp. Med. 127, 915, 1968.
21. Landsteiner, K. The Specificity of Serological Reactions. Harvard University Press, Cambridge, 1945.
22. Gell, P. G. H., and Benacerraf, B. J. Exp. Med. 113, 571, 1961.
23. Paul, W. E., and Benacerraf, B. Science 195, 1293, 1977.
24. Gell, P. G. H., and Benacerraf, B. Immunol. 2, 64, 1959.
25. Schirrmacher, V., and Wigzell, H. J. Exp. Med. 113, 1635, 1974.
26. Ishizaka, K., Kishimoto, T., Delespesse, G., and King, T. P. J. Immunol. 113, 70, 1974.
27. Katchalski, E., and Sela, M. Adv. Protein Chem. 13, 243, 1958.
28. Levine, B. B., Ojeda, A., and Benacerraf, B. J. Exp. Med. 118, 953, 1963.
29. Bluestein, H. G., Green, I., and Benacerraf, B. J. Exp. Med. 134, 458, 1971.
30. Lieberman, R., Paul, W. E., Humphrey, W., and Stimpfling, J. H. J. Exp. Med. 136, 1231, 1972.
31. Vaz, N. M., de Souza, C. M., and Maia, L. C. S. Int. Arch. Allergy Appl. Immunol. 46, 275, 1974.
32. Green, I., and Benacerraf, B. J. Immunol. 107, 374, 1971.
33. McDevitt, H. O., and Sela, M. J. Exp. Med. 122, 517, 1965.
34. Martin, W. J., Maurer, P. H., and Benacerraf, B. J. Immunol. 107, 715, 1971.
35. Gunther, E., Rude, E., and Stark. O. Eur. J. Immunol. 2, 151, 1972.
36. Amerding, A., Katz, D., and Benacerraf, B. Immunogenetics 4, 340, 1974.
37. Dorf, M. E., Balner, H., and Benacerraf, B. J. Exp. Med. 142, 673, 1975.
38. McDevitt, H. O., and Chinitz, A. Science 163, 1207, 1969.
39. Benacerraf, B., and McDevitt, H. O. Science 21, 273, 1972.
40. Benacerraf, B., and Katz, D. H. Adv. Canc. Res. 21, 121, 1975.
41. Ellman, L., Green, I., Martin, W. J., and Benacerraf, B. Prof. Natl. Acad. Sci. USA 66, 322, 1970.
42. Snell, G. D. J. Genet. 49, 87, 1948.

43. McDevitt, H. O., Deak, B. D., Shreffler, D. C., Klein, J., Stimpfling, J. H., and Snell, G. D. J. Exp. Med. 135, 1259, 1972.
44. Dorf, M. E., and Benacerraf, B. Prof. Natl. Acad. Sci. 72, 3671, 1975.
45. Dorf, M. E. Springer Sem. Immunopath. 1, 171, 1978.
46. Shreffler, D. C., and David, C. S. Adv. Immunol. 20, 125, 1974.
47. Klein, J., and Hauptfeld, B. Transplant. Rev. 30, 83, 1976.
48. McDevitt, H. O., Delovitch, T. L., Press, J. L., and Murphy, D. B. Transplant. Rev. 30, 197, 1976.
49. Schwartz, B. D., Paul, W. E., and Shevach, E. M. Transplant. Rev. 30, 174, 1976.
50. Dorf, M. E., and Unanue, E. R. In: Ir Genes and Ia Antigens, H. O. McDevitt, ed., Academic Press, New York, p. 171, 1978.
51. Cullen, S. E., David, S. C., Shreffler, D. C., and Nathenson, S. G. Proc. Natl. Acad. Sci. USA 71, 648, 1975.
52. Cullen, S. E., Freed, J. H., and Nathenson, S. G. Transplant. Rev. 30, 236, 1976.
53. Schwartz, B. D., and Cullen, S. E. In: The Role of the Histocompatibility Gene Complex in Immune Response. D. H. Katz and B. Benacerraf, eds., Academic Press, New York, 1976.
54. Springer, T. A., Kaufman, J. F., Siddoway, L. A., Mann, D. L., and Strominger, J. L. J. Biol. Chem. 252, 6201, 1977.
55. Niederhuber, J. E., and Frielinger, J. A. Transplant. Rev. 30, 101, 1975.
56 Meo, T., David, C. S., Rijnbeek, A. M., Nabholz. M., Miggiano, V. C., and Shreffler, D. C. Transplant. Rev. 7, 127, 1975.
57. Livnat, S., Klein, J., and Bach, F. H. Nature 243, 42, 1973.
58. Klein, J., Geib, R., Chiang. C., and Hauptfeld, V. J. Exp. Med. 143, 1439, 1976.
59. Levine, B. B., and Benacerraf, B. Science 147, 517, 1965.
60. Green, I., Paul, W. E., and Benacerraf, B. J. Exp. Med. 123, 859, 1966.
61. Dunham, E. K., Unanue, E. R., and Benacerraf, B. J. Exp. Med. 136, 403, 1972.
62. Shevach, E. M., and Rosenthal, A. S. J. Exp. Med. 138, 1213, 1973.
63. Sredni, B., Matis, L. A., Lerner E. A., Paul, W. E., and Schwartz, R. H. J. Exp. Med. 153, 677, 1981.
64. Germain, R. N., and Springer, T. unpublished data.
65. Shevach, E. M., Paul, W. E., and Green, I. J. Exp. Med. 136, 1207, 1972.
66. Schwartz, R. H., David, C. S., Dorf, M. E., Benacerraf, B., and Paul, W. E. Proc. Natl. Acad. Sci. 75, 2387, 1978.
67. Barcinski, M. A., and Rosenthal, A. S. J. Exp. Med. 145, 726, 1977.
68. Rosenthal, A. S., Lin, C. S., Hansen, T., Thomas, J. W., Danho, W., Ballesbach and Fohles, J. In: Proceedings of Basic and Clinical Aspects of Immunity of Insulin International Workshop, K. Keck and P. Erb. eds., in press, 1981.
69. Berzofsky, J. A., Schecter, A. N., Shearer, G. M., and Sachs, D. H. J. Exp. Med. 145, 123, 1977.
70. Kipps, T. J., Benacerraf, B., and Dorf, M. E. Eur. J. Immunol. 8, 415, 1978.
71. Green, I., Paul, W. E., and Benacerraf, B. J. Exp. Med. 126, 959, 1967.
72. Miller, J. F. A. P., Vadas, M. A., Whitelaw, A., and Gamble, J. Proc. Natl. Acad. Sci. 73, 2486, 1976.
73. Miller, J. F. A. P., Gamble, J., Mottram, P., and Smith, F. I. Scand. J. Immunol. 9, 29, 1979.
74. Katz, D. H., Hamaoka, T., Dorf, M. E., and Benacerraf, B. Proc. Natl. Acad. Sci. 70, 2624, 1973.
75. Kindred, B., and Shreffler, D. C. J. Immunol. 109, 940, 1972.
76. Katz, D. H., Hamaoka, T., Dorf, M. E., Maurer, P. H., and Benacerraf, B. J. Exp. Med. 138, 734, 1973.
77. Sprent, J. Immunol. Rev. 42, 108, 1978.
78. Kappler, J. W., and Marrack, P. J. Exp. Med. 148, 1510, 1978.
79. Singer, J., Hathcock, K. S., and Hodes, R. J. J. Exp. Med. 149, 1208, 1979.
80. Jones, P. P., Murphy, D. B., and McDevitt, H. O. J. Exp. Med. 148, 925, 1978.
81. Cook, R. G., Vitetta, E. S., Uhr, J. W., and Capra, J. D. J. Exp. Med. 149, 981, 1979.

82. Silver, J., Russell, W. A., Reis, B. L., and Frelinger, J. A. Proc. Natl. Acad. Sci. 74, 5131, 1977.
83. Lerner, E. A., Matis, L. A., Janeway, C. A., Jr., Jones, P. P., Schwartz, R. H., and Murphy, D. B. J. Exp. Med. 152, 1085, 1980.
84. Kimoto, M., and Fathman, C. G. J. Exp. Med. 152, 759, 1980.
85. Cantor, H., and Boyse, E. A. J. Exp. Med. 141, 1390, 1975.
86. Binz, H., and Wigzell, H. J. Exp. Med. 142, 197, 1975.
87. Eichmann, K. Adv. Immunol. 26, 195, 1978.
88. Greene, M. I., Bach, B. A., and Benacerraf, B. J. Exp. Med. 149, 1069, 1979.
89. Germain, R. N., Ju, S-T., Kipps, T. J., Benacerraf, B., and Dorf, M. E. J. Exp. Med. 149, 613, 1979.
90. Weinberger, J. Z., Greene, M. I., Benacerraf, B., and Dorf, M. E. J. Exp. Med. 149, 1336, 1979.
91. Longo, D. L., and Schwartz, R. H. Fed. Proc. 39, 1127, 1980.
92. von Boehmer, H., Haas, W., and Jerne, N. K. Proc. Natl. Acad. Sci. 75, 2439, 1978.
93. Rosenthal, A. S. Immunol. Rev. 40, 135, 1978.
94. Benacerraf, B. J. Immunol. 120, 1809, 1978.
95. Schlossman, S. F. Transplant. Rev. 10, 97, 1972.
96. Yaron, A., Dunham, E. K., and Schlossman, S. F. Biochemistry 13, 347, 1974.
97. Kapp, J. A., Pierce, C. W., and Benacerraf, B. J. Exp. Med. 140, 172, 1974.
98. Debré, P., Waltenbaugh, C., Dorf, M. E., and Benacerraf, B. J. Exp. Med. 144, 272, 1976.
99. Adorini, L., Harvey, M. A., Miller, A., and Sercarz, E. E. J. Exp. Med. 150, 293, 1979.
100. Schwartz, M., Waltenbaugh, C., Dorf, M., Cesla, R., Sela, M., and Benacerraf, B. Proc. Natl. Acad. Sci. 73, 2862, 1976.
101. Murphy, D. B., Herzenberg, L. A., Okumura, K., Herzenberg, L. A., and McDevitt, H. O. J. Exp. Med. 144, 699, 1976.
102. Tada, T., Taniguchi, M., and David, C. S. J. Exp. Med. 144, 713, 1976.
103. Tada, T., Taniguchi, M., and David, C. S. Cold Spring Harbor Symp. Quant. Biol. 41, 119, 1976.
104. Tada, T., Taniguchi, M., Hayakawa, K., and Okumura, K. In: T and B Lymphocytes: Recognition and Function E. S. Vitetta and C. F. Fox, eds., Academic Press, New York p. 293, 1979.
105. Thèze, J., Kapp, J. A., and Benacerraf, B. J. Exp. Med. 145, 839, 1977.
106. Waltenbaugh, C., Thèze, J., Kapp, J. A., and Benacerraf, B. J. Exp. Med. 146, 970, 1977.
107. Zinkernagel, R. M., and Doherty, P. C. J. Exp. Med. 141, 1427, 1975.
108. Shearer, G. M., Rehn, T. G., and Garbarino, C. A. J. Exp. Med. 141, 1348, 1975.
109. Bevan, M. J. J. Exp. Med. 142, 1349, 1975.
110. Simpson, E., and Gordon, R. D. Immunol. Rev. 35, 59, 1977.
111. Zinkernagel, R. M., Althage, A., Cooper, S., Kreeb, G., Klein, P. A., Sefton, B., Flaherty, L., Stimpfling, J., Shreffler, D., and Klein, J. J. Exp. Med. 148, 592, 1978.
112. Gorer, P. A., Lyman, S., and Snell, G. D. Proc. Roy. Soc. Lond. [Biol.] 135, 499, 1948.
113. Jerne, N. K. Eur. J. Immunol. 1, 1, 1971.
114. Lemonnier, F., Burakoff, S. J., Germain, R. N., and Benacerraf, B. Proc. Natl. Acad. Sci. USA 74, 1229, 1977.
115. Burakoff, S. J., Finberg, R., Glimcher, L., Lemonnier, F., Benacerraf, B., and Cantor, H. J. Exp. Med. 148, 1414, 1978.
116. Burakoff, S. J., Ratnofsky, S. E., and Benacerraf, B. Proc. Natl. Acad. Sci. USA 74, 4572, 1977.
117. Zinkernagel, R. M., Althage, A., Cooper, S., Callahan, G., and Klein, J. J. Exp. Med. 148, 805, 1978.
118. Bevan, M. J. Nature 269, 417, 1977.
119. Finberg, R., Burakoff, S., Cantor, H., and Benacerraf, B. Proc. Natl. Acad. Sci. 75, 5145, 1978.
120. von Boehmer, H., Turton, K., and Haas, W. Eur. J. Immunol. 9, 592, 1979.
121. Sredni, B., and Schwartz, R. H. Immunol. Rev. 54, 187, 1981.

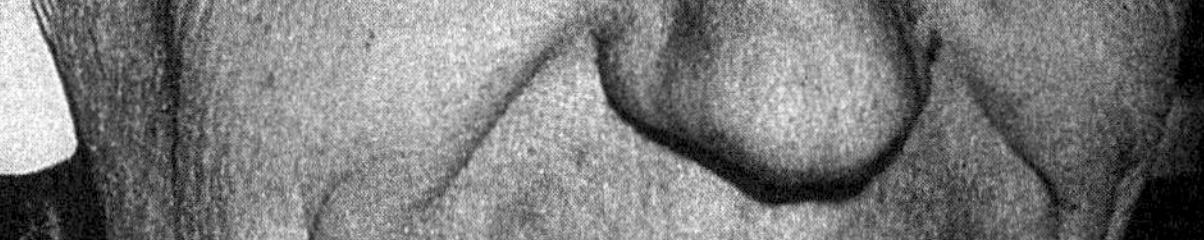

JEAN DAUSSET

His mother originated from Lorraine, his father from the Pyrénées, two French provinces very distant from one another and with vast cultural differences. His parents met in Paris. During the First World War, his father, a doctor and captain in the army, sent Jean Dausset's mother and the first three children to Toulouse. It was there that Jean Dausset was born, on 19th October 1916, and this region has held a strong attachment for him ever since.

After the war, his father worked as a physiotherapist and radiologist, dividing his time between Paris and the spa towns. Jean Dausset spent his early childhood in Biarritz, until the age of secondary school. Then, when he was 11 years old, his family came to settle permanently in Paris. He pursued his secondary studies at the Lycée Michelet and obtained his baccalaureate in mathematics.

His choice of career was almost dictated by that of his father, Henri Dausset, who pioneered Rheumatology in France. His medical studies progressed without incident until the advent of the Second World War, when they were interrupted. He was mobilized in 1939 and returned from the French Campaign in 1940 to a Paris occupied by the Germany Army. He began to devote his time ardently to the preparation of a competitive examination for the title of Intern of the Paris Hospitals. Upon receiving this title, he immediately left to join the fighting forces in North Africa. During the Tunisian Campaign, he performed blood transfusions in the army. This was his first introduction to immuno-haematology.

While training in Algiers, he performed his first laboratory experiments and carried out his first scientific study on blood platelets.

On his return in 1944, to a liberated Paris, he was given the responsibility for collection of blood samples in the Paris area, working from the Regional Blood Transfusion Centre at Hôpital Saint-Antoine.

As soon as the war was over, he undertook his first real research study, in collaboration with Professor Marcel Bessis. Professor Bessis had just developed exchange-transfusion in new-born babies and adults. It is impossible to say how much time he spent treating, with this method, women who had become anuric following abortion manoeuvres resulting in septicaemia due to Clostridium perfringens — this was his first contact with kidney failure!

His clinical years oriented towards haematology and pediatrics, with a constant attraction to the laboratory. In 1948, he was sent, as a French trainee, to the Children's Hospital in Boston (Professors L. K. Diamond and Sydney Farber) where he worked in one of the Harvard Medical School laboratories.

On his return to France, he took up work againat the regional Blood Transfusion Centre, where he immediately became interested in the new immuno-haematology techniques for red blood cells. He decided to transpose these techniques to white blood cells and platelets.

The principal obsrvation of leuco-agglutination and thrombo-agglutination was made in 1952. Since that time, he has retained a constant interest in the immunogenetics of blood cells.

In 1958, while Head of the Immuno-haematology Laboratory at the National Blood Transfusion Centre, he described the first leucocyte antigen, MAC, which has become known as HLA-A2.

Preoccupied with the state of medical research in France, he undertook, with Professor Robert Débre, to institute radical reforms in the hospital and university structures. This work as Advisor to the Cabinet of the National Ministry of Education spanned three consecutive years and culminated in the introduction of a law which established full-time employment in French hospitals, introducing to the hospitals Professors of Basic Sciences, who were given hospital responsibilities. This reform permitted a soar in French biology and brought a new lease of life to French medical research.

Despite the administrative struggles which ensued during this period, he never abandoned his laboratory work. In 1958, he was named Assistant Professor of Haematology at the Faculty of Medicine in Paris, then Professor of Haematology in 1963 and was appointed Head of the Immunology Department at Hôpital Saint-Louis. Again, he devoted his time entirely to research and, in 1965, described the first tissue group system (Hu-1, later named HLA).

Thanks to the admirable volunteer blood donors, skin donors and skin recipients, grafted under the care of Professor F. T. Rapaport, correlations were established between graft survival and tissue incompatibility.

He participated in the creation of the Research Institute in Blood Diseases, directed by Professor Jean Bernard, and was Assistant Director there until 1968. One of the departments under his direction was the Research Unit on Immunogenetics of Human Transplantation, an INSERM (National Institute of Health and Medical Research) unit of which he has been director since 1968.

In 1977, the Collège de France called him to the Chair of Experimental Medicine, a position held by Claude Bernard from 1958 to 1978, but his research laboratory remained at Hôpital Saint-Louis.

In 1963, he married Rose Mayoral from Madrid who gave him two children, Henri and Irène.

In addition to his scientific interests, he has only two passions in life: his family and modern plastic art.

N.B. This biography was written in 1980 when Jean Dausset received the Nobel Prize. Since that time, he has created, in Paris, the Human Polymorphism Study Center (Centre d'Etude du Polymorphisme Humain, CEPH) and set up an intensive international collaboration to establish a genetic map of the human genome.

Professor Honoris Causa:

University of Brussels	1977
University of Geneva	1977
University of Liège	1980

Membership of Academies:

Honorary foreign member of the Belgian Royal Academy of Medicine	1969
Académie des Sciences de l'Institut de France	1977
Académie de Médecine (Paris)	1977
Hon. Member, American Academy of Arts and Sciences (Boston)	1979
Hon. Member, Yugoslavian Academy of Arts and Sciences (Zagreb)	1979

Prizes

Grand Prix des Sciences Chimiques et Naturelles (Académie des Sciences)	1967
Médaille d'Argent du Centre National de la Recherche Scientifique	1967
Grand Prix Scientifique de la Ville de Paris	1968
Prix Cognac-Jay (Académie des Sciences de l'Institut de France)	1969
Stratton Lecture Award (U.S.A.)	1970
Landsteiner Award AABB, San Francisco (U.S.A.)	1970
Gairdner Foundation Prize (Canada)	1977
Koch Foundation Prize (Germany)	1978
Wolf Foundation Prize (Israel)	1978

THE MAJOR HISTOCOMPATIBILITY COMPLEX IN MAN — PAST, PRESENT, AND FUTURE CONCEPTS,

Nobel Lecture, 8 December, 1980

by

JEAN DAUSSET

University of Paris VII

Institute of Research in Blood Diseases, Paris, France

As George Snell (*1*) so rightly said, the supergene, the major histocompatibility complex (MHC), is like a page from the nature book read outside of the context.

Today this context is beginning to be better understood. We would here like to recall the evolution of concepts regarding these molecular structures found in the membrane of cells. First, attention was centered on the almost botanical description of their genetic polymorphism. Then the spotlight was turned, for several years, on their importance in transplantation. More recently, their role in the immune response has become more and more apparent. This, however, is probably not the last stage in our search. We, as well as others (*2–5*), have suggested that the essential function of these structures resides in self-recognition. These structures are, in fact, the identity card of the entire organism.

We will discuss these four viewpoints successively. Far from being mutually exclusive, they are landmarks in the stages of our thought process as we have gained deeper knowledge of the subject.

First concept: Polymorphism and Linkage Disequilibrium

Polymorphism: Since Landsteiner's discovery of the first genetic polymorphism in man, knowledge of polymorphic genes has not ceased to increase and will continue to increase with DNA hybridization techniques. Most of these systems, however, are pauci-allelic and more often than not have one very frequent allele, one that is more infrequent, and a few variants. None of these can be compared with the extreme polymorphism of genes in the MHC of vertebrates, and particularly in the human lymphocyte antigen (HLA) complex

The definition of this polymorphism began to emerge in three laboratories: in ours where the first antigen Mac (HLA-A2) was defined (*6*), in Van Rood's laboratory (*7*) with the 4a 4b series (Bw4, Bw6), and Rose Payne's and W. Bodmer's (*8*) with the two alleles HLA-A2 and -A3. Then, thanks to an intense international effort that has spanned more than 15 years and included eight workshops, the web began to be disentangled. The importance of this international effort, launched by Amos in 1964 [see (*9*)], and followed by other workshops directed by Van Rood (*10*), Ceppellini (*11*), Terasaki (*12*), Dausset (*13*), Kissmeyer-Nielsen (*14*), Bodmer (*15*), and again Terasaki (*16*), cannot be over-emphasized and is to the credit of the whole histocompatibility community. The four presently well-defined, closely linked loci, HLA-A, HLA-B, HLA-C, and HLA-D/DR, have each from 8 to 39 codominant alleles, and the number of haplotypical or genotypical combinations already amounts to several million. It is very likely that other closely linked polyallelic loci will be discovered, similar, for example, to the various loci in the I region of the mouse. If one adds to this complexity the polymorphism of other genes in the HLA region, coding, for example, for factors C2, $C4^S$, $C4^F$, and B^f of complement, one reaches such levels of complexity that virtually every human has a different gene combination. If one considers all the genes of the human genome, it can be said that there is not and will never be on earth, apart from true twins, two identical people: every person is unique.

A question that immediately comes to mind is: Why is the MHC so complex? It is clear that a particular pressure was exerted on these genes to make them different and to maintain this differentiation. If it is true that these structures play a role in self-recognition and that they derive from primitive genes coding for surface molecules, then one can conceive of this diversity quickly becoming a necessity when living matter passed from the unicellular stage — or from a syncytium of identical cells able to fuse without harmful consequences — to the organized multicellular organism whose tissues must coexist and even cooperate and whose cells therefore cannot even merge with the cells of an organism of the same species.

Maintenance of this polymorphism is undoubtedly aided by the selective advantage given to the heterozygotes, possibly through the immune functions attributed to the MHC molecules in a subsequent stage of evolution.

The HLA system is now known to have two types of products that are very different from each other (their single nomenclature sometimes obscures this difference).

The products of the HLA-A, -B, and -C loci [Klein's class I (*17*)] are ubiquitous, being present at the surface of all (or almost all) cells of the organism. This wide distribution would suggest that they play a *very general* biological role.

In contrast, the products of the D/DR locus — and probably of the "future" DR loci (class II) — exist only at the surface of certain specialized cells, essentially immunocompetent cells, a valuable piece of information with respect to their functions.

We must not neglect another valuable piece of information afforded by the major similarities between the class I products and immunoglobulins: the light chain (the b2-microglobulin) as well as one of the domains (a3) of the heavy chain have significant similarities with the immunoglobulins and therefore suggest the possibility of a common ancestral gene (*18, 19*).

The light and heavy chains of the class II products bear no similarity either with the class I products or with the immunoglobulins. It can be said, therefore, that the products of the HLA complex are *bipolar* and are probably derived, by duplication and successive mutation, from two very distinct genes, the function and origin of which go far back in the evolution of the species.

At present, at least three variable regions on the heavy chain of the class I (*20*) products are known. In the most distal domain (a1) is the main variability zone (between amino acids 60 to 80), which probably corresponds to the serologically defined allelic epitope (individual antigenic determinant). This same domain also has (between amino acids 30 to 40) an apparently variable zone that is responsible for interaction with the influenza virus. In the median domain (a2) is a third variable zone (between amino acids 105 to 114).

The extreme frequency of cross-reactions between the various allelic molecules of each HLA locus is well known. Two interpretations that are not mutually exclusive are possible: the similarity in the structure of the allelic epitopes [see Colombani *et al.* (*21*)] or the existence of determinants common to two molecules but different from the epitopes; determinants which I, and Ivanyi (*22*), have called "antigenic factors" (also known as supertypical antigens or public antigens).

According to our hypothesis, the molecules having an identical epitope—let us say, for example, A2—are not identical in their composition. Some may have one or more different antigenic factors. This variability may be found in the same population but is more often found in different populations (*23*).

This concept suggests that the various parts of an HLA molecule might have different functions (as is the case for the different portions of the immunoglobulin molecule). It has recently been shown that the interaction between the HLA molecule and the influenza virus does not take place at the level of the serologically distinguishable determinant since this virus has a different interaction with molecules, which, nevertheless, have a serologically identical A2 determinant (*24*).

The same hypothesis could be applied to the DR molecules and could perhaps make it possible to solve the problem of the relations between the D and DR series. In effect, D might be no more than a variable part of the DR molecule having a stimulating function, since disassociated haplotypes do appear to exist—that is, where the determinant D is not the determinant usually found with the DR antigen (*16, 25*).

A second possibility, which should not be excluded, is that D in itself does not exist but is defined by an average of allostimulation due to a certain combination of alleles in the loci of region D, the linkage disequilibrium involving not only the DR locus but other loci as well, equivalent to IA, IJ, IC, IE of the mouse.

A second series of DR molecules is already in the process of being defined both by serological and cellular procedures. In particular an allelic SB series, centrometric in relation to the DR locus, has just been described (*26, 27*) through the use of mixed secondary lymphocytic cultures.

With regard to the genetic organization, we cannot yet grasp this in precise terms, but with the aid of modern DNA hybridization techniques it will not be long before we do understand it (*28*).

Linkage disequilibrium. The four loci of the HLA complex are closely linked on the short arm of chromosome 6 (21p). They are, however, sufficiently distant for relatively frequent recombinations to occur (0.8 percent between A and B and 1 percent between B and D/DR in man). This special situation seems presently to be virtually unique to human genetics. It is, moreover, accompanied by a particular phenomenon, which is the preferential gametic association between alleles of several loci of the same complex. A linkage disequilibrium is said to exist between these alleles. The phenomenon has given rise to numerous speculations.

Is this merely reminiscent of ancestral combinations (when populations were isolated some 2000 to 4000 years ago) that were revealed or increased by human migration? Could a mixture of populations temporarily set a certain haplotypical formula that would survive for only so long as was necessary for it to be dispersed through segregations occurring in successive generations (*29*)? Or is this linkage disequilibrium really a preferential association, that is to say, selected for the biological advantage or advantages it confers in a certain environment?

Of course, these two mechanisms can operate simultaneously.

One might ask whether the feature of linkage disequilibrium is specific to the MHC or whether it is a very general feature that recurs at other points of the genome.

HLA polymorphism and its linkage disequilibrium are valuable tools for anthropologists and epidemiologists. They allow the former to characterize a population; to discern its origin and draw up its genetic history. They allow the latter to compare HLA formulas and HLA haplotypes with the particular susceptibility of a population or groups of populations to certain diseases; and perhaps in the future they will be able to reconstitute major diseases and epidemics that have occurred in the past by observing the selections that have operated.

Finally, in formal genetics, the HLA complex is certainly the segment of the human genome that is best known and is a major example of our relating a human product to the sequence of the corresponding gene.

Second Concept: Transplantation Antigens

It is in such terms that these membrane structures are most often defined, because at the same time that genetic polymorphism was being elucidated, transplantation in humans was assuming great importance in therapeutics.

Our understanding of allogenic response in man has evolved rapidly. When only the HLA-A and -B antigens were known, allogenice response amounted to cytotoxicity on targets A and B. Thanks to admirable volunteers, the correlation between the survival of skin grafts and the number of HLA-A and -B incompatibilities was clearly demonstrated (*30–33*). The same correlation was seen in recipients of related or non-related donor kidneys. This correlation, which was long debated, is no longer challenged; however, the benefit of compatibility limited to these two loci is very variable depending on the categories of patients.

When locus D was discovered (*34, 35*), its importance in transplantation was immediately suspected (*36, 37*). In fact, it was shown in vitro that lymphocytic proliferation in allogenic culture was only possible when there was incompatibility at the D locus (*38*). Clinically, a correlation has been found between the intensity of proliferation during the mixed lymphocytic reaction, that is, between the recipient and the related donor, and the survival of the graft. However, it has not been possible to apply this observation to the transplantation of nonrelated organs because of the time required for its elaboration.

In contrast, as soon as DR antigens (*14, 15*) could be detected by serological means—and thus rapidly—it became possible to use this new method of selection successfully. A DR incompatibility is accompanied by a drop in the survival rate of grafts, both skin (*39, 40*) and organ transplants (*41, 42*). In both cases there is a clear additive effect with those of the A and B incompatibilities (Fig. 1).

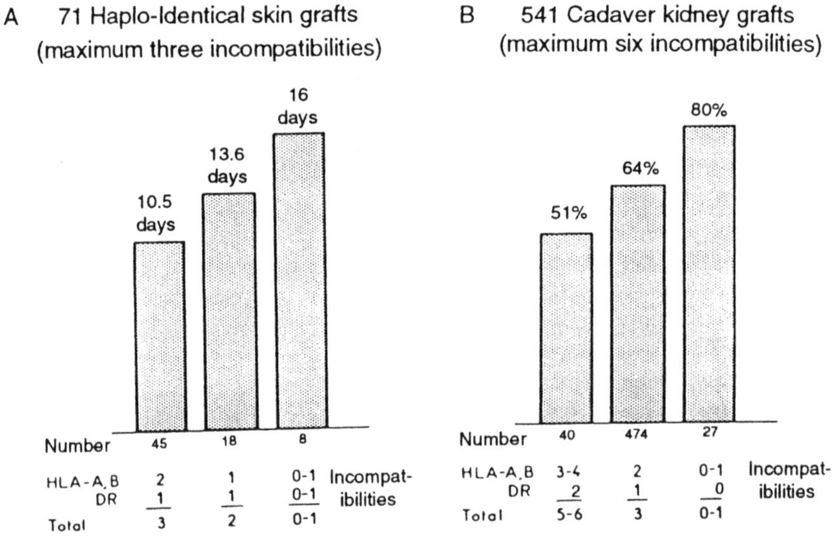

Fig. 1. Additive effect of HLA-A, -B and -DR incompatibilities. (A) Survival time (in days) of skin grafts done between HLA-haplo-identical (most often child to father). Survival of the grafts increases progressively as the total number of HLA-A, -B plus -DR incompatibilities diminshes (30, 39). (B) Percentage survival (at 2 years) of kidney transplant (done in the France-Transplant Network). The same tendency is observed, that is, survival is much improved where there are fewer total HLA-A, -B plus -DR incompatibilities.

Incompatibility is in most cases both DR and D, and thus has two consequences:

1) With DR incompatibility it provides a *target* for cytotoxic cells. DR antigens are true targets: they do not behave like minor antigens because they do not need the HLA-A and -B identity between the killer cell and the target (*43*).

2) It induces, by the D disparity, the appearance of auxiliary cells, some of which are helper cells and others suppressor cells. In the normal state, helper cells dominate suppressor cells. However, it must be made clear that in certain circumstances, the suppressor cells dominate the helper cells and the scale then tips in favor of tolerance.

It is possible that the indisputably beneficial effect of preoperative transfusions is due to the development of suppressor cells or factors in the recipient (*44*). In fact, with Sasportes and others (*45, 46*) we have shown that hyperimmunization against DR is accompanied by the in vitro appearance of a suppressive factor capable of producing a specific feedback inhibition on its own cells. The exact circumstances that cause the suppression in vivo to sometimes dominate immunity are unknown. However, it is known that in the monkey the beneficial effect of transfusions has been observed only where the animals are also immuno-suppressed (*47*). Recipients of kidneys are, so to speak, always immunosuppressed due to their renal insufficiency; this would explain why, in the course of transfusions, the immune balance leans in favor of suppression.

On the basis of the preceding considerations we have proposed (*48*) a theoretical plan, of necessity provisional, for the choice of blood donors for transfusion and organ donors. Without going into detail, it is based on the following principles:

1) Before transplantation, a state of tolerance must be developed in the recipient and at the same time immunization against HLA-A and B-targets must be avoided. Thus, transfusions should be made with DR incompatible blood that is A, B compatible. The same DR incompatibility should be used constantly in order to increase the chances of the appearance of suppressive cells and factors of the allogenic proliferation.

2) In selecting the organ donor one must avoid providing targets against which the recipient could be immunized. Thus, priority should be given to HLA-DR compatibility in patients who have not produced antibodies against HLA-A and -B antigens in the coarse of transfusions (nonresponder recipients). On the other hand, priority should be given to HLA-A and -B compatibility in those who have been sensitized in the course of preoperative transfusions (responder recipients). Indeed, for the former there is little chance of immunization occurring against A or B antigens, but the appearance of helper cells and a supply of DR targets should be avoided. Conversely, in the responders, helper cells are already present and their action must be neutralized by not contributing incompatible A and B targets.

Very precise and detailed treatment protocols will be necessary to verify or disprove the validity of this plan.

Third Concept: Role in the Immune Response

This third concept is essentially based on our knowledge of animals since systematic experiments are ethically difficult and thus rare in man (*49*). Nonetheless, to date, the parallel with the H-2 complex is striking. Here again we find the *bipolar* division of the functions of the products of the HLA complex:

1) Class I products appear to serve as targets when a cell is either infected by a virus or covered with a hapten.

2) Class II products appear to serve as a regulator between the various cell subgroups involved in the immune response.

In both cases, a phenomenon of restriction is most often observed, that is to say, an identity with class I or II products is apparently necessary between the cooperating cells.

Thus a phenomenon of restriction exists in the cytotoxicity of a killer T lymphocyte cell against a cell carrying a virus (*50, 51*), a hapten (*52*) such as the

DNP, or a normal antigen such as the H-Y antigen (*53*). The killer cell must have at least one identity with the HLA-A and -B (class I) antigens of the target cell in order for the lysis to be effective, or else the killer must have matured in the presence of the histocompatibility antigens of the target.

Similarly, when an antigen such as PPD is presented by human macrophages (*54*), the presence of a DR identity (class II) is apparently necessary for the presentation to be effective and for lymphocytic proliferation to occur. This restriction is not absolute, however, and a certain number of proliferative reactions that can be explained by cross-reactions between DR antigens has been observed.

As in the mouse, where soluble factors carry antigens from the I region that convey a specific message to another T or B cell population, so in man there is evidence of a certain number of soluble factors of this type (*55, 56*). Undoubtedly, when we have a better understanding of the various products of region D in man, numerous specific and aspecific factors, either restricted or unrestricted, will be described.

The restriction phenomenon is probably the most direct proof of the role of the products of the HLA complex in the immune response of man.

Indirect proof has been sought in the numerous associations between HLA and diseases. Based on the murine model, the first study of associations between HLA and disease was done in our laboratory on acute lymphoblastic leukemia (*57*). A slight but definite increase of A2 has now been demonstrated in numerous worldwide studies (*58*). Likewise, the A1 antigen is slightly increased in Hodgkin's disease. These two observations would suggest that a gene acting on hematopoiesis may exist near locus A (*58*).

However, most of the diseases indisputably associated with HLA are neither tumorous nor obviously infectious (*59*). These are chronic or subacute diseases having a definite familial though mild character, that are of unknown etiology and are not included in any of the major classifications. For a good number of these there is an obvious autoimmune component.

We will give only two examples to illustrate once again the bipolar nature of the functions of HLA products. In the first example, class I products may still be considered as possible targets; in the second example, class II products may be considered as regulators of the immune response.

Articular and more especially sacropelvic disorders that are strongly associated with B27 seem to require the molecule HLA-B itself to play an essential role. In fact, the same B27 antigen is found in a series of disorders (Reiter's syndrome, ankylosis spondylarthritis, psoriatic rheumatism) which tend to affect the articulations of the sacrum and pelvis. Further, this same predisposition is found in all populations of the globe. However, it has now been clearly demonstrated that the same pathological manifestations can also affect a small number of individuals who are B27 negative. The B27 epitope is therefore not indispensable. At least two hypotheses may be advanced: either the responsible gene in all populations is strongly linked with the B27 antigen, or molecule B27 has a variable part (an antigenic factor) that is responsible for susceptibility; this antigenic factor would not always be present on all B27 molecules and could be found on other HLA molecules probably having cross-reactions with B27 (in keeping with our concept explained above).

It seems, moreover, that for these diseases there is a factor that triggers infection. In fact, it is known that some acute intestinal infections caused by Gram-negative bacteria such *Shigella, Salmonella,* and *Yersinia* are complicated by ankylosing spondylarthritis mainly in patients who are B27 positive. Recently, a direct relationship was suggested between the B27 antigen and a type of *Klebsiella* (*Klebsiella* B43). The antibodies against *Klebsiella* would be capable of recognizing specifically an antigen present on the lymphocytes of B27 positive patients affected by ankylosing spondylarthritis. Further, lysates from this infectious agent would be capable of transforming lymphocytes in B27 normal individuals and of making them sensitive to the antibodies against *Klebsiella* (*60, 61*). Although clinically the link between a *Klebsiella* infection and ankylosing spondylarthritis is still unclear, this observation, not yet confirmed, suggests that certain infectious agents would be capable of modifying HLA antigens and of apparently making them similar to those in patients. It is not impossible that this type of mechanism may one day explain the linkages with other microorganisms referred to above. These microorganisms would be capable of modifying the B27 antigen and perhaps certain other HLA molecules as well (to take into account B27 negative patients), and of making them privileged targets for T-immune autolymphocytes. Although this hypothesis is enticing, it cannot yet explain the very special localization of lesions, since the B27 antigen, like all HLA antigens, is practically ubiquitous.

If we now consider disorders associated with HLA-D/DR, one is at the outset struck by the large number of them that are associated with Dw3/DR3 and, more especially in Caucasians, with the A1, B8, DR3 haplotype. For the most part, there are diseases with a strong autoimmune component and a low family penetrance, such as myasthenia gravis, Graves' disease, Addison's disease, Sjögren's syndrome, disseminated lupus erythematosus, and active chronic hepatitis. It appears that this haplotype, in strong linkage disequilibrium, has a gene or perhaps a series of genes that are conducive to autoimmunization.

Insulin-dependent juvenile diabetes (IDD), itself associated with A1, B8, DR3 and also with B18, DR3, and B15, DR4 is in this respect very instructive (*62, 63*). The viral etiology of IDD is highly suspected: experimental models of the disease do exist and specific observations in some cases in man incriminate the B4 Coxsackie virus. One can therefore infer that the virus has destroyed a certain number of islets of Langerhans and triggered a process of cellular autoimmunity where cytotoxic lymphocytes persist in the organism against antigens modified by or associated with a virus. The disease would thus be self-sustained. In this hypothesis the D region products would have been incapable of inducing an adequate immune reaction against the virus causing the disease. In contrast, individuals who are DR2 positive (most frequently A3, B7, DR2), who appear to be "protected" against IDD would have a more effective immune response against the responsible agent.

This is only a working hypothesis which would have the advantage of applying to other diseases associated with HLA-DR such as multiple sclerosis (DR2) and chronic polyarthritis (DR4) or juvenile rheumatism (DR5).

The reality, however, is certainly far more complex. In fact, in no case is the association complete with a DR antigen. This is generally explained by a linkage disequilibrium between a simple susceptability gene for the disease and a DR allele. Here too, however, one might think there are polymorphic parts to DR molecules other than the epitopes presently known and that these might have a particular immune function. Even better, it could be assumed that the interaction of two (or more than two) genes from the D region would be conducive to an adequate immune response. This interaction could take place in the *cis* position (*64*) between genes of the same haplotype (as in the interaction between $I\text{-}A^b$ and $I\text{-}E^b$ to form an I-E molecule in the mouse) or in the *trans* (*65*) position between two haplotypes (such as $I\text{-}A^k$ and $I\text{-}A^b$ complementation in the mouse).

As a corollary to these gene interactions we propose that each HLA haplotype, and especially those in linkage disequilibrium that are found most frequently in the numerous diseases associated with HLA, has its own gene configuration that confers on it a particular capacity for immune response, which may be favorable in certain environmental conditions and unfavorable in others (for example the A3, B7, DR2) haplotype gives a susceptibility to multiple sclerosis and protects against IDD). Thus, each HLA complex would be composed of a set of genes that have subtle interactions among one another (such as gene C2 with the two C4 genes) thereby giving them a specific identity in immunological terms.

Likewise, every individual possessing two HLA haplotypes has his or her own immunological capacity which is conferred on him by the two particular haplotypical formulas inherited from his two parents, but which is also the result of the genetic interaction or complementation between the these two complexes. Thus each individual has a personal immune response that makes him either susceptible or resistant to certain diseases. Here again each haplotypical combination may be beneficial or harmful depending on the type of challenge to which the individual is subjected.

In terms of the population, we can thus conceive that certain individuals are or were more exposed and thus are or were more easily eliminated than other individuals more resistant to past and present epidemic or endemic diseases. But it is not the same individuals who are susceptible or resistant to the different attacks; this is what makes the survival of a population possible, and thus the perpetuation of the human species.

Fourth Concept: Self-Recognition

MHC products are distributed on the surface of cells. Those of class I are virtually ubiquitous. Class II products may be found on immunocompetent cells but also on endothelial and other specialized cells. Their location suggests that they play a role in the social organization of cells of the same organism.

This assumption is strongly supported by the following observation: it appears to be necessary for MHC products to share an identity in order for cooperation to be etablished between two populations of cells in the same organism or in two different organisms. This is the restriction phenomenon which we discussed above, and which is valid for the two classes of products. Results confirming this apparent need for identity are accumulating very rapidly in both animals and in man and are no longer limited solely to immunocompetent cells. The same is true, for example, in the adhesion phenomenon between fibroblasts and especially in the "homing" phenomenon in ganglia. Degos and others (*66*) have observed that splenocytes injected intravenously must share an identity with the cells of the capillary endothelium with regard to class I products so that homing can occur. The identity of class II products does not intervene (Fig. 2).

	I								
K	A	B	J	E	C	S	G	D	Mean r values
0	0	0	0	0	0	0	0	0	0.23
•	•	•	•	•	•	•	•	•	0.92
•	0	0	0	0	0	0	0	0	
•	•	0	0	0	0	0	0	0	
•	•	•	•	•	0	0	0	0	0.76
•	•	•	•	•	•	•	0	0	
•	•	•	•	•	•	•	•	0	
0	0	0	0	0	0	0	0	•	
0	0	0	0	0	•	•	•	•	
0	0	0	0	•	•	•	•	•	0.86
0	0	0	•	•	•	•	•	•	
0	•	•	•	•	•	•	•	•	
0	•	•	•	•	0	0	0	•	
•	0	0	0	0	0	0	0	•	0.87
0	•	•	•	•	0	0	0	0	0.33

Fig. 2. Homing in lymph nodes according to identities (•) or differences (0). Labeled lymphocytes were injected intravenously into mice under different genetic conditions: allogenic (first line), syngenic (second line), congenic where the difference involves only one or several genes of the H-2 complex (all other lines). The average value of r (percentage of homing from which the control value has been deducted) is high wherever there is identity (•) with H-2D or H-2K; H-21 identity has no influence (*66*).

We are thus faced with a very general phenomenon with such a consistent record that it is difficult to escape the conclusion that the two types of molecules have a common function. They could serve as a recognition signal (recognizers) among cells of the same organism; a signal necessary but probably insufficient to permit effective cooperation between the two subpopulations of cells because it would be the same, at least at class I, for all cells of the organism (*2–5*).

The passive or negative discrimination of self implies that the cells "ignore" one another. This seems improbable in view of the cohesion of tissues and of their interaction. Without self-recognition, each specialized cell and each tissue would be isolated and incapable of surviving. These considerations thus suggest that self-recognition is an active phenomenon.

The subjacent mechanism of self-recognition is still unknown. At least three possibilities come to mind. The most orthodox is the complementarity between two different molecules. But one cannot exclude recognition by identity, whether that recognition takes place between two identical molecules or through a ligand. This fascinating problem has been discussed in detail elsewhere (*67*). Suffice it to mention here just two remarks in relation to complementarity:

- If a second molecule (receptor) existed with the same immunogenicity as the HLA determinant, a second allelic system as complex as the former would have been found. To date no such system has been found. However, one should remain aware of the weak immunogenicity of the idiotypes which could represent these receptors.
- If the receptor and the determinant were coded by two different genets, any mutation and selection of one should correspond to the mutation and selection of the other. This is unlikely. Here again, however, one might envisage, according to Jerne's theory, that each individual has all possible receptors, the appropriate receptor being selected in early life, perhaps in the thymus. This is probably what occurs, for example, in the growth of allophenic mice in which cells carrying different H-2 antigens coexist.

Whatever the mechanism, the fact remains that self-recognition is a general and active phenomenon of any cell that is at least partially, linked to the MHC. We suggest that class I products are responsible for individuality, for integrity, and perhaps for the general cohesion of the being and that class II products are an example of cellular cooperation, thanks to self-recognition at the level of the differentiated cells of the immune system, allowing the immune system to function harmoniously.

Future Prospects

Thus the increasingly deep understanding of the MHC in man opens up exhilarating prospects both in public health and in basic science.

With regard to organ transplantation, we do not feel that the choice of the most compatible donor will be the last word. On the contrary, our aim must be to provoke in the recipient a specific tolerance to incompatible donor antigens without at the same time diminishing his or her immunological defenses. It seems that with preoperative transfusions the way has been opened to this type of preparation. We must now attempt to unravel its detailed mechanism so that the method can be used more generally. This will be the objective of the years to come, and we have no doubt that it will be achieved.

Although organ and bone marrow transplantations mark a milestone and have already brought help to numerous patients, they should not be considered an end in themselves. Etiological treatment should progressively replace them.

The discovery of more than 50 diseases associated with or linked to HLA is perhaps still more promising, and although the diagnostic or prognostic benefits to practising physicians are still limited, physicians recognize the validity of this approach. They know that correction of the abnormality which provokes a disease is close at hand when the gene responsible has been located and its function defined. Thanks to the astounding possibilities offered by genetic engineering it will henceforth be possible to know the exact DNA sequence in the vicinity of the HLA genes. The latter will serve as markers and will make it possible to discern anomalies of susceptibility genes. We must emphasize the possibility that in some cases no anomaly will be found because a gene (or combination of genes) may be perfectly active in the defense against certain antigens but totally inactive in the defense against others. Thus an inventory of the immunological capacities of each individual will need to be drawn up. This inventory will show the weaknesses (susceptibility), the excesses (autoimmunization), and the good capacities (protection) afforded by each type of gene combination. In this way, preventive medicine of high precision will be possible; a personalized medicine that will be more efficient and less burdensome for the community than the present mass system.

At the same time, researchers now have the means with which to approach the crucial problem represented by the subtle organization of man's immune system. The cascade of interrelationships, and the language, between different immunocompetent cells will be clarified; the place of the "specific" and the "nonspecific" will be recognized; the role of HLA products in messages between well-defined cells will be determined. This deeper understanding of man's immune response will quickly have major repercussions in pathology. It will perhaps provide the key to the irritating problem of the treatment of cancer, and may also provide a simple means of inducing graft tolerance at will. It will also perhaps lead to an immunological treatment of the major parasitic diseases that still afflict such a large part of mankind.

Finally, the discovery of the primary function of the molecules of the MHC found at the surface of all, or almost all cells of the organism will be a decisive step in our understanding of the differentiation and social organization of cells.

The way already trod is but a simple introduction. They are still many marvellous pages to be written.... .

References and Notes

1. G. D. Snell, *Havey Lect*, **74**, 49 (1979).
2. ______, *Folia Biol.* (Prague) **14**, 335 (1968).
3. N. K. Jerne, *Eur. J. Immunol.* **1**, 1 (1971).
4. J. Dausset, A. Lebrun, M. Sasportes, *C. R. Acad. Sci. Paris* **275**, 2279 (1972).
5. W. F. Bodmer, *Nature (London)* **237**, 139 (1972).
6. J. Dausset, *Acta Haematol.* **20**, 156 (1958).
7. J. J. Van Rood and A. Van Leeuwen, *J. Clin. Invest.* **42**, 1382 (1963).
8. R. Payne, M. Tripp, J. Wiegle, W. Bodmer, J. Bodmer, *Cold Spring Harbor Symp. Quant. Biol.* **29**, 285 (1964).

9. P. S. Russell, H. J. Winn, D. B. Amos, Eds., *Histocompatibility Testing* (Publ. No. 11229, National Academy of Sciences, Washington, D. C., 1965).
10. H. Balner, F. J. Cleton, J. G. Eernisse, Eds., *Histocompatibility Testing 1965* (Munksgaard, Copenhagen, 1965).
11. E. S. Curtoni, P. L. Mattiuz, R. M. Tosi, Eds., *Histocompatibility Testing 1967* (Munksgaard, Copenhagen, 1967).
12. P. I. Terasaki, Ed., *Histocompatibility Testing 1970* (Munksgaard, Copenhagen, 1970).
13. J. Dausset and J. Colombani, Eds., *Histocompatibility Testing 1972* (Munksgaard, Copenhagen, 1973).
14. F. Kissmeyer-Nielsen, Ed., *Histocompatibility Testing 1975* (Munksgaard, Copenhagen, 1975).
15. W. F. Bodmer, J. R. Batchelor, J. G. Bodmer, H. Festeinstein, P. J. Morris, Eds., *Histocompatibility Testing 1977* (Munksgaard, Copenhagen, 1978).
16. P. I. Terasaki, Ed., *Histocompatibility Testing* 1980 (University of California, Typing Laboratory, Los Angeles, 1980).
17. J. Klein, *Science* **203**, 516 (1979).
18. H. T. Orr, D. Lancet, R. J. Robb, J. A. Lopez de Castro, J. L. Strominger, *Nature (London)* **282**, 266 (1979).
19. J. L. Strominger, *Immunology* **80**, 541 (1980); M. Fougereau and J. Dausset, Eds., *Fourth International Congress on Immunology, Paris, 1980* (Academic Press, London, 1980).
20. H. T. Orr, J. A. Lopez de Castro, P. Parham, H. Ploegh, J. L. Strominger, *Proc. Natl. Acad. Sci. U.S.A.* **76**, 4395 (1979).
21. J. Colombani, M. Colombani, J. Dausset, in (*12*), pp. 79–92.
22. P. Ivanyi and J. Dausset, *Vox Sang.* **11**, 326 (1966).
23. J. Dausset, *Transplant. Proc.* **3**, 1139 (1971).
24. W. E. Biddison, M. S. Krangel, J. L. Strominger, F. E. Ward, G. M. Shearer, Shaw, *HumImmunol* **1**, 225 (1980).
25. M. Sasportes, A. Nunez-Roldan, D. Fradelizi III, *Immunogenetics*, **6**, 55 (1978).
26. C. Mawas, D. Charmot, M. Sivy, P. Mercier, M. M. Tongio, G. Hauptmann, *J. Immunologenet.* **5**, 383 (1978).
27. S. Shaw, M. S. Pollack, S. M. Payne, A. H. Johnson, *Hum. Immunol.* **1**, 177 (1980).
28. H. L. Ploegh, H. T. Orr, J. L. Strominger, *Proc. Natl. Acad. Sci. U.S.A.* **77**, 6081 (1980).
29. L. Degos and J. Dausset, *Immunogenetics* **1**, 195 (1974).
30. J. Dausset, F. T. Rapaport, P. Ivanyi, J. Colombani, in (*10*), pp. 63–72.
31. J. J. Van Rood, A. Van Leeuwen, A. Shippers, M. J. Vooys, E. Frederiks, H. Balner, J. G. Eernisse, in (*10*), pp. 35–50.
32. R. Ceppellini, E. S. Curtoni, P. L. Mattiuz, G. Leighes, M. Visetti, A. Colombi, *Ann. N. Y. Acad. Sci.* **129**, 421 (1966).
33. D. B. Amos, H. F. Siegler, J. G. Southworth, F. E. Ward, *Transplant Proc.* **1**, 342 (1969).
34. F. H. Pach and D. B. Amos, *Science* **156**, 1506 (1967).
35. E. J. Yunis and D. B. Amos, *Proc. Natl. Acad. Sci. U.S.A.* **68**, 3031 (1971).
36. J. Hamburger, J. Crosnier, B. Descamps, D. Rowinska, *Transplant Proc.* **3**, 260 (1971).
37. K. C. Cochrum, H. A. Perkins, R. O. Payne, S. L. Kountz, F. O. Belzer, *ibid.* **5**, 391 (1973).
38. V. P. Eijsvoogel, J. J. Van Rood, E. D. DuToit, P. H. A. Schellekens, *Eur. J. Immunol.* **2**, 413 (1972).
39. J. Dausset, L. Contu, L. Legrand, A. Marcelli-Barge, T. Meo, F. T. Rapaport, *J. Clin. Invest.* **63**, 893 (1979).
40. M. Jonker, J. Hoogeboom, A. Van Leeuwen, C. T. Koch, D. B. Van Oud Alblas, J. J. Van Rood, *Transplantation* **27**, 91 (1979).
41. A. Ting and P. J. Morris, *Lancet* **1978-I**, 575 (1978).
42. T. Moen, D. Albrechtsen, A. Flatmark, A. Jakobsen, J. Jervell, S. Halvorsen, B. G. Solheim, E. Thorsby, *N. Engl. J. Med.* **303**, 850 (1980).
43. C. F. Feighery and P. Stastny, *Immunogenetics* **10**, 39 (1980).
44. G. Opelz, M. R. Mickey, P. I. Terasaki, *Lancet* **1972-I**, 868 (1972).
45. M. Sasportes, D. Fradelizi, J. Dausset, *Nature (London)* **276**, 502 (1978).
46. M. Sasportes *et al.*, *J. Exp. Med.* **152** (No. 2), 270s (1980).
47. A. A. Van Es and H. Balner, in *Immunogenetic and Transplantation Studies in the Rhesus Monkey*, A. A. Van Es, Ed. (J. H. Drukkerij, B. V. Pasman's, s'Gravenhage, 1980), p. 117.
48. J. Dausset and L. Contu, *Transplant. Proc.* **13**, 895 (1981).
49. ______________, *Immunology* **80**, 513 (1980); M. Fougereau and J. Dausset, *Fourth International Congress on Immunology, Paris, 1980* (Academic Press, London, 1980).

50. A. J. McMichael, A. Ting, H. J. Zweerink, B. A. Askonas, *Nature (London)* **270**, 524 (1977).
51. S. Shaw, G. M. Shearer, W. E. Biddison, *J. Exp. Med.* **151**, 235 (1980).
52. E. Dickmeiss, B. Soeberg, A. Svejgaard, *Nature (London)* **270**, 526 (1977).
53. E. Goulmy, J. D. Hamilton, B. A. Bradley, *J. Exp. Med.* **149**, 545 (1979).
54. H. Hirschberg, O. J. Bergh, E. Thorsby, *ibid,* **150**, 1271 (1979).
55. F. B. Mudawwar, E. J. Yunis, R. S. Geha, *ibid.* **148**, 1032 (1978).
56. S. M. Friedman, O. H. Irigoyen, D. Gay, L. Chess, *J. Immunol.* **124**, 2930 (1980).
57. F. M. Kourilsky, J. Dausset, N. Feingold, J. M. Dupuy, J. Bernard, in *Advances in Transplantation, First International Congress of the Transplanation Society, Paris, 1967* (Munksgaard, Copenhagen 1967), pp. 515–522.
58. L. P. Ryder, E. Andersen, A. Svejgaard, Eds., *HLA and Disease Registry, Third Report* (Munksgaard, Copenhagen, 1979).
59. J. Dausset and A. Svejgaard, Eds., *HLA and Disease* (Williams & Wilkins,Baltimore,1977).
60. A. F. Geczy, K. Alexander, H. V. Bashir, J. Edmonds, *Nature (London)* **238**, 782 (1980).
61. C. Druery, H. Bashir, A. F. Geczy, K. Alexander, J. Edmonds, *Hum. Immunol.* **1**, 151 (1980).
62. A. Svejgaard, P. Platz, L. P. Ryder, in (*16*), pp. 638–656.
63. I. Deschamps, H. Lestradet, F. Clerget, C. Bonaiti, M. Schmid, M. Busson, A.Benajam, A. Marcelli Barge, J. Dausset, J. Hors, *Diabetologica,* **19**, 189 (1980).
64. P. P. Jones, D. B. Murphy, H. O. McDevitt, *J. Exp. Med.* **148**, 925 (1978).
65. W. P. Lafuse, J. F. McCormick, C. S. David, *ibid.* **151**, 1709 (1980).
66. L. Degos, M. Pla, J. Colombani, *Eur. J. Immunol.* **9**, 808 (1979).
67. J. Dausset and L. Contu, *Hum. Immunol.* **1**, 5 (1980).

George Snell

GEORGE D. SNELL

My parents were both New Englanders, though my father was born in Minnesota where his father had moved from Massachusetts to join a frontier community. My father moved east as a young man, and for a number of years was YMCA secretary in Haverhill, Massachusetts. Subsequently he invented and worked in the application of a device for winding induction coils used in ignitors for the motorboat engines of that day. I was born in Bradford, Massachusetts, a suburb of Haverhill, in December 1903, the youngest of three children. My parents moved when I was four to the home built by my great grandfather in Brookline, Massachusetts, and it was in the excellent Brookline public schools that I received my pre-college education.

Science and mathematics were my favorite subjects. In spare time I read books on astronomy and physics as well as the usual boyhood classics. But I also enjoyed sports, and a group of five or six youngsters used to gather at our house to play touch football or scrub baseball in our yard or a neighboring vacant lot. Imaginative stories and games also were very much a part of my childhood.

In 1900, three years before I was born, my mother's parents had purchased a run-down farmhouse and 70 acres of land in South Woodstock, Vermont. The house was gradually restored and furnished, and the summers I spent at "the farm" were among the delights of my childhood and youth. An interest in gardening, farming, and forestry have been a permanent legacy of the experience this home provided.

Music was a major interest of the whole family. My mother played the piano, and we did a great deal of family singing in which friends often joined. It has been a source of great pleasure that my wife is also a pianist.

I entered Darmouth College in 1922 and again found science and mathematics my favorite subjects. A course in genetics taught by Professor John Gerould proved particularly fascinating, and it was that course that led me to the choice of a career. When the decision was finally made to enter graduate school, it was on Professor Gerould's advice that I enrolled as a graduate student with Harvard's Professor Castle, the first American biologist to look for Mendelian inheritance in mammals.

My thesis work on linkage in mice largely determined my future work. Two years spent teaching and two years as a postdoctoral fellow under Herman Muller studying the genetic effect of x-rays on mice served to convince me that research was my real love. If it was to be research, mouse genetics was the clear choice and the Jackson Laboratory, founded in 1929 by Dr. Clarence Cook Little, one of Castle's earlier students, almost the inevitable selection as a place to work. The Laboratory was a small institution when I joined the staff of seven in 1935, but under the talented leadership of Dr. Little and his successor, Earl Green, it has grown into the world center for studies in mammalian genetics. I owe a great deal to it for providing the ideal home for my subsequent research.

It was in Bar Harbor that I met and married Rhoda Carson, and where we raised our three sons, Thomas, Roy and Peter.

I have always enjoyed sports, with skiing, which I learned at Dartmouth, and tennis perhaps being my two favorites.

While for 25 years I concentrated almost exclusively on studies of histocompatibility genes and especially of the *H-2* complex, and for 35 years have pursued these subjects to some degree, I also have become involved in other areas. While working under Dr. Castle, I spent parts of two summers at Woods Hole with Dr. Phineas Whiting, an earlier student of Castle, studying the genetics of the parasitic wasp, *Habrobracon*. An outcome of this work was a paper on The Role of Male Parthenogenesis in the Evolution of the Social Hymenoptera. The problems of social evolution have remained a continuing interest, to which I am now returning in a more active way in retirement. The two years with Muller at the University of Texas resulted in the first demonstration of the induction by x-rays of chromosomal changes in mammals. My first several years at the Jackson Laboratory were spent in continuation of this work, and especially in the detailed genetic analysis of two of the induced reciprocal translocations. In the late 1930s, I became involved in problems of gene nomenclature in mice, and this, together with problems of strain nomenclature, remained a concern for many years. The efforts of the Committee on Standardized Nomenclature for mice have led to the universal acceptance of a well organized and convenient nomenclature system for this species. Some experiments, which I carried out at about the same time that I was becoming interested in histocompatibility genetics, led to the discovery of immunological enhancement, the curious inversion of the expected growth inhibition seen with certain tumors when transplanted to pre-injected mice. I soon found that I was not the first person to have seen this phenomenon, but the mouse system proved very amenable to further exploitation. I had to drop this topic in favor of the genetic studies, but it has been interesting to see it grow through the work of Dr. Nathan Kaliss and many others into a major area of research with possible implications for organ transplantation in man. A final interest, developed jointly with Dr. Marianna Cherry during my last few years at the Jackson Laboratory, concerned serologically demonstrable alloantigens of lymphocytes.

Much of the work sketched above was carried out on a collaborative basis. I cannot here give names, but I owe a great debt to the many wonderful people with whom it has been my privilege to work in these studies.

STUDIES IN HISTOCOMPATIBILITY

Nobel lecture, 8 December, 1980

by

GEORGE D. SNELL

The Jackson Laboratory, Bar Harbor,
Maine 04609, U.S.A.

The major histocompatibility complex (MHC) is a group of closely linked loci present in remarkably similar form in all mammals and perhaps in all vertebrates. It plays a still imperfectly understood but clearly important role in immune phenomena. Because of the unusual concentration of similar genes, I referred to it in 1968 as a supergene (1). Bodmer (2) has gone me one better, calling it a super supergene. The term is not inappropriate, because we now know that the MHC contains at least four gene clusters, each with its own type of end product and its own specific effects on the immune response.

The MHC was originally discovered because of its role in the rejection of transplants made between incompatible individuals. Genes competent to play this role in the appropriate experimental or surgical context are called *histocompatibility* or *H genes*. An influence on transplants probably is entirely irrelevant to the true function of such genes, but the influence does give the geneticist a handle by which to study them. It was by this route that, over a period of a good many years, I became involved first in immunogenetics and then in the new and fascinating area of cellular immunity.

That susceptibility and resistance to transplants are influenced by multiple genes showing Mendelian inheritance was demonstrated in the pioneering studies of Little and coworkers (3, 4, 5). Dr. Little became interested in this subject as a graduate student at Harvard because of experiments with tumor transplants in mice carried out by Tyzzer at the Harvard Medical School. Little suggested (3) a Mendelian interpretation of Tyzzer's data, and worked jointly with him in experiments to test this hypothesis. After he founded the Jackson Laboratory, he and his associates returned again to an investigation of transplant genetics. While these studies revealed the existence of multiple histocompatibility genes, they did not provide any means of identifying the individual loci. Any individuality was masked by the non-discriminatory nature of a test based on only one variable—the success of transplant growth.

Were there any methods by which the individuality of these genes could be revealed? A project in radiation genetics which I had pursued during my first few years at the Jackson Laboratory was winding down in the late 1930's, and in examining histocompatibility genetics as one of several potential new undertakings, I thought I saw possibilities for new openings. Two methods of locus identification appeared possible. The first was the use of visible marker genes to tag chromosome segments carrying a single or, at most, a few *H* genes. The second was the transfer, by appropriate crosses, of an *H*-bearing chromosome segment from one strain onto the inbred background of another. The result would be a new strain, appropriately referred to as *congenic resistant* or CR relative to its inbred partner. Dr. Cloudman, at the Jackson Laboratory, was at the time carrying a number of transplantable tumors, and he generously made these available for typing purposes. These plans were formulated in 1944, though not published until considerably later (6).

The linkage method promised faster results than the CR strain method, and was the first undertaken. Starting in 1945, extensive crosses were set up involving a total of 18 marker genes (6, 7). These markers were assembled in six stocks which I had either produced or acquired (6). This grouping substantially simplified the testing process. One of the first crosses set up, utilizing the three dominant marker genes , *Ca, Fu,* and *W,* revealed a linkage between a histocompatibility gene and *Fu* or fused tail.

Shortly after this linkage was established, Dr. Peter Gorer of Guy's Hospital, London, came to spend a year at The Jackson Laboratory. Gorer had previously identified a blood group locus in mice, and shown that blood type segregated with susceptibility and resistance to a transplantable tumor (8). This was the first case of individual identification of a histocompatibility locus. During Dr. Gorer's stay in Bar Harbor, he tested our backcross animals segregating for the

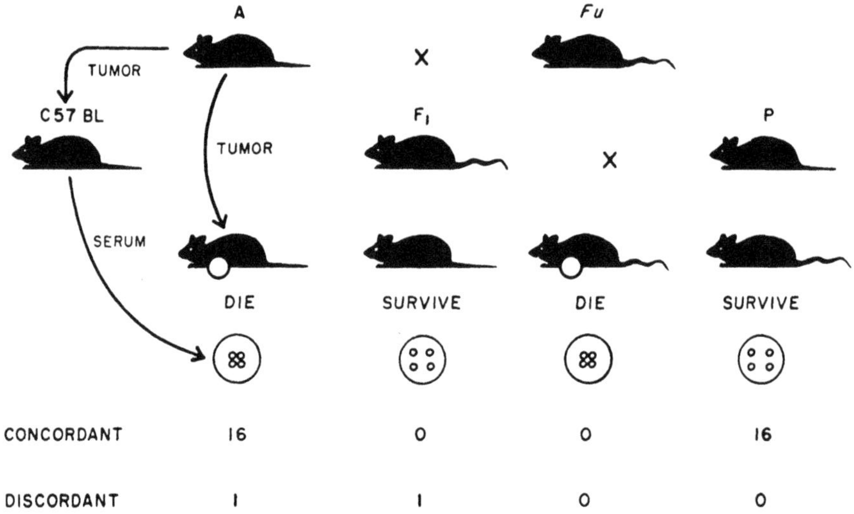

Figure 1. Association of normal tail, tumor susceptibility, and the H-2 antigen of strain A as indicated by red cell agglutination, in the cross (A × *Fu*) × P. A and P are inbred strains, Fu a strain carrying the dominant marker gene *Fu* which produces kinks in the tail. Death from the strain-A tumor and agglutination of red cells with the C57BL anti-A antiserum are tests for the *H-2* allele of strain A, whereas survival and non-agglutination are tests for the allele from strain Fu. Both tests were concordant for all except two of the 34 tested backcross mice. The two discordant mice are listed according to their blood type, presumably the more reliable indicator for *H-2*. The discordant mouse showing non-agglutination but dying from the tumor was a probable recombinant. All other 33 mice were non-recombinants. [From Snell (9), courtesy the Journal of the National Cancer Institute.]

Fu gene and transplant resistance, and found that his blood group antigen also segregated with *Fu* (Fig. 1). His locus and mine were one and the same. Because Gorer was using an antiserum reactive with that he had called antigen II, the locus was called *H-2*. The existence of three alleles was indicated by the serological data (9).

Marker genes, in appropriate circumstances, can be powerful genetic tools. Using the linkage of *Fu* and *H-2*, it was possible to show whether a previously untested inbred strain carried a previously known or a new *H-2* allele. Using this method, the number of known alleles was raised to seven (10).

One by-product of the linkage study was the finding that the F_1 from any cross giving an $H\text{-}2^{d}/H\text{-}2^{k}$ heterozygote was susceptible to tumors from $H\text{-}2^{a}/H\text{-}2^{a}$ donors (11). Some sort of complementation was taking place. This suggested that the $H\text{-}2^{a}$ genotype carried two *"components"* present separately in $H\text{-}2^{d}$ and $H\text{-}2^{k}$. In recognition of this, $H\text{-}2^{a}$ at the time was called $H\text{-}2^{dk}$. The suggestion that there were actually two *H-2* loci was not made at the time, but evidence was soon forthcoming that this indeed was the case.

In independent studies, crossing over was shown to occur within *H-2*. Amos et al (12) used red çell typing and Sally Allen (13) used the linked marker method with tumor transplants as a typing tool. Each study yielded one intra-*H-2* recombinant. Additional recombinants were soon added, including seven in a study in the Jackson Laboratory (14, 15). A recent review (16) lists a total of 90 recombinants identified by 1980.

The occurrence of crossing over proved that there are no less than two loci within *H-2*. The symbols assigned were *H-2K* and *H-2D* (sometimes abbreviated *K* and *D*). The loci are listed in this order because *H-2K* is proximal to the centromere. The crossover percent between *K* and *D* has varied in different studies, influenced certainly by the sex of the heterozygous parent and perhaps other factors, but is usually given as 0.5 percent or less.

The newly found complexity required a change of nomenclature. What had been called an *H-2* allele was actually a linked pair or cluster of alleles. Because the linkage was close, the alleles tended to be inherited as a unit, and a name for this unit was necessary. This problem ultimately was solved by borrowing the term *haplotype* from the HLA terminology (Fig. 2).

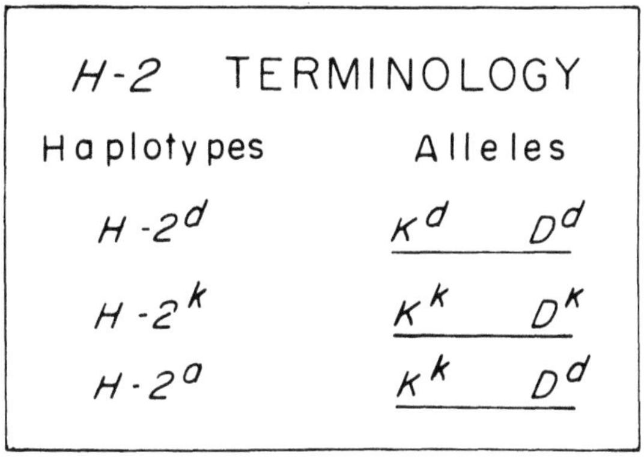

Figure 2. Figure showing the relationship of the terms *haplotype* and *allele*, and their symbols. *H-2* allelic symbols may be written either $H\text{-}2K^{d}$ or, where appropriate, abbreviated to K^{d}.

The second method for the identification of *H* loci, the production of congenic resistant lines, was by its nature a much longer project than the linkage study. The series of crosses initially used is shown in Fig. 3. An absolute minimum of 14 generations was necessary. The crosses were first set up in 1946 about the time that Gorer came to Bar Harbor. This group was lost in the Bar Harbor fire of 1947, but replaced with new crosses in 1948. In 1953, an entirely new group was set up with some refinements in method (17). All told, the first set of crosses led to the establishment of 32 CR lines, the second to 22.

Once a CR line was established, all that was known about it was that it carried a gene (or possibly a group of closely linked genes) that made it resist transplants from its inbred partner strain. Some method was still necessary to tell whether the introduced locus was the same as or different from the introduced locus of any other CR strain. Two methods were available. The first required a substantial element of luck. If a visible marker gene happened to be present on the introduced segment, and if this marker had been placed on the linkage map, the accompanying *H* gene was thereby given an individual identity. The second was a complementation test, involving a cross between a known and an unknown. This was more laborious and in some ways less informative, but it was much more widely applicable (18).

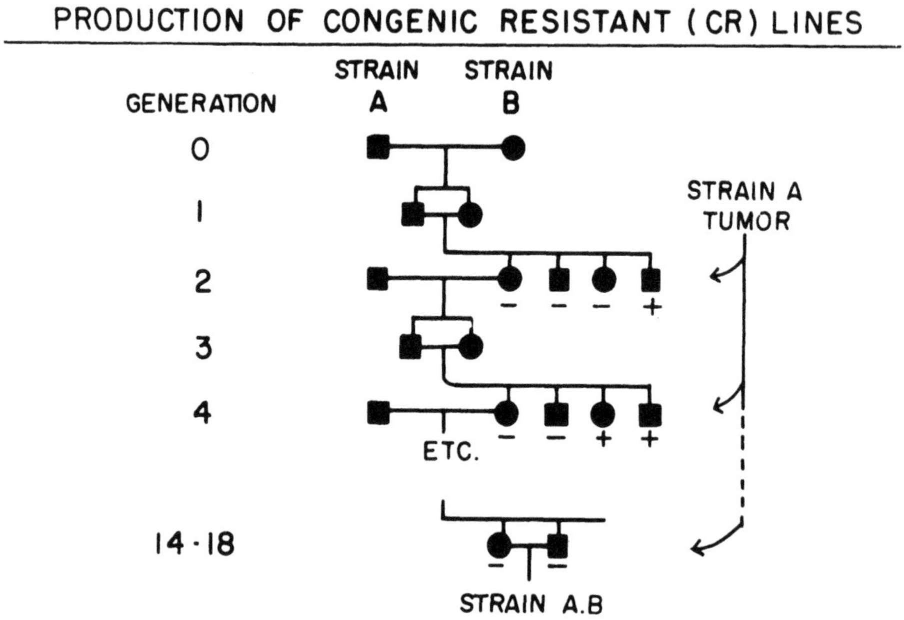

Figure 3. The cross-intercross system for the production of congenic resistant (CR) strains. Mice of every even numbered generation are challenged with a tumor from A, a highly inbred strain, and a survivor (indicated by a -) mated back to strain A. Each mating to strain A increases the proportion of strain A and reduces the proportion of strain B genes. However, the tumor challenge insures that the mouse selected for mating is homozygous for a histocompatibility gene inherited from B and foreign to A. At generation 14 or later, two resistant mice are mated, giving the CR strain A.B. [From Snell (17), courtesy the Journal of the National Cancer Institute.]

By good luck, three of the congenic strains developed in the first set of crosses carried introduced visible markers. Tests confirmed that these were linked with the introduced *H* gene. The three strains are shown in Fig. 4. The marker introduced into the first strain in the figure was albinism in Chromosome 1. The accompanying histocompatibility locus was called *H-1*. The marker in the second strain was *Fu* or fused in Chromosome 17. This suggested that the accompanying locus was *H-2,* and this was confirmed by appropriate tests. Strain three was marked by agouti on Chromosome 5. The accompanying *H* locus was called *H-3.*

Application of the complementation test ultimately led to the identification of eight more loci. In two cases, the loci were in loosely linked pairs that had to be separated by crossing over. Subsequent studies by other investigators have raised the number of known non-*H-2* loci to at least 50 (19). By far the largest contribution has been made by Dr. Donald Bailey through the use of an extensive group of congenic lines started at the University of California and completed and tested in Bar Harbor (20). Bailey used skin grafts instead of tumor transplants and a variety of other refinements.

As the congenic resistant lines became available and new histocompatibility loci were identified, certain properties of the loci became apparent. A test of the survival of skin grafts made across *H-1, H-2,* and *H-3* barriers showed that an *H-2* difference caused a much more rapid rejection than differences at the other loci. *H-2* seemed to be a uniquely "strong" locus (21). This result has been confirmed and extended in many subsequent studies (22, 23). Some representative data are shown in Table 1. The uniqueness of *H-2* became further apparent as the analysis of the first group of congenic resistant lines neared completion. Of the 32 completed and tested lines in this group, 26 differed from their inbred partner at *H-2* (7, 24).

Besides the first two groups of congenic lines, two other groups of lines of a more specialized nature were produced in collaboration with Drs. Ralph Graff and Marianna Cherry. All told, 206 lines were started and 92 carried through

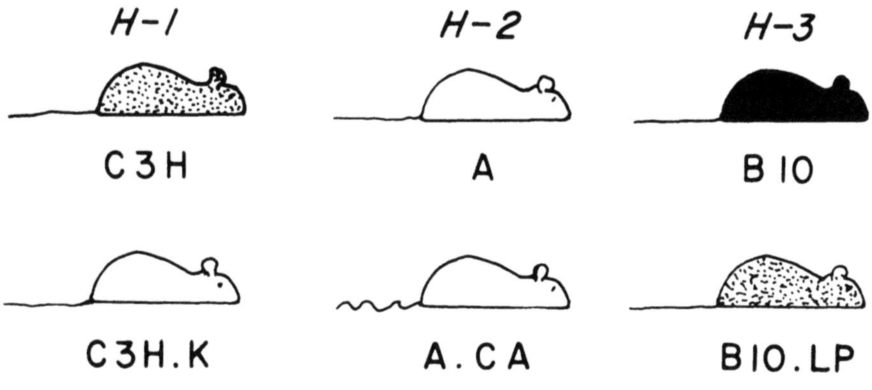

Figure 4. The identification of histocompatibility loci through the introduction into CR lines of chromosome segments bearing marker genes producing visible effects as well as histocompatibility genes. [Snell et al. (18), courtesy Academic Press.]

Table 1. Differences in "strength" of histocompatibility loci*

Locus	♀ → ♀ B10 donor	Median survival time of skin grafts (days) ♀ → ♂ B10 recipient
H-2	12	12
H-1	25	∞
H-4	120	25
H-11	78	164

* Data from Graff, Hildemann, and Snell, 1966.

to completion and tested. Sixty-one of these are still maintained. In addition, Dr. Jack Stimpfling when at the Jackson Laboratory produced seven widely used *H-2* lines (15) and Dr. Bailey 25 non-*H-2*-lines (20). Many lines have been produced in other laboratories (25).

I shall speak later of some of the uses to which the congenic resistant strains have been and are being put. It is sufficient to note here that in a one-year period ending in 1980, 87,000 of the ten most widely used congenic strains and 31,000 of their three inbred partner strains (C57BL/10Sn, A/WySn, and C3H/DiSn) were supplied by the Jackson Laboratory to other laboratories. This includes mice of four strains developed by Stimpfling. The number of congenic mice sent out has increased every year.

While the linkage testing and production and analysis of CR lines was going on at the Jackson Laboratory, there was also a growing use by Gorer and his students of the serological approach. While this was limited to *H-2,* it was a powerful tool for the study of this complex. Two basic methods were used, red cell agglutination (26) and the cytotoxic action of isoantibody plus complement on lymphocytes (27). As time went on, more and more researchers entered this field and the methods were refined and diversified (18, 28).

One of the first uses to which the CR lines were put was in the production of antisera which, because donor and recipient differed only at *H-2,* could contain only *H-2* antibodies. Hoecker, a student of Gorer's, spent a year at the Jackson Laboratory developing and applying these simplified antisera (29). This study by Hoecker and coworkers will serve to illustrate some of the serological findings.

As in the case with virtually all aspects of *H-2,* the antisera revealed an extraordinary complexity. The antisera, even if made in CR lines, could usually be simplified by absorption with mice of appropriate *H-2* type. Each antiserum so prepared showed a characteristic strain distribution in its reactions. Table 2 presents the reaction patterns known at the time of Hoecker's 1954 paper. Each reaction pattern defined a *specificity* to which a number could be assigned. It will be seen that four specificities in the table, 2, 16, 17, and 19, are confined to a single haplotype. These are *private specificities.* The specificities with a wider distribution are referred to as *public,* though with the qualification that a private

Table 2. A chart of H-2 specificities based on 1954 data*

H-2 haplotype	2	3	4	5	11	16	17	19
a	–	3	4	5	11	–	–	–
b	2	–	–	5	–	–	–	–
d	–	3	4	–	–	–	–	–
k	–	3	–	5	11	–	–	–
p	–	3	–	5	–	16	–	–
q	–	3	–	5	–	–	17	–
s	–	3	–	5	–	–	–	19

* Data of Hoecker, Counce, and Smith, 1954.

specificity may appear in two haplotypes if one is derived by recombination from the other. H-2.4 (Table 2) is a case in point. The number of identified H-2 specificities has grown steadily since these observations and now stands at 113 (16).

Along with the growth in known specificities there has been a growth in known haplotypes and alleles. Thirty-seven haplotypes have been described in inbred strains and 34 more have been added in extensive analyses by Jan Klein of mice caught in the wild (16, 30). Klein estimates that there are not less than 200 alleles at both the *K* and the *D* loci. This extraordinary polymorphism is further emphasized by a high mutation rate at *H-2* revealed in studies by Bailey, Egorov, and Kohn and their coworkers (31, 32; reviews in 18, 33). Curiously, there seems to be much less MHC polymorphism in the rat and hamster (19).

Serological methods were used in all the *H-2* recombination studies except that of Allen. The crossovers that were found made it possible to assign specificities to particular regions or loci. Thus, specificity 11 was identified with the *K* region and 2 with the *D* region of *H-2*. As data accumulated, it appeared to be necessary to postulate additional loci or regions, and the number ultimately grew to five (12, 34). However, problems began to appear with this interpretation of *H-2* structure. These centered around the discovery that there are specificities that map at both ends of *H-2* (34, 35, 36). It finally became clear that return to a two locus model of *H-2* would resolve the problems and provide an interpretation entirely consistent with the facts (37).

Prior to the onset of this debate, Drs. Démant, Cherry, and I had turned to serotyping, using the quantitatively precise chromium label method of lymphocyte cytotoxicity as well as red cell typing. One of the findings from these studies was that the private *H-2* specificities can be arranged in two mutually exclusive series, one mapping at the *K* and one at the *D* end of *H-2* (Table 3) (38). The existence of two allelic series had already been established for HLA (39). This is one of the few cases in which mouse studies of the MHC lagged behind those of other species.

One of the most interesting results of the serological studies of Démant, Cherry, and myself was the finding of a third but quite unique series of allelic

Table 3. The initial assignment of most H-2 private specificities to either the K or the D regions*

Located			Unlocated	
Haplotype	K specificity	D specificity	Haplotype	Specificity
b	33	2	*f*	9
j	15	2	*p*	16
d	31	4	*r*	18
u	20	4	*v*	21
k	23	32		
m	23	30		
q	17	30		
qp	17	12		
s	19	12		

* Based on Snell, Cherry, and Démant, 1973. In current charts, the K specificity 23 is replaced by 11, and the unlocated private specificities have been located.

specificities determined by the *H-2* complex. Whereas the original *K* and *D* series were composed of clearly distinct private specificities, the new series were composed of two families of related specificities, the 1-family and the 28-family (40–42). Members of each family showed similar but not identical strain distributions. In the original studies, it was not suggested that the new series identified a new locus, but Démant and coworkers, in further investigations, found that this was indeed the case (43, 44). The new locus, called *H-2L*, is close to and so far has not been separated by crossing over from *H-2D*. The existence of *H-2L* has been confirmed in numerous tests. One of the most interesting confirmations was the discovery of a mutant of $H\text{-}2^d$ in which *H-2L* is lost but *H-2D* remains (45). Specificities 27, 28, 29, all members of the 28 family characteristic of the *D* end of $H\text{-}2^d$ (41, 46), are all lost in the mutant. There remain several puzzling aspects of the 1 and 28 families—for example, some form of 1 or 28 has been found on the K and D as well as the L molecules—but this does not invalidate the evidence for a separate H-2L product.

Démant (47) has recently identified two additional *H-2* products, one determined by a locus close to *K*, the other by a locus, in addition to *L*, close to *D*.

Studies of the murine *H-2K, D,* and *L* loci played a seminal role in the growth of our conception of the major histocompatibility complex, but the conception in its current form is the result of an explosion of information that followed the initial discoveries of Drs. Dausset and Benacerraf. Dausset's recognition that HLA is similar to *H-2* (48) was the first hint that a gene or genes with especially strong histocompatibility effect may be common to all mammals. *H-2*-like complexes have now been identified in at least eight other mammals and in poultry (49). The demonstration by Benacerraf and coworkers (50) of an association in guinea pigs between a major histocompatibility locus and immune capability and by McDevitt and coworkers (51) of a similar association in mice, and the further demonstration that the murine immune response genes were in a distinct (*I*) region (52), was the beginning of our conception of the true complexity of this remarkable system. The expression *major histocompatibility complex* became fully justified.

This is not the place for anything approaching a full description of the complex, but a brief summary is necessary. Some of the essential facts are set forth in Fig. 5. The complex consists of five main regions, *K*, *I*, *S*, *D*, and *Tla* (or *Qa*), producing four classes of antigens, class I associated with the *K* and *D* regions, II associated with the *I* region, III associated with the *S* region, and IV associated with the *Tla* region. Present evidence points to a total of 22 loci, but this number is likely to grow. The principal bases of the classification are the tissue distribution and the chemical properties of the end products. Information is much more complete concerning some products than others, and some changes in the classification may be necessary, but the general outline seems likely to stand. The tissue distribution is summarized in Fig. 5. The chemical properties of the class I and the class IV products seem to be similar (review in 67), but the two classes are clearly differentiated by the wide distribution of the K, D, L antigens and the restriction to lymphocytes of Tla and its relatives. With this exception, each class shows distinct chemical characteristics. The growing evidence that all the products play fundamental roles in the immune response is a common bond that unites them and justifies the conception of all 22 loci as part of a complex or system.

THE *T* AND *H-2* COMPLEXES

Map distance	Region	Class of loci	No. of loci	Tissue distribution
	T		6 ?	Early embryo
17				
	K	I	2	Most or all cells except early embryo
0.2	*I*	II	8	Each antigen on specific classes of lymphocytes and macrophages or macrophage-related cells
	S	III	2	Serum
0.3				
	D	I	3	Same as *K*
1.5	*Tla*	IV	6	Specific lymphocyte classes

Figure 5. Diagram showing in condensed form some of the genetics of the *T* and *H-2* complexes and the tissue distribution of the antigens wich they determine. The principal sources of data on the genetics are: map distance (16, 18); *T* complex (53, 54); *K* and *D* region loci (18, 47, 55); *I* region loci (56, 57); *S* region loci (56, 58); *Tla* region loci (59, 60). The principal sources of information on the tissue distribution are: *T* complex products (61–63); *K* and *D* region products (18); *I* region products (56, 64–66); *S* region products (56, 58); *Tla* region products (59, 60).

A major reason for the current interest in the MHC is the extraordinary range of processes, both immunological and non-immunological, on which it has an influence (18, 68). Iványi (68) lists 35 quantitative traits affected by it. Particularly important from the medical point of view is its influence on immune processes. I cannot begin to cover this subject thoroughly. The few examples I select reflect my judgment of importance or interest or some degree of personal involvement. I start with immunological areas and move to those apparently without immunological connection.

The first reports suggesting a possible role for *H-2* in immune processes concerned the phenomenon of hybrid resistance. I came across this phenomenon in the course of producing and analyzing the first two groups of congenic resistant lines (17, 69). The phenomenon consists of a resistance of F_1 hybrids to tumors indigenous to the parental strains, a resistance which, according to the accepted laws of transplantation, should not occur. It was noted that major resistance required heterozygosity at *H-2,* but that non-*H-2* heterozygosity could have some effect. The major role of *H-2* with, however, a minor role for one or more non-*H-2* loci, was established in much greater detail by Cudkowicz and coworkers, using marrow transplants, and the added observation made that the *D* end of *H-2* was the active region (70, 71). A recent *in vitro* study suggests a possible role for natural killer cells in the phenomenon (72), but this may not be the only mechanism of hybrid killing (73).

Another group of early studies suggesting a role for *H-2* in immune processes concerned viral leukemogenesis. In 1954–56, Gross reported that cell-free filtrates from AKR or C58 leukemias caused early leukemia development when injected into newborn C3H/Bi or C57BR/cd mice (74, 75). All these mice are $H\text{-}2^k$, a fact probably noticed by most or all of the *H-2* students of that time. Could *H-2* be involved? Lilly et al (76) answered this question in the affirmative by inoculating leukemic extracts into *H-2*-typed mice from segregating generations. In a study started independently but published somewhat later, Tennant (77) and Tennant and Snell (78, 79), using congenic resistant mice and an agent prepared by Tennant from BALB/c ($H\text{-}2^d$) leukemias, showed preferential leukemia induction in other $H\text{-}2^d$ mice and also a difference in the degree of resistance engendered by other haplotypes. Table 4 shows some of the results. These data demonstrate very nicely the power of congenic strains. Definitive results could be obtained without the production of segregating generations. The data (including some not in the table) also show that non-*H-2* loci play a significant role.

While these results suggested an immunological role for *H-2,* the real breakthrough came with the demonstration, already mentioned, that the *H-2* complex contains immune reponse (*Ir*) genes. Since this discovery, an ever-expanding effort has gone into the unravelling of the processes involved. I shall mention only two of the major findings.

H-2 linked immune response genes in the mouse were originally mapped in the *I* region of the *H-2* complex (52) and it was assumed for some years that all such genes were confined to this region. It now appears that this was an oversimplification occasioned by the use of tests in which the active cells were T

Table 4. H-2 type and percent virus-induced leukemia as seen in congenic strains differing at H-2 (Tennant and Snell, 1968).

Strain	Haplotype	Percent leukemia	Significance of difference from	
			B10.D2	B10
BALB/c*	*d*	100		
B10.D2	*d*	83		<.001
B10.A	*a*	73	>.05	<.001
B10.BR	*k*	62	<.05	<.01
B10	*b*	39	<.001	
A	*a*	100		
A.BY	*b*	75		

* Strain of origin of inducing virus.

helper cells. Most of the early studies involved antibody production to pauci-determinant antigens, and in this context the T lymphocytes are T_H or helpers. When studies turned to *in vitro* cell mediated lysis (CML), new complexities appeared; the K and D antigens as well as *I* region products seemed to be influencing the response (80, 81; review in 67). Further studies tended to confirm this (82–85). From this and other evidence it appears that *I* region products regulate helper lymphocyte activity, including immune response activity, whereas *K* and *D* region products, including the K and D antigens themselves, play the corresponding role for effector lymphocytes.

Another link between *H-2* and immune capability is the phenomenon of *H-2* restriction. The response of T lymphocytes to antigens other than H-2 requires a simultaneous response to H-2. Apparently, the recognition structures of these cells are so constructed that they react not only with a specific antigen on the surface of a foreign or altered cell, but also and simultaneously with an H-2 antigen on that cell. The H-2 antigen typically "seen" by helper cells is apparently an Ia (class II) antigen, the antigen "seen" by effector cells is a K, D or class I antigen. Whether there are two receptors on the T cell to account for this dual capability, or one receptor with two reactive sites, or a single receptor reacting with a fusion product, is still debated, but the phenomenon itself is firmly established (reviews in 18, 63). A corollary of *H-2* restriction is that T cells, when confronted with any of a number of cultured cell lines that have no *H-2* products on their surface, not only are barred from reacting with H-2, but also with any non-H-2 products on the cell surface (86–88).

A curious and, from the point of view of basic mechanisms, certainly important exception to the phenomenon of *H-2* restriction is the reaction of T cells with the products of *H-2* itself. Cell mediated attack against an *I* region target, form example, does not require concomitant recognition of a K or D target on the same cell (89). Similar results have been reported with Qa (class IV) antigens (90, 91).

I now turn to some manifestations of *H-2* which, so far as we know have no relation to immune processes.

H-2 or a gene closely linked with it can influence mating preference in mice. This was demonstrated by Boyse and coworkers using congenic strains, one carrying $H\text{-}2^b$, the other $H\text{-}2^k$. If males of one genotype were presented with esterus females of both genotypes, a statistically significant proportion of the males preferred females of the opposite type (92). The active agent maps at the right (*D-Tla*) end of *H-2* (93, 94). This surprising finding presumably reflects an ability of mice to smell an *H-2* product or products.

Another manifestation of *H-2* demonstrated with congenic strains is the degree of susceptibility to cortisone-induced cleft palate. When B10.A ($H\text{-}2^a$) and B10 ($H\text{-}2^b$) pregnant females were treated with cortisone, the incidences of cleft palate in the offspring were 81% and 21% respectively (95). Strain A mice with the $H\text{-}2^a$ genotype that favors cortisone-induced cleft palate also show a relatively high spontaneous rate of the defect.

It is interesting to note that Reed and Snell in 1931, using crosses with the A strain as one parent, found evidence for one locus with a major role in determining cleft palate susceptibility (96). Could this have been the first identification of *H-2*?

The two non-immunological manifestations of *H-2* which I have described seem largely irrelevant to any basic function of this complex. Are there any non-immunological manifestations that do suggest a basic function? The answer is yes. There is a growing body of experimental data that suggests a role for *H-2* in cell interactions.

I cannot begin to summarize the literature here. Reviews will be found in Snell (63) and Dausset and Contu (97). Perhaps the best clue as to a possible role of the MHC in cell interactions comes from some recent studies by Curtis and coworkers (98, 99).

In one experiment (98), mouse kidney tubule epithelium was grown in cultures so designed that there would be confrontation between outgrowths either matched for mismatched at *H-2*. Contact inhibition, as measured by lack of overlapping growth, was increased in mismatched cultures. In other studies, low molecular weight diffusible glycoproteins were demonstrated which, in a variety of mixed cultures, would reduce adhesion among cells of the type to which they were foreign. The active molecules were called *interaction modulation factors* or IMFs. They were demonstrated in mixtures of cells ranging from those of unrelated sponges to mouse T cells. The important finding in the present context is that T cell IMFs were active in *H-2*-disparate cultures. The active region of *H-2* appeared to be *H-2D*.

Since the T cel IMFs act only on *H-2*-disparate cells, they must have a degree of *H-2* specificity. It is interesting to note that McKenzie and coworkers (100, 101) have reported the presence in mouse serum of low molecular weight glycoproteins with Ia specificity. They appear to be of T cell origin. The nature of the interaction between IMFs and the disparate cells whose capacity to adhere they inhibit is obviously a problem of great importance which only future studies can clarify.

I find attractive for a number of reasons the concept that the ancestral function of MHC products is the regulation of cell interactions. It seems to me that the early appearance and wide tissue distribution of the K and D (class I) antigens is difficult to reconcile with a purely immunological role, whereas it is quite in keeping with a developmental one. The association on the cell surface of H-2 and other alloantigens in specific configurations demonstrated by Boyse et al (102) and Flaherty and Zimmerman (103) also seems to me to fit best with this role. Finally, the Ia (class II) antigens seem clearly to be involved in cell interactions, thoung, in keeping with their presence only or primarily on lymphocytes and macrophages, interactions restricted to these cells. Perhaps the class II genes evolved from the class I genes along with the evolution of the immune system.

A cell interaction role for the MHC has been suggested by a number of authors (104–107, and others). The major problem in this thesis is its reconciliation with the proven role of apparently all MHC products, including the class I antigens, in immune phenomena. I have proposed a possible route that such an evolution could take (63). The proposal is admittedly speculative; only time and much more research will provide firm answers. Whatever the ultimate conclusion, the fascination of *H-2* is unlikely to diminish.

Science is like a web, growing by interactions that reach out in time and space. My own place in this web was made possible by strands from the past and the help of contemporaries. To them, my deep appreciation. Dr. Clarence Little provided the background for my studies of histocompatibility and, as founder of the Jackson Laboratory, the environment that made them possible. Dr. Peter Gorer was the original discoverer of *H-2* , and although my own identification of the complex was independent, our studies, once united, reinforced each other. Dr. Gorer's untimely death was a tragic loss to his many friends and to this field of science. I was aided in the work by many wonderful associates. It is unfair to single out a few but I feel I must note the contributions of Jack Stimpfling, Bill Hildemann, Ralph Graff, Marianna Cherry, Peter Démant, and Ian McKenzie. One of the greatest satisfactions of the work has been to see it develop in directions I could not possibly have foreseen. In these developments, the contributions of Jean Dausset and Baruj Benacerraf have been outstanding. Finally, I would like to express the appreciation of my family and myself to the members of the Nobel Foundation and to Mrs. Ingela Johansson of the Royal Ministry for Foreign Affairs for their wonderfully kind and helpful hospitality that has contributed so much to the pleasure of this occasion.

REFERENCES

1. Snell, G. D., Folia biol. (Praha) 14, 335 (1968).
2. Bodmer, W. F., Harvey Lectures 72, 91 (1978).
3. Little, C. C., Science 40, 904 (1914).
4. Little, C. C., and Tyzzer, E. E., J. Med. Res. 33, 393 (1916).
5. Little, C. C., in The Biology of the Laboratory Mouse, Snell, G. D., Ed. (Blakiston, Philadelphia, 1941), p. 279.
6. Snell, G. D., J. Genet. 49, 87 (1948).
7. Snell, G. D., in Origins of Inbred Mice, H. C. Morse III, Ed. (Academic Press, New York, 1978), p. 119.
8. Gorer, P. A., J. Pathol. Bacteriol. 44, 691 (1937).
9. Gorer, P. A., Lyman, S., and Snell, G. D., Proc. Roy. Soc. London B 135, 499 (1948).
10. Snell, G. D., Smith, P., and Gabrielson, F., J. Natl. Cancer Inst. 14, 457 (1953).
11. Snell, G. D., J. Natl. Cancer Inst. 14, 691 (1953).
12. Amos, D. B., Gorer, P. A., and Mikulska, Z. B., Proc. Roy. Soc. London B 144, 369 (1955).
13. Allen, S. L., Genetics 40, 627 (1955).
14. Gorer, P. A., and Mikulska, Z. B., Proc. Roy. Soc. London B 151, 57 (1959).
15. Stimpfling, J. H., and Richardson, A., Genetics 51, 831 (1965).
16. Klein, J., in The Mouse in Biomedical Research, Vol. 1, Foster, H. L., Small, J. D., and Fox, J. G. Eds. (Academic Press, New York, 1981), p. 119.
17. Snell, G. D., J. Natl. Cancer Inst. 21, 843 (1958).
18. Snell, G. D., Dausset, J. and Nathenson, S., Histocompatibility (Academic Press, New York, 1976).
19. Snell, G. D., in Mammalian Genetics and Cancer, Russel, E. S., Ed. (Alan R. Liss. New York), 1981, p. 241.
20. Bailey, D. W., Immunogenetics 2, 249 (1975).
21. Counce, S., Smith, P., Barth, R., and Snell, G. D., Annals of Surgery 144, 198 (1956).
22. Graff, R. J., Hildemann, W. H., and Snell, G. D., Transplantation 4, 425 (1966).
23. Graff, R. J., and Brown, D. W., Tissue Antigens 14, 223 (1979).
24. Snell, G. D., J. Natl. Cancer Inst. 21, 843 (1958).
25. Klein, J., Transplantation 15, 137 (1973).
26. Gorer, P. A., and Mikulska, Z. B., Cancer Res. 14, 651 (1954).
27. Gorer, P. A., and O'Gorman, P., Transplant. Bull. 3, 142 (1956).
28. Klein, J., Biology of the Mouse Histocompatibility-2 Complex (Springer-Verlag, New York, 1975).
29. Hoecker, G., Counce, S., and Smith, P., Proc. Natl. Acad. Sci. U.S.A. 40, 1040 (1954).
30. Wakeland, E. K., and Klein, J., Immunogenetics 8, 27 (1979).
31. Bailey, D. W., and Kohn, H. I., Genet. Res. 6, 330 (1965).
32. Egorov, I. K., Genetica (Moscow) 3, 136 (1967).
33. Kohn, H. I., Klein, J., Melvold, R. W., Nathenson, S. G., Pious, D., and Shreffler, D. C., Immunogenetics 7, 279 (1978).
34. Shreffler, D. C., Amos, D. B., and Mark, R., Transplantation 4, 300 (1966).
35. Démant, P., Snell, G. D. and Cherry, M., Transplantation 11, 242 (1971).
36. Murphy, D. B., and Shreffler, D. C., Transplantation 20, 443 (1975).
37. Klein, J., and Shreffler, D. C., J. Exp. Med. 135, 924 (1972).
38. Snell, G. D., Cherry, M., and Démant, P., Transplant. Proc. 3, 183 (1971).
39. Dausset, J., Iványi, P., and Feingold, N., Ann. N.Y. Acad. Sci. 129, 386 (1966).
40. Snell, G. D., Démant, P., and Cherry, M., Transplantation 11, 210 (1971).
41. Snell, G. D., Cherry, M., and Démant, P., Transplant. Rev. 15, 3 (1973).
42. Snell, G. D., Démant, P., and Cherry, M., Folia biol. (Praha) 20, 145 (1974).
43. Démant, P., Snell, G. D., Hess, M., Lemonnier, F., Neauport-Sautes, C., and Kourilsky, F., J. Immunogenet. 2, 263 (1976).

44. Démant, P. Iványi, D. Neauport-Sautes, C. and Snoek, M. Proc. Natl. Acad. Sci. U.S.A. 75, 4441 (1978).
45. McKenzie, I. F. C., Morgan, G. M., Melvold, R. W., and Kohn, H. I., Immunogenetics 4, 333 (1977).
46. Stimpfling, J. H., and Pizarro, O., Transplant. Bull. 28, 102 (1961).
47. Démant, P., personal communication (1980).
48. Dausset, J., Rapaport, F. T., Iványi, P., and Colombani, J., in Histocompatibility Testing 1965, Balner, H., Cleton, F. J., and Eernisse, J. G., Eds. (Munksgaard, Copenhagen, 1965), p. 63.
49. Götze, D., Ed., The Major Histocompatibility System in Man and Animals (Springer-Verlag, New York, 1977).
50. Ellman, L., Green, I., Martin, W. J., and Benacerraf, B., Proc. Natl. Acad. Sci. U.S.A. 66, 322 (1970).
51. McDevitt, H. O., and Tyan, M. L., J. Exp. Med. 128, 1 (1968).
52. McDevitt, H. O., Deak, B. B., Shreffler, D. C., Klein, J., Stimpfling, J. H., and Snell, G. D., J. Exp. Med. 135, 1259 (1972).
53. Gluecksohn-Waelsch, S., and Erickson, P. P., Curr. Topics Dev. Biol. 5, 281 (1970).
54. Klein, J., and Hammerberg, C., Immunol. Rev. 33, 70 (1977).
55. Démant, P., Neauport-Sautes, C., Iványi, D., Joskowitz, M., Snoek, M., and Bishop, C., in Current Trends in Tumor Immunology, Reisfeld, R., Herberman, R. B., Gorini, S., and Serrone, S., Eds. (Garland STPM Press, New York, 1979), p. 289.
56. David, C. S., in The Major Histocompatibility Complex in Man and Animals, Götze, D., Ed. (Springer-Verlag, New York, 1977), p. 255 (1980).
57. David, C. S., personal communication (1980).
58. Shreffler, D. C., Transplant. Rev. 32, 140 (1976).
59. Flaherty, L., in Role of the Major Histocompatibility Complex in Immunobiology, Dorf, M., Ed. (Garland Press, New York, 1981), p. 33.
60. Hämmerling, G. J., Hämmerling, V., and Flaherty, L., J. Exp. Med. 150, 108 (1979).
61. Artz, K., and Bennett, D., Nature (London) 256, 545 (1975).
62. Klein, J., and Hammerberg, C., Immunol. Rev. 33, 70 (1977).
63. Snell, G. D., Harvey Lectures 74, 49 (1980).
64. Murphy, D. B., Okomura, K., Herzenberg, L. A., Herzenberg, L. A., and McDevitt, H. O., Cold Spring Harbor Symp. Quant. Biol. 41, 497 (1977).
65. Cowing, C., Schwartz, B. D., and Dickler, H. B., J. Immunol. 120, 378 (1978).
66. Rowden, G., Phillips, T. M., and Delovitch, T. L., Immunogenetics 7, 465 (1979).
67. Snell, G. D., Adv. Genet. 20, 291 (1979).
68. Iványi, P., Proc. Roy. Soc. London B 202, 117 (1978).
69. Snell, G. D., and Stevens, L. C., Immunology 4, 366 (1961).
70. Cudkowicz, G., and Stimpfling, J. H., Science 147, 1056 (1964).
71. Cudkowicz, G., and Rossi, G. B., J. Natl. Cancer Inst. 48, 131 (1972).
72. Harmon, R. C., Clark, E. A., O'Toole, C., and Wicker, L. S., Immunogenetics 4, 601 (1977).
73. Nakano, K., Nakamura, I., and Cudkowicz, G., Nature (London) 289, 559 (1981).
74. Gross, L., Proc. Soc. Exp. Biol. Med. 86, 734 (1954).
75. Gross, L. Cancer 9, 778 (1956).
76. Lilly, F., Boyse, E. A., and Old, L. J., Lancet 2, 1207 (1964).
77. Tennant, J. R., J. Natl. Cancer Inst. 34, 633 (1965).
78. Tennant, J. R., and Snell, G. D., Natl. Cancer Inst. Monograph 22, 61 (1966).
79. Tennant, J. R., and Snell, G. D, J. Natl. Cancer Inst. 41, 597 (1968).
80. Gordon, R. D., and Simpson, E., Transplant. Proc. 9, 885 (1977).
81. von Boehmer, H., Fathman, G. G., and Haas, W., Eur. J. Immunol. 7, 443 (1977).
82. Maron, R., and Cohen, I. R., Nature (London) 279, 715 (1979).
83. von Boehmer, H., Haas, W., and Jerne, N. K., Proc. Natl. Acad. Sci. U.S.A. 75, 2439 (1978).

84. Wettstein, P. J., and Frelinger, J. A., Immunogenetics 10, 211 (1980).
85. Zinkernagel, R. M., Althage, A., Cooper, S., Kreeb, G., Klein, P. A., Sefton, B., Flaherty, L., Stimpfling, J., Shreffler, D., and Klein, J., J. Exp. Med. 148, 592 (1978).
86. Bevan, M. J., and Hyman, R., Immunogenetics 4, 7 (1977).
87. Dennert, G., and Hyman, R., Eur. J. Immunol. 7, 251 (1977).
88. Zinkernagel, R. M., and Oldstone, M. B. A., Proc. Natl. Acad. Sci. U.S.A. 73, 3666 (1976).
89. Klein, J., Chiang, C. L., and Hauptfeld, V., J. Exp. Med. 145, 450 (1977).
90. Forman, J., and Flaherty, L., Immunogenetics 6, 227 (1978).
91. Stanton, T. H., and Hood, L., Immunogenetics 11, 309 (1980).
92. Boyse, E. A., Lab. Anim. Sci. 27, 771 (1977).
93. Yamaguchi, M., Yamazaki, Y., and Boyse, E. A., Immunogenetics 6, 261 (1978).
94. Andrews, P. W., and Boyse, E. A., Immunogenetics 6, 265 (1978).
95. Bonner, J. J., and Slavkin, H. C., Immunogenetics 2, 213 (1975).
96. Reed, S. C., and Snell, G. D., Anat. Rec. 51, 43 (1931).
97. Dausset, J., and Contu, L., Human Immunol. 1, 5 (1980).
98. Curtis, A. S. G., Dev. Comp. Immunol. 3, 379 (1979).
99. Curtis, A. S. G., and Rooney, P., Nature (London) 281, 222 (1979).
100. Parish, C. R., Chilcott, A. B., and McKenzie, I. F. C., Immunogenetics 3, 113 (1976).
101. Jackson, D. C., Parish, C. R., and McKenzie, I. F. C., Immunogenetics 4, 267 (1977).
102. Boyse, E. A., Old, L. J., and Stockert, E., Proc. Natl. Acad. Sci. U.S.A. 60, 886 (1968).
103. Flaherty, L., and Zimmerman, D., Proc. Natl. Acad. Sci. U.S.A. 76, 1990 (1979).
104. Snell, G. D., in Immunogenetics of the *H-2* System, Lengerová A. and Vojtísková, M., Eds. (S. Karger, Basel, 1971), p. 352.
105. Bodmer, W. F., Nature (London) 237, 139 (1972).
106. McDevitt, H. O., Fed. Proc. 35, 2168 (1976).
107. Medawar, P. B., Nature (London) 272, 772 (1978).